Handbook of Common Polymers

Fibres, Films, Plastics and Rubbers

Handbook of Common Polymers

Fibres, Films, Plastics and Rubbers

Compiled by

W. J. ROFF,
B.SC., M.SC.

formerly at the Shirley Institute, Cotton, Silk and Man-made Fibres Research Association, Manchester

and

J. R. SCOTT,
PH.D., F.R.I.C., F.INST.P., F.I.R.I.

Rubber and Plastics Research Association, Shawbury, Shropshire (Director 1940–58)

with the assistance of

J. PACITTI,
M.A.

Rubber and Plastics Research Association

CRC PRESS

A DIVISION OF
THE CHEMICAL RUBBER CO.
CLEVELAND, OHIO

INTERNATIONAL SCIENTIFIC SERIES

English edition published in 1971 by
Butterworth & Co. (Publishers) Ltd
88 Kingsway, London WC2B 6AB

First published in the USA by
The Chemical Rubber Co.
18901 Cranwood Parkway
Cleveland, Ohio 44128

ISBN 0–87819–933–0
Library of Congress Catalog Card Number 74–173090

Printed in England by C. Tinling and Co. Ltd, London and Prescot

Preface

An earlier book* was intended, largely for use by persons in the textile industry, as a handbook of ready reference not only on fibres and polymeric substances employed in their treatment, but also on related constructional materials, i.e. plastics and rubbers employed in plant and equipment, and in research and development work.

The response to the book was so gratifying that it was realised that, whether in the textile industry or elsewhere, *most* persons—including students—concerned with polymeric materials have interconnected interests in them, and would often find it useful to have general information on them immediately to hand. Therefore, as in the course of time new developments caused the earlier book to become out of date and incomplete, an approach was made to friends at the *Rubber and Plastics Research Association of Great Britain* for co-operation in the present project. The authors are pleased to record that this has led to a happy collaboration, which—though they were 60 miles apart and the matter necessitated many visits, much very lengthy correspondence, and still more drafting and re-drafting—has resulted in the present volume.

The authors hope it will be found widely acceptable, because perhaps never before has so much information on **fibres, films, plastics** and **rubbers** been presented in a handy form in a single volume.

W.J.R.
J.R.S.

Fibres, Plastics and Rubbers, by W. J. Roff; Butterworths Scientific Publications, London, 1956.

Acknowledgements

One of us, recalling the compilation of the earlier book* from which the present volume originates, is pleased to acknowledge afresh his indebtedness to the many people from whom he received assistance and encouragement, especially to the following by name:

Dr. P. W. Cunliffe, Mr. B. Farrow, Dr. G. L. Gibbons, Dr. F. Holmes, Dr. F. O. Howitt, Dr. F. Howlett, Mr. H. Kennedy, Professor R. Meredith, the late Mr. T. H. Messenger, Mr. R. C. W. Moakes, Mr. D. B. Moore, Dr. F. C. Toy, Dr. A. R. Urquhart, Dr. W. C. Wake and Dr. J. O. Warwicker.

Concerning the present work, our first thanks go to Mr. J. Pacitti for considerable assistance in compilation of several Sections and most of the Comparison Tables. For encouragement, information supplied, and for much valued constructive criticism of various parts of the manuscript, we acknowledge with pleasure our indebtedness to the following friends and associates:

At the Shirley Institute: Dr. J. R. Holker, Dr. F. H. Holmes, Dr. J. D. Owen, Mr. S. C. Simmens and Dr. J. O. Warwicker.

At the Rubber and Plastics Research Association: Mr. G. M. Gale, Mr. D. I. James, Mr. J. MacLachlan, Mr. J. Matthan, Mr. R. H. Norman, Mr. J. R. Pyne, Dr. W. C. Wake,† Mr. R. R. Whisson and Mr. A. G. Williamson.

In addition, for confirmation of points relating to particular items our thanks are due to Mr. J. Knightley (polyvinylidene chloride), Professor A. Robson (proteins), also to several other persons who contributed information on minor topics, by attention to which we trust that the work has benefited. Finally, with regard to the large amount of typing and photo-copying involved in the project, we gratefully acknowledge the early help of the Shirley Institute and the very considerable assistance given throughout by the Rubber and Plastics Research Association.

W.J.R.
J.R.S.

---

*Fibres, Plastics and Rubbers, by W. J. Roff; Butterworths Scientific Publications, London, 1956.
†Now Consultant and Visiting Professor, The City University, London.

Contents

How to Use this Book

As listed on pp. 1 to 482, *Sections 1–42 contain data on individual polymers* and *Sections 51–84 deal with specific properties and related information.*

To find items pertaining to a particular **polymer**—e.g. specific gravity or dielectric constant of polyethylene—refer to the Section on that polymer (i.e. Polyethylenes, Section 1).

To obtain information relating to a particular **property**—e.g. materials with a given specific gravity or dielectric constant—look up the appropriate Section (i.e. Specific gravity, Section 58, or Dielectric constant, Section 70).

For other purposes refer to the index.

UNITS

Values of properties have mostly been given in metric units rather than SI units, because the choice of the most suitable multiples and sub-multiples of the latter—especially the unit for stress—has not hitherto been finalised. The 'Note on SI Units' (pp. 659–660) gives the factors required for converting values in normal metric units to those in SI units.

1. List of Sections on Individual Polymers

In each of Sections 1–42, listed below, the data commence with *synonyms* and *trade names* and *general characteristics*; then appear in the following order:

1	**Structure**	5	**Serviceability**
2	**Chemistry**	6	**Utilisation**
3	**Physics**	7	**History**
4	**Fabrication**	8	**Additional Notes**
		9	**Further Literature**

Olefin and vinyl-type polymers

Section 1. Polyethylenes
With notes on chlorinated and chlorosulphonated polyethylenes, ethylene/ vinyl acetate copolymers, and ionomers.

Section 2. Polypropylene
With notes on polybutene-1 and poly-4-methylpentene-1.

Section 3. Polytetrafluoroethylene
With notes on tetrafluoroethylene/hexafluoropropylene copolymers, poly-chlorotrifluoroethylene, polyvinyl and polyvinylidene fluorides, and fluoro-rubbers.

Section 4. Polystyrene
With notes on blends (inc. ABS plastics), copolymers of styrene (with acrylonitrile, divinylbenzene, and maleic anhydride), related materials (polymethylstyrenes, polychlorostyrenes, polyvinyl-naphthalene and poly-acenaphthylene), polyvinylpyridines, poly-N-vinylcarbazole, and poly-p-xylylenes.

Section 5. Indene and coumarone/indene resins
With notes on hydrogenated resins and allied resins (olefin, cyclopentadiene, terpene, and cyclohexanone resins).

Section 6. Polyvinyl acetate
With notes on partially hydrolysed polyvinyl acetates.

Carbohydrate-type polymers

Proteins

Synthetic condensation-type polymers

Section 23. Polyamides

With notes on nylons by number, modified polyamides (including polyimides), polynonamethyleneurea and polyaminotriazoles.

Section 24 A. Polyesters: alkyd resins.

With notes on drying oils, rosin ester gums, shellac, polyester rubbers, and factice.

Section 24B. Polyesters: fibre-forming types (principally as polyethylene terephthalate).

Section 24C. Polyesters: polycarbonates

With notes on polyphosphonates and polyanhydrides.

Section 24D. Polyesters: unsaturated types (principally as glass-reinforced resins)

With a note on moulding compounds.

Section 25. Epoxy resins

With notes on phenoxy resins and epichlorohydrin rubbers.

Section 26. Polyformaldehyde

With notes on polyethylene oxide, polypropylene oxide and polyphenylene oxide, and chlorinated polyethers.

Section 27. Amino-formaldehyde resins (principally as urea- and melamine-formaldehyde resins)

With notes on benzoguanamine-, aniline-, and toluenesulphonamide-formaldehyde resins.

Section 28. Phenol-formaldehyde resins.

Natural and synthetic rubber-type polymers

Section 29. Natural rubber

With notes on latex, Para rubber, plantation rubbers, anti-crystallising rubber, iso-rubber, graft copolymers, gutta percha and related substances, cyclised rubber, cellular rubber, reclaimed rubber, and liquid rubber.

Section 30A. Natural rubber derivatives: oxidised rubber.

Section 30B. Natural rubber derivatives: chlorinated rubber.

Section 30C. Natural rubber derivatives: rubber hydrochloride.

Section 31. Ebonite.

Section 32. Isoprene rubber (synthetic).

With notes on synthetic polyisoprene latex and synthetic gutta percha.

Section 33. Butadiene rubber

With notes on carboxylated rubbers, latex, and thermoplastic polybutadienes.

Section 34. Styrene/butadiene rubbers

With notes on latex, carboxylated rubber, styrene/butadiene/vinyl(or methylvinyl)pyridine rubber, and α-methylstyrene/butadiene rubber.

Section 35. Nitrile rubbers

With notes on latex.

Section 36. Chloroprene rubbers.

2. List of Sections on Specific Properties and Related Information

Structure

Section 51. Simplest fundamental unit.
Section 52. Molecular weight and degree of polymerisation.
Section 53. X-Ray data.

Chemistry

Section 54. Preparation of high polymers.
Section 55. Solvents.
Section 56. Plasticisers.
Section 57. Identification.

Physics—General Properties

Section 58. Specific gravity and density.
Section 59. Refractive index.
Section 60. Birefringence.
Section 61. Water absorption.
Section 62. Moisture regain (fibres).
Section 63. Water retention (fibres).
Section 64. Permeability.

Thermal Properties

Section 65. Specific heat.
Section 66. Thermal conductivity.
Section 67. Coefficient of linear expansion.
Section 68. Physical effects of temperature.

Electrical Properties

Section 69. Electrical resistivity.
Section 70. Dielectric strength.

PART 1
Data on Individual Polymers

POLYETHYLENES

With notes on chlorinated and chlorosulphonated polyethylenes,
ethylene/vinyl acetate copolymers, and ionomers

For ethylene/propylene rubbers, *see* Section 37.

For ethylene/propylene rubbers, *see* Section 37.

SYNONYMS AND TRADE NAMES

Poly(ethylene), Polythene. Alathon, Alkathene, Carlona, Dylan, Hi-Fax, Hilex, Hostalen, Lupolen, Marlex, Petrothene, Riblene, Rigidex, Rotene, Staflen, Vestolen. Courlene (Fibres).
Note: The term *polymethylene* is applied to products resembling commercial polyethylenes but with truly linear (i.e. unbranched) molecules. Such polymers have so far been prepared only experimentally, e.g. by decomposition of diazomethane.

GENERAL CHARACTERISTICS

Translucent, relatively soft and waxy but tough solids. Two main types: *low density polymers* and the more opaque and harder *high density polymers*. They find extensive use as general-purpose plastics and packaging films, and are employed to a less extent for monofils and fibres. Polymers of lower molecular weight are waxes and can be dispersed as latices.

1.1 STRUCTURE

Simplest Fundamental Unit

$$-CH_2- \qquad CH_2, \text{ mol. wt.} = 14.$$

Usually, however, the structure is considered as $[-CH_2CH_2-]_n$ with occasional short branches, each 2–3 C-atoms or more in length and terminated by a methyl group, which occur at intervals of 25–100 chain C-atoms. For every 1000 chain C-atoms, *low density polyethylenes* may have 8–40 (usually 20–30) branches, some exceeding 10 C-atoms in length, while *high density polyethylenes* may have less than 5 short branches and are therefore nearly linear in structure. Usually, too, there is some small degree of unsaturation present (e.g. from 0·1 to 2·5 double bonds per 1000 C-atoms) according to the type of polymer. Cross-linked structures, produced chemically or by irradiation, are also encountered in commercial polymers.

Molecular weight Commercial polyethylenes with number-average molecular weights of 1–3 million have been produced but 20000–200000 is more usual. Typical values: low density polyethylenes, up to 50000; high density polyethylenes, 20000–40000 (Phillips process) and up to 200000 (Ziegler process); waxes, 1000–10000; latices,

2000–30000; fibres, minimum *c.* 20000. *Note:* For practical purposes, the processing properties of polyethylenes are indicated by the *Melt Flow Index.* This (measured in g/10 min) is the weight of polymer extruded under specified conditions, e.g. at 190°C (*see* BS 2782, Part 105 C); being inversely related to melt viscosity, it is inversely related to molecular weight. The value of the MFI at a given molecular weight varies with polymer density.

Degree of polymerisation Plastics: up to 200000, but usually 1500–7000; waxes, *c.* 70–700.

X-ray data Polycrystalline, but branching lowers the degree of crystallinity (which can be as low as 15%). Crystallinity of commercial polyethylenes: low density, *c.* 60%; high density, up to 95%. The crystalline regions are similar in structure to those of low molecular weight solid paraffins. Unit cell: orthorhombic with $a = 7.40$ Å, $b = 4.93$ Å, $c = 2.53$ Å (identity period). The crystallites associate to form three-dimensionally ordered structures (spherulites) whose growth is promoted by annealing. Shock-cooling (quenching) produces a more random arrangement with shorter crystalline segments and smaller spherulites, resulting in a high degree of optical clarity in films of the polymer. Cross-linking reduces crystallinity, and almost entirely amorphous low density polymers are obtained at 25% cross-linking. Drawn polyethylene exhibits a characteristic fibre x-ray diffraction pattern.

1.2 CHEMISTRY

1.21 PREPARATION

Ethylene ($CH_2 = CH_2$, b.p. -104°C) is obtained industrially by cracking natural gases, such as propane or light oil fractions. The monomer needs to be pure (free of other hydrocarbons, water, oxygen, etc.) before polymerisation, which is usually effected continuously, by one of the following methods.

HIGH-PRESSURE POLYMERISATION

This depends on free radical catalysis and gives *low density polyethylenes.* The monomer is passed over heated copper to remove adventitious oxygen and is then compressed to 30000–50000 lbf/in². The catalyst (oxygen, peroxides, azo-compounds) and a chain modifier (hydrogen, carbon tetrachloride, aldehydes, ketones) are introduced in controlled amounts before or after compression. Polymerisation, which is exothermic and requires careful control, takes place at temperatures up to 200°C; propagation of the polymer radical may be terminated by combination, disproportionation or chain transfer reactions. The polymer is extruded as a ribbon which is cooled and granulated. Bulk polymerisation is commonly used, but solution processes (using benzene or methyl alcohol) have been developed to assist heat transfer and to give a lower molecular weight; suspension and emulsion polymerisation have also been tried, the last yielding latices (though these are mostly made by re-dispersing solid polymers).

A high density product (sp. gr. 0·955) has been obtained with a high-pressure process developed by Du Pont, in which the pressure is over 100 000 lbf/in^2 and the catalyst is α, α'-azobisisobutyronitrile.

Copolymers (e.g. with acrylates or vinyl acetate) may be prepared by high-pressure processes.

LOW-PRESSURE POLYMERISATION

This gives *high density polyethylenes*, for the commercial production of which there are two principal processes differing mainly in the type of catalyst used. In the *Ziegler process,* in which gaseous ethylene is fed into an inert hydrocarbon solvent containing a mixed metal alkyl/metal halide catalyst (e.g. triethyl aluminium and titanium tetrachloride), polymerisation takes place at or near atmospheric pressure and at 50–75°C. Chain terminators (e.g. oxygen, hydrogen, acetylene) are included to control molecular weight. The resulting slurry, containing up to 30% of polymer, is freed from solvent and then washed to remove the catalyst residues. In the *Phillips process,* in which the catalyst is partially-reduced chromium oxide supported on activated silica or alumina and suspended in cyclohexane, poly-merisation takes place at 400–500 lbf/in^2 and 100–175°C. The polymer is obtained as a 10% solution, and is recovered by first removing the catalyst by centrifuging, then either stripping the solvent or cooling to cause precipitation. The molecular weight of the product may be increased by using a fluoride in the preparation of the catalyst. Properties such as resistance to stress-cracking may be improved by copolymerising 2–5% of butene-1 with the ethylene, thus producing a controlled amount of branching.

In the less used *Standard Oil process* the catalyst is either a mixture of nickel and cobalt supported on activated charcoal, or reduced molybdenum oxide supported on alumina, titania, or zirconia; polymerisation takes place at *c.* 100°C and 500 lbf/in^2 and the end-products range from waxes to high molecular weight polymers. Copolymers may also be produced by this process.

After any of these processes, dry-blending may be carried out to ensure homogeneity; low- and high-density polymers are commonly blended together, along with stabilisers and anti-static agents.

1.22 PROPERTIES

Solvents Generally insoluble below *c.* 50°C (low-density types) or *c.* 60°C (high-density types), but at higher temperatures soluble in hydrocarbons and chlorinated hydrocarbons. Solubility increases with melt flow index, decreases with rise in density and crystallinity; dissolution of highly crystalline polyethylene may require a temperature of up to 170°C (but solutions have been made at room temperature by first shock-cooling a molten polymer to obtain a minimum degree of crystallinity). Solvents suitable at elevated temperature include toluene, xylene, tetralin, carbon tetrachloride, trichloroethylene, and perchloroethylene, chlorinated solvents being the more effective at lower temperatures. **Plasticised by** microcrystalline waxes, polyisobutylene, rubbers and related materials, but these are little used since flexibility can be adjusted by polymerisation conditions and blending. **Relatively unaffected by** polar solvents (alcohols, phenols, esters, ketones), vegetable oils, water, alkalis, most conc. acids (including HF) at room temperature, ozone (in absence of ultra-violet light). *Note:* (i) *Environmental stress cracking* can take place when a polymer subjected to strain (or internal stress) is exposed to polar liquids or vapours. The probability of this occurring is greatest with high density and high melt flow index polyethylene. Liquids giving rise to this phenomenon include carbon disulphide, chloroform, ethylene dichloride, diethyl ether, decalin, tetrahydrofuran, cyclohexanone, xylene, paraffin oil, silicone fluids and detergents. (ii) *Exposure to ozone* or a corona discharge changes the surface properties of polyethylene, making it more readily wettable (and more acceptable of inks, adhesives, etc.). Although this reaction is confined to the surface layers, oxidative degradation is accelerated in presence of ozone. In general, chemical resistance is highest in linear (high density) polyethylene and in cross-linked material. **Decomposed by** strong oxidising agents (fuming HNO_3 and H_2SO_4), slowly attacked by halogens and chlorinating agents (chlorosulphonic acid, phosgene, thionyl chloride).

Weight increase after immersion for 1 month in various liquids at 25°C (%; first value low-, bracketed value high-density material): acetone, 1·2 (0·8); paraffin oil, 3·8 (0·8); benzene, 14·6 (5·0); carbon tetrachloride, 42·4 (13·5). Negligible change (less than $\pm 1\%$) in 50% H_2SO_4, HNO_3, NaOH, or conc. HCl.

1.23 IDENTIFICATION

Thermoplastic. Cl, S, and P absent; N may be present in antioxidant.
Combustion Burns with a yellow-tipped blue flame and little smoke; continues to burn (may drip) on removal from flame and when

extinguished gives off smell like that of hot paraffin wax. Can be distinguished from polypropylene by specific gravity and melting point. **Pyrolysis** Melts and becomes transparent; in the absence of air it is stable up to 280–300°C when it decomposes to yield low molecular weight hydrocarbons. Cross-linked types do not melt. **Other tests** There are no simple chemical tests. The low density, lack of chemical reactivity and limited solubility, however, are indicative in themselves (but *see also* §2.23).

1.3 PHYSICS

1.31 GENERAL PROPERTIES

Specific gravity Low density types, 0·915–0·94. High density types, 0·94–0·95 (Ziegler); 0·96–0·97 (Phillips, Standard Oil). Polymethylene, 0·98. Low molecular weight types (waxes), 0·87–0·93. For some purposes three classes are recognised: low density (0·910–0·925); medium density (0·926–0·940); high density (0·941–0·965). Density is normally related linearly to crystallinity but cross-linking produces an increase in density with falling crystallinity (*see* p. 60, BOENIG, H. V., *Polyolefins,* as §1.9). **Refractive index** 1·51–1·52, increases with density. Amorphous component, 1·49; crystalline components (biaxial), α, $\beta = 1·52$, $\gamma = 1·582$. Fibres: low density, $n_{\parallel} = 1·54$, $n_{\perp} = 1·50$; high density, $n_{\parallel} = 1·57$, $n_{\perp} = 1·52$. **Birefringence** (varies irregularly with stress in specimen), 0·001–0·004 (*see*, for example, HOSHINO, S. *et al., J. Polym. Sci.,* **58**, 185 (1962)). Fibres: $(n_{\parallel}-n_{\perp})$, low density 0·04; high density 0·05. **Moisture relations** being paraffins, polyethylenes are very resistant to water. Increase in weight after immersion at 20°C for a year or more is generally less than 0·2%; at higher temperatures oxidation may result in polar groups and increased water absorption. Incorporation of carbon black tends to increase absorption. Water absorption after 28 days at 70°C (mg/cm^2 of surface): low density, 0·22; high density, 0·15; cross-linked low density polymer with 100 parts carbon black (per 100 polymer), 0·6. Water retention of polyethylene fibres, c. 2%. **Permeability** Transmission of gases and vapours decreases with increase of crystallinity and is thus least in high density polyethylenes. With organic vapours it is lowest with strongly polar materials and rises according to the following general order: alcohols, acids, nitro-derivatives, aldehydes and ketones, esters, ethers, hydrocarbons and halogenated hydrocarbons. Irradiation has little effect until the dose reaches 10^7 roentgens, at which level gas permeability is reduced by half. Modification by other means may also alter characteristics, e.g. grafting on acrylonitrile can decrease permeability to benzene by as much as tenfold. Examples:

Specific gravity of polymer	Temp. (°C)	Permeability*			Transmittance
		Nitrogen	Oxygen	Carbon dioxide	Water vapour
0·92	25	1·2–2·0	5·6	17–35	0·7–1·7
	50	9·1	2·1	71	—
0·92†	25	1·1	3·2	14	—
0·94	25	0·65	2·1	7·4	0·20
	50	2·51	7·1	19·4	—
0·95	25	0·45	1·5	4·3	—
	50	1·75	4·65	11·8	—
High density	20–30	0·18–0·3	0·7–1·1	3–4·3	0·1–0·2

*Units. Gases: $(10^{-10}$ cm$^2)$/(s. cmHg)
  Water vapour: $(10^{-10}$ g)/(cm^2 s. cmHg) at 90% r.h., through films 1 mm thick
†Irradiated (10^8 roentgens)

Water vapour through cellular polyethylene (sp. gr. 0·04, 2·5 cm thick, units as above), 0·01–0·02.

1.32 THERMAL PROPERTIES

Specific heat (similar values for low and high density polyethylenes) 0·55 at 20°C rising to 0·70 at 120–140°C. **Conductivity** $(10^{-4}$ cal)/ (cm s°C). Low density types, 8·4 at 0°C (7 at 50°C); high density, 10 at 0°C (8·7 at 50°C); cellular (sp. gr. 0·05), 1·1 at 10°C. **Coefficient of linear expansion** $(10^{-4}/°C)$. Low density 1·7–2·2; high density, 1·3–2·0. About 15% shrinkage on cooling from melt to room temperature. **Physical effects of temperature** Low and high density types become brittle only below −70°C; low molecular weight polymers (waxes) are brittle in the range −30°C to room temperature. There are second order transition temperatures at −125 to −100°C and at −21°C, and low density polyethylene shows an additional second order transition at 70°C. Heat distortion points (at 0·185 kgf/mm^2 stress): low density, 40–50°C; high density, 55–60°C. Softening points: waxes, 20–60°C; low density, 60–80°C; high density, 110–125°C. **Melting point** (loss of crystallinity) low density, 109–125°C; high density, 130–135°C; polymethylene, 136°C. Theoretical m.p. of perfectly crystalline polyethylene, 142·6°C. Transition points are slightly lower in cross-linked polymers; these do not melt, even at 300°C, but become rubbery above 115°C. Ageing data: retained tensile strength (TS) and extension at break (EB) after thermal ageing in air (%):

Polymer	7 days, 70°C		42 h, 127°C, 0·06 kgf/mm²	
	Percentage		Percentage	
	TS	EB	TS	EB
Low density	115	107	*Melts*	
High density	108	86	117	86
Low density*	110	102	105	71

*Cross-linked, 100 parts carbon black per 100 polymer

1.33 ELECTRICAL PROPERTIES

Resistivity (ohm cm) Low density, 10^{17}–10^{19}; high density, 10^{15}–10^{16}; low molecular weight waxes, $> 10^{14}$. Conductive polyethylenes, with resistivities of 30–50 ohm. cm, have been prepared by incorporating 100 parts of channel black (per 100 polymer) and cross-linking with organic peroxides. **Surface resistivity** (ohms) Low density, 10^{14}–10^{17}; high density, 10^{14}–10^{15}. **Dielectric strength** (kV/mm) 20–160 (e.g. 40 at 0·5 mm thickness; thin films up to 320); increases with density and molecular weight but is more affected by thickness and temperature. **Dielectric constant** (polyethylene sp. gr. 0·92) 2·28, over wide range of frequency; rises slightly with increase of density, falls slightly with increase of temperature. Cellular polyethylene (used in cable-making), effectively *c.* 1·5. **Power factor** (polyethylene sp. gr. 0·92) usually under 0·0002, over wide range of frequency; little affected by temperature, but in high density polyethylenes traces of catalyst may raise value to *c.* 0·0005. Cellular polyethylene (with residues of blowing agent), up to 0·0006. Oxidation or other modification raises the power factor, e.g. cross-linked polyethylene, up to 0·002. The presence of occluded air and voids (due to high coefficient of expansion and changes in crystallinity) has an adverse effect on dielectric properties, increasing the risk of corona discharge at high voltage. Oxidation and the presence of mechanical stresses in the polymer accelerate dielectric breakdown.

1.34 MECHANICAL PROPERTIES

It is difficult to correlate the stress-strain properties of polyethylenes with basic characteristics such as density, crystallinity and melt flow index (*see* §1.1) since the conditions of preparation of the specimen and the test itself greatly affect the results. In general, however, properties involving small deformations (e.g. modulus, creep) depend upon crystallinity, and thus upon density; while for large deformations (e.g. tensile strength, creep rupture) molecular weight and branching (as indicated by melt flow index) appear to be the determining factors. **Viscoelastic behaviour** Deformation is initially largely elastic and recovery is virtually complete with 5% strain, which may be taken as a safe operating limit for design purposes. Under load, polymers with high melt flow indexes tend to break fairly rapidly although not necessarily at high strain. Radiation cross-linked polymers exhibit lower creep values but time to break is reduced.

Table 1.T1 TIME TO REACH 5% STRAIN, TIME TO BREAK, AND STRAIN AT BREAK UNDER VARIOUS LOADS (60°C; TIME IN HOURS, STRAIN AS %)

Specific gravity	0·92		0·93		0·95
Melt flow index	1·2	0·3	9·5	1·87	0·03
At 0·14 kgf/mm²					
Time to 5% strain	100	900	>20000	>20000	—
Time to break	15000	†	†	†	—
Strain at break	8	6‡	3‡	2‡	—
At 0·28 kgf/mm²					
Time to 5% strain	*	*	300	*	>5000
Time to break	15000	†	10000	†	§
Strain at break	35	22‡	7	6‡	3‖
At 0·42 kgf/mm²					
Time to 5% strain	*	*	*	*	1000
Time to rupture	1500	15000	200	6000	4000
Strain at break	>200	62	9	12	7

*Initial strain ⩾ 5%
†Not determined during period of test but > 20000 h
‡Strain at 20000 h
§No break at 5000 h
‖Strain at 5000 h

Elastic modulus (Young's) (kgf/mm²).

Density	Plastics	Monofils
0·92	15·4	80 (10 gf/den.)
0·94	63·0	
0·95	77·0	400 (50 gf/den.)
0·96	126·0	
	171·5*	

*Cross-linked high-density polymer

Tensile strength and extension at break Low stresses produce recoverable strains up to a critical yield point above which cold drawing takes place and, eventually, rupture. Shock-cooling (as may occur in injection-moulded specimens and extrudates) tends to lower yield point and ultimate strength, and to increase extension at break, but slowly-cooled compression moulded and annealed specimens have higher yield points. During cold-drawing (in monofilaments) the crystalline regions are orientated and high ultimate strength is obtained. The strain rate appreciably affects the results, e.g. high density polymer:

Strain rate (mm/min)	50	500
Yield strength (kgf/mm^2)	2·28	2·36
Ultimate tensile strength (kgf/mm^2)	3·43	1·33

In view of the above it is difficult to quote truly comparable values for different grades of polyethylenes; however, some examples of fairly typical values follow:

Polymer type	Specific gravity	Melt flow index	Tensile strength (kgf/mm^2)	Extension at break (%)
Waxes*	0·90	2000	0·5	40
Commercial injection-moulding polymers†	0·92	10	0·7	300
	0·92	1·0	2·5	900
	0·94	10	1·2	150
	0·94	1·0	3·2	800
	0·96	10	1·96	20
	0·96	0·7	3·2	50
Monofils*,‡	0·92	—	8–12 (1–1·5 gf/den.)	25–50
	0·95–0·96	—	35–50 (4·5–6 gf/den.)	20–30
Cross-linked polymers*	low	—	1·6–1·8	150–250
	high	—	3·0	170
Polymethylene†	0·98	—	3·5	500

*Strain rate not known
†Strain rate 500 mm/min
‡Wet characteristics same as dry

Flexural and Compression properties, Impact strength, Hardness

	Low density	High density	
Flexural modulus (kgf/mm²)*	c. 35	c. 77	
Flexural strength (kgf/mm²)†	11–14	17–24	
Compression modulus (kgf/mm²)	—	c. 50‡	
Compression strength (kgf/mm²)	1·1–1·7	1·8–3§	
Hardness (Shore D)	50–60	65–70	
		sp. gr.	
Impact strength‖	0·92	0·95	0·96
Compression moulded	>20	2·5	5
Injection moulded	>20	6	20

*Strain rate 5 mm/min
†Strain rate 1·3 mm/min
‡36–94 when cross-linked
§1·5–3·8 when cross-linked
‖Izod, ft lbf/in notch. Adversely affected by crystallinity but rises with molecular weight. Low density polymers too flexible to fracture at room temperature

Friction and abrasion 0·6 may be taken as representative of μ static and dynamic for polyethylene against polyethylene or polished steel. However, the value tends to fall as the load is increased and depends on the type of polymer, being highest in the softer low density types, e.g. μ at 22°C and 0·07 kgf/mm², for samples containing carbon black/chromium-plated steel: low density, c. 0·5; high density, c. 0·3.

1.4 FABRICATION

Processing temperatures vary with the type of application and the melt flow index of the polymer. Extrusion: pipes, 140–170°C; films, coatings, 200–340°C. Injection moulding: 150–370°C, 8000–30000 lbf/in² (moulds, c. 50°C). Compression moulding: 130–230°C, up to 800 lbf/in². Polyethylenes are capable of being blow-moulded, extrusion-moulded, extrusion-coated (on to paper or fabric), or processed in powder form (e.g. rotational moulding, fluid bed coating, flame-spraying). They are, however, difficult to vacuum-form because they are slow to heat up and, as a result of their semi-crystalline structure, the forming temperature tends to be close to the melting point. Calendering is seldom employed since it is difficult to obtain smooth dimensionally stable sheets, and film is more readily produced by extrusion blowing or chill roll casting. Fabrication by welding is possible, and strong bonds to metals can be obtained with epoxy or nitrile-phenolic adhesives following treatment of the polymer surface to render it polar, e.g. oxidising acids, flame, ozone, or electrical discharge. Similar treatment is required before printing. Cross-linking can be effected by incorporation of peroxides or similar agents during compounding, or by irradiation of end-products (e.g.

cables). Fillers impair mechanical properties and are seldom used, but carbon black reinforces cross-linked polymers and is also used, in small proportions, as a pigment and stabiliser (ultra-violet light absorber) especially in films for outdoor use. Anti-oxidant stabilisers (e.g. di-β-naphthyl-p-phenylenediamine) are used mainly to preserve electrical properties over long periods. Anti-blocking or slip additives (e.g. oleamide) are often incorporated to prevent sticking between layers of low density polyethylene films; antistatic agents (e.g. glycol derivatives, long chain amines) serving a similar purpose, also reduce dust pick-up. Monofilaments are melt spun, followed by quenching and drawing; multifilament yarns and staple fibre have also been produced. Because of its low wettability and inertness, the material is usually mass coloured before spinning.

1.5 SERVICEABILITY

Outdoor applications are limited by the susceptibility to photo-oxidation by ultra-violet light, necessitating the use of protective agents such as carbon black; oxidative degradation at high tempera-tures and susceptibility to creep (even at room temperature) are further limitations. Although high density polyethylenes are theoretically more resistant to oxidation than low density types, in practice they behave similarly, probably because of the presence of catalyst residues. The material is normally inert to microbiological attack but unless specially compounded may not inhibit surface growth, and the relatively low hardness renders it liable to attack by insects and rodents. Exposure to high energy radiation in the absence of air produces cross-linking, with a consequent improvement in certain mechanical properties (tensile strength, modulus) and a marked temporary rise in electrical conductivity; small amounts of gases (mainly hydrogen) are evolved. In the presence of oxygen, radiation cross-linking proceeds much more slowly and an overall effect of chain degradation may be observed at low dose-rates.

1.6 UTILISATION

The first uses exploited the excellent dielectric properties and water resistance of polyethylene (e.g. for radar components, submarine cables). Cables still constitute an important application, and both cellular and radiation-cross-linked polyethylenes have been developed for this purpose. The packaging industries, however, are the largest consumers of both low and high density polymers in the form of films, coated paper, and blow-moulded bottles. Injection mouldings of various types (e.g. buckets, food boxes, chemical engineering components), and extruded tubing (e.g. domestic cold water pipes) are further important outlets. Other significant but small applications include abrasion-resistant coatings and fibres; the latter find use mainly in ropes, upholstery fabrics, fishing nets, and fusible interlinings. Low molecular weight polymers are used mainly in paper coatings and polishes; other uses include lubricants for thermoplastics (e.g. PVC), dielectrics and oil additives. Latices (often prepared from emulsifiable waxes) are used in floor polishes and in textile finishes to improve tear strength and resistance to needle cutting. Special purpose papers, e.g. for filters or battery separators, are prepared from polyethylene-encapsulated cellulose fibres, the polymer being formed directly on the fibre surface. In addition to cables, the major applications of cross-linked polyethylenes are in shrinkable packaging films (radiation processes) and tubing (chemically cross-linked, carbon black filled polymers).

1.7 HISTORY

Polymethylenes were prepared, by decomposition of diazo compounds, at the turn of the century (H. von Pechmann, 1898; E. Bamberger, F. Tschirner, 1900). It was not, however, until 1933 that R. O. Gibson and E. N. Fawcett, studying high pressure reactions (at ICI), unexpectedly obtained polyethylene of high molecular weight. A pilot plant for the new material was in operation by 1937, and full scale production commenced about 1942, most of the output being allocated to radar insulation. By 1943 production had been started in U.S.A. (under ICI licence) and Germany (IG Farbenindustrie). All three low pressure processes (Ziegler, Phillips, and Standard Oil) were disclosed around the same time in 1954, and of these the Phillips product was the first to become available commercially. Production of low and high density polymers in the U.K. now exceeds that of any other plastics material, the nearest rival being PVC.

1.8 ADDITIONAL NOTES

1.81 CHLORINATED POLYETHYLENES

These were patented (by ICI) in 1938 and were later produced under the name Halothene. Apart from applications where their flame-resistance has proved especially valuable, they have attracted relatively little interest (until recently when vulcanisable types were introduced) being overshadowed by the much greater success of PVC and chlorosulphonated polyethylene (§1.82). The structure varies with the method of preparation. The presence of chlorine atoms reduces crystallinity, resulting in more flexible products with lower softening points; at 25–40% chlorination rubbery polymers are obtained, higher levels of chlorination giving more rigid materials resembling PVC.

Chlorination can be carried out in bulk, solution, aqueous and non-aqueous suspension, and emulsion. Typically, gaseous chlorine is passed through a hot solution of polyethylene in carbon tetrachloride, in the presence of a catalyst (e.g. oxygen, peroxides, azo-compounds, light, high-energy radiation) until the desired chlorine content is achieved, whereupon the polymer is precipitated by dropping the solution into boiling water. Bulk systems, e.g. fluid bed, tend to be used with high density polymers on account of their low solubility.

Rubbery chlorinated polyethylenes are characterised by good resistance to attack by ozone, heat (epoxy or organometallic stabilisers required) and oils, and by high tear strength; they can be compounded with fillers (e.g. carbon black) and plasticisers (e.g. mineral oils, which also assist processing) and are vulcanised by means of metal oxides, peroxides, or amines. Summary of properties:

Unvulcanised polymer (25% Cl): sp. gr. 1·06; refractive index, 1·52; brittle point, −103°C; hardness (IRHD), 25; tensile strength (kgf/mm^2), 1·54; extension at break (%), 1700; dielectric constant (1 MHz), 4·26; power factor (1 MHz), 0·134.

Vulcanised polymer (40% Cl) containing 75 parts oil and 60 parts carbon black (per 100 polymer): hardness (IRHD), 53; tensile strength (kgf/mm^2), 1·54; extension at break (%), 550.

Applications are as yet relatively undeveloped; suggested uses include cables, flooring, films, and blends (e.g. with polyethylene, to impart flame-resistance; with PVC, to improve impact strength). Polymers with chlorine contents of $c.$ 60%, soluble in esters, ketones, and aromatic hydrocarbons, are used in flame-retardant, chemically-resistant paints.

1.82 CHLOROSULPHONATED POLYETHYLENES
HYPALON

Chlorinated polyethylenes containing a small proportion of chlorosulphonyl (—SO$_2$Cl) groups provide useful synthetic rubbers. The chlorine content is generally 25–35% and the sulphur content

0·9–1·7%, thus out of every 100 ethylene units, some 25 to 42 are chlorinated, i.e. as —$CH_2CH(Cl)$— units and 1·0 to 2·2 are chlorosulphonated, i.e. as —$CH_2CH(SO_2Cl)$— units.

The chlorosulphonation process resembles that of chlorination, gaseous chlorine and sulphur dioxide being introduced simultaneously into a solution of polyethylene (e.g. in hot CCl_4), possibly in the presence of ultra-violet light and/or azo-initiators. Introduction of chlorosulphonyl groups supplies reaction sites for cross-linking, and the polymers can be compounded in much the same way as general purpose rubbers, the curing system consisting of metal oxides (MgO, PbO), an organic acid (hydrogenated rosin), an accelerator mercapto benzthiazole; *see* §29.4 and a trace of moisture. Organic curing systems based on low molecular weight epoxy resins, TMTDS, and a guanidine accelerator can also be used. Antioxidants are not normally required, but enhance heat resistance; fillers (carbon black, clay, $CaCO_3$, etc.) may be used to improve high temperature mechanical properties or processing characteristics. **Identification** A dil. solution of chlorosulphonated polyethylene in pyridine gives an orange or red colour when treated with a little 2-fluoreneamine (ESPOSITO, G.G., *Off. Digest Fed. Soc. Paint Technol.*, **32** (1960, Jan.) 67–70). **Chemical resistance** The rubbers are outstanding in resistance to ozone, weathering, heat ageing, and general chemical attack; they do not, however, withstand hot oxidising acids (e.g. HNO_3), and they swell in aromatic and chlorinated solvents, and in esters (e.g. butyl acetate). **Volume swelling** of vulcanisates containing 50 parts carbon black (per 100 polymer), after 12 months' immersion at 25°C (%): toluene, 188; perchloroethylene, 106; methyl ethyl ketone, 86; hexane, 25; 10% hydrochloric acid, 16; alcohol, water, 10% sodium hydroxide, under 10. **Permeability** of unfilled vulcanisate (as % relative to natural rubber) at 25°C: H_2, 22; CO_2, 12; N_2, 11; O_2, 9. Absolute values $(10^{-10}$ cm²)/(s cmHg), N_2, 1·16; CO_2, 21; O_2, 2·8.

Physical Properties

Specific gravity 1·11–1·28. **Thermal conductivity** 2·7 $(10^{-4}$ ca)/(cm s °C). Brittle point, down to —60°C (depending on compounding). Maximum service temperature, *c.* 120°C. Retention of mechanical properties of vulcanisate containing 30 parts clay filler per 100 polymer, after ageing for 7 days at 150°C (%): tensile strength, 110; extension at break, 35; hardness increased by 15 IRHD.

Electrical Properties

Volume resistivity (ohm cm) 10^{14}. **Surface resistivity** (ohms) 10^{13}. **Dielectric strength** (kV/mm), 20 (short term, film 2 mm thick). **Dielectric constant** at 25°C and 1 kHz, 6–7. **Power factor** at 1 kHz, 0·03.

Mechanical Properties

Examples of typical vulcanisates of comparable hardness:

Filler (parts per 100 polymer)	Curing system	Hardness (IRHD)	Tensile strength (kgf/mm²)	Extension at break (%)	Stress at 100% extension (kgf/mm²)
Clay (30) Process oil (10)	Metal oxide	60	1·26	400	0·35
Calcium carbonate (25) Titanium dioxide (35)	Metal oxide	65	0·93	500	0·33
Carbon black (40)	Organic	67	2·15	270	0·53

Applications

Being white (colourless in thin films) and possessing a high level of mechanical properties without needing reinforcing fillers, chlorosulphonated polyethylenes lend themselves to many uses in coloured articles; also where resistance to weathering, corrosion, or abrasion is required, e.g. flooring, linings for chemical plant, roofing, shoe soling, proofed fabrics (clothing, collapsible fuel tanks), rollers, cables, and white sidewalls for tyres. Paints for corrosive and marine environments, and lacquers to impart ozone-resistance to rubber articles are important applications; ketones and aromatic hydrocarbons are suitable solvents. The polymers are sometimes used in unvulcanised form, generally with high proportions of fillers and plasticisers, in flooring. The ozone-resistance of natural rubber and some synthetic rubbers is improved by blending with chlorosulphonated polyethylene.

1.83 ETHYLENE/VINYL ACETATE COPOLYMERS (EVA)

Copolymerisation with vinyl esters was studied in the early days of high pressure polymerisation of ethylene, using rather lower temperatures and pressures than for homopolymerisation; free-radical initiators commonly used include azobisisobutyronitrile, organic peroxides, and persulphates. Suspension and emulsion systems have since been developed. The presence of vinyl acetate reduces crystallinity, and at 20–30% concentration produces rubbery copolymers; products containing above 75% vinyl acetate are, however, hard and rigid.

Commercial copolymers fall into two main categories; products of relatively low molecular weight and low softening point, commonly used as wax additives; and rubbery materials resembling low density polyethylene in appearance. In both instances the vinyl acetate

17

content is 20–30 % and the simplest repeat unit may be represented as:

$$-(CH_2CH_2)- \qquad CH_2CH-^{7-12}$$
$$\underset{|}{\qquad\qquad\qquad} CH_3COO$$

Wax Additives (e.g. Elvax)

These impart toughness, flexibility, and adhesive properties to paraffin waxes, and find application in paper coatings, hot melt adhesives, and plasticisers. They are soluble in aromatic and chlorinated hydrocarbons. Summary of properties:

Melt flow index	125–175	12–18
Specific gravity	0·95	0·95
Refractive index	1·485	1·482
Softening point (°C) (Ring and ball)	90	135
Brittle point (°C)	Flexible down to −58°C	
Tensile strength (kgf/mm²)	0·32	1·4
Extension at break (%)	650	750

Rubbers (e.g. Alathon, Levapren, Montothene, Ultrathene)

These, recently introduced, appear to have good prospects as general purpose rubbers. They can be compounded with fillers for cheapness or to increase stiffness and reduce flow properties. They weather better than polyethylene but ultra-violet stabilisers or carbon black are required for best results. The main disadvantage is poor heat resistance. This can be improved by cross-linking with an organic peroxide and, optionally, triallyl cyanurate (which is grafted on to the copolymer during the reaction and provides further cross-linking sites). In practice, however, the copolymers are usually handled as uncross-linked thermoplastics. Since their flexibility does not depend upon the use of plasticisers (which may be toxic or tainting), they have advantages in surgical goods and food packaging.

General Properties

Specific gravity 0·926–0·95. Chemical resistance is similar to that of low density polyethylene. **Soluble in** aromatic and chlorinated hydrocarbons at elevated temperature. **Permeability to gases** $(10^{-10}$ cm²)/(s cmHg) at 25°C: O_2, 6·7; N_2, 3·4; CO_2, 8·8. Transmittance of water vapour through 1 mm films $(10^{-10}$ g)/(cm² s cmHg) at 25°C, 90 % r.h.: 0·4 to 0·6.

18

Thermal Properties

Specific heat 0·55. **Coefficient of linear expansion** 1·6–2·0 $(10^{-4})/°C$. Brittle point below $-70°C$. Softening point (Vicat): 60–80°C.

Electrical Properties

Volume resistivity (ohm cm) *c.* 10^{14}. **Dielectric strength** (kV/mm), 21 (short term). **Dielectric constant** (at 1 MHz) 2·8. **Power factor** (at 1 MHz) 0·03 to 0·04.

Mechanical Properties of Typical Compounds

Filler (parts per 100 polymer)	Melt flow index	Hardness (IRHD)	Tensile strength (kgf/mm²)	Extension at break (%)
None	1·3–4	90–95	1·3–2·0	500–700
Clay (50)	2·3	—	0·91	560
Calcium carbonate (100)	1·0	—	0·57	230
Carbon black (40)	1·0	—	1·4	220
Carbon black (20) Clay (30)*	—	—	1·68	530

*Peroxide cross-linked compound

1.84 IONOMERS (e.g. Surlyn A)

In 1964 the Du Pont and Union Carbide companies introduced ethylene copolymers having an unusual type of cross-linkage based on ionic intermolecular bonds between oppositely-charged groups. The structure has not yet been fully elucidated but has been likened to that of certain inorganic crystals. Pendant anionic carboxyl groups are attached to the ethylene main chain by copolymerisation, and metal cations (Mg^{2+}, Zn^{2+}, also Na^+ and K^+) are also introduced. The cross-linked structure established by the interaction of these groups does not impair flow properties at processing temperatures and the polymers possess some characteristics of both thermoplastics and thermosets. They can have a very high degree of transparency, and abrasion resistance and adhesive properties are good. The polymers can be processed on conventional equipment and may be compounded with fillers. Possible important applications include packaging films and coatings, blow moulded bottles, vacuum formed goods, and wire covering.

General Properties

Specific gravity 0·93–0·97. Chemical resistance is similar to that of unmodified polyethylene, but, being polar materials, the resistance to water is lower, and to strong acids relatively poor. **Permeability to gases** $(10^{-10} \text{ cm}^2)/(\text{s cmHg})$ at 25°C: O_2, 5; N_2, 0·9; CO_2, 6. Transmittance of water vapour through 1 mm films $(10^{-10} \text{ g})/(\text{cm}^2 \text{ s cmHg})$ at 25°C, 90% r.h.: 0·3.

Thermal Properties

Specific heat 0·61. **Conductivity** $(10^{-4} \text{ cal})/(\text{cm s °C})$ 5·3–6·2. **Coefficient of linear expansion** $1·0–1·2 \times 10^{-4}/°\text{C}$. Brittle point, below −100°C. Second order transition temperatures (*see* §1.32), −125°C, −20°C, and 60–70°C. Softening point (Vicat): 70–90°C.

Electrical Properties

Volume resistivity (ohm cm) 4×10^{16}. **Dielectric strength** (kV/mm), 20 (short term). **Dielectric constant** (at 1 MHz): 2·48. **Power factor** (at 1 MHz): 0·004.

Mechanical Properties

Elastic modulus (Young's) (kgf/mm^2), 19·7–52·7. **Tensile strength** (kgf/mm^2), 2·46–3·87 (yields, inflection in stress–strain curve, 1·40–1·75). **Extension at break** (%), 200–600. **Hardness** (Shore D), 60–65.

1.9 FURTHER LITERATURE

BOENIG, H. V., *Polyolefins: Structure and Properties*, Amsterdam (1966)

KRESSER, T. O. J., *Polyethylene*, New York and London (1957)

RENFREW, A. and MORGAN, P. (Eds), *Polythene*, London (1960)

SMITH, W. M. (Ed), *Manufacture of Plastics* (Chapter 2, H. D. ANSPON Ed), New York and London (1964)

RAFF, R. A. V. and DOAK, K. W. (Eds), *Crystalline Olefin Polymers I and II (High Polymers, Vol. XX)*, New York and London (1965)

COOK, J. G., *Handbook of Polyolefin Fibres*, London (1967)

Specification
BS 3412 (low density polyethylene materials for moulding and extrusion).

POLYPROPYLENE

With notes on polybutene-1 and poly-4-methylpentene-1

For propylene-modified vinyl chloride polymer, *see* §11.82, and for ethylene/propylene rubbers, *see* Section 37.

SYNONYMS AND TRADE NAMES

Poly(propylene), polypropene, polmethylethylene. Carlona P, El Rex, Marlex, Napryl, Novolen, Oleform, Pro-fax, Propathene, Propylex (sheets), Vitraline. Fibres: Herculon, Meraklon, Ulstron.

GENERAL CHARACTERISTICS

Translucent thermoplastic material, somewhat resembling stiffened polyethylene but usually more glossy in appearance; also available as clear biaxially-oriented films and as fibres. It has high tensile strength, and, compared with polyethylene, a higher softening temperature; it becomes brittle at only moderately low temperatures, but is free from environmental stress cracking.

Polypropylene is used as a general purpose moulding material (and provides an 'integral hinge' effect) and the films and fibres find packaging and textile applications respectively.

2.1 STRUCTURE

Simplest Fundamental Unit

$$-CH_2-\overset{*}{C}H- \qquad C_3H_6, \text{mol. wt.} = 42$$
$$|$$
$$CH_3$$

Branched chains also occur. The $\overset{*}{C}$-atom is asymmetric and the commercial polymers are some 95% isotactic, i.e. stereoregular in structure, each repeat unit in a given chain being unidirectionally arranged and having the —CH$_3$ group in the same spatial configuration as illustrated below:

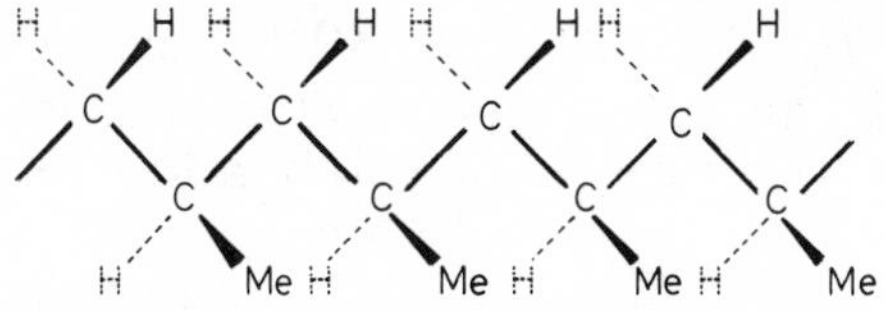

— = C—C bonds, lying in plane of the paper
◢ = bonds protruding *above* plane of paper
... = bonds protruding *below* plane of paper
Me = the methyl group, CH$_3$

The crystallinity and rigidity of the commercial isotactic polymer arise from the above regularity of structure.*

Molecular weight Number average, 25000–500000. Commercial polymers, over 80000 (degree of polymerisation over 2000). The melt flow index is used to characterise commercial polymers; it is indirectly related to molecular weight and is usually the weight of polymer extruded per minute at 230°C under a load of 2·16 kg from a cylindrical die as employed for polyethylene (*see* §1.1).

X-ray data Commercial polymers, 95% isotactic in structure (*see above*) with degree of crystallinity of 60–70% (higher values from infra-red measurements). In oriented polymer the unit cell is monoclinic, $a = 6·6$ Å, $b = 20·8$ Å, $\beta = 99$ degrees, c (i.e. identity period) $= c.$ 6·5 Å (value of 6·42 Å obtained from biaxially oriented film). The molecular chains have a helical configuration with three repeat units per turn; right- and left-handed helices are possible and both fit into the same crystal structure.

2.2 CHEMISTRY

2.21 PREPARATION

Monomer Propylene (CH_2=$CHCH_3$, b.p. $= -47·8°C$) is obtained from natural gas or the light oil fraction from petroleum. The gas or oil is cracked at temperatures above 800°C, and propylene is separated from C_3 fractions by successive distillation.

Polymer For continuous polymerisation, the highly purified monomer is fed, in the liquid state under low pressure, into an inert solvent (e.g. naphtha) in which is suspended a stereospecific catalyst of the Ziegler-Natta type. The catalyst consists essentially of a titanium halide and an aluminium alkyl; usually it is titanium tetrachloride with triethylaluminium or diethylaluminium monochloride. The two catalyst components are reacted in naphtha under nitrogen to form a slurry (10% catalyst, 90% naphtha). A third component (co-catalyst) and a chain transfer agent (e.g. hydrogen) may be employed to control molecular weight. Polymerisation occurs between 30–100°C at pressures below 20 lbf/in^2. At 60°C, reaction time is $c.$ 8 h and leads to 80–85% conversion. After completion of the process, in addition to isotactic polymer the reaction mixture contains atactic polymer together with solvent and catalyst. Monomer is flashed off,

---

*Syndiotactic polypropylene, where the Me groups lie alternately above and below the plane of the carbon chain, is also crystalline but less dense, more soluble and of lower m.p. than the isotactic form; see, for instance PETERLIN, A., *et al.* (Eds), *Macromol. Reviews*, vol. 2 (YOUNGMAN, E. A. and BOOR, J., 33–69), New York and London (1967). It is not available commercially. *Atactic* polypropylene, where the Me groups are randomly arranged above and below the plane of the carbon chain, is non-crystalline, more readily soluble and fusible.

and the solvent (together with dissolved atactic polymer) is removed by centrifuging; the remainder is treated with a catalyst-destroying agent (e.g. H_2O or CH_3OH) and the polymer is washed, centrifuged, and dried at 80°C.

2.22 PROPERTIES

Polypropylene, as a linear hydrocarbon containing little or no unsaturation, in several of its characteristics resembles polyethylene (Section 1); however, the —CH_3 groups, attached isotactically to alternate C-atoms in the molecular chain, modify the properties in certain ways, e.g. causing stiffening (which raises the m.p.) and rendering it *less* stable than polyethylene with regard to oxidation. In particular, the —CH_3 groups lead to asymmetry and the possibility of obtaining products of different tacticity, though (as explained in §2.1) it is only the isotactic structure that is important in commercial polymers. The latter are characterised by an *isotactic index*, i.e. the percentage of polymer remaining insoluble in boiling heptane (98·4°C). **Solvents** Polypropylene is dissolved only at elevated temperature, by aromatic hydrocarbons (e.g. xylene, tetralin, decalin), and chlorinated hydrocarbons (e.g. chloroform, trichloroethylene) above 80°C; amorphous or atactic polymer dissolved by hot aliphatic hydrocarbons (e.g. kerosene, molten paraffin wax). **Plasticised by** elastomers (e.g. nitrile rubbers; *see* Section 35) and ester plasticisers (e.g. dioctyl phthalate), but best plasticised internally by copolymerisation. **Swollen by** aromatic hydrocarbons (e.g. xylene) and chlorinated hydrocarbons (e.g. chloroform) at room temperature; also by esters (e.g. ethyl acetate, dibutyl or dioctyl phthalate), ethers (e.g. diethyl ether, tetrahydrofuran, dioxan) and by various aqueous oxidising agents (e.g. 10% HNO_3, 10% $KMnO_4$, dilute H_2O_2, NaOCl). **Relatively unaffected by** many organic liquids at room temperature (e.g. alcohols, glycols), and aqueous solutions including moderately conc. acids and alkalis (but less satisfactory above 60°C especially with mild oxidising agents). **Decomposed by** strong oxidising agents (e.g. HNO_3, oleum, chlorosulphonic acid, bromine) especially when warm (e.g. attacked by hot concentrated H_2SO_4); slowly weakened (e.g. fall in fibre strength) by conc. alkali. Polypropylene is more readily oxidised than polyethylene, the rate of attack and range of reagents increasing with rise of temperature.

2.23 IDENTIFICATION

Thermoplastic. Cl, N, S and P normally absent. **Combustion** Burns with blue-based non-smoky candle-like flame with flaring molten drops, leaving little or no residue; waxy, slightly acrid, odour on extinction. **Pyrolysis** Becomes transparent, melts (c. 170°C) and decomposes (c. 350°C) with evolution of neutral white fumes, i.e. hydrocarbons of low molecular weight, including c. 2% monomer, with only small proportion, or none, left as charred residue. **Other tests** Because of lack of chemical reactivity, physical tests are preferred. (i) The low density is characteristic but not unique, since polyethylene (PE) and poly-4-methylpentene-1 (PMP, see §2.82) also float in water; however, PP floats in isophorone (PE sinks) and sinks in 80/20 alcohol/water by volume (PMP floats).

2.3 PHYSICS

2.31 GENERAL PROPERTIES

Specific gravity depends on degree of crystallinity; e.g. cast films (chilled from melt), 0·89; general purpose and impact grade mouldings (slow cooled and/or oriented), 0·90–0·91; fibres, 0·90–0·92, Talc-filled polymer, 0·98–1·3; glass-filled polymer, 1·04–1·22; asbestos-filled polymer, 1·11–1·36. In unfilled polymer, increase of density (i.e. crystallinity) raises softening temperature, stiffness and yield strength, but decreases permeability, low temperature brittleness, tensile strength, and extension at break. **Refractive index** 1·49 Fibres ($\lambda = 5790$ Å), $n_{\parallel} = 1\cdot530$; $n_{\perp} = 1\cdot496$. **Birefringence** $n_{\parallel} - n_{\perp}$, 0·034. **Haze** (%) Cast unoriented films 1–2; tubular (oriented) films, 10–25; biaxially oriented films, 1 (packaging do., surface treated, 2–3). **Specular transmission** (percentage of light incident at 45 degrees reflected at 45 degrees) Cast films, 85–90; tubular films, 20–40. **Water absorption** (discs 3 mm thick; wt. % in 24 h), 0·01–0·03. Increase in weight after 6 months in aqueous environment: at 20°C, under 0·5%; at 60°C, under 2%. **Moisture regain** (percentage at 65% r.h.) 0·0–0·15. **Water retention** 5%. **Permeability** to gases (10^{-10} cm^2)/(s cmHg), at 20–30°C:

	O$_2$	N$_2$	CO$_2$
(Type not defined)	0·4–2·3	0·35–0·5	3·0–9·2
Cast film	0·6–2·5	0·4	1·8–9·0
Tubular film	0·45	—	1·2
Uniaxially oriented film	1·4	0·3	4·5
Biaxially oriented film	0·35–0·8	0·2	1·8
Irradiated oriented film	1·1	—	3·3

Transmittance of water vapour through 0·025 mm films (10^{-10} g)/(cm^2 s cmHg) at 38°C, 28; cast, unoriented, 30; biaxially oriented, 0·013 mm films, 25.

2.32 THERMAL PROPERTIES

Specific heat At 40°C, 0·46; 100°C, 0·5. **Conductivity** $(10^{-4}$ cal)/ (cm s °C), c. 5. **Coefficient of linear expansion** $(10^{-4}/°C)$ At 20–60°C, 1·1; 60–100°C, 1·5–1·7; 100–140°C, 2·1. **Physical effects of temperature.** Second order transitions occur between 4 to -12°C and -20 to -35°C. The polymer thus embrittles more readily than polyethylene (which is brittle only below -70°C) but onset of the change, e.g. as shown by loss of impact toughness, can be depressed by incorporation of additives or by copolymerisation. Heat distortion temperature (°C) at 66 lbf/in², 105; 264 lbf/in², 55–60. Maximum service temperature, 110°C. Vicat softening point (5 kg load), 85–105°C; increases with isotactic index and melt index. Fibres shrink 6–8% at 100°C (e.g. in boiling water); soften 145–150°C. Films can be heat-sealed at 170°C. Crystalline m.p. 165–175°C. The polymer is subject to oxidation at 100°C but, when suitably protected, the molten material is stable up to 250°C and does not decompose in full until over 300°C.

2.33 ELECTRICAL PROPERTIES

Volume resistivity (ohm cm) Over 10^{16}. Talc- or glass-filled, 10^{16}; asbestos-filled, 10^{15}. Resistivity falls off rapidly above 100°C. **Surface resistivity** Over 10^{15} ohm. **Dielectric strength** (kV/mm) short term, 3 mm sheet >32; stepwise 14 V. (Higher values in short term tests on thin sections). Non-tracking and arc-resistant. **Dielectric constant** 2·25 (largely independent of frequency from 10^2 Hz to 50 MHz and of temperature from 20 to 80°C). **Power factor** (at 20°C) 0·0005; (at 80°C) 0·0004 (largely independent of frequency from 10^2 to 10^6 Hz).

2.34 MECHANICAL PROPERTIES

Young's modulus (kgf/mm²) 120. Fibres: as cont. fil. yarn, 650 (80 gf/den.); monofil, 325 (40 gf/den.). **Elastic recovery** Fibres: cont. fil. yarn, measured immediately after loading and unloading at 8 gf/den. per min. e.g.,

	Stress (gf/den.)		Strain (%)		
	2	5	2	5	10
Recovery (approx. %)	>90	85	95	90	85

Tensile strength (kgf/mm²) General purpose grades, 2·8–3·5. Film

25

(unoriented 0·025 mm) 3·85–4·6; (biaxially oriented, 0·013 mm) 17·5. Fibres: cont. fil. yarn, 60 (7·4 gf/den.), 3·3 gf/den. at 95°C; monofil, 35 (4 gf/den.). **Wet strength** as dry strength. **Extension at break** (%) General purpose, 20–300 (at yield, *c.* 15); talc- and glass-filled, 2–4; asbestos filled, 3–20; films (unoriented 0·025 mm), 200–1000 (biaxially oriented 0·013 mm) 70; fibres 17–19. **Flexural yield strength** (kgf/mm^2) General purpose, 4·2–4·9; high impact, 2·9; glass-filled, 5·6–7·75; asbestos-filled, 5·25–6·3. **Compressive yield strength** (kgf/mm^2) General purpose, 3·85–4·6; high impact, 3·1; glass-filled, 4·5–5·0; asbestos-filled, 5·0. **Flexural modulus** (kgf/mm^2) 110. **Impact strength** (Izod; ft lbf/in of notch) General purpose, 0·4–2·2; high impact, 1·5–12; glass-filled 0·5–2; asbestos-filled 0·5–1·5. The polymer is notch sensitive. Polypropylene has poor impact strength at sub-zero temperatures as the glass temperature lies between 4 and −12°C. **Hardness** (Rockwell) General purpose, R80–100; high impact, R28–95; glass-filled, R90–115; talc-filled, R94–99; asbestos-filled, R90–110. Hardness increases with isotactic index and molecular weight. **Friction** μ_{dyn}: polymer/polymer, 0·1–0·3. **Abrasion** Tabor, CS-17 wheel; 18–28 mg/kilocycle.

2.4 FABRICATION

Processing follows closely that of polyethylene (Section 1) but polypropylene has certain special characteristics. The melt viscosity remains relatively high up to 230°C, then falls slowly with rise of temperature to 275°C, and thereafter falls off rapidly; thus for moulding and extrusion the preferred working range lies between 230–275°C. **Injection moulding** Temperatures 230–260°C; pressures 10 000–20 000 lbf/in^2; mould temperature 35–90°C. Mould shrinkage *c.* 1·8%. Durable integral hinges, produced as narrow web (0·25–0·38 mm thick) connecting thicker parts, can be satisfactorily moulded in polypropylene; flexing induces molecular orientation normal to the axis of the hinge. **Extrusion** Temperature 210–270°C; screw L/D ratio 24/1; high compression screws giving 4/1 compression as a minimum. Cast films are extruded from a slot-die into water or on to a chill-roll; clear films require to be extruded at 270–300°C. Biaxially oriented films of high strength and clarity are obtained by bubble-blowing extruded tubing. Extrusion coating is effected by a technique similar to that used for polyethylene with slightly lower temperatures of 300–320°C. *Vacuum forming* A short heating cycle is needed because of low specific heat, and the possibility of fabricating thin-walled articles of adequate stiffness, makes polypropylene a useful material for vacuum forming. Heater temperatures are between 540–650°C, for periods of 12 s/mm. *Other techniques* employed with polypropylene include pressure forming, skin packaging, blow moulding, hot gas welding, ultrasonic welding, machining, powder techniques such as flame or electrostatic spraying, fluidised bed dipping. *Fibres* are melt spun at temperatures up to 270°C, then hot-drawn to 5–10 times their length; wet-spinning and dry-spinning from hot solutions have also been used, and linearly oriented films can be fibrillated.

2.5 SERVICEABILITY

Because of the tertiary carbon atoms polypropylene is more susceptible to oxidation than polyethylene (which in turn is less resistant than polyisobutylene or polystyrene) and requires incorporation of appropriate antioxidants. These are especially important when the polymer is heated during processing, and if it is exposed to the degradative action of sunlight. Thus, to protect it from exterior weathering, ultra-violet radiation acceptors are required, although best protection is afforded by the use of small proportions of carbon black, e.g. 2 parts per 100 parts of polymer.

Unlike polyethylene, polypropylene is not subject to environmental stress cracking. It will withstand the temperatures used in sterilisation. It is, however, subject to creep and its use under continuous loading of high stress is not recommended.

2.6 UTILISATION

Polypropylene is a versatile material, a cross-section of end-uses being as follows—(i) *Injection moulded components* such as automobile parts, sanitary equipment, domestic appliances, soil pipe systems, hospital equipment, footwear, etc.; (ii) *Blow moulded articles* (made essentially from copolymers) such as air-ducts, expansion tanks in sealed engine cooling systems, large capacity tanks for chemical plant; (iii) *Extruded products* such as sheets for chemical plant and lining tanks, pipes for hot water; (iv) *Films,* unoriented, tubular and biaxially oriented, for packaging; (v) *Fibres,* monofils and multifils for ropes, cordage, netting, blankets, carpets, brushes.

2.7 HISTORY

Early attempts to polymerise propylene resulted in soft rubbery materials. However, in 1954, G. Natta discovered that catalysts of the metal alkyl/metal halide type (developed by K. Ziegler for the polymerisation of ethylene) have a stereospecific effect during polymerisation of propylene, yielding largely isotactic polymer. Commercial production of polypropylene started in 1959 and in 1963 the usage of polypropylene in the U.K. was 10500 tons. Since 1959 the major developments in material properties have been the improvement of heat stability and of shock resistance. An increasing number of grades is marketed. Copolymers with particular balances of properties have been produced in order to provide a full range of properties. Filled polypropylenes using glass-fibre, asbestos, mica, etc., are also widely employed.

2.8 ADDITIONAL NOTES

2.81 POLYBUTENE-1 (Polyethylethylene. Bu-Tuf)

Butene-1 (CH_2=$CHCH_2CH_3$, b.p. $-6°C$), a dimer of ethylene, is obtained mainly as a by-product from petroleum cracking. In the presence of a catalyst of the Ziegler-Natta type (§§1.21 and 2.21) it yields a polymer with the following repeat unit.

$$-CH_2-CH-$$
$$\underset{C_2H_5}{|} \qquad C_4H_8, \text{mol. wt.} = 56$$

Properties

Polybutene-1, obtained as above, has a weight-average mol. wt. of 10^5-10^6 and is largely isotactic and highly crystalline (76–79%, with three crystalline forms). It resembles polyethylene and polypropylene in being inert to many common reagents, but it dissolves in *n*-alkyl acetates and (at an elevated temperature) in ligroin, aromatic and chlorinated hydrocarbons (e.g. chloroform, toluene and decalin). It is a strong, rigid—but flexible—thermoplastic material, with excellent resistance to creep and environmental stress cracking (*see* §1.22). According to the crystalline form, the **specific gravity** ranges from 0·87 to 0·95 (amorphous, 0·85–0·86) and the m.p. from 107° to 141°C. **Coefficient of linear expansion** 1·5 ($10^{-4}/°C$). **Elastic modulus** Young's, flexural, 18 kgf/mm². **Tensile strength** 1·7–3·2 kgf/mm² (up to 12 for drawn fibres); *yield strength* 0·4–2·5 kgf/mm², according to crystalline form. **Extension at break** *c*. 350%. **Hardness** (Shore Durometer D) 65.

Utilisation

Uses and proposed uses of this relatively new material include: pipes and tubes, gaskets and diaphragms, heavy-duty bags, base for pressure sensitive tapes, and agricultural binder fibres. For **further literature,** *see* RUBIN, §2.9.

For polyisobutylene (polyisobutene) *see* §38.83.

2.82 POLY-4-METHYLPENTENE-1
(Polyisobutylethylene, TPX)

This is a rigid stereoregular (predominantly isotactic) polymer distinguished from other polyolefins by exceptionally low density, transparency and high softening point, the last property making it suitable for moulded products that require to be sterilised.

Structure

Simplest Fundamental Unit

$$—CH_2—CH— \qquad C_6H_{12}, \text{mol. wt.} = 84$$
$$|$$
$$CH_2CH(CH_3)_2$$

X-ray data Moulded articles are largely isotactic and *c*. 40% crystalline; annealing raises crystallinity to *c*. 60%.

Chemistry

Preparation The monomer is obtained by catalytic dimerisation of propylene (*see* §2.21), using alkali metals complexed on graphite or anhydrous potassium carbonate,

$$2CH_2{=}CHCH_3 \xrightarrow{\text{Na complex}} CH_2{=}CHCH_2CH(CH_3)_2$$

The product, dissolved in an inert hydrocarbon solvent and treated with a Ziegler type (anionic co-ordinated) catalyst yields the 1,2-polymer.*

Properties The polymer is relatively unaffected by many aqueous solutions, acids and alkalis (including concentrated H_2SO_4 and concentrated NaOH) but it is attacked by strong oxidising agents (e.g. HNO_3, H_2CrO_4, Br_2); it is also relatively unaffected by several organic liquids (including mineral oils and ester plasticisers) but swells, with loss of rigidity, in others, e.g. in aromatic and chlorinated hydrocarbons (xylene, CCl_4), some medium-boiling aliphatic hydrocarbons (such as petroleum ether, white spirit), glacial acetic acid, and low boiling esters (ethyl and hexyl acetate). Polar solvents give rise to environmental stress cracking.

---

*Cationic polymerisation, acid initiated at low temperature yields a rubbery 1,4-product, which is an ethylene-isobutene copolymer, i.e. with repeat unit $—CH_2CH_2CH_2C(CH_3)_2—$.

Identification This inert hydrocarbon is distinguished from poly-propylene by its transparency and infra-red spectrum. The low density is unique, the material floats in a mixture of 80/20 ethanol/water by volume (in which polypropylene sinks). Like polystyrene, a moulded sample of this polymer makes a metallic 'tinkle' when struck or dropped.

General properties

Specific gravity 0·83 (lightest commercial polymer). **Refractive index** (n_{20}^D) 1·465. **Birefringence** very low (assists transparency). **Optical transmission** up to 90%. **Water absorption** (3 mm moulded discs; percentage in 24 h) 0·01. **Permeability** Poly-4-methylpentene-1 is more permeable than polypropylene. Permeability to gases (10^{-10} cm^2)/(s cmHg) at 25°C: O_2, 27; N_2, 6·5; CO_2. Transmittance of water vapour through 0·025 mm films (10^{-10} g)/(cm^2 s cmHg) at 38°C, 90% r.h., *c.* 230.

Thermal

Specific heat 0·52. **Conductivity** 4.10^{-4} cal/ (cm s °C). **Coefficient of linear expansion** $1·17.10^{-4}/°C$. **Effects of temperature** Second order transition (T_g) *c.* 40°C Vicat softening point, 179°C. Crystalline melting point, 240–245°C (fairly sharp). Useful mechanical properties are retained up to 200°C (useful life—at 200°C, *c.* 1 day; at 125°C, *c.* 1 year).

Electrical

Resistivity $> 10^{16}$ ohm cm. **Dielectric strength** (kV/mm) 3 mm moulded sheets, 28. **Dielectric constant** (10^2–10^6 Hz) 2·1. **Power factor** Depends on frequency (min. value $< 0·0001$ at 25°C, 50 kHz) and more particularly on temperature, e.g. at 200 Hz or 10 kHz:

Temperature °C	0	25	60–70	100
Power factor	$< 0·0001$	0·0002	$> 0·0007$	0·0002

Mechanical

Young's modulus (kgf/mm^2) (100 s value at 0·2% str.), 112; at 200°C *c.* 10. **Tensile strength** (kgf/mm^2) (at yield at 50% str/min.), 2·8. **Extension at break** (%) 15. **Impact strength** (Izod test) 0·8 ft lbf/in. **Hardness** (Rockwell), L67–74; (pencil) HB.

Fabrication

Extrusion is best carried out at 260–275°C on high compression screws having a sharp transition into the metering zone and a L/D ratio of 20:1 or greater. Small bore tube extrusion is possible;

diameters above 13 mm require vacuum or pressure sizing dies; below 13 mm free extrusion with sizing plates gives good results. Wire coating by vacuum tubing or pressure coating method, with careful control of cooling to prevent voids. *Injection moulding* at 270–300°C, 15000–20000 lbf/in^2, with moulds at 60–70°C; mould shrinkage 1·5–3 %. It can be blow moulded at 275–290°C under high pressure. *Compression moulding* is normally uneconomical but may be used to prepare blocks or sheets, and a temperature of 280°C is required. Vacuum forming is restricted because of the relatively sharp m.p. Rotational casting and fluidised bed coating may be employed. *Fibres* can be melt-spun at 260–350°C. Fibres suitable for cordage, binder twine, etc. can be made by fibrillation of linearly oriented films. Adhesion (e.g. in printing, painting, metallising) is improved after the surface has received an oxidative treatment ($KMnO_4$ or H_2CrO_4).

Stability

Poly-4-methylpentene-1 resembles polypropylene in needing stabilisation against oxidative degradation when hot, i.e. antioxidants are incorporated. It is attacked by oxidising agents and degraded by ultra-violet radiation and it shows environmental stress-cracking. It is not recommended for outdoor use; on the other hand, it maintains the general chemical inertness of a hydrocarbon up to high temperatures. It withstands repeated sterilisation at 160°C and remains clear at that temperature.

Applications

Lighting, e.g. light fittings for vehicles, blow-moulded lamp-shades, moulded lamp reflectors for high intensity light sources. Chemically resistant apparatus, e.g. laboratory ware including beakers, measuring cylinders, burettes, connector pieces for general chemical purposes, catch-pots in laboratory plumbing systems, injection moulded bobbins for the dyeing of polyester fibres. Electrical applications, e.g. in high frequency and microwave applications including co-axial connectors, spacing discs and bushes, low-loss cables, and as a rigid base for printed circuits. Medical applications (awaiting results on toxicological inertness) for non-disposable syringes, tubing and vessels. Packaging applications for sterile water and other fluids which require sterilisation in the container; also for prepacked foods which need to be reheated or cooked in the container at 200°C.

History

Poly-4-methylpentene-1 was first obtained by G. Natta in 1955, and has been commercially available since 1965.

2.9 FURTHER LITERATURE

KRESSER, T. O. J., *Polypropylene*, New York (1960)

GALANTI, A. V. and MANTELL, C. L., *Polypropylene Fibres and Films,* New York (1965)

WIJBA, P. W. O., *Soc. Chem. Ind. Monograph No. 5,* London, 35 (1959) (Structure and Properties of Polypropylene)

SHOOTEN, J. V. and WIJBA, P. W. O., *Soc. Chem. Ind. Monograph No. 13,* London (1961) (Degradation of Polypropylene)

CLARK, K. J. and PALMER, R. P., *Soc. Chem. Ind. Monograph No. 20,* London (1966) (Transparent polymers from 4-methylpentene-1)

COOK, J. G., *Handbook of Polyolefin Fibres,* London (1967)

FRANK, H. P., *Polypropylene*, London (1968)

RUBIN, I. D., *Poly-1-butene*, London (1968)

Specifications

    ASTM D 2146 (propylene plastic moulding and extrusion materials)

    ASTM D 2530 (nonoriented propylene plastic film)

    ASTM D 2673 (oriented polypropylene film)

POLYTETRAFLUOROETHYLENE

With notes on tetrafluoroethylene/hexafluoropropylene copolymers,
polychlorotrifluoroethylene, polyvinyl and polyvinylidene fluorides, and fluoro-rubbers

SYNONYMS AND TRADE NAMES

Poly(tetrafluoroethylene), PTFE. Fluon, Hostaflon TF, Teflon.

GENERAL CHARACTERISTICS

White or bluish-white, translucent to opaque, smooth material, resembling polyethylene in appearance but denser. It possesses excellent electrical characteristics together with exceptional chemical inertness (being resistant to acids, alkalis and solvents) and thermal resistance (withstanding temperatures from below $-100°$ to $350°C$), and in addition has low-frictional and non-adhesive properties.

3.1 STRUCTURE

Simplest Fundamental Unit

$$—CF_2—CF_2— \qquad C_2F_4, \text{ mol. wt.} = 100$$

The simplest possible unit is $—CF_2—$ but it is usual to double it, as shown, in view of the ethylenic structure of the monomer (as with polyethylene, § 1.1).

Molecular weight Uncertain, but high (500 000 to 5 000 000) in commercial polymers; e.g. degree of polymerisation of a specific sample, as estimated from radioactive end-groups, $c.$ 10 000.

X-ray data Highly crystalline, but becomes amorphous above $327°C$. Identity period (oriented polymer) $2 \cdot 5$–$2 \cdot 6$ Å. Since there are two fluorine atoms on each carbon atom, the C—F bond length is reduced from the normal value of $1 \cdot 42$ to $1 \cdot 35$ Å.

3.2 CHEMISTRY

3.21 PREPARATION

Tetrafluoroethylene (C_2F_4; b.p. $-76°C$) can be obtained, as it was initially, by dechlorination of 1,2-dichlorotetrafluoroethane,

$$CClF_2—CClF_2 \xrightarrow[\text{EtOH at boil}]{\text{Zn dust}} CF_2 = CF_2 + ZnCl_2$$

However, industrial syntheses usually make use of the reaction between chloroform and anhydrous hydrogen fluoride to yield monochlorodifluoromethane as an intermediate, e.g.

$$2CHCl_3 + 3HF \xrightarrow[\text{SbCl}_5]{70°C} CHClF_2 + CHCl_2F + 3HCl$$

$$2CHCl_2F \xrightarrow{\text{AlCl}_3} CHClF_2 + CHCl_3 \text{ (re-cycled)}$$

The chlorodifluoromethane (b.p. $-41°C$) is then cracked thermally, and the resulting monomer is purified by scrubbing and fractionation,

$$2CHClF_2 \xrightarrow[\text{Pt-lined tubes}]{600-800°C} CF_2 = CF_2 + 2HCl$$

The monomer is a colourless, odourless, non-toxic, but reactive gas (it can be burnt in air) which slowly polymerises on storage, and more rapidly—can be violently, as the reaction is highly exothermic— when a radical initiator is present, e.g. under pressure in an aqueous solution of a persulphate. Commercially, using highly purified monomer in order to obtain a product of high molecular weight, the polymer is manufactured either as a fine granular powder or in a very finely-divided state (particle size $c. 10^{-4}$ mm) as an aqueous suspension.

3.22 PROPERTIES

Soluble in no known liquids. **Plasticised by** no common plasticisers. Prolonged heating in a fluorocarbon oil (e.g. $C_{21}F_{44}$) has some effect. **Relatively unaffected by** conc. alkalis and acids (including *aqua regia*) and all common liquids even at the boil. **Decomposed by** molten alkali metals, by alkali metal complexes (such as sodium naphthalene in tetrahydrofuran), and by prolonged exposure to fluorine (can be ignited in fluorine).

3.32 IDENTIFICATION

Thermoplasticity extremely limited. Cl, N, S and P absent. **Combustion** Will not burn in air; melts only slowly, with decomposition, when held in a flame. **Pyrolysis** Assumes transparent highly viscous state above 327°C; carbonises on further heating, producing white sublimate and volatile fluoro-derivatives (*N.B. Inhalation dangerous*). The vapours give a strongly acid reaction in the presence of moisture. Heating *in vacuo* above 600°C yields largely monomer. **Detection and estimation of fluorine** The element can be determined by (i) combustion with oxygen in the presence of silica, to yield silicon tetrafluoride (e.g. $C_2F_4 + O_2 + SiO_2 \rightarrow 2CO_2 + SiF_4$) which, when hydrolysed by water, can be titrated as hydrofluosilicic acid, or (ii) bomb-ignition with sodium peroxide followed by gravimetric or volumetric estimation of sodium fluoride. **Distinction between related fluoro-compounds** 1. Fluorine content (%): C_2F_4, 76·0; C_2ClF_3, 48·9; C_2H_3F, 41·3; $C_2H_2F_2$, 59·4. 2. Fluoro-polymers containing chlorine are distinguished by the Beilstein copper wire test (*see* Section 57; fluorine does *not* respond). 3. Fluoroacrylates are saponified by alcoholic alkali. 4. Fluoro-silicones decompose in hot 80% sulphuric acid (most other fluoro-polymers resist).

3.3 PHYSICS

3.31 GENERAL PROPERTIES

Specific gravity 2·1–2·3. **Refractive index** 1·37–1·38. **Water absorption** 0·00% even after prolonged immersion. **Permeability** to water vapour (10^{-10} g)/(cm s cmHg) at 20–30°C: 0·03; transmittance through 0·025 mm films (10^{-10} g)/(cm^2 s cmHg) at 38°C, 90% r.h.: 26.

3.32 THERMAL PROPERTIES

Specific heat 0·25. **Conductivity** $5·8.10^{-4}$ cal/(cm s °C). **Coefficient of linear expansion** (10^{-4}/°C) At 25–100°C, 1·2; 25–200°C, 1·5; 25–300°C, 2·2. **Physical effects of temperature** Flexible at —70 to −80°C and not completely brittle even at −180°C; dimensionally stable up to at least 250°C (but can creep under load). Maximum service temperature, 200–300°C. Embrittles when aged in air at 300°C. Undergoes second-order transitions at −113 and +127°C, and a 1 to 2% volume change between 19 and 23°C; first-order transition at 327°C (crystalline m.p., loses opacity and strength but does not soften appreciably). Above 400°C slowly decomposes, e.g. with 10% weight loss in 2 h at 460–480°C; above 600°C reverts largely to monomer.

3.33 ELECTRICAL PROPERTIES

Volume resistivity Over 10^{17} ohm cm, up to 100% r.h. **Surface resistivity** 10^{12}–10^{15} ohm. **Dielectric strength** Over 20 kV/mm at 1 mm thickness (thin films, 80 kV/mm); non-tracking. **Dielectric constant** 2·0–2·1 (over wide range of frequency and temperature). **Power factor** 0·0002 or lower (over wide range of frequency and temperature). Increases between −40 and −80°C (doubtful, due to impurities?); loss peak at $c.$ $10^{8 \cdot 5}$ Hz (but absent in highly crystalline samples).

3.34 MECHANICAL PROPERTIES

Elastic modulus (Young's) $c.$ 40 kgf/mm^2. **Elastic recovery** Some delayed elasticity, but from sustained or heavy loading, fairly poor. **Tensile strength** (kgf/mm^2) Approx. 1·5–2·5 (yields at 1·0–1·5), e.g. extended at 30 mm/min at 20–25°C, 2·1–1·75; rises at low temperature (e.g. up to $c.$ 4), falls with temperature rise (e.g. 0·7 at 100°C). Oriented films or tapes, 10·5–17·5. **Extension at break** 50–450%, depending on fabrication, temperature, and speed of testing; e.g. extended at 30 mm/min at 20–25°C, 350%, falling with both decrease and increase of temperature (e.g. 200% at 0° and 100°C). **Flexural strength and compression strength** Low at room temperature; $c.$ 0·1% deformation at 1 kgf/mm^2 (but polymer thereafter subject to creep under load). **Impact strength** (Izod test; ft lbf/in of notch) $c.$ 4 (2, at −60°C; 6, at 80°C). **Hardness** Brinell, 1–2; Shore D, 50–70; Rockwell, R20–25. **Friction and abrasion** Friction is unusually low and not much affected by lubrication. At low speeds, the coefficient decreases with decrease in speed (minimising 'stick-slip' effects, the static tending to be lower than the dynamic value) and at low loads it decreases with increase in load, but it shows little change with temperature (until 327°C, when the value rises appreciably). Minimum friction when sliding in direction of orientation. Some typical values, μ_{stat}: polymer/self, 0·1–0·2; polymer/steel, 0·02–0·3. μ_{dyn}: polymer/self, 0·04; polymer/steel, 0·04–0·1; grey cotton yarn (1 m/s)/polymer, $c.$ 0·2. For dry bearings/steel the polymer is preferably incorporated in porous bronze to facilitate removal of heat. The polymer shows moderately good abrasion resistance but tends to wear at a faster rate than polychlorotrifluoroethylene (§3.82); glass fibres and laminar fillers such as mica assist abrasion resistance.

3.4 FABRICATION

Commercial polymers are of such molecular weight that they do not readily flow under pressure and cannot be moulded or extruded in the conventional ways. Instead, mouldings are cold-pressed from powder (e.g. pre-formed at 2000 lbf/in^2) and then sintered—out of the mould, without loss of shape—at a temperature above the crystalline m.p. (e.g. at $c.$ 380°C); extrusion with special plant is also possible, slow passage through a die at 400–450°C effecting the necessary sintering. Heat-resistant fillers may be included. Aqueous dispersions of the polymer, as employed for surface coating, etc., can also be coagulated and then extruded. The bulk polymer can be cut, punched, and machined. Although no substance sticks directly to PTFE, adhesives can be used (*a*) if the polymer is first etched with molten sodium— or with sodium in liquid ammonia, or with a solution of sodium naphthalene in tetrahydrofuran—and then metallised, or (*b*) if a vinyl-type monomer (e.g. styrene) is graft-polymerised on to the surface.

3.5 SERVICEABILITY

The polymer is unaffected by long exposure to exterior weathering and sunlight (cf. polyethylene), it is not attacked by micro-organisms or moulds, and it possesses the following further merits: excellent electrical properties (maintained over a wide range of frequency, temperature and humidity), chemical inertness, thermal resistance (serviceable to 250–300°C) and non-adhesive properties. However, the material suffers from creep and should not be subjected to continuous stress.

3.6 UTILISATION

Because of its properties, some of which are unique, this polymer finds a variety of uses, e.g. (i) as a high-temperature and/or low-loss insulator, as used in cables, as a dielectric in electronic components, and as a film-base for printed circuits, (ii) in chemical plant, for gaskets, diaphragms, rings, tubing, taps, etc., employed under hot and/or corrosive conditions, and (iii) as an engineering material, where its release and sliding characteristics make it valuable for non-stick surfaces, expansion bearings and low-friction applications. The dispersions are used for spray coating, impregnation, and (coagulated) for extrusion.

3.7 HISTORY

Following extensive research by earlier workers with organic fluorides and chlorofluorides (the second being developed by A. L. Henne and T. Midgley in 1930), the monomer was described by O. Ruff and O. Bretschneider in 1933, and a patent for obtaining the polymer catalytically under pressure was granted to R. J. Plunkett, in the U.S.A., in 1941 (the polymer having been discovered unexpectedly in 1938, when a cylinder in which tetrafluoroethylene had been stored was cut open). Pilot scale production commenced in the U.S.A. (du Pont) in 1943 and in the United Kingdom (ICI) in 1944–47, the polymer becoming commercially available in 1948. Related fluoro-polymers appeared at roughly the same period; *see*, for example, §3.82 (PCTFE, first investigated by F. Schoffer and O. Scherer in Germany in 1934), §3.83 (PVF) and §3.84 (PVDF).

3.8 ADDITIONAL NOTES

3.81 TETRAFLUOROETHYLENE/ HEXAFLUOROPROPYLENE COPOLYMERS.
(Fluorinated ethylene/propylene copolymers. FEP resins.)*

These fully fluorinated copolymers are represented by the following repeat unit,

$$\left[CF_2CF_2 \right] CF_2CF-$$
$$\underset{5-25}{} CF_3$$

The commercial materials are translucent (transparent in thin films) and are 40–50% crystalline in the quenched form or 50–70% crystalline when annealed. They resemble PTFE, except that being less heat-resistant they can be moulded and extruded. Some typical properties are given below.

General

Specific gravity Quenched, 2·15–2·16; annealed, 2·15–2·17, **Refractive index** 1·338. **Water absorption** (disc, immersed 24 h, 23°C) <0·1%. **Permeability** to gases and vapours, roughly comparable with chloroprene and butyl rubbers.

Thermal

Specific heat 0·28. **Conductivity** $6 \cdot 10^{-4}$ cal/(cm s °C). Heat distortion temperature (*ASTM* D648) at 0·05 kgf/mm^2 stress, 70°C; at 0·18 kgf/mm^2 stress, 51°C. Serviceable from -268 to $+205$°C.

Electrical

(Much resembles PTFE, *see* §3.33.)

Mechanical (at 23°C)

Flexural modulus 60–65 kgf/mm^2. **Tensile strength** 2·27 kgf/mm^2. **Extension at break** 320%. **Impact strength** (Izod) No break; superior to PTFE, but inferior in *fatigue resistance*. **Hardness** Rockwell R25; Shore D 55. **Friction** against steel: μ_{stat}, 0·11; μ_{dyn}, 0·42.

*For another tetrafluoroethylene copolymer, *see* §3.85 ii.

The copolymers were introduced in the late 1950s, as materials largely resembling PTFE but much easier to fabricate. Their chief uses are as engineering materials, with low friction and good anti-stick properties, and in heat- and fire-resistant applications such as tubing and wire-covering, e.g. as electrical insulation in space-craft where temperatures from below -260 to above $200°C$ may be encountered.

3.82 POLYCHLOROTRIFLUOROETHYLENE
(PCTFE. Fluorothene, Hostaflon, Kel-F)

This material, approaching PTFE—but not equalling it—in chemical and thermal resistance, retains greater ductility and tensile strength at low temperatures; in addition, it can be dissolved in a few solvents (at elevated temperatures) and, being thermoplastic, is more tractable and amenable to fabrication. It can also be obtained in a transparent state, and is available as polymers of low molecular weight in the form of oils and waxes.

Repeat Unit

$$-CF_2-CF-$$
$$|$$
$$Cl \qquad C_2ClF_3, \text{ mol. wt.} = 116.5$$

Molecular weight $100000–200000$ (deg. of polymerisation, 850–1700).

X-ray data Moderately crystalline (restricted by asymmetry); extent depends on rate of cooling (shock-cooled amorphous form stable).

Preparation

The monomer, chlorotrifluoroethylene (C_2ClF_3; b.p. $-28°C$) is obtained by dechlorination of 1,1,2-trichloro-1,2,2-trifluoroethane (a refrigerant fluid),

$$CCl_2F-CClF_2 \xrightarrow[\text{hydrogenation (Ni cat.)}]{\text{Zn + EtOH, or vap. phase}} CClF = CF_2 + ZnCl_2$$

The colourless gas readily polymerises at room temperature but the reaction requires several hours for completion even with activated persulphate initiators. The high polymers are insoluble in the liquid monomer. The low polymers are oils and waxes, obtained either by carefully controlled polymerisation of the monomer or by pyrolysis of polymers of higher molecular weight.

Chemical Properties

Soluble in hot fluorinated solvents (e.g. 2,5-dichloro-α-trifluorotoluene at 130°C); **swollen by** aromatic and chlorinated hydrocarbons, esters and ethers; **resistant to** conc. acids, conc. alkalis and strong oxidising agents including ozone. **Identification,** *see* §3.23.

Physical Properties

Normally the polymer is opalescent (crystalline) but quenching from temperatures above 200°C provides a stable transparent (amorphous or microcrystalline) form, which reverts to the crystalline state only on heating above 150°C. **Specific gravity** 2·11 (quenched), 2·13 (crystalline; crystallites, 2·15). **Refractive index** 1·43 (transparent). **Water absorption** 0·00 %. **Permeability** to most gases and vapours is very low. Examples, as $(10^{-10}$ cm$^2)/$(s cmHg) the lower figures being for highly crystalline material, N_2, 0·003–0·13; O_2, 0·01–0·54; CO_2, 0·048–1·25; water vapour $(10^{-10}$ g)/(cm s cmHg), 0·00023––0·03. **Specific heat** 0·28. **Thermal conductivity** 4·8 $(10^{-4}$ cal)/(cm s °C). **Effects of temperature** Tougher than PTFE at very low temperatures, but maximum service temperature *c.* 150°C (under only light stress). Second-order transition, 210–218°C; softens, 270–310°C; decomposes above 300°C (approx. 100° below decomp. temp. of PTFE); at 475°C reverts to monomer and dimer.

Volume resistivity *c.* 10^{17} ohm cm; **surface resistivity,** *c.* 10^{14} ohm. **Dielectric strength** (kV/mm) 24; thin films 100. Non-tracking. **Dielectric constant** 2·4–2·7 (not greatly affected by change of frequency or temperature). **Power factor** depends somewhat on frequency; tends to fall (though not universally) with rise of temperature, e.g. 0·01–0·02 at 25°C, 0·0007–0·004 at 200°C.

The mechanical properties depend on the degree of crystallinity, amorphous material being tough and ductile while crystalline polymer is rigid (though softer than PTFE). **Elastic modulus** (Young's), *c.* 130 kgf/mm^2. **Tensile strength** (kgf/mm^2), 3·2–4·0, yields at 1·5–2·5; amorphous material weaker. **Extension at break,** *c.* 125 %.

The polymer can be moulded (230–345°C) and extruded, and is readily machined. It has good resistance to ultra-violet radiation and exterior weathering. It finds uses in hot and/or corrosive environments, for gaskets, seals, valves, etc., and as packaging films. Polymers of low molecular weight are inert oils and waxes (also employed, filled, as greases) with a comparatively high specific gravity (1·8–2·1) and low surface tension (20–30 dyn/cm) and they find uses as lubricants and seals at high temperature and/or in corrosive situations.

3.83 POLYVINYL FLUORIDE (PVF, Tedlar)

Repeat Unit

$$—CH_2—CH—$$
$$|$$
$$F$$

C_2H_3F, mol. wt. = 46

Vinyl fluoride (CH_2=CHF, b.p. $-88°C$) was obtained by F. Swarts, in 1901, by the action of zinc dust on difluorobromoethane, but nowadays it is synthesised from acetylene and hydrogen fluoride,

$$CH \equiv CH + HF \xrightarrow[\text{on charcoal}]{HgCl_2} CH_2 = CHF$$

or obtained by pyrolysis of 1,1-difluoroethane (which is also derived from acetylene and hydrogen fluoride),

$$CH \equiv CH + 2HF \longrightarrow CH_3 - CHF_2$$

$$CH_3CHF_2 \xrightarrow[\text{or } 400°C, \text{ small amount } O_2]{725°C, \text{ Pt} + \text{Cr fluorides}} CH_2 = CHF + HF$$

The monomer can be polymerised under high pressure, or in acetone solution, with a radical initiator such as benzoyl peroxide; it can also be co-polymerised, e.g. with ethylene.

Polyvinyl fluoride was first marketed in 1942 as sheets or films, the form in which it is usually employed. It is resistant to concentrated acids and alkalis, and dissolves only at elevated temperatures, e.g. in hot cyclohexanone, or dimethylformamide. Some of its properties are as follows: **molecular weight,** 60000 to 180000; **specific gravity,** 1·38; **refractive index,** 1·46; **moisture absorption,** under 0·5 % (and low water-vapour transmission). **Permeability to gases** at 20–30°C (10^{-10} cm^2)/(s cmHg), O_2, 0·02; N_2, 0·004; CO_2, 0·09. It is a partially crystalline material, resembling polyethylene rather than polyvinyl chloride. It will burn slowly in air, it softens only above 175°C and is mouldable under pressure at 200–210°C.

Films of polyvinyl fluoride when cold-drawn yield a sharp x-ray pattern, and can reach a tensile strength of 20 kgf/mm^2. They show good resistance to creasing or cracking even at $-80°C$, and they are much superior to films of polyethylene or polyvinyl chloride in resistance to sunlight and exterior weathering (thus, one of the main uses is as a protective surface-laminate on wood or metal).

3.84 POLYVINYLIDENE FLUORIDE (PVDF)

Repeat Unit
$$—CH_2—CF_2— \qquad C_2H_2F_2, \text{ mol. wt.} = 64$$

Vinylidene fluoride or 1,1-difluoroethylene (CH_2=CF_2, b.p. $-82°C$) can be obtained by reacting trichloroethylene with anhydrous hydrogen fluoride, followed by dechlorination of the dichlorodifluoroethane so formed,

$$CHCl = CCl_2 + 2HF \xrightarrow[200°C]{-HCl} CH_2Cl - CClF_2 \xrightarrow[\text{EtOH}]{Zn}$$

$$CH_2 = CF_2 + ZnCl_2$$

or by pyrolysis of 1,1-difluoro-1-chloroethane, which is made from acetylene and hydrogen fluoride,

$$CH \equiv CH + 2HF \longrightarrow CH_3CHF_2$$

$$CH_3CHF_2 + Cl_2 \longrightarrow CH_3CClF_2 + HCl$$

$$CH_3CClF_2 \xrightarrow{725°C} CH_2 = CF_2 + HCl$$

The monomer is polymerised by radical initiators (e.g. peroxides), under pressure, to yield a highly crystalline product which—except for its high **specific gravity** (1·75)—resembles polyethylene rather than polyvinylidene chloride.

The polymer resists acids, alkalis and several solvents, but is less inert than PTFE or PCTFE, being attacked by fuming sulphuric acid and by strong amines, and dissolving or swelling in polar solvents, e.g. dimethylformamide (but highly oriented polymers insoluble). Other properties: **refractive index,** 1·42; **water absorption,** 0·04%; **brittle** at −40°C, serviceable to *c.* 150°C; melts, 170°C; mouldable, 200–275°C (and resistant to oxidation at 200°C); **tensile and impact strengths** are higher than for PTFE and PCTFE; like PVF this material shows good resistance to outdoor weathering. Copolymerisation of the monomer can reduce the rigidity and provides important rubbers (*see* below).

A monomer closely related to vinylidene fluoride is vinylidene chlorofluoride or 1-chloro-1-fluoroethylene ($CH_2=CClF$, b.p. −24°C) which can be obtained from vinylidene chloride (§12.21) as follows,

$$CH_2 = CCl_2 \xrightarrow{Br_2} CH_2Br—CCl_2Br \xrightarrow[(HgO)]{HF} CH_2BrCCl_2F \xrightarrow[EtOH]{Zn}$$
$$CH_2 = CClF$$

This monomer readily polymerises in the presence of radical initiators, and, being asymmetrical in structure, gives rubbery products.

3.85 FLUORO-RUBBERS

Outstanding advantages offered by most of the fluoro-rubbers are that, even at relatively high temperatures (150–200°C), they are oil-resistant and chemically inert. Thus they are not only non-swelling in many common organic solvents but are also little affected by strong acids and alkalis or by powerful oxidising agents such as fuming nitric acid and ozone. This superior thermal stability and chemical resistance is attributed in part to the high energy of the F—F bond (107 kcal/mol, compared with 87 for the C—H bond and 70–80 for the C—C bond).

Various types of fluoro-rubbers available commercially are indicated below.

(i) **Vinylidene fluoride copolymers** (for homopolymers, *see* §3.84).

The best-known of these are the 30/70 hexafluoropropene/vinylidene fluoride copolymer (Fluorel, Viton), and the 30/70 and 50/50 chlorotrifluoroethylene/vinylidene fluoride copolymers (Kel-F Elastomer 3700 and 5500 respectively, the first of these providing the better tensile strength and low-temperature flexibility). The molecular weight of the fluoropropene copolymer is *c.* 60000 and that of the other copolymers is *c.* 750000 to 10^6.

The unvulcanised materials, translucent and off-white in colour, have a comparatively high specific gravity (1·80–1·86) and will dissolve in such solvents as hexafluorobenzene, methyl ethyl ketone, and certain esters. Vulcanisation is effected with organic peroxides (e.g. benzoyl peroxide, with ZnO present), amines (e.g. tetramethylenepentamine, hexamethylenediamine carbamate) or di-isocyanates; cure in the mould, for 15–30 min at 150°C, needs following with an oven cure for several hours at 150–250°C to improve the properties at elevated temperatures.

The cured rubbers can be used over a wide range of temperature, the thermal resistance of the hexafluoropropylene copolymer being exceptionally good (e.g. 10% weight loss in 2 h at 360°C), but the low-temperature flexibility is relatively poor (brittle point, -20 to -50°C) and at elevated temperatures—although serviceable at 200°C (for short periods up to 350°C)—they show poor resilience. The rubbers are very resistant, even at elevated temperatures, to most oils (though swollen by aromatics and some chlorinated hydrocarbons) and to chemical reagents, resisting even conc. oxidising acids (e.g. fuming HNO_3) and conc. alkalis (although embrittled and decomposed by anhydrous hydrazine). Permeability to gases is relatively high, e.g. Viton at 20–30°C $(10^{-10}$ cm^2)/(s cmHg), O_2, 1·5; N_2, 0·44; CO_2, 7·8.

(ii) **Nitroso rubber**

This is a 1:1 copolymer, of high molecular weight, with the repeat unit

$$-CF_2-CF_2-N-O-$$
$$|$$
$$CF_3$$

It is obtained by radical-initiated copolymerisation of tetrafluoroethylene* and trifluoronitrosomethane at or below 0°C (at *c.* 100°C a cyclic oxazetidine, $\overline{CF_2CF_2N(CF_3)O}$, is formed).

Nitroso rubber is incombustible even in oxygen, resists solvents and chemical attack, and is serviceable to low temperatures (T_g, -51°C). However, it is inferior to other fluoro-elastomers in thermal resistance;

---

*For another tetrafluoroethylene copolymer, *see* §3.81.

e.g. 20 h at 200°, negligible weight loss; 2 h at 230°, 10–25% loss; at 270°C, rapid disintegration.

(iii) **Butadiene-based fluoro-elastomers**

Examples of these are:

(*a*) **polyfluoroprene,** a fluorine equivalent of neoprene, in which the principal repeat unit, arising from 1,4-polymerisation, is represented as

$$-CH_2-CF = CH-CH_2-$$

(also some 1,2- and 3,4- structures).

and (*b*) **a copolymer of octafluorocyclohexa-1,3-diene with butadiene,** with repeat unit represented by

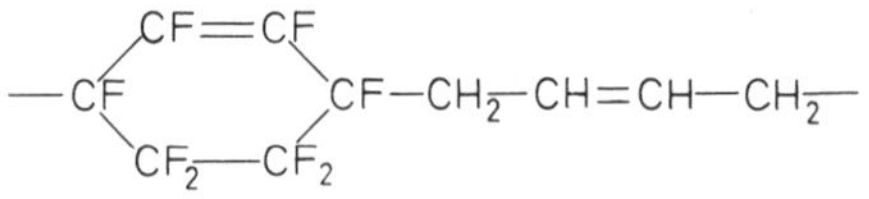

These show good oil and chemical resistance but their thermal stability is not exceptional, and an increase of the F-content is detrimental to low-temperature serviceability.

(iv) **Polyfluoroacrylates** *see* §9.82.

(v) **Fluoro-silicones** *see* Section 41.

3.9 FURTHER LITERATURE

BARSON, C. A. and PATRICK, C. R., *Brit. Plas.* **36**, 70 (1963) (review of chemistry of fluoro-organic polymers)

KENNEDY, J. P. and TÖRNQUIST, E. (Eds.), *Polymer Chemistry of Synthetic Elastomers* (High Polymers, Vol. XXIII), New York, 273 (1968) (COOPER, J. R., fluorine-containing elastomers)

POLYSTYRENE

With notes on blends (including ABS plastics), copolymers of styrene (with acrylonitrile,
divinylbenzene and maleic anhydride), related materials (polymethylstyrenes,
polychlorostyrenes, polyvinylnaphthalene and polyacenaphthylene), polyvinylpyridines,
poly-N-vinylcarbazole, and poly-p-xylylenes

SYNONYMS AND TRADE NAMES
Polyvinylbenzene, polyphenylethylene. Afcolene, Bextrene, Carinex, Distrene, Dylene, El Rex, Erinoid, Lacqrene, Lorkalene, Luran, Lustrex, Piccolastic (low molecular weight resins), Restirolo, Starex, Styron.

GENERAL CHARACTERISTICS
Transparent (frequently pigmented), colourless, glass-like material. A moulded sample produces a metallic 'tinkle' when struck or dropped.*

4.1 STRUCTURE

Simplest Fundamental Unit

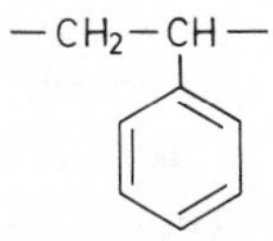

C_8H_8, mol wt = 104

Molecular weight Up to 10^6 or more; commercial moulding grades, 200000–300000 (degree of polymerisation up to 3000).

X-ray data Commercially available polymers (prepared by radical initiation) are substantially amorphous, even when stretched. A crystalline isotactic form can be obtained by stereospecific polymerisation.

4.2 CHEMISTRY

4.21 PREPARATION

Styrene or vinylbenzene ($C_6H_5CH=CH_2$; m.p. $-30°C$, b.p. $145°C$, sp. gr. 0·91) was originally obtained from storax balsam (oriental sweet gum), a fragrant viscid exudate from *Liquidambar orientalis*, a tree of Asia Minor. The balsam contains cinnamic acid, and styrene may be prepared in the laboratory by thermal decarboxylation of the acid.

*Poly-4-methylpentene-1 (§2.82) makes a similar sound.

$$C_6H_5CH = CHCOOH \xrightarrow{-CO_2} C_6H_5CH = CH_2$$

Industrially it is obtained by reaction of ethylene (from petroleum or natural gas) and benzene (from coal or synthetically from petroleum), in the presence of a catalyst of the Friedel-Crafts type, to yield ethylbenzene, which is then dehydrogenated,

$$C_6H_6 + C_2H_4 \xrightarrow{AlCl_3} C_6H_5C_2H_5$$

$$\xrightarrow[\text{H}_2\text{O and Fe at 600–650°C)}]{-H_2 \text{ (catalytic action of}} C_6H_5CH = CH_2$$

Variations of the route include oxidative dehydrogenation of the ethylbenzene. Styrene may also be synthesised directly from acetylene via vinylacetylene, or from acetylene and benzene; some styrene is formed during the aromatisation of aliphatic hydrocarbons, and it is a minor component of coal tar.

Styrene polymerises on heating or exposure to ultra-violet radiation, or in the presence of free radicals; oxygen inhibits polymerisation and should be absent, but to prevent polymerisation during storage an inhibitor such as sulphur, quinol or *p-tert.* butylcatechol is added. The main commercial processes, using radical initiators, employ mass (batch or continuous), solution or suspension polymerisation. Mass polymerisation yields a transparent product but a broad molecular weight distribution (which leads to difficulties during moulding). Solution and suspension processes permit greater control of molecular weight but the clarity of the polymer is impaired. Emulsion polymerisation is also used, particularly for the preparation of copolymers with butadiene (*see* §34.21). In all forms of polymerisation the temperature is important in determining the molecular weight and physical properties of the product. Low temperatures, resulting in a slow rate of reaction and few side effects, yield tough polymers of high molecular weight which are difficult to process; high temperatures give brittle polymers of lower molecular weight.

4.22 PROPERTIES

Soluble in aromatic and chlorinated hydrocarbons (e.g. toluene, perchloroethylene, carbon tetrachloride), carbon disulphide, tetralin, dioxan, methyl ethyl ketone, pyridine, cyclohexanone, ethyl acetate. **Plasticised by** common ester and hydrocarbon plasticisers (cf. §11.22); however, except for certain applications, plasticisers are seldom used as they are detrimental to mechanical properties (but *see* §4.81). **Swollen by** oils, ketones, esters, some aliphatic hydrocarbons (e.g. *n*-hexane), cyclohexane, nitrobenzene. **Relatively unaffected by** water, alcohols*, alkalis, non-oxidising acids, aq. solutions of inorganic salts, aliphatic hydrocarbons (some)*. **Decomposed by** prolonged contact or boiling with oxidising agents (e.g. conc. HNO_3 or H_2SO_4).

4.23 IDENTIFICATION

Thermoplastic. Cl, N, S and P absent. **Combustion** Softens and burns fairly readily, with an extremely sooty flame; characteristic odour of styrene noticeable on extinction. **Pyrolysis** Becomes rubbery at *c.* 80°C, then fluid; decomposes at 310–350°C and distils as monomer, dimer, 1,3-diphenyl-1-butene, and other products; neutral condensate, with the characteristic smell of styrene pervading (derivative, dibromo-styrene, m.p. 73–74°C).

Colour test A small sample (e.g. 0·1 g) is refluxed with conc. nitric acid (5 ml), and the clear solution that results (after *c.* 1 h) is poured into water (20 ml), producing a pale yellow precipitate. The aq. suspension, which contains *p*-nitrobenzoic acid, is extracted with ether (10 + 10 ml successively) and the combined ether extracts are washed with water (5 + 5 ml), then extracted with dil. sodium hydroxide (5 + 5 ml), and again with water (5 ml). The alkaline extracts and final aq. extract are combined, and the nitro-derivatives present in them are reduced by acidifying with conc. hydrochloric acid (5 ml), adding granulated zinc (1 g), and warming on a steam bath until effervescence ceases (*c.* 30 min). The acid solution, which contains *p*-aminobenzoic acid, is then filtered, cooled in ice, and diazotised by slow addition of dil. sodium nitrite, after which it is poured into an excess of alkaline *β*-naphthol solution. Styrene (and its copolymers) produces a rich red colour. *N.B.* An acid extract from an aniline/formaldehyde resin (§27.83) will produce the same result, but without requiring nitration and subsequent reduction.

4.3 PHYSICS

The values quoted below are for typical samples moulded from general-purpose grades; data for high-impact grades and ABS plastics (§4.81) are in some instances included for comparative purposes.

4.31 GENERAL PROPERTIES

Specific gravity 1·04–1·06; h. impact, do.; ABS, 1·12. **Refractive index** 1·59. *Optical transmission* of visible light through 2·5 mm

---

*Stressed polystyrene can exhibit *environmental stress cracking* (cf. §1.22) in the presence of liquids or vapours which otherwise have little or no effect. Such substances include the lower alcohols, paraffins, white spirit, fats, and milk products. High-impact polystyrene and ABS plastics (*see* §4.81) are less susceptible to this form of deterioration.

sheet ($\%$), 80–90; h.impact, 55–60. *Dispersion*, $(n_D - 1)/(n_F - n_C) = 31$.
Water absorption 24 h immersion at 23–25°C ($\%$), 0·03–0·05; h.impact, 0·05–0·08; ABS, 0·2–0·35.

Permeability The permeability to gases is comparable with that of low density polyethylene, polypropylene, and butyl rubber. Permeability of films at 20–23°C (10^{-10} cm^2)/(s cmHg): O_2, 0·6–1·7, N_2, <0·1–0·4 (oriented, 0·7); CO_2, 12 (25); H_2, 16. Permeability to water vapour at 20–23°C (10^{-10} g)/(cm s. cmHg), 0·8.

4.32 THERMAL PROPERTIES

Specific heat 0·32; h.impact, do.; ABS, 0·38. **Conductivity** (10^{-4} cal/cm s°C)) 1·9–3·3; h.impact, 1–3; ABS, 4–7. **Coefficient of linear expansion** (10^{-4}/°C), 0·7; h.impact, do.; ABS, 0·8. **Physical effects of temperature** For reduction of brittleness at low temperature, *see* §4.81. Second order transition* to rubbery state (i.e. glass temperature; *see* Section 68) above 80°C (fibres and films shrink and disorient); soft above 100°C; above 150°C in air, darkens, but with little further change until *c*. 250°C when it partly reverts to monomer; in a vacuum (absence of O_2) there is no appreciable liberation of styrene until above 300°C. Typical heat distortion temperatures under stress of 0·17 kgf/mm^2 (°C), 75–95; h.impact 65–90; ABS, 75–110. Maximum continuous service temperatures, unstressed conditions (°C), 65–75; h.impact, 50–75; ABS, 70–105.

4.33 ELECTRICAL PROPERTIES

Volume resistivity (ohm cm) 10^{15}–10^{19}; h.impact, *c*. 10^{16}; ABS, 2·10^{14}–10^{15}. *Surface resistivity* (ohm) 10^{15}: ABS, 2·10^{12}. **Dielectric strength** (kV/mm) >30; h.impact, 18–27; ABS, 35.

Dielectric Constant (DC) and Power Factor (PF)

Frequency (Hz):	60	10^3	10^6
DC Polystyrene	2·5–2·6	*c*. 2·4	2·4–2·65
H. impact do.	2·5–3·5	—	*c*. 2·6
ABS	2·9–4·9	—	3·7–4·1
PF Polystyrene	0·0001–0·0005	0·0001–0·0003	0·0001–0·0004
H. impact do.	0·003–0·005	—	0·0009–0·001
ABS	0·005–0·007	—	0·07–0·08

*Other transitions at 40, 97 and 160°C. (SIMPSON, W., *Chem. and Ind.*, 215 (1965)).

4.34 MECHANICAL PROPERTIES

The values given below are typical, but as the rigidity of polystyrene restricts the release of internal stresses induced during processing (particularly during injection moulding) the mechanical properties of fabricated components can vary considerably (*see* SCHMITT, B., SCHUSTER, R., ORTHMANN, H. J., *Kunststoffe*, **54**, 643 (1964); **55**, 779 (1965)).

Elastic modulus (Young's) (kgf/mm^2) 280–350; h.impact, 270–320; ABS, 190–270. **Tensile strength** (kgf/mm^2) 3·2–5·0; h.impact, 2·1–3·5; ABS, 3·9–5·2, at yield. **Extension at break** (%) 2–3; h.impact, 20–40, with yield at 1·5; ABS, 3–10, at yield. **Flexural strength** (kgf/mm^2) 5–10·5; h.impact, 3·2–4·2; ABS, 5·7–8·0, at yield. **Compression strength** (kgf/mm^2) 8–11; h.impact, 5·6–11; ABS, 4·2–8. **Impact strength** Izod, notched (ft lbf/in) 0·25–0·5; h.impact, 0·8–1·7; ABS, 2·5–7·5. **Hardness** Rockwell M 65–80; h.impact, M 40–60; ABS, R 90–105. **Friction** μ_{stat}, polymer/polymer, 0·5; μ_{dyn}, polymer/steel, 0·46; h.impact, 0·50; ABS, 0·50. Grey cotton yarn (1 m/s)/polymer, *c*. 0·25. The coefficient of friction is not much affected by the presence of lubricants.

4.4 FABRICATION

Polystyrene is a tractable material, with excellent flow characteristics and thermal stability at processing temperatures, and relatively low shrinkage on cooling. Injection moulding is carried out at 150–280°C; moulding stress or orientation (cf. §4.34) is minimised by appropriate design of moulds, the use of heated moulds, a high filling rate, or annealing of the finished product. Other processing temperatures: extrusion, 150–260°C; blow-moulding, 190–235°C (die) and 150–215°C (barrel). Unmodified grades of polystyrene tend to be brittle and do not machine well; some improvement is obtained on annealing. Extruded high-impact sheets are extensively used for thermoforming. Polystyrene can be bonded to itself by the use of solvents, and to other substrates by means of rubber-type and related adhesives. Care is required to avoid the risk of environmental stress cracking (*see* §4.22) when using solvent-based adhesives or lacquers.

4.5 SERVICEABILITY

Prolonged exposure to sunlight (or fluorescent lighting) causes yellowing of clear polystyrene and fading of pigmented types, often with crazing and impairment of mechanical properties, and in high-impact grades the presence of unsaturated rubbers accelerates degradation; ultra-violet stabilisers (e.g. amines, phenols, benzo-phenones) give some protection. On exposure to high energy radiation some cross-linking takes place, but polystyrene is essentially one of the most radiation-resistant of polymers due to the protective action of the benzene ring. Micro-organisms and insects have little effect, although surface growths of fungi can impair electrical properties.

4.6 UTILISATION

As one of the main general-purpose thermoplastics, polystyrene is surpassed in usage only by PVC and polyolefins. Its most important applications are in packaging (including disposable cups for vending machines), domestic electrical appliances (e.g. casings for radio sets, refrigerator liners), toys, electronic equipment (particularly as low-loss insulation for high-frequency apparatus) and building (e.g. wall tiles, light-diffusing panels). Expanded polystyrene is extensively used as thermal insulation in buildings and refrigerated vehicles, acoustic tiles for ceilings, and shock absorbing packaging. Latices and solutions of low molecular weight polymers find uses in surface coating (e.g. as paints and on paper).

4.7 HISTORY

Styrene was obtained by M. Bonastre in 1831, by distillation of storax balsam (*see* §4.21); it was also examined in 1839, by E. Simon, who observed that after storage for several months the liquid became a solid (which he wrongly supposed to be an oxide). In 1845, by heating styrene, J. Blyth and A. W. Hofmann obtained—even in the absence of oxygen—a similar solid (which they termed metastyrene). Practical uses of the polymer were considered by A. Kronstein in 1902 and F. E. Matthews in 1911, but the problem of preventing premature polymerisation of the monomer impeded development until C. Dufraisse and C. Moureu used aromatic amines and phenols as inhibitors in 1922. Important studies of the mechanism of the poly-merisation of styrene were made by I. I. Ostromislenskiï (*c.* 1911–25) and H. Staudinger (*c.* 1929–35).

Large scale commercial development dates from 1935, when the Dow Chemical Co. commenced production of high-purity monomer based on a synthesis devised by M. Berthelot in 1851 (earlier attempts to market polystyrene having met with little success because of poor ageing characteristics). During the Second World War the production of styrene was greatly increased, in the U.S.A. and Germany, for use in the manufacture of synthetic rubbers, and in the post-war period the availability of the low-cost monomer led to rapid growth in the production of polystyrene. High-impact grades were introduced in the U.S.A. in 1948. Isotactic polystyrene was obtained by G. Natta in 1955. It is estimated that current world consumption of styrene-based plastics exceeds 500000 tons per annum.

4.8 ADDITIONAL NOTES

4.81 HIGH-IMPACT POLYSTYRENE (INCLUDING BLENDS, COPOLYMERS AND COPOLYMER BLENDS)

Incorporation of an elastomer such as natural or styrene/butadiene rubber makes polystyrene less brittle; however, the compatibility of such additives is limited and maximum toughness is developed only if styrene is first graft-copolymerised on to the elastomer, e.g. *see* §29.85. A blend, consisting of polystyrene, ungrafted rubber and a graft-copolymer, which exhibits high impact strength, is considered to do so because of (i) good adhesion—promoted by the presence of the graft copolymer—between the two homopolymers and (ii) the presence of very small particles of rubber that prevent or disperse any high concentration of stress and thus lower the possibility of crack propagation. Some materials of this kind are listed below. For physical properties *see* §4.3.

Styrene/acrylonitrile copolymers (SA or SAN Resins. Restil, Terluran, Tyril). Polystyrene shows poor resistance to organic liquids and is readily moulded (softens *c.* 80°C) whereas poly-acrylonitrile is solvent-resistant and difficult to mould. Hence, copolymers of styrene and acrylonitrile—e.g. in the respective proportions of 70/30 or 76/24 by weight—are found to be rigid transparent thermoplastics resembling polystyrene but superior in certain ways, e.g. harder, less prone to crazing, more resistant to heat (soften *c.* 95°C) and more resistant to reagents and organic liquids (less swollen by oils); impact strength is also improved but is still relatively low (under 0·8 ft lbf/in, Izod, notched).

These materials are injection-moulded and extruded as 'up-graded' forms of polystyrene. A further development is that of moulding compounds based on styrene, acrylonitrile and an acrylic ester graft-copolymerised on to an elastomer; the most important characteristic of these products (ASA) is improved resistance to light ageing.

Styrene/copolymer blends (ABS Plastics. Abson, Blendex, Cycolac, Kralastic). Since *styrene/butadiene* and *acrylonitrile/butadiene* copolymers constitute important rubbers (Sections 34 and 35 respectively) inclusion of butadiene might be expected to increase the toughness of a styrene/acrylonitrile copolymer. Terpolymers of the three monomers are disappointing, being tough only at the expense of rigidity, but—

as with high-impact polystyrene, above—good results are obtained where an elastomeric phase (lightly cross-linked) is discretely dispersed in a continuous phase of styrene/acrylonitrile copolymer, to which it exhibits good adhesion.

Of such mixtures there are two main types as follows.

(i) *Polyblends* (*Type B*). These consist of a butadiene-based rubber (usually nitrile rubber) physically dispersed in a styrene/acrylonitrile copolymer.

(ii) *Graft copolymer mixes* (*Type G*). These consist of a butadiene-based rubber (usually polybutadiene) graft-copolymerised with styrene/acrylonitrile chains, which—along with ungrafted poly-butadiene—is physically dispersed in a styrene/acrylonitrile copoly-mer. Typical composition (%): acrylonitrile, 20–30; butadiene, 20–30; styrene, 40–60.

In both instances light cross-linkage (usually effected during the initial polymerisation) restricts dissolution of the rubbery particles, while the graft copolymerisation of polybutadiene improves its adhesion to the continuous phase of the copolymer.

ABS plastics, tough and resilient, and chemically resistant but readily processed, are employed in the form of injection mouldings, extruded pipes and profiles, and thermoplastic sheets that can be vacuum formed; the main applications are of an engineering kind, and include domestic impact-resistant goods for use at room temperature. For physical properties *see* §4.3.

An analogous material (MBS), in which methyl methacrylate replaces acrylonitrile, combines some of the advantages of ABS (e.g. impact strength) with improved transparency.

Further Literature

C. H. BASDEKIS, *ABS Plastics*, New York and London, 1964.

4.82 FURTHER COPOLYMERS OF STYRENE*

Styrene/divinylbenzene copolymers Styrene or vinylbenzene is essenti-ally difunctional and normally yields only linear polymers, but inclusion of a tetrafunctional compound such as *p*-divinylbenzene,

produces a three-dimensional (cross-linked) network. As little as

---

See also copolymers with drying oils, §24A.81; with unsaturated polyesters, §24D.21; with butadiene, Section 34.

0·01 % divinylbenzene can render the product insoluble but swellable in benzene, while 1 % renders it glass-like and virtually unaffected by organic liquids. These copolymers find uses as matrices for ion exchange resins (e.g. can be sulphonated) and as a machinable dielectric serviceable at a temperature of 120°C or more.

Styrene/maleic anhydride copolymers. It is very difficult to induce maleic anhydride,

$$
\begin{array}{c}
CH\!\!=\!\!CH \\
\diagup \qquad \diagdown \\
CO \qquad CO \\
\diagdown \quad \diagup \\
O
\end{array}
$$

to polymerise but it will readily copolymerise in solution with styrene. The product, soluble in water as an ammonium or alkali metal salt, with a molecular weight of 10 000 and a softening temperature of 190°C, finds uses in floor polishes, emulsification, and as a protective colloid (Lytron 822) also as a textile size (Stymer S).

4.83 MATERIALS RELATED TO POLYSTYRENE

(i) The following polymers of styrene derivatives show greater thermal resistance than polystyrene (softens c. 80°C) but have attained little importance commercially.

Poly-α-methylstyrene*

The monomer, α-methylstyrene ($CH_2\!\!=\!\!C(CH_3)C_6H_5$, b.p. 163°C) differs from styrene in that by itself it does not respond to free radical initiation, though in the presence of styrene copolymerisation occurs; however, when initiated cationically (e.g. with conc. H_2SO_4 or a Lewis acid) at low temperature—as with isobutene (§38.83)—it yields polymers of high molecular weight. Poly-α-methylstyrene and styrene/α-methylstyrene copolymers have appreciably higher softening temperatures than polystyrene (the homopolymer softening at 160–170°C or higher—increasing inversely with the temperature of polymerisation) and the monomer and its polymeric products have been used when an improvement in thermal resistance is required, e.g. in methylstyrene/acrylonitrile copolymers and ABS-type plastics. α-Methylstyrene is also used as a co-monomer (more compatible than styrene) for the modification of drying oils (§24A.81).

Poly-o-, m- and p-methylstyrenes (Polyvinyltoluenes)

o-Methylstyrene ($CH_2\!\!=\!\!CHC_6H_4CH_3$, b.p. 169°C at 752 mmHg) yields polymers with a softening temperature c. 20°C above that of

---

See also copolymers with butadiene, §34.84.

polystyrene. The other isomers are of less interest, *m*-methylstyrene yielding polymers softening at *c*. 60°C while the polymers of *p*-methylstyrene resemble those of styrene, but a 3/1 mixture of *m*- and *p*-methylstyrenes (b.p. 171°C, known commercially as *vinyltoluene*) readily copolymerises with drying oils (§24A.81). *o*-Methylstyrene, and α-*p*-dimethylstyrene (or 2-*p*-tolylpropene, $CH_2{=}C(CH_3)C_6H_4{-}CH_3$, b.p. 184°C) have been used to overcome the inadequate heat resistance provided by styrene in certain copolymers, e.g. with butadiene.

Polychlorostyrenes

2,5-Dichlorostyrene ($CH_2{=}CH$ $C_6H_3Cl_2$, b.p. 72°C at 2 mmHg) yields polymers that have been investigated for their hardness, impact and flame-resistance, and relatively high softening temperature (130–150°C).

(ii) *Polyvinylnaphthalene and Polyacenaphthylene*

These amorphous substances resemble polystyrene and, although of no commercial importance, are of interest in that they illustrate the influence of aromatic groups other than phenyl.

α-Vinylnaphthalene ($CH_2{=}CH$ $C_{10}H_7$, b.p. 125°C at 15 mmHg)

can be obtained by dehydrogenation of α-ethylnaphthalene, and by other methods, and in the polymer obtained from it (repeat unit I) the large pendant group raises the softening temperature to 120–130°C. Acenaphthylene ($CH{=}CHC_{10}H_6$; m.p. 93°, b.p. *c*. 270°C) is obtained by vapour-phase oxidation of acenaphthene ($CH_2CH_2C_{10}H_6$; m.p. 95°, b.p. 278°C) a reactive hydrocarbon occurring in the heavy oil fraction (265–290°C) of coal tar and in a similar cut from cracked petroleum. The monomer can be polymerised (repeat unit II) by heat or radical initiators, also cationically at low temperature, and the products again resemble polystyrene except that they are yellow or amber-coloured, and the presence of the 5-membered ring structure in the main chain raises the softening temperature to *c*. 250°C. The monomer has been copolymerised with styrene to raise the thermal resistance.

4.84 POLYVINYLPYRIDINES

These materials resemble polystyrene but changing the phenyl group to a pyridyl group (i.e. CH replaced by N in either the *o, m* or *p* regions) makes them harder, raises the softening temperature, and introduces the possibility of ionisation along the polymer chain.

The three isomeric monomers are obtained by reaction of 2-, 3- or 4- picoline (= methylpyridine) with formaldehyde, to yield the appropriate β-hydroxyethylpyridine ($HOCH_2CH_2C_5H_4N$) which on dehydration with alkali gives 2-, 3- or 4-vinylpyridine ($CH_2=CH\ C_5H_4N$); these monomers, which are lachrymatory and vesicant liquids, are polymerisable with radical initiators to give the corresponding polymers. The monomers are sometimes used in minor proportions in copolymers, e.g. with acrylonitrile (*see* §10.21), also in co- and ter-polymers, e.g. with butadiene and/or styrene (notably to improve adhesion between tyre cords and rubber, *see* §34.83).

The presence of the pyridyl group means that the polymers can be quaternised to provide water-soluble (cationic) polyelectrolytes, e.g.

When dissolved in water these products yield viscous solutions but addition of a salt (e.g. KCl) suppresses the ionisation—and, with it, the mutual repulsion along each chain—and can markedly *lower* the viscosity.

4.85 POLYVINYLCARBAZOLE
(Poly-N-vinylcarbazole. Luvican, Polectron)

This contains nitrogen in the substituent group. It resembles polystyrene but is harder, and superior in thermal resistance, it is amorphous in structure and the **simplest fundamental unit** is as follows,

$$C_{14}H_{11}N, \text{ mol wt} = 193$$

Preparation

Carbazole ($C_{12}H_8NH$; m.p. 246°, b.p. 355°C) is extracted from the anthracene oil fraction of coal tar and can be purified by reaction with an alkali metal, usually potassium; it can also be synthesised by reaction between cyclohexanone and phenylhydrazine, followed by dehydrogenation of the tetrahydrocarbazole so produced. Potassium carbazole can be reacted with vinyl chloride, in an inert solvent (white spirit, hexahydroxylene) under heat and pressure, to yield the monomer and potassium chloride; but in industrial practice carbazole is usually vinylated with acetylene, in solution as previously and in the presence of an alkaline catalyst, e.g.,

Removal of the solvent and recrystallisation from methanol yields the monomer, N-vinylcarbazole ($C_{12}H_8N\ CH{=}CH_2$; m.p. 67°C), which in the presence of radical initiators polymerises on heating (in the molten state in aq. emulsion); polymerisation by fast electron radiation of the monomer has also been used. The polymer can be moulded at 210–270°C; copolymerisation with a minor proportion of a monomer such as isoprene reduces brittleness in the product.

Principal Properties

Soluble in aromatic and chlorinated hydrocarbons, and in tetrahydrofuran. **Relatively unaffected by** aliphatic hydrocarbons (incl. mineral oils), water, dil. acids and alkalis, alcohols, esters. **Decomposed by** strong oxidising agents (conc. H_2SO_4, conc. HNO_3, sodium chlorite solutions). **Specific gravity** 1·2. **Refractive index** 1·69. **Softening temperature** *c.* 200°C (glass temperature, as change in rate of thermal expansion, at 85°C). **Volume resistivity** *c.* 10^{16} ohm cm. **Dielectric strength** (thin films) 40–50 kV/mm. **Dielectric constant** *c.* 3·0. **Power factor** *c.* 0·001 (some variation with frequency and small increase above 180°C).

Applications

The polymer was developed because the volume resistivity, along with resistance to oil, water, and elevated temperatures, render it suitable for high temperature electrical insulation. It can be injection moulded (at 220–280°C, giving strong machinable products), and the monomer has been used as a condenser impregnant that is polymerised *in situ*. Spun fibres have been employed as an asbestos substitute (though not for use at high temperatures).

4.86 POLY-*P*-XYLYLENE (PPX)

The thermal stability of linear polymers tends to be raised by the presence of cyclic structures linked diametrally in the chain (e.g. 1,4-linked aromatic rings; *see* Sections 24B, 24C and 25, also §23.82b) since the alignment of the links—equivalent to a valency angle of 180 degrees—restricts chain flexibility (although the rings themselves may rotate).

Poly-*p*-phenylene* (repeat unit I) and poly-*p*-methylenephenylene (II) are in this category and have been shown to have high melting points, but a more promising substance is poly-*p*-xylylene (III).

(I)   (II)   (III)

First described by JACOBSON, R. A., (*J. Am. Chem. Soc.,* **54**, 1513 (1932)) and later by SZWARC, M., (*Disc. Farad. Soc.,* **2**, 46 (1947)), poly-*p*-xylylene can be prepared in several ways, e.g. by (i) Wurtz reaction between a *p*-xylylene dihalide and sodium, (ii) a Hofmann degradation of a *p*-methylbenzyltrimethylammonium halide refluxed with alkali, (iii) vacuum pyrolytic dehydrogenation of *p*-xylene, or (iv) vac. pyrolysis of cyclic di-*p*-xylylene, as obtained from pyrolytic dehydrogenation of *p*-xylene in steam at 950°C. The reaction product polymerises spontaneously, i.e. on quenching from high temperature or when radical-bearing vapour, as obtained from (iv), condenses on cold surfaces.

The polymers, which are white or cream coloured, show a second order transition at 60–70°C, but on heating the amorphous form becomes converted to a crystalline state (α-form, soluble in boiling amylnaphthalene, and a more stable β-form) and the ultimate melting point is over 400°C. Certain of the polymers can be moulded at 200°C; they are mostly insoluble but above 300°C dissolve in certain esters (benzyl benzoate), chlorinated diphenyls, and terphenyl (b.p. 387°C). The **specific gravity** is reported as 1·14; oriented samples show negative birefringence. The **volume resistivity** and **dielectric strength** are high (10^{18} ohm cm and 40 kV/mm respectively); the **dielectric constant** and **power factor** are low (*c.* 1·7 and 0·0002 respectively, over a wide frequency range) but, depending on impurities according to the method of preparation employed, may be higher (up to 4 and 0·002). In air, under high voltage stress or electron bombardment, the polymer exhibits a much higher stability than polystyrene.

*For polyphenylene oxide, *see* §26.83.

4.9 FURTHER LITERATURE

BOUNDY, R. H. and BOYER, R. F., *Styrene*, New York (1952)
GIBELLO, H., *Le Styrène et ses Polymères*, Paris (1956).
TEACH, W. C. and KIESSLING, G. C., *Polystyrene*, New York and London (1960)

Specifications
BS 1493 (moulding materials)
BS 3126 (toughened moulding materials)
BS 3241 (toughened sheets)
BS 4041 (mouldings)
ASTM D703 (moulding and extrusion materials)
ASTM D1788 (ABS plastics)
ASTM D1431 (SAN plastics)

INDENE AND COUMARONE/INDENE RESINS

*With notes on hydrogenated resins and some allied resins (olefin,
cyclopentadiene, terpene, and cyclohexanone resins)*

SYNONYMS AND TRADE NAMES
Catarex, Cumar, Epok C, Piccoumaron.

GENERAL CHARACTERISTICS

These polymers, of relatively low molecular weight, range from viscous liquids to hardish brittle resins and from clear pale yellow to dark brown in colour. The resins are also available as aq. dispersions.

5.1 STRUCTURE

Simplest Fundamental Unit

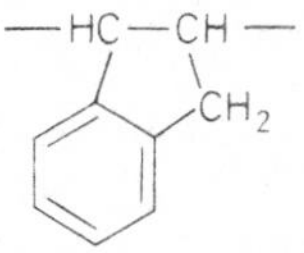

Polyindene: C_9H_8,
mol. wt. = 116

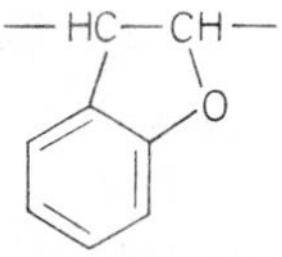

Polycoumarone: C_8H_6O,
mol. wt. = 118

N.B. More complex structures may be present. *See also* §5.61.

Molecular weight Commercial resins, up to *c.* 1000 (degree of polymerisation under 10). Experimental resins from purified indene, several thousand (d.p. up to *c.* 35).

X-ray data Substantially amorphous substances.

5.2 CHEMISTRY

5.21 PREPARATION

Indene (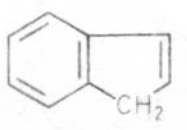 , m.p. −2°C, b.p. 182°C) occurs in the naphtha

fraction of coal tar, being obtained for instance from the coking industry. Also present, in smaller proportion, is coumarone or benzofuran (⌬ , m.p. $-18°C$, b.p. $174°C$), and because of the closeness of the boiling points—and the similarity of the end-products—the monomers are not usually separated from each other (some methyl derivatives, dicyclopentadiene and traces of styrene may also be present). If it is required, indene can be either frozen out or separated as sodium indene following reaction with sodamide. Indene, free from coumarone but containing other resin-forming hydro-carbons (*see* §5.82), is also available from high temperature cracking of petroleum.

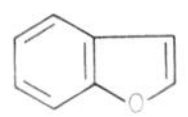

The purified indene or the indene/coumarone fraction is polymer-ised cationically, in solution (e.g. 30% in naphtha) at $0°C$ or room temperature, by addition of an initiator of the Friedel-Crafts or Lewis-acid type (e.g. 1% of conc. H_2SO_4; or $AlCl_3$, $SnCl_4$ or BF_3 ethyl etherate). Heat and ultra-violet light, in conjunction with a radical initiator, have a similar effect with indene; coumarone is less reactive.

5.22 PROPERTIES

Dissolved by aromatic and chlorinated hydrocarbons (also by aromatic/aliphatic hydrocarbon blends), ketones (hydrogenated resins are insoluble in acetone), esters, dioxan, pyridine, drying oils. **Plasticised by** common ester-type plasticisers, castor oil, drying oils; compatible with alkyd resins and some waxes. Swollen by aliphatic hydrocarbons (hydrogenated resins dissolve). **Relatively unaffected by** water, lower alcohols, most acids and alkalis; incompatible with polyethylene and paraffin wax. **Decomposed by** conc. oxidising acids.

5.23 IDENTIFICATION

Thermoplastic. Cl, N, S and P absent (or N, S in trace amount). Unsaponifiable, but unsaturated (*see* §5.81). **Combustion** Resins melt, burn with a smoky flame, produce tarry odour on extinction. **Pyrolysis** Resins melt, ultimately depolymerise, yielding neutral distillate, with tarry odour (commercial resins may yield acid or phenolic distillate); treatment with aqueous picric acid yields indene picrate (m.p. $96–98°C$) and—if present—coumarone picrate (m.p. $102–103°C$).

N.B. Hazard, picrates liable to explosion.

Passing indene vapour through a hot tube causes condensation, with dehydrogenation, to complex polycyclic hydrocarbons, e.g. chrysene, $C_{18}H_{12}$. Gas-liquid chromatography of the pyrolysate identifies the components in a commercial resin. **Colour tests** 1. *Storch-Morawski* (*Liebermann-Storch*) *test.* When an indene or coumarone/indene resin is boiled with acetic anhydride, and 1 drop of the extract is treated with 1 drop of conc. sulphuric acid, a deep and permanent rosy-red colour is obtained.

N.B. Natural rosin produces a similar colour which changes to blue or violet; some alkyd and phenolic resins and polysulphide rubbers give a red-brown colour.

2. *Hirschsohn test.* As an alternative to 1, the sample can be heated in a conc. aq. solution of trichloroacetic acid, producing the same order of colours. 3. Dissolve a small sample in 10 ml chloroform, add 1 ml glacial acetic acid and 1 ml of a 10% solution of bromine in chloroform; indene polymers produce a red colour. **Other tests** The comparatively high density (sp. gr. over 1·0) and high refractive index (*c.* 1·6) distinguish these polymers from related resins (*see* §5.82). For iodine number *see* §5.81. The infra-red spectra are characteristic. The resins can be determined polarographically.

5.3 PHYSICS

5.31 GENERAL PROPERTIES

Specific gravity Commercial resins, 1·10–1·14. Polyindene, 1·04–1·10. Polycoumarone, 1·20. **Refractive index** 1·60–1·65. **Water absorption** Practically zero. Example: $0·4\%$ increase in weight after 6 months in $3·5\%$ sodium chloride solution at 20°C.

5.32 THERMAL PROPERTIES

Physical effects of temperature The commercial polymers, thick liquids or rosin- or pitch-like solids, soften over a range from *c.* 10 to *c.* 150°C. They do not decompose until above 250°C, except when kept at a high temperature (e.g. 200°C) for several days.

5.33 ELECTRICAL PROPERTIES

Resistivity High (above 10^{15} ohm cm). The insulating properties, by virtue of the low water absorption, are good even under humid conditions. **Dielectric strength** *c.* 25 kV/mm (55, for thin films). **Dielectric constant** *c.* 3·0. **Power factor** 0·005 (50 Hz), 0·0005 (1 MHz).

5.34 MECHANICAL PROPERTIES

The polymers are too soft, or brittle, for mechanical applications, but are excellent as surface coatings, a field in which they find considerable use (*see below*).

5.5 SERVICEABILITY

The polymers are largely inert, but as surface coatings they darken on prolonged exposure to light and air; hydrogenated resins (§5.81) remain clear. Biological attack, nil.

5.6 UTILISATION

The polymers display good adhesion to metal, glass, and wood, and are used for high gloss surface coatings, including varnishes and floor polishes; also in chemically-resistant paints, and as non-tarnishing media for metallic paints; also in certain printing inks, and as adhesives (e.g. for bonding flooring compositions, linoleum and roofing felt, and in bookbinding). They are compatible with drying oils, for varnishes, and compoundable with various rubbers as softeners or tackifiers.

5.7 HISTORY

Coumarone was isolated by Fittig, in 1883. The mixed resins were first prepared, by Kraemer and Spilker in 1890, from coal tar fractions. They were developed in Germany, as a rosin substitute during the 1914–18 War, and have been available commercially since about 1920.

5.8 ADDITIONAL NOTES

5.81 HYDROGENATED RESINS

Polyindene may be represented thus, in its simplest form,

Since one end of the molecule contains a double bond and the degree of polymerisation is low, the resins show appreciable unsaturation (the iodine number of commercial resins being *c.* 20–40 g iodine/ 100 g polymer). The unsaturation causes the resins, although otherwise inert, to discolour on prolonged exposure, but this can be prevented by hydrogenation under pressure at *c.* 350°C, using a Ni or Pt catalyst. The two end-units are then almost alike, i.e. they are units of 2- or 3-substituted hydrindene. Coumarone polymers behave in a similar manner.

5.82 RESINS ALLIED TO COUMARONE/INDENE RESINS

Several unsaturated compounds derived from natural sources or industrial by-products can be polymerised to low molecular weight resins with physical properties similar to those of polyindene, though they are not necessarily related in chemical structure. The comparatively low values of specific gravity (usually under 1·0) and refractive index (well under 1·6) help to distinguish these polymers from coumarone/indene resins, with which they share the two merits of relatively low price and excellent water-resistance. The principal products of this kind are listed below; they are available as solid resins and sometimes as aq. dispersions compatible with latices.

Petroleum Resins

Pale yellow or amber-coloured low-molecular weight resins of this type can be of two kinds.

(*a*) *Olefin-type resins* (Piccopale). These are polymers of unsaturated aliphatic, alicyclic, and substituted aromatic hydrocarbons of low molecular weight derived from high temperature cracking. They have greater solubility in aliphatic hydrocarbons than resins of the aromatic type (e.g. they dissolve in white spirit) and they are chemically neutral, though they retain some unsaturation. They are used in high-gloss surface coatings (including paper coating) and for hardening waxes, with which they are compatible. Commercial resins have specific gravity, 0·93–1·05; refractive index, *c.* 1·525; softening temperature, 70–115°C.

(*b*) *Cyclopentadiene resins.* Cyclopentadiene (⬠ , b.p. 42°C),

also available from petroleum, readily dimerises to 4,7-methylene-4,7,8,9-tetrahydrindene (cf. relation to indene, §5.1), and in the presence of an ionic catalyst yields low-molecular weight resins which are complex in structure. Two forms of repeat unit, resulting respectively from 1,2- and 1,4-polymerisation, are possible, namely,

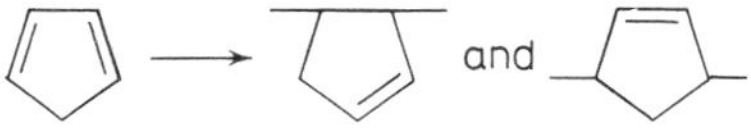

Commercial resins melt at 100–120°C and become insoluble when stoved as surface films.

Terpene resins (BX Resin, Piccolyte)

The α- and β-pinenes, two of the hydrocarbons of turpentine (the β-isomer predominating), yield low-molecular weight polymers of a rosin-like kind. These are of interest not only industrially but also because of their relation to natural rubber, through the common basic isoprene unit, C_5H_8. The structures of these polymers are thought to arise as follows.

The bicyclic structure of α-pinene (I) probably opens at the bridge and isomerises to dipentene (dl-limonene, II) which then polymerises (repeat unit, III).

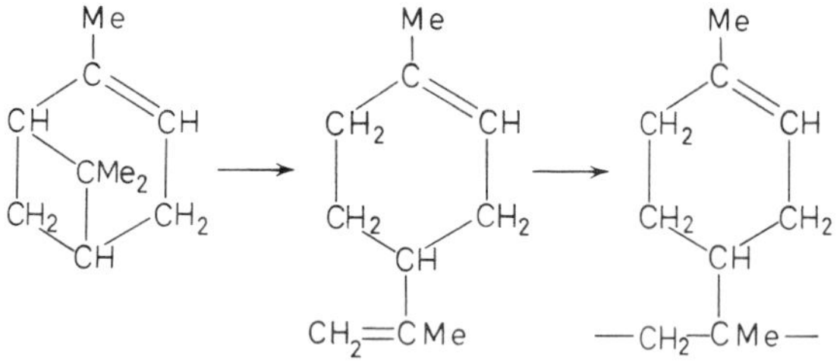

I II III

where Me = CH_3

With β-pinene (IV) it appears that the bridge opens and the structure rearranges and polymerises directly (repeat unit, V), without involving dipentene as an intermediate. However, because of the unsaturation remaining in the structures shown, the polymers are likely to be rather more complex than has been indicated.

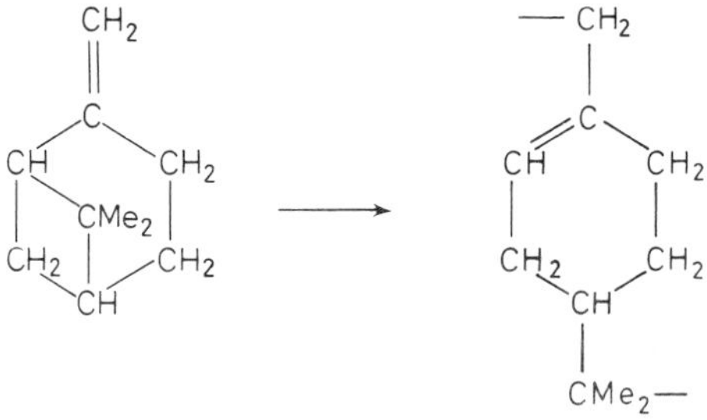

IV V

Polymerisation is effected with cationic initiators, and commercial resins reach only a low molecular weight (e.g. degree of polymerisation of 10–15). The products range from partially-fluid to clear, pale yellow, brittle resins that are light-fast and chemically neutral; they are resistant to acids and alkalis but dissolve in many organic liquids, including aliphatic hydrocarbons (e.g. white spirit), and they are compatible with waxes, bitumens, drying oils and several rubber derivatives. Commercial resins have: specific gravity, 0·98–1·0; refractive index, *c.* 1·535; and softening points ranging from 10 to 135°C.

The resins find many uses; e.g. in printing inks, paints and varnishes, floor coverings and polishes, hot-melt adhesives, also as modifiers, plasticisers and tackifiers for other polymers (including rubber and rubber-based adhesives).

Cyclohexanone (or ketone) resins (Resin AW2, Resin MS2, Setanon 600).

These are low molecular weight resins obtained by polymerisation of cyclohexanone (and methylcyclohexanone) in the presence of a basic catalyst. The structure arises in part from self-condensation of the aldol type, e.g.

but polycyclic structures are undoubtedly formed as well, since a typical product (comprising 6 units) has been found to possess 1 carbonyl-, 1 methoxy-, 1 other ether-group, and 3 hydroxy-groups.

The clear, almost colourless, light-fast and water-resistant brittle resins are inert to many reagents but dissolve in hydrocarbons (including white spirit) and are compatible with drying oils and many polymers (e.g. improving adhesion, gloss, and hardness in cellulose nitrate lacquers). Commercial resins have sp. gr. 1·08; refractive index, *c.* 1·505; and a softening range from 75–110°C. They find uses in paints, varnishes, printing inks, and adhesives, also in paper coating and the stiffening or retexturing of textiles.

For other types of ketone resins, *see* §9.86.

65

POLYVINYL ACETATE

With notes on partially hydrolysed polyvinyl acetates

SYNONYMS AND TRADE NAMES

Poly (vinyl acetate). PVA*, PVAc, PVAC.
Elvacet, Gelva, Mowilith.
Dispersions: Epok V, Emultex, Calatac, Texicote V, Texilac, Vandike, Vinalak, Vinamul, Vinnapas.

GENERAL CHARACTERISTICS

A clear colourless thermoplastic substance, available in several grades ranging from soft and brittle to hard and tough; also marketed as viscous solutions in organic solvents and more especially as aq. dispersions of low viscosity but high solids-content. It is light-resistant and chemically rather unreactive, but as it exhibits creep under load its applications are restricted mainly to adhesives and surface coatings, particularly as 'emulsion paints'.

6.1 STRUCTURE

Simplest Fundamental Unit

$$-CH_2-CH-$$
$$|$$
$$OCOCH_3$$

$C_4H_6O_2$, mol. wt. = 86

Molecular weight Commercial products, 5000 to over 500000 (degree of polymerisation, approx. 60–6000).

X-ray data Substantially amorphous (diffuse pattern) even when stretched.

6.2 CHEMISTRY

6.21 PREPARATION

Vinyl acetate (CH_2=$CHOCOCH_3$, b.p. 73°C), a colourless mobile liquid with a characteristic odour, can be prepared by passing acetylene into acetic acid in the presence of a mercuric compound as

---

*Unfortunately, the abbreviation PVA tends to be used for both polyvinyl acetate and polyvinyl alcohol; it is preferable to distinguish the two as PVAC (or PVAc) and PVAL (or PVAl) respectively.

catalyst, conditions being chosen to minimise the yield of ethylidene diacetate from a side reaction,

$$CH \equiv CH + CH_3COOH \xrightarrow[\text{(HgO)}]{\text{Catalyst}} CH_2 = CHOCOCH_3$$

or from the same components in a catalytic vapour-phase process. In a more recent process it is derived inexpensively from reaction of ethylene, oxygen and acetic acid (the last also obtained from controlled oxidation of ethylene),

$$CH_2 = CH_2 + CH_3COOH \xrightarrow[\text{(Metallic cat.)}]{O_2} CH_2 = CHOCOCH_3 + H_2O$$

Alternatively, the monomer can be obtained from the catalysed reaction of acetic anhydride and acetaldehyde, via ethylidene diacetate,

$$(CH_3CO)_2O + CH_3CHO \xrightarrow{\text{Heat}}$$

$$(CH_3CO \cdot O)_2CHCH_3 \xrightarrow[\text{+cat.}]{\text{Heat}} CH_2 = CHOCOCH_3$$

Vinyl acetate polymerises under the influence of ultra-violet light or with radical-type initiators (peroxides, persulphates, redox systems). With increasing conversion, transfer reactions involving the α or methyl H atoms can lead to some chain branching. Polymerisation may be effected in bulk (i.e. in the pure monomer), in solution (e.g. in toluene), or in aq. suspension or emulsion. The last two methods allow good control of the temperature, and yield the polymer respectively as small spheres or dispersed as a latex (particles under 1 μm in diameter, as in 'emulsion paints').

Polymerisation is adversely affected by contamination with acetaldehyde (a possible impurity) and is prevented by inhibitors such as hydroquinone, diphenylamine, sulphur compounds and copper salts, which are used to stabilise the monomer. Copolymerisation of vinyl acetate is used for various purposes, e.g. with vinyl propionate or an acrylic ester to improve water resistance and flexibility, and with crotonic acid or maleic anhydride to obtain alkali-soluble products. For copolymers with ethylene, *see* §1.83.

6.22 PROPERTIES

Dissolved by aromatic and chlorinated hydrocarbons (toluene, chloroform), ketones (acetone), lower alcohols (preferably with a little water present), esters (ethyl or butyl acetate, also monomeric vinyl acetate), and solvent mixtures such as 20/80 ethylene dichloride/alcohol (by vol.). Non-solvents such as white spirit or butanol are tolerated as diluents. **Plasticised by** conventional high-boiling esters (dibutyl phthalate, tritolyl phosphate), tributyrin, acetamide, diphenyl. Plasticised internally by copolymerisation of vinyl acetate with, for example, vinyl propionate, $CH_2{=}CHOCOCH_2CH_3$, or vinyl caprate, $CH_2{=}CHOCO(CH_2)_8CH_3$. **Swollen by** lower alcohols (anhydrous) and aq. solvent mixtures causing incomplete solution; slightly swollen (and whitened) by water alone. **Relatively unaffected by** aliphatic hydrocarbons, most oils and fats, turpentine, ether, carbon disulphide, polyhydric alcohols, dil. acids and dil. alkalis (effects similar to water, *see below* and §6.31). **Decomposed by** conc. alkalis (especially on heating) and conc. acids, decomposition being accentuated in alcoholic solution (*see* §7.21).

6.23 IDENTIFICATION

Thermoplastic. Cl, N, S and P absent (in unplasticised polymers); saponifiable. **Combustion** Liquefies (viscous 'melt'), burns feebly with a yellow smoky flame, yields acetic acid and has a characteristic 'vinyl' odour on extinction; the hot residue can be drawn to brittle threads. **Pyrolysis** Liquefies, chars, evolves acid fumes; no monomer is liberated. **Colour tests** 1. Solid polyvinyl acetate stains brown on contact with dil. iodine solution (*see* §7.23 for suitable reagent) and the depth of colour increases, rather than decreases, on rinsing in water. *N.B.* Polyvinyl acetals (Section 8) also respond but to a less extent. 2. Polyvinyl acetate in alcoholic solution is readily hydrolysed by boiling under reflux in the presence of acid, or by warming with alcoholic alkali (Saponification No. up to 650 mg KOH/g). The polyvinyl alcohol that precipitates is dissolved in water and tested as in §7.23; the acetate anion can be detected as in §14.23.

6.3 PHYSICS

6.31 GENERAL PROPERTIES

Specific gravity 1·17–1·19. **Refractive index** 1·46–1·47. **Water absorption** 1–3%; up to 8% on prolonged immersion (films become white and opaque but regain clarity when dried). Some copolymers (e.g. with an acrylic ester) are more resistant to water. **Moisture regain** (films, 65% r.h.) c. 1%. **Permeability** Only moderately impermeable to water vapour; films highly impermeable to most oils and greases.

6.32 THERMAL PROPERTIES

Specific heat 0·4. **Conductivity** 4.10^{-4} cal/cms°C). **Coefficient of linear expansion** $0·85.10^{-4}$/°C. **Physical effects of temperature** Second-order transition temperature (change in expansion rate) at 28°C; becomes rubber-like above 55°C; softens (viscous melt) from 70 to 190°C according to molecular weight; tends to degrade above 130°C; *in vacuo*, above 190°C decomposes to acetic acid and an unsaturated acetylene-type polymer.

6.33 ELECTRICAL PROPERTIES

N.B. Electrical applications are limited by the small but appreciable moisture absorption.

Volume resistivity Solution-cast films, 10^{13}–10^{14} ohm cm. **Dielectric strength** Thin films 16–32 kV/mm (lower at high humidity). **Dielectric constant** Films, dry, 3·0–3·5 (rising to 10–10·5 at high humidity). **Power factor** *c*. 0·03.

6.34 MECHANICAL PROPERTIES

N.B. Cold flow, particularly in polymers of low or moderate molecular weight, restricts mechanical applications.

Tensile strength 1·5–3·5 kgf/mm². **Extension at break** Low, for rapid deformation, but plastic under sustained stress. **Impact strength** Low molecular weight polymers, brittle; toughness increases with molecular weight. **Hardness** Increases, from soft and pitch-like to hard and horny, with increase of molecular weight. **Abrasion** In general, durability and resistance to abrasion are good and increase with molecular weight.

6.4 FABRICATION

Compression moulding: 95–120°C, *c*. 1000 lbf/in² (however, difficulties with adhesion, and cold flow in the products, restrict moulding applications). The polymer is mostly used for surface coatings and as an adhesive, when it is applied from solution or aq. dispersion; it can also be used as a melt-adhesive, e.g. coated joints can be cooled under pressure from about 20°C above the softening temperature. The adhesion of the polymer is sometimes improved by baking or stoving.

6.5 SERVICEABILITY

The properties of this polymer do not deteriorate at low temperatures or in sunlight, and it is unaffected by bacteria, fungi or insects; but it will not sustain a prolonged stress, particularly at elevated temperatures. It swells slightly and becomes opaque on long immersion in water, but recovers on drying; appropriate copolymers show greater water resistance.

6.6 UTILISATION

Vinyl acetate is an inexpensive monomer, and its polymers and copolymers—inert and non-toxic—are much used in 'emulsion' paints. Aqueous dispersions also find uses, along with solvent solutions, as coatings or adhesives in other fields; e.g. as adhesives and lacquers for leather, for the stiffening, grease-resisting, sealing and packaging of paper, and for stiff finishes (permanent starches), back-filling, and adhesive-bonding in the treatment of textiles (including non-wovens). Additional to application from solution or dispersion, plasticised polyvinyl acetate can be used as a hot-melt adhesive; aq. dispersions find additional uses as cement and plaster additives, improving both cohesion (strength) and adhesion.

6.7 HISTORY

The polymer was first prepared by F. Klatte in 1912, and was subsequently put into large scale production in Germany and in Canada (by Shawinigan Ltd.) in 1917.

6.8 ADDITIONAL NOTES

6.81 PARTIALLY-HYDROLYSED POLYVINYL ACETATES
(Solvar, also several of the trade names given in Section 7).

These materials result from graded hydrolysis of polyvinyl acetate (*see* §7.21). Products having a low degree of hydrolysis resemble the parent acetate and show good adhesion to metals; but the more important products are those in which hydrolysis has proceeded so far that they more resemble the ultimate product, polyvinyl alcohol.
Commercially, moderately to highly hydrolysed materials are classed along with the fully hydrolysed product as 'polyvinyl alcohol',

the precise 'grade' being specified with reference to (i) the molecular weight, and (ii) the degree of hydrolysis, as expressed respectively by viscosity* and by either the percentage hydrolysis (i.e. the extent to which acetyl groups have been converted to hydroxyl groups) or the percentage of polyvinyl acetate remaining in the product†.

Partially-hydrolysed polyvinyl acetates (which can be further hydrolysed by acids and alkalis), being less regular in structure and less extensively hydrogen-bonded than fully hydrolysed (99–100% hydrolysed) polyvinyl acetate, yield films that are slightly weaker and show slightly less resistance to solvents. Some solubility data are given below (*see also* §7.22).

% Hydrolysis	% Residual PVAC	Insoluble in:	Soluble in:
Under 50	Over 65	Water	Organic solvents
c. 70	c. 45	Hot water	Cold water; aq. solvents*
75†–80	33–39	Hot water; aq. solvents*	Cold water
85–90	18–25	—	Cold and hot water
98–100	0–4	Cold water (swells)	Hot water‡

*Aq. alcohol or aq. acetone
†Above c. 75% hydrolysis (i.e. under c. 40% residual polyvinyl acetate) the products can be dissolved either in cold or hot water, or in both.
‡Resists cold water; remains in solution on cooling a hot preparation but tends to set to a reversible gel on storage (*see also* §7,22)

The partially-hydrolysed acetates give reddish brown to violet colours with dil. solutions of iodine (*see* colour test in §7.23) but the colour changes to deep blue on addition of borax.

6.9 FURTHER LITERATURE

SCHILDKNECHT, C. E., *Vinyl and Related Polymers* (Chapter 6), New York and London (1952)
SMITH, W. M. (Ed.), *Manufacture of Plastics* (Chapter 4, VONA, J. A., *et al.*), New York and London (1964)
BRYDSON, J. A., *Plastics Materials* (Chapter 11), London and Princeton (1966)

*The viscosity in cP, at 20°C, of a molar solution (86 g/l) of the initial polyvinyl acetate in benzene.
†The percentage of the residual polyvinyl acetate is given by

$$\frac{4300(100-x)}{43(100-x)+22x}$$

where x is the percentage hydrolysis.

SECTION 7

POLYVINYL ALCOHOL
Including insolubilised fibres
With notes on polyvinyl ethers and polyvinylpyrrolidone

SYNONYMS AND TRADE NAMES

Poly(vinyl alcohol). PVA*, PVAL, PVAl.
Alcotex, Brax, Cipoviol, Elvanol, Gelvatol, Gohsenol, Moviol, Pevalon, Polyviol, Vinarol, Vinol.
Fibres: Cremona, Kanebiyan, Kuralon, Mewlon, Vinal, Vinylon.

GENERAL CHARACTERISTICS

A white or cream-coloured, slightly hygroscopic powder, that swells and eventually dissolves in water (some grades dissolve only over a certain range of temperature). Evaporation of the water yields tough transparent films, resembling sheet gelatin; tough, flexible, plasticised films are available commercially. The polymer, which can be extruded, or cast on to paper or fabric, possesses outstanding resistance to hydrocarbons (e.g. petrol, mineral oils) and to many other organic liquids. Fibres—rendered insoluble in water by a cross-linking process—in certain characteristics, though not necessarily in appearance, usually resemble cotton, e.g. in moisture regain.

7.1 STRUCTURE

Simplest Fundamental Unit

$$-CH_2-CH-$$
$$\underset{\displaystyle OH}{\vert}$$

C_2H_4O, mol. wt. = 44

Molecular weight Dependent on and approximately half that of the polyvinyl acetate (Section 6) from which the polymer is derived. Commercial products are 'graded' by their degree of hydrolysis and the viscosity of a molar solution of the initial polyvinyl acetate in benzene (*see also* §6.81); commonly the degree of polymerisation of low-viscosity grades is 200–1000 and that of high-viscosity grades is 1500–3000.

X-ray data Amorphous to polycrystalline material, dependent on mechanical treatment. When stretched as fibres it becomes highly oriented and up to 60% crystalline, yielding a sharp diagram; identity period, 2·55 Å. Hydrogen atoms and —OH groups can replace one another without disturbing the crystal structure, hence

---

*Unfortunately, the abbreviation PVA tends to be used for both polyvinyl alcohol and polyvinyl acetate; it is preferable to distinguish the two as PVAL (or PVAl) and PVAC (or PVAc) respectively.

isotactic polyvinyl alcohol is not much different from the common atactic material.

7.2 CHEMISTRY

7.21 PREPARATION

The monomer, vinyl alcohol (theoretical formula, $CH_2{=}CHOH$; tautomeric with acetaldehyde, CH_3CHO), is unknown, but the polymer is a stable substance obtained by hydrolysis or alcoholysis of polyvinyl acetate (§6.22),

$$-CH_2-CH- \quad \xrightarrow{\text{H}_2\text{O (or ROH)}} \quad -CH_2-CH- \ + \ CH_3COOH$$
$$\underset{\displaystyle OCOCH_3}{|} \qquad\qquad\qquad \underset{\displaystyle OH}{|} \qquad \text{(or } CH_3COOR\text{)}$$

The hydrolysis can be carried out—preferably in the absence of oxygen—under (i) alkaline conditions, such as by boiling under reflux with alcoholic alkali, or (ii) acid conditions, such as leaving to stand or boiling in an excess of methanol containing c. 1 % of sulphuric acid. The reaction proceeds substantially to completion, especially in (i), unless arrested intermediately (for partially-hydrolysed polyvinyl acetates, *see* §6.81) and the product, after washing with methanol, can be purified by precipitation from aq. solution with acetone. The products from (i) are usually pale yellow, while those from (ii) are nearly colourless.

The properties of the end-products depend on the 'grade', which is determined by the initial molecular weight and the final degree of hydrolysis (*see also* §6.81).

INSOLUBILISATION OF FIBRES

Aq. solutions of polyvinyl alcohol can be dry spun, or spun into a coagulating bath (e.g. saturated aq. sodium or ammonium sulphate), as fibres. Soluble fibres of polyvinyl alcohol have certain uses. Fibres that are insoluble—notably when held under tension—by reason of high orientation and high crystallinity are also available, being produced, for instance, by treatment of fully-hydrolysed polymer with superheated steam, or by heating in dil. mineral acid (along with sodium sulphate to suppress swelling). However, more frequently, polyvinyl alcohol fibres are insolubilised by a treatment producing intermolecular cross-linkages. Usually the fibres after spinning are stretched to some 5–10 times their length, stabilised at 180–230°C, then passed into a bath containing an aldehyde (commonly formaldehyde, but sometimes benzaldehyde or furfural), mineral acid (as catalyst) and sodium sulphate (to suppress swelling),

$$2 -CH_2-CH- + HCHO \xrightarrow[(-H_2O)]{\text{Dil. } H_2SO_4} \quad \begin{array}{cc} CH_2 & CH_2 \\ | & | \\ CH-OCH_2O-CH \\ | & | \end{array}$$
$$\quad | \\ OH$$

Cross-linkage can also be effected by esterification, e.g. with maleic anhydride. Dyeing properties of the fibres may be improved by starting with a copolymer of vinyl acetate (e.g. with vinylphthalimide) or by blending polyvinyl alcohol (e.g. with casein) before spinning.

7.22 PROPERTIES

Soluble in water, especially on warming to 70–80°C; but for more detailed characteristics *see* §6.81. Aq. solutions are markedly thickened (gelled) by addition of a little borax; conc. solutions of the fully hydrolysed polymer tend to gel on standing, like those of starch. The water-insoluble fibres will usually dissolve in formic acid and to some extent in *m*-cresol. **Plasticised by** moisture and water-soluble polyhydric alcohols (glycerol, glycol, triethylene glycol, sorbitol), hydroxy-esters (glyceryl lactate), amides (urea), and also calcium chloride, acting as a humectant. **Swollen by** polyhydric alcohols, especially when hot. **Relatively unaffected by** almost all organic liquids, i.e. hydrocarbons (including petrol) and chlorinated hydrocarbons, alcohol, acetone, esters, oils and fats. Only moderately degraded by periodate oxidation (hence 1,2-glycol structure largely absent) and resistant to ozone. The polymer is insoluble in conc. solutions of certain salts (e.g. NaCl, Na_2SO_4); it can be insolubilised by cross-linkage with an aldehyde or an amino-aldehyde precondensate in the presence of an acid catalyst, also by reaction with di-isocyanates; polymer films containing a dichromate become insoluble on heating or exposure to light. Fibres are insolubilised as described in §7.21. **Decomposed by** conc. acids, especially oxidising acids. Boiling with hydrogen peroxide reduces the viscosity of polyvinyl alcohol solutions, lowering the molecular weight and introducing acid groups. Fibres decompose in conc. mineral acids and in warm dil. nitric acid; when warmed with *c*. 25% H_2SO_4 they yield formaldehyde (from the cross-links) and dissolve; stronger acid destroys the polymer completely.

7.23 IDENTIFICATION

Limited thermoplasticity. Cl, N, S and P absent. **Combustion** Softens, burns with slightly smoky flame, leaves black residue. **Pyrolysis**

Softens (fibres shrink), chars, evolves vapour with an unpleasant odour and a mildly acid reaction; aldehydes, particularly formaldehyde, may be detected in the pyrolysate from insolubilised fibres (*vide infra*). **Colour test** Addition of iodine to solutions of polyvinyl alcohol produces a deep blue colour (cf. starch, §19.23), which is fugitive to heat but reappears on cooling. The conditions under which this test can be applied are more critical than for starch (appearing to require the presence of both an acid and iodide ions) and the sensitivity is much increased by addition of borax. A suitable spot-plate procedure is as follows: to 1 drop of test solution (containing *c.* 0·5% PVAL) add 1 drop of test reagent (0·1 g I_2 + 1·0 g KI, dissolved in 20 ml 1/1 alcohol/water, made up to 100 ml with 2 N HCl); the blue colour appears either immediately or on addition of 1 drop of a saturated solution of borax (or solid borax). Too much polyvinyl alcohol gives an almost black coloration; too little (under 0·05%) produces a green coloration. *N.B.* Positive results are also obtained with partially hydrolysed polyvinyl acetates (but *see* §6.81, as these initially may give other colours), and from PVAL fibres and polyvinyl acetal (Section 8) after acid hydrolysis. **Additional tests** When a drop of a moderately conc. solution of polyvinyl alcohol is merged into an adjacent one of saturated aq. borax, a characteristic ropiness is immediately apparent.

Insolubilised fibres The solution obtained when these are warmed with 20–25% sulphuric acid responds to the above tests and to the CTA test for formaldehyde (§26.23a), of which commercial fibres yield *c.* 3%; stronger acid destroys the polymer and discolours the solution.

7.3 PHYSICS

7.31 GENERAL PROPERTIES

Specific gravity 1·25–1·35, depending on moisture content; fibres, 1·26–1·32 (crystalline phase, 1·348; amorphous phase, 1·265). **Refractive index** 1·50–1·53, depending on moisture content. Fibres, variable according to type: $n_{||}$, 1·532–1·550; $n_{\perp}$, 1·505–1·526. **Birefringence** $(n_{||}-n_{\perp})$: 0·022–0·037; highly oriented fibres, up to 0·04. **Water absorption** When conditioned to 65% r.h. the polymer contains 6–9% water; it swells when wet and ultimately dissolves (*see also* §6.81). Insolubilised fibres show only limited swelling and withstand boiling water. **Water retention** Staple fibres, 30–35%. **Moisture regain of fibres** (%, 20–25°C) At 20% r.h., 1·3–1·8; 65% r.h., 4·5–5; 95% r.h., 10–12. **Permeability** Highly impermeable to most organic vapours and to inert gases and hydrogen (much less permeable than natural rubber); relatively impermeable to alcohol vapour but permeable to water vapour (1·9–11 (10^{-10} g)/(cm s cmHg) at 20–30°C) and ammonia.

7.32 THERMAL PROPERTIES

Specific heat c. 0·4. **Conductivity** c. 5 (10^{-4} cal)/(cm s°C). **Coefficient of linear expansion** c. $1.10^{-4}/°C$. **Physical effects of temperature** Suitably plasticised polymer remains flexible at $-40°C$ or lower. Second-order transition at 85°C. The polymer is serviceable up to 120–140°C though it slowly yellows (degrades) above 100°C, and darkens (with evolution of water and conversion to an unsaturated polymer) if kept at 150–200°C. Thermal treatment of stretched fully hydrolysed material (e.g. in superheated steam) increases the resistance to water. The polymer does not properly melt; however, films— preferably containing plasticiser or moisture—are heat sealable at c. 150°C. Insolubilised fibres retain 70% of normal strength at 100°C, and show changes associated with decomposition and melting between 200–260°C.

7.33 ELECTRICAL PROPERTIES

N.B. Electrical applications are very limited, by reason of the absorption of moisture under normal atmospheric conditions.

Volume resistivity (ohm cm) Films: dry, 10^{10}–10^{11}; at 65% r.h., 10^8–10^9. **Dielectric strength** (kV/mm) Over 40 when dry, but under humid conditions can fall to 0·4. **Dielectric constant** (at 1 kHz) 3·5 to over 10 (65% r.h.). **Power factor** (at 1 kHz) 0·03 to over 0·1 (65% r.h.).

7.34 MECHANICAL PROPERTIES
(65% r.h., 20–25°C, unless otherwise stated)

Elastic modulus (Young's) (kgf/mm²) Fibres: 300–1200 (25–100 gf/ den.); up to 2000 (180 gf/den.) in high-tenacity filaments. **Elastic recovery** (%) Fibres: varies with type; example, from 1% short-term strain, 75; from 5%, 30; from 10%, 20. Values of 70–80 (from 3% strain) and 45–60 (from 5% strain) quoted for special fibres. **Tensile strength and extension at break** *see table on p. 77.*

Other mechanical properties (chiefly of films). *Impact strength* is high. *Hardness* increases with molecular weight and varies inversely with moisture content. *Abrasion resistance* is good, as is resistance to repeated flexure; abrasion resistance of fibres and fabrics is at least as good as that of cotton.

	Tensile strength (kgf/mm^2)	Extension (%)
Extruded material		
Fully hydrolysed	3·5	225
*Cast films**		
With 2% water	4–12	2–15
10%	2–5	100–280
25%	0·4–1·2	200–550
Insolubilised fibres†		
Staple	50–75 (4·5–6·5 gf/den.)	17–26
Do. high-tenacity	up to 95 (8 gf/den.)	*c.* 13
Cont. filament	*c.* 50 (4 gf/den.)	*c.* 15
Do. high-tenacity	up to 105 (9 gf/den.)	*c.* 10

*Maximum cohesion is obtained with fully hydrolysed material of high molecular weight, which as dry unplasticised films is quoted as exhibiting a tensile strength of 10–15 kgf/mm^2
†The wet strength of fibres is *c.* 80% of the air-dry strength

7.4 FABRICATION

Plasticised or moist polymer can be extruded (70–115°C) or compression moulded (e.g. 110–155°C, up to 1000 lbf/in^2). Fibres are prepared as described in §7.21.

7.5 SERVICEABILITY

Polyvinyl alcohol is much affected by moisture, and at high humidity will creep under load; however, it is resistant to sunlight, showing only slight loss in strength after prolonged exposure, and normally it is unaffected by bacteria, fungi or insects (though solutions need a preservative to prevent mould growth). The insolubilised fibres are resistant to biological attack, are unaffected by burial in soil or immersion in sea water, and show only slight shrinkage even when boiled in water.

7.6 UTILISATION

The polymer is employed, plasticised or otherwise, for flexible sheeting, tubing, protective aprons, gloves, etc., to withstand petrol, chlorinated solvents, fats and oils, and it is used in the preparation of polarising screens; also for gas-proof diaphragms, and as a coating to render fabric impermeable to many gases and vesicant vapours. Aqueous solutions find uses as adhesives, pigment binders, and textile sizes (for nylon and polyester warps); also in the sizing, greaseproofing and gas- or vapour-proofing of paper, and as thickening and emulsion-stabilising agents. Stretched and oriented material has been used for oil- and flex-resistant driving belts (substitute for leather or cotton), also for surgical sutures; water-insoluble fibres are suitable for a variety of applications, e.g. dress and industrial fabrics (including tyre cords), carpets, rot-resistant netting and filter cloths, cordage, and bristles.

7.7 HISTORY

The polymer was first prepared in 1924 by W. O. Herrmann and W. Haehel and was investigated by H. Staudinger, K. Frey and W. Starck in 1926 (*Ber.*, **60** (1927) 1658 and 1782). It has since been developed commercially, notably in Germany, Japan, and the U.S.A. Insoluble polyvinyl alcohol fibres were pioneered in Japan (and later in America), commencing in 1939–40, with pilot-scale production in 1942 and expansion to commercial scale by 1950.

7.8 ADDITIONAL NOTES

7.81 POLYVINYL ETHERS

Vinyl alkyl ethers yield useful products when polymerised alone and when copolymerised with, for example, vinyl chloride or maleic anhydride. The homopolymers (e.g. Igevin, Lutonal), the simplest fundamental unit of which is represented by

$$—CH_2—CH—$$
$$|$$
$$OR$$

(where commonly R = $—CH_3$, $—C_2H_5$ or $—C_4H_9$), are largely inert materials unsuitable for structural purposes, since they are mostly soft and sticky; but they are soluble in various solvents and are employed as tackifiers (e.g. for pressure-sensitive adhesives and for improving the adhesion of coatings) and as non-migratory plasticisers. The copolymers may be soft or relatively hard, the vinyl ether component either acting as an internal plasticiser or being modified by the other component; thus, with vinyl chloride it provides very flexible products (e.g. Vinoflex), while with maleic anhydride it gives very versatile, soluble and reactive products (e.g. Gantrez AN). The repeat unit of a 1:1 copolymer of the last kind is represented by

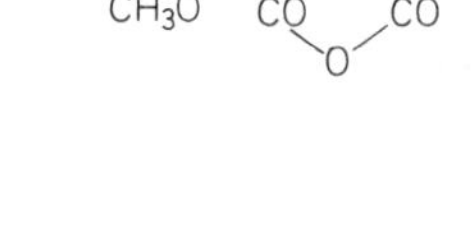

Vinyl alkyl ethers are obtained in various ways, an industrial process employing a Reppe synthesis involving addition of an alcohol to acetylene, using an alkaline catalyst at high pressure and elevated temperature,

$$CH \equiv CH + ROH \xrightarrow[160-175°C]{KOH} CH_2 = CHOR$$

Polymerisation is effected in bulk with an ionic catalyst (e.g. BF_3); the properties of the product depend upon the method employed, slow polymerisation yielding the highest molecular weight—and hardest—polymers. Emulsion polymerisation in a neutral or alkaline medium, with iodine as catalyst, is also used.

Polyvinyl alkyl ethers tend to age by oxidative degradation in light, or on milling or long storage, unless stabilised with antioxidants. A solution in 1/1 acetic anhydride/toluene is said to give a blue colour when treated with a trace of conc. sulphuric acid.

Polyvinyl methyl ether is soluble in water (below 35°C) and more readily in many organic liquids, including alcohol and benzene, but insoluble in alkalis, aq. solutions of salts, and aliphatic hydrocarbons. It is compatible with polyvinyl acetate, shellac, and cellulose nitrate, but not with polyisobutylene.

Polyvinyl ethyl ether somewhat resembles the methyl polymer but is insoluble in water.

Polyvinyl isobutyl ether dissolves in many organic liquids, including aliphatic hydrocarbons, but is insoluble in water and alcohol. It is compatible with polyisobutylene, but not with shellac or cellulose nitrate.

Copolymers of vinyl methyl ether and maleic anhydride (1/1) dissolve in water, aq. solutions of salts (with some exceptions), and some polar organic liquids, but not in hydrocarbons or chlorinated hydro-carbons. Aqueous solutions contain free acid groups, and addition of alkali causes a large increase in viscosity (cf. polyacrylic acid §9.81). Dissolution in alcohols is accompanied by the formation of partial esters; ammonia yields an ammonium salt of a half amide. The properties of these polymers make them useful as thickeners and stabilisers in emulsion polymerisation, as film formers (non-tacky), and as textile and paper sizes; compounding with polymers such as polyvinyl acetate or polyvinylpyrrolidone provides adhesives suitable for special surfaces, e.g. polyethylene.

7.82 POLYVINYLPYRROLIDONE
(Poly(*N*-vinyl-2-pyrrolidone). PVP. Albigen A, Kollidon, Periston).

Polyvinylpyrrolidone, a material similar to polyvinyl alcohol in its solubility and neutrality in water, is represented by the following repeat unit,

$$-CH_2-CH-$$

The polymer, developed in Germany in 1942, is obtained via an addition reaction between acetylene and formaldehyde to form 1,4-butynediol, which is then hydrogenated to 1,4-butanediol. This product is next oxidised to γ-butyrolactone (I), which by autoclaving with ammonia is converted to α-pyrrolidone (II); finally, vinylation with acetylene (or vinyl chloride) under pressure yields the required monomer (III),

$$HO(CH_2)_4 OH \xrightarrow{O}$$

I   II   III

N-Vinylpyrrolidone can be polymerised in bulk or in aq. solution under the influence of heat and radical catalysts (e.g. hydrogen peroxide activated by ammonia). The polymers are soluble in water, in acid solutions, and in many organic liquids (including chlorinated hydrocarbons); they are swollen by aromatic hydrocarbons, but are unaffected by aliphatic hydrocarbons. Strong alkalis cause precipitation from aq. solution. A dil. solution in $2N$ hydrochloric acid gives a dark brown precipitate or brown coloration on addition of a weak solution of iodine.

The polymers have been employed in aq. solution as extenders, or substitutes, for blood plasma (e.g. a 3·5 % solution of 25 000 molecular weight polymer, made isotonic by addition of appropriate salts); they find uses also as adhesives, and as thickening or emulsifying agents. The material shows a pronounced affinity for certain dyestuffs, and is used in alkaline solution, with sodium hydrosulphite, for stripping sulphur colours, vat dyes, or prints, and for partial stripping of direct dyes from cellulosic textiles.

Copolymers of vinylpyrrolidone of some commercial importance include those with vinyl acetate, ethyl acrylate, and styrene. These are used as textile and paper sizes (including sizing of glass fibres) and in adhesive formulations (including pressure-sensitive and re-wettable kinds).

A formulation said to give good adhesion to polyethylene is based on polyvinylpyrrolidone and a partial butyl ester derived from the copolymer of vinyl methyl ether and maleic anhydride (*see* §7.81).

7.9 FURTHER LITERATURE

DAVIDSON, R. L. and SITTIG, M. (Eds), *Water-soluble Resins* (Chapter 5, ARGANA, C. P.), New York and London (1962)

SMITH, W. M. (Ed), *Manufacture of Plastics* (Chapter 5, DICKSTEIN, J. and BOUCHARD, R.), New York and London (1964)

POLYVINYL FORMAL, ACETAL AND BUTYRAL

SYNONYMS AND TRADE NAMES
Polyvinyl formal: Formvar, Pioloform F. Polyvinyl acetal: Alvar. Polyvinyl butyral: Butacite, Butvar, Pioloform B.

GENERAL CHARACTERISTICS
Clear or translucent, tough, and inert materials, employed mainly for adhesive, abrasion-resistant, high gloss surface coatings. The butyral is also used (plasticised) as an interlayer in safety glass, for flexible sheeting, and for textile coatings.

8.1 STRUCTURE

Simplest Fundamental Unit

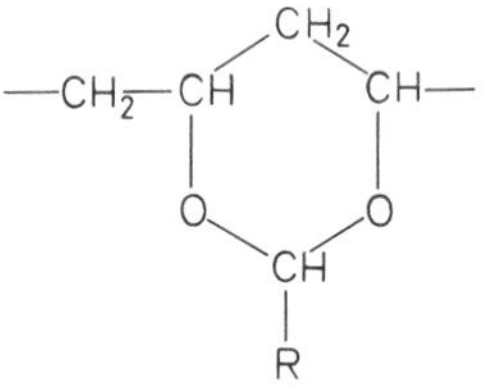

Molecular weight

Formal, $C_5H_8O_2 = 100$
Acetal, $C_6H_{10}O_2 = 114$
Butyral, $C_8H_{14}O_2 = 142$

where the main chain derives from polyvinyl alcohol, and R = H, CH_3 or C_3H_7.

Molecular weight Dependent on that of the original polyvinyl acetate and alcohol, and on the degree of acetalisation (*see* §8.21); approximate range, commercial polymers, 50 000–500 000.

X-ray data Substantially amorphous materials.

8.2 CHEMISTRY

8.21 PREPARATION

By hydrolysis of polyvinyl acetate, and treatment of the resulting polyvinyl alcohol with an aldehyde, under acid conditions,

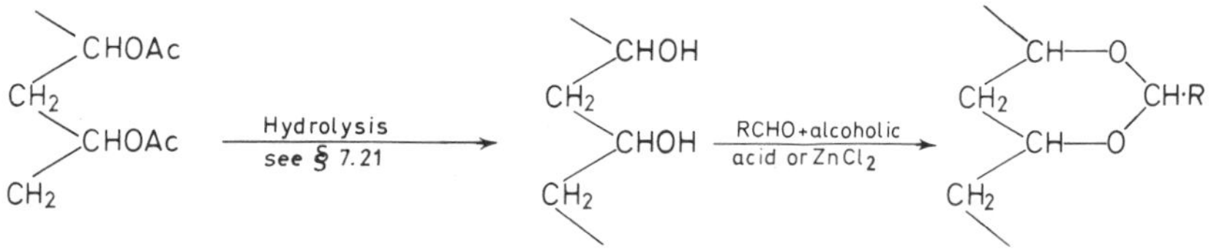

The product can be obtained in a one-stage operation by treatment of a solution or suspension of polyvinyl acetate with a slight excess of the requisite aldehyde in the presence of alcohol and a mineral acid, e.g. formaldehyde or paraformaldehyde is reacted, in the presence of hydrogen chloride, with polyvinyl acetate dissolved or suspended in dioxan or acetic acid, and the resulting formal is precipitated by water and washed free from acid.

In commercial products the degree of hydrolysis is usually high, and formals and acetals are *c.* 90% acetalised. A typical butyral might contain 2% residual acetyl, 22% hydroxyl, and 76% *n*-butyral groups; material of this approximate composition would be soluble in organic liquids but sufficiently hydrophilic to have good adhesion to glass or cellulose.

For partially formalised (cross-linked) insoluble fibres of polyvinyl alcohol, *see* §7.21.

8.22 PROPERTIES
(Commercial polymers, normally highly acetalised)

Solvents *Formal* Ethylene dichloride, dioxan, isophorone, tetrahydrofuran, glacial acetic acid, phenols; mixed solvents, such as 50/50 ethylene dichloride/diacetone alcohol, 40/60 alcohol/toluene, and 30/70 alcohol/carbon tetrachloride. Formals with high acetate content dissolve in ketones and esters. *Acetal and butyral* Largely as for formals, but also soluble in lower alcohols, acetone, aromatic hydrocarbons (toluene) and esters. Polyvinyl acetal is compatible with several other resins (e.g. cellulose nitrate, with improvement in weather resistance); butyral of high hydrolysis but low acetalisation is soluble in water. **Plasticised by** castor oil (especially for butyral), common ester-type plasticisers (triacetin, phthalates, phosphates, adipates, sebacates), polyethylene glycol ethers, certain epoxy resins. Increase in the length of the side-chain (R) facilitates plasticisation. **Swollen by** aromatic hydrocarbons and ketones (butyral may dissolve), pyridine. **Relatively unaffected by** aliphatic hydrocarbons (petrol, mineral oils), turpentine, vegetable oils, water (some absorption), cold alkalis and dil. acids. **Decomposed by** hot alkalis and hot dil. acids, which tend to regenerate free aldehydes and to hydrolyse any residual acetate groups.

8.23 IDENTIFICATION

Thermoplastic. Cl, N, S, P normally absent. **Combustion** The polymers burn fairly readily, with a yellow flame and some smoke; on extinction, residue has 'sweetish' odour (butyraldehyde detectable if present). **Pyrolysis** The polymers liquefy (viscous 'melt') and char; volatile products are acid, and contain the free aldehyde. *N.B.* Pyrolysis of polyvinyl alcohol also produces aldehydes. **Colour tests** Distillation with dil. sulphuric acid yields the respective aldehyde in the distillate and polyvinyl alcohol (*see* §7.23) in the residue. Each aldehyde colours Schiff's reagent, but only formaldehyde responds to the CTA test (§26.23a); butyraldehyde can be recognised by its odour. *N.B.* Polyvinyl alcohol gives traces of aldehydes when distilled with dil. acid.

8.3 PHYSICS

8.31 GENERAL PROPERTIES

Specific gravity Formal, 1·20; acetal, 1·14–1·16; butyral, 1·11 (up to 1·5 when compounded and plasticised). **Refractive index** Formal, 1·50; acetal, 1·45; butyral, 1·49. **Water absorption** (%) Formal, c. 1; acetal, c. 2; butyral, 3–5 (c. 2·2 when compounded and thermoset). **Moisture regain** (thin films, unplasticised, unoriented; 65% r.h.; %) Formal and acetal, c. 1·5; butyral, c. 2·0. (Moisture relations depend on proportion of free —OH groups and on residual acetate groups present.) **Permeability** Unless of high acetal content the polymers are relatively permeable to water vapour and relatively impermeable to organic vapours and hydrogen. The butyral (moderately impermeable to water vapour) has been used as a barrier for protection from vesicants, e.g. mustard gas.

8.32 THERMAL PROPERTIES

Specific heat c. 0·3. **Conductivity** Formal and acetal, $c.\ 4\,(10^{-4}\ \text{cal})/$ (cm s °C). **Coefficient of linear expansion** $(10^{-4}/°C)$ Formal, 0·8; acetal, 0·7; butyral, 1·5. **Physical effects of temperature** Second order transition temperature (°C): acetal, 82; butyral, 49. Formal and acetal soften at 110–160°C; butyral softens at 80–140°C (unplasticised); all three decompose c. 200°C. Plasticised butyral is soft, flexible, and shock-resistant at least down to −20°C.

8.33 ELECTRICAL PROPERTIES

Volume resistivity $c.\ 10^{14}$ ohm cm. **Dielectric strength** (kV/mm) Formal and acetal (thin films), c. 40; butyral, c. 16. **Dielectric constant** c. 3. **Power factor** c. 0·01.

8.34 MECHANICAL PROPERTIES

Extensometric data	*Formal*	*Acetal*	*Butyral*	*Butyral (plasticised)*
Elastic modulus (Young's) (kgf/mm^2)	over 200	over 200	over 200	(Low)
Tensile strength (kgf/mm^2)	4–7	3·5–6·5	5–6	Under 2
Extension at break (%)	4–10*	3–5	c.5†	Up to 400
Flexural strength (kgf/mm^2)	c. 12	—	c. 8	—

*Up to 300% for films with 50% plasticiser (tritolyl phosphate)
†Experimental films, unplasticised, yielded at 4·5 kgf/mm^2 (3·7% extension), extended c. 100% then hardened to break at 11 kgf/mm^2 (over 150% extension)

Impact strength (Izod test) Formal and butyral, up to c. 1 ft lbf/in of notch. **Hardness** Formal, Rockwell M80–90. Plasticised butyral is soft and rubber-like. **Friction and abrasion** Grey cotton yarn, running at 1 m/s against smooth polyvinyl formal, $\mu_{dyn} = 0.22$. Compounded polyvinyl butyral shows high abrasion resistance (but low dynamic flexural resistance).

8.4 FABRICATION

Films can be cast from solution or extruded; heat sealable at c. 200°C. For surface coatings the polymers are applied from solution, often as blends with other resins (e.g. shellac, coumarone/indene resins, cellulose nitrate). For maximum mechanical and thermal endurance, surface coatings and structural adhesives are based on polyvinyl formal and a cross-linking component, e.g. 2 parts formal and 1 part thermosetting phenol-formaldehyde resin provides a durable stoving enamel. Plasticised polyvinyl butyral can be applied (e.g. to textiles) from aq. dispersion. Approximate moulding data (formal and butyral): compression moulding, 120–165°C, 500–2000 lbf/in^2; injection moulding or extrusion, 150–185°C, 20000 lbf/in^2.

8.5 SERVICEABILITY

The polymers are only slightly affected by prolonged exposure to sunlight and otherwise remain unchanged by exterior weathering. They are not attacked by bacteria, fungi, or insects.

8.6 UTILISATION

Polyvinyl formal Lacquers, adhesives, and priming coats for metals. Used for non-cracking and oil-resistant electrical insulation on copper wire (*see* BS 1844). With a thermosetting phenolic resin it provides a thermally-resistant high-strength light alloy adhesive (Redux). It has been employed as a thin film, for bonding high-grade plywood, and in expanded form as a low-density constructional material. The formal and *polyvinyl acetal*, alone or compounded with other resins, are used for high-gloss enamels and furniture lacquers. *Polyvinyl butyral* Used in wash primers for improving adhesion of paint to metal. Plasticised polymer is employed for bonding laminated safety glass, also as a protective coating on fabrics and as unsupported sheeting, e.g. for rain capes.

8.7 HISTORY

The polymers have a comparatively short history, being first developed in Canada in the late 1920s. Cross-linking was first noted, by Haas, in 1942.

POLYACRYLATES AND POLYMETHACRYLATES
Acrylic resins

With notes on polyacrylic acids and their salts, polyacrylate rubbers,
reactive acrylic copolymers, cyanoacrylate and anaerobic adhesives,
polyacrylamide, and ketone-formaldehyde resins

For an allyl resin, resembling hard acrylic resin but cross-linked, *see* §24C.81.

SYNONYMS AND TRADE NAMES

PMMA (Polymethyl methacrylate), PMA (Polymethyl acrylate), etc. *Plastics:* Acrylite, Asterite, Diakon, Lucite, Oroglas, Perlac, Perspex, Plexiglas, Plexigum. *Aqueous dispersions or solutions:* Breon 2671E2, Plex, Plexileim, Plextol, Primal (Rhoplex), Revacryl, Revertex A, Texicryl, Texigel, Vinacryl. *Solvent-based solutions:* Acronal, Epok D, Larodur, Plexisol, Scopacron, Scopacryl.

GENERAL CHARACTERISTICS

Normally thermoplastic substances, soluble in organic liquids, acrylic resins range from PMMA—the hard glass-clear material (e.g. Perspex)—to softer, rubbery, tacky, or wax-like products. Included in this section, however, are water-soluble polymers and certain reactive copolymers that can be set to insoluble products, e.g. as surface coatings.

9.1 STRUCTURE

Simplest Fundamental Unit

$$-CH_2-CH-$$
$$|$$
$$COOR$$

(Acrylate polymers)

$$\overset{\displaystyle CH_3}{\underset{\displaystyle COOR}{-CH_2-\overset{|}{\underset{|}{C}}-}}$$

(Methacrylate polymers)

where R = an alkyl or aryl radical, or (in water-soluble polymers) a monovalent cation. The hard glass-clear material, polymethyl methacrylate, is thus represented by

$$-CH_2-C(CH_3)-$$
$$|$$
$$COOCH_3 \quad C_5H_8O_2, \text{ mol. wt.} = 100$$

Molecular weight Plastics type, of the order of 5.10^5 to over 10^6 (degree of polymerisation approx. 5000–10000). Surface-coating types: organic-solvent-based, 3.10^4–15.10^4; aq. dispersions, 10^5 to over 5.10^6.

X-ray data Substantially amorphous materials.

9.2 CHEMISTRY

9.21 PREPARATION

(i) *Acrylate monomers* These can be obtained from ethylene, by aerial oxidation and conversion of the resulting oxide to ethylene cyanohydrin followed by dehydration and hydrolysis in the presence of an alcohol,

$$CH_2{=}CH_2 \xrightarrow{O} \overline{CH_2CH_2O} \xrightarrow[\substack{(NaOH\ or \\ amine)}]{HCN} CH_2(OH)CH_2 \underset{CN}{|} \xrightarrow[(H_2SO_4)]{ROH}$$

$$CH_2{=}CH\ (+NH_4{\cdot}HSO_4)$$
$$\underset{COOR}{|}$$

In an alternative route the cyanohydrin is obtained from ethylene via the chlorohydrin, $CH_2(OH)CH_2Cl$. Substitution of water (as steam) for the alcohol yields acrylic acid, $CH_2{=}CH{\cdot}COOH$.

The esters (or acid) can also be made by a Reppe synthesis from acetylene, carbon monoxide, and the appropriate alcohol (or water), in the presence of a catalyst such as nickel, or nickel carbonyl with hydrogen chloride,

$$CH \equiv CH + CO + ROH\ (or\ H_2O) \xrightarrow[(heat,\ pressure)]{catalyst} CH_2 = CH{\cdot}COOR$$

$$(or\ COOH)$$

Acrylic acid can be synthesised directly from propylene, by a two-stage oxidation process, via acrolein, using metal oxide (Mo-type) catalysts at 400–500°C,

$$CH_2 = CHCH_3 \xrightarrow[(Cat.)]{O} CH_2 = CHCHO \xrightarrow[(Cat.)]{O} CH_2 = CHCOOH$$

Another process depends on pyrolytic dehydration of acetic acid to ketene, and its reaction with anhydrous formaldehyde to yield

88

β-propiolactone, which is converted by acid catalysis to the acid (or, in the presence of an alcohol, to an ester),

$$CH_3COOH \xrightarrow[\text{(hot metal)}]{-H_2O} CH_2 = CO \xrightarrow[\text{(ionic cat.)}]{HCHO} CH_2\!-\!CH_2 \xrightarrow[\substack{\text{(or ROH} \\ +H_2SO_4)}]{(H_3PO_4)}$$

$$\underset{O\!-\!-\!-\!CO}{}$$

$$CH_2 = CH$$
$$\underset{COOH \text{ (or COOR)}}{}$$

A further method of preparation is by hydrolysis of acrylonitrile (§§10.23(4) and 10.82).

(ii) *Methacrylate monomers* Methyl α-methacrylate, monomer for the hard glass-clear polymer, is usually derived from acetone, via acetone cyanohydrin, followed by dehydration, hydrolysis and esterification,

$$CH_3COCH_3 \xrightarrow{HCN} CH_3C(OH)CH_3 \xrightarrow[\text{(H}_2SO_4)]{(-H_2O)(H_2O)}$$
$$\underset{CN}{}$$

$$CH_2 = CCH_3 \xrightarrow[\text{(H}_2SO_4)]{CH_3OH} CH_2 = CCH_3$$
$$\underset{COOH}{} \qquad \underset{COOCH_3}{}$$

Another process depends on oxidation of isobutylene with nitric acid, followed by dehydration,

$$CH_2 = C(CH_3)_2 \xrightarrow{O} \text{(various intermediates)} \xrightarrow{-H_2O}$$

$$CH_2 = CCH_3$$
$$\underset{COOH \text{ (or COOR)}}{}$$

(iii) *Polymerisation* Acrylate and methacrylate esters of lower alcohols are mobile liquids (examples of boiling points: methyl acrylate, 80°C; ethyl do., 99·5°C; *n*-butyl do., 147°C; methyl methacrylate, 100°C; *n*-butyl do., 163°C) which in general show only slight solubility in water (though methyl acrylate dissolves to some 5%). The pure monomers are stable in the absence of light and air, or when containing an inhibitor (e.g. hydroquinone) but heat, ultra-violet light, oxygen, and peroxides initiate radical polymerisation. Sheets and small blocks of the hard plastics material are cast (e.g. between glass sheets) from a monomer in which has been dissolved some polymer—to thicken it—and an initiator, such as benzoyl peroxide, following which the assembly is slowly heated for several hours up to a moderately elevated temperature. Also employed are: solution polymerisation, yielding products for surface coating lacquers; suspension or granular polymerisation, for moulding powders; and emulsion polymerisation, yielding latices of importance for water-based paints and the impregnation or coating of leather, paper and textiles.

The molecular weight and rate of polymerisation can be controlled by the use of a suitable concentration of an initiator, such as a peroxide, azo-compound, or redox system (e.g. persulphate-bisulphite). Copolymerisation of mixed monomers greatly widens the range of the properties of the products, particularly when one monomer is other than an acrylate, e.g. acrylamide (yielding cross-linkable products) or vinylidene chloride (giving high gloss surface finishes).

For further information *see* §§9.81 and 9.82.

9.22 PROPERTIES

Data are for the hard glass-clear material (PMMA) unless otherwise stated.

Soluble in aromatic and most chlorinated hydrocarbons (toluene, ethylene dichloride, chloroform), esters (ethyl acetate), ketones, tetrahydrofuran; 80/20 toluene/methanol gives low-viscosity solutions. Polymers of butyl and higher esters are soluble in aliphatic hydrocarbons (e.g. white spirit, also in molten waxes). Hard polymers yield the more viscous solutions; cross-linked polymers are insoluble, but swell in chlorinated hydrocarbons. **Plasticised by** some ester-type plasticisers (tritolyl phosphate, dibutyl phthalate), castor oil; also by copolymerisation or by blending with softer grades. Cast sheets and blocks are usually unplasticised. **Swollen by** alcohols, phenols, ether, carbon tetrachloride. **Relatively unaffected by** aliphatic hydrocarbons (white spirit, paraffin, oil—except butyl and higher polymers), conc. alkalis (including 0·880 NH_4OH), most dil. acids (also conc. HCl), aq. solutions of salts and oxidising agents (NaOCl, H_2O_2), ozone. **Decomposed by** conc. oxidising acids (HNO_3, H_2SO_4, H_2CrO_4), alcoholic alkalis.

9.23 IDENTIFICATION

Data are for the hard glass-clear material (PMMA) unless otherwise implied.

Thermoplastic. Cl, N, S, P nominally absent; copolymers may contain N or Cl. **Combustion** Once ignited, burns slowly with almost smoke-free, blue-based, candle-like flame, leaving negligible residue; does not smoulder on extinction. Insoluble acrylics (§9.82) usually burn in a similar way but leave a charred residue. **Pyrolysis** Distils largely as monomer (formaldehyde may be detected, §26.23a). Vacuum pyrolysis at 350°C yields largely pure monomer (characteristic fruity odour; infra-red spectrum; n_D^{25}, 1·4120; Saponification No., 560 mg KOH/g). Acrylic esters can be estimated by treatment with an excess of methanolic mercuric acetate and subsequent titration with methanolic hydrogen chloride (MALLIK, K. L. and DAS, M. N., *Chem. and Ind.* 162 (1959)). **Other tests** Saponification of a polyalkyl acrylate, e.g. by boiling a solution in acetone with 2N alcoholic alkali, yields a precipitate of a salt of polyacrylic acid (tested in neutral aq. soln.: $BaCl_2 \rightarrow$ ppt. insol. in NH_4OH; $AgNO_3 \rightarrow$ ppt. sol. in NH_4OH). Polymethacrylates are less easily saponified, but treatment with alkali in ethylene glycol yields the alcohol along with complex products.

9.3 PHYSICS
Data are for the hard glass-clear material (PMMA) unless otherwise stated.

9.31 GENERAL PROPERTIES

Specific gravity 1·18–1·19. Other methacrylate polymers: ethyl, 1·10–1·11; *n*-butyl, 1·05; isobutyl, 1·02. Polymethyl acrylate, 1·22. **Refractive index** *c.* 1·49. Other methacrylate polymers: ethyl, *n*-butyl, 1·48; isobutyl, 1·475. Polymethyl acrylate, 1·47. *Dispersion*(n_D-1)/ (n_F-n_C)*, 49–58. *Transmission*: (i) *ultra-violet*—cut off below 2800–3000 Å; (ii) *visible*—very low absorption, but transmission of light incident normal to surface of sheet material suffers 8% loss, due to 4% reflection at each interface; rapid increase in reflection above 60 degree angle of incidence, makes average transmission of daylight *c.* 85%; (iii) *infra-red*—cut off above 23000 Å (*see also* Fabrication, below). **Water absorption** (%) Immersed 24 h, 0·3–0·4; much prolonged, *c.* 2. Initial moisture content (at 65% r.h.), up to 0·8; polymethyl acrylate, 1·5. Absorption decreases with ascent of both ester series. **Permeability,** as $(10^{-10} \text{ cm}^2)/(\text{s cmHg})$ at 23°C: hydrogen, 3·3; nitrogen, oxygen, 0·1. Water vapour at 40°C, 3·0 $(10^{-10} \text{ g})/$ (cm s cmHg).

9.32 THERMAL PROPERTIES

Specific heat 0·35. **Conductivity** *c.* 4·5 $(10^{-4} \text{ cal})/(\text{cm s °C})$. **Coefficient of linear expansion** $(10^{-4}/°C)$ 0·75 (20°C), 1·05 (80°C). **Physical effects of temperature** Maximum service temperature, *c.* 80°C (heat-

---

*As inverse of *dispersive power*, see Notes, Section 59.

shaped sheet tends to return to flat state above *c.* 90°C). Softens, 100°C (loaded beam), 90–110°C (Vicat). Cast sheet becomes rubbery and flexible, 120–150°C (*see* Fabrication, below). Decomposition evident, 180–190°C; readily decomposes (little residue, reverts largely to monomer), 250–300°C. Other *n*-alkyl ester polymers have lower second-order glass temperatures and soften before PMMA (and acrylates soften before corresponding methacrylates); this is reflected in the brittle temperature, e.g. (°C):*

	Methyl	*Ethyl*	n-*Butyl*	n-*Octyl*	n-*Dodecyl*	iso-*Butyl*
Acrylate polymers	*c.* 6	−24	−55	−65	−3	−24
Methacrylate polymers	*c.* 105	65	20	−20	−65	53

9.33 ELECTRICAL PROPERTIES

Volume resistivity 10^{15}–10^{17} ohm cm; rises with polarisation time. *Surface resistivity, c.* 10^{14} ohms. **Dielectric strength** (kV/mm) 3 mm sheet in oil: 16 (conditioned at 75% r.h.), 15 (immersed 24 h in water); higher (e.g. 20) for thinner samples. Non-tracking. **Dielectric constant** and **power factor** The first property slowly rises with temperature rise and falls with frequency rise; second property exhibits similar relations but above 50°C behaves in a more complex way. Examples, approximate values:

Temperature	°C	−10	20	50	80
Dielectric constant	50 Hz	3·0	3·3	3·8	4·3
	10^5 Hz	2·7	2·75	2·85	3·0
Power factor	50 Hz	0·04	0·06	0·075	0·05
	10^5 Hz	0·01	0·02	0·04	0·07

9.34 MECHANICAL PROPERTIES

Note on physical state of acrylic resins at room temperature

Commencing with the strongest and most rigid, PMMA (to which most of the data below apply), the *n*-alkyl polymethacrylates increase in softness with ascent of the series until waxy characteristics supervene, poly(*n*-dodecyl methacrylate), i.e. polylauryl methacrylate, being the softest methacrylate polymer; the corresponding acrylate polymers are initially softer (headed by PMA, which is already rubbery) and increase in softness until poly(*n*-tetradecyl acrylate). Hardness increases with bulky or branched-chain substituents, and is related to brittle temperature (*see* earlier).

*Lowest values of which are shown by the *n*-octyl acrylate and *n*-dodecyl methacrylate polymers, beyond which waxy crystallinity causes hardening.

Elastic modulus (Young's) (kgf/mm^2) c. 300 (200 at 70°C, 170 at 80°C). **Tensile strength** and **extension at break** Some typical examples, values approximate:

| | Methacrylate polymers | | | | Acrylate polymers | | |
	Methyl	ethyl	n-butyl	iso-butyl	Methyl	ethyl	n-butyl
Tensile strength (kgf/mm^2)	7*	3·5	1	2·4	0·7	0·2	0·02
Extension (%)	4†	7	200	2	750	1800	2000

*Cast sheet up to 8·5. Falls to 50% at 50°C
†Yields at low rate of strain or above 25°C

Flexural strength (kgf/mm^2) 10–11, cast sheet up to 14. **Compressive strength** (kgf/mm^2) 8–13. **Impact strength** (Izod, ft lbf/in of notch) 0·3–0·4; falls at low temperature (-40°C). Higher methacrylate (e.g. n-butyl) and flexible acrylate polymers are tougher. **Hardness** Brinell (25 kgf load, 2·5 mm ball), c. 25; Rockwell, M60–120; Vickers diamond pyramid (5 kgf load, 15 s), 17 (hard sheet 22); Mohs, 2–3; Pencil, 9H. Methacrylate polymers are harder than corresponding acrylates (*see* note at commencement of this section). **Friction and abrasion** (polymethyl methacrylate, approximate values) μ_{stat} : polymer/polymer, 0·8; μ_{dyn}: polymer/mild steel, 0·45–0·5; grey cotton yarn (1 m/s)/smooth rod, c. 0·25. Compares with aluminium in surface hardness and abrasion.

9.4 FABRICATION

Compression moulding, c. 160°C, 2000 lbf/in^2. Injection moulding, 160–220°C, 20000 lbf/in^2. Cast sheet is readily heat-shaped when softened at 140°C; can be blow moulded and vacuum formed (e.g. domes) at 150–170°C. Though slightly brittle when cold, and rubbery if overheated, the material can be satisfactorily cut, machined, and polished; it is cementable (with chloroform, or with a solution of the polymer in methylene chloride). In addition to the preparation of sheets by bulk or cast polymerisation, it can be fabricated by a dough technique, i.e. ⎡monomer mixed with finely-divided polymer and 0·5% catalyst sets hard on heating for some hours at 70°C (or at 70°C then 100°C, for dental casting).

9.5 SERVICEABILITY

Acrylic resins, as the hard plastics material and the varieties used as surface coatings, possess excellent fastness to light and resistance to exterior weathering. They are not subject to microbiological attack.

93

9.6 UTILISATION

Clear resins Sheets, mouldings, extrusions and formed shapes, particularly for windows and lighting fittings; also (dyed or pigmented) for decorative displays, illuminated signs, bathroom fittings, telephones, etc. *Solvent solutions* Varnishes and printing inks for rubber, leather, PVC leathercloth, and paper. *Aq. solutions* (§9.81) for thickening and sizes. *Aq. dispersions* These have numerous uses as emulsion paints, adhesives, printing pigment binders (e.g. in screen printing) and textile finishes (e.g. non-slip, soft or stiff 'texturising', backcoating). The reactive kinds (§9.83) are particularly useful in adhesive and binding applications.

9.7 HISTORY

Duppa and Frankland prepared ethyl methacrylate in 1865, and Fittig and Paul prepared ethyl methacrylate polymers in 1877; acrylic acid (first synthesised in 1843, and its polymerisation noted in 1872) and its esters and polymers thereof were also investigated about this period, transparent polymethyl acrylate being prepared by Kahlbaum in 1880. The main study of these polymers was, however, made by Röhm, commencing *c.* 1901. Commercial production of polyacrylates began in Germany *c.* 1927 (Röhm and Haas), and following discovery of the relative hardness of methacrylate polymers, by W. Chalmers in Canada; these were developed a few years later (e.g. from 1931–38, by J. W. C. Crawford of ICI Ltd.).

In 1912 Röhm vulcanised ester polymers with sulphur, obtaining rubbery products. 'Unmodified' polyacrylate rubbers were developed (B. F. Goodrich Co.) in the early 1940s. The U.S. Department of Agriculture and Akron University pioneered rubbers containing a small proportion of a cyano- or chlorine-containing monomer, which facilitates vulcanisation, and 'modified' rubbers of this type are now the more widely used. Commercial production of both 'unmodified' and 'modified' rubbers began in 1948 (B. F. Goodrich Chemical Co.).

9.8 ADDITIONAL NOTES

9.81 POLYACRYLIC ACIDS AND THEIR SALTS
(Plexileim, Syncol, Texigel)

Acrylic acid ($CH_2{=}CHCOOH$, m.p. 13·5°C, b.p. 142°C), obtained as mentioned in §9.21, and methacrylic acid ($CH_2{=}C(CH_3)COOH$, m.p. 15°C, b.p. 163°C)—also the related itaconic acid ($CH_2{=}C(CH_2COOH)COOH$, m.p. 167°C), obtained by a fermentation process and of importance in reactive copolymers (§9.83)—can be readily polymerised, in bulk or in solution, with radical initiators to yield the respective polymeric acids. More commonly, a salt of poly-acrylic acid is made by alkaline hydrolysis of polyacrylamide (§9.85) or polyacrylonitrile (Section 10); alternatively, saponification of a polyacrylic ester, such as polymethyl acrylate, is employed (saponification of polymethacrylic esters is more difficult).

Polyacrylic acids and their salts with monovalent cations are hard, brittle, water-soluble solids (the salts more soluble than the acids). The viscosities of the solutions vary greatly with pH, reaching a maximum (due to chain uncoiling) under neutral or mildly alkaline conditions. The dissolved acids and their salts give precipitates with polyvalent cations, e.g. Ca^{2+}, Cu^{2+}. Because of their solubility, thickening and dispersing power, and film-forming capacity, the acids and their sodium or ammonium salts have many applications, e.g. for thickening aq. latices (in paints and in the dipping and spreading of textiles), for creaming natural rubber latex, for textile sizes (particularly polyacrylic acid on nylon) and backfilling, in soil conservation by aggregation, and in drilling muds.

9.82 POLYACRYLATE RUBBERS
(Cyanacryl; Hycar PA, PA-21, PA-31, and 4021; Krynac 880; Lactoprene BN and EV; Poly FBA; Paracril OHT; Thiacril 36 and 44).

Polyacrylates can be cross-linked by reaction with various reagents (diamines, peroxides), and a fluorine-containing monomer 1,1-dihydroperfluoro-n-butyl acrylate (CH_2=$CHCOOCH_2CF_2CF_2CF_3$) yields a vulcanisable elastomer (Poly FBA) that shows good resistance to hot oils and lubricants. More commonly, however, reactive sites are introduced by copolymerising an acrylic ester with a small proportion of a monomer containing an appropriate substituent (—Cl, —CN, —COOH) or providing a residual double bond. Typical elastomers of this kind are 95/5 ethyl acrylate/2-chloroethyl vinyl ether ($ClCH_2CH_2 \cdot O \cdot CH$=$CH_2$) and 87·5/12·5 n-butyl acrylate/ acrylonitrile (CH_2=$CHCN$). Polymerisation, radical-initiated (by a peroxide or persulphate), is usually carried out in aq. emulsion followed by coagulation, or by a suspension method yielding the elastomer directly as granules.

The ethyl acrylate copolymer has sp. gr. c. 1·1, n_D^{20} 1·4345, Mooney viscosity (100°C, 4 min reading) 42–58; Poly FBA has sp. gr. 1·54, n 1·367. The ethyl and butyl acrylate copolymers are soluble in benzene, toluene, chlorinated hydrocarbons, ketones and esters; the 'ethyl' dissolves in methyl and ethyl alcohols but is insoluble in aliphatic hydrocarbons, the 'butyl' is soluble in the latter but insoluble in the alcohols.

Polyacrylate rubbers are normally vulcanised with amines, e.g. triethyltrimethylenetriamine, triethylenetetramine, hexamethylene-diamine carbamate (sulphur acts as an anti-ager rather than vulcanising agent). The normal press (mould) cure is often followed by 'tempering' for (say) 12–24 h at 150°C to reduce compression set. Reinforcing fillers (carbon black, silica, calcium silicate) are needed to get the best physical properties, up to tensile strength $1 \cdot 7$ kgf/mm^2, breaking elongation 500 %.

The poor room-temperature resilience (5–30 % rebound) and low-temperature resistance of the ethyl acrylate copolymer (glass temperature $-24°C$, brittleness temperature $-13°C$) can be improved by ester-type plasticisers. Glass temperature: butyl acrylate copolymer, $-55°C$; Poly FBA, $-30°C$.

Useful properties of polyacrylate rubbers are good resistance to flex-cracking, atmospheric oxidation, ozone, ultra-violet radiation and oils (notably sulphur-containing lubricating oils up to 180°C; fluorine-containing rubbers (e.g. Poly FBA) are exceptionally resistant to aliphatic and aromatic hydrocarbons and ester type lubricants, and exhibit low permeability to gases with a wide service temperature range ($-40°$ to $+200°C$) and retention of light colours in sunlight. Resistance to water, steam, methyl alcohol, glycols and alkalis is poor; e.g. water absorption (100°C, 48 h), 37–38 %. Butyl acrylate copolymers have better resistance to water, but poorer heat and oil resistance, than those of ethyl acrylate. Electrical properties are fairly good.

Uses *Solid rubbers*: gaskets, seals, coatings for heat-resistant textiles (glass, asbestos), printing rolls, air bags, sun-resistant temporary coatings (e.g. on tyres and de-icers), wire and cable insulation (e.g. transformer leads), white and pastel-shade articles. *Latex*: non-skid rug backings, non-woven textiles, warp size, pigment binder for textile and paper printing, water- and grease-proof and high strength papers, leather finishes, adhesives.

9.83 REACTIVE ACRYLIC COPOLYMERS

Copolymerisation of an alkyl acrylate or methacrylate (e.g. ethyl or *n*-butyl acrylate and/or methyl methacrylate) with a small proportion —say 5 %—of a monomer containing a reactive substituent yields film-forming products that (as with polyacrylate rubbers, above) can be rendered insoluble by covalent cross-linkage. The reactive groups may be supplied by acrylic acid, methacrylic acid, or itaconic acid (§9.81), or by acrylate esters containing hydroxyethyl ($-CH_2CH_2OH$), aminoethyl ($-CH_2CH_2NH_2$), or glycidyl ($-CH_2\overset{\frown}{CHCH_2O}$) groups; acrylamide (§9.85) serves a similar purpose, and the amide groups can be after-methylolated by addition of formaldehyde. Inclusion of a monomer such as styrene reduces the cost.

These products are used as solutions or aq. latices. The latices can possess high solids content and low viscosity, but can if desired be readily thickened, e.g. with hydroxyethylcellulose (Section 16), or—if acid groups are present—with ammonium hydroxide. The films are cured by heating, the final product being obtained in various ways, e.g. by stoving for several minutes at 200°C (butanol-based lacquers) or a few minutes at 130–150°C (films from latices containing —COOH or —CONH$_2$ groups), but more usually—with films from aq. latices—by prior incorporation of a supplementary cross-linking agent such as a melamine-formaldehyde precondensate (Section 27) particularly when acid or amide groups are present, or by prior incorporation of an acid catalyst when methylolated amide groups or epoxy groups are present.

Reactive acrylic copolymers can be designed to cure to soft or hard products resistant to ageing and weathering. They find important uses (from solution in organic solvents) as stoving lacquers and paints for metal, and (as water-based latices, subsequently curing to water- and solvent-resistant products) for bonding non-woven fabrics and for coating and bonding textiles—the thickened latices being very suitable for knife coating.

9.84 CYANOACRYLATE AND ANAEROBIC ADHESIVES

Two special acrylic-based adhesives are as follows:

(A) CYANOACRYLATE ADHESIVES (Eastman 910)

These are highly polar liquids, based on alkyl α-cyanoacrylates (e.g. $CH_2{=}C(CN)COOCH_3$) together with a plasticiser, thickener and stabiliser (polymerisation inhibitor). They polymerise and rapidly set solid at room temperature simply on pressing into a thin film between metals, glass, ceramics, rubbers, or certain plastics having a mildly basic reaction or carrying a film of adsorbed water. The bonds are strong, particularly to polar substances and after ageing (at room temperature or up to 80°C); the polymerisation mechanism is probably ionic, and the monomer needs to be kept away from moisture, but when stabilised (e.g. with SO_2) it can be stored in an inert container, e.g. of polyethylene. Polymerisation is retarded by acids but accelerated by alkalis (hence by glass surfaces, as being both polar and alkaline). The polymerised material resists many common solvents, including oils, but dissolves in dimethylformamide and is slowly attacked by dil. acids, alkalis, hot water or steam. Bonds are serviceable at least down to −17°C, and up to 80°C or above, but strength is permanently impaired by heating above 100–120°C; the softening point of the polymer is 165°C. The refractive index is 1·46. At 1 MHz the dielectric constant is 3·34 and the power factor is 0·02.

(B) ANAEROBIC ADHESIVES (Loctite)

Acrylic esters of aliphatic diols, e.g. tetramethylene glycol dimethacrylate $(CH_2=C(CH_3)CO\cdot O(CH_2)_4O\cdot CO(CH_3)C=CH_2)$, have the property, when in contact with air, of remaining freely fluid in the presence of a free radical catalyst and accelerator, but of polymerising at room temperature when penetrating between closely-fitting surfaces of metal, glass, ceramics, and certain polymers. Since the monomer is doubly unsaturated, the product is cross-linked, and the particular features of the reaction enable it to be used for locking thread grooves, the resulting bond being stronger than can be obtained with locknuts.

9.85 POLYACRYLAMIDE

Acrylamide $(CH_2=CHCONH_2$, m.p. 85°C), obtained by acid hydrolysis of acrylonitrile and soluble in water and other polar liquids, readily polymerises in the presence of free radicals and yields polymers of high molecular weight (alternatively, heated with alkalis, it yields poly-β-alanine or nylon 3, *see* §23.81a). The polymers are amorphous and dissolve in water*, to produce viscous solutions but, being non-ionic, their viscosity is much less affected by alkali than that of polyacrylic acids (§9.81); above pH 10, however, polyacrylamide undergoes further hydrolysis with conversion of the amide to carboxyl groups. In addition to its reactions with formaldehyde (*see also* §9.83), polyacrylamide can be rendered insoluble by treatment with glyoxal under alkaline conditions, or by partial hydrolysis followed by reaction with a polyvalent salt. The polymer softens at 210°C but finds most uses in aq. solution; e.g. for flocculation or suspension of powdered minerals and ores, as a textile size, and for increasing the strength of paper and retention of paper fillers.

9.86 KETONE-FORMALDEHYDE RESINS

The simplest product from the reaction of acetone with formaldehyde (1:1) under alkaline conditions is monomethylolacetone (or γ-ketobutanol), which when dehydrated by distillation with zinc chloride yields methyl vinyl ketone (or methylene acetone),

$$CH_3COCH_3 + HCHO \xrightarrow[\text{alkali}]{\text{dil.}} CH_3COCH_2CH_2OH \xrightarrow{-H_2O}$$

$$CH_3COCH = CH_2$$

---

*Certain forms of ionic polymerisation, and solid-state polymerisation initiated by exposure to γ-radiation, yield crystalline polymers that are water insoluble.

Substitution of methyl ethyl ketone for acetone yields the corresponding methyl isopropenyl ketone. Methyl vinyl ketone can also be made by reaction of vinylacetylene and water, with a mercuric salt as catalyst,

$$CH_2{=}CHC{\equiv}CH + H_2O \xrightarrow{Hg^{2+}} CH_2 = CHCOCH_3$$

These unsaturated ketones polymerise with radical catalysts to transparent thermoplastic resins with the simplest of repeat units as follows (cf. those from methyl acrylate and methacrylate, §9.1):

$$
\begin{array}{cc}
-CH_2-CH- & -CH_2-C(CH_3)- \\
\quad | & \quad | \\
\quad COCH_3 & \quad COCH_3
\end{array}
$$

The linear polymers are inferior to those from acrylates, being subject to discoloration and shrinkage; however, occasionally acetone-formaldehyde applied in the form of a curable precondensate of di- or tetramethylolacetone is used as a textile finish.

For further reference, *see*: WHITE, T. and HAWARD, R. N., *J. Chem. Soc.* 25 (1943).

9.9 FURTHER LITERATURE

RIDDLE, E. H., *Monomeric Acrylic Esters,* New York (1954)
HORN, M. B., *Acrylic Resins,* New York and London (1960)
PIGGOTT, K. E., *J. Oil and Col. Chemists Assn.,* 46 (1963) 1009 (thermosetting acrylic resins)

POLYACRYLONITRILE AND COPOLYMERS
*Principally as Acrylic Fibres**

With notes on 'modacrylic'† and related fibres, and the degradation of acrylonitrile polymers

SYNONYMS AND TRADE NAMES
PAN, polyvinyl cyanide. *Fibres:* Acrilan (various types), Courtelle, Orlon (various types), and many other names; e.g. Acribel, Creslan, Crylor, Dolan, Dralon, Leacril, Nitron, Nymcrylon, Redon, Tacryl, Wolcrylon, Zefran.

GENERAL CHARACTERISTICS
Rigid transparent material, little developed for plastics purposes but used in great variety as soft, light-weight, wool-like, textile fibres.

10.1 STRUCTURE

Simplest Fundamental Unit

$$—CH_2—CH—$$
$$\underset{CN}{|} \qquad C_3H_3N, \text{ mol. wt.} = 53$$

N.B. This unit comprises at least 85% of an acrylic fibre, and is present in lower proportion in 'modacrylic' fibres—*see* §10.81.

Molecular weight Number-average, of the order of 5.10^4–10^5 (degree of polymerisation 1000–2000); weight-average values approx. 2·5 times greater.

X-ray data Substantially amorphous, partially crystalline when stretched and oriented as fibres; prominent chain-spacing of 5·3 Å normal to fibre axis; identity period uncertain (pattern diffuse, probably *c.* 4·5 Å). The limited solubility probably arises from dipole interaction and hydrogen bonding involving the —CN groups.

10.2 CHEMISTRY

10.21 PREPARATION

Acrylonitrile or vinyl cyanide (CH_2=CHCN, b.p. 78°C) can be obtained by:

---

*Acrylic fibres here implies consisting largely of acrylonitrile units, cf. *Acrylic plastics*, Section 9, which are based on acrylic esters.
†Fibres containing less than 85% acrylonitrile, *see* §10.81.

(i) direct combination of *acetylene* and hydrogen cyanide in the presence of a catalyst,

$$CH \equiv CH + HCN \xrightarrow[(Cu_2Cl_2)]{Catalyst} CH_2 = CHCN \ (+ \ tars)$$

(ii) catalytic oxidation of *ethylene* to ethylene oxide (or to acetaldehyde) followed by reaction with hydrogen cyanide and dehydration of the intermediate cyanohydrin,

$$\overset{\frown}{CH_2CH_2O} \ (or \ CH_3CHO) + HCN \longrightarrow CH_2(OH)CH_2CN$$

$$(or \ CH_3CH(OH)CN) \xrightarrow[(Al_2O_3, \ 300°C)]{-H_2O} CH_2 = CHCN$$

or (iii) reaction of *propylene*, ammonia, and atmospheric oxygen in the presence of a molybdenum- or phosphate-based catalyst,

$$CH_2 = CHCH_3 + NH_3 + 3/2O_2 \xrightarrow[H_3PO_4, \ 500°C]{Cat. \ (Mo \ or} CH_2 = CHCN + 3H_2O$$

The last method is now of major importance. In a variation nitric oxide replaces ammonia and oxygen,

$$4CH_2 = CHCH_3 + 6NO \xrightarrow[(Ag, \ 500°C)]{Catalyst} 4CH_2 = CHCN + 6H_2O + N_2.$$

Polymerisation of the monomer, inhibited by air (O_2), can be radical-initiated (by azo-compounds, peroxides, persulphates, various redox combinations, etc.) and is carried out in aq. solution, suspension or emulsion*, or in solution in organic solvents. Ionic-polymerisation can also be used. To improve the affinity of fibres for acid dyes, acrylonitrile is commonly copolymerised with a basic monomer, such as 2-vinylpyridine or *N*-vinylpyrrolidone, while for basic dyes an acid co-monomer such as acrylic acid or methallylsulphonic acid may be used. For other copolymers of acrylonitrile, *see* §10.81; for fibre production *see* Fabrication, §10.4.

10.22 PROPERTIES

Dissolved by organic liquids such as *N,N*-dimethylformamide, γ-butyrolactone, dimethyl sulphoxide, hot tetramethylene sulphone, hot propylene carbonate; also by fairly conc. mineral acids (e.g. 60% HNO_3, 70% H_2SO_4), conc. aq. solutions of highly soluble salts (e.g. LiBr, NaCNS, $ZnCl_2$; particularly with 20% alcohol present), and aq. or molten quaternary ammonium salts.

N.B. Aqueous solutions discolour if left exposed to air.

---

*At room temperature acrylonitrile is 7·5% soluble in water, and water is 3·5% soluble in acrylonitrile.

Plasticised by castor oil (can be added to spinning bath). **Relatively unaffected by** common organic liquids (including dry-cleaning fluids), oils, greases, aq. solutions (including sodium carbonate and hypochlorite), moderately strong acids—but see above—and dil. alkalis. The polymer is insoluble in the monomer. **Decomposed by** conc. alkalis, hot dil. alkalis, warm 50% sulphuric acid. *See also* §10.82.

10.23 IDENTIFICATION

Limited thermoplasticity. N present (up to 26·4%); Cl, S, P absent. Infra-red spectrum characteristic ($C \equiv N$ band at *c.* 2240 cm^{-1}) and distinctive for different kinds of acrylic fibres. **Combustion** Fibres shrink from flame, ignite (burning freely, as cotton); odour somewhat resembles that of burnt hair; hard black residue. **Pyrolysis** (*see also* §10.82). Fibres shrink, darken, and melt; evolve vapour (poisonous) with unpleasant odour and alkaline reaction (no monomer, principal products NH_3 and HCN; copolymers may produce an *acid* reaction), leave carbonised residue. **Colour tests** on the pyrolytic products (cyanide detection): 1. Test paper previously impregnated with cupric acetate and dried is freshly moistened with a dil. solution of benzidine (or a benzidine salt) in dil. acid and held in the pyrolytic vapour; the presence of hydrogen cyanide produces a bright blue coloration. 2. A little of the condensed pyrolysate is made just alkaline, then boiled with ferrous sulphate and finally acidified; cyanide ions yield a precipitate of Prussian Blue. *N.B.* Tests 1 and 2 give positive results with all pyrolysates from polymers of acrylonitrile and (sometimes less strongly) from copolymers thereof; also from cyano-compounds and (less strongly) proteins, and (still less strongly) amino-formaldehyde resins. Polyamides, however, do not respond. **Other tests:** 3. A solution of an acrylonitrile polymer in dimethylformamide becomes orange-red when warmed with sodium hydroxide, but turns yellow or colourless on acidification. 4. Acrylonitrile polymers are hydrolysed by moderately strong acid (e.g. 12N H_2SO_4) at the boil; when made alkaline, the hydrolysate evolves ammonia, changes from yellow to red-brown and ultimately becomes nearly colourless. 5. Staining tests. Treatment with mixed dyes serves to distinguish the various kinds of acrylic fibre by differences in receptivity, e.g. when dyed for a few minutes at the boil in a bath containing Sevron Orange L, Lissamine Green SFS (or Kiton Green V) and dil. H_2SO_4, acid-fibres (§10.21) stain orange-red and basic-fibres stain green. For combined staining and solvent tests see, for instance, *Acrylic Fibre Identification* (*Chemstrand Tech. Inf. Booklet A*12).

10.3 PHYSICS

10.31 GENERAL PROPERTIES

Specific gravity Fibres, $1\cdot17$–$1\cdot18$. **Refractive index** ($\lambda = 5790$ Å) Fibres: $n_{||}$, $1\cdot511$; $n_{\perp}$, $1\cdot514$–$1\cdot515$. **Birefringence** ($n_{||} - n_{\perp}$) $-0\cdot003$ to $-0\cdot004$ (some copolymers $+$ve). **Water absorption** (moulded plastics), 2–3%. **Moisture regain** (fibres, 65% r.h., %) $1\cdot2$–$2\cdot0$ (absorption), $1\cdot6$–$3\cdot0$ (desorption). **Water retention** (fibres) 10–20%. **Permeability** Films of the polymer show moderate permeability to water vapour, e.g. $1\cdot0$ (10^{-10} g)/(cm s cmHg) at 40°C, but lower transmission of organic vapours.

10.32 THERMAL PROPERTIES

Specific heat $0\cdot36$. **Coefficient of linear expansion** $1\cdot6$–$2\cdot0 . 10^{-4}$/°C. **Physical effects of temperature** Second order transition temperatures at approx. 87–95°C and 140°C. Fibres can be ironed at 150°C but continued heating or higher temperatures cause darkening (*see also* §10.82). Fibres stick at 210–245°C, decompose above 250°C. Fibres shrink up to 5% in boiling water, and 5–10% in dry heat at 250–265°C unless previously stabilised, e.g. by treatment for a few seconds at 250–350°C.

10.33 ELECTRICAL PROPERTIES (Films, air-dry)

Volume resistivity (ohm cm) 10^{14}; (purified fibres, 10^{12}–10^{14}). **Dielectric constant** $6\cdot5$ (50 Hz), $4\cdot2$ (1 MHz). **Power factor** $0\cdot11$ (50 Hz), $0\cdot03$ (1 MHz).

10.34 MECHANICAL PROPERTIES (Fibres)

Elastic modulus (Young's) Staple fibres, 400–700 kgf/mm^2 (40–70 gf/den.). The modulus of acrylic fibres (particularly the wet modulus) falls rapidly with rise of temperature (e.g. *c.* 1 gf/den. at 95°C) but severe heating causes permanent stiffening (and discoloration, *see* §10.82). **Elastic recovery** Staple fibres (Courtelle), similar results from a modacrylic fibre (Dynel, *see* §10.81);

Extension (%)	1	2	5	10
Recovery (%), immediate*	75	51	28	20
delayed†	90	81	59	44

*Extended at rate of 25% of initial length/min; held for 1 min at maximum extension; reversed; measured immediately at zero stress.
†As * but measured after 1 min at zero stress

Tensile strength c. 30–40 kgf/mm^2 (3–4 gf/den.), varies with type of fibre. Wet strength, c. 85% of air-dry strength. **Extension at break** c. 20–40%, varies with type of fibre. **Friction and abrasion** Fibres are soft but resilient (lofty) and of high resistance to dynamic flexure; abrasion resistance generally superior to that of cellulosic fibres but less than half that of nylon. Experimental value for coefficient of friction, μ_{dyn} (1 m/s): cont. fil. yarn (normal or lubricant-free)/mild steel, c. 0·3.

10.4 FABRICATION

For production of fibres the earliest process is still much used, being based on dry-spinning a solution of the polymer in an organic solvent (commonly dimethylformamide). Wet-spinning, however, is also employed, using a variety of solvents (e.g. dimethylformamide, dimethylacetamide, ethylene carbonate, dimethyl sulphoxide, nitric acid, sodium thiocyanate solution, or zinc chloride solution) and an appropriate coagulating bath (e.g. a liquid hydrocarbon or—for aq. systems—calcium chloride solution); in some processes wet-spinning is effected directly following solution-polymerisation of the monomer. In all instances spinning is followed by hot-stretching and annealing. Bi-component fibres, that crimp when heated, are made by spinning two solutions (two acrylonitrile copolymers, or one copolymer with another polymer) simultaneously through the same orifice. The polymer can be moulded (with difficulty) at 200–300°C under considerable pressure.

10.5 SERVICEABILITY

Acrylic fibres in general are unaffected by bacteria, fungi, or insects, and they exhibit good resistance to sunlight and ageing but tend to discolour when heated, e.g. on prolonged exposure at 100°C (*see* §10.82).

10.6 UTILISATION

Staple acrylic fibres, being soft and resilient, are used as a substitute or diluent for wool, and fabrics made from them show good crease resistance and crease retention (e.g. in permanent pleats); they are also made into resilient bonded batting.

The rot- and light-resistant properties suggest numerous outdoor applications for acrylic fibres in heavy-duty and finer fabrics, and in netting and filter fabrics.

10.7 HISTORY

Acrylonitrile was prepared by C. Moureu in 1893. Polyacrylonitrile was developed for fibres in Germany and the U.S.A. in 1940, after suitable solvents had been found (patents registered by Bayer—for conc. aq. solutions—and by Du Pont—for organic liquids—in 1942); pilot-scale production commenced in the U.S.A. in 1943, and the first acrylic fibre (Orlon) was made in 1948 and put into commercial production in 1950. Production of acrylonitrile in the U.K. began in 1959.

10.8 ADDITIONAL NOTES

10.81 'MODACRYLIC' AND RELATED FIBRES

As noted in §10.1, 'modacrylic' fibres contain under 85% (say 35–84%) of acrylonitrile copolymerised with another vinyl monomer, polymerisation of the mixed monomers being effected, by radical initiation, in aqueous emulsion or more particularly in an organic solvent (*see* §10.21). In the second instance, the solution of polymer (e.g. in acetone, ethylene carbonate, tetrahydrofuran) can be spun directly into—for example—hot water, followed by stretching (e.g. in hot aq. $CaCl_2$) which may render the fibre insoluble. Most 'modacrylic' fibres are resistant to sunlight and water (e.g. moisture regain, 0·5–1% at 65% r.h.) and moderately resistant to acids and alkalis (though not to hot alkalis), and they are cheaper than products containing a higher proportion of acrylonitrile; however they are mostly more readily swollen or dissolved by common organic solvents, soften at a lower temperature (fibres readily shrink unless stabilised), and have slightly lower strength. Fibres containing chlorine have relatively high densities, but have the advantage of being flame-proof (and therefore suitable for curtains, children's nightwear, etc.).

The principal kinds of 'modacrylic' fibres, identifiable by infra-red spectroscopy and by staining (*see* ref. in §10.23), are indicated below.

(i) *Acrylonitrile/vinyl chloride* copolymers (Dynel, Vinyon N). Simplest fundamental unit:

$$-CH_2-CH-\left(CH_2-CH\right)_n$$
$$\qquad\quad | \qquad\quad\quad | $$
$$\qquad\quad CN \qquad\quad Cl$$

The vinyl chloride content is 60% by weight ($n = c.$ 1·3). Specific gravity, 1·30. Refractive index, 1·53; birefringence, $c.$ 0·003.

(ii) *Acrylonitrile/vinylidene chloride* copolymers (Saniv, Teklan,

Verel). These somewhat resemble (i) both giving flame-proof fabrics. Specific gravity, 1·34.

(iii) *Acrylonitrile/vinyl acetate* copolymers. Fibres of this material have been made but tend to show poor thermal stability. *Vinylidene dinitrile* (or vinylidene cyanide, $CH_2=C(CN)_2$) copolymerised with *vinyl acetate* has been used to produce fibres related to the 'modacrylics', e.g. Darvan (a 1:1 copolymer), Travis.

10.82 DEGRADATION OF ACRYLONITRILE POLYMERS

Polymers and some copolymers of acrylonitrile decompose in a characteristic way, forming insoluble coloured products. The colour changes—which occur on heating in air, or in a vacuum or an inert gas, and are accelerated by the presence of organic acids or bases— progress through yellow to orange-red or brown, and ultimately to black; e.g. fibres become yellow in 30 min at 150°C, and blacken in 5 min at 250°C. Similar changes (yellow, orange, brown) occur when acrylic polymers, and more particularly solutions of them in dimethylformamide, are warmed with aq. sodium hydroxide, the changes in this instance being accompanied by a fall in molecular weight though not with development of insolubility.

The black fibres obtained as the ultimate result of thermal treatment ('graphitised' or 'carbon' fibres) show exceptionally high values of Young's modulus (*c.* 35000 kgf/mm^2) and tensile strength (*c.* 200 kgf/mm^2) and are finding engineering uses in resin-bonded composites, e.g. for aircraft components with a high stiffness/weight ratio. As produced by heating in air the initial discoloration is unaccompanied by changes in weight or strength, but an increase in basicity is ascribed (with some uncertainty) to intramolecular cyclisation and the formation of a structure containing conjugated double bonds, i.e.,

Further heating, at 200–300°C in air, brings about oxidation (with some liberation of NH_3, HCN and vinylacetonitrile—but no acrylonitrile) to yield a semi-conducting material considered to be a mixture of structures, i.e.,

and

Drastic heating (1000°C in an inert atmosphere, with a final 'graphitising' treatment at 1500–3000°C) ultimately leads to the crystalline end-product.

POLYVINYL CHLORIDE

With notes on chlorinated polyvinyl chloride, on copolymers of vinyl chloride and vinyl acetate or of vinyl chloride modified with vinylidene chloride or propylene, and on other types of modified PVC

SYNONYMS AND TRADE NAMES

Poly(monochloroethylene), poly(vinyl chloride), PVC. Breon, Carina, Corvic, Darvic, Ekavyl, Exon, Geon, Gobanyl, Halvic, Hefa, Hostalit, Kanevinyl, Lonza, Lutofan, Marvinol, Novon, Opalon, PCU, Pechiney, Pevikon, Pliovic, Policloro, Rhodopas, Sicron, Solvay, Solvic, Vestolit, Vinnol, Vipla, Vybak, Vygen. *Fibres:* Clevyl, Fibravyl, Isovyl, Khlorun, Movyl, Rhovyl, Thermovyl (heat stable).

GENERAL CHARACTERISTICS

(i) *Rigid PVC* Hard, horny and normally amber-coloured material; precise properties dependent on formulation; can be colourless and transparent, translucent, opaque; used for rods, sheets, tubes, mouldings, fibres and stiff films.

(ii) *Plasticised PVC* Can be made as flexible as required by selection of type and proportion of plasticiser. It can be transparent, translucent or opaque, colourless or pigmented, and is used for sheets, tubes, mouldings, coatings and unsupported films.

11.1 STRUCTURE

Simplest Fundamental Unit

$$-CH_2-CH- $$
$$\overset{|}{\underset{Cl}{}}\qquad C_2H_3Cl, \text{ mol. wt.} = 62{\cdot}5$$

Molecular weight Commercial grades, 50 000–120 000 (degree of polymerisation, approx. 800–2000); experimentally, 25 000–220 000 (number average), 30 000–520 000 (weight average).

X-ray data Substantially amorphous; commercial polymers, 5–15 % crystalline. Fibres under tension show an identity period of *c.* 5 Å with two fundamental units per identity period. Polymerisation by γ-radiation yields polymers of higher crystallinity (up to 25 %).

11.2 CHEMISTRY

11.21 PREPARATION

The monomer, vinyl chloride ($CH_2{=}CHCl$, b.p. $-13{\cdot}5°C$), which easily liquefies under pressure, can be obtained by various routes:

(i) Addition of hydrogen chloride to acetylene (from carbide or cracked petroleum)

$$CH \equiv CH + HCl \xrightarrow{HgCl_2 + BaCl_2} CH_2 = CHCl$$

(ii) Chlorination of ethylene to give 1,2-dichloroethane, which is then dehydrochlorinated by treatment with alkali or by cracking at 300–600°C (Woolfe process*),

$$CH_2 = CH_2 + Cl_2 \xrightarrow{catalyst} CH_2Cl - CH_2Cl \xrightarrow{-HCl} CH_2 = CHCl$$

(iii) Oxyhydrochlorination of ethylene,

$$C_2H_4 + HCl + air \xrightarrow[500°C]{CuCl_2 + KCl} CH_2 = CHCl \text{ (major yield)}$$

If mixed acetylene and ethylene are available, process (i) removes acetylene; ethylene is then chlorinated by (ii), supplying at the same time hydrogen chloride for (i).

Vinyl chloride can be polymerised by free-radical initiators or high-energy radiation. The polymerisation can be effected by bulk, solution, suspension or emulsion techniques. Most commonly used are suspension and emulsion methods.

Bulk polymerisation of liquefied vinyl chloride yields the purest product, the polymer being precipitated from the monomer. With an initiator such as benzoyl peroxide polymerisation can be effected in c. 17 h at 55°C to 80% conversion, or with azobis*iso*butyronitrile, in 12·5 h at 62°C, to 62·5% conversion.

Suspension polymerisation employs water-soluble suspension agents (polyvinyl alcohol, methylcellulose) and a monomer-soluble initiator (benzoyl or lauroyl peroxide), the system being stirred to maintain the suspension and heated to effect polymerisation.

Emulsion polymerisation follows a similar pattern except that emulsifying agents are employed, the initiator system is initially contained in the aq. phase, and the polymer is obtained in a very finely-divided state (e.g. particle diameter under 10^{-3} mm compared with c. 0·1 mm in suspension polymerisation). Emulsion polymers can be produced with higher molecular weights and greater surface area/weight ratio (for plasticiser absorption) than suspension polymers, but are more expensive, and the presence of residual emulsifiers confers inferior water resistance and electrical properties.

---

*From which the monomer is obtained in 95% yield; the hydrogen chloride can be reacted with acetylene as in (i).

11.22 PROPERTIES

Dissolved by tetrahydrofuran, cyclohexanone, methyl ethyl ketone, dimethylformamide; also by mixtures of solvents, such as acetone/ (carbon disulphide, carbon tetrachloride, or benzene), ethyl acetate/ carbon tetrachloride, and 4/1 trichloroethylene/nitromethane. Low molecular weight polymers are also soluble in warm tetrahydro- furfuryl alcohol, acetone, dioxan, ethylene dichloride and *o*-dichloro- benzene. Polymer of mol. wt. below 15000 is soluble in warm toluene. **Plasticised by** *primary plasticisers* which are normally high-boiling esters of C_{8-10} alcohols (phthalates, phosphates, sebacates and various fatty acid derivatives); polymeric plasticisers are also used (e.g. adipates, sebacates, or azelates of propylene glycol). *Secondary plasticisers*, of limited compatibility, include esters of some fatty acids, petroleum residues, alkyl and/or aryl hydrocarbons and their nitrated or halogenated derivatives (e.g. chlorinated paraffin wax). Phthalates are general-purpose plasticisers, phosphates offer some degree of fire-retardance, and sebacates are low-temperature plasti- cisers. The plasticisers most commonly employed are dioctyl (or 2-ethylhexyl) and di-iso-octyl phthalates, dialphanyl (i.e. C_{7-9}) phthalate, trixylyl phosphate, dioctyl sebacate, polypropylene sebacate, epoxidised soya bean oil. *See also* Table 56.T1. **Swollen** (particularly when plasticised) by aromatic and chlorinated hydro- carbons, nitroparaffins, acetic anhydride, aniline, acetone. **Relatively unaffected by** water, conc. alkalis, non-oxidising acids, hypochlorite solutions, aliphatic hydrocarbons (can leach out soluble plasticisers), oils, ozone. **Decomposed by** conc. oxidising acids (H_2SO_4, HNO_3, H_2CrO_4) which slowly attack the polymer. The rate of decomposition may be increased in the presence of zinc or iron or their compounds. *Stabilisation* All PVC polymers are degraded by heat and light, hydrogen chloride is eliminated and oxidation generally occurs. With heat, release of hydrogen chloride precedes oxidation; with photo-oxidation the reverse is believed to occur. The degradation is considered to proceed as a chain reaction initiated and propagated by activated chlorine ($Cl^\bullet$), e.g.

$$Cl^\bullet + -CH_2-\underset{\underset{Cl}{|}}{CH}-CH_2-\underset{\underset{Cl}{|}}{CH}- \longrightarrow$$

$$-CH_2-\underset{\underset{Cl}{|}}{CH}-C^\bullet H-\underset{\underset{Cl}{|}}{CH}- + HCl$$

$$-CH_2-\underset{\underset{Cl}{|}}{CH}-C^\bullet H-\underset{\underset{Cl}{|}}{CH}- \longrightarrow -CH_2-\underset{\underset{Cl}{|}}{CH}-CH=CH- + Cl^\bullet$$

The newly-released Cl˙ combines with further H (reactivating the chain and yielding further HCl) and is in turn again regenerated as the chain deactivates and forms another double bond, and so on . . ., so that the polymer decomposes by an 'un-zipping' mechanism, to produce a system of highly chromophoric conjugated double bonds (a polyene structure); colour develops when five to seven conjugated double bonds are formed. Common stabilisers (which interfere with the chain reaction and accept or scavenge hydrogen chloride) preventing degradation by heat include white lead (basic lead carbonate); tribasic lead maleate or sulphate; dibasic lead phosphite, phthalate or stearate; lead silicate, salicylate or stearate; lithium, calcium, barium, cadmium or zinc stearate; barium-cadmium laurate, dibutyl tin dilaurate, dibutyl tin dimaleate, tin mercaptides (e.g. dibutyl or dioctyl thiotin), esters and epoxidised oils (as mentioned under plasticisers; the epoxy ring opens to absorb H and Cl), some epoxy resin intermediates, triphenyl or tri*iso*nonyl phosphite, glycerol mono-oleate, diphenylurea and diphenyl oxide. Mixed stabilisers are often used and frequently the effect is synergistic. Tribasic lead sulphate is tending to replace white lead in opaque plasticised (especially electrical) PVC products, and is used in opaque rigid PVC. Barium/cadmium systems are used for clear and brightly coloured plasticised PVC, and organo-tin compounds for clear rigid PVC. Octyl tin compounds or calcium/zinc systems are used for non-toxic grades, depending on regulations.

11.23 IDENTIFICATION

Thermoplastic; Cl present (approx. 56%), N, S, P, normally absent but may be present in additives. **Combustion** Softens and chars, fibres shrink; burns but is self-extinguishing unless combustible additives are present; when kept burning (e.g. in a bunsen flame) produces a very sooty yellow flame with a green tip and leaves a black residue. **Pyrolysis** Chars, evolves hydrogen chloride (at 300°C *in vacuo*) and various hydrocarbons (400°C). No monomer is liberated. **Other tests** 1. To a warm solution, made by boiling a sample in 5 ml pyridine, add 0·5 ml of 2% w/v methanolic sodium hydroxide. A brown colour and eventually a brown precipitate are produced. 2. *Chlorine-content* The chlorine can be determined by (i) measurement of the hydrogen chloride evolved on pyrolysis at red heat, (ii) combustion in oxygen or (iii) ignition with sodium peroxide followed by acidification and titration with silver nitrate. (For semi-micro method of (iii), *see* HASLAM, J. and HALL, J. I., *Analyst*, **83**, 196 (1958).)

For colour tests for plasticisers, *see* §24A.23 (phthalates), §28.23

(phenols), and §57 2.3 and 2.5 (phosphorus). For distinction between PVC and polychloroprene, *see* §36.23.

11.3 PHYSICS

A wide range of properties can be obtained depending on formulation, compounding and processing, *Rigid PVC* usually implies that no plasticiser has been added; however, the term also covers very lightly plasticised PVC, where the rigidity is not drastically affected. *Plasticised PVC* is softer, the flexibility being dependent on the type and amount of plasticiser added. Compounding ingredients include stabilisers, lubricants, fillers, pigments, processing aids and impact modifiers; paste techniques are also employed (*see* §11.4(iv)).

11.31 GENERAL PROPERTIES

Specific gravity *Rigid PVC* Normal impact, 1·39; high impact, 1·34. Fibres, 1·4. *Plasticised PVC* 1·1–1·7 (dependent on plasticiser). **Refractive index** *Rigid PVC*, 1·53–1·56. Fibres, 1·54. *Plasticised PVC* 1·55–1·6. **Birefringence** of fibres practically zero, e.g. Fibravyl ($\lambda = 5790$ Å): $n_{\parallel}$, 1·541; $n_{\perp}$, 1·536; $n_{\parallel} - n_{\perp}$, 0·005. **Water absorption** Surface absorption after 32 days immersion at 20°C, rigid PVC, 5 g/m^2; plasticised do., up to 200 g/m^2. Absorption by plasticised PVC, after 48 h immersion at 50°C: without filler, 0·6% by weight; with mineral fillers, up to 11%. Absorption increases greatly with temperature, e.g. rigid PVC, 0·4 mm thick, 32 days immersion at 20°C, 0·95%; do. at 60°C, 11·7%. Absorption is somewhat greater for emulsion-polymerised than for suspension-polymerised (unplasticised) PVC, e.g. 0·4–0·6% compared with 0·2–0·3%. **Water retention** of fibres at 65% r.h., *c.* 8%. **Moisture regain** of fibres at 65% r.h., 0·10–0·15%. **Permeability** *Gases* (10^{-10} cm^2)/(s cmHg) at 31–32°C

	O$_2$	N$_2$	CO$_2$	He
Rigid PVC	0·34	0·12	1·6	4·6
PVC with 20% plasticiser	1·8	—	7·1	10·1

Water vapour (10^{-10} g)/(cm s cmHg) at 30°C. Rigid PVC, 0·2; PVC with 25% tritolyl phosphate, 0·25; PVC with 25% dioctyl phthalate, 1·1.

11.32 THERMAL PROPERTIES

Specific heat Rigid PVC, 0·25; plasticised PVC, 0·3–0·5. **Conductivity** 3–4 (10^{-4} cal)/(cm s °C) for rigid and plasticised PVC. Falls with increasing plasticiser content; rises with temperature up to the glass temperature, then decreases (*see* SHELDON, R. P. and LANE, K., *Polymer*, **6**, 77 (1965)). **Coefficient of linear expansion** (10^{-4}/°C) Rigid PVC, 0·7–0·8; plasticised PVC, *c.* 2·5. **Physical effects of temperature** *Rigid PVC* Brittle below -40°C, hard and tough at room temperature, softens at 80–85°C (second order transition). Fibres contract on heating, thermally stabilised types (e.g. Thermovyl) not shrinking until heated above 100°C. *Plasticised PVC* Flexible down to -50°C (with suitable plasticisers) and serviceable up to 75°C. In air above 100°C both rigid and plasticised PVC decompose, autocatalysed by release of hydrogen chloride, unless stabilisers are incorporated (*see* §11.22). Full decomposition occurs above 180°C.

11.33 ELECTRICAL PROPERTIES

Rigid PVC (23°C) **Volume resistivity** $>10^{14}$ ohm cm. **Dielectric constant** 3·0 (high impact, 3·3) at 800 Hz. **Power factor** 0·02 at 800 Hz. *Plasticised PVC* The above properties are highly dependent on plasticiser type and concentration and test temperature. **Dielectric strength** (kV/mm) Rigid PVC, 14–16; plasticised PVC, 20–28. *Resistance to tracking*, good (grade 4 on 1–5 scale of increasing resistance).

When electrical properties are important it is essential to use suspension or bulk polymer and not emulsion polymer.

11.34 MECHANICAL PROPERTIES

Elastic modulus (Young's) (kgf/mm^2) Rigid PVC, 280. Fibres: cont. fil., *c.* 500 (40 gf/den.); staple, relaxed, *c.* 100 (8 gf/den.). Plasticised PVC (stress at 100% elongation), 0·28–1·90. **Elastic recovery** (%) Fibres: cont. fil., from 1% extension, 100; 2%, 95; 4%, 85; 9%, 75; staple, relaxed, over much of extensible range, *c.* 10. **Tensile strength** (kgf/mm^2) Rigid PVC, 4·25–5·6 (falls with rise of temperature, e.g. 2 at 60°C, 1·5 at 70°C). Fibres: cont. fil., *c.* 34 (2·7 gf/den.); staple, relaxed, *c.* 15. Wet strength practically as dry strength. Plasticised PVC (for 60 parts plasticiser per 100 polymer), *c.* 2. **Extension at break** (%) Rigid PVC, 5–25. Fibres: cont. fil., 20; staple, relaxed, 180. Plasticised PVC, 200–450. **Flexural strength** (kgf/mm^2) Rigid PVC, 9·5; elastomer-modified PVC, 8·5. **Compression strength** (kgf/mm^2) Rigid PVC, 5·6–6·7. **Impact strength** (Izod test) Rigid PVC, 0·8 ft lbf/in notch; 'impact-modified' PVC (*see* §11.83), up to 15 ft lbf/in notch. **Hardness** Rigid PVC (Rockwell), R110–120. Plasticised PVC (Shore A), 50–100. **Friction** μ_{stat} Rigid PVC, 0·4–0·45. As the plasticiser content is increased the coefficient of friction increases and may pass through a maximum. μ_{dyn} Grey cotton yarn (running at 1 m/s, 65% r.h.)/rigid PVC, 0·23; do./plasticised PVC, 0·45. Suitably compounded plasticised polymers possess better **abrasion resistance** at room temperature than vulcanised rubber, but at elevated temperatures soon become too soft.

11.4 FABRICATION

Rigid PVC (i) Processing temperature, normal impact, 150–195°C; high impact, 140–180°C. (ii) Mould temperature for injection moulding, normal and high impact, 20–60°C. (iii) Mould shrinkage, normal and high impact, 0·7–1·0%. (iv) Suitable for injection moulding, compression moulding, blow moulding, extrusion, vacuum or thermo-forming, calendering, high-pressure laminating, coating. *Plasticised PVC* (i) Processing temperature, 150–200°C. (ii) Mould shrinkage 1–5% dependent on plasticiser. (iii) Suitable for injection moulding, compression moulding, blow moulding, extrusion, casting, rotational moulding, calendering, and laminating (especially on sheet steel). (iv) Paste techniques can be used where plasticised PVC needs first to be spread (e.g. on fabric) or dip-coated. *Plastisols* for this purpose consist of very finely-divided (high surface area) unplasticised PVC dispersed (along with stabilisers) in a plasticiser to form a creamy paste. The dispersion remains spreadable and free-flowing, and the polymer in it remains unplasticised until the temperature is raised; at *c.* 80–110°C it sets to a low strength product, but when raised above 150–160°C it becomes similar to plasticised PVC as obtained by other means, though inferior in some physical properties. The method is much used, possessing the advantages that no solvents need be employed (where they are used the mixture is known as an *organosol*), while the initially fluid paste can be deposited where required and then gelled *in situ.** *Fibres* of unplasticised polymer are wet-spun (from tetrahydrofuran or cyclohexanone) or dry-spun from mixed solvents (*see* §11.22).

11.5 SERVICEABILITY

(*See also* §11.22). *Rigid PVC* Processing affects end-use stability markedly, but generally chemical stability is good and weatherability

*For further reference to PVC pastes, *see* CLARKSON, D. L. and MACLEOD, N. D., *Chem. and Ind.*, 751 (1949); also WELLING, M. S., *Plastics*, 21, 121, 161, 194 (1956).

excellent; high impact polymer has good stability but is adversely affected by ultra-violet light. *Plasticised PVC* Stability almost entirely controlled by the type and amount of plasticiser and stabiliser, for both of which *see* §11.22.

11.6 UTILISATION

Rigid PVC Pipes, profiles, rods, tubes, sheets for uses such as guttering, corrugated roofing, ventilation ducts, wall cladding, lining for chemical tanks, and general chemical engineering. Cellular PVC (rigid) is used for battery separators and thermal insulation. Garments made from fibres of unplasticised PVC have been claimed to have therapeutic effects in arthritic conditions. The fibres have also been used as blends with wool. PVC powder is employed in pyrotechnics to deepen the colour produced by barium or strontium salts.

Plasticised PVC Used in sheet or tube form for flexible hose and ducting, curtaining, cover-sheets and tarpaulins, clothing (aprons, raincoats). Widely used for wire and cable insulation, footwear, fabric-backed wall covering, vinyl and vinyl/asbestos flooring, etc. Cellular plasticised PVC bonded to woven or knitted fabrics finds garment and upholstery uses. *Plastisols and organosols* are used to produce fabric-backed leathercloth for the furniture and automobile trades, heavy duty industrial clothing, handbags, shoes, gloves, etc.; also applied by dip-coating where a degree of cushioning and anti-corrosion is required.

11.7 HISTORY

H. V. Regnault first produced vinyl chloride in 1835 and observed its polymerisation in sunlight. E. Baumann examined the polymer in 1872, and I. Ostromislenskii investigated it more thoroughly (and plasticised it) in 1912. F. Klatte obtained the polymer by peroxide initiation in 1917, and in 1930 H. Staudinger, by fractionating the polymer from solution, showed the relation between molecular weight and solution viscosity. About 1930 plasticised PVC was introduced commercially, and was extensively developed for unsupported sheeting by about 1939. Internal plasticisation by copolymerising vinyl chloride with vinyl acetate was patented by E. W. Reid in 1928 (*see also* §11.82i). The paste technique originated in Germany in 1931 and was developed commercially in Great Britain in 1942. In Germany Klatte made fibres, and bristles and films from unplasticised material as early as 1913, and again in 1931 (PCU fibres), but modern production of unplasticised fibres originated in France and dates from 1941.

Due to the shortage of rubber during the Second World War, plasticised PVC was used for cable covering. PVC is the second most widely employed synthetic polymer, polyethylene taking first place.

11.8 ADDITIONAL NOTES

11.81 CHLORINATED POLYVINYL CHLORIDE
(Genclor S, Rhenoflex)

(*a*) *PeCe Fibres.* The polymer for these is made by chlorinating PVC to a chlorine content between 62 and 65%. The product, freely soluble in acetone and chlorinated hydrocarbons, has been used in lacquers and to provide textile fibres (stretch-spun into hot water from 30% acetone solution). The fibres have a specific gravity of 1·44, and resemble those of unplasticised polyvinyl chloride; they resist acids and alkalis, have practically zero moisture regain, are rotproof and flameproof, and possess good thermal and electrical insulating properties. They are, however, of relatively low strength, up to 24 kgf/mm^2 (1·8 gf/den.) with extension at break up to 45%. These fibres were developed in Germany 1934, and have been applied to non-rotting fishing nets, filter fabrics, tarpaulins, and belting. The molecular structure is represented as:

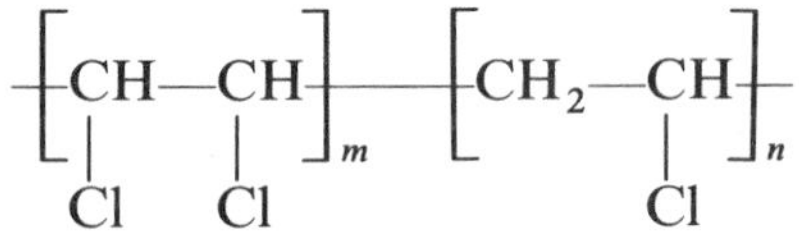

$$\left[\begin{matrix} CH-CH \\ | \quad | \\ Cl \quad Cl \end{matrix}\right]_m \left[\begin{matrix} CH_2-CH \\ | \\ Cl \end{matrix}\right]_n$$

where *m* and *n* are approximately equal; the dichloro-unit is mainly as shown, with $-CH_2-CCl_2-$ present to a less extent.

(*b*) *Heat-resistant chlorinated PVC* ('high-temperature PVC', Hi-Temp Geon, Welvic). Made by chlorinating PVC dissolved in (e.g.) a chlorinated solvent; structure of this is as shown above but with *m* much greater than *n*, i.e. a higher chlorine content than the material described under (*a*).

Chemical properties

Dissolves in tetrahydrofuran, cyclohexane and tetrachloroethylene. **Swollen by** *n*-butanol, conc. hydrochloric and nitric acids. **Relatively unaffected by** dil. acids and alkalis, conc. sulphuric acid, glycerol and petroleum.

Physical properties

General

Specific gravity 1·54. **Water absorption** 0·75 in pipe, 24 h at 23°C, 0·048–0·08%; unspecified test piece, 28 days at 23°C, 0·49%; at 100°C, 14·3%.

Thermal

Specific heat 0·29. **Conductivity** 3·9 (10^{-4} cal)/(cm s °C). **Coefficient of linear expansion** (10^{-4}/°C), 25–30°C, 0·57; 30–40°C, 0·65; 40–50°C, 0·80; 50–70°C, 0·94; 70–80°C, 0·69; 80–90°C, 0·57; 90–100°C, 0·40. **Physical effects of temperature** *Heat distortion temperature* (°C) normal grades, 98–105 at 0·185 kgf/mm^2; transparent grade, 83·2 at 0·185 kgf/mm^2 and 90·6 at 0·046 kgf/mm^2. *Vicat needle penetration temperature* (°C) Normal grades, 110–128; transparent grade, 96. *Softening temperature* (°C) Normal grades, 102–107; transparent grade, 87. *Heat shrinkage* (%) For transparent grade sheet varies from 0 at 80°C to −3 at 140°C in the longitudinal direction, and from −0·5 at 80°C to +1 at 140°C in the transverse direction.

Electrical

Volume resistivity (30°C, 65% r.h.) 5·8.10^{15} ohm cm. **Dielectric constant** (30°C, 55% r.h.) At 60 Hz, 2·93; 10^3 Hz, 2·92; 10^6 Hz (20°C), 2·0. **Dielectric strength** at 30°C, 39 kV/mm. **Power factor** (30°C, 55% r.h.) At 60 Hz, 0·0109; 10^3 Hz, 0·0110; 10^6 Hz, 0·0092.

Mechanical

Elastic modulus (Young's) (kgf/mm^2) Tensile, 290–330; compression, 190–200. **Yield stress** (kgf/mm^2) Tensile, 6·0–6·7; compression, 8·2–8·8. **Yield strain** (%) Tensile, 8·3–8·9; compression, 8·6–9·1. **Impact strength** (for material containing 'impact modifiers', since without them impact strength is lower than for ordinary PVC) (kgf mm/mm^2) Notched Charpy, 0·48–1·5; notched Izod, 0·4–1·1. **Hardness** (Rockwell) 117–119R.

Fabrication

Chlorinated PVC is slightly hygroscopic, hence drying for 1–2 h at 90–100°C, or hopper drying, before fabrication is recommended. Heat stability during processing is inferior to that of ordinary PVC.

Extrusion Screw type, constant pitch with gradually reducing groove depth full-flight screw with 25/1 L/D ratio and compression ratio of 2·5/1. Cooling must extend to the screw tip. Die pressure should be *c*. 6000 lbf/in^2; temperature 150–185°C. *Calendering* Reverse

L or Z-type calenders are suitable. The feedstock obtained from the Banbury should be at 160–170°C and should pass through the warm-up roller and calender is 3–4 min. Calender roll temperatures should be 175–195°C for the side roll and upper roll, and 180–195°C for the middle and bottom rolls. Transparent grades should be processed at temperatures about 10°C lower than normal grades. *Compression moulding* 180–195°C at 400–850 lbf/in^2 pressure for 5–10 min with full pressure cooling is recommended. *Injection moulding* 200–210°C, at 9000–17000 lbf/in^2 pressure, mould temperature of 75–120°C, in-line type screw, vacuum hopper, and a screw back-pressure of 700 lbf/in^2 is recommended. *Thermoforming* is generally carried out at 145–155°C preferably with indirect heating. *Welding* can be carried out with a hot-air gun operating at 240–250°C about 1 cm from the nozzle, gun pressure being over 6 lbf/in^2. *Jointing* can be effected with a solution of chlorinated PVC in tetrahydrofuran. *Printing* generally requires a tetrahydrofuran-based ink solution.

Applications

Because of its superior temperature resistance (45–60°C higher as compared with ordinary PVC), heat-resistant chlorinated PVC is used where PVC is unsuitable, e.g. hot-water pipes, waste-pipes, etc., also for chemical plant.

11.82 VINYL CHLORIDE COPOLYMERS

The following copolymers are regarded as modified forms of PVC, vinyl chloride being the major component. For copolymers of acrylonitrile with vinyl chloride *see* §10.81(i), and for copolymers of vinylidene chloride where vinyl chloride is the minor component, *see* Section 12.

(i) *Vinyl chloride/vinyl acetate copolymers and related products* (Vinylite, Vinyon, Vinyon HH).

These materials are colourless, water- and flame-resistant, flexible (at least as thin films), and more soluble than unmodified PVC. They are also more adhesive (i.e. when applied from solution or the melt) and this property can be enhanced by inclusion of carboxyl groups. Copolymerisation of vinyl chloride with vinyl propionate yields soft products, especially suited to coating sheet metal (e.g. from ethyl acetate solution) and for heat-sealing at *c*. 100°C. Ter-polymers of vinyl chloride, vinyl acetate, and—for instance—ethyl acrylate are also made. The chief uses of these products are as surface coatings and bonding agents, as fibres (of low softening point), unsupported films and flexible sheeting, and transparent dimension-ally-stable mouldings.

The **repeat unit** of the vinyl chloride/vinyl acetate copolymers may be represented as follows:

$$\left(\!\!-CH_2\!-\!CH\!-\!\!\right)_{\!n}\!\!-CH_2\!-\!CH\!-$$
$$\underset{Cl}{|}\qquad\underset{CH_3COO}{|}$$

where the average value of *n* for fibres (*c.* 85/15 vinyl chloride/vinyl acetate by weight) is approx. 8, or, for plastics (*c.* 90/10 vinyl chloride/ vinyl acetate), approx. 12·5. **Molecular weight** 20000–50000 (degree of polymerisation, 300–800 mixed units). **X-ray data** Substantially amorphous, even when stretched.

Chemistry

The monomers are copolymerised by the emulsion technique, using a free-radical initiator. Vinyl chloride polymerises at the faster rate, and by the use of rate-determining additives block copolymers are obtained. In general 6–20 chloride molecules polymerise before addition of an acetate molecule. The products are acetone-soluble unless above 95 % in chloride content. They can be plasticised, additional to the flexibility conferred by internal plasticisation, by common ester plasticisers. They are virtually as unaffected as PVC by most chemical reagents, but are swollen by aromatic hydrocarbons and decomposed by alcoholic alkalis and hot oxidising acids. The chlorine content (48·4 % for 15 % acetate) is, of course, lower than that of PVC (56·7 %).

Physics

General

Specific gravity 1·37–1·45; fibres, 1·34–1·36. **Refractive index** 1·55; fibres, 1·53. **Birefringence** Nil. **Water absorption** 0·08–0·15 %. **Water retention** Fibres, *c.* 12 %. **Moisture regain** Fibres (65 % r.h.), *c.* 0·1 %. **Permeability** Highly impermeable to vapour vapour.

Thermal

Specific heat 0·2–0·5. **Conductivity** 3·5–4·0 (10^{-4} cal)/(cm s °C). **Coefficient of linear expansion** $0·7.10^{-4}$/°C. **Physical effects of temperature** Plasticised material is flexible to very low temperatures (-50°C).

Unplasticised material becomes rubbery in boiling water; fluid at 125–150°C; decomposes at higher temperatures with evolution of hydrogen chloride. Fibres shrink at 60–65°C; become tacky above 85°C; fluid at 135–150°C.

Electrical

Volume resistivity (ohm cm) 10^{13}–10^{16}; fibres (90% r.h.), *c.* 10^{11}. **Dielectric strength** (kV/mm) Unplasticised, 55; plasticised, 16–24. **Dielectric constant** *c.* 3·0–3·5 (50 Hz and 1 MHz); increased by plasticisers. **Power factor** 0·01–0·02 (50 Hz and 1 MHz).

Mechanical

Elastic modulus (Young's) (kgf/mm^2) Plastics, 240–425; fibres, *c.* 350–450 (30–40 gf/den.). **Elastic recovery** Continuous filament, 70–90% for 1–2% extension; drops to about 30% beyond 5% extension. Staple, 10% from 5% extension; 5–10% from more than 10% extension. **Tensile strength** (kgf/mm^2) 5·3–6·0. Fibres: cont. fil. 25–30 (2·5 gf/den.); staple, *c.* 8·5 (0·7 gf/den.). **Extension at break** (%) Plasticised, 200–450. Fibres: cont. fil., 20–30; staple, up to 100. Wet strength of fibres is little different from air-dry strength. **Flexural strength** 8·5 kgf/mm^2. **Impact strength** (Izod test) 0·5–1·0 ft lbf/in of notch. **Hardness** Unplasticised material (Brinell, 2·5 mm ball, 25 kg load), *c.* 15; softened by plasticisers and increase of acetate content.

Fabrication of the copolymers is similar to that of PVC, though processing is easier and possible at lower temperatures (115–165°C) and hence with less degradation than with PVC, although the copolymers are *less* stable to heat (at fixed temperature) and light. Fibres are dry-spun from acetone solution, stretched at 65°C, and stabilised while under tension at 90–100°C. The copolymers resemble PVC in many respects. Their uses as mentioned earlier include dimensionally-stable clear or coloured mouldings; flame- and rot-resistant fibres suitable for fishing nets and filter fabrics, also for thermo-bonding of fibrous webs or fabrics in which they are included; flexible (additionally plasticised) tubing, sheets, etc. (e.g. for curtaining, rainwear, packaging); and from emulsion or solution they provide protective surface coatings (for metal, plaster, and brickwork) and adhesive primers (e.g. promoting adhesion between PVC and textiles). Solutions have proved useful in archaeological conservation work, e.g. transportation of ancient mosaics (the tesserae are bonded *in situ* to a thin skrim on which, after undercutting, the mosaic can be rolled up).

History The copolymers were developed in Germany and U.S.A. in the late 1920s. Fibres were made in 1928 and were developed in the U.S.A. in 1935, with commercial production commencing from

1937–39, a low-stretched fibre being introduced in 1942. Plastics with a range of acetate content were introduced in the early 1950s.

(ii) *Vinyl chloride/vinylidene chloride copolymers*
For high vinylidene copolymers, *see* Section 12. At the other end of the range copolymers with 95% vinyl chloride are available. These modified forms of PVC have greater solubility and better flow properties in processing; they are used in lacquers and impregnating compositions and for mouldings.

(iii) *Vinyl chloride/propylene copolymers** (Airco Series 400 Resins; Cumberland do.)
Propylene, employed as a co-monomer in the free radical suspension polymerisation of vinyl chloride (the vinyl chloride being added gradually to maintain a fairly homogeneous copolymer composition), yields a product having approximately the following repeat structure:

$$-\left[CH_2-\underset{\underset{Cl}{|}}{CH}\right]_{6-20}CH_2-\underset{\underset{CH_3}{|}}{CH}-$$

The propylene content usually lies between 3 and 10%. Propylene provides high chain transfer action which lowers the molecular weight and leaves propylene units at ends of the chain, because of which the copolymer is more stable than PVC (degradation by an 'unzipping' mechanism—*see* Stabilisation, §11.22—is less likely, both at the chain ends, and also in the chain). **Specific gravity** $1\cdot286–1\cdot445$ (decreases with increasing propylene content). Lower melt viscosity means lower processing temperatures and hence less need for stabilisers. Lower content of liquid stabilisers improves rigidity at high temperatures (also lowers permeability). **Tensile strength** (kgf/mm^2) 3–4·9. **Yield strength** (kgf/mm^2) 4·7–6·1. **Extension at break** (%) 100–200. **Elastic modulus** (Young's) (kgf/mm^2) 240–320. **Flexural strength** (kgf/mm^2) 6·1–7·75. **Flexural modulus** (kgf/mm^2) 270–327. **Impact strength** (Izod; ft lbf/in of notch at 23°C) 0·3–30. **Hardness** (Rockwell) R105–118.

Injection moulding Temperature 150–165°C. Pressures (lbf/in^2), (i) booster, 800–1200, (ii) holding, 700–1100, (iii) back, 50–75; mould temperature, 40–60°C. *Extrusion* Temperatures 150–175°C. The material can also be blow-moulded. **Applications** include lighting domes, television cabinet facings, refrigerator and vacuum cleaner parts, sheets and bottles, threaded enclosure caps, and flame-resistant films.

---

*Vinyl chloride/ethylene copolymers, with generally similar properties and uses, are also available (trade names: Bakelite QSQH, QSQL, QSQM).

11.83 OTHER TYPES OF MODIFIED PVC

Acrylonitrile-based rubbers introduced into PVC by blending or copolymerisation produce a material ('impact-modified' PVC) with better impact strength than normal PVC (Izod test up to 15 ft lbf/in notch, unaged), though the strength is gradually lost on weathering; remainder of properties are similar to unplasticised PVC. Copolymers of vinyl chloride with butadiene are soluble and have been used for lacquers, films and impregnation.

11.9 FURTHER LITERATURE

KAINER, F., *Polyvinylchlorid und Vinylchlorid-Mischpolymerisate*, Berlin (1951)

SCHILDKNECHT, C. E., *Vinyl and Related Polymers*, New York (1952)

BRITISH GEON LTD., *Geon PVC Resins: Technical Manual No. 1*, London (1959)

PENN, W. S., *PVC Technology*, 2nd edn., London (1967)

THOMASS, H., *PVC-Folien*, Munich (1963)

CHEVASSUS, F. and BROUTELLES, R. DE, *The Stabilization of Polyvinyl Chloride*, London (1963)

Anon., Society of Plastics Engineers, *Guide to PVC Technology*, 2nd edn., Stamford, Conn. (1964)

RITCHIE, P. D. (Ed), *Physics of Plastics*, London, 323 (1965). LANNON, D. A. and HOSKINS, E. J., effects of plasticisers and fillers, especially in PVC)

KAUFMAN, M., *The History of PVC*, London (1969) (Chemistry and industrial production)

SARVETNICK, H. A., *Polyvinyl Chloride*, New York (1969)

Standards
BSS 1763, 1774, 2739, 3757, 3878, 4023 and *4203* (sheeting)
BSS 2571 and *3168* (extrusion compounds)
ASTM D2383, D2396 and *D2538* (test methods)

SECTION 12

POLYVINYLIDENE CHLORIDE
*Principally as high-vinylidene copolymers**

SYNONYMS AND TRADE NAMES

Poly(vinylidene chloride), poly(1,1-dichloroethylene), PVDC, PVdC. Lumite, Polidene, Saran, Tygan (monofil), Velon, Viclan, Zetek.

GENERAL CHARACTERISTICS

Hard, tough, translucent to transparent material, sometimes faintly yellow in colour; produced also in brightly-coloured forms (e.g. as pigmented filaments) and as clear highly moisture-impermeable unsupported films (for packaging, etc.). Latices of both hard and soft copolymers find various uses, e.g. for barrier films, surface coatings, and gloss emulsion paints.

12.1 STRUCTURE

Simplest Fundamental Unit (homopolymer)

$$-CH_2-\overset{\displaystyle Cl}{\underset{\displaystyle Cl}{C}}- \qquad C_2H_2Cl_2 \ \text{mol. wt.} = 97$$

Molecular weight Commercial plastics, usually above 20000 (degree of polymerisation, over 200); emulsion copolymers, 200000.

X-ray data The homopolymer is too crystalline to be readily fabricated. However, when this material or a crystallisable copolymer of high vinylidene content is quenched from a melt, it is obtained temporarily in an amorphous state which is soft and formable, although ultimately it hardens and crystallises, e.g. on warming or cold-drawing. Oriented material is highly crystalline; fibre identity period, 4·7 Å. Copolymers prepared in aq. dispersion are usually amorphous but can become crystalline when converted to films.

---

*For copolymers where vinylidene chloride is present in a lower proportion, *see* §10.81ii (acrylonitrile) and §11.82ii (vinyl chloride).

12.2 CHEMISTRY

12.21 PREPARATION

The monomer, 1,1-dichloroethylene ($CH_2{=}CCl_2$, b.p. 31·9°C), is obtained by chlorination of ethylene, ethylene dichloride, or vinyl chloride to yield 1,1,2-trichloroethane, which is then dehydrochlorinated by heating with lime or alcoholic alkali, e.g.

$$CH_2 = CH_2 \xrightarrow{Cl_2} CH_2Cl - CHCl_2 \xrightarrow{-HCl} CH_2 = CCl_2$$

Alternatively, trichloroethylene can be cracked thermally (at 400°C) to yield a mixture containing also *cis* and *trans* 1,2-isomers, from which the monomer can be separated by distillation.

1,1-Dichloroethylene readily polymerises under the action of ultra-violet radiation or with radical-type initiators. It is only slightly soluble in water (can be stored under aq. alkali) and commonly it is polymerised in aq. suspension or emulsion, with exclusion of oxygen. To facilitate processing of the final product, commercial preparations are usually copolymers of vinylidene chloride (e.g. with 10–15 % acrylonitrile or vinyl chloride, or up to 50 % ethyl acrylate) or terpolymers (e.g. with 13 % vinyl chloride and 2 % acrylonitrile).

The yellow or brown colour which the polymer may acquire, especially under alkaline conditions, is probably attributable to conjugated unsaturation caused by dehydrochlorination, i.e.

$$[-CH_2-CCl_2-]_n \xrightarrow{-HCl} [-CH{=}CCl-]_n$$

To avoid this form of decomposition stabilisers of the acid-acceptor type are incorporated (e.g. amines, phenyl glycidyl ether, certain organo-metallic compounds) and, in the preparation of copolymers, precautions may be taken to minimise the occurrence of sequences containing more than 4 or 5 $-CH_2CCl_2-$ units; this can only be achieved, however, by lowering the vinylidene chloride content towards 50 %. In addition, since the decomposition is catalysed by common metals (Cu, Fe, Sn, Zn) it is necessary to use stainless or glass equipment in the preparation of these polymers; similarly, in their fabrication (§12.4), it is necessary to avoid contact of hot polymer, or solutions or dispersions of the polymer, with these metals.

12.22 PROPERTIES

Solubility The crystalline homopolymer resists organic liquids (dissolves in trichlorobenzene at 170°C) but polymer in the amorphous state will dissolve in tetrahydrofuran. Solubility is improved by copolymerisation, e.g. 90/10 vinylidene chloride/vinyl chloride copolymers dissolve in tetrahydrofuran, and 85/15 copolymers tend to be soluble in liquids such as ketones or ketone/aromatic hydrocarbon mixtures (for copolymers with only a minor proportion of vinylidene chloride, also soluble in ketones, *see* §11.22). **Plasticised by** diphenyldiethyl ether (facilitates fabrication; graphite has also been used). Common plasticisers ineffective, but highly chlorinated aromatics (e.g. hexachlorodiphenyl oxide) are compatible; flexibility improved by copolymerisation. **Swollen by** oxygen-containing liquids (dioxan, cyclohexanone, dimethylformamide), some chlorinated hydrocarbons. **Relatively unaffected by** hydrocarbons (mineral oils), alcohols, phenols, common organic solvents (but see above), moderately conc. acids and alkalis (except ammonia and strong amines; some yellowing in the presence of other alkalis). The polymer is almost insoluble in the monomer.

N.B. Chemical resistance declines with rise of temperature.

Decomposed by prolonged contact with ammonia or related compounds (*see* Colour test, below). Attacked by chlorine; slowly by hot conc. sulphuric acid. Copper and iron catalyse decomposition of the polymer, *see* §12.21.

12.23 IDENTIFICATION

Thermoplastic; Cl present (approx. 70%), N present in some copolymers; S and P absent. **Combustion** Softens, melts, does not readily burn—self-extinguishing; leaves swollen carbonaceous residue. Copolymers with high vinylidene chloride content are flame-resistant. **Pyrolysis** Melts, evolves strongly acid fumes (HCl and monomer). **Colour test** Films or shavings of PVDC darken when left for some days in contact with conc. ammonium hydroxide. Immersion in morpholine causes both sample and liquid to blacken within a few hours.

N.B. Chlorinated PVC (§11.81) *dissolves* in morpholine, yielding a brown solution; polymers and copolymers of vinyl chloride may swell and dissolve, but impart no colour.

12.3 PHYSICS

12.31 GENERAL PROPERTIES

Specific gravity 1·67–1·71. 90/10 vinylidene chloride/ethyl acrylate copolymer: amorphous, 1·67; crystalline, 1·875. **Refractive index** 1·60–1·63. Monofils (0·01 mm diam.): $n_{||}$, 1·603; $n_{\perp}$, 1·611. **Birefringence** $(n_{||} - n_{\perp})$ −0·008. **Water absorption** (plastics), 0·1%. **Moisture regain** (fibres; 65% r.h., 25°C), 0·0%.

N.B. Fibres may show slight permanent loss in weight if dried at 105°C.

Permeability Films with high vinylidene chloride content are highly impermeable to most gases and vapours; transmission increases with fall in vinylidene chloride content and with rise of temperature. Examples, (i) As $(10^{-10} \text{ cm}^2)/(\text{s cmHg})$: O_2, 0·005; N_2, 0·0009; CO_2, 0·03. (ii) Water-vapour, as $(10^{-10} \text{ g})/(\text{cm s cmHg})$: 0·0004.

12.32 THERMAL PROPERTIES

Specific heat 0·32. **Conductivity** c. $2\cdot5.10^{-4}$ cal/(cm s °C). **Coefficient of linear expansion** $0\cdot8-1\cdot75.10^{-4}/°C$. **Physical effects of temperature** Commercial copolymers: not brittle at $-40°C$; second-order transition, -17 to $-18°C$ (change in expansion rate); serviceable up to c. 75°C (95° intermittently); softens (fibres may first shrink), 115–135°C; melts, 170–175°C. Note: crystalline homopolymer melts c. 210°C (decomp. c. 250°C); but too high a proportion of vinylidene chloride introduces instability at c. 175°C, with darkening and loss of hydrogen chloride (*see also* §12.21).

12.33 ELECTRICAL PROPERTIES

Volume resistivity $10^{14}-10^{16}$ ohm cm. **Dielectric strength** (kV/mm) plastics, 16; thin films, 120–200. **Dielectric constant** c. 3 (50–10^6 Hz). **Power factor** between 0·03 and 0·1 (50–10^6 Hz).

12.34 MECHANICAL PROPERTIES

Elastic modulus, tensile strength, and extension at break

	Elastic modulus (Young's) (kgf/mm²)	Tensile strength (kgf/mm²)	Extension (%)
Plastics	20–60	3–4	10–40
Monofils, fibres	100–150 (7–10 gf/den.)	30–40 (2–2·6 gf/den.)*	20–40†
Films	—	5–10	25–35

*Higher in highly stretched material
†Unstretched material, 200–300%

Elastic recovery Stretched filaments are moderately elastic to short-term deformation (e.g. 65% recovery from 20% extension) and slowly recover from long-term loading (e.g. sagging of seat coverings).

Flexural strength Up to 12 kgf/mm². **Compression strength** *c.* 6 kgf/mm². **Impact strength** (Izod test) Up to 2 or more ft bf/in of notch, dependent on orientation. **Hardness** Rockwell M50–65. **Friction** Grey cotton yarn (1 m/s, 65% r.h.)/moulded polymer or film, μ_{dyn} 0·24.

12.4 FABRICATION

Compression moulding: *c.* 150°C, 2000 lbf/in². Injection or extrusion: 170–175°C, 20000 lbf/in²; moulds should be hot (80–90°C) to assist recrystallisation. For precautions against contact with metals, *see* §12.21. Soft amorphous polymer, obtained by shock-cooling, can be shaped at room temperature before reversion to the hard crystalline form. Plastics material can be cut, punched, machined, and hot-welded (300°C); copolymer films can be heat sealed. Filaments are melt extruded into warm water and immediately stretched up to 400%.

12.5 SERVICEABILITY

The polymers resist all but a few liquids and are unaffected by all but a few reagents (darkened by ammonia and related substances, and by hot conc. sulphuric acid). They are also unaffected by bacteria, fungi and insects, and resist mechanical abrasion, but the homopolymer shows poor resistance to sunlight and ultra-violet radiation (this can be overcome in some of the copolymers by addition of stabilisers).

On prolonged exposure to light, or to heat (especially in the presence of Cu, Fe, etc.) the polymers may darken and lose strength, but with properly compounded materials the deterioration is not serious.

12.6 UTILISATION

Moulded or extruded products, e.g. pipes resistant to water or solvents. Monofils and yarns, for heavy-duty readily-cleanable seat coverings, deck-chair canvas, blinds and awnings, insect screens, filter and bolting cloths; fibres, for dolls' non-flam hair. Highly impermeable gas- and vapour-resistant packaging films; heat-sealable and heat-shrinkable. Latices, for moisture-excluding all-enclosing webs (cocoons, i.e. films built up by a spraying technique); also for coating purposes (e.g. on paper and some textiles, also—in conjunction with Sb_2O_3—as a flame-retardant paint, and—as acrylate copolymers—for high-gloss paints).

12.7 HISTORY

V. Regnault prepared the monomer (from trichloroethane and alkali), and observed its polymerisation, in 1838; E. Baumann prepared the monomer (from vinyl chloride) in 1872, and I. Ostromislenskii examined its polymerisation in 1916, but the polymer was not produced commercially until 1939–40. It appeared as melt-extruded monofils in 1940 and as packaging films in 1942, staple fibres were available later but have not found many uses.

12.9 FURTHER LITERATURE

WOODFORD, D. M., *J. Soc. Chem. Ind.*, 316 (1966) (Emulsion polymerisation and properties of vinylidene chloride copolymers)

ASTM D729 (moulding compounds)

CELLULOSE
Principally as Fibres

With notes on natural, regenerated, standard, and α-, β- and γ-forms of cellulose, hydro- and oxycelluloses, cross-linked cellulose, and chitin

SYNONYMS AND TRADE NAMES

Polyanhydroglucose, poly(1,4-anhydro-β-D-glucopyranose). *Natural forms* (of high cellulose content) Pure cotton, flax, ramie; purified wood pulp. *Regenerated cellulose* Viscose and cuprammonium rayons. Cellophane. For additional forms of cellulose, and names of common fibres, *see* §§13.81 and 13.82.

GENERAL CHARACTERISTICS

Cellulose is the chief structural material of the vegetable world. The purest forms occur naturally in a fibrous state, as exemplified by cotton, flax, ramie, and hemp; extracted filter paper and regenerated forms such as viscose rayon also consist of almost pure cellulose (although reduced in molecular weight).

13.1 STRUCTURE

Simplest Fundamental Unit

$C_6H_{10}O_5$ mol. wt. = 162

N.B. Structurally cellulose is best represented with a repeat unit consisting of two 1,4-anhydro-β-glucose units,* i.e. as poly-1,4'-anhydrocellobiose, depicted as follows:

*See footnote p. 130.

or shown conformationally, in perspective

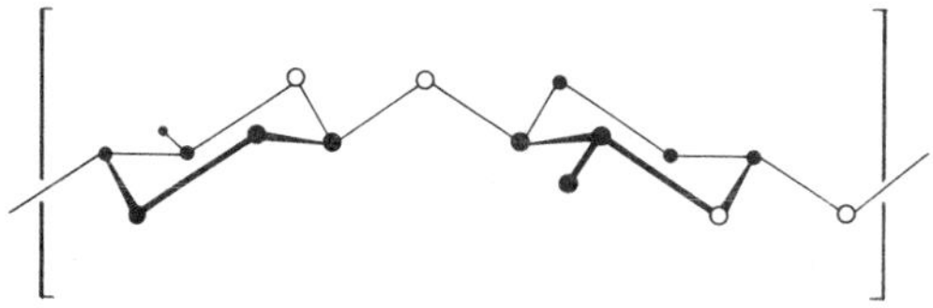

One of the two terminal units of the molecule contains a hemi-acetal group that exhibits aldehydic properties (as shown to the right, p. 129). In addition, particularly in regenerated cellulose, the molecule may contain a small proportion of modified units.

Molecular weight By light-scattering and sedimentation methods: native cellulose (cotton, flax, wood cellulose, etc.), 300 000 to over 10^6, initial high values falling on processing; regenerated cellulose, 50 000–150 000.

Degree of polymerisation Native cellulose, 2000 to over 6000; regenerated cellulose, *c.* 350 (normal viscose rayon), *c.* 700 (polynosic fibres), 500–1000 (high tenacity yarns).

X-ray data Polycrystalline. There are four anhydroglucose residues to the unit cell, in which the identity period is constant at 10·3 Å (cellobiose unit) but the lateral and angular dimensions vary, those of native cellulose (Cellulose I) being different from those of mer-cerised and regenerated forms (Cellulose II); additional specialised modifications (Cellulose III and IV) are of academic interest. Native fibres are composed of highly crystalline elementary fibrils, some 50–100 Å wide, aggregated in microfibrils *c.* 250 Å wide; the regions of maximum crystallinity (crystallites) appear to be 300–600 Å long. Regenerated fibres generally have smaller microfibrils and fewer crystallites, orientation of which increases on stretching; in polynosic fibres the crystallites are particularly large and highly oriented.

---

*Footnote to p. 129.
The difference between a 1,4-polymer of the β-configuration of anhydroglucose and one of the α-configuration (*see* Starch, Section 19) is illustrated below.

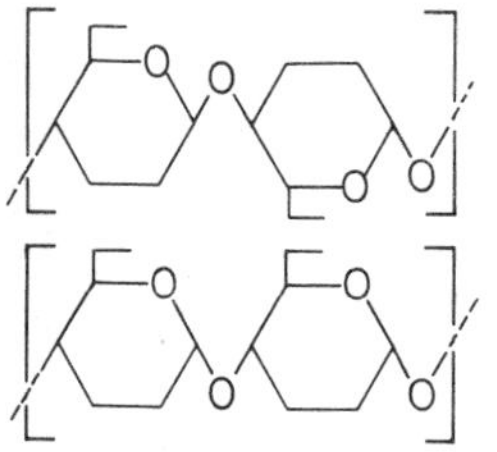

β-configuration

Cellulose (cellobiose unit)

α-configuration

Amylose (maltose unit)

The cellulose chain is essentially linear; the amylose chain is curved and assumes a helical structure.

13.2 CHEMISTRY

13.21 PREPARATION

Cellulose is present, as the chief constituent of the cell wall, in all plants (and, in association with lignin*, composes the structural material of wood), but it has not been synthesised artificially. For regenerated cellulose (i.e. cellulose precipitated from solution) the chief sources are cotton linters, spinners' waste, and especially wood pulp (e.g. from spruce, eucalyptus, etc.) which is first freed from lignin by treatment with alkali or acid bisulphite.

The Cuprammonium Process utilises a solution of purified cotton linters in cuprammonium hydroxide, from which cellulose is regenerated by spinning into a setting bath of warm water, followed by an acid wash.

The Viscose Process is most commonly used for the manufacture of fibres (viscose rayon), and the principle employed is as follows. Soda cellulose, from high grade α-cellulose treated with 18 % (w/w) sodium hydroxide and expressed to about 15 % retention, is 'aged' under controlled conditions, in which the combined degradative action of alkali and air yields a cellulose more suitable for the subsequent spinning. The product is then treated with carbon disulphide, forming sodium cellulose xanthate (degree of substitution, approx. 0·5):

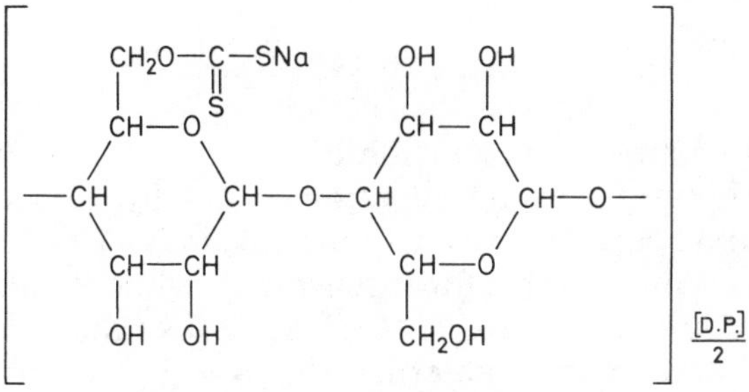

The xanthate is dissolved in dil. sodium hydroxide to give a viscous orange-coloured solution (viscose) which is filtered and allowed to 'ripen' (i.e. undergo a complex series of reactions characterised by first a fall and then an increase in viscosity) until thickened to the required degree. For the production of fibres the viscose is then spun into a bath (principal components dil. H_2SO_4, $NaHSO_4$, and $ZnSO_4$) in which the extruded filaments of xanthate undergo coagulation and regeneration to filaments of cellulose; these are wound off and stretched while wet and ductile, in order to orient the

*A highly complex polymeric substance present in woody materials; for composition see, for instance, BAYLIS, P. E. T., *Sci. Prog.* **48**, 409 (1960) and PEARL, I. A., *The Chemistry of Lignin*, London (1967).

chain molecules in the direction of the fibre axis and improve strength. For the production of films (e.g. Cellophane) the solution is extruded through a linear or annular slit; to soften the product, the final washing bath contains glycerol.

Fibres prepared under the normal conditions have a 'skin/core' structure. Developments of the viscose process are as follows: *High tenacity rayons* (e.g. tyre cords and tough staples), having an 'all skin' composition are obtained by addition of modifying substances to the viscose or the coagulating bath followed by application of high stretch in hot dil. acid. Other strong rayons have been made by stretching in 60% sulphuric acid (Lilienfeld fibres). An alternative process consists in stretching cellulose acetate fibres in steam, followed by deacetylation (Fortisan). *Polynosic rayons*, possessing a high molecular weight (D.P. twice the usual value), high crystallinity and high orientation, are characterised in particular by a high wet modulus and low swelling in water, so that they resemble cotton more than rayon in several properties (including improved dimensional stability in fabrics). These differences from ordinary rayons are achieved by omission of the 'ageing' and 'ripening' stages (*see* above), by using less alkali in the xanthate solution, and by spinning at high stretch into a bath of low acid concentration.

For natural forms of cellulose, and proprietary names of regenerated fibres, *see* §§13.81 and 13.82.

13.22 PROPERTIES

Solvents No simple solvents. Soluble in aq. solutions of metal complex compounds, such as cuprammonium hydroxide (Schweizer's reagent) and cupriethylenediamine (cuene). Less effective with natural forms but suitable for regenerated cellulose, are tris(ethylenediamine) cadmium hydroxide (Cadoxen), alkaline iron/tartaric acid complexes, and certain quaternary bases, e.g. 2N dimethyldibenzylammonium hydroxide (Triton F). In addition, ordinary regenerated cellulose dissolves in an aq. solution of an alkali and the oxide of a metal such as zinc (e.g. by aq. sodium zincate) and in ice-cold alkalis (e.g. 2N sodium hydroxide at $-5°C$); polynosic rayons are more resistant.

N.B. Cellulose in alkaline solutions is particularly susceptible to degradation by atmospheric oxygen.

Plasticised by Water; polyhydric alcohols, e.g. glycerol. **Relatively unaffected by** many organic liquids, e.g. most types of hydrocarbons, chlorinated hydrocarbons, oils, esters, high alcohols. **Swollen by** water, and especially by moderately conc. alkalis. At 20°C, maximum swelling of cotton, forming a more reactive structure, occurs in 4·5 N sodium hydroxide, but commercially (i.e. for Mercerisation) it is usually carried out in 6–7 N solutions, where the process is less temperature dependent; at $-10°C$, 2 N solutions can be used. Cellulose is also swollen by conc. solutions of certain highly soluble salts, e.g. calcium thiocyanate or zinc chloride (*see* solvents above, and vulcanised fibre §13.82iii), and it can be 'parchmentised' by brief immersion in approximately 20 N sulphuric acid. **Decomposed by** hot acids of sufficient strength, and most oxidising agents (intermediate products, *see* §13.85). In the presence of an alkaline medium, atmospheric oxygen causes some degradation. Ordinary regenerated cellulose is more prone to chemical attack than stretch-spun fibres or natural forms.

13.23 IDENTIFICATION

Non-thermoplastic. Cl, N, S, P, normally absent. **Combustion** Chars, burns without melting; when pure, with free access of air, leaves practically no ash. **Pyrolysis** Chars above 200°C without melting; 'woody' odour; carbonaceous residue. Main products: water, carbon dioxide, carbonyl compounds, acids, and a tar, of which the major component is laevoglucosan (1,6-anhydro-β-D-glucopyranose). **Colour reactions** The following sensitive tests distinguish cellulose and its derivatives, and related carbohydrates, from practically all other polymers. *Anthrone test* To a small sample suspended in 0·5 ml water, add some 2 ml of a 0·2% solution of anthrone (9, 10-dihydro-9-ketoanthracene) in conc. sulphuric acid, until the precipitate first formed redissolves. The formation of a distinctive sea-green or dark bluish-green colour, together with a greenish-yellow precipitate on careful dilution with water, occurs in the presence of cellulose, alginates, starch, or furfural, and to a less extent with cellulose esters and ethers. *Molisch test* Shake a small sample with 1 ml water and 2 drops of a 10% solution of α-naphthol in chloroform, then add slowly at least 2 ml conc. sulphuric acid so as to form a lower layer. When carbohydrates are present, a violet ring is produced at the interface, and on careful mixing a purple solution is obtained that yields a violet precipitate on dilution with water. The colour is due to formation of 5-hydroxymethylfurfural and its interaction with α-naphthol, and is given by all carbohydrates (including wood, paper, regenerated cellulose), alginates, and furfural. Ester and ether derivatives of cellulose respond less readily (nitrate hardly at all) but the purple colour may appear on warming. For quantitative colorimetric determination, *see* SMITH, B. F., *et al., Text. Res. J.* **32**, 29 (1962).

Other tests 1. Acetolysis, with acetic anhydride in the presence of sulphuric acid, converts cellulose in good yield into cellobiose octa-acetate (insoluble in water, m.p. 228–229°C). 2. Distinctions between natural and regenerated cellulose:

Iodine test Cellulose is deeply stained by contact with a 10–15% solution of iodine in aq. potassium iodide, but on prolonged immersion in water natural cellulose (cotton, linen) washes clear while regenerated kinds (e.g. viscose, cuprammonium, and polynosic rayons) retain a deep blue or blackish blue colour. Mercerised cellulose behaves intermediately; secondary cellulose acetate is coloured brown; partial ethers (methylcellulose, hydroxyethylcellulose) also stain deeply with iodine.

Cross-section of Fibres

In natural cellulose fibres the outline of the transverse section and the shape of the central canal, or lumen, are often characteristic, and in many of the bast fibres the interstices in the fibre bundles are filled with lignin (stains red-violet with acid phloroglucinol). In fibres of regenerated cellulose, lumen and lignin are absent, but the cross-sectional outline is usually characteristic, e.g. as commonly produced, cuprammonium rayon is smooth and circular but viscose rayon is deeply corrugated.

13.3 PHYSICS

13.31 GENERAL PROPERTIES

Specific gravity Varies with moisture content and method of measurement, e.g. cotton (dry, in He), c. 1·56; do. (in water), c. 1·60. Typical values, room-conditioned samples (measured, e.g., in toluene): hemp, jute, 1·48–1·49; flax, 1·50; viscose rayon, 1·51–1·52; cotton, ramie, 1·53–1·55; regenerated film, 1·53. **Refractive index and birefringence** Typical mean values,

	n	$n_{\parallel}$	$n_{\perp}$	$n_{\parallel}-n_{\perp}$
Cotton	1·555	1·577	1·532	0·045
Flax, ramie	1·563	1·595	1·531	0·064
Jute	1·536	1·55	1·51	0·04
Viscose rayon	1·534	1·547	1·521	0·026
Polynosic rayon	1·532	1·551	1·513	0·038

Optical activity Cellulose is optically inactive in solution, but the complex it forms with cuprammonium hydroxide exhibits laevorotation. **Water absorption** (%) Regenerated foil (air-dry), over 100; do. (undried gel), 300. Vulcanised fibre, c. 25. **Water retention of fibres** (%) Cotton, 45–50; hemp, c. 65; flax, jute, c. 70; ordinary regenerated rayons, 80–>100; do., high tenacity and polynosic types, 65–80. **Moisture regain** Typical values, percentage at 25°C (desorption values in brackets):

R.h., %	10	30	65	90
Cotton:				
Grey	2·1 (2·5)	3·9 (4·5)	6·8 (8·2)	11·1 (14·3)
Kier-boiled	1·9 (2·6)	3·8 (4·7)	7·1 (9·0)	12·3 (15·3)
Kiered and bleached	2·0 —	3·7 (4·6)	6·8 (8·4)	11·8 (14·9)
Mercerised	2·4 (3·0)	4·6 (5·4)	8·5 (10·2)	14·3 (17·6)
Viscose rayon:				
Normal	3·4–4·0	6·5–7·5	12·0–13·5	20–23
	(4·5–5·0)	(8·0–9·0)	(14–16)	(24·5–27·5)
High-tenacity	3·5 (4–5)	6·5–7·5	11·5–14·5	18·5–22·5
		(7·5–8·5)	(13–15)	(22–26)
Polynosic	—	—	12 (14)	—
Regenerated film:				
Washed	4 (5)	7 (8·5)	11·5–13 (15)	22 (25)
Bast and related fibres:				
Flax	—	—	9·0	—
Hemp	—	—	9·8	—
Jute	2·8 (3·1)	5·3 (6·1)	10·12 (12–14)	>18
Manila hemp	—	—	10·3 (11·3)	—
Sunn hemp	—	—	8·8 (9·9)	—

Permeability (*see also* §13.82ii) Dry films of regenerated cellulose show low permeability to nitrogen, hydrocarbon vapours, etc. Permeability of 'cellulose' to gases, at 25°C (10^{-10} cm²)/(s cmHg): H_2, 0·0065; He, 0·0005. However, unless specially coated, cellulose films are permeable to water vapour and moderately permeable to alcohol vapour. Permeability, in general, increases with moisture content, e.g.,

r.h. %*	0	40	48	80	87	90	100
O_2	0·0006	0·006	0·0065†	0·020	0·075†	0·021	8·0†
CO_2	0·002	0·020	—	0·038	—	0·50	—

*Temperature not stated
†Data from a different source (value at 100% r.h. is for never-dried film)

13.32 THERMAL PROPERTIES

Specific heat 0·32–0·35. **Conductivity** (10^{-4} cal)/(cm s °C) Cotton (compressed), *c.* 5·5; 'cotton wool' (loose), *c.* 1. **Physical effects of temperature** Loses strength when maintained at elevated temperatures.* Time, in air at 150°C, for tendering to half-strength (typical

*Hemp ropes, exposed to hot dry conditions, lose moisture and suffer temporary embrittlement with loss of strength.

values): Cotton 70 h, viscose rayon 50–120 h; hemp and manila ropes begin to char about this temperature. Decomposes in air (chars and ignites), without melting, *c.* 270°C (decomp. in absence of air >350°C). **Heat of wetting** Average results, fibres wetted with water (cal/g dry cellulose),

Initial moisture regain (%)	0	2	7	10	20
Cotton, scoured and bleached	11·0	7·1	2·4	—	—
mercerised	18	13·3	6·2	3·8	—
Viscose rayon	20–25	18–21	—	6–10	—
High-tenacity rayon, stretched	23	18	12	8	2·5
saponified acetate	18·2	13·9	6·9	4·1	0·9
Flax	13	8	2·5	—	—

See also GUTHRIE, J. C., *J. Text. Inst.* **40**, T489 (1949)

The above measure is approximately proportional to the accessibility. The heat of adsorption of all cellulosic fibres at zero moisture regain is 280–300 cal/g of water absorbed. Heat of wetting of cotton with alcohols decreases, in the series from methyl to amyl, from 7·3–1·3 cal/g dry cellulose. **Heat of combustion** 4·1–4·2 kcal/g.

13.33 ELECTRICAL PROPERTIES

Resistivity Cellulose fibres have a resistivity of approx. 10^6–10^7 ohm cm at 65% r.h. and 25°C, but the figure depends on moisture content and the purity of the sample. Experimental values, comparative but not necessarily representative (ohm cm, at 25°C):

R.h., %:	65	90
Cotton, grey yarn	5.10^6	3.10^4
Cotton, purified	10^9	5–20.10^4
Viscose cont. fil. yarn	7.10^6	$2·5.10^4$
Viscose, purified	13.10^6	5–20.10^4

Red vulcanised fibre (*see* §13.82iii) at 65% r.h. and 25°C, 2–3.10^9 ohm cm.

Dielectric strength (kV/mm) Cotton (wire covering), *c.* 4 (declines with increase of humidity); vulcanised fibre, under 8 (electrical grades, up to 12).

Dielectric constant and **Power factor** (20–25°C)

	Dielectric constant Frequency, Hz, as in ( )	Power factor 50 Hz	10^6 Hz
Cotton*, dry	3·2 (10^3), 3·0 (10^5)	—	—
40% r.h.	—	0·1	0·03
65% r.h.	18 (10^3), 6 (10^5)	—	—
Acetylated do., 40% r.h.	—	0·005	0·01
Viscose rayon*, dry	3·6 (10^3), 3·5 (10^5)	—	—
65% r.h.	8–15 (10^3), 5–7 (10^5)	—	—
Regenerated film, dry	7·7 (50), 6·7 (10^6), 4 (3.10^9)	0·009	0·06
Paper (cotton), dry	2·6 (10^3)	0·002 (10^3)	
Vulcanised fibre	4·5 (10^6)		0·05
Wood†, 10–15% moisture	3–8 (10^6), 3–7 (10^8)	0·4 (10^6), 0·8–0·9 (10^8)	

*Data for dielectric constant are minimum values. Approximate value for various raw cottons (20°C, 200 Hz): 4 (45% r.h.), 5 (55% r.h.), 6 (65% r.h.)
†Values variable with type and moisture content

13.34 MECHANICAL PROPERTIES
(at 65% r.h., 20°C unless otherwise stated)

Elastic modulus, elastic recovery, tensile strength and **extension at break** Examples; extension and recovery in percentage, other values in kgf/mm² and (gf/den.), values wet shown in [ ].

		Initial elastic modulus		Elastic recovery, from gf/den. stress / % strain							Tensile strength		Extension
		Range	Examples	0·5	1	2	3	2	5	10	Range	Examples	Examples
Staple fibre	Cotton*	600–1100	750(55)	79	60	41	34	74	45	—	25–80	50(3·6)[60(4·3)]	6–8[c. 10]
	Cotton†(6)		415(30)[(10)]	70	55	35	30	70	45	—		50(3·6)[55(4·0)]	9[10]
	Cotton‡		900(65)	—	76	55	40	73	46	34‖		57(4·1)	8
	Cotton§		910(66)	—	61	38	31	68	41	26		62(4·5)	10
	Flax*	2500–>5000	2740(200)	—	78	71	66	65	—	—	50–100	85(6·3)[90(6·7)]	2(2·2)
	Flax†(4)		760(55)[(27)]	65	60	50	—	60	55	—		37(2·7)[41(3·0)]	2·6[3·5]
	Hemp*	2500–>5000	3260(245)	—	64	55	50	50	—	—		70(5·2)	2·2
	Jute*	2500–>5000	2600(195)	—	72	74	75	75¶	—	—	40–80	45(3·4)	1·8
	Viscose‡		825(60)	88	50	33	—	74	50	37		35(2·5)	18
	Viscose§		900(65)	—	36	—	—	54	38	34		30(2·2)	20
	Polynosic†(3)		2100(150)[(12)]	95	85	50	—	65	45	—		41(3·0)[29(2·1)]	7[8·5]
	Polynosic§		1900(140)	—	65	41	—	45	41	—	40–70	45(3·4)[35(2·5)]	7–10[9–12]
Cont. fil.††	Viscose*	600–900	890(65)	87	45	32	—	82	52	40	23–34	30(2·2)[15(1·1)]	20–25[c. 35]
	Viscose†(1)		1400(100)[(5)]	95	60	25	—	80	45	30		28(2·0)[14(1·0)]	19[38]
	High tenacity† (5%/min)		1400(100)[(2)]	95	85	65	55	75	60	50		65(4·7)[41(3·0)]	12[24]
	Cuprammonium		2100(150)[(12)]	90	65	40	—	65	45	35	23–32	25(1·8)[15(1·1)]	10[17]
Regenerated film**		400–1200	—		—		—		—			9[4]	15–30[30]

*–§For elastic recovery:
*Loaded at rate of 10 gf den^{-1} min^{-1}, load held for 0·5 min. recovery measured 1 min. after start of unloading
†Loaded at rate shown in parenthesis (gf den^{-1} min^{-1}) then reversed to zero load and measured
‡Extended at rate of 25% of initial length/min, then reversed to zero load and measured 0·5 min later
§Extended as ‡, extension held for 1 min then reversed to zero load and measured 1 min later
‖At 8%
¶At 1·5%
**Measured along length; strength across film c. 30% lower, extension increased up to fourfold
††Tested as low-twist yarn

Tensile strength of vulcanised fibre (fabricated from cellulose, *see* §13.82iii), 3·5–6 kgf/mm². **Flexural strength** (kgf/mm²) Vulcanised fibre, *c.* 10. **Compressive strength** (kgf/mm²) Vulcanised fibre, 15–20. **Impact strength** (Izod test, ft lbf/in of notch) Vulcanised fibre, up to *c.* 5. **Hardness** Vulcanised fibre: Brinell 10; Rockwell, R50–80. **Frictional data** Coefficients of friction (approximate values for yarns) μ_{stat}; cotton/cotton, 0·3–0·6; viscose rayon/viscose rayon, 0·2. μ_{dyn} (mostly at 1 m/s, 65% r.h.): grey or mildly scoured cotton or linen, or viscose rayon/mild steel, *c.* 0·3; do. (lubricant free)/do., 0·6–0·8; do. (lubricated)/do., 0·15–0·35. Grey cotton/red vulcanised fibre, *c.* 0·25. Cotton/hard glass or glazed porcelain, *c.* 0·3; viscose rayon/do., *c.* 0·5. For grey cotton yarn, running against other materials, *see also* this sub-section (.34) in Sections 1, 3, 4, 8, 9, 11, 12, 14, 21, 23, 24D, 27, 28, 29 and 42.

13.4 FABRICATION

Although a linear polymer, cellulose does not soften or melt before it chars, and therefore cannot be moulded as a thermoplastic material. It can, however, be fabricated as 'vulcanised fibre' (*see* §13.82iii) and sheets of this product can be cut, drilled, and machined satisfactorily. Most forms of cellulose can be stuck with glues of the types employed for wood or paper.

The spinning of fibres is dealt with under §13.21.

13.5 SERVICEABILITY

Ageing effects under normal conditions are very slight. Reduction in molecular weight, with loss of strength, occurs on exposure in moist air to sunlight, the effect being greatly accelerated by certain dyes (e.g. red and yellow anthraquinonoid vat colours). For ageing or tendering at elevated temperatures, *see* §13.32; for data on photochemical degradation of fibres, *see* Section 82.

Fungal (mildew) or bacterial attack is favoured by dampness and the presence of appropriate nutrients. Attack can be either on non-cellulosic materials associated with the fibre, causing discoloration and odour, or on the cellulose itself, leading to loss of fibre properties as well. The β-linkage (*see* earlier footnote for distinction from α-linkage of starch) prevents animals from digesting cellulose directly, but ruminants are able to utilise it via rumen micro-organisms. Cellulose fibres are not attacked by moth larvae but occasionally are damaged by insects such as silver fish (*Lepisma saccharina L.*) or spider beetles (e.g. *Niptus hololeucus*).

13.6 UTILISATION

Too widespread for simple enumeration. *Natural fibres* Textiles and related materials, paper, fillers for imparting added strength to thermosetting plastics; also largely composing the universal structural material, wood. *Regenerated forms* Rayons, for general purposes and industrial uses (e.g. tyre cords); clear films for packaging. *See also* §§13.81 and 13.82.

13.7 HISTORY

The utility of *flax* and *hemp* twines was recognised even in the Stone Ages. Flax fabrics were used in pre-dynastic Egypt (fine linen cloth being woven in 3800 B.C.) and in Neolithic lake dwellings in Switzerland. Hemp was used for coarse textiles in ancient China (Shang period, eighteenth–twelfth century B.C. ?). Both flax (λινον) and hemp (καννaβις) are clearly indicated in Greek literature, and flax is frequently mentioned in the Old Testament and in Babylonian records. The development of these fibres, along with the use of leather, made possible the maritime exploits and commerce of the classical world. Fine linen is still a justly prized material: strong coarse linen provided the elegant smocks worn by English country labourers in the seventeenth–nineteenth centuries.

Excavation at Mohenjo-Daro has shown that a coarse *cotton* featured in the Indus Valley civilisation (3000–2500 B.C.). The fibre is mentioned in early Hindu literature (1500 B.C.) and the 'tree wool' of India is referred to by Herodotus. The Indian cottons were introduced to Europe by Alexander the Great (356–323 B.C.); however, the cotton industry in Great Britain did not originate until towards the end of the eighteenth century, and commercial growing of cotton in Egypt dates from *c.* 1820. In the New World, cotton appears to have been known in Mexico 7000 years ago, and to have been used for coarse textiles (at Huaca Prieta, in northern Peru) before 2000 B.C.; again, however, the cotton industry in America began only at the close of the eighteenth century (in 1793, after the invention of the cotton gin).

References to hemp, *sunn hemp*, and *ramie* (fibres cultivated in temperate latitudes, and unknown to the ancient Egyptians and Hebrews or to the early Greeks) occur in Chinese writings dating to 2800 B.C., and in early Sanskrit literature. *Jute* appears to have been cultivated in Bengal from the most ancient times.

The first scientific approach towards elucidating the composition of the structural matter of the vegetable kingdom was made by A. Payen, who investigated it over the years 1837–42 and named it cellulose. Some of the more prominent steps from that period are as follows. Solubility in cuprammonium hydroxide noted by Schweizer, 1857. Chain formula of condensed glucose molecules proposed by B. Tollens, 1883. X-ray diagrams obtained by Herzog and Jancke, 1920, and dimensions of the elementary cell calculated by Polanyi, 1921. In 1926, as a logical extension of Emil Fischer's work on simple sugars, W. N. Haworth adopted a cyclic structure for the β-D-glucose units in cellulose. This last step in particular, together with the development of the concept of polymeric substances in general, led to the present picture of the cellulose molecule.

On the more technical side Mercer, commencing c. 1844 (first patent 1850), developed a process for treating cotton fabric with alkali to improve the dyeing properties and increase the strength and weight.

N.B. Mercerisation under tension, as used today to impart lustrous effects, is an 1889 modification introduced by Lowe; much more recently (1945) slack mercerisation, where yarn swelling causes fabric shrinkage, has been used to obtain 'stretch cottons'.

The earliest regenerated filaments were produced by J. Wilson Swan, in 1878, in connection with the development of the first practical electric lamps (for which purpose the filaments were carefully carbonised to render them conductive) and although the textile possibilities were not entirely overlooked they were not followed up until 1885, when Chardonnet produced practical textile fibres from saponified collodion, and exhibited commercial products at the Paris Exhibition of 1889. Despeissis, utilising Schweizer's reagent, produced cuprammonium rayon in 1890 (now the Bemberg process). Cross, Bevan, and Beadle invented the viscose (xanthate) process in 1892, and it was developed for the spinning of fibres by Stearn (1898), Topham (c. 1900) and Courtaulds Ltd. High-tenacity regenerated fibre (Lilienfeld rayon, from fibre stretched in 60% H_2SO_4) was first developed c. 1930; deacetylation of stretched cellulose acetate fibres (for the same purpose) followed c. 1940. Regenerated cellulose in the form of thin transparent films, now much used for packaging purposes, was first introduced in 1924.

13.8 ADDITIONAL NOTES

13.81 NATURAL FORMS OF CELLULOSE

Cellulose appears in nature in the following forms:

(i) *WOOD*

Available as a variety of hard and soft timbers, this contains 50–70% cellulose, which is embedded in hemicelluloses, lignin, and rosin, along with traces of protein and mineral substances. The softer grades, when pulped and purified, are important in paper manufacture and in the preparation of cellulose derivatives.

(ii) *TEXTILE AND RELATED FIBRES*

The hairs surrounding the seeds of certain plants, the long fibres comprising the bast (inner bark) of others, and the fibres extracted from certain leaf stalks, contain—when cleaned and purified—a high proportion of high-grade cellulose, and are important for cordage, textile, and packing purposes. The principal kinds are given below.

SEED HAIRS:

Cotton, from hybrids of the genus *Gossypium* (fam. *Malvaceae*) grown in tropical and sub-tropical regions, contains (air-dry) 85–90% cellulose, 6–8% water, and the rest protein, wax, and *c*. 1% mineral matter (for specially purified cotton, *see* §13.83). The textile fibres are single cells, flattened and twisted along their length; according to source, they range from *c*. 15–35 mm in length, and inversely from *c*. 20–10 μm in diameter. *Cotton linters* are the short coarse fibres (under 5 mm long) remaining on the seeds after removal of the textile fibres; they are used for the preparation of cellulose derivatives.

Kapok, from *Ceiba pentandra* (fam. *Bombaceae*) of Java, is a resilient lightweight floss consisting (air-dry) of *c*. 65% cellulose, 12% water, 15% lignin, and the rest pentosans, wax, protein, etc. The fibres, tubular with an average diameter of 32 μm and length of 18 mm, are used for stuffing mattresses, cushions and life-jackets, and for thermal and acoustic insulation.

Coir, obtained from the husk surrounding the hard shell of the fruit of the coconut palm (*Cocos nucifera*), comes mainly from Ceylon and the coast of S.W. India. The husks are 'retted' in brackish mud then beaten to separate the fibres, which range from 15–35 cm in length and from 0·1–1·5 mm in diameter. Coir, consisting mainly of cellulose (40%) and lignin (40%) is extensible, springy, and resistant to rotting by sea water, and is used for matting, sacking, upholstery, and native cordage.

BAST FIBRES:

Flax (linen), from *Linum usitatissimum* (fam. *Linaceae*), a plant requiring more temperate conditions than cotton, contains when air-dry 80–90% total cellulose (65% α-cellulose), 8–10% water, and the rest lignin, wax, and traces of mineral substances. The fibres are freed from enveloping matter by a fermentation process ('retting') then separated by beating or crushing ('scutching'); they consist of overlapping cells (ultimates) ranging from *c*. 6–70 mm in length by 15–20 μm in diameter, and effect a total fibre length of from 30 to 100 cm according to source. The long yellow fibres are called *line flax*, and the shorter ones are known as *tow*. Flax (and ramie) fibres are oriented in left-hand helices, and wet fibres twist on drying so as to tighten the helix (cf. hemp and jute fibres, which are oriented in right-hand helices, and on drying twist accordingly). *New Zealand flax*, from *Phormium tenax*, is a leaf fibre resembling Manila hemp (*see* later).

Hemp, obtained by 'retting' the bast of *Cannabis sativa* (fam. *Moraceae*) a shrub of Asian origin (now more widely distributed), contains (air-dry) *c.* 78% total cellulose, 9% water, and the rest lignin, pectin, wax, and mineral matter. The individual cells average *c.* 20 mm in length and *c.* 20 μm across, and effect in overlap a total fibre length (in *line hemp*) of 120–200 cm. Hemp is a soft fibre, much used for twine, ropes, netting, and sacking. *Sunn* or *Indian hemp* (from *Crotalaria juncea*) and *Ambari hemp, Bimlipatam hemp*, or *kenaf* (from *Hibiscus cannabinus*) are fibres more resembling coarse jute than hemp. Manila and sisal hemps are leaf fibres (*see* below).

Jute, obtained by 'retting' the bast of *Corchorus capsularis* and *C. olitorius* (fam. *Tiliaceae*), comes mostly from India and Pakistan. The fibres, brown-coloured, contain (air-dry) *c.* 65% α-cellulose, 6–12% water, and 10% or more lignin; the ultimates, polygonal in cross-section, are *c.* 1·5–5 mm long by 20–25 μm across, and the fibre bundles may reach 250 cm in length. Jute, second to cotton in importance as a textile fibre, is used for cordage, sacking, hessians (including linoleum backing), and for carpet and tarpaulin foundations. For technology of jute, *see* §13.9.

Ramie is obtained on washing (degumming) the bast of tall subtropical nettles (*China grass* from *Boehmeria nivea*, and *rhea* from *B. nivea* var. *tenacissima*). The bleached fibre consists of almost pure α-cellulose in a highly oriented and crystalline state. The individual cells, circular or oval in cross-section, with characteristically thick walls, range from *c.* 50–250 mm in length and *c.* 17–64 μm across. The fibres are lustrous, soft, and very strong, but variable in length (1–50 cm, which restricts applications); they are used in special papers, gas mantles, rot-resistant cordage, and occasionally as substitutes for cotton or flax.

LEAF FIBRES:

Manila hemp (*Abaca*), a leaf-stem fibre from *Musa textilis*, a plantain originating in the Philippine Islands, contains (air-dry) *c.* 65% α-cellulose, 12% water, 22% pectin, and the rest as wax, siliceous matter, etc. The ultimate cells are *c.* 3–15 mm long, *c.* 15–30 μm across. The fibres (100–200 cm or more in length) are exceptionally strong, and are much used for high-strength paper, marine cordage, and haulage ropes; also for local fine muslins.

Sisal hemp, from *Agave sisalana* of Central America (and *henequen*, from *A. fourcroydes*), are coarse leaf fibres used for ropes and binder twine. *Bowstring hemp* (from species of the genus *Sanseveria*) and *Mauritius hemp* (from *Furcraea gigantea*) are cellulosic fibres resembling sisal.

(iii) *BACTERIAL CELLULOSE*

Certain bacteria, notably *Acetobacter xylinum*, can synthesise cellulose as lengthy fibrils growing outwards from the cell wall, and from suitable nutrient media (e.g. glucose solutions) it is possible to obtain the product as a continuous film, for which commercial applications have been proposed.

(iv) *ANIMAL CELLULOSE*

A substance closely resembling vegetable cellulose is secreted as a covering mantle by certain marine tunicates, e.g. the sea squirts. It has also been found, as a sheath on protein fibres, as a minor component of mammalian tissue including human skin.

13.82 REGENERATED AND ARTIFICIAL FORMS OF CELLULOSE

(i) *FIBRES*

The more important cellulose rayons are represented by the following (proprietary) names:

Cuprammonium rayons—Bemberg, Cuprama, Cupresa.

Ordinary viscose rayons—Fibro, Triple A. Evlan, Sarille (crimped staple). Textile Tenasco. Corval, Topel (cross-linked). Rayolanda (dyeing properties modified by treatment with an amino-aldehyde polymer).

High tenacity rayons—Avron, Durafil, Tenasco Super, Tyrex. Fortisan (*see* §13.21).

Polynosic rayons—Avril, Polyflox, Toramomen, Vincel, Zantrel.

(ii) *FILMS*

Cellophane and *Cuprophane* are regenerated as sheets or films from viscose and cuprammonium solutions respectively. Normally they contain hydrophilic plasticisers (glycerol) but some grades are coated (e.g. with cellulose nitrate) to render them less permeable to water vapour. Permeability of dry plasticised film (Cellophane) to gases, at 25°C (10^{-10} cm^2)/(s cmHg): O_2, 0·002; N_2, 0·003; CO_2, 0·005. Transmittance of water vapour through 0·025 mm nitro-cellulose-coated films (MSAT) (10^{-10} g)/(cm^2 s cmHg) at 38°C, 90% r.h.: *c.* 10.

(iii) *VULCANISED FIBRE*

Sheets, rods and tubes are fabricated from fibrous cellulose (e.g. paper) that has been swollen to gelation in zinc chloride solution, washed out, and dried to shape under compression. The toughness and relatively low cost of the product make it a useful constructional material, despite its swelling when immersed in water. It is employed for electrical insulation, printing rollers, and such utilitarian items as containers, protective corners, and washers. The name is an unfortunate one, since the substance contains no rubber and is not vulcanised in the usual sense of the term.

For quality specifications *see* BSS 216, 934, 2768, 3964 and F64, and ASTM D710.

13.83 STANDARD CELLULOSE

Pure cotton scoured under pressure with alkali, freely washed in dil. acid and water, and bleached in dil. sodium hypochlorite or hydrogen peroxide, yields a high-purity cellulose, the characteristics of which are as follows. α-Cellulose content (dry), *c.* 99·8 %. Ash, 0·1 % or less. Fluidity (0·5 % in standard cuprammonium solution), <5 poise^{-1}. Copper number (Braidy's solution), 0·005 to $<0·2$ g Cu/100 g dry cellulose. Methylene blue absorption (pH 7), *c.* 0·5 mmoles/100 g dry cellulose. Soluble in boiling 0·1 N sodium hydroxide, <1 %.

For purification methods, *see* p. 3 of M. L. Wolfrom (Ed) as under §13.9.

13.84 α-, β-, AND γ-CELLULOSES

Natural and regenerated celluloses are heterogeneous in chain length, and it is convenient (especially in the wood-pulp industry) to use the following classification for the component portions.

α-*Cellulose* The portion of high molecular weight; present in purified material, up to 95 % or more. It is resistant to, and insoluble in, 17·5 % sodium hydroxide at 20°C. β-*Cellulose* The fraction soluble in 17·5 % alkali but precipitated on neutralisation. γ-*Cellulose* The fraction remaining in solution even after neutralisation.

β- and γ-celluloses are polyanhydroglucoses of low molecular weight (D.P. <200), together with oxidation products.

13.85 HYDRO- AND OXYCELLULOSES

These are modified forms of cellulose, produced respectively by limited hydrolysis or oxidation. The decrease in molecular weight on hydrolysis is reflected in a progressive loss of strength, and, although the products at first still resemble cellulose in appearance, excessive degradation gives rise to powdery or soluble materials. Most oxidative treatments also decrease the molecular weight and at the same time may increase the sensitivity to degradation, particularly by alkalis. Control of the degradation of cellulose is important in the production of spinnable viscose, in the prevention of the weakening of fabrics by sunlight, in the prevention of yellowing, in the safe use of bleaching and laundering agents, etc., and extensive studies have been made to elucidate the mechanisms involved.

Hydrocelluloses, formed by hydrolysis of 1,4′-glucosidic linkages (with consequent shortening in average chain length and loss of strength), result from the action of dil. mineral acids or hydrolysable salts (e.g. $AlCl_3$) on cellulose (e.g. boiling cotton with 1 % HCl until disintegration occurs):

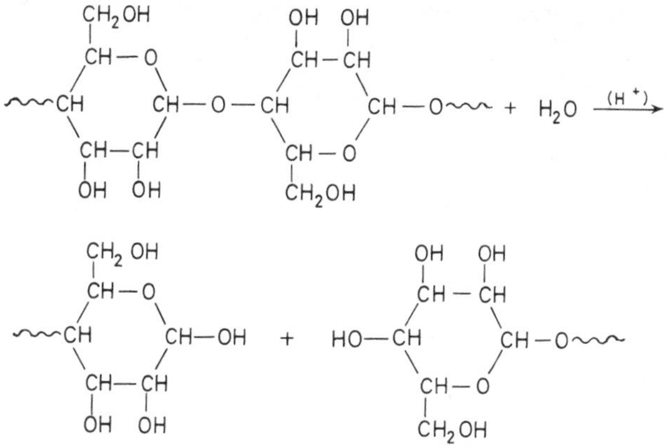

Hydrocelluloses are materials of variable composition and in part resemble β-cellulose; they give less viscous solutions than cellulose, and they show greater reducing power (copper number increases) because of the increase in the number of aldehyde end-groups (*see* §13.1). The rate of hydrolysis levels off towards zero as accessible material decreases, ultimately leaving a highly crystalline and resistant residue (*limit hydrocellulose* or *microcrystalline cellulose*) for which commercial uses have been found, e.g. as a non-assimilable dietary substance.

Oxycelluloses result from the photo-catalysed reaction of cellulose or hydrocellulose with atmospheric oxygen, and from the action of oxidising agents (hypochlorites, periodates, hydrogen peroxide with ferrous salts, etc.). Commonly the process occurs randomly, yielding aldehyde, ketone, and carboxyl groups but some forms of oxidation can give more specific products, as noted below.

(i) Oxidation under acid or neutral conditions generally brings about a marked increase in aldehyde groups and the products are characterised by exhibiting, when subsequently boiled in aq. alkali, an increase in carboxyl groups and a further loss in strength (*reducing oxycellulose*: copper number increases; degraded by alkali).

(ii) Oxidation under alkaline conditions causes more particularly an increase in carboxyl groups, and subsequent treatment with alkali has little effect (*acidic oxycellulose*: methylene blue absorption increases, affinity for direct dyes decreases).

Examples are given below of two particular oxycelluloses obtained by the use of reagents largely specific in their action (note—in these instances even at a high degree of substitution, the products retain their initial fibrous form, although most of the strength is lost).

(*a*) *Periodate oxycellulose* (*dialdehyde cellulose*). Treatment with periodate solutions (specific for oxidation of α-glycols) results almost exclusively in a type of reducing or aldehyde oxycellulose arising from cleavage of the 2,3-glycol linkages with oxidation of the secondary alcohol groups to aldehyde groups (some of which convert to a hemi-acetal structure),

(*b*) *Nitrogen peroxide oxycellulose* (*celluronic acid*). Exposure to dry gaseous nitrogen peroxide, producing reactions resulting largely in oxidation of the primary alcohol groups, yields an acidic oxycellulose of the glucuronic acid type,

Oxycellulose of the above kind is used as a haemostatic surgical dressing, soluble in body fluids.

13.86 CROSS-LINKED CELLULOSE

Fabrics of cotton or viscose rayon staple may be given a 'resin' treatment to improve the crease recovery, smooth-drying properties, crease retention, etc. It is applied by impregnating the fibres with an appropriate solution and carrying out a reaction *in situ*; the 'resin' is not visibly present in the finished product, and usually the aim is not to affect the stiffness although the 'weight' of the fabric increases. The mechanism of the action is controversial, but even at a very low degree of substitution the cellulose becomes insoluble and the process consists essentially in introducing covalent cross-linkages between adjacent cellulose chains (somewhat analogous to vulcanisation of rubber).

146

The simplest substance employed for this purpose is formaldehyde, which in acid solution yields a cellulose methylene ether, Cell-O·CH₂·O-Cell (*see also* Section 16.21) or a polyoxymethylene ether, Cell-O(CH₂O)ₙ-Cell (*see also* §26.81). Di- or polymethylol amino-compounds give rather better results, and a large number of these has been examined (*see also* §27.82). Many other di- and poly-functional compounds have been examined for this purpose, e.g. those with two or more vinyl, epoxy, chloro, isocyanate, or carboxyl groups, and some have been employed industrially.

The properties imparted to the cellulose are less dependent upon the chemical nature of the cross-links than upon their physical dimensions and the sites they occupy; thus cross-linkage effected when the cellulose is in a dry state restricts subsequent swelling in water (hence resistance to shrinkage improves) whereas reaction in swollen cellulose tends to keep it in that state (which improves the crease recovery of wet fabrics).

For further reference to this subject, on which there is extensive literature, *see* MARSH, J. T., *Self-smoothing Fabrics*, London (1962) and FETTES, E. M. (Ed.), *Chemical Reactions of Polymers: High Polymers*, Vol. XIX (REEVES, W. A. and GUTHRIE, J. D.), New York and London, 1165 (1964).

13.87 CHITIN

This is a constituent of the hard parts of the arthropoda, being present along with sclerotin (a protein) in the horny exoskeletons of insects and spiders and in certain fungi and marine organisms; similarly, it is left as a residue after extraction of calcium carbonate and protein from the shells of crabs, crayfish, lobsters, etc.

It is not a true carbohydrate but is largely a poly-N-acetyl-D-glucosamine, having the following structural unit:

$$C_8H_{13}O_5N, \text{ mol. wt.} = 203$$

i.e. the unit resembles that of cellulose but with a $CH_3CONH—$ group substituted for the $—OH$ group at the 2-position. X-ray identity period, 10·46 Å.

Chitin is less reactive than cellulose, being less readily esterified or etherified. It is insoluble in water, cuprammonium hydroxide, alkalis, dil. acids, and organic solvents, but dissolves (with slow degradation) in conc. hydrochloric acid, sulphuric acid, phosphoric acid, and formic acid.

Although alkali-insoluble, conc. potassium hydroxide (40%, at 150°C) deacetylates it to the base, *chitosan* (other polysaccharides dissolve). This material, being acid-soluble but insolubilised by alkalis, and showing good adhesion to textiles, has been proposed as a semi-permanent size or water-repellent finish, and as a pigment-binding agent. Fibres, having ion-exchange properties, have been produced experimentally.

Chitin was recognised by H. Braconnot in 1811, and was investigated by A. Odier—who named it (from χιτον = a coat or covering)—in 1823. It was examined by G. Ledderhose in 1876–8 and shown to yield glucosamine and acetic acid on hydrolysis, but its precise structure has only recently been elucidated. An obstacle in the development of this widely distributed material of the animal world lies in finding suitable commercial sources of supply.

For further reference, *see* WOLFROM, M. L. (Ed.), *Advanc. Carbohyd. Chem.*, Vol. 15 (FOSTER, A. B., and WEBBER, J. M., *Chitin*), New York, 371 (1960).

13.9 FURTHER LITERATURE

MARSH, J. T. and WOOD, F. C., *An Introduction to the Chemistry of Cellulose*, London (1938) (revised edns 1942, 1945)

OTT, E., SPURLIN, H. M. and GRAFFLIN, M. W. (Eds), *Cellulose and Cellulose Derivatives* (*High Polymers*, Vol. V; revised edn, 3 parts), New York and London (1954)

GRANT, J., *Cellulose Pulp*, London (1958) (paper, rayon) and *A Laboratory Handbook of Pulp and Paper Manufacture*, London (1960)

HONEYMAN, J. (Ed), *Recent Advances in the Chemistry of Cellulose and Starch*, London (1959)

WHISTLER, R. L. (Ed), *Methods in Carbohyd. Chem.*, Vol. III, *Cellulose*, New York and London (1963)

WOLFROM, M. L. (Ed), *Advanc. Carbohyd. Chem.*, Vol. 19 (JONES, D. M., structure and reactions of cellulose), New York and London, 219 (1964)

STAMM, A. J., *Wood and Cellulose Science*, New York (1964)

ATKINSON, R. R., *Jute–Fibre to Yarn*, London (1964) (technology of jute)

CELLULOSE ACETATE

With notes on fibres, cellulose propionate, acetate-propionate, and acetate-butyrate

SYNONYMS AND TRADE NAMES

Acetylcellulose. Safety Celluloid. Bexoid, Celastoid, Cellastine, Cellomold, Clarifoil, Dexel, Kodapak, Rhodoid, Tenite.
For trade names of some common fibres *see* §14.81.

GENERAL CHARACTERISTICS

White fibrous material, available partially acetylated (*secondary acetate*) and more completely acetylated (*triacetate*). Both kinds are produced as films and textile fibres; plasticised secondary acetate, often dyed or pigmented (commonly black), is used for moulding and for clear, tough, flexible sheets and foils.

14.1 STRUCTURE

Simplest Fundamental Unit

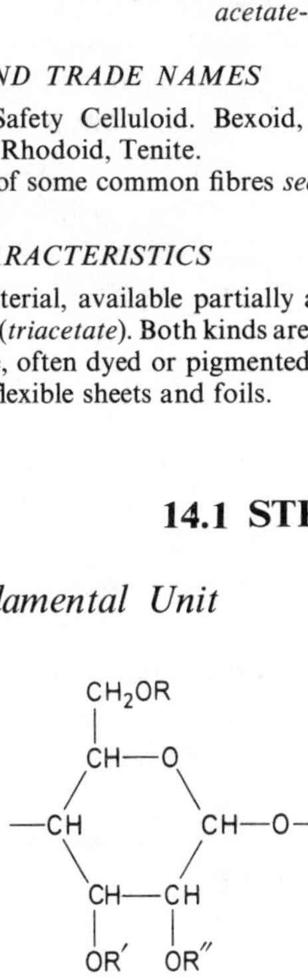

Mol. wt.,
secondary acetate = 260–275
triacetate = 288

where R, R′, and R″ (= H atoms in cellulose) are partly or completely replaced by acetyl (CH_3CO-) groups.

Molecular weight Plastics, *c.* 80000. Fibres, over 100000. *Degree of polymerisation* Lacquers, 200 or less. Plastics, *c.* 300. Fibres, up to 400.

X-ray data *Secondary acetate* Amorphous if shock cooled from the melt; partially crystalline when cooled slowly or annealed, but in general the crystallinity is very low. Fibre identity period, 10·3 Å; also quoted as 5·15 Å or multiples thereof according to the degree of substitution. *Triacetate* Amorphous if shock-cooled from the melt but crystalline when cooled slowly or annealed at a temperature near the melting point. Can be highly crystalline if appropriately prepared; occurs in two crystalline forms, the usual one has a fibre identity period of 10·4 Å.

14.2 CHEMISTRY

14.21 PREPARATION

Cellulose in the form of cotton linters or wood pulp, swollen and dehydrated by pretreatment with acetic acid, is esterified with acetic anhydride in the presence of further acetic acid and a catalyst (conc. H_2SO_4 or $HClO_4$). The reaction, which is exothermic and requires cooling to restrict degradation, is taken to completion since intermediate products are heterogeneous in composition. A limited amount of water is added to the solution of cellulose triacetate so obtained, which is then allowed to 'ripen' (i.e. hydrolyse) homogeneously until the required composition is obtained; thereafter, precipitation with an excess of water, followed by washing and drying, results in a partially-hydrolysed acetone-soluble product (*secondary acetate*). Milder hydrolysis, sufficient only to stabilise by removing combined catalyst, provides an almost fully acetylated product (described broadly as *triacetate*).

Cellulose acetates are also obtained by variations of the above process in which the reaction is controlled either (i) by replacing some of the diluent acetic acid by an inert solvent (methylene chloride), or (ii) by the use of an inert non-solvent (benzene, cyclohexane), the last permitting full acetylation with retention of the original fibrous form and minimum degradation in molecular weight.

Catalyst-impregnated cellulose fibres can be esterified in vaporised acetic anhydride, with retention of much of the original form and strength. *See* §14.81 (Alon, Cotopa).

For plastics use, secondary cellulose acetate is compounded with plasticisers; textile fibres are unplasticised.

Composition

The compositions of the theoretical esters, and of the commercial products finding various uses and available in various grades according to molecular weight, appear below.

Theoretical Esters

Average degree of substitution	1	2	3 *(triacetate)*
Acetic acid yield (CH_3COOH), %	29·3	48·8	62·5
Acetyl content* (CH_3CO—), %	21·1	35·0	44·8

*Acetyl content = 0·717 × acetic acid yield. The former term is mostly used in the U.S.A.; acetic acid yield is employed in Great Britain especially in the textile industry

Commercial Esters

The following ranges of acetic acid yield (%) are commonly met with: plastics, 52–56 (high acetyl, up to 59); films, 53–59 or more; fibres, lacquers, 53–55 (triacetates, approx. 61·5–62·5). Equivalent values of degree of substitution and acetyl content are given below:

	Secondary acetates					Triacetates		
Degree of substitution*	2·14	2·28	2·41	2·56	2·71	2·79	2·87	2·96
Acetic acid yield, %	51	53	55	57	59	60	61	62
Acetyl content, %	36·6	38·0	39·4	40·9	42·3	43·0	43·7	44·5

*Degree of substitution $= (27 \times \text{acetic acid yield})/(10 - [7 \times \text{acetic acid yield}])$

14.22 PROPERTIES

Solvents (*see also* §14.23) *Secondary acetate* Ketones (acetone, cyclohexanone; acetates $> 57\%$ in acetic acid yield are usually acetone-insoluble, 80/20 methylene chloride/methanol, low molecular weight esters (methyl acetate, ethyl lactate), ethylene glycol derivatives such as 2-methoxyethyl acetate (Methyl Cellosolve acetate, Methyl Oxitol acetate), pyridine, dimethylformamide, phenols, glacial acetic acid.

Triacetate Restricted range of solvents. Soluble—especially with addition of a little alcohol—in certain chlorinated hydrocarbons (methylene chloride, chloroform, trichloroethane, trichlorodifluoroethane); also in nitromethane, pyridine, dimethylformamide, formic acid or glacial acetic acid. Insoluble (swells) in acetone or trichloroethylene. **Plasticised by** high-boiling esters such as dimethyl phthalate (DMP), diethyl and dibutyl do., triphenyl phosphate (TPP), dibutyl tartrate, triacetin; certain glycollates and sulphonamides, e.g. methylphthalyl ethyl glycollate and *N*-ethyl *o*- and *p*-toluene sulphonamides. Secondary acetate is more readily plasticised by ester-type plasticisers than the triacetate. **Relatively unaffected by** hydrocarbons, most oils and greases, carbon tetrachloride (sec. acetate swells), perchloroethylene, alcohol (plasticised sec. acetate partly soluble), ether, weak acids, weak alkalis (borax). Triacetate fibres are not delustred by soap-phenol solutions as are those of secondary acetate. **Decomposed by** moderately conc. acids, alkalis above pH 9·5. Triacetates more resistant, except to conc. acids and alkalis or to strong oxidising agents; surface saponification (e.g. by treatment of 95°C with 0·25% sodium hydroxide) is accelerated by the presence of aliphatic quaternary ions.

14.23 IDENTIFICATION

Thermoplastic. Cl, N, S nominally absent (P may be present in plasticisers). Saponifiable. **Combustion** Melts, burns (not very readily, and non-inflammable plasticised forms are available), slightly smoky flame and a hissing sound; woody 'acetic' smell on extinction, charred

151

residue; fibres leave hard black bead. **Pyrolysis** Chars; woody smell; yellow, acid distillate. **Colour reactions** *Cellulose component See* Anthrone and Molisch tests (§13.23). *Acetate radical* Acetic acid can be distilled from a dil. solution of the hydrolysis product that results when a sample is left to stand in 1/1 sulphuric acid/water, or in N alcoholic potash (followed by acidification). To detect the acid, mix 1 drop of the distillate with 1 drop 5% aq. lanthanum nitrate, then add 1 drop 0·1% alcoholic iodine followed by 1 drop N ammonia solution. Acetic acid produces a blue to blue-brown colour (may appear only slowly, and is given also by propionic acid).

Solubility distinctions between secondary and triacetates (*see also* §14.22).

	Conc. hydrochloric acid, or 80/20 acetone/water	Methylene chloride, or chloroform
Secondary acetate	Soluble	Insoluble (swells)
Triacetate	Insoluble (swells)	Soluble

Analysis Acetic acid yield is determined by saponification. For references to this and determination of hydroxyl content, *see* §14.9.

14.3 PHYSICS

14.31 GENERAL PROPERTIES

	Plastics	Fibres	
		Sec. acetate	Triacetate
Specific gravity	1·28–1·32	1·32–1·33	1·30–1·32
Refractive index	c. 1·5	$n_{\parallel}$ 1·477, $n_{\perp}$ 1·473	1·469
Birefringence $(n_{\parallel} - n_{\perp})$	—	c. 0·004	nil or very low

Water absorption Plastics, c. 3%, depends on plasticiser. *N.B.* acetate-butyrate lower, *see* §14.82. **Water retention** (fibres, %) Sec. acetate, 25–30; triacetate, 13–18 (10–16 heat-set). **Moisture regain** (fibres) Shows an initial rise reaching a maximum at a degree of substitution of approx. 0·8, then falls with increase in esterification. *Sec. acetate* (typical values, %, at 25°C):

R.h., %	10	30	65	90
Absorption	0·8–0·9	2·4–2·5	5·9–6·0	11·0–11·6
Desorption	1·0–1·2	2·9–3·3	7·5–8·0	14–15

Triacetate (%, at 65% r.h., 25°C) Untreated, up to 4·5 (typical sample, cont. fil.: absorption 3·2, desorption 3·9); heat-set fibres, 2·5–3·3.

See also BEEVER and VALENTINE, *J. Text. Inst.*, **49**, T95 (1958). *Heat of wetting* (sec. acetate fibres at zero moisture regain): 8·2 cal/g dry material. **Permeability** Fairly permeable to water vapour; decreases with increase of substitution, thus permeability of cellulose > sec. acetate > triacetate. Permeability of sec. acetate to gases at 20–30°C $(10^{-10}$ cm^2)/(s cmHg): N_2, 0·16–0·5; O_2, 0·4–0·78; CO_2, 2·4–18 (the higher values are for plasticised material).

14.32 THERMAL PROPERTIES

Specific heat Sec. acetate, *c.* 0·4; triacetate, 0·32 (20–100°C), 0·34 (20–200°C). **Conductivity** Sec. acetate, *c.* $4·5–7·5.10^{-4}$ cal/(cm s °C). **Coefficient of linear expansion** Sec. acetate, *c.* $0·8–1·6.10^{-4}/°C$; triacetate, $1·5.10^{-4}$ (below 60°C), $3·5.10^{-4}$ (above 120°C). **Physical effects of temperature** *Sec. acetate* Plastics lose impact strength, $-50°C$; soften, 75–120°C, lowered by high moisture content; maximum service temperature, *c.* 75°C; flow temperature, 120–160°C. Second order transitions reported to occur at 15, 50, 90 and 114°C (DAANE, J. H., *J. Pol. Sci.* B2, 343 (1964)). Easily softened by infra-red radiation or by immersion in hot brine (minimum softening temp. at *c.* 53% acetic acid yield); readily formable above 200°C; melts, 235–260°C, with decomposition. Fibres lose strength when maintained at elevated temperature; time, in air at 150°C, for tendering to half strength (typical value), *c.* 220 h. Safe ironing temperature, 135°C. *Triacetate* Much more resistant than sec. acetate. Second order transitions reported at 105 and 157°C, also (fibres) 180°C. Softens, 220–225°C; melts, 290–300°C. Triacetate fabrics can be 'heat set' by fixation in open width in the dry state (e.g. 10–20 s at 240–220°C) or in steam (e.g. 30 min at 125°C) and can be durably pleated; setting raises the softening temperature (safe ironing, 200–240°C), lowers water imbibition, and improves dimensional stability.

14.33 ELECTRICAL PROPERTIES

Volume resistivity (ohm cm) Plastics, $10^{10}–10^{13}$; secondary acetate films, plasticised, $>10^{12}$ (65% r.h.), *c.* 10^{10} (85–90% r.h.). *Surface resistivity*, up to 10^{13} ohm. Resistivity of fibres varies with moisture content and history of sample; approximate values (ohm cm):

R.h., %	65	85–90
Secondary acetate	$10^{11}–10^{12}$	$10^{10}–10^{11}$
Triacetate	$>10^{14}$	10^{13}

Dielectric strength Usually upwards from 14 kV/mm (minimum 8 kV/mm for mouldings), increases with degree of acetylation; up to 80 kV/mm for thin foils. **Dielectric constant** and **Power factor** (values at 50 Hz and 1 MHz approximately similar).

	Dielectric constant	Power factor
Secondary acetate	c. 4–5	c. 0·01–0·05
Triacetate	c. 4	0·015–0·02

Power factor falls slightly with rise of temperature from -25 to 150°C. For curves of tan δ/frequency, up to 100°C and > 1 MHz, *see* KATO, N., *et al., Chem. High Polymers,* 19 (1962) 95 (Japanese, English summary).

14.34 MECHANICAL PROPERTIES

N.B. Properties of plastics depend on compounding (plasticisation) and on temperature. Secondary acetate and triacetate fibres are approximately alike in mechanical characteristics.

Elastic modulus (kgf/mm^2) Plastics, 100–200 (and higher). Fibres and films, 300–450 (25–40 gf/den.). **Elastic recovery** *Plastics* Recovery from very small strains, good. Fibres, recovery from short-term deformation (%, at 65% r.h., 20°C):

	Stress (gf/den.)			Strain (%)		
	0·5	0·75	1	2	5	10
Secondary acetate						
Cont. fil. yarn*	98	82	15	95	75	35
Staple fibre†	65	14	—	54	24	13
Do.‡	85	27	—	76	43	25
Triacetate						
Cont. fil. yarn*	98	75	20	93	89	25
Staple fibre†	50	13	—	44	22	12
Do.‡	68	24	—	61	38	22

*Loaded at 1 gf den.$^{-1}$ min^{-1}, measured immediately after reversal
†Extended at rate of 25% of initial length per min, held 1 min, measured immediately after reversal
‡As † but measured 1 min after reversal

Tensile strength and Extension at break

	Tensile strength (*approx.* kgf/mm^2)	Extension (*approx.* %)
Plastics, 65% r.h.	2–6	50–25*
Fibres, secondary acetate, normal†		
65% r.h.	16 (1·3 gf/den.)	25–30
Wet	9 (0·75 gf/den.)	35–45

*c. 5% if unplasticised
†High tenacity, up to 75 (6 gf/den.). Triacetate fibres similar to normal secondary acetate fibres, but slightly stronger when wet.

Flexural strength 2·5–10 kgf/mm². **Compression strength** From about 5 or 10, up to 20 kgf/mm². **Impact strength** (Izod test, ft lbf/in of notch) *c.* 1; resistant types, 3–7. **Hardness** Brinell (2·5 mm ball, 10 kg load), 6–15. Rockwell, up to M80 or R120. **Friction** Coefficient of friction, approximate values, μ_{stat} : Fibre/fibre, 0·3. μ_{dyn} (1 m/s, 65% r.h.): Sec. acetate filament yarn, lubricant-free/steel, 0·7; lubricated/ do., 0·3–0·5. Cont. fil. yarn/copper, 0·25; /hard glass, 0·4; /glazed porcelain, 0·6. Grey cotton yarn/sec. acetate plastics, 0·23–0·24. Cellulose acetate is relatively poor in **abrasion** resistance.

14.4 FABRICATION

Compression moulding: 100–200°C, up to 5000 lbf/in². Injection moulding: *c.* 200°C, up to 20000 lbf/in². Blow-moulding and vacuum-forming are employed. Plastics rod and sheet can readily be sawn, blanked, machined or polished, and can be bent to shape when hot; component parts can be cemented together (with acetone, or a solution of scrap polymer in a mixture of acetone and an ester such as ethyl acetate or lactate). In addition to mouldings, forms available include films cast from solution, extruded sheets, and sheets cut from blocks. Secondary acetate fibres (dry spun from 25% solution in 95/5 acetone/ water) and triacetate fibres (dry spun from solution in methylene chloride) can be dope-dyed, or coloured with dispersed dyes at elevated temperature (*c.* 75°C, higher for triacetate); triacetate fabrics are capable of being set by heat and can be durably pleated, *see* §14.33.

14.5 SERVICEABILITY

Stabilised material undergoes only slight reduction of strength with age, exposure to sunlight, humidity changes, or exterior weathering; dimensional stability improves with increasing degree of substitution (*see also* §14.82). Material in which the residual acid catalyst has been incompletely neutralised may become unstable and insoluble on storage. For data on photochemical tendering of fibres, *see* Section 82. Unaffected by bacteria, fungi, or insects (dependent on plasticiser and filler; insects have been known to eat through it), and by most oils and greases.

14.6 UTILISATION

Plastics Injection-moulded, extruded, and sheet-fabricated articles in great variety (e.g. telephones, lamp shades, spectacle frames, machinery guards, buttons, transparent cartons). Tough but clear and flexible cast or extruded films are used for packaging, photographic bases, horticultural glazing, and colour filters. Used in solution for adhesives and as a finish for yarns, fabrics, and starchless collars.

Typical formula for lacquer for coating leather or fabric (parts by weight; can be thinned by addition of toluene or naphtha): cell. acet. (lacquer grade), 12; acetone, 55; ethylene glycol mono-methyl ether, 10; plasticiser (with resin if required), 20.

Fibres Secondary acetate fabrics have a soft handle and excellent draping qualities but possess a low m.p. and relatively low strength. A resilient bonded wadding has been used for moulded articles and upholstery padding. Triacetate fabrics resemble those of secondary acetate in softness and drape but possess lower water imbibition and a higher softening temperature; by thermal treatment they can be rendered dimensionally stable and permanently smooth or pleated.

14.7 HISTORY

First prepared by P. Schützenberger in 1865, and investigated by Cross and Bevan, 1894. The acetone-soluble secondary acetate was introduced commercially by Miles in 1903, employed for safety film 1909, and used extensively (from solution) for aircraft dope *c*. 1917. The main plant in Great Britain was later developed in 1921 for the dry-spinning of fibres by the brothers H. and C. Dreyfus and British Celanese Ltd. Produced commercially as sheets and rods in 1927, and as a moulding powder in 1929–30 (with high acetate content, 1933). Triacetate, produced in a small way as fibres in 1914 but generally considered too intractable, was re-examined in the 1940s and re-introduced as fibres in 1954.

14.8 ADDITIONAL NOTES

14.81 TYPES OF CELLULOSE ACETATE FIBRES

Some of the commoner proprietary names are as follows:
Arnel, Tricel—triacetate rayons.
Celafibre, Fibroceta—sec. acetate staple fibres.
Dicel, Lansil—sec. acetate cont. filament.
Cotopa—partially acetylated cotton, rot-resistant and having better thermal and electrical resistance than cotton.
Alon—a Japanese fibre made by vapour-phase acetylation of catalyst-impregnated high-strength viscose rayon staple; it resembles secondary acetate fibre (giving an acetic acid yield of about 50%) but is stronger (2·8 gf/den., 2·2 gf/den. wet).

14.82 CELLULOSE PROPIONATE, ACETATE-PROPIONATE AND ACETATE-BUTYRATE

These products are prepared by methods similar to those used for the acetates, by reacting pre-treated cellulose in esterification mixtures containing the appropriate anhydrides. Commonly the first may have a propionic acid yield of approximately 50% (propionyl content 38·5%; degree of substitution, *c.* 1·8), while cellulose acetate-butyrate (CAB) appears to be produced in various compositions ranging upwards from 7% acetic acid yield and upwards from 18% butyric acid yield, e.g. a moulding compound may have an acetic acid yield of 19% and a butyric acid yield of 44% (acetyl content 13·6%; butyryl content 35·5%; total degree of substitution, *c.* 1·8) but a low-viscosity product intended for coating might have less acetic and more butyric acid. Unsubstituted hydroxyl groups, introduced as in secondary acetate, assist solubility in a range of solvents.

Suitable solvents: acetone, cyclohexanone, methylene chloride, chloroform, esters of low molecular weight (ethyl or butyl acetate), tetrahydrofuran, pyridine, dimethylformamide, acetic acid. The esters are swollen by alcohol and aromatic hydrocarbons, which can be used as diluents, and are insoluble in aliphatic hydrocarbons, carbon tetrachloride, and most oils; they are inert to dil. acid and alkalis but decomposed by the conc. reagents. Ester-type plasticisers are compatible (but less plasticiser is needed than with cellulose acetate).

CAB evolves a butyric odour on heating, and particularly on leaving to stand in alcoholic alkali; it responds to tests for cellulose (§13.23) but the presence of butyric acid interferes with the lanthanum nitrate test for acetic acid.

Physical properties (CAB, propionate somewhat similar)

N.B. Principal advantages over cellulose acetate are low *water absorption*, better *resistance to chemical attack*, and better *dimensional stability*.

Specific gravity 1·2–1·25. **Refractive index** 1·47–1·48. **Water absorption** (1–2%; max. *c.* 3%) and **permeability** to water vapour are lower than in cellulose acetate; moderately permeable to gases. **Thermal** and **electrical properties:** resemble approximately those of cellulose acetate; m.p. 150–200°C; decomposes (chars), 280–300°C; electrical properties less affected by humidity changes. **Mechanical properties** (depend appreciably on grade and plasticiser content) Generally slightly lower in strength but higher in extension at break and impact strength, otherwise similar to cellulose acetate (propionates, approximately intermediate between acetate and acetate-butyrate, show high flexural strength combined with high impact strength).

Applications

The esters are used chiefly as moulding powders where a material tougher and more weather-resistant than cellulose acetate is required. The propionate (Forticel) is available as clear or coloured sheeting (Celadex) suitable for general fabrication or vacuum forming. The acetate-butyrate is also used for lacquers, hot-melt applications, and dip-coating powders (e.g. giving tough and weather-resisting enamel-like coatings on metals); monofils have been used for upholstery.

14.9 FURTHER REFERENCES

LIPSCOMB, A. G., *Cellulose Acetate*, London (1933)
STANNETT, V., *Cellulose Acetate Plastics*, London (1950)
YARSLEY, V. E., *et al.*, *Cellulosic Plastics*, London (1964)
British Celanese Ltd, *'Tricel' Technical Service Manual*, London
FORDYCE, C. R. and MEYER, L. W. A., *Ind. Engng Chem.*, **32** (1940) 1054 (**40** plasticisers for CA and CAB)
HOWLETT, F. and URQUHART, A. R., *Chem. and Ind. (Rev.)*, 82 (1951) (solubility and structure of CA)
See also §13.7.

Analysis:
GENUNG, L. B., *et al.*, *Ind. Engng Chem. (Anal. Edn)*, **13**, 369 (1941) (acyl content); **14**, 292 (1942) (mixed esters); **14**, 935 (1942) (hydroxyl content); *Anal. Chem*, **22**, 401 (1950) (review of methods)
HOWLETT, F. and MARTIN, E., *J. Text. Inst.*, **35**, T1 (1944) (acetic acid yield). Committee Report, *Anal. Chem.*, **24**, 400 (1952) (examination of Eberstadt method)

Specifications:
BS 1524 (moulding material)
BS 2880 (testing flake, includes acetic acid yield)
BS 2610 (fluidity test)
BS 3186 (sheet for spectacle frames)
BSS 2792, 3334, 3344 (analysis of fibre mixtures)
ASTM D706 (CA moulding material)
ASTM D707 (CAB moulding material)
ASTM D786 (plastic sheets)
ASTM D871 (method of testing CA)
ASTM D1562 (cellulose propionate)

CELLULOSE NITRATE

SYNONYMS AND TRADE NAMES

Nitrocellulose. Gun-cotton (highly nitrated form). Collodion (solution). Celluloid, Pyroxylin, Xylonite.

GENERAL CHARACTERISTICS

White fibrous or clear colourless material, frequently pigmented. Hard and tough but flexible in thin sheets and films. Employed also as a surface coating.

15.1 STRUCTURE

Simplest Fundamental Unit

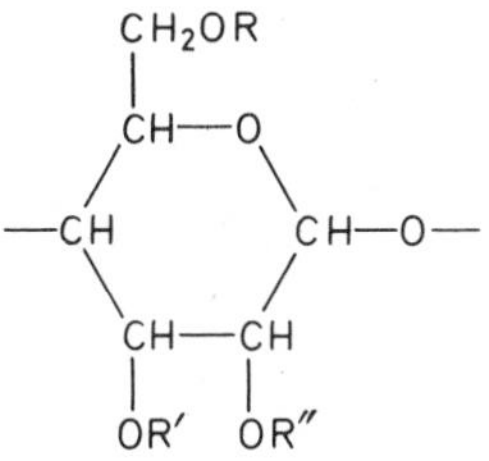

$$
\begin{array}{c}
CH_2OR \\
| \\
CH-O \\
/ \qquad \backslash \\
-CH \qquad CH-O- \\
\backslash \qquad / \\
CH-CH \\
| \qquad | \\
OR' \quad OR''
\end{array}
$$

where R, R′ and R″ (= H atoms in cellulose) are partly or completely replaced by nitro (—NO$_2$) groups.

Molecular weight Lacquer grades, 50000 (or lower). Celluloid-type plastics, up to 150000 (explosives higher).

Degree of polymerisation Lacquer grades, 50–500; plastics, *c.* 250–600. Other forms up to 1000 or more.

X-ray data Highly nitrated fibres, identity period 25·6 Å (5 anhydroglucose units), but affected by formation of various addition compounds with solvents. Plastics, frequently amorphous, appearing when stretched to be of random structure with identity periods that are multiples of the anhydroglucose unit; camphor-plasticised material exhibits identity periods of 25·6 and 38·3 Å.

15.2 CHEMISTRY

15.21 PREPARATION

From dried cotton linters, or paper pulp, steeped in a mixture of conc. nitric and sulphuric acids (the latter may be replaced by a mixture of phosphoric acid and phosphoric anhydride, or by acetic acid and acetic anhydride), at 20–40°C. The concentration of the acid governs the degree of nitration obtainable; nearly anhydrous conditions and a low temperature must be maintained if minimum degradation is required. After washing the product, to ensure that traces of acid do not at a later stage initiate its decomposition, it is stabilised by boiling in alcohol or very dil. acid, to remove residual sulphate groups.

According to the degree of substitution—and the degree of polymerisation (adjusted by treatment in water at elevated temperature and pressure)—different uses are found for the practical products. These, though not necessarily uniformly substituted, may be compared for nitrogen-content with the theoretical esters as below:

Practical products, %N		*Theoretical esters*	
		Deg. of substitution	%N
Plastics, lacquers	10·5–11·5	1	6·76
Lacquers, films	11·5–12	2	11·11
Explosives (high mol. wt.)	12–13·5	3	14·14

15.22 PROPERTIES

Solvents Plastics, fibres, and lacquer grades: alcohol, alcohol/ether; ketones, esters, and mixtures thereof, e.g. 1/1 acetone/amyl acetate; ethylene glycol monoethyl ether (Cellosolve, Oxitol) and the monoalkyl ether acetates (2-methoxy- and 2-ethoxyethyl acetate); diethylene glycol; glacial acetic acid. Products with 12% nitrogen or more: ketones, esters; alcohol/ether mixtures (trinitrate insoluble in the last). **Plasticised by** camphor (e.g. 15–25%), castor oil, aryl phosphates (triphenyl or tritotyl phosphate), alkyl phthalates (diethyl, dibutyl, or dioctyl phthalate), diphenyldiethylurea, sulphonamides and alkyl polyvinyl ethers (§7.81). **Swollen by** higher alcohols, ethylene dichloride, aromatic hydrocarbons (solutions tolerate dilution with toluene). **Relatively unaffected by** cold dil. acids and mild alkalis (Na_2CO_3), aliphatic hydrocarbons (petrol and many oils), chloroform (leaches out plasticiser). **Decomposed by** conc. acids (H_2SO_4) and alkalis, alkali hydrosulphides; surface attack by dil. alkalis (NaOH). Unless a stabiliser (diphenylamine) is incorporated, decomposed by prolonged exposure to ultra-violet light.

15.23 IDENTIFICATION

Thermoplastic. N present (but not detectable by the soda-lime test); Cl, S, and P absent. **Combustion** Inflammable, with exceptionally high burning rate, little smoke, and little residue. **Pyrolysis** Camphor, if present, can be smelt on warming (also on scraping). On further heating the material chars, evolves strongly acid yellow fumes, and may ignite or (at c. 180°C) explode.

Colour reactions 1. Add a few crystals of diphenylamine to 0·5 ml of approx. 90% sulphuric acid, and apply 1 drop of the mixture direct to the sample; cellulose nitrate produces an intense blue colour. Alternatively, if the reagent is poured carefully below the neutralised saponification liquor the blue colour develops at the interface. Other oxidising agents (e.g. dil. HNO_3) must be absent; the presence of hydrochloric acid increases the sensitivity. Cellulose acetate gives no immediate colour with this test. 2. Carbazole gives a deep green colour with cellulose nitrate, and with carboxymethylcellulose (*see* §16.23).

N.B. Cellulose nitrate responds to the anthrone test (§13.23) but not to the Molisch test or the soda-lime test (§§57/2.2).

Cellulose nitrate may be determined, in the presence of other cellulose derivatives, by a characteristic infra-red absorption band at 11·92 μm (ROSENBERGER, H. M. and SHOEMAKER, C. J., *Analyt. Chem.,* **31**, 1315 (1959)). Nitrogen may be estimated by the yellow colour produced with phenoldisulphonic acid (GARDON, J. L and LEOPOLD, B., *Analyt. Chem.,* **30**, 2057 (1958)).

15.3 PHYSICS

15.31 GENERAL PROPERTIES

Specific gravity 1·35–1·5 (trinitrate). Plastics, c. 1·38. Films, c. 1·45. **Refractive index** c. 1·50. **Birefringence** Oriented polymer slightly negative. **Water absorption** Plasticised sheet, 1–3%. **Permeability** to water vapour at 25°C: 2·1–5·0 $(10^{-10}$ g)/(cm s cmHg). To gases at 20–25°C $(10^{-10}$ cm^2)/(s cmHg): H_2, 2·0; O_2, 1·95; N_2, 0·12; CO_2, 2·12; He, 6·9.

15.32 THERMAL PROPERTIES

Specific heat c. 0·35. **Conductivity** $3–5.10^{-4}$ cal/(cm s °C). **Coefficient of linear expansion** $1·0–1·6.10^{-4}/$°C. **Physical effects of temperature** Plastics serviceable to 60°C, soften 80–90°C. Flow temperature c. 150°C. Material ignites in air c. 160°C (starts to decompose c. 90°C).

15.33 ELECTRICAL PROPERTIES

Volume resistivity *c.* $10^{10}-10^{11}$ ohm cm. **Surface resistivity** *c.* 10^9 ohm. **Dielectric strength** *c.* 25 kV/mm (50 for thin foil or short periods). **Dielectric constant** *c.* 7 (50 Hz), *c.* 6 (1 MHz). **Power factor** 0·1 (50 Hz), *c.* 0·09 (1 MHz).

15.34 MECHANICAL PROPERTIES (Plastics and films)

Elastic modulus (Young's) 200–300 kgf/mm². **Tensile strength** up to *c.* 7 kgf/mm². **Extension at break** 30% or more (typical clear plasticised sheet). **Flexural strength** 5–10 kgf/mm². **Compression strength** 14–21 kgf/mm². **Impact strength** (Izod test) *c.* 2–6 ft lbf/in of notch. **Hardness** Brinell (2·5 mm ball, 10 kg load), 8–15; Rockwell, M25–60, or up to R120.

15.4 FABRICATION

Sheet material can be cut or punched satisfactorily, can be welded by hot air, cemented (e.g. with 1/1 acetone/amyl acetate containing scrap polymer), bent to shape in hot water or moulded at 100–110°C; swollen in a solvent/non-solvent mixture, it can be stretched over formers to dry as a tough covering. Clear sheet is obtained by slicing uniform blocks made from compressed rolled sheets. Film is solvent-cast. Extrusion (rods, etc.) or compression moulding: 85–100°C, 2000–5000 lbf/in².

15.5 SERVICEABILITY

Satisfactory in the dark or for interior purposes, but becomes yellow and brittle with moderate or prolonged exposure to sunlight, or if stored at elevated temperature. Deteriorates when heated, and is highly inflammable (phosphate plasticisers reduce danger). Highly nitrated forms are explosive if insufficiently stabilised. All forms are unaffected by bacteria, fungi, and insects.

15.6 UTILISATION

Highly nitrated forms Explosives (gun-cotton; cordite). *Plastics* Wear-resistant moulded articles (knife handles, piano keys). Clear, coloured or pigmented sheets. Cinematograph film-base (now largely obsolete). Surface coating for leather, artificial leather, and bookbinding cloth. Moisture-proof coating for regenerated cellulose foil. General adhesive, pigment binder, waterproof finish, and lacquer. Intermediate (now obsolete) for cellulosic rayons.

15.7 HISTORY

The action of conc. nitric acid on carbohydrates was investigated by H. Braconnot in 1833, and by T. J. Pelouze in 1838. Relatively pure material was first prepared in 1845 by C. F. Schönbein, who (incorrectly assuming it to be a nitro-compound) developed it commercially as 'guncotton'; in 1846 he sent a sample of a transparent product, 'capable of being shaped into all sorts of . . . vessels', to Faraday. Pelouze re-investigated the subject at this time and produced 'pyroxylin'. Cellulose nitrate was employed in alcohol/ether solution as 'collodion' by W. Maynard in 1846, and was used by F. Scott Archer for the collodion wet-plate process in photography in 1851. A. Parkes developed it as a versatile surface-coating and moulding material ('Parkesine', patents 1855–1861) and succeeded in plasticising it (with camphor) in 1864. Independently it was plasticised in the U.S.A. by J. W. Hyatt, *c.* 1869; Hyatt manufactured the product as 'artificial ivory', which in turn led to its employment for other purposes—e.g. dental plates—under the names Celluloid (1871) and Xylonite (1877).

Audemars made experimental fibres from cellulose nitrate in 1855; and in 1878 J. Wilson Swan prepared filaments (by extrusion of collodion) which he carbonised for use in the first practical electric lamps. As a result of Swan's process, cellulose nitrate filaments, spun on a commercial basis for textile purposes, were put into production by Chardonnet from 1880 onwards. (In 1884 the filaments were partially saponified, giving fibres of regenerated cellulose, but Chardonnet's method is now entirely superseded by the viscose and cuprammonium processes. *See under* Cellulose, §13.21.) Cellulose nitrate was used as a 'dope' for coating the fabric of early aircraft and after the discovery (*c.* 1920) of its progressive degradation by superheated water it was much developed as a lacquer, e.g. for car bodies.

15.9 FURTHER LITERATURE

MILES, F. D., *Cellulose Nitrate,* London (1955)
YARSLEY, V. E., *et al., Cellulosic Plastics,* London (1964)
See also §13.9.

Specification:
ASTM D301 (methods of test, including determination of nitrogen)
ASTM D701 (sheets, rods, and tubes)

SECTION 16

CELLULOSE ETHERS (I)
*Alkali- and water-soluble types**

SYNONYMS AND TRADE NAMES

Methylcellulose MC, Celacol M and MM, Methocel, Methofas M. Mixed (modified) ethers: Cellofas A (methyl/ethyl), Celacol HEM (methyl/hydroxyethyl), Celacol HPM, Methofas P (methyl/hydroxypropyl). Carboxymethylcellulose (as the salt, *sodium carboxymethylcellulose*) CMC or SCMC, cellulose gum, cellulose glycollic acid (sodium salt). Cellofas B, Courlose, Tylose. *Hydroxyethylcellulose* HEC, glycolcellulose, Celacol HE, Cellosize. Mixed ethyl/hydroxyethyl ether: Ethulose, Modocoll.

GENERAL CHARACTERISTICS

These materials appear in commerce mostly as fibrous granules or powders that give viscous solutions capable of film formation. The solubility depends upon the type of substituent and on the degree and homogeneity of substitution, and these factors together with the degree of polymerisation govern the viscosity of the solutions. They are not commonly used as fibres or structural materials.

16.1 STRUCTURE

Simplest Fundamental Unit

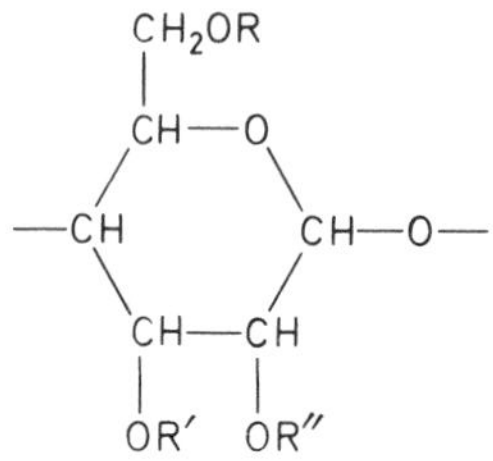

$$\begin{array}{c} CH_2OR \\ | \\ CH-O \\ / \qquad \backslash \\ -CH \qquad CH-O- \\ \backslash \qquad / \\ CH-CH \\ | \quad | \\ OR' \quad OR'' \end{array}$$

The ethers retain the basic unit of cellulose (§13.1) but with the —OH groups partly replaced by —OR groups, in which for example R = methyl (—CH_3), carboxymethyl (—CH_2COOH, or —CH_2-COONa for the sodium salt), or β-hydroxyethyl (—CH_2CH_2OH).

Molecular weight Varied during manufacture to obtain products graded over a wide range of viscosities.

Degree of polymerisation of commercial CMC, 500–2000.

X-ray data The water-soluble ethers give mainly amorphous-type diagrams, with a tendency towards crystallinity (especially in methylcellulose).

*For alkali- and water-insoluble ethers *see* Section 17.

164

16.2 CHEMISTRY

16.21 PREPARATION

The ethers are commonly prepared by treatment of alkali-swollen cellulose (cotton linters or wood-pulp) with esters of inorganic acids, e.g. alkyl chlorides or related substances, the alkali catalysing the reaction and removing hydrogen chloride. The general reactions are indicated below, cellulose being represented as Cell—OH.

$$\text{Cell—OH} \,(+\text{NaOH}) \xrightarrow[\text{methyl chloride, CH}_3\text{Cl*}]{\text{Autoclaved at 40–80°C with}}$$

Cell—OCH$_3$
(methylcellulose)

$$\text{Cell—OH} \,(+\text{NaOH}) \xrightarrow[\text{monochloroacetic acid, CH}_2\text{ClCOOH†}]{\text{Treated at 20–100°C with}}$$

Cell—OCH$_2$COOH
(carboxymethylcellulose)

$$\text{Cell—OH} \,(+\text{NaOH}) \xrightarrow[\text{oxide, } \overline{\text{CH}_2\text{CH}_2\text{O}}‡]{\text{Treated at 50°C with ethylene}}$$

Cell—OCH$_2$CH$_2$OH
(hydroxyethylcellulose)

Water-soluble mixed ethers (e.g. containing methyl groups together with ethyl, hydroxyethyl, or hydroxypropyl groups) are made for special applications.

16.22 PROPERTIES

Solubility depends on the degree of substitution (D.S.). Lightly etherified products are alkali-soluble, intermediate ethers (D.S. roughly 0·5–2) are water-soluble, and those with a high D.S.—which are commercially unimportant—dissolve only in organic solvents.

Methylcellulose is remarkable in being soluble in cold water but insoluble in *hot* water (solutions form reversible gels on warming); mixed methyl/ethylcellulose behaves similarly, methyl/hydroxy-propylcellulose tolerates higher temperatures before precipitating from solution. There is some overlap in the solubility distinctions;

*Or treated with dimethyl sulphate, (CH$_3$)$_2$SO$_4$, at 50°C.
†Sodium monochloroacetate may be used. In either instance the product is obtained as the sodium salt.
‡Or treated with ethylene chlorohydrin, ClCH$_2$CH$_2$OH.

the more highly methylated but still water-soluble forms are also soluble in organic solvents, such as methylene chloride/alcohol mixtures or formic or acetic acid. Aq. solutions of methylcellulose on evaporation yield strong oil-resistant films, which can be plasticised (glycerol, polyethylene glycols, sorbitol) or rendered insoluble (by treatment with glyoxal or an amine-formaldehyde precondensate).

Carboxymethylcellulose (always sold as the water-soluble sodium salt, SCMC) is ionic, the others are non-ionic. Unlike methylcellulose, SCMC and hydroxymethylcellulose are soluble in both cold and hot water.

The compositions of the theoretical ethers and the commercial products are as follows:

	Theoretical ethers			*Practical ethers (approx. composition)*		
	Mono-	*Di-*	*Tri-*	*Alkali-soluble*	*Water-soluble*	*Organic solvent soluble**
Methylcellulose						
Degree of substitution	1	2	3	0·3	1·3–2·0	Above 2·5
—OCH$_3$, %	17·61	32·63	45·59	5	22–33	Above 40
Carboxymethylcellulose						
Degree of substitution	1	2	3	0·1–0·2	0·4–0·9	—
—OCH$_2$COOH, %	34·09	53·96	66·96	5–10	15–30	—
Hydroxyethylcellulose						
Degree of substitution	1	2	3	1	Above 1·2	Above 2·1
—OCH$_2$CH$_2$OH, %	29·61	48·80	62·24	Up to 30	Above 35	Above 50

*Not developed commercially

16.23 IDENTIFICATION

Thermoplastic (at high degree of substitution). Cl, S, and P nominally absent; technical SCMC contains NaCl. **Cellulose component** *See* Anthrone and Molisch tests, §13.23. **Alkoxyl component** This can be demonstrated, or determined, by means of the Zeisel procedure; the alkoxyl groups are split off by the action of boiling conc. hydriodic acid, and the alkyl iodides liberated are detected, or estimated, by various methods. For quantitative determination (modified Zeisel method) *see* SAMSEL, E. P. and MCHARD, J. A., *Industr. Engng Chem.* (*Anal.*), **14**, 750 (1942), also COBLER, J. G. *et al.*, *Talanta*, **9**, 473 (1962) (modified Zeisel method and analysis by gas chromatography). **Other characteristics** (*see also* references to analytical procedures, §16.9): *Methylcellulose* As this material is insoluble in hot water it may be separated from impurities or mixtures with other ethers (e.g. SCMC) by warming the solution. *Carboxymethylcellulose* (*sodium salt*) (i) A 0·1% solution of carbazole in conc. sulphuric acid, added in excess to solid or dissolved polymer, produces a deep green colour; alginates and aldehydes (Sections 18 and 26) also give colours. (ii) SCMC stains deeply with 0·5% methylene blue, and exhausts the dye from weak aq. solution. (iii) An excess of a strong acid added to a solution of SCMC precipitates the free acid; polyvalent metal salts, such as alum (*see also* test for alginates, §18.23) and copper sulphate, cause precipitation of metal salts of CMC and can be used for gravimetric determination. (iv) The degree of substitution can be determined by boiling a sample with acetic acid and estimation of the resulting sodium acetate.

16.3 PHYSICS

The ethers are not commonly used as structural materials and not much information is available on their physical properties.

Specific gravity Methyl ether, 1·3–1·4; SCMC, 1·59. **Refractive index** Methyl ether, 1·4; SCMC, 1·5. **Moisture regain** Methylcellulose (%, at 25°C):

—OCH$_3$, %	18*	26	32†	43·5
Regain at 65% r.h.	11·9	10·2	7·7	2·5
Regain at 80% r.h.	16·0	13·7	11·1	4·4

*Degree of substitution = 1
†Degree of substitution = 2

Physical effects of temperature The ethers become brown at approx. 250–300°C and eventually char; methylcellulose liquefies during decomposition. **Mechanical properties** Films of methylcellulose: tensile strength, up to 7 kgf/mm^2; extension at break, 10%.

16.5 SERVICEABILITY

Exposure of aq. solutions to air, under alkaline conditions, causes chemical degradation with reduction of viscosity. The ethers are less susceptible to microbiological attack than starch but are not wholly resistant; solutions requiring to be stored need a preservative (e.g. formaldehyde, 8-hydroxyquinoline, sodium pentachlorophenate) and should not be exposed to excessive light or heat.

16.6 UTILISATION

Methylcellulose Permanent textile finish (alkali-soluble type). Size and starch substitute. Thickening and emulsifying agent; printing ink component; adhesive. Purified methyl/ethylcellulose is used in foodstuffs for its emulsifying properties; an aq. solution can be whipped to a 'solid' (pseudoplastic) foam.

Carboxymethylcellulose (*sodium salt*) Thickening and emulsifying agent (stabiliser in emulsion paints and oil-drilling muds; size, glue, and agar-agar substitute). Laundry assistant (prevents re-deposition of soil loosened by synthetic detergents; soil-resistant starch or starch additive, facilitates dirt removal on subsequent washing). Textile warp size; printing ink thickener; paper size and paper-making additive (reduces beating time, improves strength). Adhesive (high viscosity, non-staining, non-tacky, for wallpaper). Binder for pigments, also to improve strength of unfired ceramics and foundry cores. Purified forms are used in pharmaceuticals, dietetic foods, and ice-cream. Fibres of the aluminium salt have been proposed for fire-resisting fabrics.

Hydroxyethylcellulose Thickening agent (e.g. in water and emulsion paints), adhesive, paper size and additive (in conjunction with glyoxal, improves wet strength), textile size, etc. It is non-ionic, but unlike methylcellulose is not precipitated on heating. A mixed ethyl/hydroxyethylcellulose, resembling methylcellulose, is available.

16.7 HISTORY

Cellulose ethers were first investigated by W. Suida, in 1905, and subsequently were developed commercially by Lilienfeld from 1912 onwards. The early work on these materials has been summarised by E. C. Worden in Vol. I of *Technology of Cellulose Ethers* (New Jersey, 1933).

16.9 FURTHER LITERATURE

DAVIDSON, R. L. and SITTIG, M. (Eds), *Water-soluble Resins,* New York and London (1962). *See also* §13.9.

Analytical procedures
General GENUNG, L. B., *Analyt. Chem.,* **22**, 401 (1950) (review of methods)
Methylcellulose SAMSEL, E. and LAP, R. DE, *Analyt. Chem.,* **23**, 1795 (1951) (colorimetric determination with anthrone)
Carboxymethylcellulose BLACK, H., *Analyt. Chem.,* **23**, 1792 (1951) (anthrone); GHOSH, K. G., *et al., J. Sci. Industr. Res. B* **19**, 323 (1960) (precipitation and iodimetric estimation of the free acid); TIMOKHIN, I. M. and FINKELSHTEIN, M. Z., *Zhur. Priklad Khim.,* **36**, 415 (1963) (survey of methods, in Russian with English summary)
Hydroxyethylcellulose MORGAN, P. W., *Ind. Engng Chem. (Anal. Ed),* **18**, 500 (1946) (modified alkoxyl method); QUINCHON, J., *C.r. Acad. Sci. Paris* **248**, 225 (1959) (degree of substitution by selective reaction of phthalic anhydride with hydroxyls substituted by ethylene oxide)

Specifications
ASTM D1347 (testing methylcellulose)
ASTM D1439 (testing SCMC)

CELLULOSE ETHERS (II)
Alkali- and water-insoluble types (mainly ethylcellulose)*

With notes on benzylcellulose and cyanoethylcellulose

SYNONYMS AND TRADE NAMES
Cellulose ethyl ether. Ethocel, Ethulon.

GENERAL CHARACTERISTICS
Thermoplastic ethylcellulose, soluble in organic solvents and corresponding approximately to a diether in composition, is employed for tough, hard, translucent to transparent mouldings, sheets, and coatings.

17.1 STRUCTURE

Simplest Fundamental Unit
As in §16.1, the ethers retain the basic unit of cellulose but with the —OH groups partly or completely replaced by —OR groups, where most commonly R=ethyl (—C_2H_5).

Molecular weight Varied during manufacture to obtain products graded over a wide range of viscosities.

Degree of polymerisation of ethylcellulose *c.* 150–500.

X-ray data Intermediate products are largely amorphous but crystallinity increases (diagram sharper) as etherification approaches completion.

17.2 CHEMISTRY

17.21 PREPARATION

Cellulose treated with sodium hydroxide (i.e. alkali cellulose) is autoclaved with ethyl chloride (C_2H_5Cl, b.p. 12°C) for several hours at 80–140°C. An alternative treatment with alkali and diethyl sulphate at 50–80°C can be used. A diluent, such as toluene, may be added to facilitate control of the process.

*For water-soluble ethers see Section 16.

17.22 PROPERTIES

As with other cellulose ethers (§16.22), the properties depend on the degree of substitution (D.S.), the products that can be obtained being indicated below.

	Theoretical ethers			Practical ethers* (approx. composition)		
	Mono-	Di-	Tri-	Alkali-soluble	Water-soluble	Organic solvent soluble
Degree of substitution	1	2	3	0·2–0·6	0·8–1·3	1·9–2·6
—OC$_2$H$_5$, %	23·68	41·28	54·88	5–15	20–30	40–50

*Only the thermoplastic products, degree of substitution approximately 2·15–2·6 (43–50% —OC$_2$H$_5$), are commercially important. The solubility of the alkali- and water-soluble products is more dependent on composition than it is in methylcellulose, and is lower than that of hydroxyethylcellulose

Solvents The choice depends on the degree of substitution. Widest range of solubility occurs at D.S. approx. 2·4–2·5 (47–48% —OC$_2$H$_5$), solvents being alcohol, acetone, aromatic or chlorinated hydrocarbons, certain esters (e.g. ethyl or butyl acetate), dioxan, pyridine, and polar/non-polar mixtures (e.g. low viscosity solutions from 20/80 alcohol/toluene). Solutions can be diluted with aliphatic hydrocarbons. The tri-ether is soluble in aromatic hydrocarbons. **Plasticised by** high-boiling esters (e.g. dibutyl or dioctyl phthalate, tritolyl phosphate), castor oil. Compatible with resins (e.g. *p*-toluenesulphonamide resins), drying oils and waxes. **Swollen by** hot mineral oils. **Relatively unaffected by** water, weak acids, conc. alkalis. **Decomposed by** conc. acids.

17.23 IDENTIFICATION

Thermoplastic. Cl, N, S, P nominally absent. **Combustion** Melts, darkens, burns slowly with a clear flame; smells both 'woody' and waxy on extinction. **Pyrolysis** Melts and chars; acid distillate, with sweetish oily smell.

Other characteristics
Cellulose component See Anthrone and Molisch tests, §13.23. In the Molisch test the colour may appear only on warming the mixed liquids. *Ethoxyl component See under* §16.23 and §16.9. Chromic acid oxidises the ethoxy groups to acetic acid.

17.3 PHYSICS

17.31 GENERAL PROPERTIES

Specific gravity 1·13–1·15. **Refractive index** 1·47. **Water absorption** *c.* 1·5% (decreases with rise in substitution). **Moisture regain** (films, 65% r.h.)

Ethoxyl ($-OC_2H_5$) content, %	44	46	48	50
Moisture, approx. %	4·5	3·6	2·7	1·8

Permeability Ethylcellulose, plasticised, at 20–30°C (10^{-10} cm^2)/ (s cmHg): O_2, 26·5; N_2, 8·4; CO_2, 41; do. (not stated if plasticised), H_2, 87; He, 400. Water vapour: 1·1–10 (10^{-10} g)/(cm s cmHg).

17.32 THERMAL PROPERTIES

Specific heat *c.* 0·4. **Conductivity** *c.* 4–6.10^{-4} cal/(cm s °C). **Coefficient of linear expansion** 1·0–1·4.10^{-4}/°C. **Physical effects of temperature** Shock-resistant down to at least −30°C (films flexible to −70°C). Maximum service temperature, *c.* 80°C; softens *c.* 150°C, melts *c.* 160°C; minimum softening temperature at 48–49% $-OC_2H_5$.

17.33 ELECTRICAL PROPERTIES

Volume resistivity 10^{13}–10^{15} ohm cm. **Surface resistivity** up to 10^{13} ohm. **Dielectric strength** *c.* 20 kV/mm; for thin foil or short periods, up to 60 kV/mm. **Dielectric constant** *c.* 2·5–3·5 (up to 1 MHz). **Power factor** 0·03 or lower (up to 1 MHz).

17.34 MECHANICAL PROPERTIES

Elastic modulus (Young's) Up to 350 kgf/mm^2 (higher for tough films). **Tensile strength** and **Extension at break**

	Tensile strength (kgf/mm^2)	Extension at break (%)
Plastics	Up to 6*	5–50†
Fibres (experimental)	10–15 (1–1·5 gf/den.)	*c.* 10

*Films, up to 7·5 kgf/mm^2
†Varies with grade and plasticisation; in general slightly greater than for cellulose acetate. Films, up to 30%

Flexural strength 2–8·5 kgf/mm^2. **Compression strength** 5–14 kgf/mm^2. **Impact strength** (Izod test) *c.* 2–6 ft lbf/in of notch. **Hardness** Brinell (2·5 mm ball, 10 kgf load), up to 10; Rockwell, *c.* M65 or up to R110. Films in general slightly softer than those of cellulose acetate; minimum hardness at 48–49 %—OC_2H_5.

17.4 FABRICATION

Compression moulding: *c.* 170°C, 2500 lbf/in^2. Injection moulding: *c.* 220°C, up to 20 000 lbf/in^2. Can be cut, punched and machined satisfactorily; can be employed in adhesive compositions and applied as coatings from hot-melt formulations.

17.5 SERVICEABILITY

When suitably stabilised against oxidative photo- or thermal degradation by light or heat, ethylcellulose is satisfactory for exterior use. Dimensional changes in hot water are less than those of secondary cellulose acetate. Unaffected by bacteria, fungi, or insects.

17.6 UTILISATION

Alkali-soluble ethers have been used as permanent textile finishes, and the water-soluble ethers have been used as paper and textile sizes, but only the plastics-type ethers are important. These are employed similarly to cellulose acetate but where toughness and flexibility at low temperature are required. Plasticised compositions can be calendered on to fabric, and can be applied from solution or hot melt (containing mineral oil) for protective lacquers or adhesives.

17.7 HISTORY

See §16.7.

17.8 ADDITIONAL NOTES

17.81 BENZYLCELLULOSE

Etherification of alkali-swollen cellulose with benzyl chloride ($C_6H_5CH_2Cl$, b.p. 175°C) yields benzyl(or phenylmethyl)cellulose. This is at no stage alkali- or water-soluble, but at high degrees of substitution (60–65% —OCH$_2$C$_6$H$_5$) dissolves in organic solvents. It is very resistant to water and ozone and was at one time commercially available as a moulding material, but it is susceptible to photo- and thermal degradation. Partially benzylated cotton, thermoplastic and with increased rot resistance, has been described (*Text. Res. J.,* **28**, 659 (1958)).

The volatile products from benzylcellulose heated to 100°C in the presence of oxygen consist largely of benzaldehyde. Left standing in 1/1 acetic anhydride/sulphuric acid, benzylcellulose yields benzyl acetate, which can be saponified after separation by steam distillation from the neutralised mixture.

17.82 CYANOETHYLCELLULOSE

Treatment of alkali-cellulose with acrylonitrile (*see* §10.21) yields cyanoethyl ethers of cellulose (represented by Cell —OCH$_2$CH$_2$CN). Cotton can be cyanoethylated with retention of the fibrous form and is much improved in rot-, heat- and acid-resistance (*Text. Res. J.,* **25**, 58 (1955); **26**, 67 (1956)). A more highly substituted ether has found applications for electrical purposes. Cast as a film, or moulded, this material possesses an unusually high dielectric constant and a low power factor. The commercial product (Cyanocel) has the following properties. Degree of polymerisation, 300–400; degree of substitution, 2·6–2·8. **Soluble in** acetonitrile, acetone, dimethylformamide and similar polar solvents. **Specific gravity** 1·2. **Water absorption** (65% r.h.), *c.* 2·5%. **Coefficient of linear expansion** 1–2.10^{-4}/°C. **Volume resistivity** 10^{10} ohm cm; *surface resistivity* 10^{11} ohm. **Dielectric strength** (thin films, short time), 60 kV/mm. **Dielectric constant** and **power factor** (approximate values),

Frequency, Hz	10^3	10^6
Dielectric constant	13	11
Power factor	0·002	0·01

Elastic modulus (Young's) 225–240 kgf/mm^2. **Tensile strength** and **extension at break** 1·3 kgf/mm^2, 1·1% (films: 3·8 kgf/mm^2, 9%).

17.9 FURTHER LITERATURE

See §13.9

Specifications
ASTM D787 (moulding and extrusion materials)
ASTM D914 (testing methods)

SECTION 18

ALGINIC ACID AND ALGINATES

With notes on alginate esters, carrageenin, agar-agar, pectic acid,
and natural gums (g. arabic, g. tragacanth and g. karaya)

SYNONYMS AND TRADE NAMES
Poly(1,4-anhydro-β-D-mannuronic acid) together with L-guluronic acid. Sodium salt: Manucol, Manutex.

GENERAL CHARACTERISTICS
The free acid swells, but is practically insoluble, in water. The sodium salt is soluble, and is available commercially as a pale yellow powder, aq. solutions of which are viscous and leave transparent films on evaporation. Calcium alginate is available as textile fibres characterised by unusually jagged cross-sections, and by their solubility in dil. alkalis.

18.1 STRUCTURE

Simplest Fundamental Unit

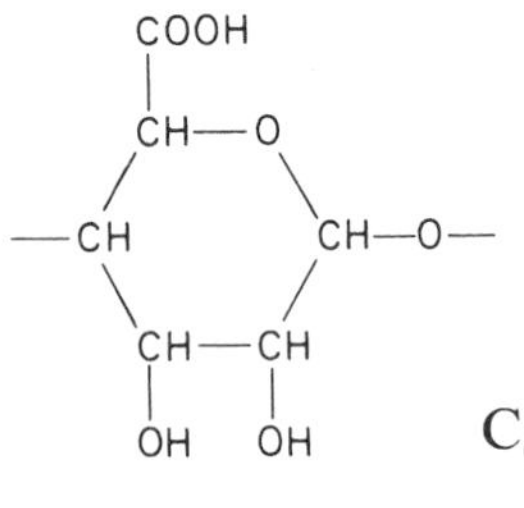

$C_6H_8O_6$, mol. wt. = 176

The unit is related to that of cellulose (§13.1) and the structural configuration is better represented by two 1,4-linked residues of β-D-mannuronic acid,

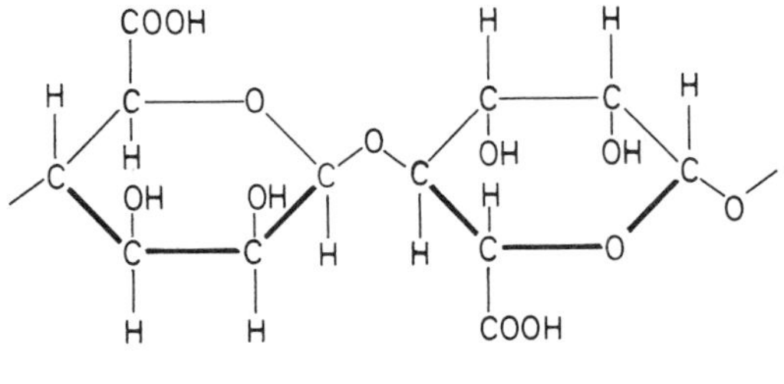

In addition 1,4-linked residues of L-guluronic acid are present.*

*That is with a configuration similar to that of the β-D-mannuronic acid residue except that on C atom 5 —H and —COOH are interchanged.

174

Molecular weight Natural acid, at least 150000 (degree of polymerisation *c.* 1000). Regenerated forms and fibres, dependent upon preparation and age, usually at least 30000–60000 (degree of polymerisation 200–400) for low viscosity grades and over 100000 (degree of polymerisation 600) for high viscosity types.

X-ray data The identity period (I.P.) of sodium alginate fibres (5 Å) closely approaches the theoretical length of a mannuronic acid residue (cf. cellulose), but in stretched alginic acid the I.P. is 8·7 Å (considered to represent the projection of two fundamental units inclined to each other; two more fundamental units, contained in a different chain and reversed in direction from the other two, complete the unit cell). The I.P. of calcium alginate fibres is slightly less than for the free acid; both calcium and beryllium alginates show only low orientation and crystallinity.

18.2 CHEMISTRY

18.21 PREPARATION

The cell walls of brown seaweeds (notably the various species of *Laminaria* common on the N. Atlantic and Japanese coasts) contain 15–40% alginic acid, which is obtained as a solution of the sodium salt on extraction with dil. alkali or sodium carbonate; the free acid can then be liberated by addition of dil. mineral acid (H_2SO_4). Alginic acid is insoluble in water; its alkali and ammonium salts are soluble but addition of salts of polyvalent metals (e.g. $CaCl_2$, $FeCl_3$) precipitates insoluble alginates. This is the basis of alginate fibre production, ripened alginate solutions being stretch-spun into an appropriate coagulating bath.

N.B. Although alginates yield viscous solutions, the slippery coating on seaweeds is composed of a different substance, fucoidin, which is not exploited commercially. For carrageenin, derived from red seaweeds, *see under* §18.82.

18.22 PROPERTIES

Solvents *Alginic acid.* Very sparingly soluble in water, freely dissolved by alkalis. *Alkali and ammonium alginates.* Water, alkalis. Magnesium and ferrous alginates also are soluble in water. *Water-insoluble alginates.* Calcium alginate readily disperses in weak alkalis (e.g. warm soap or sodium carbonate solutions) being converted to the sodium salt. *N.B.* Calcium alginate fibres cross-linked by di-iso-cyanate treatment are insoluble in alkalis. Copper, zinc, and alu-

175

minium alginates dissolve in ammonium hydroxide. Beryllium, beryllium-calcium, chromium, and ferric alginates are alkali-resistant. **Plasticised by** humectant polyhydric alcohols (glycerol), urea. **Swollen by** water and hydrophilic compounds (swelling most pronounced in alginic acid and calcium alginate). **Relatively un-affected by** hydrocarbons, organic solvents, and hydrophobic compounds in general. Soluble alginates applied to textiles may become insoluble when cured with a resin finish. **Decomposed by** acids and alkalis, especially on warming. The free acid degrades in molecular weight on heating, and soluble alginates readily degrade when warmed in solution.

18.23 IDENTIFICATION

Non-thermoplastic. Cl, N, S, and P absent. **Combustion** Metallic alginates are non-inflammable, they char without melting when ignited and leave an ash in which the original shape of the sample is sometimes retained. The ash can be raised to incandescence without further change. **Pyrolysis** Alginates char without melting; emit a 'woody' odour; produce acid fumes and an acid distillate. Metallic radicals, if present, are retained in the residue. **Characterising re-actions** 1. Alginates respond to the Anthrone and Molisch tests (*see* §13.23) but the colour in the Molisch test is redder than for cellulose. 2. When boiled with dil. sodium hydroxide, a bright yellow colour is imparted to the solution. 3. Solutions of soluble alginates precipitate alginic acid (almost colourless) on acidification. Alternatively, addition of polyvalent cations to an aq. solution of a soluble alginate causes double decomposition and coagulation; e.g. with the aid of a needle, or glass rod drawn to a point, merge 1 drop of a neutral test solution (which should be concentrated to contain approx. $0 \cdot 5 - 1 \cdot 0 \%$ polymer) into an adjacent drop of saturated alum solution; a charac-teristic ropiness of the bridge between the two drops is at once apparent. If the alum solution is replaced by one of copper sulphate or ferric chloride, the ropy product is coloured.

N.B. The same effect is obtained with soluble derivatives of pectic, polyacrylic and cellulose glycollic (CMC) acids. Pectates are precipitated by magnesium sulphate, alginates are not.

The carbazole derivative (pink) may be used for colorimetric estimation of alginic acid. For quantitative estimation of uronic acids in alginates, *see* HAUG, A. and LARSEN, B. *Acta chem. scand.,* **16**, 1908 (1962). For other distinctions of alginates and analytical methods, *see* SCHULZEN, H., *Textil—Praxis,* **19**, 67 (1964).

18.3 PHYSICS

18.31 GENERAL PROPERTIES

Specific gravity Free acid (dry), *c.* 1·6; calcium alginate (fibres), 1·75–1·78. **Refractive index** Calcium alginate (fibres), 1·525. **Birefringence** Very low (0·001). *Optical activity* Specific rotation (aq. solution), $[\alpha]_D^{15} = -133°$. **Water absorption** Calcium alginate, 50% or more. **Water retention of fibres** Calcium alginate, over 100%; beryllium alginate, *c.* 40%. **Moisture regain of fibres** (65% r.h.) calcium alginate 20–30%.

18.32 and 18.33 THERMAL AND ELECTRICAL PROPERTIES

Note. Employment of alginates is restricted by their hygroscopicity and low mechanical strength.

Volume resistivity (ohm cm) Fibres (experimental values, not necessarily representative):

R.h., %	65	90
Calcium alginate	$1·5.10^7$	$0·015.10^7$
Beryllium alginate	450.10^7	150.10^7

18.34 MECHANICAL PROPERTIES

Note. The useful mechanical properties of alginates decline with increase of humidity, this being particularly so with the calcium salt used for fibres.

Elastic recovery Calcium alginate (65% r.h.), 70–80% from 1 to 2% extension. **Elastic modulus** (Young's) Calcium alginate fibres (65% r.h.), 2000 kgf/mm² (125 gf/den.).
Tensile strength and **Extension at break** (fibres):

	Tensile strength (kgf/mm²)	Extension at break (%)
Alginic acid	Under 10 (0·5 gf/den.)	—
Calcium alginate (65% r.h.)	25–30 (1·6–2·0 gf/den.)	Up to 6
Calcium alginate (wet)	Under 10 (0·5 gf/den.)	5–25
Beryllium, chromium and related alginates*	Usually low but can reach 2·4 gf/den.	Under 5

*Retain 75% of strength when wet

177

18.4 FABRICATION

Mechanical applications limited. Low-density materials can be prepared by foaming viscous solutions and adding an insolubilising substance, such as an aq. calcium salt, the solid foam so produced being subsequently washed and dried out. Production of fibres is mentioned in §18.21.

18.5 SERVICEABILITY

Humidity, rise of temperature, and prolonged storage cause degradation, with consequent loss of strength. Purified alginates are said to be unaffected by bacteria, fungi and insects. Formalin, sodium pentachlorophenate, etc., assist preservation.

18.6 UTILISATION

Alginic acid has no direct applications. Sodium alginate, yielding viscous aq. solutions at $1-2\%$ concentration, is employed as a suspending and emulsifying agent, for thickening (e.g. in water-based paints, textile printing pastes, and as a temporary gumming paste during textile printing), and for textile and paper sizing. Ammonium alginate assists creaming of rubber latex (§29.81); other uses are in dental impression materials and ice cream formulations. Fibres of water-soluble alginates have been used for military purposes, e.g. disposable parachutes.

Fibres of calcium alginate are used for fire-resistant fabrics, for assimilable surgical threads, and for auxiliary or scaffolding purposes (e.g. low-density woollens, open-work lace, temporary seams, linking in knitting, etc., also disposable laundry bags) being subsequently removed by washing in mild alkali.

18.7 HISTORY

The natural product was first recognised in 1881 when E. C. C. Stanford extracted it from the commoner marine algae (*Laminaria* and *Fucus*) and named it 'algin'. In 1929 W. L. Nelson and L. H. Cretcher showed it to be a polyuronic acid based on D-mannuronic acid, and in 1955 F. G. Fischer and H. Dorfel showed L-guluronic acid to be present as well. Fibres and other commercial possibilities of the material were investigated by L. Sarason in 1912, and in the mid 1930s work was done on it in Japan and by Bonniksen. An early production was in the form of alginate paper and transparent sheeting (Cefoil), and in 1939–40 an investigation by J. B. Speakman, N. H.

Chamberlain and others led to the first commercially successful production of textile yarns.

18.8 ADDITIONAL NOTES

18.81 ALGINATE ESTERS

Alginic acid is not acetylated or methylated without undergoing considerable decomposition, but derivatives have been obtained by reaction with epoxy compounds. Thus, propylene oxide yields propylene glycol alginate, which is both water-soluble and (having the —COOH group blocked) compatible with acids and polyvalent metal salts. It is, however, decomposed by alkalis. Insoluble alkali-resistant products have been obtained by refluxing with a diepoxide, e.g. 1,2,5,6-diepoxyhexane.

18.82 CARRAGEENIN AND AGAR-AGAR

These neutral mucilaginous substances, relatively unreactive in solution and horny when dried, are obtained from red seaweeds in the class *Rhodophyceae*.

Carrageenin, from Irish moss (*Chondrus crispus*), is a heterogeneous partially sulphated polymer, separable into λ-carrageenin, an esterified water-soluble linear component, and κ-carrageenin a gel fraction that is less esterified but branched. The predominant structural units derive from D-galactose; e.g. 1,3 (also 1,4)-linked D-galactose 2-sulphate (I) in the λ-fraction, and 1,4'-linked 3,6-anhydrogalactose units (II) in the κ-fraction.

I   II

Agar-agar (or agar) is obtained mainly from various red algae in Japanese waters. Like carrageenin, it is a complex polymeric material, and appears to consist largely of 1,3'-linked α-D-galactose (with only a small proportion of ester sulphate groups) together with 1,4'-linked 3,6-anhydro-L-galactose.

These substances swell in cold water, dissolve on boiling, and 'set' to firm gels on cooling. They find uses as thickening and suspending agents, especially in foodstuffs; carrageenin has lubricating and anti-tack properties, and has been used as a thickener in textile printing; agar-agar yields stiff sharp-melting gels, and is employed as a bacteriological culture medium.

For further literature, *see* §18.9.

18.83 PECTIC ACID

This is essentially poly-1,4-D-galacturonic acid, a polysaccharide isomeric and closely related to alginic acid, having the repeat unit:

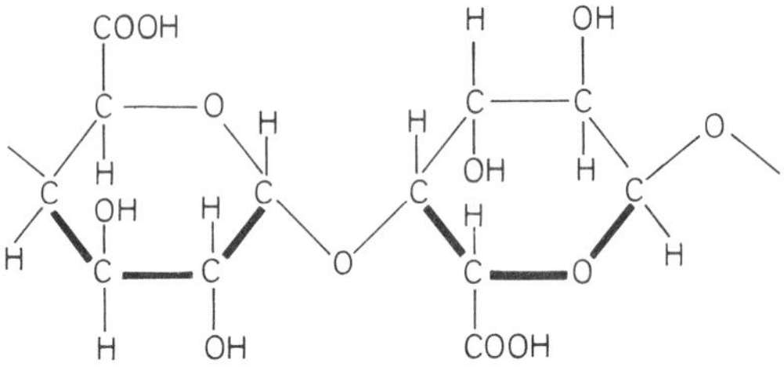

It is present in an insoluble form in green plants and particularly in fruits, from which *pectin* (largely a partial methyl ester) is extracted by hot water, e.g. from residues of apples and citrus fruits after expression of the juice. Pectin, first isolated by Vauquelin in 1790 and examined by Braconnot in 1825, is now widely used as an assimilable thickening and gelling agent in foodstuffs (optimum gelling: 0·7 % pectin, 60 % sugar, pH 3). It is degraded by alkalis and hydrolysed by mineral acid; the hydrolysed product yields a white precipitate with basic lead acetate, soluble in excess of the reagent but reappearing as a red precipitate on boiling.

Residues of D-galacturonic acid are also present in gum tragacanth (§18.84). *See also*: KERTESZ, Z. I., *The Pectic Substances,* New York and London (1951), and WORTH, H. G. J., *Chem Rev.,* **67**, 465 (1967) (chemistry of pectic substances).

18.84 NATURAL GUMS (*see also* §19.8)

The following gums contain uronic acid residues:

Gum arabic, an exudation derived from *Acacia* trees occurring mainly in the Sudan, has a molecular weight of some 250 000 (though it is easily degraded by acids) and is composed of magnesium (with calcium and potassium) salts of arabic acid, a complex branched polymer of arabinose, galactose, rhamnose and D-glucuronic acid. The free acid is precipitated when an acidified aq. solution of the gum (from which the metals may be removed by a cation-exchange resin) is poured into acetone or alcohol. As one of the most soluble of natural gums, giving clear colloidal solutions, it is used for stabilising dispersions and emulsions, for thickening or stiffening, and as an adhesive ('gum').

Gum tragacanth, an exudation deriving from various species of *Astragalus* shrubs occurring in Greece, Asia Minor, and Iran, is a complex polymer of arabinose, fucose, galactose, xylose and D-glucuronic acid, together with D-galacturonic acid which renders it mildly acid (a substituted pectic acid). The molecular weight is about 840 000. It is only partly soluble in cold water, requiring to be boiled for dispersion; the commercial material is variable in quality, and maximum viscosity is obtained only with the best grade. It is used for thickening and for stabilising dispersions and emulsions, being employed in textile printing, pharmacy and certain foods. For gum tragon, or locust bean gum, *see* §19.83.

Gum karaya, an exudate derived from *Sterculia* trees in India, is a polymer based on galactose and rhamnose with D-galacturonic and D-glucuronic acids. It is partially acetylated and hydrolyses to an acid solution in warm water. Though degraded under alkaline or sufficiently acid conditions, it is normally an acid resisting gum, providing a cheap substitute for gum tragacanth (being also known as Indian tragacanth) but distinguished from it by giving a pink colour with an ammoniacal solution of ruthenium red, or when boiled with conc. hydrochloric acid.

18.9 FURTHER LITERATURE

SMITH, F. and MONTGOMERY, R., *The Chemistry of Plant Gums and Mucilages,* New York (1959) (Amer. Chem. Soc. monograph)
WHISTLER, R. L. (Ed), *Industrial Gums,* New York and London (1959)

STARCH
and starch derivatives

With notes on dextran, locust bean gum, and guar gum

GENERAL CHARACTERISTICS

Starch comprises the principal carbohydrate reserve of the vegetable kingdom, being stored as minute white granules in the roots, bulbs, tubers, seeds, pith and fruit of plants. The granules, which vary in size (from about 0·005 to 0·02 mm across, with some up to 0·15 mm in length), have a concentric layered structure and are often very distinctive in shape. The chief source is maize (corn starch), but potato, rice, sago, tapioca and wheat starch are also important.

At room temperature natural starch is insoluble, but when heated with some fifteenfold its weight of water to 65–85°C it forms viscous dispersions that may set to gels when cold. As a result of modifying treatments, however, the properties of many commercial products differ greatly from those of the parent material.

For mention of animal starch or glycogen, *see* section below.

19.1 STRUCTURE

Pure natural starch (Latin name *amylum*) consists of two polygluco-sans, *amylose* and *amylopectin*, commonly present in a ratio of *c.* 1:3, but the proportions can vary considerably, according to the origin of the starch; for example, wrinkled pea starch is largely amylose, whereas waxy maize starch consists almost entirely of amylopectin. Glycogen, the animal equivalent of starch (i.e. glucose reserve) has the structure of a highly branched amylopectin.

Simplest Fundamental Unit

1. *Amylose* is a linear polymer of 1,4-anhydroglucose units linked through α-glycosidic bonds, as shown (for comparison with the β-configuration of cellulose, *see* §13.1):

$C_6H_{10}O_5$, mol. wt. = 162

2. *Amylopectin* is a branched polymer of anhydroglucose units. Each branch has the linear structure of amylose, but branching occurs through 1,6′ linkages.

Molecular weight Amylose: varies from 30 000 to 300 000; degree of polymerisation 200–2000. Amylopectin: over 10^6; degree of polymerisation of average branch, *c*. 25.

X-ray data Dried starch is amorphous, but undried natural starch and solidified gels show three crystalline modifications (one of which yields an identity period of 10·5 Å). Because of the α-configuration, dry amylose possesses a helical structure having about six anhydroglucose units per turn; in the blue complex formed with iodine (§19.23), the iodine atoms are positioned centrally within the helix.

19.2 CHEMISTRY

19.21 PREPARATION

Starch is synthesised by green plants and stored as a reserve carbohydrate (*see* General characteristics earlier). Seeds and roots contain 10–50% or more of starch, which is released by grinding or pulping under water, cereal starches having been previously treated to remove the endosperm, proteins, etc. The granules are then strained free of fibrous matter, allowed to settle, washed, centrifuged, and finally dried in warm air. Sometimes a slight oxidative treatment is included to yield a white and odourless product. The commercial material usually contains small residues of bound phosphates together with traces of proteins and fatty acids.

Amylopectin may be obtained from a dispersion of farina (potato starch, 80% amylopectin) or of waxy maize starch (largely amylopectin) after addition of magnesium sulphate or butanol to precipitate the amylose. Modified starches, which may be considerably changed in properties, are mentioned under §19.81.

19.22 PROPERTIES

Action of solvents Starch is soluble in dil. sodium hydroxide, anhydrous ethylenediamine, cupriethylenediamine solution, 90/10 dimethyl sulphoxide/water, and fused chloral hydrate, but only dispersions in water are of industrial significance. The granules swell reversibly in water, and in sufficient concentration form a thick slurry, but disintegrate only on heating (assisted by mechanical agitation) to give a viscous opalescent dispersion or 'paste' that may form a turbid elastic gel on cooling. For films made by spreading and drying the gels, *see* §19.3.

The properties of pastes and of gels from them depend on the proportions of amylose and amylopectin in the starch. Amylose is the less stable in solution, favouring formation of quick-setting irreversible gels and tough films; whereas amylopectin, though giving a high viscosity, yields more stable and less clouded solutions which either do not gel on storage or form only soft reversible gels.

Action of reagents When boiled with dil. mineral acids, starch is hydrolysed to products of lower molecular weight which pass into solution and ultimately yield glucose. It is also hydrolysed by enzymes of the diastase type; *α-amylase* (e.g. malt diatase) converts it to a mixture of glucose, maltose and trioses, while the less active *β-amylase* (e.g. barley diastase) converts it to maltose and a *β*-dextrin from the amylopectin. It is almost inert to aq. alkali although it swells and dissolves more readily in it than in water. It is oxidised by all common oxidising agents, and is charred by conc. sulphuric acid. Films may be plasticised, with concomitant loss of strength, by urea, ethanolamine, glycerol, castor oil or tallow.

19.23 IDENTIFICATION

Non-thermoplastic. Traces of N, P and Si present. **Combustion** On heating in air starch chars and burns, yielding products similar to those from cellulose and leaving a carbonaceous residue. **Pyrolysis** Degradation *in vacuo* commences at *c.* 220°C, with evolution of CO, CO_2, H_2O, and minor proportions of aldehydes, ketones, etc., leaving a carbonaceous residue. **Tests** Starch responds to the anthrone and Molisch tests (§13.23). The granules of natural starch (spherical, ovoid, etc., according to origin, and birefringent) may be identified with certainty under the microscope, and the hydrolytic products are characteristic (*see* action of reagents, above). With a dil. solution of iodine and a trace of iodide ion, starch and aq. solutions of starch or of amylose give a characteristic deep blue colour which disappears on heating but returns on cooling (*see also* §19.1; for a similar reaction of iodine with polyvinyl alcohol, *see* §7.23). Amylopectin and degraded forms of starch give a red or violet colour with iodine (glycogen gives a brown colour). Starch is precipitated from solution on addition of ammonium sulphate.

19.3 PHYSICAL PROPERTIES

Specific gravity 1·5. **Refractive index** (films) 1·53. Starch granules are **birefringent,** being composed of concentric layers in which amylose and amylopectin form radially-oriented mixed crystals, i.e. the granules act as double refracting spherulites. *Specific rotation* $[\alpha]_D^{15}$ amylose, $+200°$; amylopectin, $+200°$. **Moisture content** (natural starch granules, 65% r.h.) 10–20%.

The **mechanical properties** of films prepared by spreading and drying starch pastes are relatively poor, and dependent on composition (*see* §19.22). At 65% r.h., the **elastic modulus** (Young's) is c. 300 kgf/mm^2, while **tensile strength** and **extension at break** are of the order of 4·0–4·7 kgf/mm^2 and 2·6–4·2% respectively (*see* NEAL, S. M., *J. Text. Inst.*, **15**, T443 (1924)).

19.5 SERVICEABILITY

Starch is the basis of most carbohydrate foods (cereals) and is the raw material of the fermentation industries. Its gel- and film-forming properties make it useful in paper sizing and in the textile industry (size, filler, printing vehicle, stiffening and glazing agent); its adhesive applications are also considerable (flour pastes, British gum). Red iron oxide bound with a starch paste is used as a preservative on wood (Swedish red paint). Amylose fibres, which can be dissolved by the action of amylase, have been used for scaffolding threads in textile manufacture. However, unless appropriately protected, starch is unsuited to damp situations since it is subject to various forms of microbiological attack.

19.7 HISTORY

Additional to its nutritive function, starch seems early to have been employed in Egypt as an adhesive and filler in the preparation of papyrus (Pliny gives a recipe), and in China, c. A.D. 100, for sizing paper. Stiffening and filling cloth also began at an early date, finding special importance in the ruffs, wimples, etc., of the sixteenth century; however, the present industrial significance of starch commenced with technological expansion in the nineteenth century, being linked in particular with the production of maize in America.

Starch received attention from several early investigators—e.g. Leeuwenhoek described the granules in 1719—but scientific understanding of its composition began in 1811 when G. S. C. Kirchoff showed it to give a sugar (glucose) on boiling with dil. acid, and again in 1814 when Kirchoff reported the enzymic hydrolysis to another sugar (maltose). The blue colour imparted by iodine was also discovered in 1814. Practical roasting (§19.81) saw application in 1821, for British gum, and the intermediate breakdown products (dextrins) were examined by J. B. Biot and J. F. Persoz in 1833; A. Payen and J. F. Persoz isolated the starch-hydrolysing enzyme, diastase, in the same year. P. Schützenberger prepared starch acetate in 1870. Long recognised as complex in composition, starch was separated into amylose and amylopectin by L. Maquenne and E. Roux in 1906.

Names prominent in modern investigation of starch, dating from *c.* 1928, are those of W. N. Haworth, E. L. Hirst, J. K. N. Jones and others (1,4-anhydro-α-glucose structure, etc.), also those of J. P. Staudinger (branching) and K. H. Meyer among many workers in this field.

19.8 ADDITIONAL NOTES

19.81 MODIFICATIONS AND DERIVATIVES OF STARCH

Starch can be modified by the treatments indicated below. The purpose in so doing is usually to increase the solubility and lower the viscosity, but in highly degraded or substituted products none of the original characteristics remains.

Controlled hydrolysis by acids or enzymes, or *controlled oxidation* (e.g. 'chlorination' by means of sodium hypochlorite), which yield a number of commercially-important 'thin-boiling starches'. The action of periodic acid or periodates yields a reactive 2,3-dialdehyde starch, which is commercially available (e.g. Sumstar) and has potential applications in the textile, paper and leather industries.

Roasting at 180–250°C, with or without a basic catalyst (soda ash), degrades starch to products known as yellow (or canary) **dextrins** or British gums. White or light-coloured dextrins are obtained by first moistening starch with a dil. mineral acid and then heating, for shorter times, at *c.* 100°C. The molecular structure of dextrins is much deranged from that of starch, and when sufficiently modified they are completely soluble in cold water. They find uses in textile finishing and in printing pastes, and particularly as adhesives for paper and cardboard; borax increases their effectiveness.

Esterification Starch acetate (Feculose) forms clear films and can be used as a size; fibres of amylose triacetate have been prepared, resembling those of the cellulose analogue but lower in strength. Starch nitrate, of a high degree of substitution, is used as an explosive (for review, *see* CAESAR, G. V., *Advanc. Carbohyd. Chem.,* **13**, 331 (1958)).

Etherification Alkylation, hydroxyalkylation, cyanoethylation, and carboxymethylation (cf. cellulose ethers, Section 16) give useful gum-like products, some of which yield solutions that do not gel on cooling (Solvitose, Ten-o-film). An allyl ether, 35% allyl-content, insoluble in water but soluble in organic solvents, was developed as an air-drying varnish and impregnant for paper and textiles.

19.82 DEXTRAN

This name is given to water-soluble gummy or slimy polysaccharides synthesised by (enzymes produced by) bacteria, e.g. by *Leuconostoc* species from sucrose or by certain *Acetobacter* species from starch dextrins. They have very high molecular weights and are composed predominantly of 1,6'-linked anhydro-α-D-glucose units, branched, through formation of 1,3'-linkages. These substances have been used as blood plasma substitutes (with the molecular weight reduced to *c.* 75 000 by partial acid hydrolysis), in paper and textile coating, and for experimental fibres. For a review, *see* NEELY, W. B. *Adv. Carbohyd. Chem.*, **15**, 341 (1960).

19.83 LOCUST BEAN GUM AND GUAR GUM

These substances are natural polysaccharides related to starch, though rather different in properties. For other natural gums *see* §18.84 and references under §18.9.

Locust bean gum, carob gum, or *gum tragon,* is obtained from the seeds of *Ceratonia siliqua,* a tree of the Mediterranean regions. *Guar gum* comes from the seed-coat of *Cyamopsis tetragonolobus,* a bean plant cultivated in Pakistan and the U.S.A. Both products are largely polymers of D-mannose (65–80%) and D-galactose, differing in their structural arrangements. The molecular weight of locust bean gum is *c.* 300 000. Both polymers are readily degraded by acids. They are neutral substances, swelling in cold water and dissolving on warming, which at low concentrations provide viscous solutions that yield flexible films on drying. The chief uses are as substitutes for gum tragacanth (§18.84) in thickening, as stabilising agents, and in textile sizing and printing.

19.9 FURTHER LITERATURE

KERR, R. W., *Chemistry and Industry of Starch*, New York (1950)

NORD, F. F. (Ed), *Advanc. Enzymol.*, **12** (MEYER, K. H. and GIBBONS, G. C., starch chemistry), New York and London, 341 (1951)

RADLEY, J. A., *Starch and its Derivatives,* London (1953)

HONEYMAN, J. (Ed), *Recent Advances in the Chemistry of Starch and Cellulose,* London (1959)

WHISTLER, R. L. (Ed), *Methods in Carbohyd. Chem.,* Vol. IV (starch), New York (1964)

WHISTLER, R. L. and PASCHALL, E. F. (Eds), *Starch: Chemistry and Technology*, Vols. I and II, New York and London (1965)

SECTION 20

SILK AND WOOL*

With notes on types, other animal fibres, collagen, gelatin, and leather

General Characteristics
See §§20.21, 20.81 and 20.82.

20.1 STRUCTURE

Simplest Fundamental Unit
The main structural materials comprising these fibres are essentially proteins—known as fibroin (in silk) and keratin (in wool)—which are composed of mixed but unidirectional (i.e. head-to-tail) α-amino-acid residues, or polypeptide fragments, of the type

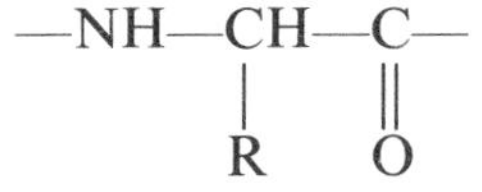

$$-NH-CH-C-$$
$$\underset{R}{|} \quad \underset{O}{\|}$$

where R = H or one of the groups listed in §22d. Hydrolysis yields the constituent amino-acids, $NH_2CH(R)COOH$, the residues of which in silk fibroin are principally those of glycine, alanine, serine, and tyrosine, and in wool those of glutamic acid, cystine, leucine, serine, and arginine. Silk fibroin is a linear polymer containing very little sulphur, but wool keratin exhibits cross-linkage largely through the dithio (—S—S—) groups in the cystine residues.

	Silk	*Wool*
Molecular weight	Minimum 84000; native fibroin probably 150000	By end-group analysis 60000 minimum; soluble fractions, up to about 80000
Degree of polymerisation	Not less than 1000 α-amino-acid residues, of at least 16 different kinds	Unknown: 18 to 20 different α-amino-acids have been isolated
X-ray data	Polycrystalline fibre. Identity period 7 Å (2 *trans*-positioned fundamental units), or a half-period of 3·4–3·5 Å (length of single α-amino-acid residues); other dimensions of unit cell, 9·2 Å × 9·4 Å	*Natural unstretched form* (α-keratin): A folded or contracted chain molecule; repeat period of fold 5·1 Å; equatorial spacing of 9·8 Å. *Stretched form* (β-keratin): Identity period 6·65 Å, and 3·3 Å (cf. silk), other dimensions of unit cell, 9·3 Å × 9·8 Å.

*Unless otherwise stated, data refer to *degummed silk* (silk fibroin) and *scoured wool* (wool keratin).

188

For further information on the composition and yield of amino-acids, and on the empirical compositions of silk and wool, *see* Section 22.

20.2 CHEMISTRY

20.21 PREPARATION

Neither silk nor wool has been synthesised artificially.

1. **Fibres of natural silk,** which are mainly derived from cultivated larvae of moths belonging to the *Bombycidae* sub-family (notably *Bombyx mori*), consist of two fine continuous filaments of fibroin coated and stuck together with a layer of silk gum, or sericin, which represents about 20% by weight of the whole and is fairly readily soluble in hot water. In commercial practice the silk gum is generally retained during weaving or knitting (acting as a size) and is later removed by hot soap solution. For types of silk, *see* §20.81.

2. **Wool hairs** (mainly from various breeds of the common sheep, *Ovis aries,* and from goats and camels) are complex structures consisting of an inner medulla or marrow of spherical cells, enclosed in a cortex of spindle-shaped cells which are aligned axially to the fibre, and protected by an external cuticle or epidermis which has an overlapping scalar structure. There may be up to 50% grease and wool-fat present, which is removed by scouring in alkaline soap solution. For types of wool, *see* §20.82.

20.22 PROPERTIES

Solvents *Silk* Soluble with little or no decomposition in conc. aq. lithium thiocyanate; soluble with slight or medium decomposition, dependent on conditions, in aq. cuprammonium hydroxide and cupriethylenediamine (also in solutions of related nickel complexes) and in conc. aq. solutions of highly soluble inorganic salts (e.g. lithium halides); soluble in liquid ammonia, with ammonolysis. *Wool* Dissolution occurs in hot conc. acids or alkalis, and in solutions of sodium sulphide or thioglycollate, accompanied by degradation. **Swollen by** *Silk* Dilute alkalis, and especially by conc. organic acids (formic and acetic). *Wool* Water, and solutions of alkalis and alkali sulphides. **Relatively unaffected by** water (but see above), organic liquids, dil. acids and alkalis (except above room temperature or over long periods, especially with silk). **Decomposed by** conc. acids and alkalis, especially when hot. Silk is readily degraded by oxidising agents, notably hypochlorites. For attack by various reagents, *see also* HOWITT, F., *J. Text. Inst.,* **51** (1960), P238 (silk) and P120 (wool).

20.23 IDENTIFICATION

Non-thermoplastic. N present; Cl absent; S (under 0·2% in silk, 3·6% in wool); P (trace in wool). **Combustion** Silk and wool burn, not over-readily unless kept hot, yielding a characteristic odour of burnt protein (like burnt hair) and leaving a swollen grey-black (coke-like) residue. **Pyrolysis** *Silk* Chars, initially evolves strongly alkaline vapours, but later the distillate is acid; swollen grey-black (coke-like) residue. Products of pyrolysis: water, carbon dioxide, acetic acid, aliphatic amines, phenolic tars. *Wool* Shrinks, chars, melts, evolves strongly alkaline vapours (but on continued heating the distillate may be acid); highly swollen grey-black residue. Products of pyrolysis include water, carbon dioxide, acetic acid, ammonia, hydrogen sulphide, and phenolic tars.

Distinction between silk and wool The first dissolves (with degradation) in conc. hydrochloric acid at room temperature, the second is insoluble. The fibres can also be identified by their physical characteristics as seen under the microscope.

For further information concerning identification of proteins, *see* Section 22.

20.3 PHYSICS

20.31 GENERAL PROPERTIES

	Silk	Wool
Specific gravity	Raw silk, 1·36; fibroin, 1·35 in benzene (1·42 in water); weighted silk, *c.* 1·6	1·30–1·32
Refractive index $n_{\parallel}$	1·591–1·595	1·556
$n_{\perp}$	1·538	1·547
Birefringence		
$n_{\parallel} - n_{\perp}$	0·053–0·057	0·009 (to 0·013)
Optical activity		
Specific rotation in aq. solution,	−53·1° water-soluble form, −58·9° denatured form	—
$[\alpha]_D^{15}$		
Water retention (%)	Raw silk, *c.* 70; degummed silk, *c.* 50	*c.* 40
Moisture regain (%, at 25°C)		
30% r.h.	6·5	8·8
65% r.h.	10·0 (11·0 desorption)	13·9 (15·7 desorption)
80% r.h.	17·4	19·0
100% r.h.	Raw silk, *c.* 38; degummed silk, *c.* 36	*c.* 33 (the cross-section increases by 38% in water)

20.32 THERMAL PROPERTIES

	Silk	Wool
Specific heat	0·30–0·35 (at 25°C)	0·26 (0°), 0·36 (10°C)
Conductivity $(10^{-4}$ cal)/(cm s °C)*	1·0–1·5	0·9 (increases with temperature)
Coefficient of linear expansion $(10^{-4}/°C)$†	−50 to −900	—
Heat of wetting (cal/g dry material)	c. 16	24–27

*Compressed to density of 0·1–0·15 g/cm³
†Along the fibre axis, at 15% r.h., 0–40°C

Physical effects of temperature *Silk* Becomes brown on prolonged drying at 120°C; decomposes (scorches) 170–200°C. *Wool* Discolours, with loss of strength, on prolonged drying above 100°C. Scorches more readily than silk, and decomposes if kept at *c.* 150°C for short period (safe ironing temperature, normally up to 165°C). The presence of unsaturated compounds such as maleic anhydride improves the thermal stability. When heated in steam, stretched wool is capable of supercontraction, i.e. contraction to a length shorter than the initial unstretched length. Note: Liquids that destroy the disulphide cross-links, such as 5% sodium bisulphite at 97°C, also cause supercontraction of wool making it rubber-like in extensibility; permanent setting of wool (and hair) depends on the cleavage and subsequent reformation of these cross-links.

20.33 ELECTRICAL PROPERTIES

	Silk (well washed)	Wool (scoured, purified)
Resistivity (ohm.cm) 65% r.h., 25°C 90% r.h., 25°C	 c. 10^{10} 10^5–10^7	 c. 10^9–10^{10} c. 10^5–10^6
Dielectric constant Dry; 120 kHz to 13 MHz	 4·2	 c. 4*

*Value rises steeply above 10% moisture regain. For detailed dielectric properties of wool-water systems, see WINDLE, J. J., and SHAW, T. M., *J. Chem. Phys.* **22**, 1752 (1954) (at 3000–9300 MHz); **25**, 435 (1956) (at 26 000 MHz)

Dielectric strength *Silk* (silk-covered wire, 65% r.h.), 14–24 kV/mm.

20.34 MECHANICAL PROPERTIES

	Silk	*Wool*
Elastic modulus (Young's) (kgf/mm^2)	*c.* 700–1000 (60–80 gf/den.)	*c.* 100–300 (10–25 gf/den.)
Elastic recovery Short-term (%)	From small strains, 100. From 5% extension, 60; 10%, 45; 20%, 30.	100–50, according to extension imposed
Tensile strength (kgf/mm^2)	35–60 (3–5 gf/den.); lowered to *c.* 80% when wet.	15–20 (1·3–1·7 gf/den.); lowered to *c.* 75% when wet
Extension at break (%)	20–25 (increased when wet)	30–40 (increased when wet; 100 in steam by α- to β-keratin change)

Coefficient of friction (values approximate only) *Silk* μ_{stat} : silk/silk, 0·2–0·3. μ_{dyn}: degummed yarn (1 m/s at 65% r.h.)/mild steel or stainless steel, 0·25–0·3; do. lubricant free/mild steel, 0·4–0·6; do. lubricated/mild steel, *c.* 0·2. *Wool* Fibres possess a surface scalar structure that causes the coefficient of friction in a root-to-tip direction to be slightly less than that from tip to root. Examples, μ_{stat} :

	Root-to-tip	*Tip-to-root*
Wool fibres/wool	0·2–0·35	0·4–0·5
Wool fibres/horn, clean	0·4–0·6	0·8–1·0
Wool fibres/horn, greasy	0·3–0·4	0·5–0·8
Wool fibres/casein plastics	0·5–0·6	0·6–0·7
Wool fibres/soft vulcanised rubber	1·5–2·0	2·0–2·5
Wool fibres/ebonite	0·5–0·6	*c.* 0·6
Wool fibres/glass	*c.* 0·6	*c.* 0·6

For a review of the frictional properties of wool, *see* LINCOLN, B., *Wool Sci. Rev.,* No. **18**, 38 (1960).

Abrasion resistance *Silk* Good. *Wool* Relatively poor.

20.5 SERVICEABILITY

Silk is susceptible to sunlight (especially if tin-weighted) but is otherwise resistant to weathering, particularly if finished in a slightly alkaline bath (pH 9–10). It is not readily attacked by bacteria or fungi (except in the gum state), or by insects. **Wool** Exposure to sunlight produces brittleness and uneven dyeing (tippy wool), otherwise it is of good stability and susceptible only to conditions favouring hydrolysis, oxidation, or reduction. It is attacked by insects (moth and beetle larvae) and, under humid conditions, by micro-organisms.

20.6 UTILISATION

Silk Thermal and electrical insulation; fine fabrics (high affinity for dyes), hosiery. **Wool** Thermal insulation (clothing); felting, carpets.

20.7 HISTORY

Silk Traditionally the culture of silk began in 2640 B.C. in China, where for many centuries its origin and the art of manipulating it were kept secret (and did not reach Japan until A.D. 300). There are Sanskrit records dating from *c.* 1000 B.C. that refer to the working of silk in ancient India, while clothing of silk is mentioned in the Old Testament and by Aristotle. Woven silk was much prized by the Romans, and the industry that they established in Byzantium (*c.* A.D. 550) ultimately spread to Italy and thence throughout medieval Europe, being introduced into England in the time of Henry VI (1422–61), and again later by weavers from Flanders.

Wool Sheep appear to have originated as mountain animals in Mesopotamia. It is known that their wool was employed for clothing at least 5000 years ago, the keeping of sheep and the use of their fleeces being frequently referred to in the Old Testament, in Assyrian and Babylonian records, and in the Odyssey (*c.* ninth century B.C.). Elaborate woollen clothing dating from *c.* 1000 B.C., and remarkably well preserved, has been found in wooden coffins in Danish peat bogs. Both the Greeks and the Romans used wool for textiles, as mentioned by Pliny. In Great Britain the domestication of sheep was understood even in pre-Roman days; while in the Middle Ages, when continental trade expanded, wool became for several centuries the premier industry contributing to the commercial prosperity of England—so much so that in the fourteenth century Edward III introduced 'wool sacks' into the House of Lords. There the symbol still survives, though today the major share of wool production falls to the newer countries (Argentina, Australia, New Zealand, and U.S.A.).

General Only in recent years, since Emil Fischer initiated his classical researches on proteins at the beginning of the twentieth century, have the fibrous proteins been investigated scientifically. The names of Astbury, Speakman, and others are connected with the elucidation of the structure of wool, and those of Howitt, S. G. Smith and colleagues, with that of silk.

20.8 ADDITIONAL NOTES

20.81 TYPES OF SILK

Most of the silk of commerce is of the cultivated type derived from the moth *Bombyx mori*. The silk thread, as secreted by the caterpillar (silk-worm), is in the form of a double filament (bave) of fibroin covered with sericin. It has a length of about 2500 m (*c.* one-third of which is reelable) with a diameter of *c.* 0·020–0·025 mm. Each single filament (brin) of degummed silk has a denier of 0·9–1·4.

Wild or tussah silks are also used to some extent (almost entirely for spun silk yarn); these have coarser filaments showing a different form of cross-section (wedge-shaped, instead of a rounded triangle) and are more resistant chemically than cultivated silk. Other types of silk, of which there are several, have so far found little commercial utilisation. Spiders' silk has been shown to be highly specialised, differing in composition and cross-section for different functions (*see* LUCAS, F., *Discovery*, **25**, 20 (1964)).

20.82 TYPES OF WOOL

Fleece wools are classified into five general types (fine, medium, long, cross-bred, and carpet or mixed) depending on the breed of sheep from which they are derived; thus, Merino sheep produce fine wools, Cotswolds long wools, and Corriedales medium wools.

Wools show great variability and their characteristics of length, diameter, strength, elasticity, lustre, colour, shrinkage and waviness change considerably with the breed of sheep. Generally wools are graded as 56's, 70's, 80's, etc., according to the finest possible count to which the wool can be spun. The staple length varies from *c*. 2·5 to *c*. 20 cm. The longer fibres (tops) are combed and spun into worsted yarns for dress fabrics while the shorter fibres (noils) are used for weft yarns. Even the single fibre is not uniform but can show significant chemical and physical variations from tip to root.

20.83 OTHER ANIMAL FIBRES

Angora and cashmere wools are derived from the hair of different varieties of the common goat, *Capra hircus,* the product of the Angora goat being known as *mohair* (as distinct from angora yarn, which is made from the fur of Angora rabbits).

The hair from certain camels, including their S. American relatives the llama, alpaca, and vicuna, is employed in the manufacture of blankets, rugs, dress fabrics, and brushes.

For identification, *see* APPLEYARD, H. M., *Guide to the Identification of Animal Fibres*, Leeds (1960).

20.84 COLLAGEN, GELATIN, AND LEATHER

Collagen is a complex sclero-protein, of which skins and hides are composed, and constitutes the white fibrous connective tissue in vertebrates. X-ray analysis shows a meridional spacing of 2·9 Å, indicating that the main chain of α-amino-acid residues (spacing

3·4 Å) is slightly contracted; there is a regular lateral spacing of about 11–15 Å (according to the degree of hydration) and a less regular spacing of *c.* 4·5 Å, corresponding to side-chain and backbone spacing respectively. The chief α-amino-acid residues in collagen are those of glycine and prolines (*see also* §22d).

Collagen, which swells in weak acids and alkalis, is dissolved by strong acids and alkalis and also by peptic enzymes. With boiling water it is the source of gelatin and glue (Gk. κολλα = glue) and with tannic acid or salts of certain heavy metals it yields leather. Of these two technically important derivatives, described below, gelatin is produced by partial breakdown of collagen while leather can be regarded as a reinforced form of collagen.

Gelatin is a heterogeneous film-forming protein, or protein-fraction, derived by extraction of the softer parts of skins and hides (collagen, as above) with boiling water. The molecular weight is of the order of 54000 and the average degree of polymerisation is *c.* 580. For the empirical composition, and α-amino-acid residues present in purified gelatin, *see* Section 22.

Gelatin is insoluble in cold water, but swells considerably in it (minimum swelling at the iso-electric point, pH 4·7–5·0) and readily passes into solution when warmed above 30°C; solutions of *c.* 1% conc. and above set to gels on standing at room temperature. It is soluble in dil. alkalis and solutions of various salts, with modification of the setting properties. It can be plasticised by polyhydric alcohols (glycerol). It is insoluble in hydrocarbons and absolute alcohol; and can be rendered insoluble in water by treatment with formaldehyde or (in the presence of light) with potassium dichromate.

Apart from the ancient and still important uses of crude gelatin in *glue* and *size*, the purified material (including that obtained from fish offal) is much employed in the sizing of paper and as a gelling and emulsifying agent, notably in the photographic industry since the invention of the gelatin dry plate in 1871. Textile fibres of gelatin were produced as early as 1894 but are now entirely superseded. Gelatin is a good adhesive for glass and paper, but it loses strength at high humidity and unless containing a preservative (such as boric acid, β-naphthol, or zinc sulphate) is subject to attack by moulds.

Leather derives chiefly from the corium or derma (the true skin) lying below the outer skin, or epidermis, of various animals. The epidermis, hairs, and subcutaneous tissue are removed from skins and hides by mechanical, bacteriological, and chemical means, leaving the corium which consists (*c.* 95% in cow hide) of fibrillar *collagen* (1 μm or less in diameter) bundled and interwoven in fibre form (each fibre about 25 μm in diameter). This material is converted by combination with tannins, chromium salts, formaldehyde, or isocyanates, to the heat-resistant and non-putrescible product known as leather. The moisture content of tanned leather, e.g. shoe leather, is *c.* 15%

at 65% r.h., rising under humid conditions to more than double this value.

Leather, used even in prehistoric times, is much employed for footwear, clothing, upholstery and travel goods; also in many industrial applications, such as belting and tapes, packing washers, spinning roller coverings and loom accessories.

20.9 FURTHER LITERATURE

Silk

HOWITT, F. O., *Bibliography of the Technical Literature on Silk*, London (1946)

ANFINSEN, C. B., *et al.* (Eds), *Advances in Protein Chemistry*, Vol. XIII (LUCAS, F., SHAW, J. T. B. and SMITH, S. G., silk fibroins), New York, 107 (1958)

BS 2804: Glossary of Terms relating to Silk (1956)

Wool

ALEXANDER, P. and HUDSON, R. F., *Wool, its Chemistry and Physics*, London (1954) (2nd edn, EARLAND, C. (Ed.), London (1963))

ONIONS, W. J., *Wool. An Introduction to its Properties etc.*, London (1963)

VON BERGEN, W., *Wool Handbook* (2 vols), New York and London (1963)

MERCER, E. H., *Keratin and Keratinisation*, London (1961)

Collagen, etc.

ALEXANDER, J., *Glue and Gelatin*, New York (1923)

ANSON, M. L. and EDSALL, J. T. (Eds), *Advances in Protein Chemistry*, Vol. IV (FERRY, J. D., gelatin gels), New York, 17 (1948)

RANDALL, J. T., *Nature and Structure of Collagen*, London (1953)

GUSTAVSON, K. H., *The Chemistry and Reactivity of Collagen*, New York and London (1956)

WOODROFFE, D. (Ed), *Standard Handbook of Industrial Leathers*, London (1949)

SPIERS, C. H., *Leather*, London (1963) (preparation, applications)

O'FLAHERTY, F., *et al.*, (Eds), *The Chemistry and Technology of Leather* (4 vols), New York and London (1956–1965) (Amer. Chem. Soc. Monograph)

BORASKY, R. (Ed), *Ultrastructure of Protein Fibres* (pp. 5–17, properties of protein fibres; pp. 19–37, collagen), New York and London (1963)

REGENERATED PROTEINS*

With notes on dissolution and regeneration, and on fibres from synthetic polypeptides

SYNONYMS AND TRADE NAMES

Fibres Fibrolane, Merinova. (Aralac, Ardil, Lanital, Soylon and Vicara are no longer made). *Plastics* Artificial horn. Erinoid, Galalith, Lactoid.

GENERAL CHARACTERISTICS

Fibres White, or cream to light brown, in colour; soft and wool-like, but without a scaly surface. *Plastics* Opaque or translucent, cream-coloured (often pigmented), tough, horny material that softens in hot water and swells considerably if left immersed.
N.B. All regenerated protein plastics and fibres contain combined formaldehyde (*see* under §21.21).

21.1 STRUCTURE

Simplest Fundamental Unit

$$-NH-CH-C-$$
$$\underset{R}{|} \quad \underset{O}{\|}$$

With additions and cross-linkages owing to treatment with formaldehyde.

The polymers are composed of mixed but unidirectional (i.e. head to tail) α-amino-acid residues, or polypeptide fragments, in which R = H or one of the groups of atoms given in §22d.

Molecular weight Of the order of 30000–50000. Casein, *c.* 33000; zein, *c.* 38000. Much higher by some methods of measurement. When regenerated, the proteins are obtained reduced in molecular weight (*see* §21.81), the molecular weight of formaldehyde-reacted proteins is very high, but cannot as yet be determined. *Degree of polymerisation* The number of units in a typical chain is *c.* 400 (or more), composed of 15–24 different types. Some 17 different α-amino-acid residues have been identified in casein (*see* §22d).

X-ray data Substantially amorphous materials. Fibres show low orientation and only slight crystallinity; identity period, uncertain, *c.* 4·5 Å?

*Unless otherwise stated, data refer to *casein-formaldehyde*.

21.2 CHEMISTRY

21.21 PREPARATION (*See also* §21.81)

Since the proteins all derive from natural sources, and those used for regenerative purposes are in a sense waste materials, the products are of economic interest. Casein is obtained from skim milk, the main sources being France and Argentina. The vegetable proteins are residual in the spent meals after the respective oils have been extracted from ground-nuts (*Arachis hypogaea*), soya-beans (*Soya max*), and the germ of maize or Indian corn (*Zea mays*)—the last substance being itself a by-product, obtained during the milling of maize seeds. The chief proteins in ground-nuts are arachin and conarachin, soya-beans yield glycinin, and maize yields a gliadin called zein. The use of other proteins (e.g. gelatin, keratin) for regenerative purposes is either obsolete commercially, or not yet developed beyond an experimental stage.

The general procedure for the preparation of casein products is as follows. Caseinogen, left in solution in skim milk (i.e. milk from which the cream has been removed) is converted to insoluble casein either by the addition of acid or by 'setting' with rennet after which the product is well washed. For conversion to fibres, it is dispersed in alkali (e.g. 20–30% in 1–2% sodium hydroxide) and regenerated, by extrusion through spinnerets, in an acid coagulating bath (e.g. 10% sulphuric acid saturated with sodium sulphate), the resulting swollen filaments being then stretched some 300% while still wet. For plastics, moistened powdered rennet casein is compressed to a dough and extruded under heat and pressure, either in the form required or in one suitable for further pressing while still flexible. In each instance the regenerated protein is finally hardened*, rendered insoluble, and made more resistant to swelling in water by treatment with aq. formaldehyde.

Regeneration of vegetable proteins follows similar lines, the materials being dispersed in alkali, spun into an acid bath, and subsequently treated with formaldehyde. Zein is extractable in, and can be regenerated directly from, an aq. alcohol; alternatively it can be spun from an alkaline solution containing urea.

The formaldehyde yield of formalised regenerated proteins is of the orde ,f 1–5%, some of which is lost on boiling with water. In certain processes formaldehyde is added at an earlier stage, together with an acid catalyst, and reaction occurs on subsequent baking. In addition to cross-linkage by formaldehyde, a mild acetylating treatment is

---

*Unhardened casein is employed as a cold-setting adhesive, e.g. as a powder containing sodium fluoride and an excess of lime (when wetted, casein dissolves in the released alkali, then forms a gel of calcium caseinate).

sometimes given. Cross-linkage by treatment with aluminium or chromium salts has also been used.

21.22 PROPERTIES
(commercial plastics and fibres unless otherwise stated)

Solvents Before treatment with formaldehyde: soluble in dil. alkalis, including aq. ammonia and borax, also on warming with dil. phosphoric or lactic acids; ground-nut and soya-bean proteins soluble in aq. urea; zein soluble in 70–90% aq. aliphatic alcohols, aq. acetone, alcohol-xylene mixtures, phenols, hot trioxan. Formalised products insoluble (except with decomposition). **Swollen (and plasticised) by** water and hydrophilic liquids. **Relatively unaffected by** most organic liquids, oils, or greases; cold dil. acids and cold dilute alkalis (swelling on prolonged immersion). **Decomposed by** conc. acids and alkalis, moderately conc. hypochlorite solutions. Formaldehyde is evolved on warming with dil. mineral acids. For attack of fibres by various reagents, *see* HOWITT, F., *J. Text. Inst.*, **51**, 321 (1960).

21.23 IDENTIFICATION

Non-thermoplastic. N, S, P present; Cl absent.
    Combustion Swells, chars, may burn with a small flame, smells strongly of burnt protein (burnt feathers, burnt milk) and of formaldehyde. **Pyrolysis** Discolours, chars, emits alkaline vapours; smells of burnt protein and formaldehyde (as for combustion).

    N.B. Owing to treatment with formaldehyde the initial distillate from regenerated proteins can have an acid reaction.

    Colour reactions 1. All commercial regenerated proteins are hardened or rendered insoluble by treatment with formaldehyde and this may be detected by warming a sample with dil. sulphuric acid and applying the CTA test (§26.23a). 2. The relatively high phosphorus content helps in the identification of casein (*see also* Section 22 and test in Section 57). 3. Most casein fibres (and to a less extent ground-nut fibres) acquire a mauve colour when left standing in warm conc. hydrochloric acid.

    For further information concerning identification of proteins, *see* Section 22.

21.3 PHYSICS

21.31 GENERAL PROPERTIES

Specific gravity Casein: plastics, *c.* 1·35; fibres, 1·29–1·30; ground-nut protein, 1·30–1·31; soya-bean protein, 1·31; zein, 1·25. **Refractive index** Plastics and fibres (all types), 1·53–1·54. **Birefringence** (fibres) Zero, or very small (zein fibres, 0·005). **Water absorption** Casein plastics, 7–14%; do. prolonged immersion, up to 30%. Zein plastics, 10%. **Water retention** Fibres (all types) *c.* 50%. **Moisture regain** Mean values, various fibres at 25°C (%):

R.h., %	35	65	95
Casein	7·6$_5$	12·3$_5$	20·8
Ground-nut protein	6·9	10·5*	16·0$_5$
Soya-bean protein	6·7	10·1	19·4$_5$
Zein	5·9$_5$	9·7	15·3

*Some types up to 15%

21.32 THERMAL PROPERTIES

(Casein-formaldehyde; other regenerated proteins assumed similar)

Specific heat 0·36. **Conductivity** 4 (10^{-4} cal)/(cm s °C). **Coefficient of linear expansion** *c.* $0·5 . 10^{-4}$/°C (shrinks on initial drying). **Physical effects of temperature** Loses strength above 90°C; softens and discolours, 100–130°C; decomposes (chars), 150–200°C; ground-nut fibres, chars *c.* 250°C. **Heat of wetting** *c.* 27 cal/g dry ground-nut fibres.

21.33 ELECTRICAL PROPERTIES

(Casein-formaldehyde, 65% r.h.)

Volume resistivity (ohm cm) *c.* 10^8; fibres, *c.* 10^7. *Surface resistivity,* 10^8 ohm (or more). **Dielectric strength** 8–28 kV/mm (much dependent upon moisture content). **Dielectric constant** *c.* 6·5 (50 Hz and 1 MHz). **Power factor** *c.* 0·05 (50 Hz and 1 MHz).

Note: When really dry, casein-formaldehyde is an excellent non-conductor, but owing to its high affinity for moisture, under normal conditions it is suitable only for low tension insulation.

21.34 MECHANICAL PROPERTIES (25°C, 65% r.h.)

Elastic modulus (Young's) (kgf/mm^2) Plastics, *c.* 350; fibres, *c.* 250 (20 gf/den.). **Elastic recovery** Except from very small strains, relatively poor. Recovery of fibres, short-term, %:

	From 1–2% *extension*	*From* 10% *extension*
Casein	70	18
Zein	80	25
Ground-nut protein	15	10

Tensile strength and Extension at break

	Tensile strength (kgf/mm^2)	*Extension at break* (%)
Plastics (casein)	5–7	c. 2·5
Fibres: casein, ground-nut	Up to 12 (1 gf/den.)	c. 60
zein	Up to 12 (1 gf/den.)	c. 25
soya-bean	c. 7 (0·6 gf/den.)	c. 40

Wet strength of fibres: approximately half air-dry strength. **Flexural strength** 7–12·5 kgf/mm². **Compression strength** *c.* 20–35 kgf/mm². **Impact strength** (Izod test) Up to 1 ftlbf/in of notch. **Hardness** Brinell (2·5 mm ball, 25 kg load), *c.* 23; Rockwell (1/16 in ball), M70–100. **Friction and abrasion** Coefficient of friction μ_{stat} : fibres/ fibres, approx. 0·15–0·2. For casein plastics/wool, *see* §20.34. μ_{dyn} : grey cotton yarn (running at 1 m/s, at 65% r.h.) against moulded casein-formaldehyde, 0·24. Regenerated fibres, being non-scaly, do not felt like wool. They have low abrasion resistance against themselves but appear to be satisfactory when blended with viscose rayon or wool, i.e. the abrasion resistance of the mixture is not lowered in direct proportion to the regenerated protein content.

21.4 FABRICATION

Compression moulding (100–150°C, 2000–4000 lbf/in²) or extrusion (below 100°C) when of high moisture content (20%), or containing a lyophilic-type plasticiser, the products being hardened for a suitable period in dil. formalin, i.e. moulding is effected before treatment with formaldehyde. For additional information on initial fabrication of plastics and fibres, *see* §21.21. Formalised material can be machined, turned, and polished very satisfactorily; it can also be cemented (e.g. with a casein or wood glue) and can be shaped to some extent when wet, but cannot be deep moulded. An alkaline bleach bath causes softening of the surface, which dries with a hard glaze (applied to buttons, etc.).

21.5 SERVICEABILITY

The chief drawback to casein-formaldehyde plastics is their high imbibition and attendant swelling under humid or wet conditions. This greatly reduces their applications, wet conditions resulting in a general loss of strength while repeated swelling and shrinkage give rise to cracking and permanent distortion. In addition, casein plastics turn yellow on prolonged exposure to sunlight and they are not immune to attack by fungi (mildew) or insects (moth larvae and beetles).

The fibres dye readily, and have good resistance to light, but can be damaged by micro-organisms. Mixtures with wool retain wool-like properties in service.

21.6 UTILISATION

Despite the foregoing disadvantages, casein-formaldehyde plastics are important materials. They are cheap, take a high polish, and can be brightly coloured, and hence are much used for decorative buttons and buckles, handles, knobs, and the like, where they will not be unduly exposed to exterior weathering or repeated changes of humidity. Unformalised casein is used as an emulsifying agent, in sizing paper and textile yarns, as a binder in water paints and as a cold-setting adhesive (insolubilised by lime). It is also used in compound adhesives (e.g. with polychloroprene), and addition of a small proportion of unformalised casein to viscose (before spinning) increases the adhesion between rayon tyre cords and synthetic rubbers.

The fibres have been developed mainly for their wool-like (although non-scaly) qualities, and because of their low wet strength they are generally used in admixture with wool, cotton, or viscose rayon.

21.7 HISTORY

Plastics Several German chemists were associated with the earlier developments in casein plastics, and as 'artificial horn', it was introduced in Europe *c.* 1885, the product hardened with formaldehyde (Galalith) appearing in commerce a few years later. Casein wood-glues were used extensively in the construction of aircraft in 1917.

Fibres A British patent for regenerated protein fibres was granted to Hughes in 1857. For a short time, in 1894–98, Millar produced a commercial fibre (Vanduara) based on formalised gelatin; experimental bristles had also been made in Germany in 1895 and F. Todtenhaupt had spun fibres (1904) but the first successful commercial development of fibres from regenerated proteins did not occur until 1935, when the production of Lanital from casein began in Italy after some years of experiment by Ferretti. Although zein (protein of maize) was first recognised by Gorham in 1821 and investigated by various workers throughout the nineteenth century, production of fibres from vegetable proteins dates only from recent years. Limited production was started in Great Britain, Japan, the U.S.A., and elsewhere in 1935–39; fibres from soya-bean protein were made in the 1940s in the U.S.A., where also larger scale production of zein fibre (Vicara) began in 1948; and the manufacture of ground-nut fibre (Ardil) was started in Scotland in 1951. However, with the exception of those from casein, fibres of regenerated proteins are no longer made, production of Vicara and Ardil being dicontinued in 1957.

21.8 ADDITIONAL NOTES

21.81 DISSOLUTION AND REGENERATION OF PROTEINS

In the natural state the molecules of casein and vegetable globulins (ground-nut protein, zein, etc.) exist in a coiled-up (globular or corpuscular) form, and in order to produce a fibrous filament it is necessary to straighten them out by a process similar to (and probably identical with) that to which the term 'denaturation' has been applied. With the exception of silk fibroin, such conversion requires rupture of dithio cross-linkages (in cystine, and possibly other residues). This requirement is reflected in the maturation process with caustic soda, as commonly used in the preparation of a spinning solution. Alkaline maturation results not only in the breakdown of cross-linkages but also in some degradation of peptide and other bonds; the derived filaments therefore almost invariably have a degree of polymerisation substantially lower than that of the parent protein, a circumstance which is reflected in the low strength of regenerated protein fibres.

21.82 FIBRES FROM SYNTHETIC POLYPEPTIDES

Synthetic polypeptides may be prepared in various ways, such as by base-catalysed polymerisation of an N-carboxyamino-acid anhydride (obtained by reaction of an amino-acid with phosgene), e.g.

$$
\begin{array}{c}
RN\text{---}CHR' \\
| \quad\quad | \\
CO \quad CO \\
\diagdown_O\diagup
\end{array}
\quad \xrightarrow{-CO_2} \quad
\begin{array}{c}
\text{---}NR\text{---}CH\text{---}C\text{---} \\
| \quad\; || \\
R' \quad O
\end{array}
$$

(cf. §21.1)

Materials of this type, e.g. poly-L-alanine, when dry- or wet-spun as fibres exhibit high crystallinity (initial α-helix configuration changing to the β-form on stretching) and have been much used in structural investigations, although not as yet developed commercially.

21.9 FURTHER LITERATURE

SUTERMEISTER, E. and BROWNE, F. L., *Casein and its Industrial Applications,* New York (1939)

ANSON, M. L. and EDSALL, J. T. (Eds), *Advances in Protein Chemistry,* Vol. V (LUNDGREN, H. P., synthetic fibres from proteins), New York, 305 (1949)

COLLINS, J. H., *Casein Plastics and Allied Materials,* London (1952) (Plastics Monograph)

WORMELL, R. L., *New Fibres from Proteins,* London (1954)

FETTES, E. M. (Ed), *Chemical Reactions of Polymers* (*High Polymers,* Vol. XIX), New York and London, 367 (1964). (WHITFIELD, F. E. and WASLEY, W. L., Reactions of Proteins)

APPENDIX ON THE COMPOSITION OF PROTEINS

22a. EMPIRICAL COMPOSITION OF SILK, WOOL, AND OTHER PROTEINS
(g/100 g dry protein)

	Silk	Wool	Gelatin	Casein	Ground-nut protein
Carbon	48·4	49·3	48·7–51·5	53	52
Hydrogen	6·5	7·6	6·5–7·2	7·0–7·5	7
Oxygen	26·7	23·7	25·1–25·3	22·5	22·5
Nitrogen	18·35	15·9	17·7–18·8	14–15	18
Sulphur	c. 0·2	3·6	c. 0·2	c. 0·7	c, 0·7
Phosphorus	0	0 (trace)	0	c. 0·8	0 (trace)

22b. NOTES ON THE POLARITY OF THE SIDE GROUPS IN PROTEINS

(i) The main polar groups present in protein molecules are given below:

Type	Polar group	Example (see §22d)
Acid	—COOH (carboxyl)	Aspartic acid
	—O·PO(OH)$_2$ (phosphoric)	Phosphoserine
	—C$_6$H$_4$OH (phenolic)	Tyrosine
	—SH (thiol)	Cysteine
Basic	=NH (glyoxalyl-imino)	Histidine
	—NH$_2$ (ω-amino)	Lysine
	—NH·C(:NH)·NH$_2$ (guanidyl)	Arginine
(Others)	—CH$_2$OH (alcohol)	Serine
	—S·S— (dithio, disulphide)	Cystine
	—CONH$_2$ (amide)	Asparagine

(ii) *Phosphoric groups* are combined with serine residues in a few proteins, notably in casein and vitellin, from which phosphoserine has been isolated.

(iii) *Thiol groups* (—SH) are formed by reduction of the dithio (—S·S—) group. It is not yet clear to what extent (if at all) this group is present in natural unmodified proteins, but it is evident that free cysteine groups in protein materials have at least partly originated from initial cystine groups.

(iv) *Free carboxyl groups*, associated with monoaminodicarboxylic acids, are present in natural proteins partly as amide groups. Analytical methods at present available for the estimation of amide groups are of doubtful accuracy, and the extent to which this group occurs in proteins whose primary structures have not yet been determined (this includes all structural proteins) is thus rather uncertain.

22c. GENERAL IDENTIFICATION REACTIONS OF PROTEINS

Non-thermoplastic. 14–18 % N.

The empirical composition of five typical proteins is given above in §22a.

The swollen residues and characteristic odours (like burnt hair, or burnt milk) deriving from proteins on combustion, or pyrolysis, are detailed in §§20.23 and 21.23. Proteins respond to the cyanide tests described under pyrolysis of polyacrylonitrile (§10.23).

Some general colour reactions for proteins are as follows: 1. *Biuret test.* Warm a sample with 1 % aq. alkali and treat a few drops of the extract with one drop of 1 % aq. copper sulphate solution. Add a *trace* of conc. sodium hydroxide; proteins and their degradation products give a violet colour (addition of sugar prevents precipitation of basic copper salts). Test detects adjacent —NH·CO— groups. 2. *Ninhydrin test.* Heat for 10–15 min at 65°C, in a dil. solution of ninhydrin (e.g. 0·5 % in 95/5 butanol/dil. acetic acid); most proteins acquire a mauve colour. 3. The following reactions apply to proteins containing phenolic groups. They are given also by certain free phenols, but these can usually be distinguished by the absence of nitrogen. (i) *Xanthoproteic reaction.* Contact with warm conc. nitric acid nitrates the phenolic groups, turning the protein yellow. Rinsing the sample, followed by treatment with dil. alkali, changes the colour to golden or orange. (ii) *Millon's reagent.* Dissolve a sample in dil. sodium hydroxide, neutralise the solution, and treat with the reagent (made by dissolving mercury in conc. nitric acid and diluting the resulting solution to half strength). On warming, proteins containing tyrosine (*see* §22d) give a pink or red precipitate.

22d. TYPES, NAMES AND APPROXIMATE PROPORTIONS OF AMINO ACIDS (NH₂CH(R)COOH) COMPOSING COMMON PROTEINS

(in which the acids are present as residues of the form —NH·CH(R)·CO—)

Type of amino-acid	Name of amino acid	Nature of R in the general formula $NH_2CH(R)COOH$	Approx. g amino-acid per 100 g dry protein*						
			Silk (fibroin)	Wool (keratin)	Gelatin	Milk-protein (casein)	Ground-nut protein (arachin)	Soya-bean protein (glycinin)	Maize germ protein (zein)
Aliphatic	Glycine	—H	41·2	5·5	25·5	1·9	2·0	0·2	0
	Alanine	—CH₃	33·0	4·3	8·7	3·5	3·0	4·1	10·0
	Serine	—CH₂OH	16·2	10·6	3·3	5·9	6·4	6·0	1·0
	Valine	—CH(CH₃)₂	3·6	5·7	2·9	6·0	3·4	4·5	4·0
	Threonine	—CH(OH)·CH₃	1·55	7·15	2·0	4·5	2·5	4·0	2·2
	Leucine iso-Leucine	—CH₂·CH(CH₃)₂ —CH(CH₃)·CH₂·CH₃	2·0	12·6	7·1	15·8	6·5	12·4	20·0
Aromatic	Phenylalanine	—CH₂·C₆H₅	3·35	4·1	3·7	6·5	4·7	5·3	7·6
	Tyrosine	—CH₂·C₆H₄OH(1, 4)	11·4	5·5	1·0	6·3	6·0	4·3	5·9
Sulphur-containing	Cystine	—CH₂·S·CH₂—	0·2	13·0	0·2	0·4	1·6	1·0	0·9
	Methionine	—CH₂·CH₂·S·CH₃	0	0·55	0·9	3·5	1·0	2·0	2·3
Heterocyclic	Tryptophan	—CH₂·C:CH·NH·C₆H₄(1, 2)	0·65	0·95	0	1·4	1·2	1·5	0·2
	Proline	—CH₂·CH₂·CH₂-(cyclic)†	0·7	6·8	19·7	10·5	5·3	3·9	9·0
	Hydroxyproline	—CH₂·CH(OH)·CH₂-(cyclic)†	0	0	14·4	—	—	—	—
Acid	Aspartic acid	—CH₂·COOH	2·75	6·8	5·6	6·7	5·0	3·9	3·4
	Glutamic acid	—CH₂·CH₂·COOH	2·15	14·5	11·2	22·0	21·0	20·0	35·6
Basic	Histidine	—CH₂·C:CH·NH·CH:N	0·4	1·2	0·8	3·2	2·0	2·5	1·0
	Arginine	—(CH₂)₃·NH·C(:NH)·NH₂	1·0	9·8	8·7	3·9	13·1	5·8	1·6
	Lysine	—(CH₂)₄·NH₂	0·5	3·3	5·9	8·3	3·0	5·4	0

*Average values assessed from various sources.

†Proline and hydroxyproline, being cyclic amino-acids, are present as the respective residues,

22e. FURTHER LITERATURE

Analysis of protein hydrolysates is much facilitated by the techniques of chromatography, and the literature on this subject is considerable. *See*, for instance:

LEDERER, E. and M., *Chromatography* (2nd edn), Holland (1957)

COLOWICK, S. P. and KAPLAN, N. O. (Eds), *Methods of Enzymology,* Vol. III, New York, 492 (1957) (chromatography of amino-acids)

LUCK, J. (Ed), *Annual Review of Biochemistry,* Vol. 28, London, 69, 97 (1959) (analysis)

STAHL, E., *Thin-layer Chromatogtaphy,* London (1963)

TRUTER, E. V., *Thin-film Chromatography,* London (1963)

ALEXANDER, P. and BLOCK, R. J. (Eds), *Analytical Methods of Protein Chemistry,* London and New York (1960)

NEURATH, H. (Ed), *The Proteins: Composition, Structure, and Function,* Vol. I (2nd edn), London and New York (1963)

SAVIDAN, L., *Chromatography,* London (1966)

For other references to chromatography, *see* Further Literature, Section 57.

SECTION 23

POLYAMIDES

*With notes on nylons by number, modified polyamides (including polyimides),
polynonamethyleneurea and polyaminotriazoles*

SYNONYMS AND TRADE NAMES

Generic name Nylon
NOTATION Numbers following the word 'nylon' refer to the number of carbon atoms in
the molecule(s) of the initial monomeric material(s); where there are two numbers (*see*
examples below) they are spoken of as nylon 'six-six', nylon 'six-ten', etc.

Particular species:
Polycaprolactam or polycaproamide Nylon 6. Akulon, Caprolan, Celon, Durethan BK,
Enkalon, Grilon, Perlon (Perlon L)
Polyundecanoamide Nylon 11. Rilsan (Rilsan B)
Polyhexamethylene adipamide Nylon 6.6. Bri-nylon, Chemstrand-nylon, Maranyl, Zytel
Polyhexamethylene sebacamide Nylon 6.10. Brulon, Perlon N
Polymerised fatty acid polyamides. Versalon, Versamid
For Perlon U (Dorlon)—a polyurethane—*see* §40.21 I et seq.

GENERAL CHARACTERISTICS

Fibres Practically colourless, transparent (usually delustred), strong and abrasion resistant;
cross-section commonly circular. *Plastics* Slightly yellow, translucent to opaque, tough,
horn-like materials.

23.1 STRUCTURE

Simplest Fundamental Unit
Fibrous polyamides Two main types: (i) from self-condensation of
ω-amino-acids or polymerisation of their cyclic anhydro-derivatives
(lactams), and (ii) from condensation of diamines with dicarboxylic
acids. Examples:

Type	Unit	Formula	SFU-weight
(i)	—NH(CH$_2$)$_5$CO— (nylon 6)	C$_6$H$_{11}$ON	113
(i)	—NH(CH$_2$)$_{10}$CO— (nylon 11)	C$_{11}$H$_{21}$ON	183
(ii)	—NH(CH$_2$)$_6$NH—CO(CH$_2$)$_4$CO— (nylon 6.6)	C$_{12}$H$_{22}$O$_2$N$_2$	226
(ii)	—NH(CH$_2$)$_6$NH—CO(CH$_2$)$_8$CO— (nylon 6.10)	C$_{16}$H$_{30}$O$_2$N$_2$	282

Non-fibrous polyamides A typical example, the polymer of diethylenetriamine dilinoleate, has an SFU-weight of 627.

209

Molecular weight For useful properties the number-average molecular weight must be above 10000. Examples (nylon 6):

	Fibres	Mouldings	Extrusions
Molecular weight	c. 20000	Up to 30000	Up to 100000
Degree of polymerisation*	100–200	150–250	450–900

*Approximate, dependent on type

Average degree of polymerisation: nylon 11, c. 100; nylon 6.6, 50–100; a typical non-fibrous polyamide (e.g. diethylenetriamine dilinoleate polymer) might have a molecular weight and degree of polymerisation of 6000–10000 and 10–16, respectively.

X-ray data Unstretched polymers: amorphous or randomly polycrystalline. Stretch-spun fibres: crystalline and oriented regions, c. 100 Å long. The general structure consists of approximately planar zig-zag C-chains, held parallel by H-bonding between $>$CO and $>$NH groups. Identity period (Å): nylon 6, 17·2 (or 8·6); nylon 11, 14·9; nylon 6.6, 17·2; nylon 6.10, 22·4.

23.2 CHEMISTRY

23.21 PREPARATION

Polyamides are obtained by self-condensation of ω-amino-acids or polymerisation of their lactams, or by condensation between diamines and dicarboxylic acids, thereby producing two types of structural unit (*see* §23.1). Typical syntheses are outlined below.

1. *Nylon* 6 (from caprolactam, m.p. 63–64°C)—Phenol (I) is hydrogenated to cyclohexanol (II) and this is dehydrogenated to cyclohexanone (III) which is then converted to its oxime (IV). Beckmann rearrangement of the oxime gives 5-caprolactam (V) which is largely converted to a linear polymer (repeat unit, VI) on heating under pressure. Alternative routes, commencing with hydrogenation of benzene to cyclohexane, depend upon aerial oxidation to yield cyclohexanol and cyclohexanone, or upon nitration followed by partial reduction to yield the oxime directly.

$$\xrightarrow[\text{(autoclave, trace }H_2O)]{>200°C} \quad [-NH(CH_2)_5 CO-]$$

VI

An industrial process developed in Japan yields cyclohexanone oxime (IV) economically in one stage by the photonitrosation of cyclohexane with nitrosyl chloride,

$$C_6H_{12} + NOCl \xrightarrow[(+HCl)]{\text{u.v. radiation}} C_6H_{10}NOH \cdot HCl$$

The oxime (as a hydrochloride) is then, as previously, converted to caprolactam by Beckmann rearrangement.

The polymerisation of caprolactam requires the presence of a trace of water or acid, which may be liberated *in situ* by a catalyst (e.g. 'nylon salt', *see below*). However, the process does not go to completion; thus, at 200°C the equilibrium mixture contains 10% monomer (some present as trimer) and the proportion increases at higher temperatures. In commercial production, unconverted lactam and aminocaproic acid are removed by washing with hot water; but on re-melting the polymer depolymerisation takes place until the appropriate monomer-polymer equilibrium is restored.

2. *Nylon* 11 (from aminoundecanoic acid, m.p. 178–179°C)— This polymer is derived from vegetable sources. Castor oil (mainly glycerides of ricinoleic acid) on methanolysis gives methyl ricinoleate (VII) which is cracked thermally into *n*-heptaldehyde and methyl undecylenate (VIII); the products are separated by fractional distillation and the ester is hydrolysed to give undecylenic acid (IX). This is reacted, in a non-polar solvent and in the presence of air, with hydrogen bromide to yield largely 11-bromoundecanoic acid (X), which on treatment with ammonia yields 11-aminoundecanoic acid (XI). This, recrystallised from hot water, when heated to 200–220°C at atmospheric pressure with removal of water polymerises in the melt, predominantly to a linear polymer (repeat unit, XII)

$$CH_3(CH_2)_5CH(OH)CH_2CH{=}CH(CH_2)_7COOMe \xrightarrow[>300°C]{\text{crack}}$$

VII

$$CH_3(CH_2)_5CHO + CH_2{=}CH(CH_2)_8COOMe \xrightarrow[\text{of VIII}]{\text{hydrolysis}}$$

VIII

211

$$CH_2=CH(CH_2)_8COOH \xrightarrow[(O_2)]{HBr} CH_2Br(CH_2)_9COOH \xrightarrow{NH_3}$$

IX
X

$$NH_2(CH_2)_{10}COOH \xrightarrow[>200°C]{-H_2O} [-NH(CH_2)_{10}CO-]$$

XI
XII

3. *Nylon* 6.6 (from hexamethylenediamine and adipic acid, m.p. 39°C and 151°C respectively)—Phenol is hydrogenated, or cyclohexane is catalytically oxidised with air, and the resulting cyclohexanol is oxidised to adipic acid (XIII). Dehydration of ammonium adipate then yields adiponitrile, which is hydrogenated to hexamethylenediamine (XIV). Equivalent amounts of XIII and XIV are used to prepare a solution of hexamethylenediammonium adipate ('nylon salt', an ω-amino acid, m.p. 183°C) which, when heated first under pressure then under reduced pressure and in the absence of oxygen, condenses to a linear polymer (repeat unit, XV). In contrast to the polymerisation of caprolactam, the condensation proceeds almost to completion.

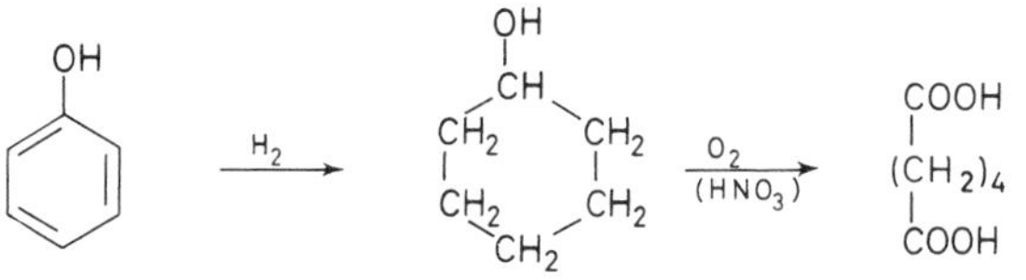

$$\xrightarrow[220° \text{ rising to } 270°C]{XIII + XIV, -H_2O} [-NH(CH_2)_6NH-CO(CH_2)_4CO-] \quad XV$$

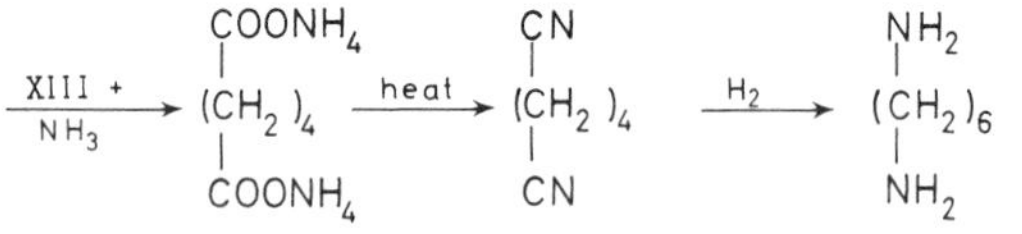

An alternative process for the preparation of hexamethylenediamine starts from furfural ($\overline{OCH_2=CHCH=C}\cdot CHO$), obtained by acid hydrolysis of plant residues (bagasse, oat husks, and straw waste). It is converted to furan, which is hydrogenated to tetrahydrofuran and this is reacted with hydrogen chloride in the presence of sulphuric acid to yield 1,4-dichlorobutane, which with sodium cyanide gives adiponitrile (*see earlier*). The same product has been obtained by chlorination of butadiene ($CH_2=CH-CH=CH_2$) to 1,4-dichloro-2-butene, followed by catalytic conversion by hydrogen cyanide to the corresponding dinitrile and its reduction to adiponitrile. *See also* 5.

4. *Nylon* 6.10 (from hexamethylenediamine and sebacic acid)—The preparation of this polyamide resembles that of nylon 6.6 but with adipic acid replaced by sebacic acid (XVI, m.p. 131–134°C). This

acid is obtained industrially as the sodium salt, together with capryl alcohol, by oxidative fusion of ricinoleic acid (*see* nylon 11, above) with alkali.

$$CH_3(CH_2)_5CH(OH)CH_2CH{=}CH(CH_2)_7COOH \xrightarrow[\text{(heat and air)}]{\text{NaOH}}$$

$$CH_3(CH_2)_5CH(OH)CH_3 + COOH(CH_2)_8COOH \quad (XVI)$$

An alternative treatment by heating with aq. alkali above 200°C under pressure, yields the same products and liberates hydrogen. Sebacic acid can also be obtained by dimerisation of butadiene with sodium and treatment of the product with carbon dioxide, followed by hydrogenation.

5. *Polyamides obtained by interfacial polymerisation**—Polymers of the nylon 6.6 and 6.10 type form at the interface when an aq. solution of one reactant (e.g. hexamethylenediamine) is brought into contact with a non-miscible solvent containing the other reactant (e.g. adipyl or sebacyl chloride dissolved in a chlorinated hydrocarbon). This method of preparation is of considerable interest since the reaction takes place almost instantaneously, without the need of heat, and it is possible to arrange for the product to be drawn off continuously.

6. *Polymerised fatty acid polyamides*—Flexible non-fibrous polyamides, used chiefly for surface coatings, are obtained by condensation of alkylene polyamines (ethylenediamine, diethylenetriamine) and polymerised fatty acids, chiefly dilinoleic acid (XVII) which is derived from linseed oil (*see* §24A.81).

$$CH_3(CH_2)_4CH{=}CH\,CH_2\,CH{=}CH(CH_2)_7COOH \longrightarrow$$

XVII

These polyamides are more readily fusible and soluble than the fibrous type, but those containing residual NH-groups (e.g. from diethylenetriamine) form cross-linked structures with certain epoxy resins (§25.21a(i)) and are used as curing agents for them.

For notes on other nylons and related polymers, *see* §§23.81–83.

23.22 PROPERTIES

Solvents (*see also* 1 and 2, §23.23) Phenols (90% phenol, *m*-cresol) and phenol/chlorinated hydrocarbon mixtures (e.g. 1/3 phenol/

*See also §54.13 (iv).

tetrachloroethane, by volume), 90% formic acid (some polyamides require heat), hot glacial acetic acid, hot formamide, hot benzyl alcohol, halogenated alcohols (2,2,3,3-tetrafluoropropanol, 1,3-dichloropropan-2-ol), 70% chloral hydrate in alcohol. Soluble in conc. mineral acids but with slow decomp. (nylon 11 insoluble in conc. HCl; *see* 2, §23.23). Polyamides that are partially substituted or of mixed composition, or derived from dimerised fatty acids, are more readily soluble than fibrous polymers; some dissolve in aq. alcohols, others in *n*-propanol or *n*-butanol or mixtures of these with toluene. **Plasticised by** some non-fibrous polyamides soft at room temperature (when blendable with harder types). *See also* §23.82a. **Relatively unaffected by** (fibrous types more resistant than non-fibrous types) hydrocarbons and chlorinated hydrocarbons, esters, ethers, oils, and many other organic liquids; conc. alkalis, dil. acids. **Decomposed by** conc. mineral acids, hot dil. mineral acids, oxidising agents, halogens (including chlorine-containing bleaching agents); hot conc. alkalis cause slow loss in weight.

23.23 IDENTIFICATION

Thermoplastic. N present; Cl, S, P absent. **Combustion** Melts, darkens (fibres fuse to bead), boils, finally burns with a small flame that is easily extinguished giving a white smoke (nylon 6.6 produces an odour like that of celery). **Pyrolysis** Decomposes 260–300°C, with alkaline distillate and characteristic odour. *ONB test for nylon* 6.6 (ROFF, W. J., *Analyst,* **79**, 306 (1954)): A test-paper soaked in a freshly prepared saturated solution of *o*-nitrobenzaldehyde in dil. sodium hydroxide is presented to the hot pyrolytic vapour; nylon 6.6 (and other derivatives of adipic acid) produces a mauve-black colour slowly fading on exposure of the paper to air; polyamides not derived from adipic acid (e.g. nylon 6) give, at most, a grey colour. Dry distillation of methoxymethyl-type nylon (§23.82a) or its distillation with dil. H_2SO_4 yields formaldehyde (identify by the CTA test, §26.23). **Additional characterisation** (*see also* specific gravity and melting point): 1. *Test for nylon* 6.6 A distinction from other polyamides (in addition to the ONB test above) is its insolubility in boiling *N,N*-dimethylformamide; nylons 6, 11 and 6.10 are readily soluble (high-tenacity nylon 6 fibre is only sparingly soluble). 2. *Solubility in hydrochloric acid solutions at room temperature.* Proceeding from most to least soluble: *nylon* 6—soluble in 4N acid (nylons 6.6, 6.10 and 11 insoluble); *nylon* 6.6—soluble in 6N acid (nylons 6.10 and 11 insoluble); *nylon* 6.10—insoluble in 6N acid but slowly soluble on warming (nylon 11 insoluble), soluble also in 90% formic acid at room temperature (nylon 11 insoluble); *nylon* 11—insoluble in conc. (11·6N) hydrochloric acid, but fibres slowly shrink to wax-like mass. 3. *Sorp-*

tion of iodine. After immersion for a few minutes in a solution of iodine (e.g. 5%) in aq. potassium iodide, followed by rinsing in water, fibres of nylons 6 and 6.6 are left stained brown to black (sorption of up to 1 atom of iodine per amide group); nylons 11 and 6.10 acquire much less colour. 4. *Tests following hydrolysis.* On hydrolysis, e.g. by heating with 2N to 6N hydrochloric acid for several hours in a sealed tube at 100°C, polyamides yield their constituent acids and amines:

Test for caprolactam. A red precipitate is obtained with potassium bismuth iodide. Quantitative paper chromatography of caprolactam by means of this reagent has been described (CZEREPKO, K., *Mikrochim. Acta.*, 417 (1962)); a subsequent spraying with dil. sulphuric acid increases the sensitivity.

Tests for adipic acid. (i) Evaporate a hydrolysate to dryness, or nearly so; mix a small sample of the product with resorcinol and a trace of zinc chloride; heat for 5–10 min at 110°C; cool, and extract with dil. alkali. Adipic acid, present in hydrolysates from nylon 6.6, produces a deep red-pink colour. (ii) Adipic acid, m.p. 151°C, can be obtained from a hydrolysate by extraction with ether, followed by recrystallisation from water. (iii) *See* ONB pyrolysis test, earlier.

Test for hexamethylenediamine. Evaporation of the aq. portion remaining after ether extraction (ii, above) yields hexamethylenediamine hydrochloride, m.p. 248°C, on recrystallising from alcohol.

More extensive qualitative and quantitative distinction between polyamides is achieved by subjecting the hydrolysates to chromatography, acids and bases being detected with indicators, and aminoderivatives producing a mauve colour on heating with ninhydrin (§22c); *see*, for instance, CLASPER, M. and HASLAM, J., *Analyst,* **74**, 224–237 (1949), **76**, 33–40 (1951), and (with MOONEY, E. F.) **82**, 101–107 (1957). For determination of amine and carboxyl end groups, *see* WALTZ, J. E. and TAYLOR, G. B., *Anal. Chem.* **19**, 448 (1947).

23.3 PHYSICS

23.31 GENERAL PROPERTIES

Specific gravity Nylon 6, 1·13–1·14; nylon 11, 1·04–1·05; nylon 6.6, 1·14; nylon 6.10, 1·09; non-fibrous polyamides, *c.* 0·98.

Refractive index and birefringence

	n_{ISO}	$n_{\parallel}$	$n_{\perp}$	$n_{\parallel}-n_{\perp}$
Nylon 6	1·54	1·575–1·58	1·52–1·525	0·05
11	1·52	1·55	1·51	0·04
6.6	1·54	1·58	1·52	0·05
6.10*	1·52	1·57	1·52	0·05

*As bristles

Heat-treatment causes a small rise in $n_{\parallel}$ and $n_{\perp}$.

Non-fibrous polyamides (Versamid 950), refractive index, 1·50.

Water absorption (plastics, %) Nylon 6.6, *c.* 1·5; nylon 6, slightly greater (*c.* 2·5); on prolonged immersion values approach 10%. Nylon 6.10, *c.* 0·4; nylon 11, slightly less; on prolonged immersion values approach 2 or 3%. In general, the sorptive capacity of polyamides is reduced by dry heating and increased by steaming. **Water retention** (fibres, %) Nylon 6, *c.* 15; nylon 11, *c.* 3; nylon 6.6, *c.* 25. **Moisture regain** (fibres, %; 25°C, 65% r.h. unless otherwise stated); nylon 6, *c.* 4; nylon 11, *c.* 1; nylon 6.10, *c.* 1·7; nylon 6.6,

R.h. %	30	65	80	100
Moisture regain, %	2	4·2	5·5	8·7
Moisture regain, %*	1·8	3·9	5·1	8·5

*Heat-stabilised sample, slightly more crystalline, at 30°C

Permeability to gases $(10^{-10} \text{ cm}^2)/(\text{s cmHg})$, nylon, O_2, 0·04; N_2, 0·01–0·02 (nylon 6, 0·01); CO_2, 0·16–0·20 (0·09); relative rates, nylon 11 film: H, CO_2, O_2, N_2 respectively 1, 0·5, 0·1, 0·01; do. (mass/time), nylon 6 film: methanol or ethanol, water or chloroform, acetone, benzene respectively > 1, 1, 0·1, 0·02. Water vapour, at 20–30°C, 0·055–1·4 $(10^{-10} \text{ g})/(\text{cm s cmHg})$. *See also* Section 64.

23.32 THERMAL PROPERTIES

Specific heat Nylons 6, 11, 6.6 and 6.10, 0·4–0·55. **Latent heat of fusion** 20–25 cal/g. **Conductivity** 4·5–7 $(10^{-4} \text{ cal})/(\text{cm s °C})$. **Coefficient of linear expansion** $(10^{-4}/°C)$ Nylon 6, 0·83; nylon 6.6, 1·0; nylons 11 and 6.10, 1·5; considerably lower values in glass-filled nylon. **Physical effects of temperature** Strength retained to *c.* −40°C. Serviceable up to 100°C for moderate periods or for short times up to 130°C; preferred maximum continuous service temperature for nylon 11, 60°C. Glass-filled nylon shows a considerably higher heat distortion temperature. Second-order (glass) transition temperature, reported as follows (°C): nylon 6, 70 (lowered by moisture); nylon 11, 46; nylon 6.6, 37–47, 57 (dry) and *c.* 80 but being substantially crystalline at room temperature these may be too high (a penetrometer method gave −65°C for nylons 6 and 6.6 (RYBNIKAR, F., *J. Pol. Sci.,* **28**, 633 (1958)). Melting point, approximate °C: nylon 6, 215–220; nylon 11, 185; nylon 6.6, 250–260; nylon 6.10, 215; non-fibrous polyamides, from *c.* 50–200.

Annealing—to relax stresses in fibres, set the twist in yarns, or stabilise the dimensions and improve crease recovery in fabrics—

is effected by treating nylon 6.6 in hot water, in saturated steam up to 150°C or in hot air or dry steam at temperatures up to *c.* 200°C, the time needed decreasing correspondingly from hours to seconds; with nylon 6, temperatures above 170–180°C tend to be detrimental. Both nylons undergo fairly rapid oxidative degradation when heated above 200°C in air; and even at 150°C are subject to degradative discoloration, e.g. the typical time for nylon 6.6 fibres exposed in air at 150°C from tender to half strength, 20–60 h. Thermal decomposition, in the absence of oxygen, commences at *c.* 300°C; a primary mechanism involves scission of —NH—CH$_2$— groups; degradation is restricted by blocking end-groups of the molecule.

23.32 ELECTRICAL PROPERTIES

Volume resistivity (ohm cm) *c.* 10^{14}; at high humidity, *c.* 10^{10}. Experimental results with nylon 6.6 fibre, not necessarily representative and somewhat higher if purified: 65% r.h., *c.* 10^9; 90% r.h., 10^8 (do., another set, 10^{10}). *Surface resistivity* (ohm) Nylons 6 and 6.6, $>10^{12}$; nylon 11, *c.* 10^{14}. At elevated temperatures polyamides show appreciable conductivity. **Dielectric strength** Up to *c.* 16 kV/mm, and increasing up to 120 kV/mm for thin films; e.g. breakdown of dry sheets (nylon 11; approximate kV/mm): 3 mm, 17; 2 mm, 20; 0·5 mm, 40; 0·1 mm, 65. **Dielectric constant** Nylon 6.6 (air-dry, 50 Hz), *c.* 4·0. Decreases with frequency rise to *c.* 3·0 (10^7–10^8 Hz), levels out from 10^8–10^{10} Hz. Above 50°C rises steeply, especially at low frequency; e.g. (at 120°C): 65 Hz, 30; 1 kHz, 16; 1 MHz, 12. Drying lowers the values slightly; increase in humidity causes rise, e.g. $>8·0$ (1 MHz) after prolonged immersion in water. A tendency for dielectric constant to decrease with frequency rise, and to rise with temperature rise particularly at low frequencies, is common to polyamides. Nylon 6 behaves similarly to nylon 6.6; nylons 11 and 6.10 show rather lower values (e.g. 3·6 at 1 kHz falling to 3·1 at 10^8 Hz) and are less affected by water. **Power factor** Nylon 6.6 (air-dry, 50 Hz), 0·01–0·02. Rises with frequency rise to 0·04 (10^5 Hz), falls again to 0·02 (10^7 Hz and above, with slight rise above 10^{10} Hz). Rises with temperature rise, to *c.* 0·08 (50°C, wide frequency range), then more steeply especially at low frequency; at 100°C maximum dielectric loss occurs around 100 kHz to 1 MHz; value decreases at high frequency and high temperature (e.g. 1 MHz and 100°C, 0·2; 1 MHz and 140°C, 0·1). Falls at low temperatures (-150°C, wide frequency range) to *c.* 0·002. Moisture causes rise, e.g. 0·2 (at 1 MHz) after prolonged immersion in water, but shifts the loss maxima to lower frequency or temperature; nylons 11 (power factor, 0·03–0·05 at 1 kHz) and 6.10 less affected than nylon 6.6.

23.34 MECHANICAL PROPERTIES
(25°C, 65 % r.h. unless otherwise stated)

Elastic modulus (Young's) (kgf/mm^2).

N.B. Values fall rapidly with temperature rise and with increase in humidity; values rise with orientation, but diminish when fibres are treated in water:

	Plastics	Fibres
Nylon 6	Up to c. 70 (> 120, dry)	Staple, up to 200 (20 gf/den.); cont. fil., up to 300 (30 gf/den.)
Nylon 11	c. 100	c. 500 (50 gf/den.)
Nylon 6.6	200–300	Staple, 200 (20 gf/den.) and higher; cont. fil., 250 (25 gf/den.) and higher
Nylon 6.10	180–200	Monofil, c. 400 (40 gf/den.)

Elastic recovery *Plastics* (unoriented) Yield, practically inelastic except to small deformations; can be cold-drawn. *Fibres* Show considerable recovery when released from short-term but relatively large deformations; however, recovery practically disappears (i.e. deformations can be permanently set) if strained fibres are relaxed in steam or hot water. Examples of immediate recovery from short-term deformation (percentage, at 25°C, 65 % r.h. *N.B.* Values may rise by 15–20 % in first half-minute after unloading, full recovery from up to 8 % extension may occur in 1 min; values decreased but rates increased by rise of temperature and/or humidity):

	Stress (gf/den.)				Strain (%)			
	0·5	1	2	4	2	5	10	20
Nylon 6 (staple)	66	60	60	55	70	61	60	53
Nylon 6.6 (staple)	79	75	70	53	80	78	73	60
Nylon 11 (staple)	75	60	60	60	80	60	60	60
Nylon 6 (cont. fil.)	100	98	95	60	100	98	95	75
Nylon 6.6 (cont. fil.)	100	98	95	85	100	98	90	70

Tensile strength and **Extension at break.** Examples:

		Tensile strength (kgf/mm^2)	Extension at break (%)
Nylon 6	plastics	6–7 (> 8, dry)	300–200 (25, dry)
	staple	35–60 (3·5–6 gf/den.)	50–35
	cont. fil.	45–65 (4·5–6·5 gf/den.)	40–30
	high tenacity	75 (7·5 gf/den.)	c. 20
Nylon 11	plastics	5–6	Up to 300
	staple	Up to c. 55 (5·5 gf/den.)	20–60
Nylon 6.6	plastics	7–8	Up to c. 100
	staple	40–70 (4–7 gf/den.)	35–20
	cont. fil.	45–60 (4·5–6 gf/den.)	30–20
	high tenacity	Up to 90 (9 gf/den.)	12–15
Nylon 6.10	plastics	5–6	Up to c. 200
	monofil	c. 40 (4 gf/den.)	c. 30

Wet strength of fibres (nylon 6 and 6.6): 85–90% of air-dry strength.

N.B. Strength falls, extension rises, with increase of temperature and/or (particularly with nylons 6 and 6.6) moisture content. Energy of tensile break of moulded nylon 6.6 shows decline below 60°C, or below 20°C in notched specimens (VINCENT, P. J., *Plastics*, **28** (1963) 107).

Flexural, compression, and impact strength (plastics):

	Flexural strength (kgf/mm²)	Compression strength (kgf/mm²)	Izod test (ft lbf/in *of notch*)
Nylon 6	c. 5	>2	3–4·5*
11	Up to c. 7	c. 5	>4
6.6	c. 10	c. 3·5	1·5–2†
6.10	Up to c. 7	c. 2	c. 2

*1·5, dry
†Copolymers higher. Nylon 6.6 withstands repeated impact. Ropes of polyamide fibre are superior in repeated shock resistance to other types of cordage

Hardness Rockwell, 25°C: nylon 6, R65–85 (>100, dry); nylon 11, R108 or α50 (50°C: R80 or α−50); nylon 6.6, L95–100, M80–90, or R110–150; (Vickers diamond pyramid No. 12); nylon 6.10, R111 (Vickers d.p. No. 11). Hardness is improved by loading with graphite or glass fibre.

Friction Approximate values for coefficient of friction (nylons 6 and 6.6 similar). μ_{stat}, plastics or cont. fil. yarn/mild steel, 0·15. Cont. fil. yarn/nylon rod, 0·2; do./yarn, 0·15–0·6. μ_{dyn}, plastics/mild steel, 0·05–0·2. Cont. fil. yarn/mild steel, 0·35–1·1 (increases with decrease of twist); do. (lubricated)/do., c. 0·25; do. (lubricant free), c. 0·5. Cont. fil. yarn/stainless steel, 0·4; do./copper, c. 0·25; do./nylon rod, 0·2; do./glass or ceramic, 0·35–0·45. Grey cotton yarn/nylon rod, 0·18. Water usually lubricates (but swells) nylon; friction is lowered (e.g. in high-load bearings and gears) by use of porous oil-absorbing nylons, or nylon filled with graphite or molybdenum disulphide.

23.4 FABRICATION

Careful temperature control is needed in injection moulding or extrusion because of the relatively sharp melting points, and the moisture content should be low; pressures up to c. 4000 lbf/in² and warm moulds (50–80°C) are used. Examples of fabrication temperatures (°C): nylon 6, injection (nozzle), 220–225; extrusion, up to 290 (blown mouldings need high viscosity grade polymer); nylon 11, above 200; nylon 6.6, rather higher than for 6 (>265); nylon 6.10, lower than for 6.6 (235–255). Articles of a cylindrical shape can sometimes be made by centrifugal casting; non-fibrous polyamides can be cast in the ordinary way.

23.5 SERVICEABILITY

Polyamides in general are dimensionally stable but undergo oxidative degradation, with discoloration and loss of strength, on prolonged exposure to sunlight or heat (*see also* §23.32). Photo-tendering is enhanced by the presence of a pigment such as titanium dioxide. Resistance to both forms of degradation is of technical importance, and is improved by incorporation of trace amounts of copper, copper compounds together with a halogen, chromium or manganese compounds, or phenolic antioxidants; black nylon is preferred for weather resistance. Polyamides are attacked by strong acids and bleaching agents but show good resistance to alkalis and appear to be immune to biological attack (although insects may eat through fibres or film, and micro-organisms may thrive on surface finishes).

23.6 UTILISATION

Nylons 11 and 6.10, which show least water absorption, are preferred for electrical uses, though nylon 6.6 is also employed; they are particularly suited to mechanical applications (e.g. oil-resistant low-friction bearings and gears) and loading with appropriate fillers (graphite, glass fibres) increases the lubrication and hardness. Soluble copolymers and derivatives are used for adhesives and varnishes. Viscous non-fibrous polyamides, hardenable with liquid epoxy resins, can be used to provide strong castings for dies, jigs, patterns, etc., also for embedding electrical parts, and as surface coatings, inks, and adhesives (including heat-sealing compositions); they have been used to reduce shrinkage of woollen fabrics, and the water of shellac as melt adhesives, and in non-drip paints.

The high strength (dry and wet), elastic recovery, abrasion resistance, and toughness of nylon fibres make them suitable for marine hawsers, driving and climbing ropes, parachute and tyre cords, and rot-resistant filter fabrics and netting. Nylons 6 and 6.6 are important for industrial fabrics (e.g. tarpaulins), carpets and apparel fabrics, including blends with wool and other fibres; nylon stockings are not only abrasion-resistant but can be heat set to shape to maintain a good fit elastically; bulked yarns (stretch-nylon) find numerous uses. Monofils, mostly made from nylons 11 and 6.10 because of low water absorption, are used for bristles, surgical sutures, and fish lines.

23.7 HISTORY

Early investigations of ω-amino-acids were made by GABRIEL, S. and MAAS, T. A. (*Berichte*, 1899), MANASSE, A. (1902), and VON BRAUN, J. (1907), but the fibrous potentialities of high molecular weight polyamides went unrecognised until the work of W. H. Carothers, begun in 1929, which led to the discovery of the 6.6 polymer in 1937. *See* MARK, H. and WHITBY, G. S. (Eds), *Collected Papers of W. H. Carothers* (*High Polymers*, Vol. I), New York (1940). Production of fibres of nylon 6.6 was commenced in the U.S.A. (by E. I. du Pont de Nemours and Co.) in 1938, and in Great Britain (by British Nylon Spinners Ltd.) in 1941.

Polymerisation of caprolactam leading to production of the German alternative (nylon 6) was patented by P. Schlack in 1938 and production of Perlon commenced in 1940; production in N. Ireland (by British Enkalon) began in 1963. The derivation of nylon 11 from castor oil was developed in France by J. Zeltner and M. Genas in 1944, production commencing in 1949. Casting-type and surface-coating non-fibrous polyamides, first described in 1944, have been made in Great Britain since 1955.

23.8 ADDITIONAL NOTES

23.81 NYLONS BY NUMBER
(for Notation *see* commencement of Section 23)

Fibre-forming polymers representing substituted forms of **nylon 1** have been obtained by low-temperature anionic polymerisation of monoisocyanates (*see* SHASHOUTA, V. E., *et al.*, *J. Amer. Chem. Soc.*, **82**, 866 (1960)). **Nylon 2,** or polyglycine, can be prepared from glycine derivatives such as anhydrocarboxyglycine and is a synthetic polypeptide (*see also* proteins, Sections 20–22). **Nylons 3, 4, and 5** have been prepared from acrylamide, 2-pyrrolidone, and 2-piperidone respectively. Nylon 3 melts at 325°C, nylon 4 at 260°C; these polymers are of interest in showing higher moisture regain and less accumulation of 'static' than nylon 6.

Nylon 7, polyenanthamide, has received much attention (e.g. by N. V. Mikhailov and others) in Russia. Production depends on telomerisation of ethylene with carbon tetrachloride to obtain 1,1,1,7-tetrachloroheptane, followed by hydrolysis to ω-chloroheptanoic acid and amination of this to the required monomer, ω-aminoenanthic acid (ω-aminoheptanoic acid), m.p. 195°C; alternative methods of preparation have employed oxidation of ricinoleic acid, or synthesis of an aminoheptanoic ester (HORN, *et al.*, *J. Appl.*

Pol. Sci., **7**, 887 (1963)). Polymerisation of the monomer, by heating under pressure with removal of water, yields the polymer, m.p. 225°C, which has been developed as a fibre (Enant); condensation with caprolactam yields a softer fibre-forming copolymer (Kapronant), m.p. 185–205°C. Nylon 7 has a specific gravity of 1·1, and is said to crystallise more readily than nylon 6 and to be more stable to heat and ultra-violet light. Melt-spun fibres (mol. wt. 15 000–30 000) have a moisture regain of 2·6–3·0% (65% r.h., 25°C), and in physico-mechanical properties appear similar to those of nylon 6 and 6.6. **Nylon 7.7,** a corresponding homologue of nylon 6.6, has been obtained by condensation of heptamethylenediamine with pimelic acid.

Nylons 8 and **10,** respectively polycaprylamide and polycapramide, have been synthesised from unsaturated cyclic hydrocarbons derivable from acetylene or butadiene; they are fibre-forming, but appear to offer no special advantages and have not attained commercial importance. **Nylon 9,** polypelargonamide, can be derived from whale oil (*see also* polynonamethyleneurea §23.83), and has been used in Russia for a fibre (Pelargon; sp. gr. 1·06, m.p. 200°C).

Nylon 12, polylauroamide, can be obtained by trimerisation of butadiene to form 1,5,9-cyclododecatriene, with subsequent conversion to laurolactam in good yield; the polymer resembles nylon 11, but, following the general effects of increase in nylon number, has a lower specific gravity (1·02) and melting point (175°C); it is used (Rilsan A, Vestamid) in applications where the slightly lower moisture absorption is advantageous.

23.82 MODIFICATIONS OF CONVENTIONAL POLYAMIDES

(a) Substituted polyamides
Polyamides derived from *N*-substituted amines (e.g. *N,N'*-dimethyl-hexamethylenediamine), and interpolymers of the components of fibrous polyamides, retain some of the strength of the simple polymers, but show less crystallinity and lower softening temperatures with greater solubility and elastic extensibility. Partial *N*-methylolation or *N*-alkoxymethylation of the amide hydrogen atoms in fibrous polymers produces similar effects, and can be brought about by reaction with formaldehyde in the presence of an acid and (in the second instance) the appropriate alcohol,

$$-NH-R-\underset{\underset{O}{\|}}{C}- \xrightarrow[R'OH]{HCHO} -\underset{\underset{(CH_2O)_xR'}{|}}{N}-R-\underset{\underset{O}{\|}}{C}-$$

N-alkoxy derivatives are the more stable but both these and *N*-methylol products tend to cross-link and become insoluble on heating. Hydroxyethylation of fibrous polyamides by reaction with ethylene oxide has also been employed.

Substituted polyamides, soluble in alcohol or alcoholic solvent mixtures, find uses as adhesives and as varnishes, e.g. for abrasion-resistant or anti-slip finishes on textile threads. In fibrous form they have been employed as 'elastic nylon'.

Many modifications of nylon fibres by graft copolymerisation (e.g. ethylene oxide, vinylpyridine or styrene) have been reported but are as yet of no commercial importance. For further information on substituted polyamides, *see* LEWIS, J. R. and REYNOLDS, R. J. W., *Chem. and Ind.*, 958–961 (1951).

(b) Modification of the conventional main chain
Polyamides with phenylene rings in the main chain have been obtained by condensation of appropriate monomers, e.g. diamino- or dicarboxy-derivatives of durene (1,2,4,5-tetramethylbenzene) with, respectively, dicarboxylic acids or diamines. They yield heat-resistant fibres of high modulus, the melting point being especially high (400°C) where both components are cyclic in structure, e.g. a diaminobenzene and iso- or terephthalic acid. Saturated rings are present in a silk-like fibre (Qiana) based on polybis(*p*-aminocyclohexyl) methane dodecanedioate, which is represented by,

Polyamides with *nitrogen-containing rings* in the main chain have been developed, particularly in Japan (as 'polyazine' fibre) to convey improved dyeing properties; typical examples are the products made by polymerisation of diamines with the dimethyl ester of isocinchomeronic acid (2,5-pyridine dicarboxylic acid, obtained by oxidation of α-collidine), and by condensation of a diamine with pyrazine dicarboxylic acid, or of 2-methylpiperazine with a dicarboxylic acid, repeat units representative of the last two polymers being shown in I and II.

I   II

Phosphorus has been introduced but conveys no particular advantage (e.g. *see* PELLON, J. and CARPENTER, W. G., *J. Pol. Sci.* A.1, 863 (1963)); *silicon* reduces fibre forming properties; *sulphur, oxygen,* and other elements have also been introduced, an example being condensation of hexamethylenediamine with thiodivaleric acid, $COOH(CH_2)_4$-$S(CH_2)_4COOH$, to give a product that yields soft elastic fibres. Nylon 6.6 fibres modified by introduction of disulphide or alkylene sulphide cross-linkages simulate the structure of keratin in wool (EARLAND, C. and RAVEN, D. J., *J. Text. Inst.,* **51**, T678 (1960); BRUCK, S. D., *J. Res. Nat. Bur. Std.,* A 65, 489 (1961), A 66, 77, 251 (1962)).

(c) Heat-resistant polyamides (including polyimides)

A demand for polymers that retain good mechanical properties at higher temperatures than usual has led to a number of interesting polyamides, among which those based on diaminoaryl terephthalates are mentioned in *(b)* above. Related materials (H-nylon, HT-1, Pyre ML) are poly-amide-imides and polyimides derived from condensation of diamines with polycarboxylic acids or their anhydrides, e.g. with pyromellitic dianhydride (obtained by oxidation of durene) to yield repeat structures represented by III, where —NRN— is the diamine residue.

III

Polypyromellitimides, fabricated as fibres and films (Kapton) via the soluble intermediate (—CONH—) stage and cured by heat, are serviceable at 250°C, retain half-strength at nearly 300°C, and decompose without melting only above 400°C (*see* U.S. Pat. 2 900 369 (1955) DU PONT ; JONES, J. I. *et al., Chem. and Ind.,* 1686 (1962)). Other heat-resistant structures are represented by a series of poly-1,3,4-oxadiazoles (repeat unit IV, derived from cyclodehydration of interfacially-obtained polyhydrazides, —RCONHNHCO—), and by polybenzimidazole (V, from condensation of 3,3'-diaminobenzidine and diphenyl isophthalate) which remains inert and unchanged up to 370°C.

IV   V

Temperatures of over 500°C can be withstood by polyamides with a 'ladder-like' structure, e.g. as results from condensation between 1,2,4,5-tetra-aminobenzene and 1,4,5,8-naphthalenetetracarboxylic acid,

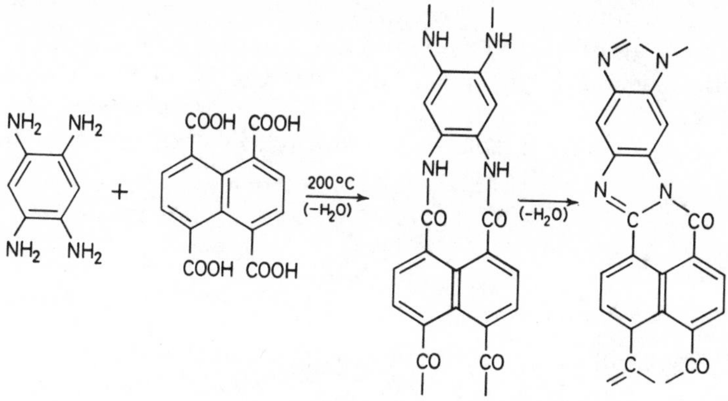

23.83 RELATED POLYMERS

Polynonamethyleneurea (Urylon) This, developed as fibres in Japan, is synthesised through ozonolysis of unsaturated fatty acids derived from rice bran oil, or whale oil, to obtain (as one of the products) azelaic acid, $COOH(CH_2)_7COOH$. This is converted via the ammonium salt and dinitrile to nonamethylenediamine which, when condensed with urea produces the polymer,

$$NH_2(CH_2)_9NH_2 + NH_2CONH_2 \xrightarrow{-2NH_3} [-(CH_2)_9NHCONH-]$$

Like polyamides, the polymer is soluble in hot glacial acetic acid and hot phenol; it has good resistance to alkalis; it most resembles nylon 11 but is distinguished from it by being more transparent, higher in nitrogen content and melting point, more resistant to acids and resistant to dissolution in 80/20 phenol/water at room temperature. Fibres, melt-spun, have the following properties: sp. gr., 1·07; moisture regain (65% r.h., 25°C), 1·4–1·7%; softening temperature, 205–210°C; m.p., *c.* 230°C; tensile strength (gf/den.), staple 3·5–4·5, cont. fil. 4·5–5·5; extension at break (%), staple 20–30, cont. fil. 10–35; wet strength resembles dry strength; initial elastic modulus is high (up to 600 kgf/mm^2) and elastic recovery is good.

High strength glass-like polyurea fibres have been prepared from di-isocyanates and piperazine derivatives (B.P. 876, 491 Du Pont), e.g. with the following repeat unit:

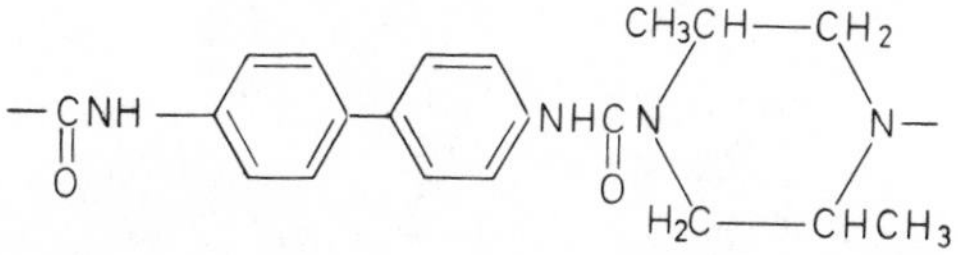

Polyaminotriazoles Aliphatic dihydrazides, obtained by reacting dicarboxylic esters with hydrazine, can be further condensed by heating

225

in the presence of excess hydrazine to yield fibre-forming polymers containing an alkylene group and a 4-amino-1,2,4-triazole ring. For example, sebacic dihydrazide, $NH_2NHCO(CH_2)_8CONHNH_2$, polymerises with loss of water (2 moles) and rearrangement to form a polyoctamethylenetriazole repeat unit (cf. IV above):

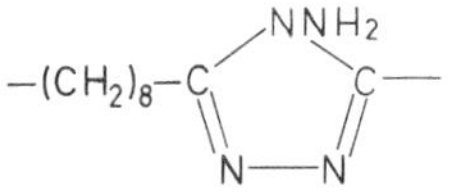

The nylon-like octamethylene polymer has a specific gravity of 1·12, refractive index 1·55, m.p. 256–260°C, moisture regain (65% r.h., 25°C) 3·9%; it is soluble in formic acid and in hot acetic acid, glycol, or phenols; it resists alkalis and mild acids but is decomposed by strong acids and bleaching agents. Melt-spun fibres have a tensile strength of 5–6 gf/den. and 20–12% extension at break, and may be readily dyed.

For further reference, *see*: MONCRIEFF, R. W., *Text. Mfr*, **79**, 559 (1953); FISHER, J.W., *J. Appl. Chem.*, **4**, 212 (1954).

23.9 FURTHER LITERATURE

HOPFF, H., MULLER, A. and WENGER, F., *Die Polyamide*, Berlin (1954) (sections on chemistry, plastics, fibres)

FLOYD, D. E., *Polyamide Resins*, New York (1958) (2nd edn 1966)

HOPFF, H. and GREBER, G., *Ciba Review*, 11 (No. 127, 1958) 2–24 (polyamide and polyester fibres)

KORSHAK, V. V. and FRUNZE, T. M., *Synthetic Hetero-chain Polyamides*, London (1965) (transl. from the Russian)

Fibre technology

KLARE, H., *Synthetische Fasern aus Polyamiden*, Berlin (1963) (mainly technology)

British Nylon Spinners Ltd., *Technical Service Manual*, Pontypool, Wales

The Chemstrand Corporation, *Technical Service Manual*, Pensacola, Fl. U.S.A.

Specifications

ASTM D789 (injection moulding material)

Textile Standard No. 54: Method for determination of relative viscosity in 90% formic acid. *J. Text. Inst.*, **49**, 734 (1958)

POLYESTERS
Sub-section 24A: ALKYD RESINS*

With notes on drying oils, rosin ester gums, shellac, polyester rubbers and factice

SYNONYMS AND TRADE NAMES

Glyptals. Alkydal, Beckosol, Beetle (BA series), Crestalkyd, Durecol, Epok, Mitchalac, Paralac, Plastokyd, Plusol, Scoplas, Scopolux, Soalkyd, Synolac, Synresate, Vilkyd, Wresinol.

GENERAL CHARACTERISTICS

Transparent to opaque, often yellow to brown, and ranging from soft to hard and rosin-like, alkyd resins constitute a wide class of materials; initially they are readily dissolved by organic solvents but when spread as films they can become insoluble and thus find extensive uses in coating systems.

An alkyd resin of the above kind may be defined as a linear polyester modified by branching or cross-linkage. A simple polyester, derived from condensation of a polyhydric alcohol and a polyfunctional acid, is sometimes described as an *alkyd* (the word comes from *alc*ohol-ac*id*), and for the present purpose one component needs a functionality† of at least 2, while the other must number at least 3. Although any 2,3-functional composition can, with sufficient condensation, yield a 3-dimensional structure, it is usual first to modify it by incorporation of a monobasic unsaturated fatty acid, a representative example being the product from glycerol, phthalic acid (or phthalic anhydride) and a drying oil (a glyceride of a monobasic unsaturated fatty acid, *see* §24A.81).

The oil-content is described in terms of 'oil-length', i.e. short-, medium-, or long-oil types, representing respectively 35–45, 46–55 and 56 to over 70% oil in the final resin.

24A.1 STRUCTURE

Simplest Fundamental Unit

The repeat-structure of a linear polyester is represented by the residue resulting from (i) self-esterification of a hydroxy-acid,

$$-\text{ORC}- \atop \underset{\text{O}}{\|}$$

or (ii) condensation of a dihydroxy alcohol and a dibasic acid,

$$-\text{OROCR}'\text{C}- \atop \underset{\text{O}}{\|} \ \underset{\text{O}}{\|}$$

---

*For polyester moulding compounds sometimes referred to as *alkyds, see* §24D.81.
†Functionality = number of reactive groups (—OH, —COOH, —NH₂, etc.) per molecule.

Inclusion of a trifunctional component introduces branching (and cross-linkage), represented by,

$$
\begin{array}{c}
-\mathrm{OR''OCR'C}- \\
\quad\ \ |\quad \| \quad \| \\
\quad\ \mathrm{R'''O}\ \ \mathrm{O}\ \ \mathrm{O}
\end{array}
$$

where R''' is an unsaturated fatty acid residue derived from a drying or semi-drying oil (*see* §24A.81).*

Molecular weight Simple fusible polyesters, *c.* 3000–6000 (higher condensation limited by increasing viscosity and difficulty in removal of water). In alkyd resins molecular weight depends on the proportions, composition, and treatment received; initial fractions of 17–20 000 and 35–39 000 have been reported. *Degree of polymerisation* Simple fusible polyesters, *c.* 7–25.

X-ray data Most alkyd resins are microcrystalline, especially if annealed.

24A.2 CHEMISTRY

24A.21 PREPARATION

The preparation of an alkyd resin—of the kind described under *General characteristics* above—involves condensation between a polyhydric alcohol (e.g. glycerol) and a polybasic acid (e.g. phthalic acid, or its anhydride), and modification of the product by incorporation of an unsaturated fatty acid or a glyceride of the acid (i.e. a drying oil, such as linseed oil, *see* §24A.81),

$$
\mathrm{HOCH_2CHCH_2OH + C_6H_4(CO)_2O} \\
\quad\quad\ \ | \\
\quad\quad\ \mathrm{OH}
$$

$$
+ \mathrm{RCH{=}CH(CH_2)_7COOH} \xrightarrow[\text{catalyst}]{\text{Heat}+}
$$

$$
\begin{array}{c}
-\mathrm{OCH_2CHCH_2OCC_6H_4C}- \\
\quad\ \ |\qquad\quad\ \|\quad \| \\
\quad\ \mathrm{O}\qquad\quad\ \mathrm{O}\ \ \mathrm{O} \\
\quad\ | \\
\quad\ \mathrm{C(CH_2)_7CH{=}CHR} \\
\quad\ \| \\
\quad\ \mathrm{O}\quad\quad + 2\mathrm{H_2O}
\end{array}
$$

The preparation is generally carried out by a batch process, of which there are four main types: (i) a fatty acid, or (ii) a fatty acid (glyceride) oil, is reacted with the other components at 200–280°C until the desired composition is achieved, or (iii) (alcoholysis process) a drying oil and a polyol are reacted together at 250°C in the presence of a catalyst (e.g. litharge) to give a monoglyceride which is then

---

*For more detailed representations of alkyd resin structures *see* SEAVELL, A. J., *J. Oil Col. Chem. Assoc.* **42**, 319 (1959).

further esterified with a dibasic acid, or (iv) resin prepared by any of the above techniques is diluted with oil at high temperature. During esterification, water may be removed directly (fusion process) or by a refluxing liquid (solvent or azeotropic process), the latter system giving lighter-coloured products. Glycerol (used mainly for short- and medium-oil resins) may be partly substituted by a difunctional glycol (complete replacement would yield only linear structures), or—particularly for long-oil resins—by pentaerythritol, which being tetrafunctional combines with twice as much fatty acid to yield fast-drying products. Phthalic acid or the anhydride can be replaced by adipic or sebacic acid to confer flexibility, or by maleic acid to increase functionality and viscosity. The choice of the unsaturated fatty acid or drying oil is dictated by local economics and the properties required in the end product; tung oil, for instance, gives faster-drying and more water-resistant products than linseed oil.

A drying oil increases the solubility and flexibility of a simple polyester, and a polyester increases the gloss and rate of drying of the oil, hence the importance of alkyd resins in paints and lacquers. The 'drying' of films takes place first by loss of added solvent, and then by radical polymerisation of the unsaturated portion, which is initiated by heating (together with further condensation) or by aerial oxidation, and renders the end-product insoluble.

Because of the presence of many functional groups, alkyd resins can be modified by combination with ester gums (§24A.82), epoxy or phenol-formaldehyde resins, or—when unsaturated—with styrene (see §24D.21). For paint manufacture the resins are normally thinned to 50% solids content with suitable solvents.

24A.22 PROPERTIES

Dissolved by hydrocarbons (petroleum ether, white spirit; toluene, naphtha) and chlorinated hydrocarbons, ketones, esters, alcohols (some alkyd resins only). Short-oil resins generally require aromatic solvents; medium- and long-oil resins, aliphatic solvents; amino-alkyd resins require blends of butanol and xylene. Many alkyd resins are compatible with rosin, ester gums, drying oils, and cellulose nitrate. **Plasticised by** phthalate and phosphate esters, but commonly plasticisation is conferred by suitable internal composition, i.e. long-chain acids and drying oils. **Swollen or decomposed by** chlorinated hydrocarbons and ketones; alkalis and conc. oxidising acids. Linear swelling of cured unpigmented coatings at equilibrium (%): aq. phenol, 12; acetone, 30; benzaldehyde, 26. **Relatively unaffected by** (as cured films) hydrocarbons, oils, weak acids, and neutral salt solutions.

24A.23 IDENTIFICATION

Thermosetting. N, Cl, S and P nominally absent. Saponifiable. **Combustion** Burns, generally with sooty flame. **Pyrolysis** Phthalic anhydride, a common constituent, sublimes. **Colour reactions and other tests** The following simple tests may be applied but it should be remembered that alkyd resins are frequently combined with other polymers which may be identified by tests described elsewhere, e.g. styrene (§4.23), polyamides (§23.23), amino-resins (§29.23) and phenolic resins (§30.23).

1. When boiled with acetic anhydride, cooled to room temperature, and treated with conc. sulphuric acid—or when heated with tri-chloroacetic acid—simple alkyd resins give a red-brown colour.

2. When fused with potassium hydrogen sulphate, glycol esters yield acetaldehyde (odour), while resins—and other compounds—containing glycerol yield acrolein (sharp biting smell).

3. *Fluorescein test for phthalates* (phthalate plasticisers also respond). Heat a small sample with a little resorcinol and a trace of conc. sulphuric acid, for 5–10 min at 120°C. Extract the cooled mass with water, then with ether, wash the ether extract with water, and finally with dil. alkali. The appearance, in the alkaline extract, of a pink to red colour and a pronounced yellow-green fluorescence indicates the presence of a phthalate initially.

4. *Phenolphthalein test** If an excess of phenol is substituted for resorcinol in the foregoing test, when the fused mass is extracted with alkali, the red colour of phenolphthalein is obtained (colourless on acidification and vice versa).

N.B. In tests 3 and 4 overheating is to be avoided, and it is advisable to run blank controls.

5. *Saponification and identification of saponification products.* On refluxing with dil. alkali (preferably, and in some instances necessarily, alcoholic alkali) alkyd resins decompose, with an appropriate absorption of alkali; most alkyd resins have saponification numbers in the range 200–300 mg KOH/g but higher values are found if low molecular weight acids are present.

The saponification residues contain polyhydric alcohols and alkali salts of polybasic acids which may be detected as follows:

Polyhydric alcohols (glycerol, glycol, etc.)†—(i) Evaporate the filtrate from the saponification liquors to dryness, and extract the residue with 1/1 alcohol/ether. Evaporate the extract to dryness and warm the residue with a little dil. bromine-water, then boil off excess bromine. To 0·5 ml of the residual solution add 2 drops 5% potassium bromide solution, 2 drops saturated guaiacol solution, and 2 ml

†*N.B.* The same colour reactions are given by soluble derivatives of cellulose, starch, and other polyhydroxy compounds, which should therefore be absent from the test solution. It is advisable also to run blank control tests, since trace colorations may be obtained even in the absence of the materials sought.

conc. sulphuric acid; polyhydric alcohols produce a blue colour, changing to purple on warming. (ii) If the filtrate is alcoholic, evaporate to dryness; dissolve the residue in water, and treat as follows: To a few millilitres of solution, add 1 drop conc. sulphuric acid and 3 to 4 drops saturated potassium periodate solution. After a few minutes, add 2 to 3 drops saturated sodium sulphite solution, followed by 0·5 ml 5% aq. chromotropic acid and 2 to 3 ml conc. sulphuric acid. Polyhydric alcohols are oxidised to formic acid and formaldehyde, the latter producing a deep violet coloration (as test in §26.23).

Polybasic acids—Acidifying the saponification residues with dil. sulphuric acid yields the appropriate acid, which in some instances precipitates from solution, e.g. saponified fibres or films yield terephthalic acid, which is identified by washing, drying, and then refluxing with methyl alcohol and a little conc. sulphuric acid, to yield the dimethyl ester (m.p. 140°C, when recrystallised from aq. alcohol).

For detailed analytical procedures *see* KAPPELMEIER, C. P. A., *Chemical Analysis of Resin-based Coating Materials*, New York (1959).

24A.3 PHYSICS

24A.31 GENERAL PROPERTIES

Specific gravity From approx. 1·0–1·4. **Refractive index** 1·47–1·57 (highest in unmodified alkyds, lowest in long-oil and natural resin modified types).

Water absorption The results of a number of studies of water resistance, though not entirely in agreement, lead to the following observations:

(*a*) Swelling after prolonged immersion depends on the polarity of the cured films and on the alkalinity of the water or any pigments present. At equilibrium, attained after several days, adhesion between the film and substrate is impaired, and at high degrees of swelling blistering may occur. Linear swelling at equilibrium, of cured alkyd resin coatings in water (%): unpigmented, 0·75; with titanium dioxide, 2; with zinc oxide, 11.

(*b*) Moisture absorption rises and tensile strength falls with rise of humidity; e.g. long-oil pentaerythritol alkyd resin (67% linseed oil):

R.h. %	20	60	100
Moisture absorption (%)	0·2	0·6	2·3
Tensile strength (kgf/mm²)	2·03	1·54	1·05

(c) Adhesion appears to be little affected by humidity until a critical value is reached (70–95 % r.h.) beyond which there is a sharp fall. On drying out, the initial adhesion is largely recovered.

24A.32 THERMAL PROPERTIES

Conductivity c. 5 $(10^{-4}$ cal)/(cm s °C). **Coefficient of linear expansion** c. 1.10^{-4}/°C. **Physical effects of temperature** Usually brittle when cooled, some types of alkyd resins are liquid or soft and plastic at room temperature, but others soften in the range 60–90°C (glycerol-phthalate types) or even as high as 140–150°C (rosin modified glycerol-maleate types). Maximum service temperature c. 100°C (higher in some types, e.g. silicone-modified, over 200°C).

24A.33 ELECTRICAL PROPERTIES

Volume resistivity c. 10^{14} ohm cm. **Dielectric strength** (kV/mm) 12–16; alkyd-bonded mica containing up to 5 % resin: 22; insulating varnishes: 100. **Dielectric constant** (at 1 MHz, 25°C) 3–5. **Power factor** (at 1 MHz, 25°C) 0·012–0·03.

24A.34 MECHANICAL PROPERTIES

The structures are so complex, and the variables of formulation and compounding are so numerous, that data on the mechanical properties of alkyd resins are not easily correlated, while the fact that they are usually tested as films on substrates introduces further variables. Little quantitative information is available on unsupported films, but some data on air-cured unsupported films of long-oil (60–70 %) alkyd resins, unpigmented except No. 5, are given below.

Composition*	Elastic modulus (Young's) (kgf/mm²)	Tensile strength (kgf/mm²)	Extension at break (%)
1. PP/soya bean oil	4	0·525	65
2. PP/linseed oil	8	0·65	65
3. GP/linseed oil, styrene-epoxy modified	16	2·065	90
4. GP/DHCO, rosin-phenolic modified	—	0·425	110
5. As 4, but with rutile (TiO₂) pigment	—	0·288	53
6. GP/linseed oil (ex fusion process)	—	0·125	85
7. As 6, but ex solvent process	—	0·175	100

*PP = pentaecrythritol phthalate. GP = glycerol phthalate. DHCO = dehydrated castor oil. Sources:
Resins 1 to 3, WRIGHT, J., *J. Oil Col. Chem. Assoc.*, **48**, No. 8, 670 (1965)
Resins 4 and 5, IVANFI, J., *Paint Manufacture*, **35**, No. 10, 37 (1965); No. 11, 46
Resins 6 and 7, BULT, R., *Off. Digest*, **33**, No. 443, 1594 (1961)

Hardness Sward rocker hardness (glass = 100): typical cured coatings, 30–40, but lower values and values up to 60 may be encountered. **Abrasion resistance** For best performance some degree of flexibility in the resin is desirable but much depends on the type of abrasive action. Superior performance is claimed for alkyd resins in road-marking paints; laboratory tests indicate that they are less abrasion resistant than epoxy or polyurethane coatings.

24A.4 FABRICATION

In the manufacture of coatings, the resins are normally handled as solutions (50% solids). Rutile (titanium dioxide) is commonly used as a white pigment and opacifier, pigment: binder ratios being in the range 0·7 or 0·8/1. For special purposes (e.g. marine paints) antimony oxide, zinc oxide, or white lead may be added. Colour is imparted by inorganic or organic pigments, e.g. iron oxide, phthalocyanines, azo compounds. Pigment wetting characteristics are influenced by oil length and acid value, and a wetting agent, e.g. a zinc soap, may be added to assist dispersion during compounding, which is carried out on a triple-roll mixer or ball mill. After dispersion of the pigments, cobalt and manganese naphthenates may be used as driers to catalyse the oxidation and accelerate the drying of applied coatings. Alkyd resins are frequently blended with other polymers (cf. §24A.21), non-fibrous polyamides (§23.21.6) producing thixotropic non-drip paints.

24A.5 SERVICEABILITY

Weathering resistance varies with composition and pigmentation, but many alkyd resins are sufficiently durable for outdoor applications. Yellowing on prolonged exposure to light is minimised by the use of oils with a high linolenic acid content (e.g. tobacco seed or soya bean oils). Organo-mercury compounds and zinc oxide give some protection against microbiological and algal growth. Alkyd resins are more resistant to high energy radiation than cellulosics, polymethyl methacrylate, or polychloroprene.

24A.6 UTILISATION

An early use of unmodified alkyd resins, as a binder for mica in electrical insulation, is now of very minor account compared with the important applications of oil-modified resins in the paint industry. In this field they are remarkably versatile, finding uses in interior and exterior finishes, both for general and special purposes, e.g. fire-retardant paints (chlorinated rubber modified), thixotropic paints (polyamide modified), crackle finishes (phenolic modified), and stoving lacquers (amino-resin or nitrocellulose modified). Water-dispersible alkyds have been developed for use in emulsion paints. Other applications include plasticisers (e.g. for polyvinyl chloride), printing ink vehicles, adhesives, foundry core-binders, floor-coverings (linoleum) and intermediates for polyurethanes (section 40).

24A.7 HISTORY

The self-condensation undergone by lactic acid on heating was observed by Gay-Lussac and J. Pelouze in 1833, and condensation of glycerol with tartaric acid was examined by J. Berzelius in 1847. From then onwards polyesters received attention from many nineteenth-century investigators, including M. M. Berthelot and J. M. van Bemmelen, but there was little practical development until 1910 when the General Electric Co. patented the use of polyesters in insulating varnishes and impregnants. By 1920 rosin- and nitrocellulose-modified types were in use. The first oil-modified type appeared in 1927 as a result of the work of R. H. Kienle, who had a fundamental part in the development of these materials, and from then on there was steady progress in the use of alkyd resins for coatings. W. H. Carothers' research on linear polyesters in the 1930s also contributed much to the understanding of alkyd chemistry. Phthalic acid and its anhydride became readily available *c.* 1930 by the vapour-phase catalytic oxidation of naphthalene. Pentaerythritol came to prominence as a starting material *c.* 1938, and about the same time phenolic- and amino-modified alkyd resins made their appearance. Styrenated types were produced in 1942. The post-war period brought further modified types incorporating silicones, polyamides, epoxies, and isocyanates, while the development of synthetic glycerol from petroleum and the introduction of isophthalic anhydride have had significant economic and technical effects.

24A.8 ADDITIONAL NOTES

24A.81 DRYING OILS

Vegetable and animal oils in this category are non-polymeric but when thin films are exposed to air, as in painting, they undergo progressive 'drying' to an insoluble and elastically-extensible phase. Drying oils are much used in association with alkyd resins (as described earlier in this section) and with certain other substances including natural rosins, whereby the rate of drying, gloss, and durability of the films are enhanced. The drying process, involving partial oxidation and polymerisation, is accelerated by sunlight and heat; also, the addition of a small proportion of catalytic 'driers' (e.g. an oxide or oil-soluble salt of a polyvalent metal, such as cobalt or manganese naphthenate) reduces the time required to obtain a hard film from several days to a few hours.

The active components of drying oils are polyunsaturated monobasic aliphatic acids, present as mixed triglycerides (glyceryl esters). Those of *vegetable* origin are mostly C_{18} acids, containing the group

$$-CH=CH(CH_2)_7COOH$$

along with additional double bonds which are either conjugated ($-CH=CH \cdot CH=CH-$) or linked through methylene groups ($-CH=CH \cdot CH_2 \cdot CH=CH-$). The acids of *animal* origin extend up to C_{24}, and appear mainly to contain the groups

$$-CH=CH(CH_2)_2COOH$$

and

$$-CH=CH(CH_2)_2CH=CH-$$

The drying power is approximately proportional to the degree of unsaturation (iodine value, *see below*); hence the presence of glycerides of saturated or mono-unsaturated acids (e.g. stearic and oleic acids, respectively) reduces the effectiveness of drying. The fastest-drying oils are those having a conjugated double bond structure, e.g. oiticica oil, and particularly tung oil (films from which also show the greatest water resistance).

The heat generated by the drying of unsaturated oils can cause spontaneous combustion, e.g. in cotton waste.

Linseed oil, obtained from flax seeds and the commonest drying oil, is employed in paints and in the manufacture of linoleum, and formerly was used for coating textiles, i.e. oil-proofing. *Boiled oil*, made by heating linseed oil in the presence of air and metallic driers, and *blown oil* made by blowing air through hot oil, are faster-drying forms. *Stand oil* is linseed oil thickened or 'bodied' by thermal polymerisation (to mol. wt. 2000) in the absence—or with limited access—of oxygen; it is used in printing inks, varnishes and high-gloss enamels. The composition of *linoxyn*, the 'dried' (i.e. oxidised and polymerised) product from linseed oil, is not fully understood; its formation is accompanied by increases in weight and density, and evolution of several volatile substances (H_2O, H_2O_2, CO_2, aldehydes, lower fatty acids). Ultimately, especially after long storage under damp conditions, superoxidation produces an alcohol-soluble component (responsible for the tackiness of old oilskins or linoleum).

Tall oil (Swedish for *pine*, and pronounced *tall* as in tallow) is a by-product of the alkali-sulphate process for pulping pinewood in the manufacture of *kraft* paper. The crude oil contains fatty acids and rosin acids (§24A.62); distillation yields refined tall oil, a source of

235

fatty acids resembling those of linseed and other drying oils, which is suitable for incorporation in alkyd resins.

The treatments mentioned above for rendering linseed oil faster-drying or increasing its viscosity may also be applied to other drying oils.

To provide products that dry rapidly and exhibit greater resistance to water, drying oils are often modified by copolymerisation with styrene or its derivatives (*see* §4.83i).

Semi-drying oils obtained from corn (maize), cotton-seed, and soya-beans contain largely linoleic and oleic acids, in approximately equal amounts, with an appreciable proportion of saturated acids. Castor, coco-nut, ground-nut, neatsfoot, olive, rape-seed*, and certain fish oils, though partially unsaturated, are not drying oils; however, by heating castor oil with a dehydrating agent (e.g. H_2SO_4), and by fractional separation of the active polyunsaturated components from certain fish and marine mammalian oils, valuable drying products are obtained. Data relating to some of the commoner drying oils are given in Table 24A.T1. For oils reacted with sulphur (factice), *see* §24A.85; for further literature *see* §24A.9.

Table 24A.T1. SOME COMMON DRYING OILS

Oil	Source	Principal polyunsaturated acids*	Iodine value†
Linseed	*Linum usitatissimum*		170–190
Hemp seed	*Cannabis sativa*	Linolenic acid,	150
Perilla	*P. nankingensis* and *ocimoides*	$CH_3CH_2CH=CHCH_2CH=CHCH_2CH=CH(CH_2)_7COOH$,	185–205
Poppy seed	*Papaver somniferum*	and linoleic acid,	135
Rubber seed	*Hevea braziliensis*	$CH_3(CH_2)_4CH=CHCH_2CH=CH(CH_2)_7COOH$	130
Walnut	*Juglans regia*		140–150
Oiticica	*Licania rigida* or *Couepia grandiflora*	Licanic acid, $CH_3(CH_2)_3(CH=CH)_3(CH_2)_2CO(CH_2)_4COOH$	140–150
Tung or Chinawood	*Aleurites cordata, Fordii* and *montana*	Elaeostearic acid, $CH_3(CH_2)_3(CH=CH)_3(CH_2)_7COOH$	145–175
Dehydrated castor	(*Ricinus communis*)	Linoleic acid, as above, and the conjugated isomer, $CH_3(CH_2)_5CH=CHCH=CH(CH_2)_7COOH$	135
Fish and marine mammalian	E.g. *Alosa menhaden*	Up to C_{20-24} acids with 2 to 6 double bonds; e.g. clupanodonic acid, $C_{17}H_{27}CH=CH(CH_2)_2COOH$	160

*Present as triglycerides
†Approx. g iodine/100 g oil

Specific gravity With the exceptions of oiticica oil and tung oil (up to 0·98 and 0·945, respectively), the specific gravities of all the oils listed above lie between 0·925 and 0·939 at 15–20°C. **Refractive index** With the exceptions of oiticica oil and tung oil (1·49–1·51 and 1·51–1·52, respectively), the refractive indices of all the oils listed above lie between 1·47 and 1·48.

*Half of the fatty acids of rape seed oil (derived from *Brassica campestris*) consists of *cis*-13-docosenoic acid or erucic acid, $CH_3(CH_2)_7CH = CH(CH_2)_{11}COOH$.

24A.82 ROSIN ESTER GUMS

Rosin or *colophony* is the involatile component, representing *c.* 68 % of the oleo-resin exuded by various species of pine trees (genus *Pinus*), which is left behind after distillation of the other major component *turpentine*. It is not a high polymer but a mixture of isomeric mono-basic partially-hydrogenated and alkyl-substituted phenanthrene carboxylic acids, i.e. *rosin acids* ($C_{19}H_{29}COOH$) principally abietic acid. Rosin softens from *c.* 80°C upwards, but, being insoluble in water while readily soluble in organic liquids (from which it may be deposited as a brittle surface coating), it finds uses in cheap varnishes, paper-sizing, rosin soaps and soldering fluxes. It is 'upgraded' (sometimes being first hydrogenated or dimerised by treatment with acid), i.e. made less acid and improved in water resistance and hardness, by partial polyesterification with glycerol or pentaerythritol, the products being known as *ester gums*. These have uses in resin-modified polyesters (§24A.21), in adhesives, and to improve adhesion and gloss in cellulose nitrate lacquers.

24A.83 SHELLAC

This is produced by a scale or coccid insect (*Tacchardia lacca, Coccus lacca*, or *Laccifer lacca* Kerr; fam. *Coccidae*) of Southern Asia; the females, which attach themselves to various host trees—notably of the *Ficus* genus—to feed parasitically on the sap, secrete the resinous excrescence as a protective covering.

The crude material (*sticklac*), after washing and sorting (as *seedlac*), is melted and filtered, and is then either allowed to fall in small drops (*button lac*) or stretched into thin sheets (*shellac*). It consists prin-cipally (30–40%) of aleuritic acid [9,10,16-trihydroxypalmitic acid, $CH_2OH(CH_2)_5(CHOH)_2(CH_2)_7COOH$], which is present as esters together with complex isoprenoid acids (hexahydromethanoazulene hydroxy-acids, e.g. shellolic acid), and is obtained on alkaline hydrolysis or on autoclaving with water. Orpiment (As_2S_3, up to 0·5%) and rosin (up to 12%) may be present as adulterants in com-mercial samples.

Refined shellac is soluble in alcohol, alkalis (including aq. solutions of borax, sodium carbonate or ammonia) and pyridine, and in formic, acetic and lactic acids; it is insoluble, or largely so, in hydrocarbons, chlorinated hydrocarbons, most esters, and drying oils. It is softened by several phthalate- and phosphate-type plasticisers, and by castor oil. A natural wax, present to the extent of up to 5% and insoluble in the usual shellac solvents, can be removed by decantation or filtration, or by extraction with aromatic hydrocarbons.

The specific gravity is 1·15–1·20. The resistance to water is poor, films blooming or whitening at high humidity or when immersed, but can be improved by extraction of a soft fraction soluble in ether. Incorporation of cellulose nitrate or non-fibrous polyamides (§23.21.6) also improves the water resistance, as does addition of cashew-nut shell liquid (§28.81). On prolonged storage shellac becomes insoluble in cold alcohol (but dissolves on heating); also, although normally thermoplastic (fluid at $c.$ 80°C), when stoved for some hours above 100°C it splits off water and becomes thermoset.

Shellac forms glossy and adhesive films (adhesion being attributed to the —OH groups) which render it useful as a spirit varnish (particularly in French polish) and as a priming barrier or 'knotting' for wood. It finds uses, too, as a laminating agent for mica and paper, a leather finish, a stiffening medium for felt hats, in printing inks and sealing wax, and in various applications for non-tracking electrical insulation; it was formerly used with slate dust for moulding gramophone records. Mixtures of shellac and ethylene glycol esters of hydrolysed shellac provide strong cements for metals, glass, etc., which after stoving will withstand boiling water.

Shellac reached Europe from India in the tenth century; a red pigment precipitated from its solutions by alum—i.e. the original *lac* or *lake* colour—was formerly an important dye that in the seventeenth century replaced cochineal.

For further literature on shellac, *see* §24A.9.

24A.84 POLYESTER RUBBERS

This term has been applied loosely to several distinct classes of materials, some now better known under other names, e.g. polyurethanes (*see* Section 40) and polyacrylate rubbers (*see* §9.82). Two other rubbery polyesters have been produced to which the term alkyd has been applied, but neither is currently available: (*a*) In 1933, R. H. Kienle described flexible alkyd resins, some of them oil-modified, which could be compounded and milled like rubber. They were tough, oil-resistant materials, remaining flexible at 5°C, but they do not appear to have been developed commercially. (*b*) During the Second World War, Bell Telephone Laboratories developed an unmodified saturated polyester (marketed as Paraplex X-100) that was processable on conventional rubber equipment and could be vulcanised with organic peroxides. It had excellent oil resistance, but proved too expensive for large-scale development.

24A.85 FACTICE

Following the discovery (C. Goodyear, 1839; T. Hancock, 1842) that heating with sulphur caused 'vulcanisation' of rubber, it was soon found that certain vegetable and animal oils given a similar treatment were converted to rubbery products, which—although mechanically inferior to rubber—might be used as partial replacements (extenders) for it; consequently these products became known as 'factice' (from *caoutchouc factice*) or 'rubber substitute' and by numerous trade names.

The principal method of preparation is by heating a drying oil (commonly rape seed oil; but linseed, whale and soya bean oils are also used—*see* §24A.81) with 10–30% sulphur for some hours at 150–170°C, which produces *hot-type* or *brown* (or *dark*) *factice*; but, as was found by A. Parkes (discoverer of 'cold vulcanisation', in 1846), treatment of fatty oils with *c.* 15% of sulphur monochloride at room temperature yields lighter products, which are known as *cold-type* or *white factice*.

Today the use of factice has declined but it is still valuable as a processing aid in compounding natural and synthetic rubbers, also for its power of absorbing—and retaining—large proportions of plasticisers in very soft rubbers (e.g. for inking rollers in printing), also in pencil-mark erasers (since the poor abrasion resistance keeps the 'rubber' clean), also to improve dimensional stability before vulcanisation and/or the surface appearance of finished goods. Treatment of an oil with less than the usual amount of sulphur gives a liquid *sulphurised oil* suitable for application to rubber as a rapid-drying varnish, which is tenaciously adhesive if applied prior to vulcanisation.

24A.9 FURTHER LITERATURE

MARTENS, C. R., *Alkyd Resins*, New York and London (1961)
PATTON, T. C., *Alkyd Resin Technology*, New York and London (1962)

Drying oils
JAMIESON, C. S., *Vegetable Fats and Oils*, New York (1943) (and, re-written by ECKEY, E. W., 1954)
HILDITCH, T. P., *The Chemical Constitution of Natural Fats*, London (1949)
MILLS, M. R., *An Introduction to Drying Oil Technology*, London (1952)
CHATFIELD, H. W. (Ed), *The Science of Surface Coatings*, London (1962) (includes drying oils)
WEXLER, H., *Chem. Rev.*, **64**, 591 (1964) (polymerisation of drying oils)
IUPAC Publication: *Recommended Methods for the Analysis of Drying Oils*, London (1966); also in *Pure and Appl. Chem.*, **10**, 190 (1965)

Shellac
PARRY, E. J., *Shellac*, London (1935)
GIDVANI, B. S., *Shellac and other Natural Resins* (*Plastics Monograph No. S1*), 2nd edn, London (1954)
HICKS, E. *Shellac*, London (1962)
BOSE, P. K., SANKARANARAYANAN, Y. and SEN GUPTA, S. C., *Chemistry of Lac*, Indian Lac Res. Inst. (1963)

Polyester rubbers
KIENLE, R. H. and SCHLINGMAN, P. F., *Ind. Engng Chem.*, **25**, 971 (1933) (flexible alkyd resins)
THE RESINOUS PRODUCTS & CHEMICAL COMPANY, *Paraplex X-100, Paraplex S-200*, Philadelphia (1943, 1944)
MORTON, M. (Ed), *Introduction to Rubber Technology*, New York (1959) (DAUM, G. A., Chapter 11, 'Nitrile and polyacrylate rubbers')
Factice
National College of Rubber Technology.
Symposium: Factice as an Aid to Productivity in the Rubber Industry, London (1962).

POLYESTERS
Sub-section 24B: FIBRE-FORMING POLYESTERS

Principally as polyethylene terephthalate. For polycarbonates, *see* Section 24C.

SYNONYMS AND TRADE NAMES

Polyethylene terephthalate (PET, PETP). Fibres: Terylene, Dacron, Diolen, Lavsan, Tergal, Terlenka, Teron, Trevira. Films: Melinex, Hostaphan, Mylar, Terphane, Videne. *Modified or related products.* A-Tell, Grilene, Kodel, Vestan, Vycron, Arnite (moulding material; *see* §24B.6).

GENERAL CHARACTERISTICS

Linear polymers of this class are represented by strong fibres and monofils, and tough, high-gloss, biaxially-oriented unsupported films. They have high softening temperatures, can be thermally set to shape, are relatively resistant to light and are little affected by moisture.

24B.1 STRUCTURE

Simplest Fundamental Unit

Polyethylene terephthalate (PET) has the following repeat unit,

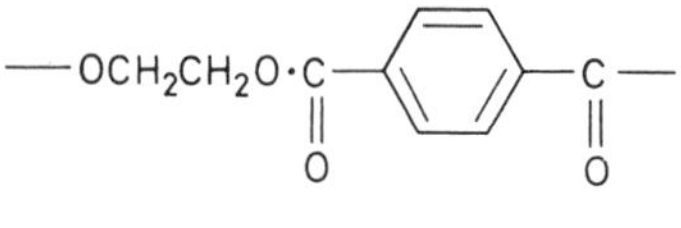

$C_{10}H_8O_4$, mol. wt. = 192

Related materials are either copolymers based on the above structure, or are derived from other aromatic or alicyclic monomers (*see* §24B.21II); so far, no completely open-chain polyesters have been developed commercially.

Molecular weight Commercial polymers: number-average, 15–20000 (degree of polymerisation 80–105); weight-average, 20–30000 (degree of polymerisation 105–155).

X-ray data Substantially amorphous when quenched from a melt, but crystalline after heating above 80°C. In oriented crystalline fibres of PET the triclinic unit cell contains a single repeat unit, identity period 10·8 Å.

24B.2 CHEMISTRY

24B.21 PREPARATION

I. Polyethylene terephthalate (PET)—The raw materials, derived from petroleum, are ethylene glycol (**EG**, b.p. 197°C), which is obtained from ethylene via ethylene oxide, and terephthalic acid (**TA**, m.p. *c.* 300°C, with sublimation), which is obtained by oxidation of *p*-xylene with nitric acid or directly with air. It is preferable, however, to prepare the polymer indirectly and commercial processes use the following stages.

(i) *Esterification* of methanol with **TA** to yield dimethyl terephthalate (**DMT**, m.p. *c.* 140°C when purified—as is readily done—by recrystallisation).

(ii) *Ester interchange* between **DMT** and an excess of **EG** at 150°C, to yield largely bis 2-hydroxyethyl terephthalate (with recovery of methanol by distillation),

$$CH_3OC\text{---}\!\langle\bigcirc\rangle\!\text{---}COCH_3 + 2HO(CH_2)_2OH \rightarrow HO(CH_2)_2OC\text{---}\!\langle\bigcirc\rangle\!\text{---}CO(CH_2)_2OH + 2CH_3OH$$

(iii) *Polycondensation* of the above product, under vacuum at 275°C (with continuous removal—and recovery—of **EG**),

$$HO(CH_2)_2OC\text{---}\!\langle\bigcirc\rangle\!\text{---}CO(CH_2)_2OH \rightarrow H\left[-O(CH_2)_2OC\text{---}\!\langle\bigcirc\rangle\!\text{---}C-\right]_n O(CH_2)_2OH + (n-1)HO(CH_2)_2OH$$

Numerous additives (metals, metal oxides, salts, etc.) have been claimed to accelerate reactions (i) and/or (ii) and minimise undesirable ones.*

When the final melt reaches the required, viscosity, the polymer is rapidly cooled to avoid degradation and discoloration; for extrusion as fibres or films it is subsequently re-melted under nitrogen (*see* Fabrication, §24B.4).

II. Related polyesters—Copolymerisation of **DMT** and **EG** with a small proportion of a third substance results in products with increased dye affinity, though usually with slight lowering of melting point and strength. Examples of suitable substances are isophthalic acid (for Vycron), *p*-hydroxybenzoic acid (for Grilene), and a polyethylene glycol, the last in particular introducing appreciable ether-linkage into the main chain. The repeat unit can be modified

---

*Even so, side reactions are not entirely avoided and the product commonly contains a small proportion (1%) of oligomers of relatively low molecular weight but high melting point (e.g. a cyclic trimer of ethylene terephthalate, m.p. 325°C). These oligomers occasionally cause difficulties industrially, by depositing on yarn guides or appearing as specks in dyed fabric.

by replacing terephthalic acid with 1,2-diphenoxyethane-4,4'-di-carboxylic acid $[(CH_2OC_6H_4COOH)_2]$; similarly, it is claimed that condensation of *p*-hydroxybenzoic acid with ethylene oxide provides silk-like fibres (A-Tell). Homopolymers different from but largely resembling PET may be obtained from *p*-hydroxybenzoic acid and from *p*-(2-hydroxyethyl)benzoic acid, while ester interchange between a dialkyl terephthalate and hexahydro-*p*-xylylene glycol (1,4-bis-hydroxymethylcyclohexane or 'cyclohexanedimethanol') yields a fibre-forming polymer with two cyclic components in the repeat unit (Kodel, Vestan)*,

A fibre-forming polyester has been prepared experimentally from lignin, a by-product in the purification of wood cellulose. Proto-catechuic acid (derived from lignin via vanillin) was treated with epichlorohydrin and alkali to yield a mixture of isomeric hydroxy-acids, and these on condensation gave polymers with repeat units of the following kind,

24B.22 PROPERTIES

Dissolved by phenols (*m*-cresol, *o*-chlorophenol) and phenol/chlorinated hydrocarbon mixtures (e.g. 1/3 (vol./vol.) phenol/tetrachloroethane), acetonylacetone, dichloroacetic acid, conc. sulphuric acid (with decomposition), hot benzyl alcohol, hot benzyl acetate. **Relatively unaffected by** hydrocarbons and most common organic liquids, including esters and dry-cleaning solvents†, and by formic, acetic, phosphoric and hydrofluoric acids. Also resistant to bleaching solutions, reducing agents, and mild alkalis (e.g. Na_2CO_3), and to moderate exposure to mineral acids, e.g. polyesters can be determined in fibre mixtures by their resistance to 75% w/w H_2SO_4 at 50°C (cotton and viscose rayon dissolve; *see* BS 3557) but *see also below*. **Decomposed by** hot alcoholic alkalis (e.g. dil. alc. NaOH), and progressive surface attack by hot aq. alkalis (e.g. dil. NaOH, particularly in the presence of quaternary ammonium compounds); also

---

*These fibres possess somewhat poorer mechanical properties than PET. It is claimed that fabrics containing them as staple fibres show less 'pilling' (production of small balls on fabric surface) than corresponding fabrics containing PET.

†Fibres and films that have not been hot-set §26B.32) may shrink in hot dry-cleaning solvents.

slowly weakened by ammonium hydroxide and decomposed by hot amines, e.g. polyesters can be determined in fibre mixtures by dissolution in 2% hydrazine in butanol at 100°C (wool remains inert). Mineral acids (*see above*) cause loss of strength, dependent on concentration and on time and temperature of exposure, e.g. tensile strength of coarse monofilament falls to 50% after 50 h at 100°C in: 7·5% HCl, 16% HNO_3, or 45% H_2SO_4. For recovery purposes scrap polyethylene terephthalate can be decomposed (trans-esterified) by boiling in a glycol or by heating under pressure with methanol (e.g. at 180°C) or water (220°C). Amorphous polymer is much less resistant to chemical attack than the usual oriented crystalline fibres or films.

24B.23 IDENTIFICATION

Thermoplastic. Cl, N, S, P absent. Saponifiable. **Combustion** Burns, with melting, with a sooty flame. **Pyrolysis** Shrinks, melts (*c.* 256°C) and chars, yielding a weakly acid distillate; principal pyrolytic product is acetaldehyde (also CO, CO_2, C_2H_4, C_6H_6, etc.). If the hot pyrolytic vapour is sampled with a test paper soaked in a freshly-prepared saturated solution of *o*-nitrobenzaldehyde in dil. sodium hydroxide, a blue-green colour results, of which the blue component (indigo) is fast to dil. hydrochloric acid (*see* same test but different result with nylon 6.6, §23.23). **Other tests** For detection of *ethylene glycol* in the saponification residues, *see* §24A.23. For detection of *terephthalic acid* the precipitate obtained on acidifying the saponification liquor is washed, dried, and then refluxed with methanol and a little conc. sulphuric acid to yield dimethyl terephthalate (m.p. 140°C, recrystallised from aq. alcohol). For reaction of polyethylene terephthalate with ethanolamine and production of a crystalline derivative (m.p. 234·5°C), *see* HASLAM, J. and SQUIRRELL, D. C M., *Analyst,* **82**, 511 (1957).

24B.3 PHYSICS

24B.31 GENERAL PROPERTIES

Specific gravity 1·38 (amorphous, 1·335; crystalline, 1·455). **Refractive index** Stretched films, 1·65 (amorphous, 1·57). Fibres: $n_{||}$, 1·70–1·72; $n_\perp$, *c.* 1·54. **Birefringence** Up to *c.* 0·18. Light transmission, 90% of visible; ultra-violet cut-off below 3150 Å. **Water absorption** (films) Up to 0·65% on prolonged immersion. **Water retention** (fibres) 4% (up to 12%, dependent on finish present). **Moisture regain** (staple fibres, 25°C)

R.h. %	35	65	95
Moisture regain %	0·25	0·45	0·70

Permeability Films have relatively low permeability to most gases and vapours. *Examples* (i) To gases, at 20–30°C (10^{-10} cm^2)/(s cmHg): O_2, 0·03–0·05; N_2, 0·005–0·006; CO_2, 0·09–0·15. (ii) To water vapour (10^{-10} g)/(cm s cmHg) at 20–30°C, 0·1–0·18. (iii) Transmittance of water vapour through 0·025 mm films (10^{-10} g)/(cm^2 s cmHg) at 38°C, 90% r.h.: 50.

24B.32 THERMAL PROPERTIES

Specific heat 0·3, rising to 0·4–0·45 at 150–200°C. **Conductivity** c. 3·4 (10^{-4} cal)/(cm s °C) **Coefficient of linear expansion** Average value (from volume expansion) (10^{-4}/°C) 0·5 at 25°C, 1·2 above 100°C. **Physical effects of temperature** Films and fibres are flexible to below $-70°C$, with increase in strength but decrease in extensibility. Second order transitions at $-40°C$ and particularly (change in expansion rate) at 70–80°C. Untreated films shrink when heated to c. 100°C (can be used for tight packaging). Heat-set films are serviceable up to 150°C, and yarns retain about half their normal strength at 180°C (but at high temperatures in air suffer oxidative degradation). Softens (sticky; can be heat-sealed), 230–240°C; m.p. 256–260°C (pure polymer, 265°C; can be raised even higher by annealing). Fabrics are stabilised dimensionally by heating in air (15–30 s, up to 220°C) or in boiling water, and can be permanently set to shape, e.g. pleated, by steaming and pressing at 130–150°C. Staple fibres are usually supplied in the thermally-stabilised form.

24B.33 ELECTRICAL PROPERTIES

Volume resistivity (ohm cm) Stretched films 10^{19} at 25°C; 6.10^{15} at 100°C; 3.10^{12} at 180°C. *Surface resistivity*, c. 5.10^{11} ohm at 100% r.h. **Dielectric strength** Films, 100–160 kV/mm. at 25°C; non-tracking. **Dielectric constant** Films, c. 3·0–3·2 (50–10^6 Hz, -25 to 80°C thereafter rising); 3·8 (10^3 Hz, 140–160°C). **Power factor** 0·003, 0·005 and 0·014 (at 25°C, and 50, 10^3 and 10^6 Hz respectively); 0·011, 0·002, 0·015 and 0·045 (at 10^3 Hz, and -40, 50, 120 and 190°C respectively).

24B.34 MECHANICAL PROPERTIES

Elastic recovery (%) Staple fibres (extended at 25% per min, held at maximum extension 1 min, then reversed to zero stress):

% Strain	1	2	5	10
Immediate recovery	68	62	39	26
Measured 1 min later	83	80	63	41

The recovery of filaments is superior to that of staple fibres, e.g. 90–100% from 3% strain. **Elastic modulus, Tensile strength and Extension at break** Approximate values:

	Staple fibres	Continuous filaments		Films	Moulded samples
		medium	*high tenacity*		
Elastic modulus (Young's) kgf/mm² and (gf/den.)	400–750 (30–60)	1250–1450 (100–115)	1400–1600 (110–130)	350	—
Tensile strength, kgf/mm² and (gf/den.)	50–70 (4–5·5)	50–65 (4–5)	75–90 (6–7)	16–18*	5·5†
Extension at break %	20–50	15–30	6–15	*c.* 70	300

*Yields at *c.* 9 kgf/mm², at 3–4% strain
†Yields at 7·3 kgf/mm²

Wet strength: practically as air-dry strength.

Friction Approximate values (cont. fil., 1 m/s), μ_{dyn}: commercial yarn/mild steel, 0·3–1·2; do. (lubricant free)/do., 0·3–0·8; do./porcelain or hard glass, *c.* 0·5. Friction against steel is lowered by matt surfaces and by increase in yarn tension. Moulded material/mild steel, 0·2. Resistance to abrasion is good though less than that of polyamides.

24B.4 FABRICATION

Melt spun fibres are drawn to four or five times their length while above 80°C, whereby they acquire orientation, increased crystallinity, and high strength; they may then be heat-set dry or in steam. Melt-extruded films are rapidly cooled to an amorphous state, then re-heated above 80°C and plane-oriented by biaxial expansion, and finally heat-set at 200°C to increase its crystallinity and stabilise it dimensionally. Adhesion to smooth polyesters is in general limited, and to obtain strong bonds it is necessary to employ special precautions; e.g. an organic isocyanate is included in a latex cement, and before bonding tyre cords to rubber they are treated with a mixture of a resorcinol-formaldehyde resin plus a vinylpyridine-containing rubber latex (§36.63). Adhesion of printing ink is improved after exposure of a polyester surface to ozone.

24B.5 SERVICEABILITY

Fibre and film-forming polyesters are little affected by moisture and/or moderate heat (except that prolonged exposure to moisture at an elevated temperature leads to hydrolysis), and they withstand many common organic liquids and reagents (including many acids, though they are less stable to alkalis and amines). They also withstand exterior weathering relatively well (showing less loss of strength than polyamides), and when exposed behind glass the resistance to photo-tendering (loss in strength by action of light) is particularly good. They are unaffected by bacteria, fungi, or insects. High-energy radiation causes some cross-linking and degradation.

24B.6 UTILISATION

Light- and moderate-weight quick-drying fabrics, including stretch-type and pleat-retaining fabrics (e.g. blends with wool, etc.). Industrial fabrics (e.g. water-resistant filter cloths, bolting (sieve) cloths, paper-making felts and 'wires', etc., fatigue-resistant reinforcing material in tyres, vee-belts, conveyer belts, etc.) and marine uses (fishing nets, tarpaulins, sailcloths). Films find numerous uses, e.g. as transparent packaging (including shrink-packs and 'boil-in-the-bag' food packs) and as electrical insulation and a capacitor dielectric, also for diaphragms, magnetic recording tape, cine film base, typewriter ribbon, hose-lining and light-weight conveyer belts; when vacuum-metallised, it is used for mirrors and decorative yarn (Lurex). Moulding material based on polyethylene terephthalate has been introduced, e.g. *see Brit. Plas.* 644 (1966).

24B.7 HISTORY

For early polyesters, *see* §24A.7. The first practicable fibre-forming heat- and hydrolysis-resistant polyester was prepared in 1939 by J. R. Whinfield and J. T. Dickson, of the Calico Printers' Association Ltd. This was developed on a commercial basis by Imperial Chemical Industries Ltd. in 1947, since when production of fibres and films (as Terylene, Melinex, etc.) has spread throughout the world.

24B.9 FURTHER LITERATURE

PETUKHOV, B. V., *The Technology of Polyester Fibres,* Moscow (1960); transl. and ed. by MULLINS, M. F. and B. P., Oxford (1963)

LUDEWIG, H., *Polyesterfasern—Chemie und Technologie,* Berlin (1965)

GOODMAN, I. and RHYS, J. A., *Polyesters (Vol. 1 Saturated Polyesters),* London (1965)

KORSHAK, V. V. and VINOGRADOVA, S. V., *Polyesters,* Moscow; translated and ed. by HAZZARD, B. J. and BURDON, J., Oxford (1965)

GAYLORD, N. G. (Ed), *Polyesters* (High Polymers, Vol. XIV), New York and London (1971) (Part I), (1971) (Part II)

POLYESTERS
Sub-section 24C: POLYCARBONATES*
With notes on cross-linked polycarbonates and polyphosphonates, and on polyanhydrides

SYNONYMS AND TRADE NAMES

PC; Lexan, Lexel (fibre), Makrolon, Merlon, Panlite.

GENERAL CHARACTERISTICS

Transparent, faintly amber-coloured, thermoplastic materials showing good dimensional stability, thermal resistance, and electrical properties, also good tensile and impact strength; suitable for mouldings, fibres, and films.

24C.1 STRUCTURE

Simplest Fundamental Unit

The commonest polycarbonate, based on diphenylolpropane, is represented by

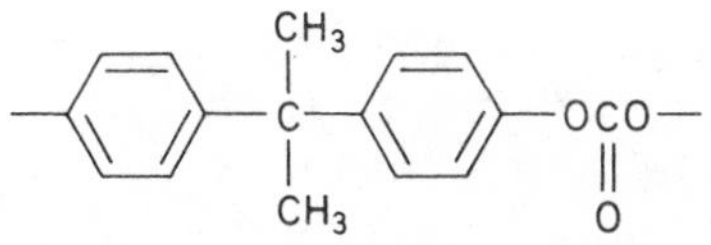

$$C_{16}H_{14}O_3, \text{ mol. wt.} = 254$$

For an example of a cross-linked polycarbonate, *see* §24C.81.

Molecular weight Commercial plastics, up to 30000 (degree of polymerisation *c.* 120) beyond which increasing viscosity limits practical processing; solvent-casting resins up to 90000; values up to 500000 experimentally.

X-ray data The commonest polycarbonate does not crystallise readily even when drawn or annealed, and crystallites tend to be very small. Unit cell contains 4 chains and 8 fundamental units; identity period 21·5 Å (PRIETZSCH, A., *Koll. Zeit.*, 156, 9 (1958)).

24C.2 CHEMISTRY

24C.21 PREPARATION

Polycarbonates are prepared as below from diphenylolalkanes, of which the commonest is 2,2-diphenylolpropane or Bisphenol A (BPA), obtained as described in §25.21.

---

*For polyvinylene carbonate, *see* §26.81(2b).

1. Ester interchange in a melt of BPA and an organic carbonate (e.g. diphenyl carbonate $(C_6H_5)_2CO_3$; m.p. *c.* 80°C), initially under reduced pressure at 200°C and ultimately with the vacuum increased and the temperature raised to 300°C to facilitate removal of phenol,

2. An extension of the Schotten-Baumann reaction, namely room-temperature treatment of BPA (dissolved in aq. alkali, plus a quaternary ammonium compound as catalyst) with phosgene, in the presence —as a dispersed phase—of an organic solvent for the polymer,

It is also possible to effect this reaction in homogeneous solution using, for example, pyridine as both base and solvent.

The alkane derivative may be replaced by diphenylol ethers or sulphides, sulphoxides, etc.; alternative methods of preparation include reaction between a diol and a bischloroformate, which may be effected directly by interfacial polymerisation (*see* §23.21.5).

24C.22 PROPERTIES (polymer from BPA)

Dissolved by certain chlorinated hydrocarbons (chloroform, methylene chloride, di-, tri- and tetrachloroethane, hot chlorobenzene), pyridine, dioxan, cyclohexanone, and hot phenols. Note: on storage conc. solutions prepared from amorphous polymer may gel or precipitate a high-melting crystalline phase. **Swollen by** acetone, benzene, carbon tetrachloride. Note: aromatic hydrocarbons, ketones, and esters may cause crazing and stress cracking. **Relatively unaffected by** aliphatic hydrocarbons, trichloroethylene, alcohols (surface crazing on long immersion in methanol), dil. acids, oxidising agents (including hypochlorite). **Decomposed by** hot alcoholic alkalis; amines and other organic bases; surface attack by aq. alkalis.

24C.23 IDENTIFICATION

Thermoplastic. Cl, N, S, P normally absent. **Combustion** The polymer from BPA burns with a bright and smoky flame, which is not readily self-supporting. **Pyrolysis** Melts, chars, evolves neutral vapours (largely CO_2). **Other tests** Aromatic polycarbonates have an unusually high C-content; e.g. polymer from BPA, over 75%. Alkaline fusion or hydrolysis yields a phenol, *see* qualitative tests under §28.23. The parent phenol may be obtained by refluxing with ethanolamine, *see* HASLAM, J. and WILLIS, H. A., *Identification and Analysis of Plastics*, London, 92 (1965). The infra-red absorption is characteristic; spectra of 13 polycarbonates are given by H. Schnell (*see* §24C.9) pp. 168–174.

24C.3 PHYSICS

24C.31 GENERAL PROPERTIES (polymer from BPA)

Specific gravity 1·20 (calculated crystalline density, 1·30). **Refractive index** 1·585. Light transmission, 3 mm sheet, 85%. **Water absorption** (%) 50% r.h., 0·15; in water, 0·35; in boiling water, 0·6. **Permeability** Cast films are moderately permeable to gases and vapours (approx. 10- to 100-fold more permeable than polyethylene terephthalate films). *Examples* (i) To gases $(10^{-10}$ cm$^2)$/(s cmHg): N$_2$, 0·23–0·3; O$_2$, 2·0–3·9; CO$_2$, 8·5–11·2. (ii) To water vapour $(10^{-10}$ g)/(cm^2 s cmHg): films 0·1 mm thick, 55–80.

24C.32 THERMAL PROPERTIES (polymer from BPA)

Specific heat 0·28–0·30. **Conductivity** 4·6 $(10^{-4}$ cal)/(cm s °C). **Coefficient of linear expansion** $(10^{-4}$/°C) 0·6–0·7 (0·76 at 60–140°C). **Physical effects of temperature** Brittle only below −135°C. Second-order (glass) transition temperature, *c.* 145–150°C. Softens, 160°C upwards; cold-drawn film shrinks. Melts, 215–230°C (crystalline m.p. 268°C). Decomposes, 310–340°C. Serviceable up to 135°C (for creep at 100°C, *see under* Elastic modulus below), but tensile strength halved at *c.* 120°C; moisture accelerates degradation when hot, resins become brown by oxidation when aged at 125°C.

24C.33 ELECTRICAL PROPERTIES (polymer from BPA)

Volume resistivity (ohm cm) *c.* 10^{16} (10^{15}, at 100°C or 80% r.h.; 10^{14} at 125°C). *Surface resistivity* 10^{12}–10^{13} ohm. **Dielectric strength** (kV/mm) Very thin films, 120; 0·05–0·125 mm films, 100; 3 mm sheets, 16. **Dielectric constant** 2·7–3·1 over the range 50–10^{10} Hz (and up to 125°C at 50 Hz). **Power factor** 0·0005–0·001 (50 Hz), 0·01 (10^6 Hz), 0·005 (10^{10} Hz). Note: electrical properties show little dependence on frequency, and are not greatly changed by heating to 140°C or by long immersion in water.

24C.34 MECHANICAL PROPERTIES (polymer from BPA)

Elastic modulus (Young's) (kgf/mm^2) *c.* 250; with incorporation of glass fibres, *c.* 900. Visco-elastic behaviour of polycarbonate resin strained at 100°C, 50% r.h.:

Stress (kgf/mm^2)	Strain (%)		Recovery (%) after 1000 h
	Initial	*after* 1000 h	
0·4	0·03	0·3	39
0·8	0·3	1·0	32
1·6	0·7	2·0	36
2·0	1·2	4·7	21

Tensile strength (kgf/mm^2) Plastics, up to 7, yielding at *c.* 6 (lower values under dynamic conditions); with incorporation of glass fibres, 14; increased to 11 by drawing, or to 17 by drawing and crystallisation. Commercial fibres, up to 40 (3·7 gf/den.). **Extension at break** (%) 60–100 (yield at 5% strain); commercial fibres, *c.* 36. **Flexural strength** 6–12 kgf/mm^2. **Compression strength** *c.* 8 kgf/mm^2 (lower under dynamic conditions). **Impact strength** (Izod test, ftlbf/in width) Notched, 0·25 in thick, 2–3; 0·125 in thick, up to 16 (1·6 at −55°C). Unnotched, over 50. **Hardness** Rockwell M70, R115–118. Polycarbonate resins are scratched by polymethyl methacrylate resin. **Friction and abrasion** Friction, normally moderate, can rise at high loads or speeds, to cause galling, i.e. surface damage. μ_{dyn} (rod/rotating cylinder, system unlubricated): polycarbonate/polycarbonate, 0·2 at low speed to 1·9 at high speed; PC/steel, 0·7–0·8; steel/PC, 0·35–0·45. Wear resistance, normally good, can rise (as with friction) at high loads or speeds.

24C.4 FABRICATION

Extrusion, 250–300°C. Injection, 275–330°C and up to 20000 lbf/in^2. Moulding pellets need pre-heating to avoid degradation by moisture at high temperatures. Films can be extruded or solution-cast. Fibres can be wet- or dry-spun, or melt-extruded at 260–280°C, then stretched five- to ten-fold, and are being extensively developed. The material can be machined satisfactorily, and to some extent can be cold-drawn; it is vacuum-formable (160–220°C), heat-sealable (220–245°C) and weldable (260°C). Cementable with epoxy resins and polyamide hot-melt adhesives, and with solutions of the polymer (in methylene chloride or ethylene dichloride).

24C.5 SERVICEABILITY

Mouldings have good dimensional stability and are both strong and tough, though strength is lowered under dynamic conditions (e.g. to maximum serviceable tensile or compressive load of *c.* 1·5 kgf/mm^2); good electrical properties are maintained at elevated temperatures. The faint amber colour increases in intensity on exposure to strong light and for outdoor applications necessitates the use of carbon black as filler, or lacquering with an ultra-violet radiation absorber. Polycarbonates, relative to other thermoplastics, show good resistance to ionising radiation; some discoloration and embrittlement may occur, but tensile and electrical properties are maintained. A high melt viscosity requires high processing temperatures, at which moisture must be excluded to avoid degradation.

24C.6 UTILISATION

Dimensionally stable transparent mouldings, particularly where impact resistance is needed and for electrical applications. The material is suitable for gears and bearings, subject to exclusion of heavy loads. Films are suitable for capacitor dielectrics serviceable up to 130°C, and as photographic film-bases.

24C.7 HISTORY

Diphenylolalkanes were obtained by A. Baeyer, in 1872, by reacting phenol with aldehydes and ketones; in 1891 A. Dianin, by condensing phenol with acetone prepared the product now known as Bisphenol A. A. Einhorn, in 1898, investigated intractable polymeric carbonates made from dihydric phenols (e.g. resorcinol) and phosgene. Thereafter, although not entirely set aside, these materials received little attention until 1953–56, when the high-melting polycarbonates derived from dihydroxydiphenylalkanes were developed by Farbenfabriken Bayer A.G. (*see* H. Schnell, §24C.9), the General Electric Company of the U.S.A. (*see* D. W. Fox), and others.

24C.8 ADDITIONAL NOTES

24C.81 CROSS-LINKED POLYCARBONATES AND POLYPHOSPHONATES

An example of a polycarbonate of this kind—a special instance of a curable polyester (Section 26D)—is obtained by heating diethylene glycol bisallyl carbonate* with a free-radical initiator, e.g. benzoyl

---

*Also termed allyl diglycol carbonate.

peroxide. The end-product (Columbia Resin 39, e.g. as cast sheets) resembles polymethyl methacrylate (Perspex) in clarity but is insoluble and harder (Rockwell M95–M100), the repeat structure being represented as

$$\underset{\overset{|}{O}}{\overset{|}{\underset{|}{CH_2}}} \qquad CHCH_2OCO(CH_2CH_2O)_2COCH_2CH \qquad \underset{|}{\overset{|}{CH_2}}$$

Polyphosphonates
Related to the above resins, hard and flame-resistant end-products have been obtained from diallyl esters of arylphosphonic acids, the repeat structure being represented (for example) as

$$\underset{|}{\overset{|}{CH_2}} \qquad C_6H_5 \qquad \underset{|}{\overset{|}{CH_2}}$$
$$CHCH_2OPOCH_2CH$$

By copolymerisation, the refractive index (*c*. 1·57) can be adjusted to match that of glass; glass-fibre laminates bonded with such resins are thus virtually transparent and also fireproof.

24C.82 POLYANHYDRIDES

Film- and fibre-forming polymers have been obtained by self-condensation of certain aromatic dicarboxylic acids, e.g.

$$HOOC-\!\!\!\bigcirc\!\!\!-R-\!\!\!\bigcirc\!\!\!-COOH \longrightarrow -\!\!\!\bigcirc\!\!\!-R-\!\!\!\bigcirc\!\!\!-COC-$$

where R is $-(CH_2)-$, $-O(CH_2)_nO-$, $-O(CH_2O)_n-$, etc. These materials, crystalline and possessing good thermal and hydrolytic stability (even to alkalis), have not as yet reached commercial importance.

An unusual polyanhydride, black and hydrogen-free, presumed to be represented by the repeat unit,

$$-\underset{\underset{O}{\overset{\diagdown}{CO}}}{C}=\underset{\overset{\diagup}{CO}}{C}-$$

is obtained when attempts are made to prepare the hypothetical carbon oxide, C_4O_3 (*see* JONES, J. I., *Chemical Communications*, 938 (1967)).

24C.9 FURTHER LITERATURE

CHRISTOPHER, W. F. and FOX, D. W., *Polycarbonates*, New York and London (1962)
SCHNELL, H., *Chemistry and Physics of Polycarbonates,* London (1964)
KORSHAK, V. V. and VINOGRADOVA, S. Y., *Polyesters* (*see* §26B.9, Chapter IX (Polyarylates)

POLYESTERS
Sub-section 24D: UNSATURATED POLYESTERS
(principally as glass-reinforced materials)
With a note on moulding compounds*

SYNONYMS AND TRADE NAMES

Contact moulding resins, low-pressure laminating resins, GRP (glass-reinforced polyester). Artrite, Crystic, Filabond, Hetron, Marco, Palatal, Paraplex, Polylite, Synres, Vibrin.

GENERAL CHARACTERISTICS

The uncured intermediates are colourless or pigmented liquids varying in viscosity from 150 to over 4000 cP; usually they are supplied as solutions in styrene, which acts both as reactive monomer and as a vehicle to assist fabrication (and confers its characteristic odour). When cured they set to hard or rubbery solids; they are most commonly employed with glass-fibre reinforcement and fabricated by contact (pressure-less) moulding techniques, but are also used in pressure moulding and as casting resins incorporating mineral fillers. Glass-reinforced polyesters possess excellent mechanical characteristics, chemical resistance, and electrical properties and are particularly suitable for large self-supporting structures produced in small numbers.

24D.1 STRUCTURE

Simplest Fundamental Unit

Although polyesters of this class may be prepared from a variety of starting materials, in the uncured state they have an essentially linear structure with unsaturation in the main chain. The repeat unit is represented by:

$$-OROCR'C-$$
$$\underset{O}{\|} \quad \underset{O}{\|}$$

where R, and/or R' (more commonly R') contain the —CH=CH— group, which on curing by copolymerisation with a vinyl monomer yields a spatial network. A simple example is glycol and maleic acid, provided by copolymerisation of styrene and the intermediate obtained by condensation of ethylene,

$$O(CH_2)_2OCCH-CHC\left[O(CH_2)_2OCCH=CHC\right]_n-$$

$$\left[CH_2CH\right]_n-$$
$$\underset{C_6H_5}{|}$$

*Sometimes referred to as *alkyds* or *allytics*. For alkyd resins of the surface-coating kind, *see* Section 24A; for drying oils *see* §24A.81.

One-component systems based on polyfunctional monomers of the allylic type, are also employed. A typical example is diallyl phthalate, which is both chain- and network-forming (*see also* §24D.8):

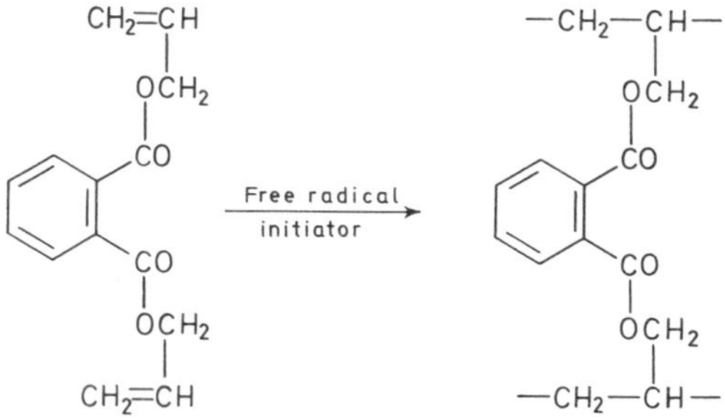

CH₂=CH   —CH₂—CH—
 | |
 OCH₂   OCH₂
 | |
 CO   Free radical   CO
[benzene ring] ——————————→   [benzene ring]
 | initiator |
 CO   CO
 | |
 OCH₂   OCH₂
 | |
 CH₂=CH   —CH₂—CH—

Molecular weight Uncured, 7000–40000.

X-ray data Generally amorphous; some crystalline forms encountered.

24D.2 CHEMISTRY

24D.21 PREPARATION

Two intermediate stages are involved: preparation of an unsaturated polyester, and blending it with a solvent/reactive monomer. Polyesterification takes place by condensation of a dihydric alcohol and one or more dibasic acids, with the elimination of water. Although unsaturation may be introduced in either the alcohol or the acid, it is usually supplied by the acid component owing to the scarcity of unsaturated dihydric alcohols. A saturated acid is often incorporated to render the polyester more soluble in the reactive monomer and to obtain the required degree of flexibility in the cured resin; phthalic acid (generally as anhydride) and its isomers are commonly used in polyesters curing to rigid end-products, while straight-chain acids (e.g. succinic, adipic, azelaic, sebacic) are used for flexible or tough end-products. Hexachloroendomethylenetetrahydrophthalic acid (chlorendic or HET acid; *see* HET anhydride §25.21 vi) is important in making fire-resistant polyesters. Of the unsaturated acids maleic acid (anhydride), and fumaric acid are the most often employed, while more expensive acids of higher molecular weight (e.g. itaconic and mesaconic acids) may be used to confer flexibility.

The alcohol is almost always a glycol (e.g. propylene, butylene, diethylene glycols); ethylene glycol tends to give crystalline end-products. Polyhydric alcohols are sometimes introduced to confer greater strength and chemical resistance; however, they introduce branching that results in increased viscosity and reduced compatibility with the solvent/reactive monomer.

Commercial practice uses a batch process. The alcohol is heated to

80–100°C and the solid ingredients (e.g. maleic anhydride) are added. A solvent (xylene) reflux system helps to remove water evolved during the condensation reaction and to control the temperature which is gradually increased to 200°C. 'Cooking' proceeds for 16 h or more under a blanket of an inert gas to prevent cross-linking of the resin. The reaction is stopped on reaching the desired acid number (usually less than 50), and the unreacted components are stripped by blowing or by applying vacuum. The resin is then cooled to 50–60°C and stabilised by addition of a polymerisation inhibitor (e.g. hydroquinone) before blending with a reactive monomer. The product is generally supplied as a syrup containing 20–50% of monomer; conversion to cross-linked structures is dealt with in §24D.4 (Fabrication).

24D.22 PROPERTIES

Solvents Uncured resins are soluble in ketones; also in styrene and in acrylic, vinyl and allyl esters, as commonly used in solvent/reactive monomer systems. Cured resins are insoluble. **Relatively unaffected by*** aliphatic hydrocarbons (petrol, mineral oils), non-polar chlorinated hydrocarbons (carbon tetrachloride, tetrachloroethylene), alcohol, non-oxidising acids (low and medium concentrations), salt solutions, organic acids. **Swollen or decomposed by*** polar chlorinated hydrocarbons (chloroform, trichloroethylene), ketones, phenol, aniline, esters (ethyl acetate); alkalis (cause hydrolysis), oxidising acids.

EFFECTS OF IMMERSION ON GLASS FIBRE REINFORCED POLYESTERS
SPECIALLY FORMULATED FOR CHEMICAL RESISTANCE (AT 25°C)

Reagent	Weight change, % after 1 month	1 year	Flexural strength retained (%) after 1 year
10% Sulphuric acid	+0·27	+0·80	60
10% Nitric acid	+0·38	−3·40	50
10% Hydrochloric acid	−4·2	−5·1	40
10% Sodium hydroxide	−0·2	−0·8	30
Acetone	+2·3	+5·0	55
Benzene	+3·4	—	60

There is a rapid deterioration in mechanical properties in the initial months of immersion until a steady state is reached; this trend

*These comments relate principally to cured general purpose polyesters (i.e. of the glycol, phthalic-maleic acid type) at temperatures up to *c.* 50°C. Chemical resistance depends considerably upon the composition of the polyester, and generally is not as good as that of epoxy resins. Better results have been obtained with polyesters based on isophthalic or fumaric acids; the alcohol component, too, has an important effect, Bisphenol A and 1,3-cyclohexanediol conferring improved resistance to acids and alkalis respectively. Of the reactive monomers N-tert.butylacrylamide improves acid resistance, and triallyl cyanurate is generally beneficial. Glass fibres used as reinforcement differ considerably in chemical resistance, A-type glass being recommended for acid-resistance, and E-type for alkali-resistance (see §42.21).

is reflected in the weight change values at 1 month and 1 year. The original strength properties, however, are generally so high (*see* §24.34) that even after deterioration a level of mechanical strength adequate for many purposes is retained.

24D.23 IDENTIFICATION

Non-thermoplastic. S and P nominally absent. **Combustion** Cured resins burn with a candle-like or sooty flame and evolve odours characteristic of the monomers. Some decrepitation may be noticed. **Pyrolysis** White, acid fumes are usually evolved; there is little charring. A strong fruity, biting odour may indicate allyl products. **Other tests** There are no simple tests for this class of polymers which embraces materials of complex and widely differing composition. E. A. Parker* has described alcoholic and aq. saponification procedures for the separation and identification of glycols and the more common acid components, including phthalic, adipic, sebacic, succinic, and fumaric acids. In most general purpose resins styrene (§4.23) may be detected, while nitrogen and chlorine can be detected in some heat-resistant types, e.g. those incorporating triallyl cyanurate, chlorostyrene, or HET acid.

24D.3 PHYSICS

24D.31 GENERAL PROPERTIES

Specific gravity Uncured resins, 1·1–1·15; cured unfilled resins, 1·25; glass-reinforced resins, 1·6–2·0. **Refractive index** Uncured resins, with 20% styrene, 1·564; do. with 20% methyl acrylate, 1·544. **Water absorption** Like chemical resistance this property is influenced by the composition of the polyester. Flexible resins are less water-resistant than rigid types. The acid component, especially the saturated acid, has an important effect, good resistance being associated with aromatic acids (e.g. phthalic) and long chain aliphatic acids (e.g. sebacic). Of the unsaturated acids maleic anhydride imparts better resistance than fumaric or itaconic acids.

In glass fabric reinforced polyesters, particularly those with A-glass fibres, immersion in water causes a rapid decline in strength to a steady minimum level. E-glass fibres are less affected, but a reactive finish, e.g. vinyltrichlorosilane (§41.83), is required on both types to enhance wet strength.†

*KLINE, G. M. (Ed), *Analytical Chemistry of Polymers I*, Chapter 11 (*High Polymers*, Vol. XII), New York (1959).
† The role of these finishes in promoting adhesion is uncertain as is the nature of the resin-glass bond itself; on drying out, the initial level of strength properties is nearly regained.

Flexural strength retained after immersion of laminates in water for 3 months at 25°C, specimens tested wt: A-glass, 40%; E-glass, 70%. Weight increase after 1 year in water at 25°C, 1·1%. **Permeability** Transmittance of water vapour through 1 mm cast films $(10^{-10}$ g)/ $(cm^2$ s cmHg) at 25°C: flexible or rigid unfilled resins, *c.* 4·2; rigid resins, 20% quartz filler, *c.* 3·6. Values rise by 60–120% as temperature is increased to 50°C. The permeability of glass-reinforced laminates appears to be influenced by the type of finish on the glass. The strength of laminates begins to deteriorate at about 50% r.h.

24D.32 THERMAL PROPERTIES

Specific heat *c.* 0·5. **Conductivity** $(10^{-4}$ cal)/(cm s °C) Cast resins, unfilled, 5; glass reinforced laminates, 6–10. **Coefficient of linear expansion** $(10^{-4}/°C)$ Cast resins, unfilled, 0·5–1·0; glass-reinforced laminates, 0·15–0·3. **Physical effects of temperature** At low temperatures glass-reinforced polyesters exhibit increases in certain properties due to stiffening. Tensile, compression, and flexural strengths are reported to rise at *c.* −180°C, to values 50–100% higher than at 25°C.

Cured polyesters do not melt, and a rough guide to their heat resistance is given by the heat distortion temperature, which varies from 80 to 130°C for unfilled rigid resins, and from 100 to 250°C for laminates of 50% glass content. In general purpose resins isophthalic acid provides the most resistant products, and still better results are obtained with silicone-modified polyesters or resins cured with triallyl cyanurate. On exposure to high temperatures some polyesters at first show an increase in mechanical properties because of an improvement in the degree of cross-linking, but after a few hours' exposure oxidative degradation leads to softening and loss of strength. Further data on heat- (and flame-) resistance appear in Table 24D.T1 and its footnotes.

Table 24D.T1. FLEXURAL STRENGTH RETAINED (%) AFTER AGEING TYPICAL LAMINATES, OF 60–70% GLASS CONTENT, IN AIR

Polyester	Reactive monomer	Ageing temperature		
		175°C	200°C	225°C
General purpose	Styrene	51	16	2
Heat-resistant*	Triallyl cyanurate	93	34	8
Flame-resistant†	Styrene	63	24	4

*Exposure to moist heat (steam) may cause hydrolytic degradation, but polyesters derived from aromatic diols or the propylene glycol diether of Bisphenol A possess a useful degree of resistance.
†Unmodified polyesters are not inherently flame-resistant, but the use of HET acid §24D.21) provides self-extinguishing grades; suitable fillers, other than glass, include antimony oxide and chlorinated waxes.

24D.33 ELECTRICAL PROPERTIES

These are influenced by filler content and water absorption. The figures below may be regarded as the approximate medians of broad ranges of values.

Product	Volume resistivity (ohm cm)	Dielectric strength (kV/mm)	Dielectric constant at 1 MHz	Power factor at 1 MHz
Unfilled resins (rigid)	10^{15}	18	3·5	0·02
Unfilled resins (flexible)	—	12	5	0·04
Glass reinforced resins (50–60 % glass)	10^{14}	24	5	0·015

24D.34 MECHANICAL PROPERTIES

The composition of the resin, the type of reinforcement, and the conditions of fabrication markedly affect the mechanical properties of cured resins.

Short-term properties The values given in Table 24D.T2 indicate properties of the commoner forms of polyesters; values up to 20 % above or below those given are often encountered, and even greater extremes are sometimes found.

Table 24D.T2. SHORT-TERM PROPERTIES

Product	Elastic modulus (Young's) (kgf/mm²)	Extension at break (%)	Tensile strength (kgf/mm²)	Flexural strength (kgf/mm²)	Compression strength (kgf/mm²)	Izod impact strength* (ft lbf/in)
Rigid castings, unfilled	350	<2	7	11	15	1·8 (0·3)
Flexible castings, unfilled	<55	<65	1	Too flexible to test		—
General purpose laminates (50–60 % glass, woven fabric)	2500	<2	35	50	25	25 (16)
General purpose laminates (30 % glass, chopped strand mat)	1000	—	15	20	15	20 (12)
High strength laminates (70 % glass, parallel roving reinforcement)	>4000	—	85	100	50	70
Dough moulding compound (60–70 % glass, chopped fibres)	1000	<2	10	12	15	(8)

*Unnotched values; notched values shown in brackets.

Long-term properties Because of creep and fatigue, only a fraction of the very high level of instantaneous or short-term mechanical properties of reinforced polyesters can be relied on where long-term stresses are involved. This is particularly true in moist environments (cf. §24D.31) where the safe stress level for periods of several years may be less than 25% of the short-term strength.

Table 24D. T3 LONG-TERM PROPERTIES. TENSILE CREEP OF POLYESTER/GLASS LAMINATES (AT 23°C)

Conditions (see below)	Stress (% of short-term tensile strength)	Strain (%)			Time to break (h)
		Initial	After 10 h	At break	
A*	50	0·29	0·34	(0·45)‡	—
	60	0·46	0·58	(0·75)‡	—
	80	0·55	0·65	(0·80)‡	—
B†	40	0·40	0·42	0·66	10 000
	50	0·51	0·60	0·72	1500
	60	0·65	0·78	0·85	100
	70	0·78	—	0·91	9

*Fibre mat laminates, conditioned and tested at 50% r.h.
†Woven fabric laminates, conditioned in water and tested parallel to the warp
‡Strain after 1000 h (not broken)

Hardness Rigid castings and laminates (Rockwell M), 70–115. Hardness generally rises with styrene content. Flexible resins (Shore A), 40–94. **Friction and abrasion** μ_{stat} , castings (35% styrene)/regen. cellulose films, 0·22–0·52. Surface abrasion resistance is generally good, especially in flexibilised resins and in polyesters based on isophthalic acid or cross-linked with an allylic monomer. Volume loss in rubbing abrasion (mild-steel = 100): glass-reinforced laminate, 1200–2800; hard natural rubber (hardness 96 IRHD): 5200; soft natural rubber (hardness 56 IRHD): negligible.

24D.4 FABRICATION

Addition of a free radical initiator (e.g. benzoyl peroxide, methyl ethyl ketone peroxide) to a polyester resin/reactive monomer syrup brings about polymerisation and formation of a cross-linked structure. Accelerators (e.g. cobalt naphthenate, dimethylaniline) speed up the reaction, particularly in room-temperature curing applications. Since no volatile matter is evolved during cross-linking, little or no pressure needs to be applied during fabrication, thus simplifying the equipment required and reducing costs. Glass fibres, in the form of non-woven

259

mats, woven cloths, or continuous filaments, are the most common form of reinforcement although other fibres (e.g. asbestos, sisal) and mineral fillers (e.g. calcium carbonate, slate) are also used. Composites may be moulded at room temperature by hand lay-up or spraying techniques, thixotropic resins being preferable for this application; higher loadings of glass are achieved with rubber bag moulding, often combined with heat-curing. Matched die moulding is used where the quantity of mouldings required justifies higher mould costs, and where high density and uniformity are important; pressures up to 500 lbf/in^2 and temperatures up to 130°C are employed. For this application glass mat is supplied pre-impregnated with viscous resins having long shelf-life, e.g. incorporating monomers of low volatility such as diallyl phthalate. Moderate shrinkage on curing (6–8%, unfilled resin) makes polyesters suitable for casting and electrical encapsulation. Continuous lamination and extrusion processes have been developed for glass-reinforced polyesters. Filament winding techniques are employed for special products of circular cross-section.

For compression moulding compounds, *see* §24D.81.

24D.5 SERVICEABILITY

Cross-linked polyesters can maintain good mechanical properties over a period of years, but the composition, type of reinforcement, and degree of cure can strongly influence ageing performance (*see* §§24D.32 and 24D.34). Discoloration from outdoor exposure is minimised by inclusion of ultra-violet absorbers; resins incorporating methyl methacrylate as reactive monomer generally show good weather resistance. Biological attack is virtually nil. Irradiation cross-links uncured resins and has been used as a method of cure; however, radiation-cure tends to lower tensile and impact strengths and sometimes causes cracking.

24D.6 UTILISATION

Glass-reinforced structures account for *c.* 85% of unsaturated polyester production. Typical large volume uses include: building panels (e.g. roofing), small boats, motor vehicle bodies, chemical plant (e.g. ducts, containers), electrical laminates, machinery covers, and aircraft components. Filament-wound structures are used in pipework, tanks, and military missiles.

Dough moulding compounds (§24D.81) are used in small electrical and motor vehicle parts. Casting resins are used for electrical encapsulation, general purpose sealants or repair compounds, mould and tool making, and some decorative purposes, principally buttons; they have also been used as concrete additives and as binders for military missile fuels. Polyester coatings (to be distinguished from those of polyester polyurethanes, Section 40) are suitable for use in industrial, corrosion-resistant applications and as varnishes.

24D.7 HISTORY

The history of this class of polyesters becomes distinguishable from the main lines of polyester development, established by W. H. Carothers in 1937, with the studies of T. F. Bradley, E. J. Kropa, and W. B. Johnston on the relation between the 'drying' or cross-linking of synthetic resins having unsaturation in the main chain. The discovery, by C. Ellis in the same year, that the rate of cross-linking is greatly accelerated in the presence of unsaturated monomers was a major step towards practical exploitation of these materials, and commercial development started in the U.S.A. during the Second World War, initially as casting resins for electrical purposes. However, the discovery of the reinforcing effects of glass fibres by the United States Rubber Co. in 1942 led to further important applications in the form of radomes and other aircraft components. Owing to the high cost of glass fibres and some technological difficulties, production of polyester laminates declined at first after the war, while casting applications grew in importance. Glass-fibre reinforced polyester moulding compounds appeared in the U.S.A. about 1942 while similar compounds incorporating natural fibres originated in the U.K. in 1947.

After 1948, with improvements in curing systems and research on glass finishes, glass-reinforced products again came into prominence and have dominated the polyester market since 1950.

24D.8 ADDITIONAL NOTES

24D.81 POLYESTER COMPRESSION MOULDING COMPOUNDS

The terminology in this field is ill-defined and it is sometimes difficult accurately to identify moulding compounds that may be encountered. They are usually supplied as putties, doughs, or ropes consisting of an unsaturated polyester resin dissolved in a reactive monomer and compounded with glass fibre or mineral fillers. When diallyl phthalate is often used as a reactive monomer (on account of its low volatility) the compounds are sometimes described as 'alkyds'.

Diallyl phthalate and cognate resins also give rise to moulding materials described as 'allylics', which have important electrical applications. The usual basis is partly polymerised diallyl phthalate (prepolymer) which cross-links on the application of heat and pressure, with no evolution of volatiles and therefore little shrinkage. Compounds with glass fibres and other fillers, together with a catalyst, can be used for moulding; alternatively, the prepolymer may be blended with monomer to form a dough moulding compound. The storage life is about 6 months. Moulding may be carried out on modified compression moulding equipment at pressures as low as 500 lbf/in^2 at c. 150°C.

Summary of properties:

Property	Unfilled resin	Glass-filled resin
Specific gravity	1·27	1·74
Heat distortion point (ASTM-D648)°C	155	180
Volume resistivity (ohm.cm)	$1·8 \times 10^{16}$	$>10^{13}$
Dielectric strength (kV/mm)	18	14
Dielectric constant at 1 mHz	3·4	4
Tensile strength (kgf/mm^2)	2·8	5·25
Compression strength (kgf/mm^2)	16·1	17·5
Izod impact strength (ft lbf/in. notched)	0·3	8

24D.9 FURTHER LITERATURE

BOENIG, H. V., *Unsaturated Polyesters,* Amsterdam (1964)

LAWRENCE, J. R., *Polyester Resins,* New York and London (1960)

PARKYN, B., LAMB, F. and CLIFTON, B. V., *Polyesters,* Vol. 2, 'Unsaturated Polyesters and Polyester Plasticisers', London (1967)

MORGAN, P. (Ed), *Glass Reinforced Plastics,* London (1961)

OLEESKY, S. and MOHR, G., *Handbook of Reinforced Plastics of the SPI,* New York and London (1964)

PENN, W. S., *GRP Technology, Handbook to the Polyester Glass Fibre Plastics Industry,* London (1966)

ROSATO, D. V. and GROVE, C. S., *Filament Winding,* New York (1964)

Allylics:

RAECH, H., *Allylic Resins and Monomers,* New York and London (1965)

EPOXY RESINS

With notes on phenoxy resins and epichlorohydrin rubbers

SYNONYMS AND TRADE NAMES
Epoxide resins, ethoxyline resins. Araldite, Epikote, Epon, Epoxylite.

GENERAL CHARACTERISTICS

The *intermediates*, commercially available, are yellow or amber-coloured syrups, or brittle fusible brown solids, containing reactive epoxy groups (commonly the glycidyl group, —CH$_2$CHCH$_2$O).

The *cured resins,* resulting from the action of heat and/or curing agents on the intermediates, are insoluble, infusible, tough and inert amber-coloured materials used as castings, surface coatings, adhesives, etc. An inert filler is often incorporated. It is characteristic of these resins that hardening of the intermediates occurs without elimination of products of low molecular weight, e.g. no water is produced.

25.1 STRUCTURE

Simplest Fundamental Unit

Intermediates These are linear polymers of relatively low molecular weight, having lateral hydroxy-groups and terminal epoxy-groups. A typical example is provided by reaction of diphenylol propane with an excess of epichlorohydrin, the repeat and terminal units of the product being represented respectively by I and II (in the complete molecule one of the glycidul end-groups needs to be preceded by an additional diphenylolpropane residue, as shown in §25.21).

I   II

Cured resins The hardened end-products are complex cross-linked networks resulting from reaction of the active groups of the intermediates with amines, or with anhydrides of carboxylic acids. *See also* §25.21.

Molecular weight *Intermediates* Low, say 400–6000. *Cured resins* Very high (cross-linked structures).

X-ray data Some of the curing agents are crystalline but the cured resins are substantially amorphous. Crystalline fillers may be present.

25.2 CHEMISTRY

25.21 PREPARATION

25.21.1 *Intermediates*

The syrups or brittle resins of relatively low molecular weight are prepared by reaction of an excess of a difunctional epoxide-type compound with a difunctional compound containing active hydrogen atoms. The epoxide-type compound is commonly *epichlorohydrin,* and the other compound is often the bisphenol, *diphenylolpropane.* The materials are synthesised initially as follows.

Epichlorohydrin, or 1-chloro-2,3-epoxypropane ($ClCH_2CHCH_2O$, b.p. 116°C), is obtained by either (a) chlorination of acrolein, or (b) high temperature chlorination of propylene, by the following sequences of reactions,

(a) $CH_2{=}CH \xrightarrow{Cl_2} ClCH_2CHCl \xrightarrow[\text{(catalyst)}]{H} ClCH_2CHCl \xrightarrow{Ca(OH)_2} ClCH_2CH{-}CH_2$ (with CHO, CHO, CH_2OH substituents; final product epoxide)

(b) $CH_2{=}CH \xrightarrow[\text{high temp.}]{Cl_2} CH_2{=}CH \xrightarrow{Cl_2 + H_2O} ClCH_2CH(OH)CH_2Cl \xrightarrow{Ca(OH)_2} ClCH_2CH{-}CH_2$ (with CH_3, CH_2Cl substituents; final product epoxide)

or directly from the allyl chloride ($CH_2{=}CHCH_2Cl$) stage, by oxidation with peracetic acid.

Diphenylopropane, or 2,2-bis(4-hydroxyphenyl)propane ($HOC_6H_4C(CH_3)_2C_6H_4OH$, m.p. 156°C) also named Bisphenol A or Diane, is obtained economically by the acid catalysed condensation of phenol and acetone.

Base-catalysed condensation between the above reactants yields the intermediates, the main reaction probably proceeding as shown,

$$ClCH_2CHCH_2 + HOC_6H_4C(CH_3)_2C_6H_4OH \xrightarrow{(NaOH)} ClCH_2CHCH_2OC_6H_4C(CH_3)_2C_6H_4OH \xrightarrow{NaOH}$$

$$CH_2{-}CHCH_2OC_6H_4C(CH_3)_2C_6H_4OH \xrightarrow[\ n \ \ \text{diphenylolpropane} + NaOH]{n+1 \ \text{epichlorohydrin} +}$$

Intermediates of this type, therefore, have a backbone structure of a polyether with —OH groups along the length and a reactive glycidyl ether group (—OCH_2CHCH_2O) at each end, i.e. they are complex diglycidyl ethers. The value of n, which ranges from 0 to c. 15, depends on the initial ratio of the reactants; where n is low the intermediates are liquids of moderately high epoxide content (and thus dermatitic)

264

but as n increases they become solids, rising—decreasingly—in hydroxyl content and falling in epoxide content (until, when $n = c.\ 12$, they are non-dermatitic).

Other difunctional compounds of the epoxide type are illustrated by vinylcyclohexene dioxide (III) and dicyclopentadiene dioxide (IV)

III   IV

Cycloaliphatic epoxides of this kind avoid the loss of chlorine accompanying the use of epichlorohydrin; also employed are phenolic novolacs (§28.21) in which some of the hydroxyl groups are epoxidised, i.e. converted to glycidyl ether groups.

Examples of compounds that can replace diphenylolpropane are 4,4′-diaminodiphenylmethane, 1,4-butanediol, resorcinol, glycerol, and pentaerythritol; phenolic novolacs may also be used.

In addition to intermediates of the glycidyl ether type it is possible to obtain reactive compounds by epoxidation of drying oils (§24A.81) e.g. epoxidised soya-bean oil. The epoxy groups in these compounds are of the internal type, i.e. —CH—CH—.

25.21.2 Cured resins

There are two principal classes of curing agents, by reaction with which intermediates are set to hard resins: (a) *polyamines*, which often react at room temperature, and (b) *carboxylic acid anhydrides*, which react less readily and usually need a high temperature but yield products with superior thermal and mechanical resistance.

The main types of reaction between intermediates and curing agents are indicated below.

(a) *Cure with amines*

(i) *Primary amines* These operate by proton transfer to the epoxide oxygen atom,

In the above manner diepoxy intermediates are coupled into longer units, while the use of primary di- or poly-amines leads to cross-linkage, e.g.

$$4 \; \text{~~CH-CH}_2 + R(NH_2)_2 \longrightarrow$$

$$\begin{array}{cc} OH & OH \\ | & | \\ \text{~~CHCH}_2\text{NCH}_2\text{CH~~} \\ & R \\ | \\ \text{~~CHCH}_2\text{NCH}_2\text{CH~~} \\ | & | \\ OH & OH \end{array}$$

Examples (some with toxic and/or dermatitic hazards): diethylene-triamine, triethylenetetramine, 4,4′-diaminodiphenylmethane, 4,4′-diaminodiphenyl sulphone, *m*-phenylenediamine, methylenediamine, and amine complexes of boron trifluoride; also acting as amine-type curing agents, especially to provide tough surface coatings, are certain polymeric substances, e.g. polyamides obtained from polymerised fatty acids (Versamids, §23.21.6), while with a high temperature cure both amine- and phenol-formaldehyde resins can be used (OH-groups reacting similarly to NH_2-groups).

(ii) *Secondary amines* Secondary diamines may be used to extend chains without cross-linkage, the reaction being partly as in (i) and partly as in (iii).

(iii) *Tertiary amines* Being devoid of labile hydrogen atoms, these do not operate by direct addition but probably act catalytically (with a trace of water as co-catalyst) opening the epoxy ring at elevated temperatures and initiating polymerisation anionically, e.g.

$$\text{~~CH-CH}_2 \xrightarrow{\text{NR}_3} \text{~~CH-CH}_2\overset{+}{\text{NR}}_3$$

$$\text{~~CH-CH}_2 \longrightarrow \text{~~CH-CH}_2 \\ \overset{|}{O} \\ \text{~~CH-CH}_2\overset{+}{\text{NR}}_3 \quad \text{etc.}$$

or by a related mechanism.

Examples: benzyldimethylamine, triethylamine, 2,4,6-*tris*(dimethylaminomethyl)phenol.

Reactions with amines are accelerated by OH-containing substances, such as phenols and furfuryl alcohol.

(b) *Cure with acid anhydrides*

The complex reactions occurring with these hardening agents may be represented as follows.

(i) Base-initiated opening of the anhydride ring and formation of a monoester by reaction with a hydroxy group of an intermediate, e.g.

followed by reaction of the free carboxy group with either (ii) a second hydroxy group (e.g. on a different molecule) to form a di-ester cross-link

or (iii) a terminal epoxy group,

(iv) In addition, some etherification occurs, catalysed by the presence of acid, by reaction of epoxy- and hydroxy-groups,

Examples: phthalic anhydride, hexahydrophthalic anhydride, maleic anhydride, endic or nadic anhydride (endomethylenetetra-hydrophthalic anhydride, a cyclopentadiene/maleic anhydride adduct, V), chlorendic or 'HET' anhydride (hexachlorocyclopentadiene/

267

maleic anhydride adduct, VI), and the particularly heat-resistant pyromellitic dianhydride (benzene-1,2,4,5-tetracarboxylic acid dianhydride, VII),

V   VI   VII

Reaction with anhydrides is accelerated by the presence of tertiary amines, such as *N,N*-dimethylaniline, benzyldimethylamine, and 2,4,6-*tris*(dimethylaminomethyl)phenol, also by pyridine and triethanolamine.

(c) Modification of cure

Modifications of some intermediates and curing agents are mentioned above; others include: (i) curing, in the presence of an accelerator, with di- or poly-functional mercaptan-ended polysulphides (Section 39) to obtain flexible products, (ii) curing with di-isocyanates, i.e. reaction with hydroxyl groups—especially after increasing their number by reacting epoxy end-groups with a dialkanolamine, e.g. diethanolamine

and (iii) reduction of syrup viscosity and limitation of cross-linkage by inclusion of mono-reactive diluents or chain-stoppers, such as phenyl (or butyl) glycidyl ether (VIII) and styrene oxide (IX).

VIII   IX

25.22 PROPERTIES

Solvents The intermediates, if required, can be thinned, e.g. with a mixture of toluene or xylene with an alcohol or ketone (e.g. butanol, diacetone alcohol, or methyl ethyl ketone) or ethylene glycol monoethyl ether; however, a merit of the resins is that normally no solvent is needed. **Plasticised,** to a limited extent, by conventional plasticisers (e.g. pine oil, polyethylene glycol, dibutyl phthalate) and by mono-functional epoxides (e.g. butyl glycidyl ether, see VIII above) but with detriment to mechanical properties. Resins used for large castings or for encapsulating articles of low thermal expansion, and subjected to appreciable variation of temperature in service, are best plasticised by partly or wholly replacing normal curing agents with specially developed diamines or polyols (flexibilisers). **Relatively unaffected,** as cured resins, by water, hydrocarbons, alcohols (may soften slightly in ketones), and moderately conc. inorganic reagents (50% H_2SO_4, 10% HNO_3, conc. HCl, 20% NaOH). Acid-cured resins are inferior to amine-cured resins in resistance to bases. **Decomposed by** conc. acids (90% H_2SO_4, 85% H_3PO_4; slowly by HCOOH, glacial CH_3COOH, 30% HNO_3).

25.23 IDENTIFICATION

Cured resins, non-thermoplastic; products cured with amines contain N and (sulphone-cured) S; anhydride-cured resins may contain Cl; polysulphide-cured resins contain S; P usually absent. **Combustion** Subject to composition (e.g. absence of Cl) the cured resins burn with a very smoky flame and leave a charred residue. **Pyrolysis** Sample evolves brown fumes (condensing to oily distillate; may be strongly alkaline, with unpleasant amine odour) and chars without melting. Cured resins commence to evolve volatile products at 200–300°C, often with rapid rise in decomposition rate above 300°C; products (from scission of phenoxy-residues) include H_2O, CO, CO_2, lower aliphatic hydrocarbons, and phenols. **Colour tests** for epoxy resins derived from diphenylolpropane. Dissolve *c*. 0·1 g powdered resin in 10 ml conc. sulphuric acid and treat 1 ml portions of the clear solution (filtered if necessary) as follows. (i) Add 1 ml conc. HNO_3 and after a few minutes pour into 100 ml dil. NaOH to obtain a bright orange or red colour. (ii) Add 5 ml Dénigès reagent (0·25 g HgO dissolved in a mixture of 1 ml conc. H_2SO_4 added to 5 ml water); the solution slowly becomes orange-coloured and may yield an orange ppt. (iii) Add 1 drop formalin to obtain an orange solution which, when poured into water, becomes green or blue. *N.B.* Polycarbonates derived from diphenylolpropane do *not* respond to the above tests. **Characterisation of intermediates** The following tests may be used. 1. Measurement of viscosity (temperature-dependent, needs good control). 2. Determination of *epoxide equivalent* (g resin/g-equivalent of epoxide) or of *epoxy value* (i.e. g-equivalent of epoxide/100 g resin, which is equal to 100/*epoxide equivalent*). 3. Determination of *total hydroxyl equivalent* (g resin esterified/g equivalent of acid, 1 epoxide group rating as 2 hydroxyl groups) or of *hydroxyl number* (g equivalents of hydroxyl/100 g resin). The values of 2 and 3 may be determined by chemical and/or infra-red techniques.

25.3 PHYSICS

N.B. The properties of the cured resins depend on the types of intermediates and hardeners used, and whether cure is effected at room temperature or elevated temperature. Therefore, the data in the sub-sections that follow must be regarded as typical but necessarily generalised.

25.31 GENERAL PROPERTIES

Specific gravity Cast resins, unfilled, *c.* 1·15–1·2; usually raised by fillers, e.g. silica-filled, *c.* 1·6–2·0. **Refractive index** Intermediates, 1·475–1·58; cast resin, *c.* 1·56. **Water absorption** ($\%$) Cast resins, *c.* 0·1–0·2 in 24 h at 20–25°C; but continues over longer periods, e.g. *c.* 1·0 (cold-set resin) or 0·5 (hot-set resin) after 10 days at 20–25°C or 1 h at 100°C. Usually lowered by inert fillers (e.g. silica) and raised by reactive diluents. **Permeability** To gases (10^{-10} cm^2)/(s cmHg): O_2, 0·05–1·6; CO_2, 0·09–1·5.

25.32 THERMAL PROPERTIES (*see N.B.* above)

Conductivity, as (10^{-4} cal)/(s cm °C). Cast resins, *c.* 3–5; fillers may increase the value, e.g. silica-filled, *c.* 15. **Coefficient of linear expansion** (10^{-4}/°C) Cast resins, general, -50 to 50°C, *c.* 0·5; but value increases above the softening temperature, e.g. resins cured with phthalic anhydride, above 70°C, *c.* 1·2; do. methylnadic anhydride, at 85°C, *c.* 0·7. Fillers usually lower the value, e.g. silica-filled, 0·3. **Physical effects of temperature** The more thermally-resistant products tend to be brittle at low temperatures. Properties decline near the softening temperature, which is lowest in resins cured with polyamides or aliphatic amines. For example, deflection temperature (approx. glass temperature in °C) of resins cured with agents named: polyamides, 50 upwards; aliphatic amines, 100–150; aromatic do., 160 (*m*-phenylenediamine) to 190–200 (diaminodiphenyl sulphone); anhydrides, 130–140°C.

25.33 ELECTRICAL PROPERTIES (*see N.B.* above)

Volume resistivity (ohm cm) Cold-set resins, 10^{13} to $>10^{15}$. Cast resins, at various temperatures: 25°C, 10^{15} to $>10^{16}$; 100°C, $>10^{14}$; 150°C, 10^{12}–10^{14}; 200°C, 10^{10}–10^{12}; 250°C, 10^8–10^{11}. *Surface resistivity* (ohms) at 25°C, 10^{12}–10^{14}; at 100°C, 10^{12}. **Dielectric strength** (kV/mm) Cast resins, 3 mm thick, 14–18; do., silica-filled, up to 21. **Dielectric constant** At 25°C and 50 Hz, slightly above 4; falls with frequency rise, to slightly above 3 at 10^{10} Hz; tends slightly to fall with moderate temperature rise, but at low frequency rises at

200–250°C, e.g. to 10–20. **Power factor** At 25°C and 50 Hz, 0·002–0·03; tends slightly to fall with frequency rise; not much changed by moderate temperature rise, but rises to high values (*c*. 0·7) when near the softening temperature.

25.34 MECHANICAL PROPERTIES

Elastic modulus (Young's) (kgf/mm²) (i) *Tensile* Cast resins, cold-set, 200–385; do., hot-set, 200–450. Lowered by rise of temperature, e.g. at 200°C, 80. Usually increased by fillers, e.g. silica-filled, 750–1750. (ii) *Compression* 100–300. (iii) *Flexural c*. 300. **Tensile strength** (kgf/mm²) Cold-set, up to *c*. 7; hot-set, up to *c*. 8. Tends to be least in products cured with aliphatic amines (e.g. 6·3), greatest with aromatic amines or anhydrides (up to 9). *Shear strength*, hot-set Al/Al adhesive, up to 3. **Extension at break** 4–7%. **Flexural strength** (kgf/mm²) 7–14, highest values from anhydride-cured resins. **Compression strength** (kgf/mm²) Cold-set, 8–12; hot-set, up to 16; silica-filled, *c*. 20. **Impact strength** (Izod, ft lbf/in of notch) 0·3–1. **Hardness** Cast resins, Shore D, 83–90; films (Persoz, s), 300–400. **Abrasion** The Taber abrasion resistance of the resins is said to approach that of aluminium alloys.

25.4 FABRICATION

A mixture of an intermediate and a curing agent will set, either at room temperature or on heating, without requiring application of pressure since no volatile matter is evolved. An accelerator may be included. The *pot life* (usable life) of the mix may vary from 30 min or less to a much longer period; the *cure* may take 24 h at room temperature or a shorter time above it, e.g. 2 h at 160–180°C. The reaction is exothermic and overheating of large bulks of resin must be avoided. To obtain maximum thermal resistance in the product, a post-cure at 100–250°C may be needed. Flexibilisers (*see* §25.22) reduce the tendency for cracks to occur in castings or potted assemblies subjected to thermal fluctuation in service. Inert fillers (silica, slate, aluminium, mica dust, etc.) reduce shrinkage and thermal expansion while increasing hardness and compression strength; they may also add an economic advantage.

Additional to casting and impregnating processes, it is possible, by using brush, spray, or fluidised-bed techniques, to build up surface coatings to appreciable thickness, there being no solvent to evaporate; however, solvent-system air-drying resins can also be used.

The good adhesion of the resins is due partly to the relatively low shrinkage (*c*. 2% by vol.) that they undergo during cure, and to the hydroxy-groups (*see* §25.21) which assist the initial wetting of polar substances; hence the resins bond strongly to glass, stone, metals, metal oxides, etc. For maximum joint-strength the surfaces to be bonded need to be properly prepared, e.g. thoroughly de-greased then pre-treated (as by etching).

HAZARD Some of the epoxide intermediates and/or the amines with which they may be cured can cause dermatitis, and persons who become sensitised may be unable to work with them. However, the cured resins are innocuous and can usually be machined or otherwise manipulated without danger.

25.5 SERVICEABILITY

The polyether structure of the cured resins resists many forms of chemical attack. The final products are therefore inert; they are also strong, of relatively low density, and usable up to moderately high temperatures, e.g. 150°C, although those most thermally resistant tend to be the more brittle at low temperatures. In epoxy electrical insulation highest tracking resistance results from resins made from alicyclic epoxides and non-aromatic hardeners.

25.6 UTILISATION

Castings for small or large electrical equipment, including outdoor-type insulators, also for tool-jigs and patterns. *Potting* (encapsulation) of electronic components, and *impregnation* of windings for sealing them against moisture and/or conferring resistance to shock and vibration. *Adhesives* for bonding metals (e.g. light alloys in aircraft), glass, ceramics, and related polar materials; *bonding* of fibre assemblies and lamination of fabrics for constructional purposes, e.g. using glass or other refractory fibres. *Surface coatings* for metals (resistant to chemical and marine corrosion) and for stone, e.g. flooring, roadways; also for can and drum lacquers, wire enamels, printed circuits, stovable finishes for metal, and air-drying drying-oil-modified varnishes.

25.7 HISTORY

A study of the reaction of epichlorohydrin with phenols (and the formation of glycidyl ethers) was reported by T. Lindeman in 1891, in 1910 E. Fourneau reacted diglycidyl ethers with amines, and in 1934 P. Schlack examined glycidyl ethers (including those of Bisphenol A) for textile purposes.* Epoxy resins did not appear commercially,

---

*Seeking strong fibres, which he found instead in nylon 6, *see* §23.7.

however, until 1939, the first applications being in dentistry (by P. Castan, of de Trey AG, Switzerland) while in 1945 they were developed as adhesives (Ciba).

25.8 ADDITIONAL NOTES

25.81 PHENOXY RESINS

Resins are commercially available made from epichlorohydrin and diphenylolpropane, with a linear structure and a repeat unit as represented by I in §25.1, but without reactive (epoxide) terminal groups. These products, having a molecular weight of *c.* 30 000 and being soluble in a number of solvents (e.g. tetrahydrofuran, dimethyl-formamide, acetone/toluene) but showing resistance to alkalis and mineral acids, find uses as surface coatings and adhesives. Films— e.g. on paper—have low permeability to water vapour and gases, can be heat-sealed at 150–175°C, and if desired can be cured by reaction of the OH-groups with polyisocyanates or amino- or phenolic-resins.

25.82 EPICHLOROHYDRIN RUBBERS (Hydrin-type rubbers)

Polyepichlorohydrin (for monomer, *see* §25.21) and copolymers of epichlorohydrin (1:1 with ethylene oxide, *see* §26.82) are elastomeric materials having the repeat units shown in X and XI respectively.

$$-CH_2-CH-O- \qquad -CH_2-CH-O-CH_2-CH_2-O-$$
$$\quad\quad\;\; | \qquad\qquad\qquad\qquad\;\; |$$
$$\quad\;\; CH_2Cl \qquad\qquad\qquad\;\;\; CH_2Cl$$

$$\text{X} \qquad\qquad\qquad\qquad \text{XI}$$

These are saturated special-purpose rubbers, curable with diamines or ammonium compounds, and resistant to ozone and organic liquids (including hot oils) and to dynamic stressing (e.g. as motor mounts showing low compression set after long running under load). The respective specific gravities are 1·36 and 1·27, and the homopolymer (Cl-content, 38·4%) is flame-resistant.

25.9 FURTHER LITERATURE

SKEIST, I. and SOMERVILLE, G. R., *Epoxy Resins,* New York and London (1964)
LEE, H. and NEVILLE, K., *Handbook of Epoxy Resins,* New York and London (1967)
BRUINS, P. F. (Ed), *Epoxy Resin Technology*, New York and London (1969)
POTTER, W. G., *Epoxide Resins*, London (1970)

SECTION 26

POLYFORMALDEHYDE

With notes on polyethylene oxide, polypropylene oxide,
polyphenylene oxide, and chlorinated polyethers

Acetal resin,* Delrin. Copolymers: Alkon, Celcon, Hostaform.

GENERAL CHARACTERISTICS
(i) Hard, tough, resilient, thermoplastic material; naturally an opaque white in colour, and with a low moisture absorption (affording good dimensional stability), it superficially resembles polyethylene but has a higher melting point and greater 'springiness'. These properties allow it to compete with die-casting metals in engineering applications, e.g. small gear wheels, bearings, handles, and snap-fit assemblies. (ii) For other kinds of formaldehyde polymers, *see* §26.81.

26.1 STRUCTURE

Simplest Fundamental Unit

$$—CH_2—O—$$

CH_2O, mol. wt. $= 30$

Molecular weight Above 15000; optimum properties at *c.* 30000 (degree of polymerisation $= 1000$).

X-ray data Highly crystalline (75–80%). Identity period, drawn film, 17·3 Å (probably helical chain configuration; 9 repeat units of 1·92 Å per hexagonal unit cell).

26.2 CHEMISTRY

26.21 PREPARATION

Formaldehyde, $CH_2{=}O$, a reactive gas (b.p. $-21°C$), is commonly made by partial oxidation of methanol, in the vapour phase, over a catalyst of metallic copper or silver,

$$CH_3OH \xrightarrow{(O)} CH_2O + H_2O$$

The reaction probably involves an endothermic dehydrogenation,

$$CH_3OH \longrightarrow HCHO + H_2,$$

*Not to be confused with polyvinyl *acetals* properly so-called (Section 8).

which is maintained by the exothermic oxidation of the hydrogen liberated.

The product is marketed as a 37% wt./wt. aq. solution (formalin, described as 40% wt./wt.), in which formaldehyde exists largely as methylene glycol, $CH_2(OH)_2$. However, in both the gaseous and dissolved states formaldehyde shows a ready tendency to polymerise, particularly when cooled—hence, to avoid the formation of a precipitate, formalin should not be exposed to low temperatures.

Formaldehyde polymers are of several kinds: (i) those, possessing important mechanical properties, the preparation and properties of which appear below, and (ii) various other types, dealt with under §26.81.

In the presence of a suitable catalyst, pure anhydrous formaldehyde undergoes polymerisation to polymethylene ethers of high molecular weight. In one method of preparation the gaseous monomer, obtained from liquefied formaldehyde or by depolymerisation of trioxan (*see* §26.81), is led into a solution or dispersion of the catalyst in an inert hydrocarbon solvent from which the polymer precipitates as it is formed; alternatively, trioxan is polymerised in the melt or in solution, either alone or with addition of a small proportion of a co-monomer such as ethylene oxide or indene. In both instances the initiating catalyst is of the ionic type (adding to the carbonyl group) and poly-merisation takes place very rapidly. Many substances have been claimed as initiators, e.g. boron trifluoride and its derivatives, phosphine derivatives, amines, quaternary ammonium salts, metal carbonyls, activated aluminium or alumina. Ionising radiation can also be used (polymerisation of trioxan in the solid state being possible by this means).

Good performance in service requires the use of very pure reactants (giving high molecular weight) and after-treatment to increase the thermal stability of the product (which has a hemiacetal structure). The stabilisation is usually achieved by conversion of the hydroxyl end-groups (*see* §26.81) into ester or ether groups, e.g. by acetylation or methylation. Copolymerisation also confers some improvement in stability. Additional protection against decomposition is conferred by incorporation of radical inhibitors (e.g. amines); compounds that absorb free formaldehyde (e.g. phenols, amino-compounds including nylon 6), by preventing its oxidation to formic acid, stop further breakdown of the polymer. For exterior exposure an ultra-violet stabiliser or carbon black filler is also essential.

26.22 PROPERTIES

Dissolves only in hot solvents, usually above 100°C; e.g. phenols, chlorophenols, dimethylformamide, dimethyl sulphoxide, benzyl alcohol. **Swollen** only slightly on long immersion at room temperature in common solvents, including phenols, etc. Linear swelling in water (%), 0·4 at 25°C; 1·5 at 95°C. Weight increase (approx. %) in: acetone, 0·95; ethyl acetate, 0·65; toluene, 0·12. **Relatively unaffected by** most common organic liquids (phenols and ethylene dichloride lower the strength). **Decomposed by** acids, including dil. mineral acids and moderately strong organic acids (rapid attack by conc. HNO_3); bases, particularly nitrogenous bases (surface discolored by NH_4OH; but copolymers more resistant); hypochlorite solutions. Preferred service range for the homopolymer, between pH 4 and 10.

26.23 IDENTIFICATION

Thermoplastic; Cl, N, P, S normally absent. **Combustion** Melts, burns with a blue smoke-free flame, leaves little or no residue. **Pyrolysis** Melts, distils partially with sublimation and partially with reversion to monomer; leaves little or no residue. **Colour tests** Formaldehyde, obtained from polyformaldehyde (and from several other materials*) on pyrolysis or on distillation with dil. sulphuric acid, may be detected with the following reagents.

(a) *Chromotropic acid* (CTA, or 1,8-dihydroxynaphthalene-3,6-disulphonic acid). This provides one of the best colour reactions for formaldehyde, which may be carried out thus: to 2 or 3 ml of the solution to be tested add a few drops of 5% aq. CTA followed by conc. sulphuric in excess, and—if necessary—warm to 100°C for 10 min. Formaldehyde produces a deep violet colour, other aldehydes having little effect. Careful addition of the sulphuric acid, to form a lower layer, causes development of the colour at the interface. Formaldehyde in resin-finished textiles may be detected in this way, after warming samples in dil. sulphuric acid (or on standing in 12N H_2SO_4) the extract being filtered or decanted free from fibres and then tested as above. (b) *Schryver's test* (adapted) Solution I: to 0·1 g phenylhydrazine hydrochloride suspended in a little water, add 5 drops conc. hydrochloric acid and dil. to 50 ml. Solution II: to 0·25 g potassium ferricyanide dissolved in a little water, add 20 ml conc. hydrochloric acid and dil. to 50 ml. Add the test solution or gas to c. 10 ml of I, and add an equal volume of II. A magenta colour develops. (c) *Schiff's test* Formaldehyde (and other aldehydes) restores the colour to a dil. aq. solution of magenta that has been *just* decolorised with sulphur dioxide. (d) *Phloroglucinol* in alkaline solution gives a red colour with formaldehyde (orange with acetaldehyde).

*For example, from polyvinyl alcohol (especially from formalised fibres) and polyvinyl formal, polymethyl methacrylate (on pyrolysis only), formalised proteins, methoxymethyl polyamides, amino- and phenol-formaldehyde resins and formal-derived polysulphide rubbers.

(*e*) *Carbazole* dissolved in conc. sulphuric acid yields a blue or blue-green colour on addition of aldehydes (test not specific).

Quantitative determination of formaldehyde may be made colorimetrically with CTA, or gravimetrically with dimedone (5,5-dimethyl-cyclohexane-1,3-dione) with solutions of which it yields a bulky precipitate; for respective methods, *see*, for instance, ROFF, W. J., *J. Text Inst.* **47**, T309 (1956), and YOE, J. H. and REID, L. C., *Ind. Engng Chem. (Anal.)* **13**, 238 (1941).

26.3 PHYSICS

26.31 GENERAL PROPERTIES

Specific gravity 1·42 commonly, as 75% crystalline (amorphous phase, 1·25; crystalline phase, 1·51). **Refractive index** 1·41. **Water absorption** (%) Discs, 24 h, 0·25; maximum, under 1·0. **Permeability** More permeable to water, less to hydrocarbons, than polyethylene. Examples: (i) Permeability to vapours (relative transmission rate through films *c.* 1 mm thick): as $g/(m^2, 24\ h)$ per 0·04 mm—i.e. 0·001 in—thickness; alcohol, (10% aq.), 1; toluene, 2·4; water, 8–16; trichloroethylene, 100 (absolute permeability): water vapour at 20–30°C, 0·4–0·8 (10^{-10} g)/(cm s cmHg). (ii) Permeability to gases, as (10^{-10} cm²)/(s cmHg), copolymer values in parentheses: O_2, 0·04–0·10 (0·03); N_2, 0·02 (0·01); CO_2, 0·19–0·30 (0·04).

26.32 THERMAL PROPERTIES

Specific heat 0·35. **Conductivity** 5·5 (10^{-4} cal)/(cm s °C). **Coefficient of linear expansion** 0·8–1·0.10^{-4}/°C. **Physical effects of temperature** Good impact resistance retained down to glass transition temperature, −40 to −60°C. Serviceable in air at 80–85°C (in water to at least 65°C) and satisfactory for short-term exposure to 120–140°C. Crystallinity increases with annealing, but at high temperatures the material degrades and discolours; melts (crystallinity disappears), 175–180°C. Copolymers soften *c.* 160°C.

26.33 ELECTRICAL PROPERTIES

Volume resistivity Over 10^{14} ohm cm. *Surface resistivity*, 10^{14}–10^{16} ohm; copolymers slightly lower. **Dielectric strength** (kV/mm) Thin films, short time, over 200; 0·5 mm thick, 48; over 1·75 mm, 20. *Arc resistance*: melts, burns; non-tracking. **Dielectric constant** *c.* 3·7 (10^2–10^6 Hz; with only slight increase at 100% r.h.). Copolymers largely similar. **Power factor** *c.* 0·003–0·005 (10^2–10^6 Hz; with small increase at 100% r.h.). Copolymers largely similar.

26.34 MECHANICAL PROPERTIES

Elastic modulus (Young's) Over 300 kgf/mm^2. **Elastic recovery** Spring-like, for small deformations (low creep, good dimensional stability); shows little fatigue in rapid cyclic loading, e.g. up to 3·5 kgf/mm^2 at room temperature or to 2 kgf/mm^2 at 65°C. **Tensile strength** (kgf/mm^2) 6·5–7 (10 at −55°C; 5 at 70°C; yields at 3·5–4 at 100°C). **Extension at break** Moulded, 15–16%; extruded film or sheet, yields (at c. 15%), extends up to 75%. **Flexural strength** 8–12 kgf/mm^2. **Compression strength** (as compressive stress, kgf/mm^2, at compression stated) 3·6 at 1%, 12·6 at 10%. **Impact strength** (Izod test, ft lbf/in) Notched, c. 1·5; the unnotched value is unusually high (c. 20). Over half of the room-temperature strength is retained even at −80°C. **Hardness** Rockwell M85–95, R120; Shore D 80–85. **Friction and abrasion** Static and dynamic coefficients of friction are similar. Examples, against steel (dry), 0·1–0·3; do. (wet), 0·1–0·2; do. (oiled), 0·05–0·1; copolymers, against metals, 0·15–0·2; against self, 0·35. The material can be run against other plastics, such as nylon. Resistance to abrasion is greater than in most plastics, but less than that of nylon 6.6.

26.4 FABRICATION

The material is readily machineable, even without cutting aids; also weldable (e.g. by hot plate technique, at 300°C) but as yet there are no strong adhesives for it, though adhesion is improved by surface etching, e.g. with p-toluenesulphonic acid in perchloroethylene, followed by exposure to air. Moulding, 195–215°C and 15–20000 lbf/in^2; mould temperatures 65–100°C. Extrusion, 200–215°C.

26.5 SERVICEABILITY

Commercial material is stabilised against thermal degradation, both from oxidative attack on the main chain and from depolymerisation, and usually it is suitable for continuous service up to 85°C (or 65°C in water). It is tough and resilient, even down to low temperatures, and is free from biological attack, but it is susceptible to ultra- · violet light and is not recommended for use in exposed situations.

26.6 UTILISATION

Tough mouldings or machined parts having good dimensional stability, particularly where some resilience is also required, e.g.

snap-fit applications. Examples include the usual injection mouldings, blow mouldings and extrusions; in some applications it can with advantage replace zinc die-casting alloys, brass, aluminium and even cast iron, being appropriate for gear wheels, valve seatings, impellors and numerous smooth-surface precision-dimensioned mechanical and electrical components.

26.7 HISTORY

Formaldehyde and its low molecular weight polymers were first reported by A. M. Butlerov in 1859–61, and in 1868 A. W. Hofmann obtained the gas by the platinum-catalysed vapour-phase oxidation of methanol. Among many nineteenth-century workers who examined the material and its products, F. Mayer, in 1888, first used the term paraformaldehyde. In the late 1920s/early 1930s H. Staudinger made a major study of the polyoxymethylenes (*see* §26.81), the results of which contributed much to the understanding of high polymers in general.

The early polymers, however, lacked the necessary thermal stability or mechanical properties; these were achieved only after work, commenced in 1947 by the Du Pont Company, which led to the commercial introduction of the first stable formaldehyde polymer (Delrin) in 1959.

26.8 ADDITIONAL NOTES

26.81 OTHER TYPES OF FORMALDEHYDE POLYMERS

1. *Polyoxymethylenes or polyoxymethylene glycols.* These, of low molecular weight, are without constructional applications, but are of interest both historically (*see* above) and as convenient sources of formaldehyde. They are obtained almost spontaneously when a trace of water is added to liquid formaldehyde; also when gaseous formaldehyde is exposed to cold surfaces, and when aq. solutions of the gas are cooled or evaporated. In these products (as in high molecular weight polyformaldehyde) all the oxygen atoms are *included* in the main chain (cf. (2) below), the polymerisation reaction being expressed as follows,

$$n\text{CH}_2{=}\text{O} + \text{H}_2\text{O} \rightleftharpoons \text{HO}[-\text{CH}_2\text{O}-]_n\text{H}$$

or $\qquad n\text{CH}_2(\text{OH})_2 - (n-1)\text{H}_2\text{O}$

The solubility of the polymers in water decreases as the degree of polymerisation rises, but they revert to formaldehyde on boiling with water or on heating. The most important representative is *para-formaldehyde*, a mixture of polymers got by evaporating an acid solution of the monomer, in which the degree of polymerisation ranges from under 10 (mostly above 12) up to 100 (formaldehyde content 94–99%). Distilling aq. formaldehyde with a trace of conc. sulphuric acid produces not a glycol but the relatively stable cyclic trimer, *trioxan*, $(CH_2O)_3$. This compound, which is soluble in water and several organic solvents and on heating sublimes unchanged (m.p. 62°C; b.p. 115°C under 1 atm) in flexible crystals, is a useful source of anhydrous formaldehyde (obtained by acid-catalysed decomposition at *c.* 300°C).

2. *Hydroxy-type polymers.* These are of two kinds: (*a*) *Sugar-type products*. In alkaline solution formaldehyde undergoes Cannizzaro disproportionation to formic acid and methanol, though under mild conditions it condenses to polyhydroxy compounds in which—as distinct from the polyoxymethylenes—most of the oxygen atoms are *lateral* to the main chain. Glycollic aldehyde is first formed.

$$2HCHO \rightarrow CH_2(OH)CHO$$

but rapidly reacts with further formaldehyde to give complex mixtures of aldoses and ketoses. With calcium hydroxide as catalyst a mixture of hexose sugars $(H(CHOH)_5CHO)$ results, but the reaction can be controlled to yield 2-, 3- or 4- carbon hydroxyaldehydes. These low molecular weight carbohydrate products are possible intermediates in the syntheses of cellulose and starch (Sections 13 and 19 respectively).

(*b*) *Polyhydroxymethylenes.* Polymers of this type, of high molecular weight, with a repeat unit represented by

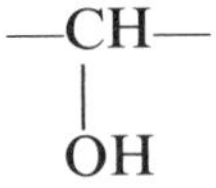

$$—CH—$$
$$|$$
$$OH$$

have not been developed commercially but can be obtained indirectly; e.g. as crystalline films, by hydrolysis of films of polyvinylene carbonate,

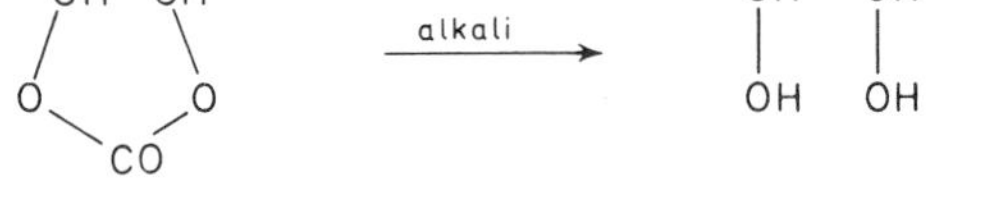

See, for instance, FIELD N. D. and SCHAEFGEN, J. R., *J. Pol. Sci.,* 58, 533 (1962).

26.82 POLYETHYLENE OXIDE AND POLYPROPYLENE OXIDE

Polyoxyethylenes or polyethylene glycols, a series of OH-terminated ethers discovered by Lourenzo in 1863, are obtained by acid- or base-catalysed polymerisation of ethylene oxide (free radicals are without effect) and have the linear repeat unit

$$—CH_2CH_2—O—$$

Although homologues of the polyoxymethylenes derived from formaldehyde, these compounds differ from them in properties, ranging from fairly mobile liquids to low-melting waxy solids (e.g. Carbowax, molecular weight c. 1000 up to 10000). They are soluble in water and in many organic solvents. They are, however, almost without useful constructional properties, except that at a sufficiently high degree of polymerisation (e.g. Polyox, mol. wt. 10^5 to $>10^6$) while remaining water-soluble they are also thermoplastic and can be extruded as films, etc., possessing appreciable mechanical strength. Ethylene oxide polymers find many uses, both as intermediates (e.g. in non-ionic detergents) and in their own right (e.g. as hydraulic and antifreeze fluids, lubricants and release agents, thickening agents, and—types with high degree of polymerisation—as textile sizes, water-soluble packaging film, etc.) and for the reduction of hydro-dynamic friction. A high molecular weight copolymer of ethylene oxide and epichlorohydrin is described under §25.82.

Polypropylene glycols, with repeat unit

$$—CH_2CH—O—$$
$$|$$
$$CH_3$$

possess somewhat similar properties to the ethylene oxide polymers, in that they are initially water-soluble, but they are less hydrophilic and become water-insoluble at a relatively low degree of polymerisation (D.P. c. 15). Polyethers of moderate D.P., derived from propylene oxide, are used in the preparation of polyurethanes (*see* §40.21 II). A low D.P. water-soluble copolymer of propylene and ethylene oxides (Oxitex) has been used for textile fibre lubrication. Copolymers having a high D.P.—e.g. propylene oxide/butadiene monoxide—are under development as synthetic rubbers (Dynagen).

26.83 POLYPHENYLENE OXIDE (PPO)

A poly-2,6-dimethyl-1,4-phenylene oxide (i.e. a polyxylenol) having the repeat unit is produced commercially by oxidative polymerisation of 2,6-dimethylphenol in the presence of oxygen and a copper-complex catalyst. It is soluble in aromatic and chlorinated hydrocarbons, and in esters, but resists aliphatic hydrocarbons, alcohols, etc., also acids and bases, and hot water (it can be sterilised). When ignited it yields a sooty flame and a charred residue, but it is not readily self-supporting in combustion. Some of the physical characteristics of this material are as follows.

Specific gravity 1·06. **Coefficient of linear expansion** $0·5.10^{-4}/°C$. It becomes brittle at $-170°C$ but is serviceable up to $160-175°C$ (heat distortion point, $c.$ 200°C, T_g $c.$ 210°C; thus, it retains its rigidity to a higher temperature than do most thermoplastics). When heated *in vacuo*, after an initial loss in weight it becomes graphitised and remains stable to above 500°C. **Volume resistivity** 10^{17} ohm cm; **dielectric strength** 16–20 kV/mm; **dielectric constant** 2·55 (up to 10^6 Hz), 2·7 (10^8 Hz), 2·6 (10^{10} Hz); **power factor,** 0·0002 (10^3 Hz), 0·0006 (10^6 Hz), 0·0008 (10^8 Hz), 0·0017 (10^{10} Hz); the electrical characteristics are maintained over a wide range of temperature. **Elastic modulus** (Young's), 230 kgf/mm²; **tensile strength,** 7 kgf/mm² (yields at 7·5); **extension at break,** 80% (yields at 9%); **flexural strength,** 10 kgf/mm²; **impact strength** (Izod, notched), 1·5–1·8 ft lbf/in; **hardness,** Rockwell R118–120; **abrasion** (Taber CS-17), 0·017 g/kc; **coefficient of friction** (μ_{dyn}, polymer/self), 0·18–0·23.

Applications
This relatively new rigid material (somewhat resembling polyformaldehyde resin) shows good chemical resistance and has good thermal, electrical, and mechanical properties; it can be moulded and extruded, or machined and fabricated from rod, sheet, etc., and it seems likely to find a variety of uses.

For other high temperature thermoplastics, *see* polycarbonates (Section 24C) and polysulphones (§39.81); for polyphenylenes, *see* §4.86.

26.84 CHLORINATED POLYETHERS
(Penton, Pentaphane)

Polymerisation of the substituted trimethylene oxide 3,3-bis(chloromethyl)oxetan = (3,3-dichloromethyloxacyclobutane) yields a product possessing good chemical stability and mechanical strength. The monomer is obtained by treatment of pentaerythritol (I) with hydrochloric and acetic acids, followed by ring closure of the intermediate trichlorohydrin acetate (II) with alkali, to provide the required product (III). Polymerisation of this is effected either cationically at low temperature (e.g. with BF_3 or its etherate, or PF_5) usually in an inert non-solvent, such as methyl chloride, from which the polymer (with repeat unit IV) separates out, or anionically at higher temperatures (e.g. with triethylaluminium).

$$
\begin{array}{cccc}
CH_2OH & CH_2Cl & CH_2Cl & CH_2Cl \\
| & | & | & | \\
HOCH_2-C-CH_2OH & ClCH_2-C-CH_2OCOCH_3 & ClCH_2-C-CH_2 & -CH_2-C-CH_2-O- \\
| & | & | \;\; | & | \\
CH_2OH & CH_2Cl & H_2C-O & CH_2Cl \\
& & & \\
I & II & III & IV
\end{array}
$$

Properties
The polymer has the following characteristics.

Molecular weight number-average, 250000–350000. Polycrystalline. **Soluble** in warm cyclohexanone, but **resistant to** most organic liquids, alkalis, moderately strong acids, and hypochlorite or chlorite solutions; stable to steam at 130°C; **decomposed by** conc. sulphuric and nitric acids, and by liquid ammonia. Self-extinguishing on combustion. Chlorine content approx. 45·5%. **Specific gravity** 1·4. **Water absorption** 0·01% (0·1% at 100°C). **Permeability** is generally low; e.g. through 0·025 mm film: water vapour at 90% r.h., 38°C, 35 $(10^{-10}$ g)/(cm^2 s cmHg). **Thermal conductivity** 3 $(10^{-4}$ cal)/ (cm s °C). **Coefficient of linear expansion** $0·8.10^{-4}$/°C. **Physical effects of temperature** Second order transition temperature at 7·5°C, rising with frequency, according to method of determination, to 32°C at 1 kHz; crystal melting commences 125°C; m.p., 180°C; decomposes in air, 290°C (in N_2, 320°C). **Electrical (volume) resistivity** c. 10^{16} ohm cm; *surface resistivity* 5.10^5 ohm; **dielectric strength,** 16 kV/mm; **dielectric constant,** 3·1 (50 Hz), 2·85 (1 MHz); **power factor,** 0·016 (50 Hz), 0·01 (1 MHz). **Elastic modulus,** 110 kgf/mm^2; **tensile strength,** 3·5–4·2 kgf/mm^2 (2·5 at 100°C); **extension at break,** 60–160% (200–250 at 100°C); **impact strength,** Izod, 0·4 ft lbf/in of notch (material tends to be brittle at room temperature but much less so above 35°C); **hardness,** Rockwell R100, Shore D 75; good resistance to abrasion.

Fabrication can be effected by machining or injection moulding; sheet or film can be heat sealed or vacuum-formed (at c. 165°C); the material can be welded (with air or N_2 at c. 350°C, also by high frequency heating above 50 MHz), can be stuck with neoprene-type adhesives, and can be powder-coated on to metal.

Applications
Solvent- and corrosion-resistant equipment (e.g. tanks, linings, bushes, gaskets, valve parts, etc., for chemical plant); also for electrical uses (e.g. where stability and inertness are paramount).

History
Polymerisation of the monomer was first reported by D. Delfs, in Germany, in 1941. Subsequently, in the 1950s, the product was developed in the U.S.A. (by Hercules Powder Co.; *see*, for instance, *Mod. Plas.*, **34**, 150 (Feb. 1957) and in the U.K. (by I.C.I. Ltd.; *see*, for instance, *J. Appl. Chem.*, **8**, 186 and 188 (1958).

26.85 POLYMERS FROM CHLOROMETHYLATED DIPHENYL ETHERS

Products obtained by reaction of diphenyl ether ($C_6H_5 \cdot O \cdot C_6H_5$) with formaldehyde and dry hydrogen chloride can condense, on heating with a Friedel-Crafts type catalyst, to provide hard non-flam foams (blown by loss of HCl). Thus, resulting from the above reaction, a simple monomer, such as *o*- or *p*-(chloromethyl) diphenyl ether, gives a polymer with the repeat unit,

$$-CH_2(C_6H_4) \cdot O \cdot (C_6H_4)-$$

and those of higher substitution yield cross-linked structures usually retaining some chloromethyl groups.

These materials can also be used as non-flam bonding agents for glass fibre laminates, etc., and have been described by DOEDENS, J. D. and CORDTS, H P., in *Ind. Engng Chem.* **53**, 59 (1961).

26.9 FURTHER LITERATURE

WALKER, J. F., *Formaldehyde*, New York (1964)

AKIN, R. B., *Acetal Resins*, New York (1962)

FURUKAWA, J. and SAEGUSA, T., *Polymerisation of Aldehydes and Oxides (Polymer Reviews, Vol. 3)*, New York and London (1963)

GAYLORD, N. G. (Ed), *Polyethers (High Polymers, Vol. XIII, Part 1)*, New York and London 1963)

MALINOVSKII, M. S., *Epoxides and their Derivatives* (trans. from the Russian), Jerusalem (1965). (Preparation and reactions of 1,2-epoxides; polymerisation, pp. 347–351; appendices on ethylene and propylene oxides.)

VOGL, O. (Ed), *Polyaldehydes (A.C.S. Symposium)*, New York and London (1967)

BARKER, S. J. and PRICE, M. B., *Polyacetals*, London (1970).

AMINO-FORMALDEHYDE RESINS
Principally as urea- and melamine-formaldehyde resins

With notes on benzoguanamine-, aniline- and toluenesulphonamide-formaldehyde resins

SYNONYMS AND TRADE NAMES

Aminoplasts, UF and MF resins. Aerolite, Beckamine, Beetle, Cascamite, Cibanoid, Cymel, Epok U, Kaurit, Melmex, Melolam, Melopas, Mouldrite, Nestorite, Paralac, Pollopas, Scarab, Setamine, Uformite.

GENERAL CHARACTERISTICS

1. Thermosetting resins translucent or opaque white (pigmentable over a wide range of colour), used for rigid *mouldings* and *laminates*. Fillers (e.g. α-cellulose fibre or woodflour) are incorporated in moulding powders to increase stability and strength.

2. Viscous syrups and water-based formulations, used as catalyst-hardenable *adhesives* and sealants for wood.

3. Water-based *pre-condensates*, used for impregnation and then reacted *in situ*, e.g. to impart dimensional stability, crease resistance, etc., to textiles, and wet-strength to paper.

4. Solvent-based *stoving enamels* and lacquers.

27.1 STRUCTURE

Simplest Fundamental Unit

Urea-formaldehyde
(UF) resins

Melamine-formaldehyde
(MF) resins

N.B. These structures represent idealised three-dimensional resins comprised of urea or melamine residues and methylene linkages ($-CH_2-$); in practice, condensation is incomplete and the end-products appear to consist of cross-linked macromolecules in which, in addition to methylene linkages, methylene ether linkages ($-CH_2OCH_2-$) and un-reacted methylol groups ($-CH_2OH$) may be present.

Molecular weight Soluble intermediates are monomers, dimers, etc., representing simple methylol or methylene derivatives, or methylol ethers (§27.81e), of urea or melamine. In cured resins the molecular weight may range from 100000 to that of a continuous network (smallest molecular weight determined by electron microscopy, 630000).

X-ray data The cured resins are substantially amorphous.

285

27.2 CHEMISTRY

27.21 PREPARATION

UF resins

Urea ($CO(NH_2)_2$, m.p. 132·7°C, obtained by heating ammonia and carbon dioxide under pressure) dissolves in and reacts with aq. solutions of formaldehyde (*see* §26.21) to give, depending on the relative concentrations of the components, a variety of products. The most important of these are *monomethylolurea* (MMU; $NH_2CONHCH_2OH$, m.p. 110°C) and *dimethylolurea* (DMU; $CO(NHCH_2OH)_2$, m.p. 126°C, also reported as up to 140°C), which can be obtained on warming or leaving to stand under neutral or mildly alkaline conditions. Under acid conditions these water-soluble intermediates undergo condensation and eventually precipitate as methyleneureas (probably $NH_2CONH(CH_2NHCONH)_{2-5}$-$CH_2NHCONH_2$) or their methylol derivatives, or as cross-linked structures; DMU, for example, ultimately forms an insoluble product (repeat unit represented ideally in §27.1), and even MMU—though theoretically capable only of linear condensation—behaves somewhat similarly, since in aq. solution it contains also DMU and free formaldehyde in equilibrium.

For the preparation of intermediates of low molecular weight (methylol derivatives and oligomers suitable for impregnation, lamination, adhesives, moulding compounds, etc.) reaction is first effected between urea (1 mol) and aq. formaldehyde (up to 2 mol) under neutral conditions, or under mildly acid or alkaline conditions followed by neutralisation. Water may then be removed—with some formaldehyde—by spray-drying or distillation under reduced pressure, and in the second method the resulting syrup if required for moulding powder may be blended with fillers (α-cellulose, woodflour, slate powder), pigments, moulding lubricant, a latent catalyst (rendered acid on heating) and a stabiliser such as hexamethylenetetramine (acid absorber during storage).

Formation of the initial intermediate and cure to the cross-linked end-product may be represented thus,

$$
\begin{array}{ccc}
NH_2 & NHCH_2OH & -N-CH_2- \\
| & | & | \\
CO \xrightarrow{CH_2O\ (neutral)} & CO \xrightarrow[\text{or with heat}]{\text{Acid catalyst, alone}} & CO \\
| & | & | \\
NH_2 & NHCH_2OH & -N-CH_2-
\end{array}
$$

It is probable, in addition, that α-cellulose (which as purified wood pulp or chopped cotton, is an excellent stabiliser and filler for UF resins) reacts with methylol groups during the final cure.

MF resins

Melamine or 1,3,5-triamino-2,4,6-triazine,

$$\overline{C(NH_2):NC(NH_2):NC(NH_2):N,}$$

(m.p. over 350°C) may be obtained by several routes. For example, cyanamide ($CN\cdot NH_2$)—obtained on hydrolysis of calcium cyanamide ($CaN\cdot CN$, produced by reacting calcium carbide and nitrogen at *c.* 1000°C), readily dimerises to dicyandiamide (or cyanoguanidine, $NH_2C(NH)NH\cdot CN$); this, heated under pressure—sometimes in an inert solvent—with anhydrous ammonia, yields the required trimer.

Melamine can combine with up to 6 mol of formaldehyde, though usually 3 mol prove ample (*see* idealised structure of end-product in §27.1). The methylol derivatives, when heated alone or with acid catalysts, condense to cross-linked end-products, the complexity of which (as with their resistance to water and heat) exceeds that of the urea analogues. Formation of the initial intermediate and cure to the cross-linked end-product may be represented thus,

(and dimethylol groups)

For additional notes on the reaction of formaldehyde with nitrogenous compounds, *see* §27.81; for textile and paper resins, *see* §27.82.

27.22 PROPERTIES

Solubility—Simple methylol derivatives of urea or melamine, also their methyl ethers, are soluble in water and in alcohol; butyl ethers, sparingly soluble in water, dissolve in butanol and in pyridine. Polymers of moderate molecular weight dissolve in pyridine, aq. formaldehyde, formic acid, and dil. mineral acids (with some decomposition). The end-products, being cross-linked, are insoluble. **Plasticised by** some polyhydric alcohols (e.g. glycerol), amines and amides, though only to a limited extent (e.g. sufficiently to facilitate processing); benzamide and *p*-toluenesulphonamide are used in moulding powders, and along with aniline and ethyleneurea have been used in UF adhesives. Butanol-based amino-resins (*see* §27.81e) are plasticised by alkyds (Section 24A). **Relatively unaffected by** common organic liquids, solvents, oils, greases, cold dil. acids or alkalis. **Decomposed by** acids and alkalis especially when heated. UF resins decompose in boiling water; fully cured MF resins are more resistant, being attacked only by conc. acids and alkalis or hot dil. acids.

27.23 IDENTIFICATION

Non-thermoplastic; N present; Cl, S, P normally absent (S present in toluenesulphonamide resins, §27.84). **Combustion** The resins crack and char in a flame but are not easily ignited; the residues smell of formaldehyde and methylamines (odour resembles that of burnt fish). **Pyrolysis** The resins crack and char, evolving alkaline vapours, formaldehyde, and the characteristic fishy odour; trace amounts of volatile cyanides are also present (detect as in §10.23). **Other tests** UF resins are slowly hydrolysed when refluxed with dil. mineral acids, giving formaldehyde and urea (and ultimately carbon dioxide and ammonia); MF resins, though more resistant, yield formaldehyde and melamine (strong acids yield cyanuric acid, *see* below). In the hydrolysate formaldehyde can be detected with chromotropic acid (§26.23); urea and melamine can be detected as follows. *Tests for urea* 1. After addition of a few drops of a sat. alcoholic solution of xanthydrol, a copious white precipitate of dixanthylurea (m.p. 265°C) slowly forms. Melamine is without effect. 2. To a small sample of hydrolysate (or resin) add a few millilitres of approximately 0·1% acetylbenzoyl (i.e. 1-phenyl-1,2-propanedione) dissolved in 50% v/v phosphoric acid, and heat in boiling water for 15 min. Urea gives a deep red-violet colour. Melamine is without effect. *Tests for melamine* 1. Addition of sat. aq. picric acid produces a yellow precipitate of melamine picrate (m.p. 350°C). Urea is without appreciable effect. 2. When melamine resins are boiled for some time in 50% phosphoric acid, the solution deposits crystals of cyanuric acid on cooling. When boiled in aniline, the resins give a small amount of a water-soluble white powder (m.p. >350°C); urea resins treated similarly give a crystalline precipitate of diphenylurea (m.p. 235–239°C; insol. H_2O, recryst. from alcohol). 3. Fabrics finished with MF resins turn yellow when treated with sodium hypochlorite.

Analysis of amino-formaldehyde resins in textiles is effected by partial acid hydrolysis followed by paper or thin-layer chromatography.

27.3 PHYSICS

a. Unless otherwise stated, the data below refer to general purpose moulded products, which contain one-quarter to one-third cellulosic filler by weight (as α-cellulose or woodflour).

27.31 GENERAL PROPERTIES (*see* note *a*)

Specific gravity *c*. 1·5; mineral- or glass-filled, 1·8–2·1. **Refractive index** UF, *c*. 1·55; MF, *c*. 1·65. **Water absorption** Disc (50 mm diam. × 3 mm) immersed 24 h at room temperature (%): up to 1·8; resistant grades, up to 0·7; mineral-filled and electrical grades, 0·1–0·2. Boiling water, immersed 30 min (%): up to 6; resistant grades, up to 1; electrical grades, up to 0·5. **Permeability** Moderately impermeable to water vapour.

27.32 THERMAL PROPERTIES (*see* note *a*)

Specific heat 0·4; asbestos-filled, 0·3. **Conductivity** $(10^{-4}$ cal)/(cm s °C) *c*. 8; mineral- or glass-filled, 10–15. **Coefficient of linear expansion** $(10^{-4}/°C)$ α-cellulose- or mineral-filled, 0·2–0·5; woodflour-filled, 0·3–0·6. **Physical effects of temperature** Maximum service temperature (without producing shrinkage or cracking): UF resins, *c*. 75°C; MF resins, *c*. 100°C. When heated in air: UF resins decompose above 100°C and char *c*. 200°C; MF resins show little change up to 150°C but become yellow (and evolve formaldehyde) from 150–200°C, and from 250–350°C they darken (with decomposition of the triazine ring and possible oxidation). Cellulose fillers char, mineral fillers are more resistant.

27.33 ELECTRICAL PROPERTIES

Data refer mainly to electrical and thermally-resistant grades, which may contain up to two-thirds filler by weight (as asbestos (*see* §42.83), mica, or other mineral powder).

Resistivity 10^{11}–10^{13} ohm cm. *Surface resistivity*, after immersion for 24 h in water, 10^{10}–10^{12} ohm. **Dielectric strength** (kV/mm) Room temperature: short time, 12–18; do., glass-filled, *c*. 8. Room temperature, step-by-step, *c*. 10; do., glass-filled, *c*. 8. At 90°C (3 mm sheet in oil, for 20 s without failure) 4–6 kV/mm. Non-tracking (particularly MF resins). **Dielectric constant** Cellulose-filled (*approximate* values), 7·5 (at 50 Hz), 7 (10^3 Hz), 6·5 (10^6 Hz); mineral- or glass-filled (values *up to*), 14 (50 Hz), 9 (10^3 Hz), 7·5 (10^6 Hz). **Power factor** Generally *c*. 0·03 (at 50, 10^3 and 10^6 Hz); glass-filled, up to 0·1 (50 Hz); mineral-filled, up to 0·3 (50 Hz), 0·07 (10^3 Hz), 0·05 (10^6 Hz).

27.34 MECHANICAL PROPERTIES (*see* note *a*)

Elastic modulus (Young's) (kgf/mm²) *c*. 700–1000; mineral- or glass-filled, up to *c.* 2500. **Tensile strength** (kgf/mm^2) Generally *c.* 5; weakened by non-fibrous filler, range *c.* 3–8. **Extension at break** *c.* 1%. **Flexural strength** *c.* 8 kgf/mm^2 (mineral powder filler may reduce value by nearly 50%). **Compression strength** up to 30 kgf/mm^2. **Impact strength** (Izod test, ft lbf/in of notch) *c.* 0·2, considerably raised by long fibrous fillers. **Hardness** Brinell (10 mm ball, 500 kg load) 48–54. Do. (5 mm ball, 125 kg load): UF, 40–45; MF, up to 50. Rockwell M 110–125. Do.: UF, E94–97; MF, E110. Mohs scale, up to 3·5. **Friction and abrasion** Grey cotton yarn (1 m/s, 65% r.h.)/ moulded UF resin, μ_{dyn}: 0·2. MF resins show high resistance to abrasion, surpassed only by certain epoxy and polycarbonate resins (*see*, for example, SIAS, C.B., Paper 7-E, 15*th Technical and Management Conf., Soc. Plas. Ind. Reinforced Plastics Division*, 1960).

27.4 FABRICATION

Compression moulding: 125–170°C, 1500–8000 lbf/in^2; up to 1 min per 0·1 in thickness. Preheating, up to 95°C, is advisable and improves dielectric strength. Loss in weight of moulded disc (50 mm diam. × 3 mm) after immersion in boiling water for 30 min should not exceed 0·6 g (0·05 g for thermally resistant electrical grades). Moulded resins are hard and not well suited to further machining, except in the form of laminated sheets. Joints made with amino-formaldehyde adhesives need clamping until set; incorporation of partially hydrolysed polyvinyl acetate (§6.81) imparts initial tack. *See also*: stoving enamels, §27.81e; textile and paper resins, §27.82.

27.5 SERVICEABILITY

Moulded UF resins, when properly filled and compounded, resist internal cracking (crazing) on ageing, and are light-fast, but are not recommended for exterior weathering (nor are UF stoving enamels); UF resins in fabrics are attacked by hot alkalis and are partially removed on boiling in water. MF products are also light-fast and are particularly resistant to heat, water, and acids; filled mouldings are highly stable, and stoving enamels of alkylated MF (blended with an alkyd) withstand severe weathering, e.g. on car bodies. Amino-resins in general (fillers, however, excluded) are unaffected by bacteria, fungi or insects.

27.6 UTILISATION

Some general applications: 1. *Mouldings* For domestic and industrial accessories, e.g. buttons, screw caps, durable 'crockery', and electrical insulation and equipment (plugs, sockets, etc.). A complete range of colours is possible. 2. *Laminates* MF in particular provides a hard upper surface for laminated boards, i.e. assemblies of cloth or paper impregnated and dried before curing by heat under pressure (for the lower layers phenolic resins—*see* Section 28—are usually used). 3. *Adhesives* Room-temperature catalyst-set or heat-set adhesives for wood/wood, e.g. in furniture manufacture and for chip board (MF-bonded joints superior to UF-bonded joints in resistance to boiling water). MF is employed for setting reactive acrylic latices (*see* §9.83), e.g. as used for fabric/fabric bonding. A cationic resin assists adhesion of cellulose nitrate in the preparation of moisture-proofed regenerated cellulose film. 4. *Textile and paper uses* For imparting dimensional stability, crease resistance, wet strength, etc. (*see* §27.82). DMU, in addition to direct applications is used to insolubilise starch and starch mixtures on paper and textiles; MF, in particular, is also used to stiffen textiles and for incorporation in flame-retardant finishes. 5. *Surface coating* Butylated stoving enamels and lacquers, e.g. for car bodies and washing machines. 6. UF *foam,* generated *in situ,* is used for thermal insulation, e.g. in cavity walls.

27.7 HISTORY

UF resins

The reaction between urea and formaldehyde first received attention from A. Hölzer (isolated methyleneurea, 1884), C. Goldschmidt (1896), and other nineteenth-century investigators; and the investigations were continued, with commercial aims in view, by workers such as A. Einhorn and A. Hamburger (studied methylolureas, 1908), A. E. Dixon (1918), H. John (impregnating syrups and adhesives, 1918–20) and F. Pollak and K. Ripper (1920–25). Various applications were envisaged, chiefly a colourless 'organic glass' (the last two investigators produced a material known as 'Pollopas').

Commercial production of moulding powders began in England *c.* 1926 and in the U.S.A. in 1928–29, thiourea being used initially in the cellulose-filled Beetle resins initiated by C. Rossiter (as much surpassing in colour possibilities the already established phenolic resins). Urea-formaldehyde glues for wood were introduced soon after, notably by I. G. Farbenindustrie (Kaurit, 1933) and N. A. de Bruyne (Aerolite, 1935–36). Lacquers (butylated UF/cellulose nitrate) were developed in 1926, and in the same year the use of UF-precondensates to improve the crease resistance of cellulosic textiles was instigated by the Tootal Broadhurst Lee Co. Ltd. (e.g. Brit. Pat. 291, 473). Applications in tub-sizing, for wet-strength paper, appeared in 1937.

MF resins

Melamine, isolated by Liebig in 1834, was not made use of until a century later when Henkel et Cie in Germany, and Ciba in Switzerland, developed MF resins for their superior resistance to heat, water and abrasion; this made it possible to produce, for example, high grade table-ware and durable decorative laminated sheet. Durable mechanical finishes on textiles (e.g. Everglaze) appeared in the 1930s. Durable lacquers (butylated MF/oil-modified alkyd resin), largely replacing those based on UF, were developed in 1938. Cationic MF derivatives for additives to paper were introduced in 1945, and MF-curable acrylic latices appeared in the early 1960s.

27.8 ADDITIONAL NOTES

27.81 THE REACTION OF AMINO-COMPOUNDS WITH FORMALDEHYDE

(*a*) *Urea* (*see also* §27.21). Urea and aq. formaldehyde react slowly and exothermically at room temperature. The rate of reaction is increased by acid and base catalysts and by rise of temperature. Under alkaline conditions an equilibrium mixture of methylolureas is produced, its composition depending on the U/F ratio. Mono- and dimethylolurea can be isolated from suitable preparations, as relatively unstable water-soluble crystalline substances; they partly dissociate in aq. solution.

Under acid conditions methylolureas condense by elimination of water between either (i) a methylol group and a free amino group, or (ii) two methylol groups, yielding products containing respectively methylene linkages (I) or methylene ether linkages (II, which by loss of formaldehyde can condense to I)

$$—CONHCH_2OH + H_2N— \qquad\qquad —CONHCH_2OH + HOH_2CHN—$$

$$\downarrow\,{-H_2O} \qquad\qquad\qquad\qquad\qquad \downarrow\,{-H_2O}$$

$$[—CONHCH_2NH—] \quad\xleftarrow{-HCHO}\quad [—CONHCH_2OCH_2NH—]$$

$$\text{I} \qquad\qquad\qquad\qquad\qquad\qquad\qquad \text{II}$$

Further condensation of formaldehyde or methylol groups with free —NH— groups of linear polymers I and II leads to branched chains and (from DMU) insoluble cross-linked products with structural units such as

$$-\text{NHCO}\overset{|}{\text{N}}\text{CH}_2- \qquad \text{and} \qquad -\text{NCO}\overset{|}{\text{N}}\text{CH}_2-$$
$$\underset{|}{\overset{|}{\text{CH}_2}}$$

A second theory (not much supported) postulates that methylol compounds dehydrate to unsaturated azomethine or methyleneimine groups from which branched or cross-linked structures build up by addition polymerisation,

$$-\text{CONHCH}_2\text{OH} \xrightarrow{-\text{H}_2\text{O}} -\text{CON}=\text{CH}_2 \longrightarrow -\text{CO}\overset{|}{\text{N}}-\text{CH}_2-$$

Yet another theory (MARVEL, C. S., *J. Amer. Chem. Soc.*, **68**, 1681 (1946)) proposes that urea, acting as both amine and amide, first condenses with formaldehyde to give a cyclic trimer, $[\text{CH}_2\text{N(CONH}_2)]_3$, which on further condensation with formaldehyde yields a spatial network.

For etherification of methylolureas, *see* (*e*) below.

(*b*) *Melamine* Although less soluble than urea, melamine reacts more readily with formaldehyde. It displays a functionality of up to six (as in hexamethylolmelamine) but in practice it is uncommon to use a M/F ratio below 1/3. Curing can be effected without catalysts, and at moderate temperatures little formaldehyde splits off. Melamine resins are generally more resistant to hydrolysis than UF resins. *See also* (*e*) below.

(*c*) *Benzoguanamine* This compound, 1,3-diamino-5-phenyl-2,4,6-triazine, $\overline{\text{C(NH}_2):\text{NC(NH}_2):\text{NC(C}_6\text{H}_5):\text{N}}$ (i.e. melamine with an —NH$_2$ groups substituted by —C$_6$H$_5$), which can be prepared by reacting dicyandiamide, aniline and anhydrous ammonia under pressure (cf. preparation of melamine, §27.21), has a m.p. of 224–227°C and by reaction with formaldehyde yields resins that are highly water-resistant. It is used mostly for hard non-staining 'crockery' and in butylated stoving lacquers; *see also* (*e*) below.

(*d*) *Aniline and Toluenesulphonamide* Polymers derived from reaction of formaldehyde with these compounds, being less related to those from (*a*), (*b*) or (*c*) are not usually classed as amino-resins but are included under §§27.83 and 27.84.

(*e*) *Amino-compounds and alcohols* Acid-catalysed reactions of formaldehyde with an amino-compound in a substantially anhydrous alcohol yields the etherified methylol derivative, e.g.

$$-NHCH_2OH + ROH \xrightarrow{-H_2O} -NHCH_2OR$$

Methyl ethers of methylolureas or methylolmelamines, water-soluble and more stable than the methylol compounds, are used as intermediates for textile resins (§27.82), formation of which involves condensation by splitting off methanol. Ethers from higher alcohols (particularly *n*-butanol) are more hydrophobic but dissolve in appropriate organic liquids and are important as surface coatings for metals, i.e. stoving enamels, cured for up to 30 min at temperatures of 120–160°C, as used on car bodies, washing machines and refrigerators. Butylated MF finishes, cured by heat and an acid catalyst, are harder and more durable than the UF counterparts, and those from butylated benzoguanamine-formaldehyde are outstanding in durable gloss and resistance to alkalis. To overcome brittleness in the final product, butyl ethers are usually plasticised with an oil-modified alkyd resin (Section 24A); during stoving, when some butanol is split off, the amino-resin is cured by acid groups (and probably reacts with hydroxyl groups) of the alkyd resin. Note—a polyhydric alcohol (e.g. glycol) has the opposite effect from butanol, yielding polymers with increased moisture absorption.

27.82 TEXTILE AND PAPER RESINS

(Aerotex, Calaroc, Fixapret, Lyofix, Uformite)

Textiles
Water-soluble precondensates, consisting largely of di- or trimethylol intermediates (*see* §27.21 and 27.81), that condense on heating under acid conditions, can be used with a deferred-action catalyst to impregnate cellulosic fabrics, the properties of which are permanently modified on drying followed by baking (e.g. for a few minutes at 150°C). Crease- and shrink-resistance are improved and durable mechanical effects can be obtained, e.g. glazing, embossing, pleating. In most instances the aim is to impregnate the *fibres* without stiffening the *fabric* by surface deposition.

Fabrics treated with simple UF or MF resins, if laundered in bleach (hypochlorite) form *N*-chloramide groups (—CO·NCl—) and on subsequent ironing become damaged and discolored, due to liberation of acid through hydrolysis of the intermediate. This defect, and a realisation that the beneficial effects of 'resin treatment' depend not so much on polymer deposition as on cross-linkage of the cellulose, led to examination of numerous other amino-formaldehyde compounds, aiming at cross-linkage without introduction of chlorine-retentive groups. A typical compound is *N,N'*-dimethylolethylene-urea,

$$\overline{CH_2N(CH_2OH)CON(CH_2OH)CH_2},$$

the parent material of which (i.e. ethyleneurea) derives from condensation of urea and ethylenediamine; though a satisfactory cross-linking agent this does not prove particularly non-chlorine-retentive (the homologous derivative of propyleneurea being more effective). Another example is the dimethyl ether of dimethyloluron, probably represented by

$$\overline{CH_2N(CH_2OCH_3)CON(CH_2OCH_3)CH_2O},$$

a derivative obtained initially from the reaction of urea with formaldehyde in the molar ratio of 1 to 4. Also used are compounds with only a single nitrogen atom per molecule, such as ethyl *N,N*-dimethylolcarbamate $(CH_2OH)_2NCOOC_2H_5$.

For additional information, see §13.86 which includes further references.

Paper
UF precondensates, modified by incorporation of bases, such as dicyandiamide and guanidine, then condensed with an excess of acid present, yield polymeric syrups having *cationic* properties, e.g. $[\sim NHRR']^+$ groups. These, substantive to cellulose, are used at the wet end of paper manufacture to improve wet strength. MF precondensates, without additional bases but partially reacted and still dissolved in aq. acid, are used for the same purpose; colloidal solutions of this kind show blue Tyndall-scattering of light, the particle size lying between 100 and 200 Å. A related colloidal product of melamine is used for durable rot-proofing of cellulose.

Modification with sodium hydrogen sulphite yields resins with *anionic* properties, e.g. $[\sim NHCH_2SO_3]^-$ groups. These products, reactive with the aluminium ion of alum, also have uses in paper-making.

27.83 ANILINE-FORMALDEHYDE RESINS

These substances, which have lost some of their former importance by reason of their colour (yellow to red-brown) and relatively high cost, are prepared by condensation of aniline with aq. formaldehyde, followed by dehydration under vacuum. Under neutral conditions, the initial product is a cyclic trimer, anhydroformaldehydeaniline, $[-N(C_6H_5)CH_2-]_3$, which on heating yields soluble resins of low molecular weight of which the linear repeat unit is considered to be

$$-N-CH_2-$$
$$\mathop{|}_{C_6H_5}$$

Under acid conditions, the amino group is protected by salt formation and nuclear substitution occurs (cf. phenol-formaldehyde resins, §28.21). Subsequent acid-catalysed condensation of a 1:1 reaction mixture then yields brittle products represented by

$$-NH-C_6H_4-CH_2-$$

while a slight excess of formaledehyde provides mouldable resins of which the cross-linked structure may be represented by

$$-NH-C_6H_3-CH_2-$$
$$\mathop{|}_{CH_2}$$
$$|$$

Linear polymers prepared with a deficiency of formaldehyde dissolve in chlorinated hydrocarbons and methylcyclohexanone, and some are soluble in acids. Mostly, however, these substances are insoluble, though they are compatible with phenolic resins and waxes. They resist attack by strong alkalis but conc. acids, and to some extent dil. acids, decompose them.

Identification
The colour, though not specific, is diagnostic. The resins burn with a sooty flame, and on extinction both aniline and formaldehyde can be smelt. These components are also obtained on pyrolysis and on refluxing a sample in 20% sulphuric acid, when they may be identified as follows. *Tests for aniline* Diazotise the acid extract by cooling in ice and slowly adding dil. solution of sodium nitrite, then after a few minutes pour into an excess of an alkaline solution of β-naphthol or R-salt. Aniline produces a bright red colour (cf. a similar result from polystyrene—*see* under §4.23—but obtained only after nitration and subsequent reduction of the derivative). *Test for formaldehyde* Distil the acid extract, and test the distillate with chromotropic acid as in §26.23a. *N.B.* The reagent is unsatisfactory for the detection of formaldehyde in the presence of aniline and resin in the undistilled extract.

Physical properties

Specific gravity (unfilled) 1·22–1·25. **Refractive index** *c.* 1·7. **Water absorption** *c.* 0·1%. **Specific heat** 0·4. **Thermal conductivity** *c.* 2·5 $(10^{-4}$ cal)/(cm s °C). **Coefficient of linear expansion** *c.* $0·55.10^{-4}$/°C. **Softens** (unfilled), *c.* 120°C; (filled), 130–150°C. Maximum service temperature, *c.* 80°C. M.p. (fusible resins) up to 200°C.

Electrical grades **Volume resistivity** 10^{16}–10^{17} ohm cm, 10^{10} at 90% r.h. *Surface resistivity* (unfilled), *c.* 10^{12} ohm. **Dielectric strength** *c.* 20 kV/mm (24 or more for short periods); non-tracking. **Dielectric constant** 3–4 (50 and 10^6 Hz). **Power factor** *c* 0·01 (50 Hz), 0·001–0·003 (10^6 Hz); filled, *c.* 0·05 (50 Hz).

Elastic modulus (Young's) (kgf/mm^2) unfilled, *c.* 300; filled, *c.* 1000. **Tensile strength** 6–8 kgf/mm^2. **Extension at break** Low (1–2%). **Flexural strength** (kgf/mm^2) 10–15. **Compression strength** (kgf/mm^2) 10–17·5. **Impact strength** (Izod) Under 0·5 ft lbf/in of notch. **Hardness** Rockwell, M100–125.

Fabrication Compression moulding *c.* 160°C, 3000 lbf/in^2 (temperature depends on A/F ratio).

Serviceability and Utilisation These materials darken in sunlight but (subject to the kind of filler present) are free from microbiological attack and show good dimensional stability. They were produced commercially in Switzerland in 1928, following the work of K. Frey, and have proved useful in electrical applications (e.g. in switch gear) but their importance has diminished in favour of newer polymers. AF condensation products are also used as accelerators in the vulcanisation of rubber (§29.4), and they have been employed to plasticise amino-resin adhesives. Fusible polymers of low molecular weight and low formaldehyde content have been used to cure epoxy resins (Section 25).

27.84 SULPHONAMIDE-FORMALDEHYDE RESINS
(Crestamide, Santolite)

These materials—developed as an outlet for *p*-toluenesulphonamide, by-product from the manufacture of saccharin—are related to aniline-formaldehyde resins but are less coloured and more readily soluble in organic liquids. A repeat unit may be represented by

$$-\text{N}-\text{CH}_2-$$
$$|$$
$$\text{SO}_2$$
$$|$$
$$\text{C}_6\text{H}_4\cdot\text{CH}_3$$

but the resins have relatively low molecular weight and are thought to be super-cooled melts containing ring structures (mainly a cyclic trimer, i.e. a derivative of hexahydrotriazine) together with unreacted *p*-toluenesulphonamide. The last (m.p. 137°C) can be extracted by shaking a benzene solution of the resin with aq. alkali, followed by neutralisation and recrystallisation from water.

The polymers range from viscous syrups to clear colourless solids resembling rosin in brittleness. They are insoluble in water, aliphatic hydrocarbons, and some chlorinated hydrocarbons (carbon tetra-chloride, perchloroethylene) but dissolve in esters and ketones. They are compatible, usually with a solvent- or plasticising action, with cellulose nitrate, acetate and acetate-butyrate, and with ethyl-cellulose, polyamides, casein, alkyd resins, ester gums and related coating materials. They are saponified by alkalis, and they lose formaldehyde on heating. The specific gravity is *c*. 1·35, the refractive index is 1·6, and the melting range of the brittle solids is around 70–80°C; the resins are not very stable to heat, and above 100°C cause decomposition when incorporated in nitrocellulose films.

The chief use is in lacquers and varnishes, which have excellent clarity and lightfastness. They have been employed as plasticisers in post-formable laminates, and have uses as adhesives (e.g. to glass) and in heat-sealing compositions (e.g. with nitrocellulose for regenerated cellulose film, reducing at the same time the moisture vapour trans-mission).

27.9 FURTHER LITERATURE

BLAIS, J. F., *Amino Resins,* New York (1959)
VALE, C. P. and TAYLOR, W. G. K., *Aminoplastics,* London (1964)

Specifications
BS 1322 (aminoplastic moulding materials, physical properties)
BS 2906 (aminoplastic mouldings)
BS 3167 (MF tableware)
ASTM D704 (MF moulding compounds)
ASTM D705 (UF moulding compounds)
ASTM D956 (recommendations for moulding aminoplastic specimens)

PHENOL-FORMALDEHYDE RESINS

SYNONYMS AND TRADE NAMES

P F resins, phenolic resins, phenoplasts, novolacs (fusible resins). Alberit, Alresins, Avis, Azolone, Bakelite, Cegeite, Chierolo, Dendrodene, Durez, Durophen, Elo, Epok R, Fabrolite, Featalak, Fiberite, Fluosite, Fudow, G. E, Gederite, Kerit, Lerite, Metholon, Moldesite, Mouldrite, Nestorite, Phenall, Progilite, Resart, Resinol, Rockite, Rogers, Sarvis, Setalict, Sirfen, Sternite, Synmold, Tessilite, Tessilplast, Trolitan. *Modified resins* Arochem, Beckacite, Caladene. *Cast resins* Catalin. *Laminates* Aroborite, Celisol, Celotex, Paxolin, Tufnol. *Ion-exchange resins* Amberlite. *Surface-coating resins* Acrophen (plywood and particle board adhesive), Amberlac, Amberol, Becophen, Compregnite.

GENERAL CHARACTERISTICS

Novolacs Transparent, almost colourless to dark red-brown, brittle, rosin-like materials; fusible and soluble; relatively simple in structure, not hardened by heat. *Cast resins* Transparent, almost colourless materials, commonly employed pigmented; infusible and insoluble after thermal hardening, but machineable: stronger and tougher than novolac resins. *Moulded resins* Shaped end-products, rendered infusible and insoluble by heat and pressure; amber to dark brown in colour, and brittle unless—as is usual—containing a reinforcing filler. *Laminates* Rigid light- to dark-brown boards composed of sheets of resin-impregnated fabric or paper bonded together under heat and pressure; readily machineable. *Foams* Rigid heat-resistant expanded products, derived from a resin syrup containing a blowing agent that reacts with a hardening acid.

28.1 STRUCTURE

Simplest Fundamental Unit

Linear polymers
C_7H_6O, mol. wt. = 106

Cured resins
(3-dimensional networks)

Molecular weight Novolacs, 500 or more (degree of polymerisation ranges from about 2 to 13). Resin intermediates, from *c*. 125 to over 1000; cured resins, indeterminate (high, but finite).

X-ray data Substantially amorphous substances (fillers may be crystalline).

28.2 CHEMISTRY

28.21 PREPARATION

The presence of the hydroxyl in phenol* (C_6H_5OH, m.p. $41°C$) causes formaldehyde (in aq. solution, as Formalin, *see* §26.21) to react with the aromatic ring at the *o*- and/or *p*-positions. This is assisted by heating but can be a slow process unless a basic or acid catalyst is included. The main reactions are of two kinds, as follows.

(*a*) Direct ADDITION of phenol and formaldehyde†, yielding *methylol* derivatives (sometimes termed phenol alcohols), e.g.,

The product shown in the above example is *o*-methylolphenol (m.p. $86°C$) known also as *o*-hydroxymethylphenol, *o*-hydroxybenzyl alcohol, or saligenin. This kind of reaction—involving the formation of methylol groups—is catalysed by strong bases and is favoured by an excess of formaldehyde. Strong acids bring about a similar reaction but make it proceed rapidly to further stages (*see below*). The use of ammonium hydroxide or an amine as a basic catalyst causes rapid production of hydrophobic compounds, e.g. alkyl-aminomethylphenols and their condensation products.

(*b*) Direct CONDENSATION of phenol and formaldehyde or of phenol and a methylol derivative‡, with loss of water, yielding *methylene* derivatives, e.g.,

---

*Additional to coal and petroleum tars, synthetic processes are now major sources of phenol; e.g. monochlorination or sulphonation of benzene followed by hydrolysis, or oxidation of toluene to benzoic acid followed by oxidative decarboxylation, or propylation of benzene to *iso*propylbenzene (cumene) which is converted to the hydroperoxide and then decomposed.

†In aq. solution formaldehyde behaves largely as methylene glycol ($HOCH_2OH$); if this substance, which is unknown in the pure state, is regarded as the active material even the initial reaction with phenol becomes one of *condensation*, e.g.,

‡Or condensation of methylol derivatives, which can yield methylene ether ($—CH_2OCH_2—$) derivatives; however, by loss of formaldehyde these may revert to methylene ($—CH_2—$) derivatives.

300

The product shown in this example is 2,2'-dihydroxydiphenylmethane, which, being a substituted phenol, is capable of further condensation with formaldehyde and/or methylol derivatives. This kind of reaction —involving the formation of methylene groups—is catalysed by bases, and particularly by acids, and is favoured by an excess of phenol, i.e. by a deficiency of formaldehyde.

If these reactions are allowed to continue, assisted by heating, hydrophobic compounds eventually separate out as a lower layer. When this is freed from the aq. layer, and further volatiles are removed by distillation under reduced pressure, it yields a resinous product of moderate molecular weight (a *resol* or a *novolac* that can be made to yield a cross-linked polymeric network, *see* (i) and (ii) below). In the initial reactions the phenolic hydroxyl group remains preserved, but in the later stages of condensation it is probably directly involved in the formation of *o*- and *p*-quinone methides, e.g. production of an *o*-quinone methide.

These reactive structures, and more complex structures produced by their polymerisation, probably account for the yellow or brown colour of cured PF resins*.

In industrial practice the following conditions are of interest.

(i) *Alkaline catalysis and a P/F molar ratio of less than unity (i.e. formaldehyde in excess).*

This, by addition reactions together with some condensation, leads first to compounds of relatively low molecular weight, which have reactive methylol groups and are both soluble in suitable organic liquids and fusible. A typical product may be represented as follows,

*The surface discoloration attendant upon ageing of PF resins in air probably arises from the production of somewhat similar structures by oxidation, e.g.,

301

A reactive resinous substance of this kind (comprising numerous low molecular weight isomers*) is known as an *A stage resin* or *resol*.

Further heating causes further condensation and yields lightly cross-linked structures that are insoluble in organic liquids (but still swell) and infusible (though they soften on heating, they are not truly thermoplastic). An intermediate material of this kind is termed a *B stage resin* or *resitol*.

Continued heating eventually leads (as in the thermosetting of a PF moulding powder) to more extensive condensation and the production of a rigid non-swelling non-thermosoftening molecular network, the ultimate structure of which is represented ideally in §28.1. This end-product is known as a *C stage resin* or *resite*.

N.B. Alkaline catalysis with a P/F molar ratio considerably greater than unity (i.e. phenol in excess) tends to give only low molecular weight methylene-linked compounds, such as diphenylolmethanes; these are necessarily less reactive than resols with free methylol groups (though ultimately with sufficient heating they may yield cross-linked structures).

(ii) *Acid catalysis and a P/F molar ratio greater than unity* (*i.e.* phenol in excess).

This, by condensation, leads to polymeric products of relatively low molecular weight, which are mainly linear in structure, soluble in organic liquids and fusible, but unreactive (having ends stoppered with hydroxyphenyl groups, and a general absence of methylol groups). A typical product may be represented by,

or by,

$$C_6H_4(OH)CH_2[C_6H_3(OH)CH_2]_nC_6H_4OH$$

where n depends on the initial P/F molar ratio, and ranges from 1 to *c.* 12. An inert resinous substance of this kind (comprising numerous low molecular weight isomers*) is known as a *novolac*.

N.B. Acid catalysis with a P/F molar ratio of considerably less than unity (i.e. formaldehyde in excess) leads—with a high concentration of a strong acid—rapidly to network structures of high molecular weight, but it is of little practical importance since it is difficult to arrest the reaction at an intermediate stage.

Of the above conditions, (i) is the basis of the preparation of a **one-stage resin,** which is first made as a *resol,* then blended with fillers, pigments, etc., converted to a *resitol,* and ultimately cured under heat

*Even for short chains the number of isomers is considerable, consequently these resins are amorphous.

and pressure to an inert *resite*; (ii) is the basis of the preparation of a **two-stage resin,** which is made by blending a *novolac* with fillers, pigments, etc., together with a catalyst or potential catalyst and a substance capable of supplying more formaldehyde (e.g. para-formaldehyde or hexamethylenetetramine). The additional formalde-hyde (released under heat and pressure during moulding) converts the fusible novolac into an infusible network structure resembling that of a *resite*, as in (i). If phenol is replaced by resorcinol, novolacs can be obtained that are sufficiently reactive with added paraformalde-hyde to cross-link under neutral conditions, even at room temperature (*see also* §28.81ii). It is possible, too, by addition of acid, to bring about novolac-type cross-linkage in an A stage resin or resol (a reaction employed in phenolic resin adhesives that set on addition of an acid hardener).

For other reactions of phenols with aldehydes, *see* §28.81.

28.22 PROPERTIES

Solvents Novolacs and resols dissolve in alcohol, acetone, esters, ethers, and (to a less extent) hydrocarbons; oleophilic phenols (e.g. *p*-tert.-butylphenol, *p*-phenylphenol) confer solubility in hydro-carbons and drying oils. Cured resins are insoluble in common solvents, but dissolve (with decomposition) in molten phenols, e.g. in α-naphthol under pressure. **Plasticisers** are not usually used, but oleophilic phenols (above) reduce brittleness; modification with rubber may also be employed. Uncured resins are compatible with furfural, glycols, castor oil, and high-boiling esters. Zinc stearate is sometimes included as a lubricant in mouldable resins. For flexible resins from cashew nutshell liquid, *see* §28.81 I(iii). **Relatively un-affected by** most aq. liquids (including conc. non-oxidising acids), organic solvents, greases and oils; cured resins may to a slight extent be swollen by water, dil. acids and alkalis, alcohols, ketones, etc. **Decomposed by** conc. oxidising acids, and hot alkalis.

28.23 IDENTIFICATION

Mostly non-thermoplastic (novolacs and uncured resins fusible). Cl, N (present in some resins), S, and P normally absent.

Combustion Novolacs and uncured resins melt; cured resins char, ultimately burn, yielding odours of phenol and formaldehyde on extinction. Woodflour and cotton-filled resins show low combusti-bility; those filled with asbestos or glass-fibre are virtually incombustible. **Pyrolysis** Novolacs melt, decompose (*c.* 500°C) and distil as phenolic alcohols; cured resins decompose and char, with evolution of phenol and formaldehyde. The volatile products have an acid reaction, particularly if cellulosic fillers are present. Vacuum pyrolysis yields various products: at moderate temperatures, scission at —CH_2— bonds yields propylene, propanol and acetone; above 800°C, ring breakdown yields carbon monoxide (with CO_2 and CH_4) and a carbon residue.

Characterisation of resin components Note: PF resins (including filled mouldings and laminates) smell of phenol and formaldehyde when filed, sawn, or scraped with a knife. For the following colour tests insoluble resins should be reduced to a finely-divided state.

1. *Formaldehyde* Warming or distillation with dil. sulphuric acid yields formaldehyde, which can be detected by the violet colour it produces with chromotropic acid (*see* §26.23a).

2. *Phenols* I. Phenols are released: (i) on dry distillation, especially if the test sample is first mixed with soda-line, (ii) on fusion with alkali, and (iii) sometimes on boiling with water or aq. alkali.

N.B. Phenols may also result from aryl ester plasticisers, such as tritolyl phosphate.

To purify a phenol an aq. solution is saturated with carbon dioxide and extracted with ether, the ether extract is then dried (with Na_2SO_4), filtered, and evaporated. The purified phenol may be identified by means of *p*-xenylcarbimide (*p*-xenyl isocyanate) with which it forms derivatives, e.g. phenyl *p*-xenylcarbamate, m.p. 173°C (*see* MORGAN, G. T. and PETTIT, A. E., *J. Chem. Soc.,* 1124 (1931)).

II. The following tests can be applied direct to a pyrolysate or alkaline extract. (*a*) *Gibbs' indophenol test* A few milligrammes of dibromoquinonechloroimide are shaken with saturated aq. borax and added to a neutral or mildly alkaline test solution (pH of resultant mixture should lie between 8 and 10). On standing, phenols produce a blue or bluish-purple coloration. This is a very sensitive test. (*b*) *Liebermann's test* A small sample, mixed on a spot plate with conc. sulphuric acid, is treated with a few crystals of sodium nitrite. Phenolic substances give a blue or green colour that becomes red on dilution with water and blue or green again when made alkaline. (*c*) *Diazo test* To an ice-cold solution of aniline (or nitroaniline) in an excess of 20% sulphuric acid, slowly add dilute sodium nitrite solution; after a few minutes add an excess of the test solution (which must be previously made alkaline). Phenols precipitate a red dyestuff. (*d*) *Phenolphthalein test* Phenols heated with phthalic anhydride and a trace of conc. sulphuric acid yield a deep red colour when the cooled melt is extracted with aq. alkali. A blank control is desirable for this test. (*e*) *Other tests* Millon's reagent and the xanthoproteic reaction (*see* §22C.3) are applicable to certain phenols; with many phenols ferric chloride gives a violet or green colour.

28.3 PHYSICS

28.31 GENERAL PROPERTIES

Specific gravity *Novolacs* Phenol, 1·25; cresol, 1·17–1·20. *Cast resins* Unfilled, 1·34. *Moulded resins* Unfilled, *c.* 1·3; cellulose-filled (wood-flour, flock, paper, pulp, chopped fabric, etc.), 1·36–1·46; mineral-filled (heat resistant), 1·54–1·75; mineral- or glass fibre-filled (electrical applications), 1·75–1·92. *Laminates* Paper or cotton fabric, 1·34–1·38; glass fibre fabric, 1·67. **Refractive index** *Novolacs* Above 1·6. *Cast resins* Unfilled, *c.* 1·55. **Water absorption** (%/24 h). *Novolacs c.* 0·2–0·5. *Moulded resins* Unfilled, 0·15–0·60; woodflour and flock-filled, 0·3–0·8; paper- or pulp-filled, 0·4–1·5; glass fibre-filled, 0·1–1·0; mineral-filled (heat resistant), 0·2–0·5; mineral-filled (electrical applications), 0·01–0·07. On prolonged immersion absorption in all mouldings with cellulosic fillers may rise to 5–10%. Walnut-shell flour has lower water absorption than woodflour. *Laminated boards* (base as stated; mg in 24 h): cotton, 60–90; paper, 15–250; asbestos, $38 \times 38 \times 3$ mm³, 60–130. **Permeability** Highly impermeable to water vapour, but dependent on fillers. Permeability to N_2 at 20–30°C, 0·095 (10^{-10} cm²)/(s cmHg).

28.32 THERMAL PROPERTIES

Specific heat *Moulded resins* Unfilled, 0·4; cellulose-filled (woodflour, flock, etc.), 0·3–0·4; mineral- or glass fibre-filled, 0·28–0·32. **Conductivity** (10^{-4} cal)/(cm s °C) *Moulded resins* Unfilled or cellulose-filled, 3–5; mineral- or glass fibre-filled, 10–15. *Laminated boards* (transversely) 5–8. **Coefficient of linear expansion** (10^{-4}/°C) *Cast resins* Unfilled, 0·8; filled, 1. *Moulded resins* Unfilled, 0·45; cellulose-filled, 0·3–0·45; mineral-filled, 0·2–0·25; glass fibre-filled, 0·15. **Physical effects of temperature** Novolacs melt at 100–150°C. *Cast resins* soften at *c.* 100°C, but do not melt; prolonged heating increases the colour (which eventually becomes ruby red) and causes internal crazing. *Cured resins* (mouldings and laminates) soften only slightly at 120–180°C, when laminated boards can be bent to simple curvatures (post-forming). Fibrous mineral fillers increase thermal stability, but prolonged exposure above the softening temperature results in slow decomposition. Maximum service temperature (°C): unfilled resins, 150–220; cellulose-filled, 150; glass fibre-filled, 220; mineral-filled (heat-resistant), 225; do. (electrical applications), 150.

28.33 ELECTRICAL PROPERTIES

Volume resistivity (ohm cm) *Cast resins* Up to 10^{14}. *Moulded resins* (filler as stated): cotton, 10^8-10^{11}; woodflour or mineral, 10^9-10^{12}; mineral (low loss), up to 10^{14}. *Laminated boards* (base as stated): cotton or asbestos, up to 10^{10} or 10^{11}; paper, up to 10^{12}. *Surface resistivity* (ohm) Values approximately one-tenth of those for volume resistivity but up to 10^{14} for electrical grade mouldings and paper based laminates. **Dielectric strength** (kV/mm) *Cast resins* 8–12. *Moulded resins* (filler as stated): cotton or mineral, 4–10; woodflour, 4–12; mineral (low loss), up to 15. Moulded materials in oil at 90°C (resin type as stated): general shock-resistant, 1·5–4; heat resistant, 6; low-loss, 4·5–11. *Laminated boards* (base as stated): asbestos, 4; cotton, 8; paper, up to 20 (thin laminates appreciably higher). The dielectric strength of cellulose-filled resins falls to three-quarters or half the above values at 100°C. Proof strength of paper laminates in oil at 90°C (kV withstood, flatwise, for 1 min; thickness as stated): 1·6–3·2 mm, *c* 25; 13 mm, 50. **Dielectric constant and Power factor** Values at 1 kHz (tend to be higher at 50 Hz and slightly lower at 1 MHz:

	Dielectric constant	Power factor
Novolacs	*c.* 4	*c.* 0·02
Cast resins	4–8	0·01–0·05
Moulded resins (filler as stated)		
Woodflour	5–10	0·03–0·15
Cotton	5–10	0·1–0·2
Mineral	5–15	0·1–0·15
Mineral, low loss	5–6	0·02–0·1
Laminated boards (base as stated)		
Cotton	6–12	0·1–0·4
Paper	4–10*	0·02–0·2*
Asbestos	10–20	0·15–0·5

*Low loss grades, at 1 to 30 MHz: dielectric constant, *c.* 5; power factor, 0·04–0·05.

Temperature effects. Dielectric constant tends to increase with temperature rise, especially at low frequencies; power factor increases similarly, very rapidly at low frequencies (e.g. at 100°C, 10 times the value at 25°C).

28.34 MECHANICAL PROPERTIES

Elastic modulus and **Tensile, Flexural, Compression,** and **Impact Strengths:**

	Elastic modulus (kgf/mm^2)	Tensile strength (kgf/mm^2)	Flexural strength (kgf/mm^2)	Compressive strength (kgf/mm^2)	Impact* strength (ft lbf/in)
Cast resins					
Unfilled	c. 280‖	c. 6·5	c. 10	10–20	0·25–0·5
Moulded resins (filler as stated)					
Woodflour†	700–1000	3–6	6–10	15–25	0·1–0·5¶¶
Cotton†	600–900	3–6	7–10‡‡	15–25	0·25–2·5¶¶
Mineral‡	1000–2000¶	2·5–6·5	6–10	10–25	0·1–1·0
Laminated boards (base as stated)§					
Cotton	500–1000**	5·5–12·5	10–15‡‡	20–25‖‖	1–5¶¶
Paper	700–1500**	5·5–15	7–15 ‡‡	17–25‖‖	0·3–3·0¶¶
Asbestos	400–1400	5–8·5††	7–15§§	20–35	0·2–1·0***

*Izod test, notched
†General purpose resins; cotton as chopped fabric
‡Heat resistant grades usually inferior to electrical grades in mechanical properties
§Parallel to laminations; values normal to lamination are lower (but vice versa for impact strength)
‖In compression, 500–700
¶2000 for mica-filled
**In compression, normal to laminations, c. 5-times greater
††Glass fibre, 15 or more
‡‡Raised by high-strength fibrous fillers, e.g. chopped tyre cord, or special fabric or paper, up to 20
§§Glass fibre, 10–30
‖‖Shear strength (kgf/mm^2): parallel to laminations, 4–5; normal to do., c. 10
¶¶Raised by high strength fibrous fillers, e.g. to c. 10
***Glass fibre, up to c. 20

Elastic recovery From short period strain, good; some permanent set after prolonged strain. **Extension at break** 1–5%. **Hardness** Cast or moulded resins: Brinell (25 kg load, 2·5 mm ball), c. 30; Rockwell, M 60–130. Novolac: Brinell (as above), 10; Rockwell, M 50–60, R 50–100. Woodflour-filled resins, laminated boards: Brinell (as above), up to 45; Rockwell, M 70–125. **Friction** Coefficients (approx. values). μ_{stat} : cotton fabric laminate/polished steel + trace lubricant, 0·2–0·3. μ_{dyn}: cotton fabric laminate (30 mm/s)/polished steel + trace lubricant, 0·15–0·2. Grey cotton yarn (1 m/s, 65% r.h.)/smooth cast resins, c. 0·2; do./moulded resin (cotton filled), 0·25–0·3. **Abrasion** The resistance of laminates compares with that of aluminium or copper, but declines at high humidity.

28.4 FABRICATION

Castings Highly viscous intermediates, prepared with an excess of formaldehyde and an alkaline catalyst which is later neutralised, can be slowly heat-hardened into transparent resins (largely superseded, however, by acrylic resins—*see* Section 9). Liquid intermediates can be hardened at room temperature or with moderate heat after addition of a strong acid, e.g. an aryl sulphonic acid. *Mouldings* To be hardened by heat, novolacs require addition of a formaldehyde-donating substance (usually hexamethylenetetramine, 10–15 parts per 100 resin by wt.). Resols cure without a hardening agent. Resins are compounded with hardener (when required), fillers, colours, etc., and moulded under heat and pressure to thermoset end-products. Common reinforcing fillers include woodflour (to over 100 parts per 100 resin), chopped cotton fabric (also high strength fibres, including glass yarns), slate, asbestos and other minerals (for heat resistance); best electrical resistance is obtained with mica powder. Release from the mould is facilitated by inclusion of stearic acid (1–3 %) as lubricant; stearates, waxes, and castor oil are alternatives. Staining of steel moulds is prevented by inclusion of calcium or magnesium oxide. *Compression moulding* (transfer moulding also used) generally employs pressures of 2000–6000 lbf/in^2, at 150–170°C, for 5–15 min. *Laminated boards* are produced by hot-pressing (e.g. 2000 lbf/in^2, 190°C) assemblies of fabric or paper sheets that—singly—have been previously dried to a given moisture content after being impregnated with a solution of a curable resin. *General manipulation* Cast resins and laminated boards, and to a less extent moulded resins, can be sawn, drilled, and machined, and they can be glued with phenolic- or epoxy-type adhesives. Laminated boards can be post-formed to shapes of simple curvature when heated to 120–180°C.

28.5 SERVICEABILITY

Novolacs can show cold flow (pitch-like). *Cast resins* may undergo slight shrinkage and darkening with time. Filled *moulded resins* and *laminated boards* are dimensionally stable, changing only slightly with changes in humidity. Thermal stability is greatest when fibrous mineral fillers are used. The resins are unaffected by bacteria, fungi, or insects, except that under appropriate conditions of temperature and humidity moulds may grow on the surface of a resin or laminate incorporating nutrient-containing fillers.

28.6 UTILISATION

Novolacs provide exposure-resistant lacquers and varnishes, including modifications with drying oils, e.g. for coating paper, metal cans, and electrical equipment (including coil impregnation). *Cast resins* (transparent or pigmented, mostly requiring after-machining) are used for handles, knobs, etc., and ornamental objects. *Moulded resins* (filled; colour usually restricted to brown or black) need little or no after-machining and have numerous general applications, e.g. knobs, bottle caps, containers, electrical fittings (of considerable complexity). *Laminated boards* are used for decorative wall panels and table tops, oil- or water-lubricated bearings and quiet-running gear wheels (meshing with bronze or steel), wear-resistant components (such as picker bushings on looms), and electrical switch and terminal panels. Phenolic resins have numerous applications as *adhesives and bonding agents,* e.g. for wood (notably plywood), brake-linings (asbestos-filled), grinding wheels, foundry sand-cores, oil-filters, and pulp preforms (for travel goods, car fascia panels, etc.). Other applications include *expanded resins* for acoustic and thermal insulation (including light-weight 'syntactic' foams made from resin-bonded microballoons of expanded PF resin); *ion-exchange resins,* e.g. for water softening, or metal- or dye-recovery; and *in rubber compounding,* as tackifiers, reinforcing fillers or cross-linking agents.

28.7 HISTORY

Resinous substances produced by the action of acids on phenol-acetaldehyde adducts were noted as early as 1843; but A. von Baeyer, investigating the reactions in 1867–72, discarded the tarry products as not being pure and well-defined compounds. Related products were obtained by A. Michael, in 1883, by the action of alkalis; around 1890–1900 L. Lederer, and O. Manasse separately, showed that the initial intermediates in the reaction of phenol and formaldehyde are *o-* and *p*-hydroxybenzyl alcohols.

Onwards from 1890, when aq. formaldehyde became available commercially (from partial oxidation of methanol, *see* §26.21), PF reactions received increasing attention, culminating in the now 'classical' study made by L. H. Baekeland from 1902–1909 (patents 1907). Baekeland found that he could get either soluble resins (shellac substitutes which he called Novolaks—now spelt *novolac* and used as a generic term) or thermosetting resins, depending on the P/F molar ratio, and that acids and alkalis catalysed the respective reactions (*see also* §§28.21 and 28.81). A contemporary, H. Lebach, showed that the soluble resins could be converted to the thermosetting kind by incorporation of a formaldehyde donor (e.g. hexamethylenetetramine, thus lowering the P/F ratio), and that acid catalysts accelerated the cure of thermosetting resins. Baekeland's work established the principles on which the moulded filler-reinforced resins industry is based; commercial production (as Bakelite) commenced in 1910–12 and has since reached industrial importance in many countries.

In 1909, Sir J. Swinburne—independently of Baekeland—produced varnish-type PF resins (Damard lacquers), and in 1910 K. Albert and L. Berend developed surface-coating resins modified by rosin and ester gum (§24A.82). These were followed, in 1926, by fast-drying lacquers incorporating tung oil, and in 1928 by oil-soluble resins, e.g. based on *p*-tert-butylphenol.

Ion-exchange properties, discovered by P. A. Adams and E. L. Holmes in 1935, have since led to extensive commercial developments in this field.

The chemistry of phenolic resins continues to receive attention, prominent among more recent investigations (since *c.* 1940) being those of H. von Euler, K. Hultzsch, and A. Zinke.

28.8 ADDITIONAL NOTES

28.81 THE REACTIONS OF PHENOLS WITH ALDEHYDES

I. Phenols
(i) *Methyl-substituted phenols*
Although in the earlier stages of the reaction with an aldehyde the phenolic hydroxyl group may not be directly involved (*see also* §28.21), it influences the direction of substitution to the *o*-, *o'*-, and *p*-positions. The presence of *m*-positioned methyl groups enhances the rate of the *o,p*-substitution reactions; but *o*- and/or *p*-positioned methyl groups reduce the rate and necessarily block reaction at these points; thus, for example, the reactivity of the phenols illustrated below (reactive positions indicated by ·) decreases from left to right in the order 3,5-xylenol > *m*-cresol > phenol > *p*-cresol > *o*-cresol.

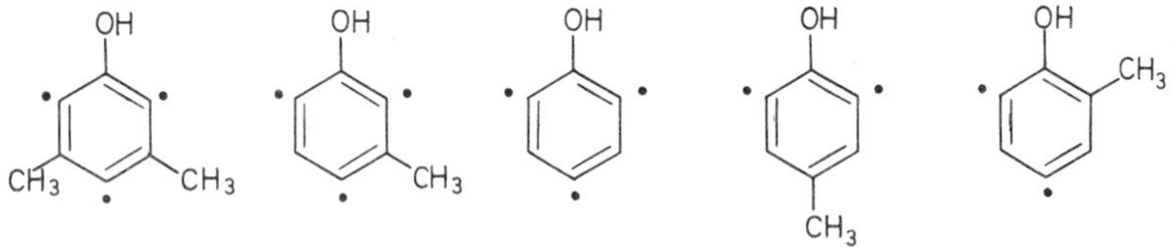

Since *o*- and *p*-cresol are bifunctional in the main reaction with formaldehyde, they tend to give only linear condensation products. Certain higher alkyl- or aryl-substituted phenols (e.g. *p-tert*-butylphenol, *p-n*-octylphenol, *p*-phenylphenol) are important in providing surface coating resins soluble in—and in some cases reactive with—drying oils (§24.81).

(ii) *Resorcinol*

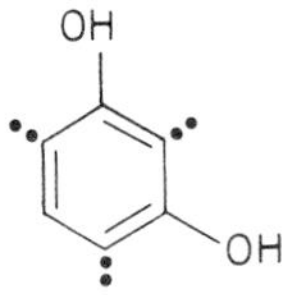

Following from above, the *m*-positioned second hydroxyl group of resorcinol, $C_6H_4(OH)_2$, reinforces the direction of substitution to *o*-, *o'*- and *p*-positions, and even under neutral conditions resorcinol-formaldehyde resins can cure at room temperature. Dark-red resorcinol-formaldehyde novolac syrups, that set and harden on stirring in additional formaldehyde (as paraformaldehyde, along with woodflour or walnut-shell flour filler), constitute valuable neutral adhesives for wood (Aerodux, Penacolite). Resorcinol-formaldehyde condensates are also employed to increase adhesion between synthetic fibres (e.g. as tyre cords) and rubbers.

(iii) *Cardanol* (cashew nutshell liquid)

The dark-coloured oil known as CNSL, which is extracted from the outer shell of the cashew nut (*Anacardium occidentale,* fam. *Anacardiaceae*), on steam distillation yields as the main product an unsaturated phenol called cardanol, which is represented as shown.

The CNSL (in which polyunsaturated phenols related to cardanol are also present) reacts with formaldehyde, and the intermediate products can be cured (either by heating or by addition of para-formaldehyde) to polymers having the usual characteristics of phenolic resins except that they are less rigid. This makes them useful where the normal resins are sometimes too brittle, e.g. for bonding brake-lining compositions and flooring compositions, flexible in-sulating oil- and water-proof varnishes, and sealing oil-filled cables. The resins are, however, relatively low in tensile strength.

II. Aldehydes

Formaldehyde (*see* §26.21) is the most reactive and cheapest aldehyde, and for these reasons—although acetaldehyde was used historically—other aldehydes are seldom employed, unless to enhance solubility in drying oils, e.g. by using acetaldehyde or butyraldehyde.

An exception, however, is *furfural* (furfuraldehyde) which is economically important as deriving from vegetable waste, e.g. oat hulls and corn cobs. With phenol it yields hard dark-coloured resins, which are used for chemical- and heat-resistant precision mouldings (Durite) and for bonding grinding wheels. The structure of the resins probably resembles that of phenol-formaldehyde resins, but with the additional possibility of vinyl-type linkages (cf. coumarone resins, §5.1), as represented below.

28.9 FURTHER LITERATURE

CARSWELL, T. S., *Phenoplasts* (*High Polymers*, Vol. VII), New York (1947)

ROBITSCHEK, P. and LEWIN, A., *Phenolic Resins: their Chemistry and Technology*, London (1950)

MARTIN, R. W., *The Chemistry of Phenolic Resins*, London (1956)

MEGSON, N. J. L., *Phenolic Resin Chemistry*, London (1958)

GOULD, D. F., *Phenolic Resins*, New York (1959)

WHITEHOUSE, A. A. K., PRITCHETT, E. G. K. and BARNETT, G., *Phenolic Resins*, London (1967)

Specifications

BS 771 (phenolic moulding materials)
BS 1137 (phenolic resin-bonded paper sheets)
BS 1314 (phenolic resin-bonded paper tubes)
BS 2076 (bonded paper insulating sheets for use at radio frequencies)
BS 2572 (phenolic laminated sheet)
BS 2907 (phenolic mouldings)
BS 2966 (phenolic resin-bonded cotton fabric sheets for electrical purposes)
BS 3927 (phenolic foam materials, thermal insulation, etc.)
BS Aerospace Standard PLI (phenolic mouldings for aircraft)
BS Code of Practice CP 3003(6) (phenolic resin linings for vessels)
ASTM D494 (acetone extraction test)
ASTM D700 (phenolic moulding materials)
ASTM D796 (recommendations for moulding phenolic resin specimens)
ASTM D834 (test for ammonia in mouldings)

NATURAL RUBBER

With notes on natural rubber latex; Para rubber; plantation rubbers;
anti-crystallising rubbers and iso-rubber; graft copolymers; gutta percha and related
substances; cyclised rubber; cellular rubber; reclaimed rubber; liquid rubbers

SYNONYMS AND TRADE NAMES

Poly(2-methyl-1,3-butadiene), *cis*-1,4-polyisoprene, caoutchouc, gum elastic, India (or india) rubber.

Dynat, Heveacrumb, Kualakep, Natcom (*see also* §§29.81 and 29.83).

GENERAL CHARACTERISTICS

Raw natural rubber is a translucent, light yellow to dark brown, soft and extensible material. It is more or less 'springy', showing elastic recovery (snap) from deformations of brief duration, but permanent deformation (set) results from a prolonged period of strain.

When kept below 0°C rubber hardens, but with rise of temperature it softens and weakens, having virtually no strength at *c.* 80°C. When softened by warming or mechanical working, it shows strong self-adhesion (auto-hesion or 'tack'). It dissolves in hydrocarbon, chlorinated hydrocarbon and some other liquids, giving useful adhesive cements. Suitably prepared raw natural rubber is very resistant to some types of abrasion (hence its use as, for example, crepe soles).

Raw natural rubber is compatible with synthetic polyisoprene, polybutadiene, styrene/butadiene, ethylene-propylene and polysulphide rubbers, but not readily with acrylonitrile/butadiene and silicone rubbers. Blending with butyl rubber is not generally practicable because the two rubbers have very different reactivities towards sulphur (cf. §38.4).

Vulcanised natural rubber shows good elastic recovery from both short- and long-period deformation, high resilience, and improved strength and resistance to temperature changes; it is insoluble, though it swells in hydrocarbon and some other liquids. When suitably compounded it has good electrical properties and resists many corrosive chemicals. The relatively high protein content of natural rubber (2–3·5%) renders it liable to attack by micro-organisms, even when vulcanised. Vulcanisates substantially taste- and odour-free can be made from pale crepe (§29.83).

Vulcanised rubbers change in properties with passage of time (ageing), due to oxidation, cross-linking and chain-scission in the molecular network, especially at elevated temperatures.

Natural rubber can be vulcanised to the hard or ebonite state (*see* Section 31).

29.1 STRUCTURE

Simplest Fundamental Unit

At least 97% consists of 1,4-isoprene units:

$$\underset{\displaystyle -CH_2-C=CH-CH_2-}{\overset{\displaystyle CH_3 \,\, | }{}}$$

C_5H_8, mol. wt. = 68

Note. 1,4-Polyisoprene exists in two stereoisomeric configurations: natural rubber has the *cis* configuration (in which the chains substituent to a double bond lie on the same side, as

313

in the diagram below), while gutta percha, balata and chicle (*see* §29.86) have the *trans* configuration (where the substituents lie on opposite sides).

Since rotation about a double bond is not possible, the two forms are not readily inter-convertible (*see*, however, §29.84).

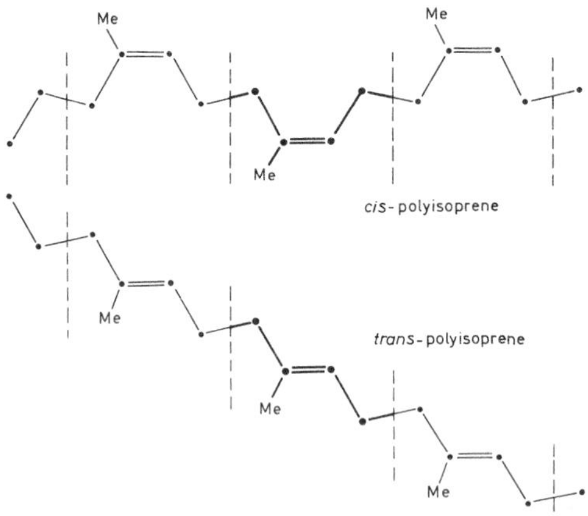

cis- polyisoprene

trans-polyisoprene

Main chain carbon atoms shown as ●, H atoms omitted. Me = CH₃

Molecular weight Ranges from $<100\,000$ to $>4.10^6$, with peak of distribution at $c.\ 1\cdot1.10^6$. Average values: natural rubber, unmilled, 680 000–840 000 (no.-av.); $1\cdot8$–$2\cdot1.10^6$ (wt.-av.); do., milled (*see* §29.4 Fabrication), 120 000–140 000 (no.-av.); 160 000 (wt.-av.). Other data (undefined averages), in fresh latex, up to $2\cdot5.10^6$, falling to 10^6 or less on standing; further reduced to $c.\ 350\,000$ when lightly milled or 100 000–150 000 in well-milled rubber.

Degree of polymerisation Lightly milled 5000; well-milled 1500–2000.

X-ray data Unstretched natural rubber, unvulcanised or vulcanised, is amorphous at room temperature, but becomes randomly crystalline when kept below its crystal melting point (T_m; *see* Section 68), i.e. $c.\ 30°C$, though crystallisation is very slow above $c.\ 15°C$. On stretching orientation causes it to become crystalline, e.g. by 40% at 1000% elongation; unit cell *either* monoclinic $a = 12\cdot46$ Å, $b = 8\cdot89$ Å, c (fibre axis) $= 8\cdot10$ Å, $\beta = 92°$, *or* orthorhombic, $a = 8\cdot97$ Å, $b = 8\cdot20$ Å, c (fibre axis) $= 25\cdot12$ Å, the cell containing 8 or 16 isoprene units respectively.

29.2 CHEMISTRY

29.21 PREPARATION

Natural rubber is obtained by cutting (tapping) the bark of *Hevea brasiliensis* (fam. *Euphorbiaceae*), a tree occurring wild in S. America and now cultivated in Ceylon, Malaya, and neighbouring countries of

S.E. Asia, and tropical Africa. It is also contained in other plants, some of which have been of commercial significance, e.g. the guayule shrub (*Parthenium argentatum*) of Mexico, and the Russian dandelion (*Taraxacum kok saghyz*)*.

Tapping—or crushing for small plants and shrubs—yields the rubber in the form of a milky suspension or latex (e.g. 32–35% rubber in *Hevea* latex; *see* §29.81).

The latex is either used as such, or the rubber is separated from it by coagulation with acid (usually acetic), the coagulum being washed, sheeted and either dried in air, giving *crepe*, or dried in wood smoke, giving *smoked sheet*. In a more recent process the coagulum is produced as a fine crumb—which facilitates washing and drying—and compressed into blocks (Heveacrumb).

For market grades of the above, and for other types of natural rubber, *see* §29.83 (also Para rubber, §29.82). For the further processing of rubber, to give finished products, *see* §29.4 Fabrication.

29.22 PROPERTIES

Solvents Raw (crude) rubber is less easily dissolved than freshly masticated rubber. Solvents include hydrocarbons (aliphatic, naphthenic, aromatic), chlorinated hydrocarbons (e.g. carbon tetrachloride, trichloroethylene), certain ketones (e.g. diethyl and ethyl propyl ketones, cyclohexanone, methylcyclohexanone), certain esters (e.g. butyl acetate and stearate, methyl benzoate, dibutyl phthalate), diethyl ether, anisole, dimethylaniline, and carbon disulphide. Addition of *n*-butanol to hydrocarbon solvents usually assists dispersal of residual matter.

Vulcanised rubber is essentially insoluble in all organic liquids although it swells in many; it will eventually dissolve in some high-boiling substances (e.g. *p*-cymene, nitrobenzene, *p*-dichlorobenzene) if heated in presence of air, but this involves complete breakdown of the molecular structure. **Plasticised by** (a) chemical peptisers, e.g. thio-β-naphthol, xylyl mercaptan, pentachlorothiophenol or its zinc salt, dibenzoyl disulphide; (b) physical softeners, e.g. fatty acids, hydrocarbon oils and waxes, petroleum jelly, bitumens ('mineral rubber'), tars, rosin, coumarone-type resins, factice. **Swollen** (when vulcanised) by many organic liquids, the swelling being greatest with hydrocarbons, their chlorine derivatives and carbon disulphide, and much less with polar liquids such as aniline, phenol and the lower alcohols and acids, *see* Table 29.T1. The vulcanised rubber swells in nearly all animal and vegetable oils except castor oil, and is slightly swollen after prolonged immersion in water (*see* §29.31). **Relatively**

*The ring of 'milk' seen on breaking the flower-stem of the common dandelion (*Taraxacum officinale*) is also a rubber latex.

unaffected by alcohol, acetone, water and dil. acids and alkalis. The vulcanised rubber is little affected by most mineral acids (e.g. conc. HCl and HF and 50% H_2SO_4 at 65°C; conc. HBr at 38°C and 85% H_3PO_4 at 65°C), alkalis (conc. NaOH and KOH at 65°C, conc. NH_4OH at 52°C), chlorine water (38°C), most salt solutions (65°C), glycol, glycerol, and C_1 to C_5 alcohols. **Decomposed by** strong oxidising agents (ozone, conc. H_2SO_4, HNO_3 unless very dil.); hardened and 'perished' by boiling conc. mineral acids.

Table 29.T1. SWELLING (% BY VOLUME, AT APPROX. EQUILIBRIUM) OF UNFILLED 95:5 RUBBER–SULPHUR VULCANISATE AT ROOM TEMPERATURE

Petrol	390 (170*)
Cyclohexane	460
Benzene	500 (220*)
Carbon tetrachloride	660 (280*)
Ethylene dichloride	225†
Diethyl ether	240
Acetone	14–27†
Ethyl acetate	33–104†
Amyl acetate	240
Carbon disulphide	580
Aniline	14

*Containing 55 parts of acetylene black (per 100 rubber)
†Rubbers not defined

29.23 IDENTIFICATION

Thermoplastic. N and P present in detectable amounts (0·3–0·6% and 0·04% respectively); P content distinguishes from synthetic polyisoprene rubber (*see* below and §32.23). Cl and S nominally absent. Presence of S with traces of N indicates a vulcanised rubber. Not saponifiable. **Combustion** Readily ignites, burns freely with a smoky flame, and (especially if vulcanised) yields a characteristic odour on extinction. **Pyrolysis** Softens to a sticky melt, decomposes to isoprene, dipentene, and other products, giving a characteristic and (in absence of sulphur) neutral distillate. The pyrolysate has a characteristic infra-red absorption spectrum, with diagnostic bands at 11·3, 7·3, 12·5, 6·1 and 11·0 μm (*see* BS 4181; also ASTM Designation D297 for colour reactions of pyrolysate. Vacuum pyrolysis yields largely isoprene.

Other tests
Colour reactions (i) and (ii) below are characteristic of isoprene polymers:

(i) *Weber reaction*—A small sample (preferably finely-divided) is extracted with acetone, then dissolved or suspended in carbon tetrachloride and treated with a little bromine in the same solvent. After shaking and warming, excess bromine and solvent are removed in a stream of air and the residue is fused with phenol. The dark-coloured product gives a purple solution with chloroform, a yellow cloudy solution with ether, and a purplish-grey precipitate (in a yellow solution) with acetic anhydride. The presence of carbon black in vulcanised rubbers darkens the solution, and the coloration may be difficult to see, but with small samples the test can still be used successfully.

A *positive* response is given by synthetic *cis*-1,4-polyisoprene (Section 32; for distinction from natural rubber see (iv) below), gutta percha (§29.86) and 'synthetic gutta percha' (§32.82), cyclised rubber (§29.87) and rubber hydrochloride (Section 30); also, when polyisoprene is the major component in a mixture a positive result is obtained, but the test does *not* detect the isoprene component of butyl rubber (Section 38). A *negative* result is also given by alkali-reclaimed natural rubber (§29.89) and chlorinated rubber (Section 30). A faint coloration given by unsaturated ethylene-propylene terpolymers (Section 37) can be distinguished from that given by polyisoprene by using controls.

(ii) *Trichloroacetic acid test*—A small sample (preferably after extraction with acetone), when warmed with trichloroacetic acid, produces a mauve-pink melt that changes to dark red or red-brown on boiling, and on dilution with water gives a grey-mauve precipitate that yellows on standing.

N.B. This test should be used with caution on vulcanisates, as some compounding resins, notably phenol-formaldehyde resins, give a positive result.

Synthetic *cis*-1,4-polyisoprene, cyclised rubber and rubber hydro-chloride respond to the test (and certain polysulphide rubbers give a pink or red colour; *see* §39.23). Chlorinated rubber, and synthetic rubbers other than those mentioned, are without effect, or produce yellow or brown colorations with hazy pale yellow or grey precipitates on dilution.

(iii) *Oxidation*—On oxidation with chromic-sulphuric acid (CrO_3/H_2SO_4, 20·5/28% by wt.) every four isoprene units of rubber yield approximately 3 molecules of acetic acid, which can be steam-distilled and titrated; however, as the isoprene/acetic acid ratio is not entirely constant, the reaction is unsuitable for quantitative work.

(iv) *Distinction from synthetic polyisoprene*. Natural rubber can be distinguished by detection of its protein content, and its higher phosphorus content (400 p.p.m. compared with 20 p.p.m. in synthetic polyisoprene—although the result with vulcanisates is unreliable, as those of synthetic polyisoprene sometimes contain lecithin). For detection of phosphorus, *see* §57.2. The protein may be detected as follows: (a) films of natural rubber from benzene solution show an infra-red absorption band at 6·5 μm, which is absent from synthetic polyisoprene; (b) the hydrolysate resulting from heating acetone-extracted rubber with dil. hydrochloric acid (to hydrolyse the protein) is subjected to 2-dimensional paper chromatography, using 75/15/10 *n*-butanol/formic acid/water in the first direction, and 80/20 phenol/water in the second. On detection with ninhydrin, the resolved amino acids (*see* Section 22) yield a characteristic pattern.

(For details of identification procedures *see*: WAKE, W. C., *The Analysis of Rubber and Rubber-like Polymers,* 2nd edn, London (1969); PARR, N. L. (Ed), *Laboratory Handbook*, London (1963), Chap. 50, EAGLES, A. E. and WILLIAMSON, A. G.)

29.3 PHYSICS

29.31 GENERAL PROPERTIES

Note. The properties of natural rubber vulcanisates are greatly dependent on the compounding ingredients, processing and vulcanising conditions used.

Specific gravity Unvulcanised, *c.* 0·93 (purified hydrocarbon 0·906–0·916); unfilled vulcanisates, 0·92–1·00. **Refractive index** n_D^{25} Unvulcanised, 1·519; unfilled vulcanisate, 1·526. For maximum *transmission of light*, a filler with a refractive index near to that of rubber must be used, e.g. 'light' magnesium carbonate (a hydrated basic carbonate) or certain silicas and silicates.

Water absorption Depends largely on water-soluble or hygroscopic components embedded in the rubber hydrocarbon matrix, which acts as a semi-permeable membrane; these components include latex serum constituents not completely removed from the crude rubber, and impurities in compounding ingredients. A high water absorption lowers electrical resistance and dielectric strength, and increases dielectric constant and power loss. For low water absorption, washed and dried rubber is preferable; deproteinised rubber shows the lowest water absorption of all. Whiting, zinc oxide, clay, and talc are suitable fillers. Absorption increases with period of immersion (sometimes, but not always, reaching an equilibrium), with rise of temperature or ambient humidity, and with ageing (oxidation) of the rubber, but is decreased by vulcanisation. Examples:

(i) Unvulcanised crepe, as strip 2·2 × 6 mm section, at 25°C:

Time immersed, days	2	3	6	11	25	58	134
Absorption, % by wt.	2·8	3·6	5·1	6·9	10·0	14·5	19·1*

(ii) Vulcanisate (95/5 smoked sheet/sulphur), as strip 2·3 × 5 mm section, at 34°C:

*Probably high due to oxidation.

Time immersed, days	1·75	8	17	37	54	132
Absorption, % by wt.	1·2	2·2	3·7	6·2	7·5	17·9*

(iii) Vulcanisates (a unfilled; b with 50 parts whiting plus 75 ZnO per 100 rubber):

Time immersed, hours	20	100	200	350
Absorption, mg/cm² of surface:				
a (at 70°C)	2·2	4·5	6	7·5
b (at 24°C)	—	2·0	2·7	4·0

(iv) Vulcanisates, as (ii), exposed to air at various humidities to give equilibrium absorption at 25°C:

Relative humidity, %	21	42	63	76	85	93	97·5	98·3
Absorption, % by wt.	0·10	0·19	0·28	0·47	0·78	1·58	2·50	2·80

(v) Temperature coefficient, per 10°C, of initial rate of absorption varies between 1·2 and 1·9.

Permeability to gases is somewhat reduced by vulcanisation; e.g. unfilled mix, permeability to H_2, with increasing time of vulcanisation:

Combined sulphur, %	0	0·8	2·2	2·4	3·0
Permeability $(10^{-10}$ cm²$)$/(s cmHg)	66	55	53	50	47

Permeability (P) increases with rise of temperature according to the equation $P = P_o \exp(-E_p/RT)$, where E_p = activation energy of permeation; e.g. data for unfilled vulcanisates:

Gas	H_2	He	N_2	O_2	CO_2
P at 25°C $(10^{-10}$ cm²$)$/(s cmHg)	45–52	29–30	7–9	20–24	130–135
E_p (kcal/mol)	6·9	6·5	9·3	6·6	6·2

Additional data 1. Permeability of unfilled vulcanisates $(10^{-10}$ cm²$)$/(s cmHg) (at 25°C): air 10–12; A, 11; CH_4, 29 (85 at 50°C); C_2H_2, 102; cyclo-C_3H_6, 192 (473 at 50°C); NH_3, 358. 2. Do., at 50°C: C_2H_6, 145; C_3H_8, 250; n-C_4H_{10}, 533; n-C_5H_{12}, 1260. 3. Transmittance of water vapour $(10^{-10}$ g$)$/(cm² s cmHg) through 0·13 mm films of raw rubber, 290; through 0·35 mm films of unfilled vulcanisate, 58.

Permeability to gases is lowered by fillers, especially if these have flattened particles disposed normally to the direction of permeation.

29.32 THERMAL PROPERTIES

Specific heat Raw, 0·45 (at 25°C; temperature coefficient 0·0012/°C); unfilled vulcanisates, 0·44; filled vulcanisates have lower values, e.g. with 50–52 parts carbon black (per 100 rubber), c. 0·36. **Heat of fusion** 16·4 cal/g. **Heat of combustion** (cal/g) Raw, 10700–10800; unfilled vulcanisates, 10400–10600; filled vulcanisates give lower values. **Conductivity** $(10^{-4}$ cal$)$/(cm s °C) Raw, 3·2; unfilled vulcanisates, 3·4–3·6; filled vulcanisates generally higher, e.g. with 50 parts carbon black (per 100 rubber), 6·8; with 95% litharge (x-ray protective rubber), above 10. **Coefficient of linear expansion** Above the glass temperature (T_g), raw or unfilled vulcanisate, $2·2–2·3 . 10^{-4}/°C$. The value increases with temperature and decreases with addition of fillers, e.g. with 50 parts carbon black (per 100 rubber), $1·5–1·8 . 10^{-4}/°C$. Below T_g, unfilled vulcanisates, $0·64–0·80 . 10^{-4}/°C$. **Physical effects of temperature** Raw natural rubber, when progressively cooled, becomes stiffer and eventually glass-hard and brittle. When cooled rapidly, it loses high elasticity at $c. -50°C$; glass temperature (T_g) -69 to $-74°C$. Crystallises and becomes relatively hard and opaque on prolonged storage at room temperature, or in a shorter time at sub-zero temperatures; crystallisation is most rapid at -25 to $-26°C$, reaching $c.$ 50% completion in 7 h. With rise of temperature raw natural rubber becomes transparent and translucent and more extensible, but at 50–60°C mechanically weak and sticky. Crystal melting point (T_m): monoclinic, 28°C; orthorhombic, 14°C. It begins to become fluid at $c.$ 120°C and decomposes with some depolymerisation at $c.$ 200°C.

Vulcanisation extends the useful temperature range, from $c. -55$ to $+70°C$. The glass temperature is -61 to $-72°C$ (unfilled vulcanisates) and the brittleness temperature -55 to $-62°C$, though this can be lowered as much as 20°C by incorporation of plasticisers, e.g. long-chain aliphatic esters such as di-2-ethylhexyl sebacate. Crystallisation at sub-zero temperatures is retarded, though not prevented, by vulcanisation; it can be greatly retarded by chemical modification of the raw rubber (*see* Anti-crystallising rubber, §29.84). With rise of temperature tensile strength and breaking elongation are not greatly changed up to $c.$ 100°C, but above this, strength may decrease greatly (*see* Table 29.T2); the vulcanisates tend to soften and become sticky at $c.$ 120°C upwards and become fluid in the region of 260°C, depending on the filler content.

29.33 ELECTRICAL PROPERTIES

N.B. Electrical properties of rubber can be greatly influenced by composition, conditions of vulcanisation, ageing, humidity, strain, temperature, the magnitude and frequency of applied potential, and the period of its action.

Volume resistivity (units $= 10^{15}$ ohm cm) Raw rubber (commercial), 0·3–5·0; do. (purified), 160–220. Unfilled vulcanisates, 2–4 (higher, e.g. 17, if dried before test); vulcanisates from purified raw rubber, 200 (20°C), 0·01–0·1 (100°C). Filled vulcanisates, 50 vol. (i.e. 50 × sp. gr.) filler per 100 parts by wt. of rubber: anatase, 1·2, china clay 4·3, halloysite clay 0·6, lithopone 1·5, rutile 4·2, talc 7·2, whiting 1·4, zinc oxide 2·7. Vulcanisate with 40 vol. clay, dried before test, 17.

Insulating rubbers up to 100. Resistivity is lowered by carbon black, e.g. tyre tread containing 50 parts HAF black (per 100 rubber), 200–600 ohm cm; hence *black rubbers are not reliable insulants.* Rubbers of very low resistivity, down to 1 ohm cm, are made by incorporating special carbon blacks, e.g. acetylene and conducting furnace (CF) blacks. Rubber products with *resistance** below 5.10^4 ohm are classed as 'conducting'; those between 5.10^4 and 10^8 ohm as 'anti-static' (BS 2050, which gives recommended values for various types of product). *See also* R. H. NORMAN, *Conductive Rubber:* its *Production, Application and Test Methods,* London (1957). *Surface resistivity* (ohm $\times 10^{15}$) Unfilled vulcanisates (50% r.h.), 11·5; do. (70–90% r.h.), 8. **Dielectric strength** (kV/mm) Raw rubber (1 mm thick, tested with a.c.), 16–17·5. Unfilled vulcanisates (0·22–0·70 mm), 34–58 (a.c., voltage raised slowly), 90–100 (a.c., voltage applied abruptly), 90 (d.c.). Filled vulcanisates give lower values, e.g. 12 (30 vol. zinc oxide per 100 parts by wt. of rubber), 15–20 (40–65 vol. whiting, 0–6 vol. gas black), 9–10 (rubber floorings). **Dielectric constant** Raw rubber (commercial), 2·4–2·6 at 1 kHz; do. (purified), 2·35–2·38 at 60–10^6 Hz. Unfilled vulcanisates, 2·5–3·3 depending on frequency (60–10^6 Hz) and temperature (-20 to $+100°C$). Dielectric constant is increased by fillers and generally by immersion in water, e.g. results at 500 Hz:

Time immersed at 70°C, days	0	7	14
No filler	2·7	3·2	3·3
Zinc oxide (50 vol.)*	7·7	10·9	10·7
Whiting (50 vol.)	3·9	4·4	4·6
Furnace carbon black (25 vol.)	28·0	21·1	19·8

*I.e. (50 sp. gr.) per 100 parts by weight of rubber

Other data: with clay (40 vol.) and rubber dried before test, at 1 kHz, 3·3; insulation rubbers, *c.* 2·8 (50 Hz). **Power factor** Raw rubber (commercial), at 1 kHz, 0·0014–0·0029; do. (purified), do., 0·0008–0·0026. (For fuller data on effects of frequency, temperature and humidity *see* NORMAN, R.H., *Proc. I.E.E.,* 2A, 100, 41 (1953).)

Unfilled vulcanisates, power factor varies greatly and in a complex manner with temperature and frequency, e.g.

Temperature °C		-20	$+25$ to 30	$+100$
Frequency, Hz,	60	0·005–0·05	0·002	0·002–0·003
	600–1000	0·02–0·07	0·002–0·004	0·002–0·010
	10^6	—	0·017–0·05	0·004–0·007

For fuller data see KITCHIN, D. W., *Ind. Engng. Chem.,* 24, 549 (1932); SCOTT, A. H., MCPHERSON, A. T. and CURTIS, H. L., *Bur. Stand. J. Res.,* 11, 173 (1933); NORMAN, R. H. and PAYNE, A. R., *Trans. Inst. Rubber. Ind.,* 41, T191 (1965).

*i.e. the resistance of the discharge path through the product.

Power factor is usually raised by fillers, e.g. clay (40 vol., 1 kHz), 0·011; whiting (50 vol., 500 Hz), 0·006; zinc oxide (50 vol., 500 Hz), 0·014 (0·021–0·022 after 1–2 weeks in water at 70°C).

29.34 MECHANICAL PROPERTIES

Raw rubber (i.e. 'crude' or unvulcanised) is in general mechanically weak, deficient in elastic recovery, sensitive to temperature (becoming stiff when cold and soft when hot) and greatly swollen or dissolved by many organic liquids. Hence, except for some limited applications, e.g. 'crepe' soles (having excellent abrasion resistance), adhesive solutions and resin/rubber blends, rubber is vulcanised (cross-linked) so as to improve strength and elastic recovery, reduce temperature-sensitiveness and improve resistance to liquids.

Vulcanisates

Note: the properties of vulcanised natural rubber are greatly influenced by the degree of vulcanisation and the amount and nature of fillers and other ingredients.

Unfilled ('*pure gum*') *vulcanisates* have high tensile strength and extensibility; recovery from deformation at room temperature is rapid and virtually complete, and resilience (rebound) is correspondingly high (typical data for vulcanisates are given in Table 29.T2).

Reinforced vulcanisates (with reinforcing fillers such as carbon black, *see* §29.4 Fabrication) show little or no increase in tensile strength over unfilled vulcanisates, but increased tear strength, stiffness and (generally) abrasion resistance, while extensibility is reduced. Fillers increase residual deformation (set) and the hysteresis (and heat build-up) in repeated cyclic deformation. However, hysteresis remains less than for similar styrene/butadiene rubber (SBR) vulcanisates (Section 34). Abrasion resistance is high, at normal or cool temperatures being at least equal to that for SBR, but it is inferior to that for butadiene rubber (BR) (Section 33). Resistance to formation ('initiation') of cracks by repeated flexing is not so good as with BR or SBR, but resistance to crack growth ('propagation') is better. Vulcanisates containing non-reinforcing (inert) fillers (§29.4 Fabrication) have lower tensile and tear strengths and especially abrasion resistance than reinforced vulcanisates. *Other properties.* Bulk modulus (kgf/mm^2): raw, or unfilled vulcanisate, 200; vulcanisate with 50 parts carbon black (per 100 rubber), 240. Poisson's ratio at small (up to 25%) deformation, vulcanisates, c. 0·50. Creep is linear with log(time); increase in strain (as percentage of initial strain, per unit of log (time), at 25°C), vulcanisates, unfilled, 1–3; do., with 50 parts carbon black (per 100 rubber), 7–12.

Table 29.T2. TYPICAL VALUES OF MECHANICAL PROPERTIES, AT *c.* 20°C UNLESS OTHERWISE STATED

	Unfilled vulcanisates	Vulcanisates with 50 parts HAF carbon black (per 100 rubber)
Elastic modulus (kgf/mm^2)		
Young's, static	0·1–0·2	0·35–0·6
shear, static	0·03–0·07	0·14–0·18
shear, dynamic	0·04–0·10†	1·1‡
Tensile strength (kgf/mm^2)	1·7–3·0§	2·2–2·8*
do. 100°C	1·75–2·1	1·8
do. 140°C	0·35	—
Breaking elongation (%)	675–900§	450–600
do. 100°C	950–1000	—
Stress at 300% elongation (kgf/mm^2)	0·14–0·21	1·0–1·65
do. 100°C	0·11–0·14	—
Tear strength (kgf/mm)	5·1	7–14
do. 100°C	4·3	8
Hardness (IRHD)	30–45	60–70††
Compression set‖	—	10
Resilience (rebound) (%)	70–93	50–62
do. −20°C	—	15
0°C	—	33
40°C	—	54
80°C	—	60
100°C	76–95	65–80**
Loss tangent (tan δ)‡‡		
60 Hz	0·03	0·10
1 kHz	0·04	0·13
Internal friction (kP)		
shear, 60 Hz	—	*c.* 7
do. 60 Hz 100°C	0·3	2·5–4
do. 140 Hz 100°C	0·15	—

*Up to 3·5 kgf/mm^2 with other carbon blacks
†Varies little with frequency, 60 Hz to 1 kHz; at 25–180 Hz and 100°C, 0·03–0·04
‡Varies little with frequency, 60 Hz to 1 kHz
§For films or threads from latex (cf. 29·81); tensile strength up to 4·5, elongation up to 950
‖Compressed 25% at 70°C for 24 h; recovery 10 min at 70°C; set expressed as percentage of compression
¶Type of carbon black not stated; minimum resilience (4%) at −35°C
**EPC carbon black
††Up to *c.* 95 with higher proportions of fillers
‡‡Other data for unfilled vulcanisates: at 1 Hz, 0·01–0·05; at 40–140 Hz and 100°C, *c.* 0·035

Other data

(1) **BSS 1154 and 1155,** for vulcanised natural rubbers of various hardness, specify the following values; the ranges indicated cover rubbers of different composition and use.

Hardness, IRHD	32–40	42–50	52–60	62–70	72–80	82–89
Tensile strength, kgf/mm², minimum	1·6–2·1	1·7–2·1	1·7–1·8	1·4	1·1	0·9
Breaking elongation, %, minimum	650–700	600	500	350–400	250	150
Compression set*, %, maximum	25–35	25–35	25–30	30–35	35	45

*Compressed 25 % at 70°C for 24 h; recovery 10 min at 70°C; set expressed as percentage of compression

(2) **Friction** Against various materials (metal, varnished and polished wood, linoleum, PVC flooring, concrete, stone, grass):

Rubber:		Raw (crepe)	Vulcanised	Resin-rubber*	Microcellular†
Hardness (IRHD)		44–48	66–85	90–95	40–55
μ_{stat}	dry	0·80–1·40	0·50–1·05	0·30–0·75	0·40–0·91
	wet	0·48–1·05	0·55–0·78	0·34–0·65	0·62–0·82
μ_{dyn}	dry	0·50–1·30	0·70–1·00	0·40–0·87	0·50–1·20
	wet	0·45–0·92	0·43–1·07	0·40–0·88	0·38–1·20

(Data from BARRETT, G. F. C., *Rubber J.*, 131, 685 (1956), *see also* BOWDEN, F. P. and TABOR, D., *Friction and Lubrication of Solids*, Oxford, 69 (1954).)

*Vulcanised mixture of rubber with high-styrene (§34·1) or other resins.
†*See* §29·88.

Against grey cotton yarn (1 m/s, 65% r.h.), soft vulcanised rubber, c. 0·5–0·9, generally increasing with softness of the rubber ebonite, 0·22.

29.4 FABRICATION

Apart from limited uses in the unvulcanised form (*see* Utilisation), natural rubber is always vulcanised and usually mixed with fillers and other ingredients.

Prior to mixing, the rubber is subjected to *mastication* (softening or plasticising) by working on a 2-roll mill (milling) or, more usually, in a 2-blade internal mixer, which reduces it to a plastic state in which it will mix with 'compounding' ingredients in powder form.

Vulcanisation, to reduce adhesiveness and improve the mechanical properties, useful temperature range and resistance to organic liquids, is effected by a vulcanising agent, usually sulphur, and the action of heat; other agents are: sulphur-containing organic substances (*see also Accelerators*, below), organic peroxides and—for thin rubber layers, e.g. proofings—sulphur monochloride (S_2Cl_2) used without heat ('cold curing'). Vulcanisation can also be effected, without sulphur or other agent, by high-energy radiation. Vulcanisation involves formation of cross-links between the rubber chain molecules; when sulphur is used these are sulphur ($-S_x-$) bridges (where $x \geqslant 1$), but a peroxide or radiation produces direct C—C

324

cross-links. In normal soft vulcanisates there is one cross-link per 500–1000 C atoms of the polyisoprene chain (contrast Ebonite or hard rubber, §31.1). Sulphur dosage is *c.* 1–3% on the rubber, depending on the accelerator (*see* below) and other factors, and vulcanisation is commonly effected in *c.* 10–50 min at 120–160°C, or even seconds at temperatures up to 200°C (e.g. continuous vulcanisation of rubber-covered cables). However, massive articles require a longer time because rubber is a poor heat conductor. (*See also* §31.4 for the special case of vulcanisation of ebonite.)

Other 'compounding' ingredients commonly added are:

Accelerators These reduce the time and/or temperature and amount of sulphur needed for vulcanisation, and often improve the mechanical properties and resistance to ageing (oxidation) of the vulcanisate. The metal oxides (MgO, PbO) formerly used are now largely superseded by organic accelerators, broadly classified as basic and acidic. Basic types, which include thiocarbanilide, diphenyl- and di-*o*-tolylguanidines, and aldehyde-amine condensation products, have, however, been largely replaced by the more active acidic types, represented by the following groups and examples:

(i) *Thiazoles,* of which mercaptobenzthiazole (MBT) and dibenzthiazyl disulphide (MBTS) are the commonest, the latter showing 'delayed action' (i.e. no activity until above a certain temperature).

(ii) *Sulphenamides,* also with delayed action, e.g. cyclohexylbenzthiazyl sulphenamide (CBS), *t*-butylbenzthiazyl sulphenamide (TBBS), and 2-(4-morpholinyl mercapto)benzthiazole:

R = cyclohexyl, *t* - butyl,
or morpholinyl — N(CH₂—CH₂)₂O

(iii) *Thiurams,* e.g. tetramethylthiuram disulphide (TMTDS):

$$(CH_3)_2N\cdot C(:S)\cdot S\cdot S\cdot C(:S)\cdot N(CH_3)_2$$

and the tetraethyl homologue, and corresponding monosulphides, all classed as 'ultra'-accelerators. The disulphides contain sufficient labile sulphur to vulcanise without need of the free element.

(iv) *Dithiocarbamates,* being metal (zinc, sodium) or alkylamine salts of dialkyldithiocarbamic acids, e.g. zinc dimethyldithiocarbamate:

$$Zn[\cdot S\cdot C(:S)\cdot N(CH_3)_2]_2$$

(v) *Xanthates*, e.g. sodium or zinc isopropylxanthate.

The last two groups ('super'-accelerators) make possible vulcanisation at low (even room) temperatures, and hence—especially the water-soluble sodium and dialkylamine salts—are particularly suitable for use in latex (*see* §29.81).

Activators: these are necessary for the effective functioning of most organic accelerators. Zinc oxide is the commonest, and a fatty acid (e.g. stearic) is often added to promote its activity. *Fillers* range from inert diluents, e.g. whiting, talc, barytes and coarse clays, to reinforcing fillers that increase tensile and tear strengths and abrasion resistance. Typical strongly reinforcing fillers are the fine-particle carbon blacks, e.g. 'high-abrasion furnace' (HAF) and 'channel' blacks; less reinforcing are lampblack, 'semi-reinforcing furnace' (SRF) and 'fine thermal' (FT) blacks*. For white or coloured products, silica, fine (e.g. china) clays, calcium and aluminium silicates and light magnesium carbonate are useful reinforcing agents, though inferior to carbon black in improving abrasion resistance. Fibrous fillers, e.g. cotton flock and asbestos, and synthetic resins, e.g. 'high-styrene' resins (§34.1), are used for special purposes. *Protective Agents* prevent or delay ageing, i.e. deterioration due largely to oxidation, which is autocatalytic. Oxidative conditions to which vulcanised natural rubber is susceptible include: natural ('shelf') ageing; oxidation catalysed by trace-metal (Cu, Mn) compounds; effects of heat and light; repeated flexing (causing cracking); atmospheric (ozone) cracking. Oxidation may be retarded by *antioxidants* and/or *sequestering agents* (for trace metals); for ozone-cracking, *antiozonants* and/or waxes (which bloom to the surface and form a protective film) are used. Important antioxidants are either amines or their derivatives (e.g. phenyl-β-naphthylamine, ketone-amine condensation products) or phenols and their derivatives (e.g. p,p'-dihydroxydiphenyl; 2,6-di*tert*butyl-4-methylphenol); in general the former are classed as 'staining' (i.e. discolouring the rubber, especially on exposure to visible or ultra-violet light, and/or staining of neighbouring materials by migration) and the latter as 'non-staining' although phenolic antioxidants may discolour on exposure to ultra-violet light. Anti-ageing effects are conferred by some organic accelerators. *Softeners and lubricants* in the form of stearic acid, waxes, mineral oils, tars, and bituminous materials ('mineral rubber', a name originally applied to the naturally occurring elaterite and gilsonite, but now to petroleum bitumens), are added to plasticise or increase the self-adhesion of rubber mixes, and to assist in their manipulation. Some softeners, especially rosin or coumarone/indene resins (Section 5), increase the tackiness of rubber, e.g. for adhesive tapes. *Other ingredients* added to rubber include *peptisers* to assist mastication, pigments, odorants and deodorants, stiffeners, fungicides, and, for cellular materials, *blowing agents* (*see* §29.88)

*Other types of blacks are: EPC (easy-processing channel), FEF (fast-extrusion furnace), MPC (medium-processing channel), MT (medium thermal), and SAF (super-abrasion furnace).

For further details on compounding and vulcanisation *see* ALLIGER, G. and SJOTHUN, I. J., *Vulcanisation of Elastomers*, New York (1964); HOFMANN, W., *Vulcanisation and Vulcanising Agents*, London (1967); WILSON, B. J., *British Compounding Ingredients for Rubber*, 2nd edn, Cambridge (1964); PIOTROVSKII, K. B. and SALNIS, K. YU., *Auxiliary Substances for Polymeric Materials*, transl. MOSELEY, R. J., Shawbury (1967).

29.5 SERVICEABILITY

The chemically unsaturated structure of natural rubber (*see* §29.1) renders it liable to oxidation, with consequent deterioration of useful properties. This occurs in all forms of the material on long keeping in air, and is aggravated by sunlight and elevated temperature. When vulcanised rubber is slightly stretched (as, e.g. in bent cables), atmospheric ozone causes cracks normal to the stretch direction. These effects can, however, be largely but not entirely mitigated by suitable protective agents (*see* **Fabrication,** above).

Natural rubber vulcanisates resist cold well, but heat only moderately; they are combustible unless specially compounded. They have good resistance to corrosive chemicals (unless strongly oxidising) but are adversely affected by hydrocarbon liquids. When appropriately compounded, they have good electrical insulating properties, or can be made electrically conducting.

The proteins, etc., normally present in natural rubber render it liable—even after vulcanisation—to attack by micro-organisms; the purified rubber hydrocarbon has been variously stated to be 'unaffected by bacteria, fungi or insects', but 'can be consumed by micro-organisms' and 'will support growth of fungi'.

29.6 UTILISATION

Uses of *unvulcanised* natural rubber are restricted by the relatively small range of temperature over which it retains its properties, by its inability to sustain a load without acquiring a permanent set, and its lack of resistance to many organic liquids. It is, however, highly abrasion resistant and hence has found use, as pale crepe, for boot and shoe soling. It is also used in pressure sensitive adhesives, adhesive solutions and rubber/resin blends, and when not subject to oxidation is a good electrical insulator.

The superior mechanical properties and resistance to organic liquids and to temperature changes make *vulcanised* natural rubber suitable for many applications. The greatest amount (*c.* 65%) is used in pneumatic tyres, especially those for heavy duty, where its relatively low heat build up—compared with competitive synthetic rubbers—is an advantage. Other uses are in conveyer and transmission belting; electrical insulation; footwear; hose and tubing; floorings, mats and road surfacings; surgical goods; sports accessories, including balls; proofed materials, e.g. clothing and inflatable articles; thread; and numerous mechanical applications including anti-vibration mountings, shock-absorbers, seals, packings, weather-strips, and valves. It is also used in cellular form (§29.88) and in ebonite (Section 31).

29.7 HISTORY

Fossilised natural rubber discovered in 1924 in lignite deposits in Germany is believed to date from the Eocene period (approx. 60 million years ago). Rubber was first seen by Europeans during the second voyage of Columbus to the West Indies (1493–6), and subsequently Spanish writers described its use by the New World natives in religious ceremonial, medicine, and ball games. It was first mentioned in print, as 'gummi optimum', in 1530; but it remained a curiosity in the West for nearly three centuries, until it was put to one of the first uses, that of erasing pencil marks (J. Priestley, 1770), whence the name India *rubber*.

In 1826 Michael Faraday showed natural rubber to be a hydrocarbon, and in 1860 Greville Williams suggested the basic C_5H_8 unit, having identified isoprene as its chief pyrolytic product. This was confirmed by W. A. Tilden, who established the general structure of isoprene. Other contributions towards the nature of rubber as polyisoprene were made by C. Harries (ozonisation) in 1904–5, H. Staudinger (catalytic hydrogenation) in 1922, and J. P. Katz and M. Bing in 1925 (x-ray crystallinity of stretched rubber).

A major use of rubber in its early applications was for waterproofing textile fabrics, a patent for this (with rubber dissolved in turpentine) being granted to S. Peal in 1791, while C. Macintosh developed double-texture proofed fabrics in 1819 (patent 1823). Solid tyres were patented by R. W. Thomson in 1867 and pneumatic tyres by J. B. Dunlop in 1888.

In 1820 T. Hancock discovered that rubber could be plasticised by purely mechanical means, such as mastication (*see* §29.4) between rotating cylinders, and in 1843—after an independent but undisclosed discovery by C. Goodyear in 1839—Hancock patented vulcanisation by heat and sulphur. Cold vulcanisation, using sulphur monochloride without heat, was discovered by A. Parkes in 1846.

Other discoveries of outstanding importance were those of the reinforcing power of fine-particle carbon black made by incomplete burning of natural gas, by S. C. Mote in 1904, and of organic accelerators of vulcanisation by G. Oenslager in 1905.

In 1876, H. Wickham conveyed seeds of the Brazilian wild rubber tree to Kew, whence, through seedlings sent to Ceylon, all the S.E. Asian plantations have developed.

For the artificial synthesis of 'natural' rubber, *see* §32.7.

29.8 ADDITIONAL NOTES

29.81 NATURAL RUBBER LATEX

(De Laval, Englatex, Hectolex, Hiltex, Lacentex, Laconvertex, Lacretex, Lanortex, Positex, Prevul, Qualitex, Revertex, Revultex, Vuljex.)

The latex is the milky fluid collected from the *Hevea* tree (*see also* §29.21). It contains 32–35% rubber and *c*. 5% non-rubber constituents (sugars, fatty acids and esters, sterols, salts, and proteins) suspended or dissolved in an aq. medium. The rubber particles average *c*. 1 μm in diameter (range 0·05–3 μm), and carry a negative charge. The specific gravity of natural latex is *c*. 0·98, and that of 60% latex (below) is 0·94–0·95.

For use in industry the latex is concentrated on the plantation, usually to about 60% rubber content (62% total solids). Concentration is carried out by centrifuging (De Laval, Hiltex, Lacentex, Laconvertex, Qualitex), creaming (Lacretex), electrodecantation, or evaporation (Revertex). All processes except the last—the standard product of which contains *c*. 67% rubber, 73–75% total solids—reduce the non-rubber constituents, thus giving a rubber with lower water absorption, an important factor in electrical products. Evaporation, however, produces a more stable latex with a higher penetrative power. Concentrated latex is usually preserved by adding ammonia; for centrifuged latex the amount is 1·6% or more NH_3 (on the water content), or 0·5% plus another preservative (e.g. sodium pentachlorophenate or boric acid)—*see* BS 4355. Potassium hydroxide plus a soap, or alternatively formaldehyde or other preservatives, can also be used.

The compounding ingredients used in latex, which are similar to those used in solid rubber (§29.4, **Fabrication**), are added either as aq. solutions or as dispersions in which the particles are of about the same size as the rubber particles and have the same polarity. The addition of a stabiliser is also necessary.

Articles such as gloves and balloons are made by dipping a former, coated with a coagulant, into the latex mixture, the coagulated and dried rubber film on the former being vulcanised in hot air. Toys and similar articles may be made in plaster or metal moulds; water filters through plaster, leaving a film of rubber, but with metal moulds the latex contains a heat-sensitising agent which causes gelation when the mould is heated.

Latex thread is made by extruding a fine stream of compounded latex into a bath of coagulant, the resulting thread of coagulum being dried and vulcanised. Latex is used also for proofing fabrics.

Latex foam rubber is mostly made by mechanically aerating a compounded latex to form a viscous foam; after adding a delayed-action gelling agent, the foam is poured into a mould, or spread on a backing (according to the end product), where it sets with or without application of heat. Further heating effects vulcanisation, and the product is finally dried.

Pre-vulcanised latex (Prevul, Revultex, Vuljex), made by heating with sulphur, zinc oxide and an ultra-accelerator (*see* §29.4, **Fabrication**) for *c.* 2 h at 70°C, allows a vulcanised film to be obtained without the need for curing ovens, etc.

A latex having *positively* charged (unvulcanised or vulcanised) rubber particles (Positex) will deposit on negatively charged textile fibres; the reversal of the latex particle charge is brought about, e.g. by pouring centrifuged ammoniated latex, diluted with water, into a solution of a quarternary ammonium salt.

For tests and specifications on natural rubber latex, *see* §29.9 **Further Literature.**

29.82 PARA RUBBER (Fine hard Para, Brazilian ball)

Named after the town at the mouth of the Amazon from which it was exported, this is derived from latex collected from wild *Hevea brasiliensis* trees. It is made by dipping a wooden paddle into the latex, holding the paddle in the smoke of a wood fire (the acid smoke coagulating the latex), and repeating the process until a ball or 'biscuit' of rubber is built up, which is cut in half and stripped from the paddle. This type of rubber is now less important than formerly.

29.83 PLANTATION RUBBERS

Within each main type (smoked sheet, crepe) there are several quality grades, which have hitherto been based largely on appearance —colour, presence or absence of dirt, bubbles, mould, etc. However, in recent years increasing amounts of plantation rubbers are being graded by technical quality (Technically Classified Rubber) or sold to a technical specification (e.g. Standard Malaysian Rubber).
Smoked Sheet The market grades of Ribbed Smoked Sheets (RSS) are: 1X, 1, 2, 3, 4 and 5; the quality falls off as the number increases.
Crepe. The following types may be sub-divided into thin or thick and/or quality grades: (i) *White* and *Pale Crepes*, made from fresh coagulum—*see* §29.21—with thorough milling and washing to remove non-rubber constituents—*see* §29.81. (ii) *Estate Brown Crepes,* from high-grade scrap—i.e. rubber coagulated spontaneously during collection or processing—generated on estates, by milling on power wash mills. (iii) *Compo Crepes*, as (ii) but from wet slab, tree scrap, lump or smoked sheet cuttings. (iv) *Thin Brown Crepes* (Remills),

as (ii) but from wet slab, unsmoked sheet, lump or other high-grade scrap on estates or small holdings. (v) *Thick Blanket Crepes*, as (ii) but thicker. (vi) *Flat Bark Crepes*, as (ii) but from all types of scrap including earth scrap. (vii) *Pure Smoked Blanket Crepes*, as (ii) but from RSS or RSS cuttings. (For full descriptions of types and grades see *International Standards of Quality and Packing for Natural Rubber Grades* (The Green Book), New York and Washington, D.C.: The Rubber Manufacturers' Association, Inc., 1969.)

Technically Classified (*TC*) *Rubber*—This is marked on the bale with a coloured circle to indicate (qualitatively) its rate of vulcanisation: yellow (normal), red (slow), blue (fast). The purpose is to supply rubber manufacturers with a raw material that can be used without the need for making tests of vulcanisation rate or blending to even out differences in rate.

Standard Malaysian Rubber (*SMR*)—This is sold to a specification based mainly on dirt content and the 'plasticity retention index' (PRI; *see* BS 1673: Part 3), a measure of resistance to oxidative deterioration. This covers pale crepe, smoked sheet, Heveacrumb (§29.21), and some other forms of plantation rubber.

Superior Processing Rubber (*SP rubber, PA 57, PA 80*)—This is made by coagulating a mixture of normal and vulcanised latices. With an 80:20 ratio the product is sheet or crepe, which, when masticated and mixed (cf. §29.4, **Fabrication**) has less tendency than normal rubber to become misshapen (due to elastic recovery) after being calendered or extruded. With a 20:80 ratio (PA 80) the product is a crumb, which can be blended with natural or synthetic rubbers for the same purpose.

Skim Rubber—Skim sheet and crepe result from coagulation of the diluted latex fraction obtained as a by-product in the centrifugal concentration of ammoniated field latex. It contains 75–85% rubber and up to 20% of protein, and is useful in hard products, although its variability and tendency to premature vulcanisation ('scorch') are disadvantages.

Rubber Powder (*Harcrumb, Mealorub, Pulvatex, Rodorub*)—This can be prepared by spraying concentrated latex, simultaneously with diatomaceous earth, into a current of hot air, so that a dehydrated mass of non-cohering particles is produced. Other methods include the preparation of lightly vulcanised powder by heating latex with substances that liberate sulphur. Rubber powder is used mixed with bitumen for road surfacings.

29.84 ANTI-CRYSTALLISING RUBBERS AND ISO-RUBBER

Anti-crystallising ('AC') rubber is characterised by greatly retarded rate of crystallisation, and consequently better maintenance of elastic properties, at sub-zero temperatures. This is achieved by destroying the regularity of the polyisoprene chain molecule of natural rubber (§29.1) by reacting it, either solid (at 40–50°C) or as latex, with an organic thiol acid, which effects some interchange from *cis* to *trans*, and may also introduce sulphur-containing side groups. A similar result is produced by reacting the rubber with butadiene sulphone (the active agent being sulphur dioxide) or cyclohexylazocarbonitrile at 170–180°C.

At *c*. −25°C (the temperature of quickest crystallisation of natural rubber) vulcanisates of 'AC' rubber crystallise up to 300 times more slowly than those from untreated rubber.

(*See*:CUNNEEN, J. I., *et al., Trans. Inst. Rubber Ind.,* **34**, 260 (1958); British Rubber Producers' Research Association, Technical Bulletin No. 4, London, *c*. 1961.)

Iso-Rubber
Removal of hydrogen chloride from rubber hydrochloride (§30C), by careful heating alone or with alkali or an organic base, gives iso-rubber, in which part of the unsaturation is positioned laterally:

$$CH_2$$
$$\|$$
$$-CH_2-C-CH_2-CH_2-$$

(cf. formula for natural rubber, §29.1).

29.85 NATURAL RUBBER GRAFT POLYMERS

Trade Names—Heveaplus MG and SG, followed by a number denoting the percentage content of grafted methyl methacrylate (MG) or styrene (SG).

General Characteristics
Natural rubber/methyl methacrylate (MMA) grafts vary, with increasing MMA, from tough 'crepe' (cf. §29.21) to granular thermoplastic solids. They can be vulcanised with sulphur, giving vulcanisates that show 'self reinforcement', being strong, tough and hard without addition of reinforcing fillers, and capable of being coloured, or more or less transparent.

Structure
Simplest fundamental unit The units are lengths of the polyisoprene chain constituting natural rubber (§29.1) and the grafted chains, normally of polymethylmethacrylate (Section 9) or polystyrene

(Section 4). These chains are attached either at the α-methylene C atom of the polyisoprene, or at a chain end of the latter having a double bond. Commercial graft polymers contain some homopolymer and unchanged rubber, e.g. 10% (of polymethylmethacrylate) and 12% in MG49.

Molecular weight $2.10^5 – 10^6$. In the usual products: main rubber chains, 200 000–330 000; grafted side-chains, 2000–5000 each.

Chemistry

Preparation Natural rubber is reacted with an unsaturated monomer (normally methyl methacrylate or styrene) in presence of a polymerisation initiator. The rubber may be (i) in latex form, using as initiator *tert*.butyl peroxide plus tetraethylene pentamine; (ii) in solution, using an organic peroxide; (iii) as solid swollen in the monomer, using, e.g. azobis*iso*butyronitrile. (For a full account *see* ALLEN, P. W. in BATEMAN, L. C. (Ed), *The Chemistry and Physics of Rubber-like Substances,* London and New York, 97 (1963).)

Properties
(*N.B.* The effects of various liquids depend on the nature and amount of grafted compound as indicated below.)

Solvents Those liquids that dissolve both natural rubber and either polymethylmethacrylate or polystyrene; e.g. for unvulcanised MG 23: benzene, cyclohexanone, carbon tetrachloride, amyl acetate; *partially soluble* in light petroleum; *insoluble* in ethyl alcohol, acetone, methyl ethyl ketone, pool rubber solvent. The graft polymer prevents precipitation of the unchanged rubber on adding a non-solvent; this makes it possible to prepare (e.g.) a 40% colloidal solution of MG49 in xylene/methyl alcohol, which is very viscous and hence suitable for spreading. According to the nature of the solvent, either the rubber chains or the grafted chains may be extended (the other being collapsed) and the solutions give respectively soft and hard dried films. **Swollen** (vulcanised unfilled MG49) by ethyl acetate (130% swelling) and acetone (74%). **Relatively unaffected** (vulcanised MG49) by water and hydrocarbon lubricating oils (cold).

Physics
Specific gravity (MG49), 1·02. **Electrical properties** (vulcanised unfilled MG49). Volume resistivity, $> 3.10^{12}$ ohm cm. Surface resistivity, $> 9.10^{11}$ ohm. Dielectric strength, 16 kV/mm. Dielectric constant, 2·9. Power factor, 0·02. **Mechanical properties** Vulcanisates

are characterised by high strength and hardness (even without reinforcing fillers), low hysteresis and exceptional resistance to fatigue and flex-cracking.

Unfilled vulcanisate of:	Trensile strength (kgf/mm²)	Breaking elongation (%)	Stress (kgf/mm²) at x% elongation	Hardness (IRHD)
MG 23	2·8	520–560	110–120 ($x = 300$)	72–75
MG 49	1·8	215	130 ($x = 100$)	96

Fabrication

Rubber/methyl methacrylate and rubber/styrene graft copolymers can be compounded, processed and vulcanised like natural rubber, being used either alone or blended with rubber. Mould flow is excellent, permitting intricate detail to be reproduced, and scorch tendency is less than for mixes reinforced with carbon black. To achieve full 'self reinforcement' vulcanisation must be at or above 120°C. The hot vulcanisates are sufficiently plastic to be shaped by post- or vacuum-forming.

Utilisation

Hard impact-resistant mouldings (e.g. electrical components), tough rubber soles and heels, engineering uses requiring low hysteresis, bonding agents (solution or latex) for rubber and polyvinyl chloride. Latex (5–10% methyl methacrylate graft) is used to make foam (cf. §29.81) with enhanced stiffness/weight ratio.

History

The 'graft' nature of the reaction products of natural rubber and vinyl monomers was demonstrated by F. M. Merrett (1954). Practical production methods were developed by the workers at the British Rubber Producers' Research Association (1954–58).

29.86 GUTTA PERCHA AND RELATED SUBSTANCES
(Balata, chicle)

Gutta percha is *trans*-polyisoprene (*see* footnote §29.1), obtained from the latex of *Palaquium oblongifolium* (fam. *Sapotaceae*) and similar trees indigenous to tropical S.E. Asia. Two temperate-zone plants, *Euonymus* sp. and *Eucommia ulmeoides* yield gutta percha in the Soviet Union. Unlike natural rubber (*cis*-polyisoprene), at room temperature gutta percha is hard and relatively inextensible, and commercial forms contain from 15–60% of resins (albane and fluavil, comprising fatty and other acid esters of α- and β-amyrin, lupeol and other phytosterols). Gutta percha first appeared in Europe as a museum curiosity acquired by J. Tradescant, senior, and described by his son (mid-seventeenth century); it was re-introduced by W. Montgomerie and (separately) J. d'Almeida, in 1843. In 1845 William Siemens suggested its use as insulation for telegraph wires, and 4 years later W. Breit laid the first experimental lengths of submarine cable at Folkestone. Gutta percha found use as submarine cable insulation because of its low water absorption and low dielectric loss.

Fundamental unit See footnote to §29.1. **Molecular weight** 37000–200000 (number-average). **X-ray data** The molecule is asymmetrical but of the same hand in any one crystal, so that, in contrast to rubber, it has a more compact structure and is highly crystalline even when unstretched. Gutta percha exists in three crystalline forms: α, β and γ. The β form is stable above *c*. 70°C, though also metastable at room temperature, so that gutta percha is normally either β or γ, depending on its previous heat treatment. On stretching, all samples change to a mixture of α and β, the latter predominating. Unit cell dimensions (Å) are:

Form	a	b	c (identity period)	angles
α	(not known)		8·76	—
β	7·8	11·8	4·75	$\alpha = \beta = \gamma = 90°$
γ	5·9	7·9	9·2	$\alpha = \beta = 90°; \gamma = 94°$

(*See* FISHER, D. G., *Proc. Phys. Soc.*, B116, 7 (1953).)

Solvents Gutta percha is less soluble than rubber but dissolves in aromatic and chlorinated hydrocarbons at room temperature, and in hot aliphatic hydrocarbons; esters are poor solvents, and it is unaffected by alcohol and water (see below). **Specific gravity** 0·945–0·955. **Refractive index** 1·523 (β modification, 1·509). **Water absorption** Purified gutta percha immersed over 2 years in O_2-free water in the dark, less than 0·2%. **Thermal properties** Glass temperature (T_g), −53 to −68°C; brittleness temperature, *c*. −60°C; crystalline melting point (T_m), 65–74°C; mouldable at *c*. 100°C.

Electrical properties

	Commercial grades	Purified material
Volume resistivity (ohm cm, at 25°C)	0·3–2·5 . 10^{15}	$>10^{17}$
Dielectric constant	*c*. 3·2	*c*. 2·6*
Power factor	*c*. 0·002–0·005	<0·002

*Decreases with rise of frequency, e.g. 2·61 (100 Hz), 2·38 (10^{10} Hz)

Balata, similar but inferior to gutta percha, since it contains 35–50 % resin, is derived from the latex of the bullet tree (*Mimusops balata* or *globosa*; fam. *Sapotaceae*) of Central and S. America. **Chicle** is a gummy resinous substance obtained from the *Achras sapota* (Sapotilla) tree of Mexico and Central America, and comprising a hydrocarbon resembling gutta percha plus a considerable proportion of resins similar to those in gutta percha (*see* above) **Utilisation** The main use of gutta percha (deresinated) is for the covers of golf balls; other uses are in dentistry, for cutting blocks (especially for gloves) and—blended with paraffin wax—as heat-sealing adhesive tissue. Balata is used mainly as an impregnant for textile belting, also for golf ball covers and cutting blocks. Chicle is used in chewing gum, adhesive plasters, waterproof varnishes, etc., though now largely superseded by synthetic products.

29.87 CYCLISED RUBBER

Synonyms and trade names
Cyclorubber, isomerised rubber; Alpex, Cyclatex, Cyclite, Detel H, Plastoprene, Plioform, Pliolite NR, Surcoprene, Thermoprene, Ty-ply.

General characteristics
Thermoplastic, off-white to amber-coloured solids, ranging from tough (gutta percha-like) to brittle resinous (shellac-like). Compatible with natural, styrene/butadiene, butyl and polychloroprene rubbers, and with many natural and synthetic resins (though not with alkyds). Resistant to alkalis and most acids. Can be vulcanised with sulphur.

Structure
Simplest Fundamental Unit

where Me = the methyl group, CH_3.

It has been shown (LEE, D. F., SCANLAN, J. and WATSON, W. F., *Proc. Roy. Soc.,* A273, 345 (1963)) that the residual double bond in the above unit permits further cyclisation to take place, to yield polycyclic structures containing an *average* of four condensed rings, i.e.

$$\begin{array}{ccccc}
-H_2C & H_2C{-}CH_2 & H_2C{-}CH_2 & \\
\diagdown & & & \\
MeC{-}CH & MeC{-}CH & CMe \\
H_2C & MeC{-}CH & MeC{-}C \\
\diagdown & & & \\
H_2C{-}CH_2 & H_2C{-}CH_2 & CH_2- \\
\end{array}$$

| 1 | 2 | 3 | 4 | 5 | C_5H_8 units |

Molecular weight 2000–14 500. **X-ray data** Gives an indefinite diffraction pattern, indicating a non-crystalline material, even when stretched.

Chemistry

Preparation Natural rubber is cyclised by catalysts of the following types: (a) RSO_2X (e.g. sulphuric acid, chlorosulphonic acid, aromatic sulphonic acids and sulphonyl chlorides), giving products used as bonding agents (e.g. Thermoprene), (b) amphoteric metal halides or their derivatives ($SnCl_4$, H_2SnCl_6, $TiCl_4$, $FeCl_3$, $AlCl_3$, $SbCl_5$, BF_3) giving moulding (Plioform) or paint (Pliolite) materials; (c) phosphorus pentoxide or oxychloride; (d) phenol plus a little sulphuric or phosphoric acid, giving soluble paint materials. Heat is generally applied, especially with RSO_2X catalysts; the rubber may be solid (milled with the catalyst), in solution (which reduces oxidation) or in the form of latex containing a cationic or nonionic stabiliser to prevent coagulation during cyclisation (by heating with sulphuric acid). Cyclised latex is often blended with untreated latex, and the mixture is coagulated to give a cyclised rubber master-batch. Cyclisation can be applied also to synthetic polyisoprene (Pliolite S-1).

Properties Dissolves in aromatic or chlorinated hydrocarbons, tetralin, carbon disulphide, some petroleum hydrocarbons. *Note* Solubility (especially in the last) depends on method and conditions of cyclisation. **Plasticised by** ester-type plasticisers, drying oils, chlorinated naphthalenes, animal and vegetable waxes, coumarone resins. **Swollen** (to a limited extent) by ether, amyl acetate, aniline. **Relatively unaffected by** some aliphatic hydrocarbons (see *Note* above), alcohols, ketones, water, alkalis and most acids (e.g. 50% H_2SO_4, conc. HCl, 50% HF and 85% H_3PO_4 all at 65°C; 10% HNO_3 at 35°C; acetic).

Identification N, Cl, S, P nominally absent (but 1–1·6% Cl may be present if made using metal halides). Thermoplastic; harder and less extensible than natural rubber. **Combustion** Softens; burns readily, smoky flame (resembling natural rubber); on extinction leaves a soft sticky residue that is hard and friable when cold. **Pyrolysis** Melts; chars; evolves yellow fumes that condense to a strong-smelling, neutral or mildly acid liquid. **Other tests** Responds to the Weber colour reaction for natural rubber (§29.23). Oxidation with chromic acid gives 25–30% of acetic acid (natural rubber, 66%). The infra-red absorption spectrum differs from that of natural rubber in showing little or no trace of the 12 μm band due to $—C(CH_3)=CH—$; spectra of pyrolysates of cyclised and untreated rubber also differ.

Physics

General properties

Specific gravity 0·96–1·12 according to method of preparation. **Refractive index** 1·525–1·545. **Water absorption** after 20 h at 70°C, 0·5–0·8 mg/cm². **Permeability** Highly impermeable to water vapour, e.g. a cyclised rubber coating (2·5–3 g/m²) reduced the permeation through cellophane (with 100% r.h. on one side and dry air on the other) from 10 000 to 55 $(10^{-10}$ g)/(cm² s).

Thermal properties

Conductivity c. 2·75 $(10^{-4}$ cal)/(cm s °C). **Coefficient of linear expansion** 0·75–0·8 $(10^{-4}/°C)$. **Physical effects of temperature** On heating, cyclised rubber softens usually between 80° and 130°C (extremes, 20° and 255°C), fluid at 110–160°C (or up to 280°C), and decomposes at 280–300°C (above 350°C for Thermoprene).

Electrical properties

Volume resistivity 10^{16}–7.10^{16} ohm cm. *Surface resistivity* (ohm) at 75% r.h., c. 5.10^{11}; at 90% r.h., 4.10^{10}–10^{11}. **Dielectric strength** 37–55 kV/mm, using c. 1 mm thick testpieces. **Dielectric constant** at 1 kHz, 2·68; at 1 MHz, 2·6–2·7. **Power factor** at 50 Hz, 0·006; at 1 kHz to 1 MHz, 0·002; at unstated frequency, 0·006–0·010; lower values are obtained by removal of hygroscopic impurities.

Mechanical properties

Tensile strength is highest for material of intermediate hardness ('hard balata' type): 2·8–3·5 kgf/mm²; lowest for brittle ('shellac') type: 0·45 kgf/mm². **Breaking elongation** decreases from 25–30% for the softer ('gutta percha') types to <1% for 'shellac' type. **Flexural strength** 1·1–6·5 kgf/mm²; usually highest for the softer types. **Compression strength** 2·7–8 kgf/mm²; usually highest for intermediate types. **Impact strength,** for testpieces not defined but probably notched, decreases from 0·9–0·025 kgf mm/mm² with increasing hardness (from 'gutta percha' to 'shellac' types); for undefined material, 3 ft lbf/in notch. **Hardness** Undefined material, probably 'gutta percha' type, 85–90 Shore A Durometer.

Fabrication
Depending on type and use, cyclised rubber may be (a) moulded by compression (125–155°C, 100–3000 lbf/in^2) or injection, after being plasticised, pigmented or mixed with fillers as required; (b) vulcanised with sulphur, e.g. giving ebonite-like products; (c) mixed with natural or synthetic rubber and vulcanised to give tough, hard, abrasion-resistant products (master-batches with natural rubber, *see* above, are commonly so used); (d) dissolved in solvents, giving adhesives or coating solutions, or added to inks, paints, etc.

Serviceability
Cyclised rubber is little affected by sunlight, except on prolonged exposure, and is unaffected by fungi, bacteria and insects. It is chemically inert, resistant to water and highly impermeable to moisture. Resistance to oxidation varies; some types used as varnish films oxidise gradually in air and become insoluble.

Utilisation
Main uses are (a) mouldings, e.g. to replace ebonite, with the advantage that colour can be varied; (b) basis or ingredient of paints, varnishes and printing inks (letterpress, offset); (c) acid- and alkali-resistant or water-impermeable coatings; (d) bonding agents for metals, rubber, wood, brick, concrete, etc.; (e) mixed with paraffin wax, as hot-melt coatings. Master-batches with untreated rubber (*see* above) are used in making hard, tough non-black rubbers (e.g. soles and heels), golf ball covers and crash-helmets.

History
J. G. Leonhardi (1781) obtained a hard brittle material by treating natural rubber with sulphuric acid. A similar cyclised rubber, obtained using rubber solution and conc. sulphuric acid, was studied by C. D. Harries (1919); F. Kirchhof (1920–22) obtained a product resembling gutta percha. H. L. Fisher (1926) found that cyclisation was produced also by sulphonic acids and sulphonyl chlorides, giving materials usable as bonding agents. H. A. Bruson and co-workers (1927–31) obtained thermoplastic mouldable or film-forming materials using stannic chloride, chlorostannic acid, or other amphoteric metal halides.

For further detail on cyclised rubber and iso-rubber *see* DAVIES, B. L.

and GLAZER, J., *Plastics derived from Natural Rubber,* London (1955); LE BRAS, J. and DELALANDE, A., *Les Dérivés Chimiques du Caoutchouc Naturel,* Paris (1950); NAUNTON, W. J. S. (Ed), *The Applied Science of Rubber,* London (1961); and on cyclised rubber, FETTES, E. M. (Ed), *Chemical Reactions of Polymers,* New York, London and Sydney, pp. 125 + ; (1964).

29.88 CELLULAR RUBBERS

Vulcanised rubbers, natural or synthetic, and including ebonites ('hard rubbers'), can be made in cellular form by: (a) incorporating a blowing agent, which evolves gas when the rubber is vulcanised by heat; blowing agents may be inorganic (e.g. sodium bicarbonate) or organic (e.g. azobisisobutyronitrile); the product is often termed 'blown sponge'; (b) saturating the unvulcanised rubber mix with nitrogen under high pressure, and releasing the pressure before and during vulcanisation ('expanded' rubber or ebonite); (c) aerating a vulcanisable latex composition and vulcanising the resulting foam ('latex foam', *see* §29.81). *See also* §31.4, Fabrication, for 'microporous' ebonite.

According to the methods of manufacture and after-treatment, the cells in cellular rubber may be either closed or open (inter-communicating). Materials with closed cells are excellent heat insulators. 'Microcellular' rubber is made from a mixture of rubber (usually natural or styrene/butadiene, Section 34) and a high-styrene resin (§34.1), containing a blowing agent and fillers. The vulcanising mould is filled, so that expansion occurs only on removal of the vulcanised article from the mould, producing a material with fine closed pores. It is used especially for footwear soles.

29.89 RECLAIMED RUBBER

This is the product obtained by treating ground scrap tyres, inner tubes, and miscellaneous rubber articles by the application of heat with or without chemical reagents (alkalis, acids), thus effecting an appreciable re-conversion or regeneration of the rubber to its original plastic state, in which it can be compounded, processed and re-vulcanised. Any textile fibre present can be removed either chemically or mechanically, but combined sulphur is not substantially removed. Reclaimed rubber is used in the manufacture of rubber goods generally in admixture with new raw rubber, but some articles not subjected to severe wear and tear may be made entirely from reclaimed rubber. Thus, first quality whole tyre reclaim (sp. gr. *c.* 1·17), containing *c.* 50% rubber hydrocarbon, gives a tensile strength *c.* 0·65 kgf/mm² and breaking elongation *c.* 350% when vulcanised with sulphur and a suitable organic accelerator. (For further detail *see* NOURRY, A. (Ed), *Reclaimed Rubber,* London (1962).)

29.8(10) LIQUID (DEPOLYMERISED)RUBBER

(Trade name: Lorival, plus grade letters and/or numbers.)

These fluid rubbers, viscosity 5000–200 000 P at 20°C, are made by heating natural rubber under controlled conditions, and used as rubber softeners, oil and bitumen additives, binders, stoving varnishes and as a basis for pourable (castable) compositions that can be vulcanised to either the soft or hard (ebonite) stage.

29.9 FURTHER LITERATURE

BATEMAN, L. (Ed), *The Chemistry and Physics of Rubber-like Substances,* London (1963)
BLACKLEY, D. C., *High Polymer Latices,* London and New York (1966)
DAWSON, T. R. and PORRITT, B. D., *Rubber, Physical and Chemical Properties,* Croydon (1935)
NAUNTON, W. J. S. (Ed), *The Applied Science of Rubber,* London (1961)
STERN, H. J., *Rubber, Natural and Synthetic,* 2nd edn, London and New York (1967)
BRYDSON, J. A. (Ed), *Developments with Natural Rubber,* London (1967)
EDGAR, A. T., *Manual of Rubber Planting (Malaya),* Kuala Lumpur (1960) (all aspects of plantation rubber production)

Specifications
BS *1154* and *1155* (standard natural rubber vulcanisates)
BS *1672: Parts 1 and 2* (methods of testing natural latex)
BS *1673: Part 2* (chemical analysis of raw natural rubber)
BS *4252* (measurement of viscosity of latex)
BS *4355* and *4556* (specifications for natural latex)
BS *4396* (specification for raw natural rubber)
ASTM *D1076* (specification and tests for natural latex)
ASTM *D1278* (chemical analysis of natural rubber)
ASTM *D2227* (specification for limits on impurities in natural rubber)

SECTION 30

NATURAL RUBBER DERIVATIVES

SUB-SECTION 30A: OXIDISED RUBBER

TRADE NAME
Rubbone.

GENERAL CHARACTERISTICS
Pale yellow to orange, viscous fluids to resinous solids, resistant to chemicals, water and heat, and with good electrical properties. Compatible with most resins, drying oils and cellulose nitrate. Can be vulcanised with sulphur.

30A.1 STRUCTURE

Simplest Fundamental Unit
No consistent repeating unit; the molecule is derived from natural rubber (polyisoprene, $(C_5H_8)_n$; *see* Section 29) by introduction of —OH (mainly), —OOH, =CO, —CO$_2$H, —CO$_2$R, and probably

$$=C-C=$$
$$\diagdown\!\diagup$$
$$O$$

and R—O—R groups. Composition varies according to grade, e.g.

	Rubbone 'A'	'B'	'C'
Percentage oxygen	5–6	10–11	15–16
Unsaturation, % of that in natural rubber	90	75	50

Rubbones 'N' and 'P' are degraded rubbers with low oxygen content.
Molecular weight Rubbone 'B', 3000–3600; 'C', 2400.

30A.2 CHEMISTRY

30A.21 PREPARATION

Natural rubber plus cobalt linoleate (oxidation catalyst) and a cellulosic filler (e.g. woodflour or chemical wood pulp, which activates the catalyst) is milled on hot rolls or in (e.g.) a Werner-Pfleiderer mixer, and the resulting oxidised rubber extracted with a solvent, e.g. benzene. Alternatively, rubber latex can be oxidised by air (in presence of manganese linoleate) or hydrogen peroxide.

30A.22 PROPERTIES

Solvents Behaviour towards solvents depends on the oxygen content (S = soluble, I = insoluble):

	Rubbone 'A'	'B'	'C'
White spirit	S	S	I
Aromatic hydrocarbons	S	S	S
Chlorinated hydrocarbons	S	S	S
Ethyl alcohol	I	I	S
Diethyl ether	S	S	S
Acetone	I	S	S

Films hardened by heat and oxidation are more resistant, e.g. those from Rubbone 'N' are relatively unaffected by water (boiling), benzene, toluene, and dil. acids and alkalis.

Plasticised by drying oils and the common plasticisers.

30A.23 IDENTIFICATION

Cl, S, N, P nominally absent. **Pyrolysis** At 350°C gives *c.* 10% volatile products, including water and dipentene. **Other tests** Gives characteristic infra-red absorption spectrum, showing bands due to —OH and =CO groups (*see* BOSTRÖM, S.(Ed), *Kautschuk-Handbuch*, Vol. 1, Stuttgart, 180 (1959) (KIRCHHOF, F.)).

30A. 3 PHYSICS

30A.31 GENERAL PROPERTIES

Specific gravity 0·97. **Refractive index** 1·519 at 28°C. **Water absorption** (films hardened by heat and oxidation) 24 h, 0·25%; 48 h, 0·45%; 168 h, 0·75%.

30A.32 THERMAL PROPERTIES

Withstands 350°C, though with some loss of weight; prolonged heating causes polymerisation.

30A.33 ELECTRICAL PROPERTIES

Volume resistivity (ohm cm) Rubbone 'B' after 3 h at 110°C, $2 \cdot 2 . 10^9$; after cooling to 20°C, $2 \cdot 8 . 10^{11}$. Rubbone 'N', 10^{15}. **Dielectric**

strength (kV/mm) Rubbone 'B', 14–19 (4 mm thickness); Rubbone 'N', 9.

30A.5 SERVICEABILITY

Oxidised rubber (Rubbone) has good resistance to thermal decomposition (better than most organic materials), water, chemicals and ageing (in bulk). In thin layers it slowly oxidises in air, giving durable films.

30A.6 UTILISATION

Oxidised rubbers have been used for the following purposes, though now largely superseded by newer materials: (a) paints, varnishes and lacquers (including electrical wire insulation), often in combination with drying oils or cellulose nitrate; (b) emulsion paints; (c) printing inks; (d) for impregnating packaging paper and cardboard (to make them impermeable to fats); (e) in sulphur-vulcanised form, as moulding powders, e.g. to replace ebonite; (f) reacted with aldehydes, to give water- and chemical-resistant resins. Oxidised rubber latex has been used as an adhesive.

30A.7 HISTORY

Formation of a resin by oxidation of natural rubber was observed by A. Adriani (1850) and J. Spiller (1865), and the chemistry of oxidation by oxygen and ozone was studied by, e.g. E. Herbst (1906), C. D. Harries (1910) and S. J. Peachey (1912). The observation (W. H. Stevens, 1930) that cobalt compounds strongly catalyse oxidation led to commercial processes for making oxidised rubbers (Rubbone; H. P. and W. H. Stevens *et al.,* working for Rubber Growers' Association and Rubber Producers' Research Association, 1933 onwards). The final process, using a cellulosic filler to promote catalysis, dates from 1937.

SUB-SECTION 30B: CHLORINATED RUBBER

SYNONYMS AND TRADE NAMES
Rubber chloride; Alloprene, Detel (paints), Duroprene, Paravar, Parlon, Pergut, Raolin, Tegofan, Tretol (paints).

GENERAL CHARACTERISTICS

Chlorinated rubber is a hard non-rubbery, pale yellow to brown solid, or a bulky, white to cream powdery or fibrous material. Not generally sufficiently thermoplastic to be moulded unless plasticised. Non-flammable and very resistant to corrosive chemicals. Stable at room temperatures but tends to decompose at moderately elevated temperatures. Available also as an aq. emulsion (latex).

30B.1 STRUCTURE

Simplest Fundamental Unit

Fully chlorinated natural rubber appears to have a saturated cyclised structure (cf. §29.87), e.g.

$$-CHCl-CMe-CCl-CHCl- \qquad C_{10}H_{11}Cl_7$$

(structural diagram with CHCl, CHCl, CHCl—CHCl, C Cl Me) $(65 \cdot 4\% \text{ chlorine})$

Commercial products contain *c.* 50–68% (usually 64–68%) chlorine. **Molecular weight** 100 000–400 000. **X-ray data** Gives amorphous rings (but *see* §30B.32, Physical effects of temperature).

30B.2 CHEMISTRY

30B.21 PREPARATION

(*a*) *Solution process*—Rubber dissolved in an inert solvent (usually carbon tetrachloride) is reacted at 80–110°C with chlorine, e.g. by counter-current spraying. The rubber must be depolymerised *either* by mastication (§29.4) before dissolution, *or* by oxidative degradation in the solution, promoted by a catalyst or ultra-violet irradiation. The chlorinated rubber is separated either by removing the solvent (by superheated steam or pouring into hot water) or adding a non-solvent. An acid-neutralising stabiliser is included in the rubber solution or added to the water to free the product from hydrogen chloride.

(*b*) *Latex process*—Latex containing a cationic or non-ionic stabiliser is acidified and chlorine passed in. The resulting chlorinated latex may be coagulated with methyl alcohol. The solid product contains up to 60% chlorine; to increase the chlorine content the product (dry, swollen or dissolved) must be further chlorinated.

(*c*) Methods using *solid rubber*, sheeted or comminuted, and gaseous or liquid chlorine have been devised.

Synthetic polyisoprene rubber (Section 32) can be similarly chlorinated (Pliochlor).

30B.22 PROPERTIES

Dissolves in aromatic or chlorinated hydrocarbons, tetralin, higher ketones (e.g. cyclohexanone), many esters, dioxan, carbon disulphide, nitrobenzene. Diethyl ether and lower ketones give partial dissolution or swelling. **Plasticised by** ester plasticisers (phosphates, phthalates, sebacates, stearates), hydrogenated rosin esters, various vegetable oils, natural and synthetic resins, camphor, amylnaphthalene, chlorinated paraffin, chlorinated diphenyl and chlorinated naphthalene. **Relative unaffected** (when unplasticised) by water (though may not resist boiling), aliphatic hydrocarbons, lower alcohols, mineral acids (97% H_2SO_4, conc. HCl and HNO_3 all at 100°C; aqua regia, chromic acid, hydrofluoric acid, 85% phosphoric acid), conc. ammonia solution, hydrogen peroxide, chlorine, ozone, sulphur dioxide. *Note.* Resistance may be lowered by plasticisers. **Decomposed by** organic bases (especially hot) and boiling alcoholic alkalis.

30B.23 IDENTIFICATION

Cl present; N, S and P nominally absent (P may be present from phosphate plasticisers). **Combustion** Non-combustible. Chars; burns only while held in a flame, leaving a black swollen residue. **Pyrolysis** Begins to decompose at 100–125°C even if stabilised; above 200°C chars, evolves hydrogen chloride and other volatile products. **Other tests** Chlorine content is unusually high but not unique (cf. polyvinyl chloride and polyvinylidene chloride, but infra-red absorption spectrum distinguishes from these). Practically unaffected by aq. alkalis but breaks down to a dark brown product when refluxed with alcoholic potash (giving apparent saponification number above 500 mg KOH/g sample). Responds to Weber colour test for natural rubber, but not to trichloroacetic acid test (§29.23).

30B.3 PHYSICS

30B.31 GENERAL PROPERTIES

Specific gravity 1·58–1·69, increasing with chlorine content (65% Cl, 1·63–1·64). **Refractive index** 1·55–1·60, increasing with chlorine content (65% Cl, 1·595). **Water absorption** 0·02% (conditions not stated); 0·27% (24 h at 80% r.h.). **Permeability** Transmittance of water vapour through 0·03 mm films $(10^{-10}$ g)/(cm^2 s cmHg) at

25°C: unplasticised, 78; with 25% of various plasticisers, 24–225.

30B.32 THERMAL PROPERTIES

Specific heat c. 0·4. **Conductivity** c. 3 $(10^{-4}$ cal)/(cm s °C). **Coefficient of linear expansion** c. 1·25 (10^{-4})/°C. **Physical effects of temperature** Softens at between 90° and 150°C, depending on chlorine content and history. Above c. 80°C can be stretched to several times its length ('racked') and then becomes crystalline and doubly-refracting. Decomposes above 100–125°C.

30B.33 ELECTRICAL PROPERTIES

Depend on amount and type of plasticiser. **Volume resistivity** $2·5 . 10^{13}$– $7 . 10^{15}$ ohm cm. *Surface resistivity* $5 . 10^{11}$–$2 . 10^{13}$ ohm. **Dielectric strength** (kV/mm) c. 0·1 mm thick films (not stated if plasticised) 50–100; thickness not stated unplasticised, >80; plasticised 16–20. **Dielectric constant** 3·0 (at 50 Hz); 2·5–3·5 (1 kHz to 1 MHz); 2·5–4·7 (frequency not stated). **Power factor** unplasticised, 0·003 (at 50 Hz to 1 kHz); 0·006 (1 MHz). Other data, 0·002–0·024.

30B.34 MECHANICAL PROPERTIES

Elastic modulus (Young's) 100–400 kgf/mm².

	Unplasticised film	Plasticised film
Tensile strength (kgf/mm²)	2·8–4·5	2·3–3·2 (25% plasticiser) 0·4–0·85 (40% plasticiser)
Elongation at break (%)	c. 3·5	Up to 250

Flexural strength 7–10 kgf/mm². **Compression strength** c. 7 kgf/mm². **Impact strength** c. 1·5 ft lbf/in notch. **Hardness** (Brinell) c. 10–15 kgf/mm². **Frictional data** Grey cotton yarn (running at 0·9 m/s, at 65% r.h.) against a film-coated cylinder, $\mu_{dyn} = 0·26$.

30B.4 FABRICATION

According to type and use, chlorinated rubber may be (a) plasticised and moulded, at 120–150°C and 6000–12 000 lbf/in²; (b) dissolved in solvents (aromatic or chlorinated hydrocarbons, esters) with or without addition of plasticisers, drying oils, resins, pigments, etc., to give paints, lacquers, inks and adhesives; (c) expanded into cellular form by heating under pressure then suddenly releasing the pressure.

30B.5 SERVICEABILITY

Chemically inert, non-inflammable, and unaffected by insects, bacteria and fungi. Sunlight or ultra-violet light causes gradual decomposition, especially in unpigmented films, with darkening and embrittlement.

30B.6 UTILISATION

Main uses are chemical- and fire-resistant paints, traffic paints and under-water paints (e.g. for ships' hulls); printing inks, impregnating (water- and flame-proofing) textiles, paper and leather; adhesives for rubber, metals, wood, plastics, laminates, ceramics and glass-fibre; packaging films; mouldings, e.g. electrical insulators; filaments; cellular products (heat- and sound-insulation). Chlorinated rubber latex is used as such, e.g. in emulsion paints.

30B.7 HISTORY

W. Roxburgh (1801) reacted a carbon disulphide solution of rubber with chlorine, though there is some doubt whether the product was chlorinated rubber. A patent by A. Parkes (1846) refers to chlorine as a 'vulcanising agent'. The first product identifiable as chlorinated rubber was that of G. A. Englehard and H. H. Day (1859), but commerical exploitation (in Duroprene paints) dates only from *c.* 1915 (S. J. Peachey). Solid chlorinated rubber was first marketed in 1930 (Tornesit). Chlorination of latex was effected by G. F. Bloomfield and E. H. Farmer (1934), but the present process using acid latex dates from 1951 (G. J. van Amerongen).

SUB-SECTION 30C: RUBBER HYDROCHLORIDE

SYNONYMS AND TRADE NAMES
Hydrochlorinated rubber; Marbon V and X, Pliofilm, Pliolam (laminate), Shrinkwrap, Tensolite, Vitafilm.

GENERAL CHARACTERISTICS
Rubber hydrochloride is a white, thermoplastic, tough solid with very low permeability to gases, water vapour and fats. Burns with difficulty; self-extinguishing. Free from taste and odour. Resistant to chemical attack, but not stable to heat or to light unless stabilised. Compatible with resins and chlorinated rubber but not with natural rubber itself.

30C.1 STRUCTURE

Simplest Fundamental Unit
Fully hydrochlorinated natural rubber consists essentially of

$$-CH_2-\underset{\underset{Cl}{|}}{\overset{\overset{CH_3}{|}}{C}}-CH_2-CH_2- \qquad C_5H_9Cl \ (33{\cdot}9\% \ chlorine)$$

Commercial materials usually have 28–33% chlorine; they may be partially cyclised (cf. §29.87) and have some residual unsaturation. **X-ray data** Unlike natural rubber, the hydrochloride (up to 105–110°C) is polycrystalline and gives a powder diagram. Unit cell monoclinic (pseudo-orthorhombic), $a = 5{\cdot}83$ Å, $b = 10{\cdot}38$ Å, $c = 8{\cdot}89$ Å, $\beta = 90°$. Above 80–110°C it can be stretched and then gives a fibre diagram, unit cell orthorhombic, $a = 11{\cdot}9$ Å, b (fibre axis) $8{\cdot}95$–$9{\cdot}1$ Å, $c = 10{\cdot}4$ Å.

30C.2 CHEMISTRY

30C.21 PREPARATION

In the usual process natural rubber, dissolved in an inert solvent (benzene, toluene, chlorinated hydrocarbon), is reacted with hydrogen chloride, e.g. at 10°C. Ketones, alcohols or terpenes (e.g. 20 volume % acetone or methyl ethyl ketone) may be added to prevent gelling; dioxan and ethyl acetate suppress cyclisation. The solvent is removed by steam distillation, or the rubber hydrochloride precipitated by a non-solvent, or the solution used directly to make films (*see* §30C.4, Fabrication). A variant of the process uses a solvent/non-solvent mixture such that the hydrochloride is precipitated as formed. In other methods (a) rubber is swollen in a liquid saturated with hydrogen chloride; (b) thin sheets of rubber are exposed to gaseous or liquid hydrogen chloride; (c) rubber latex containing a cationic or non-ionic stabiliser is reacted with hydrogen chloride, then coagulated by adding alcohol or brine. In all cases residual hydrogen chloride must be removed; stabilisers and/or plasticisers are commonly added (*see* Fabrication).

Synthetic polyisoprene rubber (Section 32) can also be hydrochlorinated.

30C.22 PROPERTIES

Dissolves in aromatic and chlorinated hydrocarbons. **Plasticised by** paraffins, amylbenzene, ester plasticisers (aliphatic stearates, adipates, phthalates, sebacates, succinates; tritolyl phosphate), castor oil, rosin, methyl ester of hydrogenated rosin, naphthalene; chlorinated paraffins and aromatic hydrocarbons. **Swollen by** some esters (hot). **Relatively unaffected by** water (boiling), alcohol, ether, acetone, petrol, most mineral acids, alkalis, and by ozone if not stressed. **Decomposed by** organic bases (hot), conc. sulphuric and nitric acids and other strong oxidising agents; slowly attacked by alcoholic alkalis.

30C.23 IDENTIFICATION

Chlorine present; N, S and P nominally absent (small amounts of N may be present from stabilisers, and P from phosphate plasticisers). **Combustion** Melts, chars, does not readily burn, but when alight burns with sooty flame and leaves black residue. **Pyrolysis** Evolves strongly acid fumes with sweetish odour; leaves black residue. **Other tests** Does not respond to Weber test for natural rubber (§29.23; contrast chlorinated rubber, §30B.23). Attacked by conc. nitric acid (contrast chlorinated rubber). Gives characteristic infrared absorption spectrum (CHECKLAND, P. B. and DAVISON, W. H. T., *Trans. Faraday Soc.*, **52**, 151 (1956)). (For distinction of rubber hydrochloride film from other transparent packaging films, *see* WAKE, W. C., *The Analysis of Rubber and Rubber-like Polymers*, 2nd edn, London, 57 (1969).)

30C.3 PHYSICS

30C.31 GENERAL PROPERTIES

Specific gravity $1\cdot11$–$1\cdot27$, increasing with chlorine content. **Refractive index** $1\cdot533$. **Permeability** to gases at 20–30°C (10^{-10} cm^2)/ (s cmHg), unplasticised, O_2, $0\cdot025$–$0\cdot54$; N_2, $0\cdot008$ (plasticised, $0\cdot62$); CO_2, $0\cdot17$ ($1\cdot8$); Pliofilm, H_2, $1\cdot4$–$4\cdot0$; O_2, $0\cdot06$–$1\cdot0$. To water vapour (10^{-10} g)/(cm s cmHg), increases with temperature from $0\cdot1$–$0\cdot13$ (25°C) to $0\cdot7$ (52°C); other values at 20–30°C, $0\cdot013$–$1\cdot5$.

30C.32 THERMAL PROPERTIES

Softens at 50–90°C, 'melts' (i.e. becomes amorphous) at 115°C. Decomposition begins at 100–130°C and is complete above 180–185°C.

30C.33 ELECTRICAL PROPERTIES

Volume resistivity $c.$ 10^{15} ohm cm. **Dielectric constant** 2·8–3·5. **Power factor** $c.$ 0·05.

30C.34 MECHANICAL PROPERTIES

Tensile strength (kgf/mm^2) 2–3 (Pliofilm 3·3–3·6). **Elongation at break** (%) 50–200 (Pliofilm 400–550). **Resistance to tearing** is good, especially in 'tensilised' films (§30C.4). **Frictional properties** Grey cotton yarn (running at 1 m/s, at 65% r.h.) against commercial foil, $\mu_{dyn} = 0.22$.

30C.4 FABRICATION

According to its end use, rubber hydrochloride may be (a) cast into films from solution (cf. §30C.21), after addition of light- and/or heat stabilisers, plasticisers, pigments, etc., as required (Pliofilm); the films may be hot-stretched ('tensilised') to increase strength (Tensolite); films can be heat sealed at 105–110°C; (b) drawn into filaments (Tensolite), used as textile material; (c) used, in solution, as adhesives; for this purpose sulphur, etc., may be added to give a vulcanisable bond (Ty-ply).

30C.5 SERVICEABILITY

Chemically inert; not readily inflammable. Not attacked by insects or rodents; does not support mould growth. Decomposes gradually at temperatures above 100–130°C, and (if not stabilised) on exposure to light. May show stress-cracking when exposed to ozone.

30C.6 UTILISATION

Packaging films; adhesives, especially for bonding rubbers to metals and other materials; textile fibres; component of laminates with paper, cellulosic films, metal foils, etc.

30C.7 HISTORY

The first experiments on the action of hydrogen chloride on rubber (R. Matthews, 1805) did not suggest any important transformation, but U.S. patents of 1881 (D. M. Lamb) covered uses of the reaction product. Identification of the product, obtained in chloroform solution, as rubber hydrochloride followed in 1900 (C. O. Weber) and 1911–13 (C. D. Harries). Its use as transparent moisture-proof packaging film dates from 1937, and as a base for adhesives, from 1939–40.

30.9 FURTHER LITERATURE

DAVIES, B. L. and GLAZER, J., *Plastics derived from Natural Rubber*, London (1955)

FETTES, E. M. (Ed), *Chemical Reactions of Polymers*, New York, London and Sydney (1964) (pp. 107–24, GOLUB, M. A., Isomerisation; pp. 125–32, SCANLAN, J., Cyclisation; pp. 142–51, CANTERINO, P. J., Halogenation; pp. 173–93, WICKLATZ, J., Hydrogenation; pp. 194–246, LE BRAS, J., PAUTRAT, R. and PINAZZI, C., Reagents with multiple bonds)

LE BRAS, J. and DELALANDE, A., *Les Dérivés Chimiques du Caoutchouc Naturel*, Paris (1950)

NAUNTON, W. J. S. (Ed), *The Applied Science of Rubber*, London (1961)

EBONITE

SYNONYMS AND TRADE NAMES

Hard rubber*, vulcanite; Dexonite, Ebonar, Keramot, Onazote (expanded ebonite), Silvonite, Stabilit.

GENERAL CHARACTERISTICS

Ebonite is a hard substance, black in bulk (translucent red-brown if very thin) and usually brown or red when pigmented. It can be machined by normal methods, but becomes flexible and shows rubber-like elasticity when above its yield temperature (*see* §31.32)— which ranges, for natural rubber (NR) ebonite up to *c.* 80°C; for synthetic rubber ebonites to *c.* 125°C.

Ebonite resists many corrosive chemicals, and its resistance to swelling by organic liquids is much better than that of soft rubber vulcanisates (aliphatic hydrocarbons having practically no action). Suitably compounded ebonite has excellent electrical properties, though the surface resistivity slowly decreases on exposure to light. Since ebonite is a cross-linked material, it is not thermoplastic; above the yield temperature, however, it can be deformed, and if then cooled, the deformation is 'frozen in' but it disappears again on warming above the yield temperature. Ebonite with a mineral filler is termed 'loaded' (otherwise it is 'unloaded'); it is then less deformable at elevated temperatures but is mechanically weaker. Ebonite can be made in cellular or expanded form.

31.1 STRUCTURE

Simplest Fundamental Unit

The structure is uncertain, but probably consists essentially of

$$
\begin{array}{cc}
\begin{array}{c}
-\text{R}- \\
| \\
\text{S}_n \\
| \\
-\text{R}- \\
\end{array}
&
\begin{array}{c}
\quad\quad \text{R}' \\
-\text{R} \diagdown \diagup \text{R}- \\
\quad \text{S} \\
\end{array}
\\
\text{I} & \text{II}
\end{array}
$$

where —R— is a segment of a chain-molecule of a highly unsaturated rubber (e.g. polyisoprene, polybutadiene, styrene/butadiene or acrylonitrile/butadiene), R′ is a short segment (probably of 2 C atoms) and n is probably either 1 or 2. Most of the sulphur is present as in II; the *total* sulphur represents *c.* 1 atom per 4–8 C atoms in the main chain. Thus, in natural rubber (polyisoprene) ebonite the sulphur is 25–50% on the rubber, and II is believed to be represented by

$$
\begin{array}{c}
\text{CH}_2-\text{CH}_2 \\
\diagup \quad\quad \diagdown \\
-\text{CH} \quad \text{C(CH}_3)- \\
\diagdown \quad \diagup \\
\text{S}
\end{array}
$$

*This term includes materials, sometimes called *pseudo-ebonites*, which are rendered hard by incorporation of resins, with or without fillers, rather than by vulcanisation.

Molecular weight Indeterminate, since there are no individual molecules.

X-ray data Ebonite gives a diffuse halo, indicating an amorphous structure, at all temperatures.

31.2 CHEMISTRY

31.21 PREPARATION

Ebonite is made from a highly unsaturated rubber (natural or synthetic) by heating (vulcanisation) with a high proportion of sulphur. The reaction is essentially addition of sulphur at the double bonds, forming *inter*molecular cross-links and *intra*molecular ring structures (*see* §31.1), but with a very high proportion of sulphur and a high vulcanising temperature some substitution, and formation of hydrogen sulphide, occurs. For other ingredients normally added to facilitate processing or modify the properties of the product, *see* **Fabrication** §31.4.

31.22 PROPERTIES

Solvents Ebonite is insoluble except under conditions (heat, oxidation) that break down the molecular structure. **Plasticisers** The hydrocarbon oils, bitumens, aromatic petroleum fractions, resins (e.g. coumarone) and linseed oil that are included in the unvulcanised mix to improve processing and 'tack' may plasticise ebonite in the sense of improving impact strength, but they lower the softening or yield temperature (§31.32). Polychloroprene (Section 36), butyl rubber (Section 38) and polyisobutylene (§38.82) act similarly and are used to make flexible and/or impact-resistant ebonites. **Swollen by** (natural rubber ebonites) aromatic, chlorinated or nitrated hydrocarbons; carbon disulphide, phenols, aniline, ethyl to butyl acetates, oleic acid. In general, swelling decreases with increase of combined sulphur and is less with synthetic than natural rubber ebonites (*see* Table 31.T1). The time to reach maximum swelling is generally much longer than for soft vulcanised rubber. **Relatively unaffected by** aliphatic hydrocarbons, alcohols (up to amyl), glycol, glycerol, vegetable oils, acetone (*see* Table), water and neutral salt solutions, most acids (including, at 65°C, conc. HCl, 85% H_3PO_4, 50% HF and H_2SO_4, and—if specially compounded—conc. HBr, formic and glacial acetic), ammonia (conc. at 40°C), caustic alkalis (conc. at 65°C), aq. solutions of chlorine or hypochlorites, and 40% hydrogen peroxide. **Decomposed by** conc. sulphuric acid, nitric acid (8–20% or above), chromic acid (50%) and other strong oxidising agents, and by dry halogens.

Table 31.T1. SWELLING OF UNLOADED EBONITES AT ROOM TEMPERATURE
(% BY VOLUME; TEST PIECES 5 mm THICK)

I. *Natural rubber ebonite:* Given in brackets is the time (days) to reach approximate maximum swelling ('+' means maximum not reached in the time shown.

Liquid	Rubber/sulphur ratio	
	100/39	100/54
Petroleum ether	10 (100+)	0
Cyclohexane	60 (200)	<1
Benzene	85 (4)	65 (15)
Carbon tetrachloride	60 (9+)	50 (90+)
Ethyl alcohol	0	0
Acetone	17 (180+)	4 (180+)
Carbon disulphide	150 (1)	80 (1)
Nitrobenzene	70 (40)	60 (150)
Aniline	75 (180)	23 (300+)
o-Cresol	70 (90+)	73 (300+)

II. *Synthetic rubber ebonites: N.B.* Immersion times were not always long enough to reach maximum swelling.

Liquid (and immersion period, days)	Initial rubber (and rubber/sulphur ratio)			
	Polybutadiene		25:75 Styrene/ butadiene	25:75 Acrylonitrile/ butadiene
	(100/35)	(100/47)	(100/47)	(100/47)
Petroleum ether (60)	3	0·2–0·4	0·3	0
Petrol (60)	17	0·2–0·9	0·2	0
Benzene (150)	55	2–12	10	0·5
Carbon disulphide (5)	55	35–45	50	4 (180 days)
Nitrobenzene (50)	10	0·4–0·7	0·6	0·2

Loaded ebonites swell less than those without filler, swelling being approximately proportional to the volume fraction of rubber-sulphur compound.

Raising the temperature, e.g. to above the 'intercept yield temperature' (§31.32), increases and accelerates swelling, though it remains very small in aliphatic hydrocarbons.

31.23 IDENTIFICATION

P, Cl normally absent (except for trace of P in natural rubber ebonite, *see* §29.23; Cl present in ebonite containing polychloroprene); N may be present in small amount from accelerators, or in large amounts in nitrile rubber ebonite; S present, 20–35% of the 'organic' part (i.e. excluding mineral fillers). Ebonite is distinguished from the only other polymers with similar sulphur content, i.e. polysulphide rubbers (Section 39), by its rigidity and by the infra-red absorption spectrum of the pyrolysate.*

Combustion When well ignited burns briskly, giving off much sulphur dioxide; leaves charred residue. **Pyrolysis** Begins to liberate hydrogen sulphide at c. 100°C (natural or nitrile rubber ebonite) or c. 140°C (styrene/butadiene ebonite); up to 250°C only traces of volatile organic matter are formed from natural rubber ebonite; on pyrolysis at 700°C a sample gave (%): liquid, 59; hydrogen sulphide, 13; water, 4; coke, 6; gases (uncondensed at −80°C) and products not accounted for, 18; the liquid portion contained (in decreasing amounts) 2-methyl-5-ethylthiophene; m-xylene and 2,3-dimethyl-thiophene; 2-methyl- and 2,4-dimethyl-thiophene; toluene; benzene.

31.3 PHYSICS

31.31 GENERAL PROPERTIES

Specific gravity Unloaded, 1·08–1·25 (increasing with S content); loaded (45% by volume of filler), up to 1·8. **Refractive index** 1·6–1·65. *Light transmission*, natural rubber ebonite 0·055 mm thick:

wavelength (Å)	≤ 5180	5450	5590	5660	5710	5815	6130	7000
transmission (%)	<0·01	0·4	0·9	2	2·9	5	10	25–30

Moisture absorption Various ebonites, unloaded, as powders at 97–98% r.h. and 20–25°C,

Initial rubber	Natural	Polybutadiene (Buna)	Styrene/ butadiene (Buna S)	Acrylonitrile/ butadiene (Buna N)
Equilibrium absorption (%)	0·4–0·8*	0·8–1·4	c. 0·6	c. 4

*At 75% r.h.: unloaded, 0·2; loaded with 50 parts of various silicas per 100 (rubber + sulphur), 0·2–1·2.

With some loaded ebonites not all of the moisture absorbed is given up on exposure to dry air.

Water absorption Natural rubber (NR) ebonite, unloaded, as sheets immersed at 25–28°C,

*Ebonites made from natural, styrene/butadiene and nitrile rubbers can be distinguished from one another by the infrared absorption spectra of liquid pyrolysates obtained in the presence of zinc dust or calcium turnings (HUMMEL, D., *Kautschuk u. Gummi*, 11, WT 185 (1958)).

Thickness (mm)	Immersion time (days)	Absorption (%)
2·5	30	0·2
1·0	315	0·25*

*Loaded, 10·6%

Surface absorption, NR ebonite, immersed 7 days at *c.* 25°C (mg/cm²): unloaded, 0·1; loaded with 75 parts precipitated silica per 100 (rubber + sulphur), 0·25–0·4. Absorption increases with rise of temperature, e.g. (i) unloaded (immersed 7 days at 70°C), 0·7 mg/cm²; (ii) absorption at 90°C compared with that at 25°C, NR ebonite, 20–40 times; styrene/butadiene or nitrile rubber ebonites, 6 times. **Permeability***To gases at 67°C $(10^{-10}$ cm²)/(s cmHg) H_2, 1·45; He, 4·7; N_2, 0·025. To water vapour at 20–25°C, 0·08–0·42 $(10^{-10}$ g)/ (cm s cmHg).

31.32 THERMAL PROPERTIES

Specific heat NR ebonite, unloaded, 0·33–0·34. **Conductivity** NR ebonite, 3·7–4·4 $(10^{-4}$ cal)/(cm s °C). **Coefficient of linear expansion** NR ebonite, varies with temperature and increases sharply at the glass temperature (T_g):

Temperature, °C	−160	−80	room temp.	up to T_g	above T_g
Coefficient $(10^{-4}$/°C)	0·41	0·55		0·65–0·8	*c.* 2

For ebonite containing mineral fillers the expansion is less, roughly in proportion to the volume fraction of rubber plus sulphur in the ebonite; if the filler particles are flat and plane-orientated, the linear expansion of a sheet is greater in the thickness than laterally. **Physical effects of temperature** Ebonite is in the glass-like state at room temperature (i.e. below T_g), e.g. NR ebonites,

Combined sulphur (parts per 100 rubber)	25	33	43	52
Glass temperature, T_g (°C)		34	66	83	*c.* 87

Above T_g ebonite becomes flexible and shows rubber-like elasticity, e.g. with Young's modulus, 0·6–2 kgf/mm² (NR ebonite), which is *c.* 10 times that of a soft vulcanisate. *Yield temperature* Heat resistance is usually expressed by a 'yield temperature', variously defined as (i) temperature at which the deformation (under a given stress) reaches a given value when temperature is raised uniformly; (ii) temperature at which the deformation—initially very small—begins to increase rapidly as shown on the temperature/deformation curve, this being obtained *either* with uniformly rising temperature *or*

*Type of initial rubber not stated; probably natural.

by plotting deformation at a succession of steady temperatures. Yield temperature defined by (ii), preferably called 'intercept yield temperature' (being read as an intercept on the temperature axis), coincides approximately with glass temperature*. Yield temperature increases with sulphur content and time of vulcanisation, and is higher for ebonites made from butadiene polymers and copolymers than for natural rubber ebonites (see Table 31.T2).

Table 31.T2. 'INTERCEPT YIELD TEMPERATURE' OF EBONITES VULCANISED 5–6 h AT 155°C (°C)

| Initial rubber | Sulphur, parts per 100 rubber | | |
	30	35	47
Natural	58	73	80–83
Polybutadiene (Buna)	—	c. 80	85–100
25:75 Styrene/butadiene	—	—	90–95*
45:55 Styrene/butadiene	90–100	—	100–110
25:75 Acrylonitrile/butadiene	—	—	125*

*Rises a further 15–20°C when vulcanised for 11 h

Mineral fillers do not substantially alter the 'intercept yield temperature', but reduce deformation at a given temperature and hence raise 'yield temperature' by definition (i) above.

At c. 200°C and above, natural rubber ebonite softens and becomes liquid.

31.33 ELECTRICAL PROPERTIES

Ebonites made from hydrocarbon rubbers have excellent electrical characteristics, but those from acrylonitrile/butadiene rubbers have a relatively high dielectric loss. Typical data are given in Table 31.T3.

Table 31.T3. ELECTRICAL PROPERTIES OF EBONITES*

| | Initial rubber | | | | |
	Natural	Natural (loaded, electrical grade)	Poly-butadiene	Styrene/ butadiene†	Acrylonitrile/ butadiene‡
Volume resistivity (ohm.cm§)	10^{16} to $>10^{17}$	10^{14} to 10^{15}	—	—	—
Surface resistivity (ohm‖)	10^{15} to $>10^{18}$	$>10^{15}$	—	—	—
Dielectric strength¶ (kV/mm)	90–150	50–115	—	—	—

*See MARK, H. F., GAYLORD, N. G. and BIKALES, N. M., *Encyclopedia of Polymer Science and Technology*, Vol. 12, New York (1970) (article by SCOTT, J. R., pp. 161–177, rubber, hard). The term 'plastic yield', often used for the deformation above the yield temperature, is incorrect, because the deformation is essentially reversible.

Dielectric constant

(800–10^6 Hz)	2·7–3·2	3·0–3·7	*c.* 3·1	2·8–3·2	3·6–3·9
(60°C; 10^6 Hz)	2·85–3·3		*c.* 3·2	*c.* 3·3	*c.* 4·0

Power factor**

(800–1000 Hz)	0·005–0·008	0·006–0·015	—	0·004–0·007	0·016–0·030
(10^5–10^6 Hz)	0·008–0·010††	0·008–0·015	*c.* 0·006	0·005–0·012††	0·020–0·035
(60°C; 10^6 Hz)	0·011–0·014	—	*c.* 0·009	0·006–0·013	0·034–0·038

*Unloaded, at room temperature unless otherwise stated
†25–45% styrene
‡25–40% acrylonitrile
§Decreases with rise of temperature
‖Decreases with increasing ambient humidity, especially after exposure to light (*see* §31.5 **Serviceability**); italic figures are for ebonite thoroughly cleaned, protected from light and tested at normal humidity; surface contamination or prolonged exposure to light can reduce resistivity by several powers of 10
¶Data are for 0·5–1·0 mm thickness; value decreases with increasing thickness, being, e.g. approximately proportional to (test-piece thickness)$^{-0.5}$
**Data for ebonite conditioned at 40–75% r.h.; value increases with moisture content, and increases rapidly with temperature above the 'intercept yield temperature' (§31.32)
††Power factor shows a maximum, for NR ebonite, at *c.* 3 MHz; for styrene/butadiene rubber ebonite, 10 MHz

BS 234 for electrical ebonites gives the following data,

	Unloaded	Loaded
Dielectric strength (min.),kV for 3·2 mm thickness	40 or 25 according to grade	30
Dielectric constant (max.) at 800–10^6 Hz	3·0 or 3·3	3·8
Power factor (max.) at 800–1600 Hz	0·006 or 0·008	0·008
Power factor (max.) at 10^6 Hz	0·009 or 0·010	0·012

31.34 MECHANICAL PROPERTIES

At room temperature ebonite is in the glass-like state and shows linear elasticity up to a stress somewhat below the breaking stress. Creep under stress is very slow, e.g. under a shear stress of 0·7 kgf/mm^2 shear deformation increased from 0·8 to 2·0% in 1 year at 16–18°C; at higher temperatures creep is more pronounced, and is commonly assessed by a 'cold flow' test at 49°C. Mechanical strength is high, except that notching greatly reduces impact strength.

The properties depend on the initial rubber, proportion of vulcanising sulphur, presence or absence of fillers, etc., and temperature and time of vulcanisation; hence the values in Table 31.T4 are not all comparable. In general, a high degree of vulcanisation, besides improving heat resistance (§31.32), increases tensile and flexural strengths but lowers impact strength.

Other properties *Bulk modulus* (kgf/mm^2). Static, 600–800; dynamic (10 kHz), 1270. *Compression strength* 7·5–8 kgf/mm^2. *Poisson's ratio* Static 0·2–0·3 (loaded), 0·39 (?unloaded); dynamic (10 kHz), 0·46 (?unloaded). *Mechanical loss tangent* (tan δ) 0·01–0·03 (varies with frequency).

Table 31.T4. MECHANICAL PROPERTIES OF EBONITES*

	Initial rubber		
	Natural (NR)	*Natural (loaded)*†	*Polybutadiene (BR) styrene/butadiene (SBR), acrylonitrile/butadiene (NBR)*‡
Elastic modulus, Young's (kgf/mm^2)§			
Static	200–300	400–600	SBR: 170–320
Dynamic (10 kHz)	335	—	—
Tensile strength (kgf/mm^2)	5·5–8	2–5	4–7·5
Breaking elongation (%)‖	3–6	2–5 (13)	1–6 (10)
Flexural strength (kgf/mm^2)‡‡	9–15	4–9	4–10
Impact strength (kgf.mm/mm^2)			
Unnotched	2·5–7	0·5–1·5 **	0·8–2·3 (up to 4·0 for NBR)
Notched¶	0·2–0·7	0·1–0·2	0·15–0·7
Hardness, Brinell (kgf/mm^2)††	11–12	*c.* 18	11–14
Rockwell, L	100	—	SBR: 90–120
Shore D Durometer	—	90–95	NBR: 70–80

*Unloaded, at room temperature unless otherwise stated
†25–80 vol. mineral filler per 100 vol. of (rubber + S); with increasing filler the modulus increases but tensile, flexural and impact strengths decrease
‡These rubbers give similar properties except where otherwise shown
§Shear modulus is *c.* one-third of Young's modulus
‖Values in brackets may be high because of inadequate vulcanisation
¶Other data (ft lbf/in notch); NR ebonite, 0·5–0·6; SBR ebonite 0·5–1·0
**With 50 parts ebonite dust as filler: (unnotched) 3·0; (notched) 0·25
††5 mm ball, 50 kgf load; 'hardness' varies with ball size and load
‡‡BSS 234 and 3164 specify proof tests corresponding to values from 4·5 to 10 kgf/mm^2 according to type of ebonite

31.4 FABRICATION

During the vulcanisation of ebonite allowance has to be made for dissipation of the heat of the reaction, which is strongly exothermic; this restricts (i) shortening of the time of vulcanisation, by raising the temperature or using accelerators, and (ii) the thickness of articles that can be vulcanised under given conditions.

In addition to the essential sulphur (25–50 parts per 100 rubber for natural and polybutadiene rubbers; rather less for butadiene copolymers), the following may be incorporated in the rubber*.

*The new raw rubber may be partly or wholly replaced by reclaimed rubber (§29.89).

Fillers. These reduce both the shrinkage during vulcanisation (*c.* 7 % by volume without fillers) and the heat build-up. Mineral fillers (e.g. clay, whiting, barytes, asbestine, graphite) also reduce deformation at elevated temperatures but lower the strength; there is no 'reinforcement' (cf. soft vulcanised rubber, §29.4); for electrical ebonite, silica or silicate fillers are preferred. Finely ground ebonite ('ebonite dust') has less effect on the properties of the ebonite and helps processing of the unvulcanised mix.

Accelerators. The most active of these are basic compounds (e.g. butyraldehyde-aniline, diphenylguanidine) but accelerators in general are less effective than in forming soft vulcanisates. Zinc oxide, commonly used to activate accelerators in the latter, is not always effective in ebonite, but magnesium oxide is more effective.

Softeners. These include polymers that do not harden by vulcanisation (e.g. polyisobutylene, butyl rubber, polyethylene); besides helping processing, they may improve impact strength, but reduce heat resistance.

Pigments. Normally inorganic, but the intense red-brown colour of the rubber-sulphur compound (*see* §31.31) makes light shades difficult to obtain.

Mixing of the ingredients and shaping of the mix by calendering, extruding or moulding are carried out generally as for soft vulcanised rubber (§29.4). Vulcanisation is relatively slow, e.g. requiring 4–5 h at 155°C without accelerator, or at 141°C with; even longer periods (at lower temperature) must be allowed for thick articles. Hence large mouldings are often given a short period of vulcanisation ('set cure') in the mould, and the process is completed in steam or inert gas. Ebonite sheets are usually vulcanised in hot water under pressure, after covering with tinfoil to prevent water discolouring the surface. Rods, tubes and extruded sections are packed in talc and vulcanised in steam or inert gas. Hand-built articles and linings of vessels are vulcanised with steam or water, under pressure if possible.

Other fabrication methods are as follows: (i) From liquid (depolymerised) rubber (§29.8(10)) which, when mixed with sulphur, etc., can be poured, e.g. at 40–80°C; used for making mouldings and for fixing paint-brush bristles. (ii) From rubber latex, especially for microporous ebonite (made by gelling the latex-sulphur-etc. mixture, vulcanising the gel, then drying out the water). (iii) Cellular ebonite is made either by incorporating a blowing agent (§29.88 and Section 81), or by saturating the unvulcanised mix with nitrogen under pressure, and releasing the pressure to produce expansion before and during vulcanisation ('expanded ebonite').

Ebonite can be bonded to most metals, except copper.

31.5 SERVICEABILITY

Ebonite has excellent electrical properties and resistance to chemical attack, good dimensional stability in moist conditions, is easy to shape by machining, and—especially in cellular form—has very low thermal conductivity. In the dark it retains its physical properties unimpaired over many years; however, exposure to light produces sulphuric and sulphurous acids, forming a conductive film that greatly reduces surface resistivity (this deterioration can be retarded, e.g. by acid-neutralising fillers, but cannot be entirely overcome).

The fact that the soft unvulcanised mix can be applied to surfaces of complex shape, and subsequently hardened by vulcanisation, makes ebonite especially suitable for chemical-resistant coverings.

The main limitation to the use of ebonite is its softening at elevated temperatures; the maximum safe working temperature depends on the stress conditions, but in general for natural rubber ebonite is not above 65°C, or exceptionally 75–80°C (styrene/butadiene ebonites, 80°C; acrylonitrile/butadiene ebonites, 110°C).

31.6 UTILISATION

Lining or covering chemical plant; chemical-resistant pipes and fittings, valves, pumps and roller-coverings; battery boxes and separators; electrically insulating components; trolley and gear wheels, steering wheel rims, textile machinery accessories, wood-wind musical instruments, pipe stems, combs, surgical appliances, bowling balls, also as intermediate layer for bonding soft vulcanised rubber to metals. *Cellular and expanded ebonites* are used for thermal insulation, flotation devices (floats, buoys), and as a core material for sandwich constructions.

31.7 HISTORY

T. Hancock obtained ebonite by immersing raw rubber in molten sulphur (British Patent 9952/1843); its first use was for making moulds for vulcanising soft rubber articles. C. Goodyear, in U.S.A., made ebonite probably before 1850, possibly before 1843 (U.S. Patent 8075/1851 granted to his brother N. Goodyear). In Great Britain the first firm to manufacture ebonite was Charles Macintosh and Co. Products made by Macintosh and by Goodyear received wide publicity at the Great Exhibition, London, 1851. In U.S.A. five companies were making ebonite in the 1850s; manufacture in France and Germany began in 1855–6.

Early products included combs, buttons, pipe stems, ornamental carvings and jewellery, denture plates (1851), electrical insulators (1855), chemical-resistant articles (*c.* 1860), fountain pens (1884). The names 'ebonite' and 'vulcanite' first appeared in print in 1860 and 1861 respectively.

In 1914–18, in Germany poly-2,3-dimethylbutadiene (Methyl Rubber H) was used to make an ebonite, and since the 1930s poly-1,3-butadiene and styrene/butadiene copolymers have been similarly used.

31.9 FURTHER LITERATURE

BOSTRÖM, S. (Ed), *Kautschuk-handbuch,* Vol. 5, Stuttgart (1962)

GENIN, G. and MORISSON, B. (Ed), *Encyclopédie technologique de l'industrie du caoutchouc,* Vol. III, Titre premier (Chapter IX), Paris (1955)

NAUNTON, W. J. S. (Ed), *Applied science of rubber* (Chapter X), London (1961)

SCOTT, J. R., *Ebonite, its nature, properties and compounding,* London (1958)

WHITBY, G. S. (Ed), *Synthetic rubber* (Chapter 16), New York and London (1954)

Specifications
BS 234 (loaded and unloaded ebonites for electrical purposes)
BS 903: Part D1–D7 (methods of test)
BS 3164 (loaded and unloaded ebonites for general purposes)
BS 3734 (dimensional tolerances for ebonite products)
ASTM D530 (tests on hard rubber)
ASTM D2135 (hard rubber materials)
ASTM D2707 (tension testing of hard rubber)

ISOPRENE RUBBER (SYNTHETIC)

With notes on synthetic polyisoprene latex and synthetic gutta percha

SYNONYMS AND TRADE NAMES

'Synthetic natural rubber', IR. Ameripol SN, Cariflex I, Cariflex IR, Coral rubber, Natysn, SKI, SKI-3.

GENERAL CHARACTERISTICS

Raw synthetic isoprene rubber is a colourless to light amber, translucent, somewhat elastic substance resembling lightly milled (masticated) natural rubber, but is not tacky and does not show cold flow. It is compatible with natural rubber and butadiene and styrene/butadiene rubbers. Available also as an artificial aq. dispersion or latex (§32.81). *Unvulcanised compounded mixes* have a good 'tack' (self-adhesion), though less with lithium-catalysed high molecular weight polymers, and show less elastic recovery than natural rubber mixes after calendering or extrusion.

Vulcanisates resemble those of natural rubber, notably in having low hysteresis and thus low heat build-up, good retention of tensile properties at elevated temperatures, and high tensile strength in unfilled mixes. Stiffness tends to be lower and breaking elongation higher than for natural rubber; ageing-, oxidation- and cracking-resistance are better, but abrasion- and tear-resistance may be poorer. Chemical-resistance is similar to that of natural rubber. Synthetic isoprene rubber can be vulcanised to the ebonite stage (*see* Section 31).

32.1 STRUCTURE

Simplest Fundamental Unit

Synthetic isoprene rubbers contain 90–99% *cis*-1,4 isoprene units (I) arranged head-to-tail as in natural rubber (*see* §29.1), with 0–8% *trans*-1,4 units (II), 0–3% 1,2 units (III) and 1–7% 3,4 units (IV), but the proportions of II and III are usually very small or nil. Synthetic *trans*-1,4-polyisoprene (*see* §32.82) is not rubber-like but resembles gutta percha and balata. (For the difference between *cis*- and *trans*-polyisoprenes *see* footnote, §29.1.)

Molecular weight Range (weight-average), 5.10^4–$5.8.10^6$ (degree of polymerisation 750–85 000). Mean, according to method of preparation, from 77 000 (number-average) or 135 000 (viscosity-average) to $2 \cdot 5 . 10^6$ (number-average) or $2 \cdot 75 . 10^6$ (weight-average). Lithium gives higher values than 'complex' catalyst. Gel content 0–20%.

X-ray data Amorphous, but when cooled or stretched the raw or vulcanised polymer gives diffraction patterns resembling those for natural rubber though with different relative intensities of crystalline, liquid and amorphous patterns. *Unit cell* monoclinic: $\beta = 92°$, $a = 12 \cdot 46$ Å, $b = 8 \cdot 89$ Å, c (fibre axis) $= 8 \cdot 10$ Å; orthorhombic: $a = 8 \cdot 97$ Å, $b = 8 \cdot 20$ Å, $c = 25 \cdot 12$ Å.

Compared with natural rubber synthetic isoprene rubber crystallises more slowly on cooling (*see* §32.32) and less completely on stretching, e.g. at 1000% extension, 25% crystalline (natural rubber, 40%).

32.2 CHEMISTRY

32.21 PREPARATION

Isoprene, or 2-methyl-1,3-butadiene $(CH_2\!\!=\!\!C(CH_3)\!\!-\!\!CH\!\!=\!\!CH_2$, b.p. 34°C), is made by the following methods:

(1) Reaction of isobutylene (§38.21) with formaldehyde (§26.21) to give 4,4-dimethyl-1,3-dioxan, which is subsequently catalytically decomposed:

$$(CH_3)_2C\!\!=\!\!CH_2 \xrightarrow[\text{acid cat.}]{2HCHO} (CH_3)_2C \overset{CH_2-CH_2}{\underset{O-CH_2}{\diagup\diagdown O}} \xrightarrow[\text{Ca phosphate cat.}]{-(H_2O \text{ and } HCHO)}$$

$$\longrightarrow CH_2\!\!=\!\!C(CH_3)\!\!-\!\!CH\!\!=\!\!CH_2$$

(2) Dimerisation of propylene to 2-methyl-1-pentene, and isomerisation of this to 2-methyl-2-pentene, which is subsequently pyrolysed:

$$2CH_2\!\!=\!\!CHCH_3 \xrightarrow{\text{Catalyst*}} CH_2\!\!=\!\!C(CH_3)CH_2CH_2CH_3 \xrightarrow[\text{catalyst†}]{\text{acid}}$$

(*e.g. tri-*n*-propylaluminium)   (†e.g. Al_2O_3/SiO_2)

$$CH_3C(CH_3)\!\!=\!\!CHCH_2CH_3 \xrightarrow[\text{HBr cat.}]{\text{pyrolysis}} CH_2\!\!=\!\!C(CH_3)CH\!\!=\!\!CH_2 + CH_4$$

(3) Dehydrogenation of isopentane or 2-methylbutenes (analogous to the Houdry process for butadiene; §33.21).

The monomer is polymerised by the following methods.

Using complex catalyst—Isoprene, in a dry de-aerated aliphatic hydrocarbon solvent (butane, pentane, hexane, heptane), is polymerised by a Ziegler-Natta catalyst. For example, the catalyst may be made by reacting an aluminium alkyl with titanium tetrachloride (approx. 1:1 molar ratio), polymerisation taking place under nitrogen, with stirring, at 45–50°C—lower temperatures give higher molecular weight but lower the rate of polymerisation; after 2–4 h the polyisoprene is precipitated as a crumb by adding acetone, then mixed with an antioxidant (or this may be added to the acetone), otherwise it oxidises readily. Use of an 'etherate' catalyst, in which the aluminium alkyl is complexed with ethyl or *iso*propyl ether, gives a higher yield in the same reaction time.

Using lithium catalyst—In this process the catalyst is an alkyl (usually *n*-butyl) lithium; the solvent for the isoprene may be an aliphatic hydrocarbon or cyclohexane.

'Complex' catalysts, compared with lithium, give lower molecular weights (hence better processing, but inferior dynamic properties in vulcanisates) and fewer *trans* units, i.e. a more stereo-regular structure, *c.* 96% *cis*-1,4 (lithium, *c.* 92%), hence vulcanisates crystallise more readily on stretching and retain greater strength when hot.

According to molecular weight, the Mooney viscosity ranges from 40–80 (lithium catalysed polymers of high molecular weight do not give correspondingly high viscosity). Other variables are (*a*) antioxidant may be either 'staining' (e.g. phenyl-β-naphthylamine), or 'non-staining'; (*b*) light-coloured grades of the polymer are available for making white or pastel-shade products; (*c*) high-molecular weight gel-free polymers can be oil-extended, e.g. with 25 parts naphthenic oil per 100 polymer.

32.22 PROPERTIES

Soluble in same liquids as natural rubber (§29.22). **Plasticised by** (*a*) *Peptisers* such as phenols, organic disulphides, β-naphthyl and xylyl mercaptans, pentachlorothiophenol, zinc salt of *o*-benzamidothiophenol; (*b*) *physical softeners* generally as for natural rubber (§29.4), e.g. pine tar and petroleum processing oils, the latter being preferred for some high molecular weight grades. **Swollen** (when vulcanised), by same liquids as for natural rubber (*see* §29.22) and to about the same extent, e.g. swelling of vulcanisates (with 50 parts carbon black per 100 rubber) in 70/30 *iso*octane/toluene at 24°C (% volume increase): syn. isoprene rubber, 197; natural rubber, 214. **Relatively unaffected by** same liquids as for natural rubber (*see* §29.22), but action may depend on the filler, e.g. dil. acids (20% CH_3COOH at 50°C, 20% HCl at 65°C, 33% H_2SO_4 at 65°C) have little effect on silica-filled vulcanisates but swell and weaken those containing lampblack. **Decomposed by** same liquids as natural rubber (*see* §29.22).

32.23 IDENTIFICATION

N, Cl, S, P nominally absent (N may be present in an antioxidant; §32.21); presence of S with traces of N indicates the vulcanised material. Not saponifiable. **Combustion, pyrolysis, other tests** as for natural rubber (§29.23). **Distinction from natural rubber** (*a*) Absence of protein (present in natural rubber and detected by hydrolysis and paper chromatography or by infra-red absorption spectrum; §29.23iv); (*b*) phosphorus content, *c*. 20 p.p.m. compared with *c*. 400 p.p.m. for natural rubber (test unsuitable for vulcanisates, as these may contain lecithin).

32.3 PHYSICS

32.31 GENERAL PROPERTIES

Specific gravity Raw rubber 0·92. **Water absorption** Lower than for natural rubber, *see* Table 32.T1.

Table 32.T1. WATER ABSORPTION OF UNFILLED VULCANISATES

	Time of immersion			
	20 h	7 days	14 days	28 days
In water at 100°C (wt. increase, %)				
Synthetic isoprene rubber	1·5	3·6	4·2	5·3
Natural rubber	3·7	9·3	13·0	19·5
In water at 70°C (wt. increase, mg/cm^2)				
Synthetic isoprene rubber	0·32	0·77	1·13	—
Natural rubber	2·2	5·8	7·6	—

32.32 THERMAL PROPERTIES

Physical effects of temperature *Raw rubber*: Hardens and eventually becomes brittle when progressively cooled; glass temperature −72°C to −68°C. At sub-zero temperatures crystallises, but more slowly than natural rubber, e.g. at −25°C requires 25–55 h as compared with 6–7 h for natural rubber to reach comparable degrees of crystallisation. On heating, softens and weakens; pyrolysis begins at 300°C and is complete at 400°C. **Vulcanisates** Brittleness temperature: −67°C to −56°C. Stiffening at low temperatures, and temperature of minimum resilience, are about the same as for natural rubber, but the tendency to crystallise is less. At elevated temperatures tensile strength and breaking elongation decrease, in much the same way as for natural rubber, depending on the type of synthetic isoprene rubber, compounding and vulcanisation (e.g. tensile strength of vulcanisates reinforced with carbon black falls by 60–65% at 100–135°C); tear strength and internal friction (damping) also decrease (*see* §32.34).

32.33 ELECTRICAL PROPERTIES

In unfilled vulcanisates these are comparable with or better than those of natural rubber (*see* comparisons below). **Volume resistivity** (ohm cm) $> 10^{14}$ (natural rubber $> 10^{14}$). **Dielectric strength** (kV/mm) 23·5 (23·0). **Dielectric constant** (at 1 kHz) Dry, 2·59; wet, 2·69 (2·64; 2·95). **Power factor** (at 1 kHz) Dry, 2.10^{-4}; wet 4.10^{-4} $(11.10^{-4};$ $11.10^{-4})$.

32.34 MECHANICAL PROPERTIES

As with natural rubber, unfilled vulcanisates have high **tensile strength**; reinforcing fillers produce little increase, but raise the **tear strength. Stiffness** in comparable mixes is generally less than for natural rubber, hence extensibility tends to be greater. **Resilience** (rebound) is already high at room temperatures (unfilled vulcanisates up to 93%; or 70–83% filled vulcanisates), and hence only little higher (by *c.* 5%) at 100°C (contrast butyl rubber, §38.34). **Abrasion resistance** of reinforced vulcanisates is comparable with that of

Table 32.T2. TYPICAL VALUES OF MECHANICAL PROPERTIES OF VULCANISATES*

	Unfilled vulcanisates	Filled vulcanisates†
Elastic modulus shear, 60 Hz; 100°C, (kgf/mm²)	0·040	0·14–0·21
Tensile strength (kgf/mm²)	1·7–3·8	2·3–2·9; up to 3·8‡
100°C	0·8–2·0	1·3–1·6
Breaking elongation (%)	720–1300	430–560
100°C	700–1100	520
Stress at 300% elongation (kgf/mm²)	0·08–0·24	0·9–1·5
Tear strength (kgf/mm)	2·0	5·5–7·7; up to 8·4‡
100°C	—	4·8–6·9
Hardness (IRHD)	43	60–67
Compression set§ (%) 70°C	—	14 (30 phr MPC black)
70°C	—	18–24 (40 phr HAF black)
100°C	—	53–62 (40 phr HAF black)
Internal friction‖, 60 Hz (kP)	—	*c.* 9
100°C	0·2	2·5 (5·0)

*At 20°C unless otherwise indicated
†50 parts HAF carbon black (per 100 rubber) except where otherwise indicated
‡Channel black
§ASTM D395, method B: constant compression for 22 h (at 70°C) or 70 h (at 100°C); recovery 30 min at room temperature; set expressed as percentage of compression
‖Measured in shear; bracketed value is for mill-mixed rubber, others for internal mixer

natural rubber vulcanisates; thus in tyres, synthetic isoprene rubber has given 75–105% the wear life of natural rubber. **Resistance to cracking** (or crack growth) in laboratory tests is better than for natural rubber. *See also* Table 32.T2.

32.4 FABRICATION

Synthetic isoprene rubber is processed in similar ways to natural rubber, except for the following differences.

(i) Lower molecular weight grades (as made with 'complex' catalyst; *see* §32.21) require no plasticisation before mixing, being similar to masticated natural rubber. Higher molecular weight grades (lithium catalyst) require plasticisation on a mill or in an internal mixer, with or without a peptiser (*see* §32.22); however, plasticisation proceeds more quickly than with natural rubber. (ii) As synthetic isoprene rubber is softer (during mixing) than natural rubber, dispersion of powdery ingredients is more difficult; master-batching of powders, and (e.g. with organic accelerators) choice of low-melting materials, are recommended to improve dispersion. (iii) Because isoprene rubber lacks the protein present in natural rubber, addition of a 95/5 lecithin/triethanolamine mixture (e.g. 1·5 parts per 100 rubber) is sometimes desirable to activate vulcanisation, prevent reversion and reduce hysteresis in the vulcanisate.

Compounded mixes of synthetic isoprene rubber calender well and can be extruded more quickly than natural mixes; calender shrinkage and extruder 'die swell' are less, tack (self-adhesion) is usually as good.

The rubber can be bonded to steel as for natural rubber.

32.5 SERVICEABILITY

Vulcanisates are susceptible to atmospheric oxidation, causing deterioration of mechanical properties, also to cracking when exposed stretched to air containing ozone; these effects can be greatly reduced by incorporating antioxidants and antiozonants respectively (*see* §29.4). Resistance to short-term rise or lowering of temperature is good, but prolonged heating aggravates oxidative deterioration while cold produces crystallisation and stiffening. Hydrocarbon and chlorinated hydrocarbon liquids cause considerable swelling, and oxidising agents attack the rubber.

32.6 UTILISATION

Synthetic isoprene rubber, being an attempt to match natural rubber, is intended primarily for uses where the latter has hitherto been indispensable because of its low hysteresis (heat build-up) and maintenance of tensile properties at relatively high temperatures, e.g. aeroplane and heavy-duty (bus, military, lorry, off-the-road) tyres—where it is generally blended with natural or styrene/butadiene rubber—or its high tensile strength in unfilled vulcanisates, e.g. rubber bands and threads. Other uses are motor and instrument mountings, footwear heels and direct-moulded soles; ebonite (*see* Section 31), insulated wires and cables (cf. low water absorption; *see* §32.31), adhesives, footwear, proofings and surgical goods (light-coloured grades; *see* §32.21). *See also* §32.81.

32.7 HISTORY

Isoprene was first obtained, by pyrolysis of natural rubber, and shown to polymerise to a 'rubber-like body' by Greville Williams (1860). G. Bouchardat (1875) concluded that natural rubber is a polymer of isoprene. W. A. Tilden (1882) observed that isoprene was converted into 'india-rubber or caoutchouc' on contact with strong hydrochloric acid or nitrosyl chloride, and deduced its structural formula. In 1884 he obtained isoprene by 'pyrogenic decomposition' of turpentine, and (1892) found that the polymer obtained on long standing could be vulcanised with sulphur. F. E. Matthews (1910), working for Strange and Graham Ltd. on production and polymerisation of dienes, found that sodium induced relatively quick and complete polymerisation of isoprene. C. D. Harries (1910) independently discovered this, and found that polymerisation could be effected by heating with acetic acid.

(For fuller details of the early history, *see* PERKIN, W. H., Jr., *J. Soc. Chem. Ind.,* **31**, 616 (1912).)

The earlier polymers, however, did not resemble natural rubber because they did not have the regular *cis*-1,4 structure (*see* §32.1). However, the discovery, in the early 1950s, of catalysts inducing stereo-specific polymerisation made possible a substantially all-*cis*-1,4 polyisoprene, production being announced by the Firestone, Goodrich, and Goodyear companies in U.S.A. in 1954–55. Similar polymers have since been developed elsewhere, including the Russian SKI and SKI-3.

32.8 ADDITIONAL NOTES

32.81 SYNTHETIC ISOPRENE RUBBER LATEX

Aq. dispersions of synthetic isoprene rubber (at least 85 %, normally 92–97 %, *cis*-1,4) are made by emulsifying a hydrocarbon solution of it with an aq. soap solution (e.g. 0·5% potassium rosinate; or a soap-forming acid may be added to the polymer solution, and aq. alkali added), removing the solvent by 'stripping', and finally concentrating to 60–65% total solids. Antioxidant (e.g. 1% 2,6-di-*tert*.butyl-4-methylphenol) is added during preparation.

The latices have sp. gr. 0·93–0·94, pH 10–10·5, average rubber particle size 0·65–0·75 μm (rather bigger than in natural rubber latex), surface tension 31–44 dyn/cm. Unlike natural latex they contain no ammonia, 'natural accelerator' or carotene (hence they have lighter colour), and the content of non-rubber constituents is low. Mechanical stability is high but chemical stability low.

Compounding and vulcanisation are broadly as for natural rubber latex, except that more accelerator is needed, and anionic stabilisers are preferable. As ammonia is absent, zinc oxide does not cause premature thickening and gelling, and gelation time does not change owing to its volatilisation. **Vulcanisates** have tensile strength similar to those from natural latex (or may be lower with Li-catalysed polymer), higher elongation, lower stiffness (especially with low gel content), lower water absorption, but poorer resistance to tearing and to ageing and ozone attack—hence protective agents against the latter must be added. **Uses:** *Foam rubber* Used alone or blended with styrene/butadiene copolymer latex (*see* §34.81), gives faster gelling, greater hot wet strength, and a softer, more resilient foam than the latter, also less shrinkage during vulcanisation than natural latex; the product can contain high filler loadings, an advantage in foam backings. *Dipped articles* Gives a high deposition rate because of large particle size and low soap content. The low stiffness of vulcanised films is advantageous in balloons and surgeons' gloves, as is also their low water absorption in, e.g. electricians' gloves. *Adhesives* Absence of fatty acid (as present in natural latex) improves adhesion, tack, and 'quick grab'. *Extruded thread* Gives better colour than natural latex, with low hysteresis and set. *Rubberised hair* Fibres 'combine' more readily because the unvulcanised rubber is more plastic under pressure. *Other uses* Carpet backings, slush mouldings.

32.82 SYNTHETIC GUTTA PERCHA (Trans-pip)

Synthetic polyisoprene having substantially an all *trans*-1,4 structure (*see* §32.1) resembles the hydrocarbon of gutta percha (*see* §29.86) in the rapidity and extent of its crystallisation (crystallinity 28–34%). It is made by solution polymerisation using a Ziegler-type catalyst (e.g. $VOCl_3 + AlEt_3$, $VCl_3 + AlEt_3$; $TiCl_3$(α form) + $AlEt_3$; $VCl_3 + AlEt_3 + Ti$-$(OC_3H_7)_4$). It has the following properties: sp. gr., 0·95; crystalline melting point 64–67°C (α form), 56·5–60·5°C (β form); vulcanisable with sulphur, giving good tensile strength in unfilled mixes. Polymerisation can be controlled to vary the molecular weight (e.g. to be greater than for natural gutta percha) and structural purity (lower *trans*-1,4 content gives a more rubbery, less crystalline polymer). It is compatible with natural, butadiene, styrene/butadiene, acrylonitrile/butadiene, and butyl rubbers, polyethylene, polyamides and ABS plastics (§4.81). **Uses** To replace gutta percha or balata in golf-ball covers, transmission belts, adhesives and film packaging applications; also in hot-melt adhesives.

32.9 FURTHER LITERATURE

HSIEH, H. L., *Polymerisation of Dienes by Means of Metallic Lithium and Lithium Alkyls, Dissertation*, Princeton University (1957) (Preparation and structure of polyisoprene)

SECTION 33

BUTADIENE RUBBER

With notes on carboxylated butadiene rubber, butadiene rubber latex and thermoplastic polybutadienes

SYNONYMS AND TRADE NAMES

BR; Ameripol CB, Astyr, Budene, Bunas 85, 115 and Buna CB, Cariflex BR, Cis-4, Cisdene, Diene, Duragen, Europrene-cis, Intene, JSR-BRO1, Plioflex 5000 and 5000S, SKB, SKBM, SKD, SKLD, SKV, Synpol E-BR, Synpol 8407, Taktene.

GENERAL CHARACTERISTICS

Raw butadiene rubber is a colourless to brown, transparent or translucent, somewhat elastic material possessing little or no tack. Most grades exhibit cold flow, but 'nonflow' grades are available. It is compatible with natural, synthetic isoprene, styrene/butadiene, nitrile and chloroprene rubbers; the solution-polymerised rubber is often blended with one of the first three to achieve satisfactory processing. Butadiene rubber is also available as a latex or aq. dispersion (§33.82).

Unvulcanised compounded mixes have less elastic recovery ('nerve') than natural rubber mixes and hence can be calendered and extruded more accurately, but are mechanically weak and deficient in tack (self-adhesion).

Vulcanisates generally resemble those of natural rubber, but are superior in resistance to heat, ageing, ozone and especially to abrasion and cold. Tensile and tear strengths (especially for unfilled mixes and 'low-*cis*' rubbers, *see* §33.21) and stiffness are lower. In laboratory tests hysteresis and heat build-up are generally greater (especially for emulsion-polymerised rubbers), but not when butadiene rubber is used as partial replacement for natural rubber in tyres. The coefficient of friction is lower than for natural rubber, especially on wet surfaces, but grip on ice or snow is better.

Butadiene rubber can be vulcanised to the ebonite stage (*see* Section 31).

33.1 STRUCTURE

Simplest Fundamental Unit

I	II	III
cis−1, 4	*trans*−1, 4	1,2 or vinyl

$$C_4H_6, \text{ mol. wt.} = 54$$

According to the method of polymerisation and catalyst used (*see* §33.21) butadiene rubbers contain I, II and III units in the following proportions (%):

Polymerisation		I	II	III
Bulk				
Catalyst:	Lithium	35–36	52–56	8–13
	Sodium	10	25	65
	Potassium	15–16	37–40	52
Solution				
Catalyst:	Lithium alkyl	32–36	53–58	8–11
	'Cobalt'	94–98	1–3	1–2
	'Titanium'	90–98	1–5	1–5
	'Nickel'	>95	1–2	1–2
Emulsion		14–18	64–69	17–18

The III units may be disposed with the vinyl groups all on one side of the main chain (isotactic) or alternately on one side then the other (syndiotactic). Polymers consisting mainly of II and/or III units are not rubber-like, but are thermoplastic materials (§33.83) resembling balata and gutta percha (§29.86).

Molecular weight *Bulk polymers* Maximum *c.* 400000; peak of distribution curve, 50000. *Solution polymers* 10^5–$1·6.10^6$ (weight-average) or 50000–650000 (number-average); usually 260000–430000 (wt. or no.-av.). *Emulsion polymers* $1·1.10^6$ (wt.-av.) or 50000–$1·2.10^6$, normally *c.* 450000 (no.-av.). The **molecular weight distribution** is wide in bulk and emulsion polymers (wt.-av./no.-av. ratio for emulsion polymers rising to 7) but in solution polymers it is increasingly narrow (ratios of $1·7$–$1·2$) in the order 'cobalt', 'titanium', and lithium alkyl. Bulk and emulsion polymers are branched; solution polymers made with 'cobalt' catalyst may also be branched, but those made with 'titanium' or lithium alkyl are substantially linear. The **gel content** of potassium-catalysed bulk polymers is *c.* 50%, but in solution and emulsion polymers is very small.

X-ray data Amorphous; but raw or vulcanised rubbers having above *c.* 90% of *cis*-1,4 units become crystalline when cooled or stretched, e.g. a polymer with 98–99% *cis* has a monoclinic unit cell, containing 4 monomer units with, $\beta = 109°$, $a = 4·60$ Å, $b = 9·50$ Å, c (chain axis, identity period) $= 8·60$ Å.

33.2 CHEMISTRY

33.21 PREPARATION

1,3-Butadiene (CH_2=CH—CH=CH_2, b.p. −4°C) is made mainly from the C_4 fractions of petroleum gases or natural gas, the following methods being used. (i) Dehydrogenation of *n*-butane using a chromic oxide/alumina catalyst gives butadiene plus butylenes (Houdry process). (ii) Fractionation of a butane/butylene feedstock to give a mixture rich in butylenes; this is treated with sulphuric acid to remove isobutylene (§38.21), and the remaining *n*-1- and *n*-2-butylenes are dehydrogenated to give butadiene by passing (along with super-heated steam) over a catalyst comprising calcium and nickel phosphates plus chromic oxide. (iii) Dehydrogenation of *n*-butane to give *n*-1- and *n*-2-butylenes, which are converted to butadiene as in (ii). (iv) Small proportions of butadiene are formed in the cracking of petroleum fractions to produce ethylene and propylene. In each of these processes the butadiene is separated in at least 98 % (commonly 99 % or more) purity, and is then stabilised by adding an antioxidant, such as *p-tert*.butylcatechol (25–50 p.p.m.).

1,3-Butadiene can be polymerised, using an ionic catalyst, in the following ways.

Bulk polymerisation Liquefied butadiene polymerises in the presence of lithium, sodium, or potassium (producing the Russian SKBM, SKB and SKV respectively). In the original Russian 'rod' process the butadiene was brought into contact with iron rods on which the alkali metal was deposited; the later 'rodless' process uses a dispersion of finely-divided alkali metal, and the German Buna 85 was formerly made (continuously, in a tubular reactor with a screw conveyer) by treating liquefied butadiene with a dispersion of potassium in low molecular weight polybutadiene. An antioxidant, such as phenyl-β-naphthylamine, is incorporated in the polymer.

Solution polymerisation Butadiene, freed from air, water and carbon dioxide, is dissolved in an inert hydrocarbon solvent and polymerised with a stereospecific catalyst, which is then deactivated by addition of antioxidant; the solvent and residual monomer are distilled off, leaving a slurry from which the polymer is separated. The catalysts used are: a *lithium* alkyl (usually butyl); *cobalt* (a dialkylaluminium chloride with cobaltous chloride); *titanium* (an aluminium alkyl with titanium tetrachloride and iodine, or with titanium tetraiodide); or *nickel* (nickel metal, oxide, carbonyl or organic salt, with a halide—e.g. BF_3 or $TiCl_4$—and an organometallic compound, e.g. $Al(C_2H_5)_3$). Solution polymers are described as 'low-*cis*' or 'high-*cis*', corresponding respectively to a content of *c*. 35 % or *c*. 90 % (or more) *cis*-1,4 units.

Emulsion polymerisation Liquefied butadiene is dispersed in water containing an emulsifier (e.g. fatty acid soap or mixed fatty/rosin acid soap) and a polymerisation catalyst (e.g. potassium persulphate plus dodecyl mercaptan, the latter acting also as a regulator of molecular weight). When polymerisation is complete, an antioxidant is added, and the polymer is separated from the suspension by an acid coagulant. Emulsion polymers made at *c*. 5°C and *c*. 50°C are called 'cold' and 'hot' respectively.

Other variable features of polybutadiene rubbers, governed by the conditions of preparation are as follows: (i) Mooney viscosity (at 100°C) ranges from 35–60 according to molecular weight; (ii) antioxidant may be either 'staining' (e.g. phenyl-β-naphthylamine, diphenyl-p-phenylenediamine) or 'non-staining' (tri-nonylphenylphosphite, 2,6-di-*tert*.butyl-4-methylphenol); (iii) Grades of lithium-alkyl catalysed polymer with reduced cold flow, to facilitate handling, are available. (iv) Light-coloured grades are available for making white or pastel-shade products. (v) High molecular weight gel-free polymers (e.g. 'cobalt'-catalysed or emulsion polymerised) can be extended with up to 80 parts (per 100 of polymer) of hydrocarbon oils, e.g. 37·5 parts of high-aromatic or 50 parts of naphthenic oil, giving products with 30–35 Mooney viscosity. (vi) 'Preprocessed' grades contain oil (e.g. 15%) and a tackifier, to make possible processing without blending with other rubbers (*see* §33.4, **Fabrication**).

33.22 PROPERTIES

Dissolves (milled raw rubber) in hexane, heptane, ligroin*, cyclohexane, methylcyclohexane*, naphtha, kerosene, pinene, turpentine, tetralin, benzene*, toluene, styrene, naphthenic petroleum oils, di-*iso*propyl and dibenzyl ethers, higher ketones (e.g. methyl ethyl ketone), tetrahydrofuran, mesityl oxide, oleic acid, higher aliphatic esters, chloro- and nitro-hydrocarbons, pyridine, carbon disulphide. **Plasticised by** physical softeners, namely 'aromatic' and 'high-aromatic' petroleum oils, pine tar, 'aliphatic' plasticisers, e.g. dioctyl phthalate; the peptisers used with other rubbers (*see* §§32.22, 38.22) are unsuitable because they cause gelation and convert *cis* to *trans* structures (*see* §33.1). **Swollen** (raw rubber) by paraffinic petroleum oils, perchloroethylene, cyclohexanone, acetone, ethyl acetate, cresol, aniline, 1-nitropropane, cottonseed oil; (vulcanised rubber) to a greater or less extent by the liquids that dissolve the raw rubber (*see* above), e.g. unfilled vulcanisates gain 80–130% by weight in *n*-heptane. **Relatively unaffected by** water, dilute acids (but not 6N HCl) and alkalis, glacial acetic acid, hypochlorite solutions, lower alcohols, glycol, phenol, furfural, castor oil, ethylenediamine. **Decomposed by** boiling nitric acid (sp. gr. 1·42).

33.23 IDENTIFICATION

N, Cl, S, P nominally absent (N may be present in an antioxidant); presence of S with traces of N indicates vulcanised material. Not

---

*May leave some gel.

saponifiable. **Combustion** Burns with a smoky flame (distinction from butyl rubber and polyisobutylene). **Pyrolysis** On heating, butadiene rubber softens and eventually pyrolyses, giving (at 325–475°C) 14% hydrocarbons volatile at 25°C (paraffins, olefins, butadiene); the liquid pyrolysate gives characteristic infra-red absorption spectrum, with diagnostic bands at 11·0, 10·4, 10·1 and 14·8 μm; absence of 'aromatic' bands, e.g. at 6·7, 12·9 and 14·3 μm distinguishes it from the pyrolysate from a styrene/butadiene rubber. **Other tests** Negative results in the Weber and trichloroacetic acid colour reactions for natural rubber (§29.23) and in reactions for styrene in styrene/butadiene copolymers (§34.23) distinguish butadiene rubber from these rubbers; dissolution in boiling nitric acid (sp. gr. 1·42) distinguishes it from butyl rubber.

Determination of proportions of I, II and III monomer units (*see* §33.1) is best effected by infra-red absorption spectroscopy (*see*, e.g. BINDER, J. L., *J. polym. Sci.*, A.**1**, 47 (1963)).

33.3 PHYSICS

33.31 GENERAL PROPERTIES

Specific gravity Raw rubber 0·88–0·91 (amorphous), 1·01 (crystalline, 98–99% *cis*-1,4). **Refractive index** Raw $n_D^{20} = 1\cdot5158$ (bulk polymer), 1·5175 (emulsion polymer); $n_D^{25} = 1\cdot5147$–1·5175 (emulsion polymers with ratios of I, II, III units ranging from 7/77/16 to 25/55/20). **Water absorption** Solution butadiene rubber (BR) shows low absorption relative to natural rubber (NR), e.g. for mixed vulcanisates:

BR/NR *ratio*	0/100	60/40	70/30	80/20	90/10
Absorption after immersion for 7 days at 75°C (%)	4·56	3·19	2·87	2·47	1·91

Permeability Vulcanisates are relatively permeable to gases (e.g. *c.* three times as much as natural, synthetic isoprene or styrene/butadiene rubber). *Examples* $(10^{-10}$ cm$^2)$/(s cmHg). (i) Unfilled vulcanisates, at 25°C: N_2, 6·5; O_2, 19; H_2, 42; CO_2, 140. (ii) Vulcanisates of oil-extended rubber with 55 parts HAF carbon black (per 100 rubber), at 30°C: air, 6·0. To water vapour at *c.* 30°C: 3·8 $(10^{-10}$ g)/ (cm s cmHg).

33.32 THERMAL PROPERTIES

Crystalline m.p. (°C) *Raw rubber* 98–99% *cis*-1,4, +2. *Vulcanised* 95·2% *cis*-1,4, −14 to −6; 91·2% do., −15 to −11; 87·3% do.,

−20 to −22. **Heat of fusion** *Raw* 940–2000 cal/mol monomer unit (*cis*-1,4). **Entropy of fusion** *Raw* 8 eu/(mol monomer unit. °K) (98% *cis*-1,4). **Coefficient of linear expansion** $(10^{-4}/°C)$ *Raw* 'High *cis*', 0·52 and 2·45 (below and above glass-transition temperature respectively); 'low *cis*' (bulk polymer), 1·0 and 2·37 (do.). **Conductivity** Rather greater than for natural rubber, e.g. *c*. 10% greater heat flow in tyre tread mixes. **Physical effects of temperature** *Raw rubber* Hardens and eventually becomes brittle when progressively cooled. Glass temperature (°C): 'high *cis*' (*see* §33.21), −110 to −95; 'low *cis*' (Na- or K-catalysed bulk polymer), −63 to −45°C; 'low *cis*' (emulsion polymer), −86 to −75. At sub-zero temperatures rubbers with 75% or more of *cis*-1,4 units crystallise; the rate of crystallisation is greatest at −57° to −52°C, and increases with molecular weight, e.g. for polymers with molecular weight 99000 and 135000 the maximum rate = 0·41 and 2·4% per hour and the half-crystallisation period = 250 and 50 minutes, respectively.

Vulcanised rubber Brittleness temperature (°C): 'high *cis*', −107 to −97; 'low *cis*' (bulk polymer), −65 to −50; 'low *cis*' (Li alkyl or emulsion polymer), −96 to −74. 'High *cis*' vulcanisates crystallise on cooling, most quickly at −55° to −50°C. Best cold-resistance is shown by a solution polymer with a ratio of I/II/III units (§33.1) of 79/16/5. On cooling, the resilience of solution and emulsion polymers is retained to a lower temperature than with natural rubber, e.g. temperature of minimum resilience (°C): 80% *cis*. −70 or lower; low *cis* (bulk polymer), −34; low *cis* (emulsion polymer), −55; natural rubber, −40. At elevated temperatures (e.g. 100°C) tensile strength and breaking elongation decrease markedly (*see* §33.34).

33.33 ELECTRICAL PROPERTIES

	Potassium-catalysed bulk polymerised rubber		
	Raw	*Unfilled vulcanisate*	*Filled vulcanisate**
Volume resistivity (ohm cm)			
at 20°C	5.10^{15}	3.10^{15}	5.10^{15}
at 100°C	5.10^{12}	$1·2.10^{12}$	$1·2.10^{13}$
Dielectric constant, 50 Hz	2·3	—	—
Power factor, 50 Hz			
at −20°C	0·002	0·0045	0·012
at 20°C	0·008	0·002	0·014
at 100°C	0·07	0·014	0·07

*70 parts clay plus 70 parts talc per 100 rubber

33.34 MECHANICAL PROPERTIES

The properties of vulcanisates depend on the relative proportions of I, II and III units in the polymer (*see* §33.1). In laboratory specimens of the polymers tensile strength, breaking elongation, resilience and abrasion resistance are highest, and tear strength, stiffness and heat build-up lowest, with 100% of *cis*-1,4 (I) units, but in commercial rubbers these relationships tend to be obscured.

In comparison with natural rubber the tensile strength of unfilled vulcanisates, especially of 'low *cis*' types, is very low, but the difference is less in carbon black reinforced vulcanisates. Compared with natural rubber vulcanisates, tear strength tends to be lower and compression set higher, but resilience greater with solution polymers at low and room temperatures, although lower with emulsion and bulk polymers at all temperatures.

According to laboratory evaluation, the hysteresis and heat build-up, compared with those of natural rubber, are the same or greater with solution polymers, but compared with styrene/butadiene rubber they are less; with emulsion polymers they are greater than for natural or styrene/butadiene rubber.

In tests of repeated flexing, butadiene rubber vulcanisates have good resistance to crack initiation, but subsequent crack growth is rapid, so that the overall flex life is poor. Abrasion resistance exceeds that of natural and styrene/butadiene rubbers, e.g. by up to 3 times in laboratory tests on solution-polymerised polybutadiene.

Table 33.T1. TYPICAL VALUES OF MECHANICAL PROPERTIES OF VULCANISATES*,

	Unfilled vulcanisates	*Filled vulcanisates (50 parts HAF carbon black per 100 rubber)*
Elastic modulus, Young's, static (kgf/mm^2)	—	0·53–0·61†
Elastic modulus, shear, 60 Hz (kgf/mm^2)	0·06	—
100°C	0·055	0·16–0·23
Tensile strength (kgf/mm^2)	0·2–0·7 (1·75)‡	1·4–2·3
100°C	—	0·7–1·4
Breaking elongation (%)	660–830	310–500
100°C	—	260–300
Stress at 300% **elongation** (kgf/mm^2)	0·06	0·55–1·4
Tear strength (kgf/mm)	0·25	3·8–5·5
100°C	—	3·1–3·7
Hardness (IRHD)	33–35	54–72
Compression set (%)‖	—	21·4
Internal friction, shear, probably 60 Hz (kP)¶	0·75–1·07	3·6–5·3
Resilience, rebound (%)§	80–84/68–74/50	55–67/47–54/39
100°C	80–84/73–80/—	60–74/55–64/—

*At 20–23°C unless otherwise indicated
†Filler: 45 parts 'CK-3' carbon black (gas black type, made from naphthalene)
‡For 100% *cis*-1,4 polymer
§Figures are for solution/emulsion/bulk polymers
‖2 h compression by 35% at 100°C: 1 h recovery
¶Unfilled, temperature not stated; filled, at 100°C

33.4 FABRICATION

Butadiene rubbers are compounded and processed like natural rubber, except for the following differences:
(i) As they are not readily plasticised by working on rolls or in an internal mixer, and peptisers cannot be used (§33.22), they are commonly made with relatively low molecular weight. (ii) Unless 'extended' with a hydrocarbon oil or when containing processing aids (§33.21), they are difficult to process and are therefore generally used in blends with natural, synthetic isoprene or styrene/butadiene rubber. (iii) Because of their great capacity for absorbing hydrocarbon oils and fillers, without serious detriment to vulcanisate properties, larger proportions of these may be used than in natural rubber. (iv) As solution butadiene rubbers contain virtually no fatty acid, this must be added to facilitate vulcanisation. Polybutadiene rubbers can be bonded to steel during vulcanisation.

33.5 SERVICEABILITY

Vulcanisates of butadiene rubber are less susceptible than those of natural rubber to atmospheric oxidation (and the consequent deterioration of mechanical properties) and to cracking when exposed stretched to air containing ozone. Resistance to short-term cooling is generally better than for natural rubber, but types having a substantially stereo-regular structure ('high *cis*') crystallise and stiffen on prolonged cooling. Loss of strength and extensibility in short-term heating is about the same as for natural rubber, but prolonged heating causes more stiffening, though less than with styrene/butadiene rubber. Resistance to organic liquids and corrosive chemicals is broadly similar to that of natural rubber.

33.6 UTILISATION

By far the largest use of butadiene rubber is as a partial replacement for natural or styrene/butadiene rubber in tyres, which results in improved wear and low-temperature performance (especially grip on ice and snow), less fatigue, rolling resistance and groove cracking, and cooler running. Other uses claimed include: belting, footwear, mountings, injection mouldings, hose, flooring, cellular rubber and ebonite; also for incorporation in polystyrene and ABS plastics to increase impact strength (*see* §4.81).

33.7 HISTORY

1,3-Butadiene was discovered in 1863 by E. Caventou (on pyrolysis of amyl alcohol—actually fusel oil, a mixture of *iso*butylcarbinol and *sec*butylcarbinol)—but it was not until 1910 that it was observed to polymerise under the action of sodium (C. D. Harries, F. E. Matthews) or of heat (S. V. Lebedev) to give a rubber; Harries (1911) stated that the sodium polymer was 'of still better quality' than that from isoprene. W. H. Perkin, Jr. (1912) developed a method for making butadiene from *n*-butyl alcohol, and commented 'it may ultimately be found that this is a cheap and convenient route to butadiene rubber'*.

In the 1920s I. G. Farbenindustrie studied polymerisation by alkali metals and produced thereby the first commercial butadiene rubbers, Buna 85 and 115. In 1926 the Russian Central Administration for State Industry offered a prize for development of butadiene rubber, and after accepting processes from B. V. Byzov ('SKA' from petroleum) and S. V. Lebedev ('SKB' from alcohol), factories were set up at Yaroslavl (1932), and Voronezh and Efremov (1933). However, further work in Germany (*c.* 1930) showed the advantages of emulsion copolymers of butadiene with styrene, and later effort was concentrated on these.

Following the introduction of 'stereospecific' catalysts and the polymerisation of isoprene to 'synthetic natural rubber' (*see* Section 32), they were applied to the cheaper monomer butadiene, and commercial production of stereospecific polybutadiene began in 1955. Meanwhile, since 1952, butadiene rubber latices (*see* §33.82) had been produced by emulsion polymerisation.

33.8 ADDITIONAL NOTES

33.81 CARBOXYLATED BUTADIENE RUBBER

The Russian 'SKD-1' is a copolymer made from butadiene and methacrylic acid (100/2), using emulsion polymerisation at 30°C with a peroxide initiator, giving a polymer with *c.* 70% 1,4-butadiene units; it contains phenyl-β-naphthylamine as stabiliser. Used as a latex (total solids at least 18%, pH 8–9, surface tension 35–40 dyn/cm) for dipping tyre cords made of viscose or synthetic fibres; the —COOH groups in the polymer give enhanced adhesion to the tyre rubber.

---

*For details of early history *see* PERKIN, W. H., Jr., *J. Soc. Chem. Ind.,* **31**, 616 (1912).

33.82 BUTADIENE RUBBER LATICES

Preparation (*a*) By polymerising butadiene in aq. emulsion, by the same general procedure as for styrene/butadiene copolymers (*see* Section 34). For example, a latex for making foam articles is obtained by using potassium oleate as dispersing agent, and effecting 'cold' polymerisation with a hydroperoxide initiator, followed by heat concentration to give a latex with a total solids content of 58–60 %, rubber particle size 0·2 μm, pH 10·5, viscosity 900 cP, and surface tension 50 dyn/cm. (*b*) By emulsifying a solution of butadiene rubber in a hydrocarbon solvent with a aq. soap solution (or a soap-forming acid may be added to the solution, followed by aq. alkali), removing the solvent by 'stripping', and concentrating the resulting dispersion; a typical latex has *cis*-1,4 content of polymer 90 %; gel content nil; total solids content 63 %; pH 10·6; and surface tension 31 dyn/cm.

Uses Foam products (mattresses, pillows, upholstery, etc.) for which its good colour and absence of odour (compared with styrene/butadiene latices) are advantageous; water paints; textile coatings (mixed with resins to increase their flexibility).

33.83 THERMOPLASTIC POLYBUTADIENES (Trans-4)

Polymers consisting mainly of *trans*-1,4 units (II, *see* §33.1) are thermoplastic and resemble gutta percha and balata. Typical products have *trans*-1,4 content 88–93 %, gel content nil, sp. gr. 0·93–0·97, Mooney viscosity (100°C) 26–130; they are soluble in carbon tetrachloride and benzene, swollen by toluene, attacked by conc. sulphuric and nitric acids but relatively unaffected by hydrochloric acid; unaffected by *n*-heptane, methyl alcohol, acetone, acetic acid and ammonium hydroxide. Glass temperature −83°C, first order transition temperature (change of crystalline modification) 70–75°C, m.p. 110–150°C. At room temperature these materials are crystalline and relatively hard (87–97 IRHD) but they soften on heating (to 20 IRHD or less at 120°C) and can be moulded at or above *c*. 80°C. They can be vulcanised with sulphur; the vulcanisates are more or less crystalline and retain deformation at room temperature but recover on heating; those reinforced with carbon black have excellent abrasion resistance. **Uses** Shoe soles (solid and cellular), floor tiles, gaskets, sponge, golf-ball covers, battery boxes.

33.9 FURTHER LITERATURE

BREUERS, W. and LUTTROPP, H., *Buna: Herstellung, Prüfung, Eigenschafften*, Berlin (1954)

GÉNIN, G. and MORISSON, R. (Ed), *Encyclopédie Technologique de l'Industrie du Caoutchouc*, Vol. I, Titre II, Paris (1958)

STYRENE/BUTADIENE RUBBERS

With notes on latex, carboxylated rubber, styrene/butadiene/vinyl(or methylvinyl)-pyridine rubber, and α-methylstyrene/butadiene rubber

SYNONYMS AND TRADE NAMES

SBR, GR-S. Ameripol, ASRC, Austrapol, Baytown, Buna OP, Buna S, Buna SS, Buna Hüls, Carbomix, Cariflex S, Copo, Duradene, Europrene, Flosbrene (liquid rubber), FR-S, Gentro, Intol, JSR, Kryflex, Krylene, Krymix, Krynol, Naugapol, Nipol, Petroflex, Philprene, Plioflex, Polysar S, S (plus a number*), Shell S, SKS, Solprene, Stereon, Synaprene, Synpol, Ugitex S.

GENERAL CHARACTERISTICS

Raw styrene/butadiene rubbers are straw-coloured, translucent, somewhat elastic substances resembling natural rubber softened by milling, but less tough and strong. They have little tack (self-adhesion) and little tendency to cold flow, are compatible with natural and most synthetic rubbers, and are available as aq. dispersion or latex (*see* §34.81).

Unvulcanised compounded mixes are deficient in tack, and show more elastic recovery (e.g. after calendering or extrusion) than natural rubber mixes.

Vulcanisates generally resemble those of natural rubber in mechanical properties, except that they have: (i) better resistance to crack formation by repeated flexing, and to abrasion especially under hot dry conditions; (ii) very low tensile strength in unfilled mixes, though with carbon black reinforcement it approaches that of natural rubber; (iii) lower resilience and inferior tear- and temperature-resistance; (iv) greater hysteresis and heat-build-up.

They also resist oxidative ageing better than similar natural rubber vulcanisates, but in resistance to swelling and corrosive liquids they broadly resemble the latter.

Styrene/butadiene rubbers can be vulcanised to the ebonite stage (*see* Section 31).

34.1 STRUCTURE

Simplest Fundamental Unit

$-CH_2-CH-$ (benzene ring)	$-CH_2 \diagdown \diagup CH_2-$ $CH{=}CH$	$-CH_2 \diagdown$ $CH{=}CH$ $\diagdown CH_2-$	$-CH_2-CH-$ $\vert$ $CH{=}CH_2$
I	II	III	IV
styrene, C_8H_8; mol. wt. $=104$	*cis*-1,4	*trans*-1,4	1,2 or vinyl
	butadiene, C_4H_6; mol. wt. $= 54$		

*Some of the trade names are used also for other synthetic rubbers; styrene/butadiene rubbers are then distinguished by adding a 4-digit number from the IISRP descriptive code, *see* §34.21.

The molar proportion of I can range from 1·8–40%, but is normally 13·5–15% (23·5–25 weight %). Copolymers with *c*. 26–40 mol % are self-reinforced stiff rubbers. Those with *c*. 55–82 mol % are 'high-styrene' resins.

Proportions of II, III and IV depend on the method and temperature of polymerisation (*see* §34.21):

	II	III	IV	
'Hot' (emulsion) polymerisation	18·3	65·3	16·3	Percentage of
'Cold' (emulsion) polymerisation	12	72	16	the butadiene
Solution polymerisation	40	54	6	units

Units I to IV are randomly arranged in emulsion copolymers, but in solution polymers may occur in uniform blocks of varying length. Such block copolymers of ABA type (A = polystyrene, B = polybutadiene) have reversible thermoplastic-elastomeric properties, being resilient and rubber-like (without vulcanisation) at room temperatures, whilst at higher temperatures they process as thermoplastics.

Molecular weight 'Hot' (*see* §34.21) rubbers, 150000–400000 (viscosity-average) or 30000–100000 (number-average); 'cold' rubbers, 280000 (viscosity-average), or 500000 (weight-average) or 110000–260000 (number-average). 'Hot' rubbers, with lower molecular weight and more low-molecular fractions than 'cold' rubbers, have more chain-branching and cross-linking. Solution polymers have a narrower molecular weight range than emulsion polymers.

X-ray data Owing to the random sequence of units I–IV, styrene/butadiene rubbers containing more than 10 mol % of styrene do not crystallise even when stretched or cooled; i.e. x-ray diffraction gives an amorphous halo. Rubbers containing *c*. 5 mol % styrene tend to crystallise and show preferred orientation when stretched.

34.2 CHEMISTRY

34.21 PREPARATION

1,3-Butadiene (*see* §33.21) and styrene (*see* §4.21) are copolymerised using a catalyst, usually in emulsion, but sometimes in solution, as described below.

Emulsion polymerisation*

The liquid monomers are emulsified in water containing a soap and/or other emulsifying agent, and are polymerised either (i) at *c.* 50°C using potassium persulphate or an organic peroxide as initiator, to give '*hot*' rubber, or (ii) at *c.* 5°C using a redox initiating system, e.g. ferrous sulphate plus a peroxy compound (*p*-menthane hydroperoxide) with or without sodium formaldehyde sulphoxylate, to give '*cold*' rubber. Normally the reaction mixture contains also a modifier or regulator (to control molecular weight), buffer salts (e.g. phosphates), and sequestering agents (to control the ferrous ion concentration when a redox system is used). Polymerisation is terminated at *c.* 70% conversion by adding a 'short-stopper' (e.g. quinol, or sodium dimethyldithiocarbamate plus an alkylene polyamine). Unreacted monomers are removed by 'flashing' followed by steam heating at reduced pressure. After addition of an antioxidant ('stabiliser'), the polymer suspension is coagulated (sometimes with prior creaming by addition of brine) and the crumb-like coagulum is separated and dried.

Solution polymerisation

The monomers, diluted with a hydrocarbon solvent, are copolymerised by a stereospecific catalyst, e.g. a lithium alkyl, generally as in the solution polymerisation of butadiene (*see* §33.21). Solution polymers contain less non-polymer constituents than emulsion polymers.

The following variables occur in emulsion polymers. (i) Monomer ratio (1·8 to 40 mol % styrene). (ii) Polymerisation temperature (normally *c.* 50° or *c.* 5°C). (iii) Emulsifying agent, normally a fatty acid soap and/or a rosin acid soap. (iv) Presence or absence of a 'modifier' (e.g. *tertiary* dodecyl mercaptan, di-*iso*propylxanthogen disulphide). (v) Whether antioxidant is 'staining' (e.g. phenyl-β-naphthylamine, dihydroquinolines) or 'non-staining' (e.g. styrenated phenol, a triaryl phosphite, 2,6-di*tert*butyl-4-methylphenol). (vi) Coagulating agent, normally $NaCl + H_2SO_4$ ('salt-acid'), glue $+ H_2SO_4$, $NaCl +$ alum, or alum $+ H_2SO_4$. (vii) High molecular weight 'cold' polymers may be 'extended' with naphthenic, aromatic or high-aromatic petroleum oils. (viii) Mooney viscosity, ranging from 20 to 130 (4-minute reading at 100°C) but mostly 42 to 58. (ix) The polymer (extended or not) may be mixed with carbon black to form a master-batch.

34.22 PROPERTIES

Soluble (raw rubber, milled or of low molecular weight) in most aromatic and chlorinated hydrocarbons, some aliphatic hydro-

The number code of the International Institute of Synthetic Rubber Producers (IISRP) is often used to denote the above variables except (iv)—'modifiers' are presumably always used:
1000 series: 'hot' rubbers.
1500 series: 'cold' rubbers.
1700 series: 'cold' rubbers, oil-extended.
1600 and 1800 series: 'cold' rubbers plus carbon black and oil.
1900 series: miscellaneous master-batches.
   Thus 1502 is a 23·5 wt.% styrene copolymer made at 6°C with mixed soap emulsifier and non-staining antioxidant, using 'salt-acid' coagulant; nominal Mooney viscosity 52.
   In the U.S.S.R., monomer ratio, polymerisation temperature, emulsifier, 'modifier' and extender are denoted by numbers and letters.

carbons, terpenes, tetrahydrofuran; *partially dissolves* in ligroin, methylcyclohexane, tetralin, diethyl ether, higher molecular weight ketones including cyclohexanone, higher molecular weight esters, diethylamine, pyridine. **Plasticised by** physical softeners, namely mineral oils and their derivatives, notably 'process oils', 'aromatic' or 'high-aromatic' fractions and bitumens; refined heavy coal-tar fractions, naphthenic acid, ester plasticisers (phosphates, phthalates); also by peptisers, e.g. β-naphthyl and xylyl mercaptans, pentachloro-thiophenol and its zinc salt, di-(o-benzamidophenyl) disulphide, though these are less effective than in natural rubber. **Swollen** (raw rubber) by mineral oils and linseed oil; (vulcanised rubber) by liquids that dissolve the unvulcanised rubber, the degree of swelling differing from that of natural rubber mainly in being relatively *less* for non-polar and *greater* for polar liquids, especially when the rubber has a high styrene content, e.g. swelling of tyre tread vulcanisates at room temperature (% by volume).

Rubber	Aniline	40/60°C Petroleum ether	Benzene
Natural	10	146	277
Styrene (*c.* 25 wt. %)/butadiene	26	85	265
Styrene (46 wt. %)/butadiene	55	38	367

Relatively unaffected by aq. alkalis and salts (except bisulphites and certain salts of Ag, Cu, Fe, Ni), most dil. mineral acids (e.g. up to 10% HCl, 50% H_2SO_4, 80% H_3PO_4, but not above 8% HNO_3), chlorine water, most organic acids, alcohols (up to octyl), ethylene glycol, glycerol, phenols, low molecular weight esters and ketones. **Decomposed by** nitric acid (above 8%), conc. sulphuric acid, 50% chromic acid, formic acid.

34.23 IDENTIFICATION

N, Cl, P, S nominally absent (some N may be present in an anti-oxidant); presence of S with trace of N indicates the vulcanised material. Not saponifiable. **Combustion** Burns with a smoky flame and distinctive odour, that of styrene being noticeable on extinction (distinction from butyl rubber and polyisobutylene). **Pyrolysis** occurs above *c.* 325°C giving (325–430°C) 11·8% of products volatile at 25°C, including 1·9% butadiene, plus styrene and other saturated and unsaturated hydrocarbons. The usually dark liquid pyrolysate has a characteristic infra-red absorption spectrum, with diagnostic bands at 14·3, 12·9, 6·7, 11·0, 10·1 and 10·4 μm; presence of the first three (aromatic) bands distinguishes from butadiene rubber (*see* BS 4181, Method for identification of rubbers). **Colour tests** *Azo dye reaction for styrene.* About 0·5 g of acetone-extracted rubber is refluxed with 5 ml conc. nitric acid and then treated as described under the colour test for styrene in §4.23. For *colour tests on the pyrolysate see* ASTM D297.

Quantitative estimation of styrene (*in the raw rubber*) (i) From the refractive index of a specimen previously extracted with ethanol-toluene azeotrope and dried:

$$\text{styrene}\ (\%) = 23\cdot50 + 1164\ (n_D^{25} - 1\cdot53456) - 3497\ (n_D^{25} - 1\cdot53456)^2$$

(for details *see* BS 1673, Part 5.1; ASTM D1416).
(ii) From the ultra-violet absorption spectrum of the pyrolysate.

34.3 PHYSICS

Data are for the rubbers with 23–25 wt. % styrene unless otherwise stated.

34.31 GENERAL PROPERTIES

Specific gravity Raw rubber 0·93; increases with styrene content, from 0·91 (8 wt.%) to 0·98–1·01 (50–55%). Vulcanised (unfilled) 0·94–1·00. **Refractive index** Raw $n_D^{25} = 1\cdot5345$; in general, for pure styrene/butadiene copolymer (solvent-extracted rubber): $n_D^{25} = 1\cdot7010 - \sqrt{0\cdot03443 - 0\cdot02861S}$, where S = wt. fraction of styrene. **Water absorption** In emulsion polymers, varies with nature of coagulant and final treatment ('finishing') of the raw rubber, e.g. vulcanisates—composition not stated—after 7 days' immersion at 70°C

Coagulant	Absorption (mg/cm^2)
NaCl + H$_2$SO$_4$	4–5 ('best') to c, 20
Alum	3–4
Glue + acid; special finishing	1–2

Other results—coagulant not stated—after immersion at 25°C (mg/cm^2): unfilled vulcanisates, 1·5–3·5 (7 days); cable mix (60 parts whiting, 50 parts bitumen per 100 rubber), 1–2 (7 days), 2–4 (31 days). Solution polymers have lower absorption than emulsion polymers, e.g. 1·3 and 2·0 mg/cm^2 respectively after 7 days at 70°C. **Permeability and Diffusivity** Unfilled vulcanisate at 25°C

	H_2	N_2	O_2	CO_2	CH_4	He
Permeability $(10^{-10}$ cm$^2)/$(s cmHg)	41	6·3	17	124	21	23
Diffusivity $(10^{-6}$ cm$^2)/$s	9	1·0	1·4	1·05	—	—

Permeability to air of vulcanisates with 50–55 parts HAF black (per 100 rubber) at 30°C, 2·1–2·6 $(10^{-10}$ cm$^2)/$(s cmHg). Permeability to water vapour at 20–30°C, 1·9 $(10^{-10}$ g$)/$(cm s cmHg).

34.32 THERMAL PROPERTIES

Specific heat Raw rubber 0·46, 0·45, 0·435 (for 9, 23–25, and 43 wt. % styrene); unfilled vulcanisate, 0·44. **Conductivity** Unfilled vulcanisate 4·6–5·9 $(10^{-4}$ cal$)/$(cm s °C). **Coefficient of linear expansion** Raw rubber $(10^{-4}/°$C) 0·9 and 2·5 (below and above glass temperature respectively; 10 wt. % styrene); 0·8 and 2·3 (do., 30–50% styrene). Other data: 2·2–2·3 (above glass temp.; c. independent of styrene content up to 56%).

Physical effects of temperature

Raw rubber When progressively cooled, hardens and eventually becomes brittle. Glass temperature (°C), 'hot' rubber, -46; 'cold' rubber, -44; in general, 'hot' $(-85+1·35\ S)/(1-0·005\ S)$; 'cold' $(-78+1·28\ S)/(1-0·005\ S)$ where $S =$ percentage styrene by weight. Unlike natural, synthetic isoprene, butadiene and chloroprene rubbers, it does not crystallise on prolonged cooling unless the styrene content is very low (*see* §34.1). On heating, the normal butadiene/styrene rubber softens and eventually pyrolyses above 325°C.

Vulcanised rubber Glass temperature -52°C (unfilled mix); brittleness temperature -50°C (cable insulation mix); Gehman 'freeze point' -51°C (mix with 42 parts EPC carbon black per 100 rubber; similar natural rubber vulcanisate -56°C). On cooling, resilience decreases more than with natural rubber, and to a greater extent the higher the styrene content, e.g. temperature of minimum rebound resilience (vulcanisates with 50 parts carbon black), natural rubber, -34°C; butadiene/styrene rubbers with 25, 31, 42 and 46 wt. % styrene, $-24°$, $-9°$, $+4°$, $+18$°C respectively. At elevated temperatures tensile strength, breaking elongation and resistance to tearing and cut growth decrease markedly, and more so than with natural rubber (*see* §34.34).

34.33 ELECTRICAL PROPERTIES

Data for rubbers containing 23–25 wt. % styrene are given in Table 34.T1.

Table 34.T1. ELECTRICAL PROPERTIES OF RUBBERS CONTAINING 23–25 WT. % STYRENE

	Raw rubber			Unfilled vulcanisates			Filled vulcanisates‡	
	−20°C	20–25°C	60°C	−20°C	20–25°C	60°C	20–25°C	60°C
Volume resistivity (ohm cm)	—	10^{14-15}*	10^{12-13}	—	10^{13-15}†	10^{12}	10^{14-15}§	10^{13}
Dielectric strength (kV/mm)	—	24–36	—	—	—	—	15–25	—
Dielectric constant‖								
50 Hz	2·6	2·6	2·65	2·6	2·9	2·95	—	—
800–1000 Hz	2·55	2·6	2·65	2·5	2·75¶	2·9	3·1–3·5¶	—
10^6 Hz	—	2·35	2·45	—	2·7**	2·6**	2·4–3·8**	2·4–3·8**
Power factor††								
50 Hz	0·004–0·009	0·001–0·004	0·003–0·008	0·007–0·11	0·004–0·022	0·005–0·085	0·015‡‡	0·024
800–1000 Hz	0·030	0·006 (c. 0·002*)	0·004	0·06–0·09	0·045 / 0·002¶	0·012–0·017	c. 0·006¶	—
10^6 Hz**	—	0·005	0·002	—	0·010–0·016	0·008	0·014–0·046	0·012–0·044

Notes

*Rubbers conditioned at 40% r.h.
†Rubbers conditioned at 40% r.h.; 1·9 mm sheets after 7 days in water at 25°C, 10^{14}; special electrical rubbers up to c. 10^{17}
‡Electrical insulation rubbers; data at 50 Hz, 800–1000 Hz and 10^6 Hz are for different rubbers
§Rubbers conditioned at 40% r.h.; after immersion as in (†), results varied greatly (down to 10^{10}) according to mix composition
‖Values from different sources vary by ±0·1
¶Rubbers conditioned at 40% r.h.; after immersion as (†) dielectric constant c. 0·1–0·2 higher and power factor increased to 0·004–0·009 and 0·006–0·010 for unfilled and filled vulcanisates respectively
**Vulcanisates conditioned at least 2 days at 75% r.h. and 25°C
††Depends greatly on humidity conditioning (usually not specified) before testing.
‡‡At −20°C, 0·007–0·025

For rubbers containing 23–25 wt. % styrene the peak of the temperature–power factor curve is at −12°C at 1 MHz or +26°C at 100 MHz (raw rubber); c. 0°C at 1 MHz (unfilled vulcanisate); −24°C at 1 kHz or −12°C at 12 kHz (cable mix containing whiting and bitumen). With higher styrene content the peak temperatures are higher.

34.34 MECHANICAL PROPERTIES

Properties of vulcanisates depend on the proportion of styrene units (*see* §34.1) in the copolymer; increasing this proportion leads to greater stiffness, especially at low temperatures, and reduced resilience at room temperatures. The following statements apply to the normal rubber with 23–25 wt. % styrene.

Unfilled vulcanisates have poor tensile strength, but vulcanisates reinforced with carbon black, especially those from 'cold' rubber, approach similar natural rubber vulcanisates and have similar or rather lower breaking elongation and stiffness (stress at fixed elongation). Tear strength is below that for natural rubber, especially at elevated temperatures. Set (permanent deformation after extension or compression) is greater, but creep during long periods at elevated

389

temperatures (e.g. 55–60°C) is less. Resilience (rebound) is lower than for natural, synthetic isoprene or high-*cis* butadiene rubbers, and hysteresis (heat build-up) correspondingly greater.

Abrasion resistance, as shown by tyre performance, is at least equal to that of natural rubber, especially under hot dry running conditions; 'cold' rubber has better abrasion resistance than 'hot' rubber (*see* §34.21).

Initial formation of cracks by repeated flexing is slower than for natural rubber, but growth of existing cracks or cuts is more rapid, especially in the 'hot' rubber and at elevated temperatures, e.g. for carbon black (tyre tread) vulcanisates, cut growth rate (arbitrary units) natural rubber, 1·0 (room temperature), 1·7 (93°C); styrene/butadiene rubber, 3–6, 40–60.

Typical values of properties of vulcanisates of styrene/butadiene rubbers, measured at *c.* 20–23°C unless otherwise stated are given in Table 34.T2.

Table 34.T2. MECHANICAL PROPERTIES OF RUBBERS CONTAINING 23–25 WT. % STYRENE

	Unfilled vulcanisates	Vulcanisates with 50 parts HAF carbon black per 100 rubber (*)	(†)	Filled vulcanisates (‡)				
Elastic modulus, Young's								
static (kgf/mm²)	0·10–0·20	0·51–0·72	—	—	—	—	—	—
shear, dynamic, 60 Hz	0·22–0·30	1·9	—	—	—	—	—	—
60 Hz, 100°C	0·10–0·13	0·63	—	—	—	—	—	—
1 kHz	0·28–0·38	2·8	—	—	—	—	—	—
1 kHz, 100°C	0·16–0·19	1·0	—	—	—	—	—	—
Tensile strength (kgf/mm²)	0·14–0·28	1·4–2·65	1·69–2·32	1·22	1·41	1·55	1·76	1·41
93°C	—	0·8–1·05§	—	—	—	—	—	—
Breaking elongation, %	400–600	400–650	300–330	500	450	400	300	200
93°C	—	480–550§	—	400	400			
Stress at 300% elongation (kgf/mm²)	—	1·0–1·55	min. 1·05–1·61 max. 1·55–2·28	—				
Tear strength (kgf/mm)	—	2·5–6·0	—	—	—	—	—	—
Hardness (IRHD)	35	56–65	—	41–50	51–60	61–70	71–80	81–88
Compression set‖ (%)	—	—	—	30	30	25	25	30
Resilience, rebound (%)	65	30–41	—	—	—	—	—	—
−20°C	—	6						
0°C	—	25 ¶	—					
40°C	—	48						
80°C	—	59						
Loss tangent (tan δ), 60 Hz	0·07–0·09	0·2						
60 Hz, 100°C	0·05–0·07	0·2	—	—	—	—	—	—
1 kHz	0·10–0·24	0·25						
1 kHz, 100°C	0·05–0·06	0·2						

Notes

*Values from the literature
†Test values (minimum tensile strength and elongation; minimum and maximum stress) specified in BSS 3472 and 3650 covering various types of raw styrene/butadiene rubbers; lowest tensile and stress values are for oil-extended rubbers (*see* §34.21)
‡From BSS 3515 and 3629 for vulcanised styrene/butadiene rubbers of various hardness ranges; minimum for tensile strength and elongation, maximum for compression set; lower values in first two columns are for extruded rubbers
§Same rubbers at 20°C gave tensile strength 2·0–2·6 kgf/mm² and elongation 600–650%
‖Compressed 25% for 24 h at 70°C; recovery 10 min at room temperature; set expressed as percentage of compression
¶Filler 50 parts EPC carbon black (per 100 rubber); resilience at 20°C = 41%

34.4 FABRICATION

Styrene/butadiene rubbers are compounded and processed like natural rubber, except for the following differences. (i) Little or no plasticisation before mixing is needed, though it can be effected by mechanical working with or without peptisers (*see* §34.22); however, styrene/butadiene (especially the 'cold') rubbers break down less readily than natural rubber. (ii) Unvulcanised mixes are deficient in tack (self-adhesion) unless they contain tack-producing additives (pine tar, coumarone/indene resins) or a substantial proportion of natural rubber. (iii) In calendering and extrusion there is more elastic recovery, but extruded sections have less tendency to collapse or deform. (iv) Less sulphur (*c.* 1·5–2·0 parts per 100 rubber) is needed for vulcanisation, but the accelerator must be increased or more powerful types used, e.g. a thiazole plus a guanidine 'booster', or a sulphenamide. (v) Reinforcing fillers are necessary to obtain good mechanical properties (*see* §34.34); carbon blacks are best, other types (e.g. silica) being less effective than in natural rubber. (vi) There is less tendency to reversion (softening and loss of strength) when vulcanisation is prolonged beyond the optimum.

Styrene/butadiene rubbers can be bonded to steel and other metals.

34.5 SERVICEABILITY

Styrene/butadiene rubber vulcanisates resemble those of natural rubber in being susceptible to atmospheric oxidation, with consequent deterioration of mechanical properties, and to cracking when exposed stretched to air containing ozone; also in being considerably swollen and weakened by (in particular) hydrocarbon and halogeno-hydrocarbon liquids and attacked by oxidising substances. However, atmospheric oxidation is slower and causes stiffening (through cross-linking) rather than softening (chain-scission) as in natural rubber. Resistance to oxygen and ozone can be improved by antioxidants and antiozonants, and in practice deterioration by ageing is less than with natural rubber. Resistance to short-term rise or fall of temperature is not so good as for natural rubber, but the absence of crystallisation eliminates the tendency to stiffen during long periods at sub-zero temperatures. Abrasion resistance is good, but heat generation under dynamic stressing, as in tyres, is greater than in natural rubber.

34.6 UTILISATION

Styrene/butadiene rubbers are general-purpose materials intended to replace natural rubber, and hence are used for substantially the same purposes as the latter, some exceptions being (i) where heat generation must be minimised, e.g. carcases of heavy-duty (bus, lorry, military) tyres; (ii) where a very soft (unfilled) vulcanisate with good strength and elastic properties is needed, e.g. thread. On the other hand, the superior abrasion resistance of styrene/butadiene rubbers, especially under hot dry conditions, and better age-resistance have caused them to replace natural rubber almost completely in tyre treads, being used either alone or blended with synthetic isoprene rubber or butadiene rubber (*see* Sections 32 and 33).

34.7 HISTORY

See also §33.7 (butadiene rubber) and §4.7 (polystyrene).

Emulsion polymerisation of olefins, discovered by F. Hofmann and co-workers (1912), was developed from *c.* 1927 by W. Bock, E. Konrad and E. Tschunkur, leading to a quicker polymerisation based on the use of synthetic emulsifiers and oxygen-yielding (e.g. peroxy) compounds as catalysts. In 1929–30 Bock and Tschunkur, using this method, found that copolymerising butadiene with styrene gave a better rubber than that from butadiene alone, and by 1934 a pilot plant was making the copolymer (Buna S), which was produced on a large scale in Germany during World War II. Further major technical advances (Germany, *c.* 1941–3) were the introduction of redox catalyst systems, making possible 'cold' polymerisation and oil-extended copolymers, and of 'modifiers' to regulate molecular weight (*see* §34.21).

In America the German process and product were being studied in 1937–8, and reached pilot plant scale production by 1939. From 1940 production was organised by the U.S. Government Rubber Reserve Co. and carried out by the major U.S. rubber companies. Production reached a peak of 720000 tons/year in 1945. After a post-war drop, it has risen steadily since *c.* 1950. In Russia styrene/butadiene rubber (SKS) was introduced in 1949, the oil-extended rubber in 1955, and 'cold, modified' rubber in 1959. Styrene/butadiene rubbers are now made also in several countries in Europe, N and S. America, Africa, Australia and Japan.

34.8 ADDITIONAL NOTES

34.81 STYRENE/BUTADIENE RUBBER LATEX

(Buna-Latex, Bunatex, Darex, Hycar 2559 and 2569, Intex, Sto-Chem, Litex, Naugatex, Pliolite, Polysar 722, Revinex 430, FR-S 200, Synthomer, Tylac, Copo.)

Preparation

The emulsion copolymerisation of butadiene and styrene (*see* §34.21) yields an aq. dispersion (latex) with total solids content *c*. 25%. Higher solids contents, as required for some uses (e.g. foam rubber, at least 60%), are best obtained by concentrating the latex, usually by evaporation under reduced pressure; to avoid unduly increasing the viscosity (which, for foam production, should not exceed 10 P) the initially very small (0·04–0·08 μm diam.) rubber particles are agglomerated, before concentration, by freezing, passage through a homogeniser, addition of electrolyte or hydrophilic colloid, or partial destruction of the soap used as emulsifier.

Commercial latices have the following characteristics:

	'Hot'*	'Cold'*
Total solids (%)	27–59	21–70
Styrene (wt. % in polymer)	23·5–48 (usually 46)	14–44 (usually 22–25)
Mooney viscosity of polymer (100°C, 4 min reading)	30–140	48–150
Average particle size (μm)†	0·06–0·22	0·06–0·30
pH	9·0–11·0	9·5–11·0
Viscosity (P)	—	5–14 (64–70% total solids)
Surface tension (dyn/cm)	—	30–40 (64–70% total solids)

*See §34.21
†Minimum 0·16 μm for foam production

The rubber particles are normally negatively charged, but 'cationic' latex, with positive charge, is also made. Other variants use, as soaps and electrolytes, only salts of volatile bases (e.g. ammonia) that evaporate on drying the latex and vulcanising, thus reducing water absorption. Styrene/butadiene latices are commonly designated by the number code of the International Institute of Synthetic Rubber Producers (IISRP) to denote type of emulsifier, total solids, and styrene content and Mooney viscosity of polymer: 2000 series ('hot' latices), 2100 series ('cold' latices).

Compounding and vulcanisation are basically as for natural rubber latex, using *c*. 2 parts sulphur (per 100 rubber) and *c*. 3 parts zinc oxide, but the amount of accelerator should be increased. Water-soluble accelerators are preferable, e.g. dimethylammonium salt of 2-mercaptobenzthiazole.

Properties of vulcanisates Unfilled vulcanised films are inferior in strength to those from natural rubber latex, e.g.

	'Hot'	'Cold'	Natural
Tensile strength (kgf/mm^2)	0·2*	1·1*	3·7
Breaking elongation (%)	400	700	850

*Varies with type of soap stabiliser, ranging up to 1·3 ('hot') or 2·7 ('cold')

Uses Foam products (giving less shrinkage during manufacture, lower load-bearing capacity at equal density, but better retention of this capacity during fatigue, than natural rubber latex); sealing compositions, tyre cord dipping compositions; carpet, rug and upholstery backings and underlays; adhesives for carpet pile; non-woven and 'combined' (footwear) fabrics; binder for jute and sisal insoles; coating and impregnation of paper; leather finishes; asbestos brake linings. For foam products the latex may be blended with those of natural, synthetic isoprene or butadiene rubber (*see* §§29.81, 32.81, 33.82). It is not suitable for making dipped articles (owing to poor wet gel strength) or thread.

34.82 CARBOXYLATED STYRENE/BUTADIENE RUBBER

(Ciago 2570 X1, Dylex K-55, Nitrex 2617, Sto Chem 6205.)

This is made by including a small proportion (e.g. 1·25–4%) of an unsaturated acid, usually acrylic or methacrylic, in the butadiene/styrene (90/10 to 50/50) mixture, which is emulsion-polymerised generally as for styrene/butadiene rubber (§34.21); e.g. the Russian SKS-30-3 uses 70:30:3 butadiene/styrene/methacrylic acid. It is used mainly as latex: total solids, 18–55%; sp. gr., 1·00; pH 3–9 (according to use); rubber particle size 0·1–0·3 μm; viscosity (cP), 2·5–4 (20–25% total solids), 300 (55%).

The latices are used for dipping tyre cord; backing carpets; making dipped gloves, coated and impregnated paper and textiles; as adhesives, e.g. for paper/aluminium foil; leather finishes; binders for tufted carpets, fibres, scrap leather and scrap foam rubber. The —COOH groups of the unsaturated acid enhance the strength and durability of the rubber/textile bond and permit cross-linking (vulcanisation) by water-soluble substances (e.g. polyamines, hexa-methylenediamine carbamate, sodium aluminate) thus avoiding the need to make aq. dispersions of sulphur, zinc oxide and accelerators; other vulcanising agents are zinc oxide, and epoxy and urea-formalde-hyde resins.

34.83 STYRENE/BUTADIENE/VINYLPYRIDINE (OR METHYLVINYLPYRIDINE) TERPOLYMER RUBBERS

(Bunatex VP, Goodyear VP-100, Pyratex.)

These are made by emulsion polymerisation of butadiene, styrene and a small proportion of vinylpyridine (§4.84) or 2-methyl-5-vinylpyridine, e.g. 3–10% of the latter replacing styrene; thus the Russian SKS-25-MVP-5 uses 70/25/5 butadiene/styrene/methylvinyl-pyridine. The terpolymers are used mainly as latices; those containing vinylpyridine have total solids, 40–42%; sp. gr., 0·98 (solids 0·95–0·97); pH 11·0; rubber particle size, 0·08 μm; viscosity, 20–27 cP; surface tension, 48 dyn/cm.

The latices are used (alone or mixed with styrene/butadiene latex) for dipping tyre cord, giving a stronger and more fatigue-resistant rubber/textile bond than the latter latex alone, thus especially useful for nylon and polyester cords; also for other uses, e.g. belts, requiring good rubber/textile adhesion.

34.84 α-METHYLSTYRENE/BUTADIENE RUBBER

The Russian SKMS rubbers are copolymers of α-methylstyrene (§4.83) and 1,3-butadiene (§33.21) in various proportions, and are made by emulsion polymerisation at 48–50°C or 4–8°C (cf. 'hot' and 'cold' styrene/butadiene rubbers; §34.21). Thus, SKMS-30ARKM-15 is made at 4–8°C with initial monomer ratio 32/68, using a mixture of disproportionated rosin and fatty acid soaps as emulsifier, and a 'modifier' to regulate molecular weight; it contain 14–17% high-aromatic extender oil and is stabilised with 1·3–2·0% phenyl-β-naphthylamine. Its properties, as given in State Standard (GOST) 11138–65, are generally similar to those of the 30/70 styrene/butadiene rubber made in the same way (SKS-30ARKM-15); its rather poorer cold-resistance can be improved by reducing the methylstyrene content to 10% (increasing it to 50% gives a tough rubber used for shoe soles and ebonite, *see* Section 31).

Methylstyrene/butadiene (30/70) rubber is available as a latex, used for dipping tyre cord.

34.9 FURTHER LITERATURE

BLACKLEY, D. C., *High Polymer Latices*. London and New York (1966)
BREUERS, W. and LUTTROPP, H., *Buna: Herstellung, Prüfung, Eigenschaften*, Berlin (1954)
WHITBY, G. S. (Ed), *Synthetic rubber*, New York and London (1954)
STERN, H. J., *Rubber, natural and synthetic*, 2nd edn., London and New York (1967)

Specifications
BS 1673: Part 5 (analysis of raw styrene/butadiene rubbers (SBR))
BS 3472 and *3650* (specifications for raw SBR, *1500* series and *1700* series respectively)
BS 3515 (vulcanised SBR)
BS 3629 (vulcanised extruded SBR)
ASTM D1416 (analysis of SBR)
ASTM D1417 (testing SBR latices)

NITRILE RUBBERS*

With notes on nitrile rubber latex

SYNONYMS AND TRADE NAMES

NBR, GR-A. Ameripol D, Breon, Buna N, NL, NN, NW, NNL and NWL, Butacril, Butakon A, AC and XA, Butaprene, Chemigum N, Elaprim, Europrene N, FR-N, Hycar, Krynac, Nipol N, Paracril, Perbunan, Perbunan N, Perbunan W, Perbunan Extra, Polysar N, SKN, Tylac, Ugitex N.

GENERAL CHARACTERISTICS

Raw nitrile rubbers are almost colourless to dark brown, tough, somewhat elastic materials, less plastic than natural rubber, and usually having an odour of acrylonitrile, though odourless grades are available. They are compatible with butadiene, styrene/butadiene, chloroprene, acrylate and polysulphide rubbers and with polyvinyl chloride, but not readily with natural or synthetic isoprene rubber, and are available as aq. dispersion or latex (*see* §35.81), as crumb (for making solutions) and as powder (for mixing with phenolic resins, e.g. in adhesives).

Unvulcanised compounded mixes are deficient in tack, and show considerable elastic recovery ('nerve') unless the rubber has first been well broken down.

Vulcanisates generally resemble those of natural rubber in mechanical properties, except that: (i) in unfilled mixes tensile and tear strengths are low, though with carbon black reinforcement tensile strength approaches that of natural rubber; (ii) resilience at room temperature is much lower, and hysteresis (heat build-up) higher; (iii) resistance to heat, light and oxidative ageing is better.

Nitrile rubber vulcanisates differ from those of natural, synthetic isoprene, butadiene, styrene/butadiene and butyl rubbers primarily in being more resistant to the swelling and weakening action of hydrocarbon (especially aliphatic) liquids, though their resistance to polar liquids is generally less; they show greater stiffening at sub-zero temperatures. These differences are greater the higher the acrylonitrile content; hence high-acrylonitrile rubbers are used when resistance to fuels, oils, etc., is all-important, but low-acrylonitrile rubbers when cold-resistance also is important.

Nitrile rubbers can be vulcanised to ebonite (*see* Section 31).

35.1 STRUCTURE

Simplest Fundamental Unit

I	II	III	IV
acrylonitrile, C_3H_3N; mol. wt = 53	cis-1.4	trans-1,4	1,2 or vinyl

butadiene, C_4H_6; mol. wt = 54

*'Nitrile rubber' was adopted by the American Chemical Society (*c.* 1938) as a general term for all rubber-like copolymers of a diene with an unsaturated nitrile; in practice these are generally butadiene and acrylonitrile respectively.

Nitrile rubbers are classified, according to content of I, in various grades from 'low-acrylonitrile' (*c.* 20%*) to 'high' (*c.* 40%) or 'very high' (*c.* 45%), but authorities differ as to the exact percentages and the number of other grades. Copolymers having 50–60% of I have been described as 'leathery plastics'. The butadiene units are largely in the *trans*-1,4 form (III).

Molecular weight Fractionation gives a range of *c.* 20000–10^6. **Degree of polymerisation** *c.* 400–20000. High molecular weight rubbers show some chain-branching and cross-linking.

X-ray data Owing to the random arrangement of units I–IV, nitrile rubbers are amorphous and do not crystallise even on stretching or cooling, and hence give a diffuse x-ray diffraction halo.

35.2 CHEMISTRY

35.21 PREPARATION†

Acrylonitrile (*see* §10.21), freed from inhibitor by passing through silica gel, and freshly distilled 1,3-butadiene (§33.21) are copolymerised in emulsion by methods analogous to those for styrene/butadiene rubber (§34.21), e.g. using as initiator a persulphate ('hot' process) or a cumene hydroperoxide/dextrose/sequestering agent redox system ('cold' process), plus an emulsifier (synthetic detergent or a soap) and a 'modifier' (a mercaptan) to control molecular weight; antioxidant is added before coagulating the resulting polymer suspension.

35.22 PROPERTIES

Note. The solvent action of liquids is less the higher the molecular weight of the rubber, and (with hydrocarbon liquids) the higher the acrylonitrile content.

Soluble or partially soluble (raw rubber) in aromatic hydrocarbons, tetrahydrofuran, tetralin, aliphatic and aromatic nitro-compounds, acetic acid esters (methyl to butyl), ketones (up to methyl *iso*butyl), aldehydes, most chlorinated hydrocarbons (especially aromatic),

---

*Wt.% and mol % of acrylonitrile are practically the same.
†In addition to monomer ratio (§35.1) the following variables exist among nitrile rubbers: (i) Polymerisation temperature, normally *c.* 50°C ('hot') or *c.* 5°C ('cold'). (ii) Mooney viscosity, ranging from 20 to 115 (4 min reading at 100°C); low values, giving easier processing, are obtained by using more 'modifier'. (iii) Type of antioxidant, whether 'staining' or 'non-staining'. (iv) A small proportion of a third monomer may be included to achieve special properties (*see also* §35.82).

aromatic amines, pyridine, benzyl alcohol, aromatic ethers. **Plasticised by** petroleum and coal tar derivatives, dibenzyl ether, organic esters (phosphates, phthalates, sebacates), liquid acrylonitrile/butadiene copolymers, octadecenenitrile; also by some peptisers (e.g. alkyl esters of an N,N-dialkylcarbamic or dialkyldithiocarbamic acid), but these are less effective than in natural rubber. **Swollen** (raw rubber) by ethyl ether, perchloroethylene, cyclohexanone, phenols, butyl alcohol, carbon disulphide, turpentine; (vulcanised) by liquids that dissolve or swell the raw polymer, and some others (*see* Table 35.T1). **Relatively unaffected** (raw and vulcanised) by aliphatic hydrocarbons including mineral oils, water and most salt solutions, ethyl alcohol, glycol, glycerol, aliphatic amines, organic acids (except acetic), vegetable oils; (vulcanised) also by caustic alkali solutions, ammonium hydroxide, hypochlorite solutions, moderately dil. sulphuric acid (cold), 30% hydrochloric acid (below 65°C), phosphoric acid (except hot conc.), hydrofluoric acid (below 65%). **Decomposed by** chlorine, bromine, nitric acid, chromic acid, chlorosulphonic acid, fuming sulphuric acid, hot 80% (w/w) sulphuric acid.

Table 35.T1. SWELLING OF NITRILE RUBBER VULCANISATES AFTER 48–72 h IMMERSION AT ROOM TEMPERATURE (% BY VOL.)

Acrylonitrile content (%)	18	25–29	35	40	40
Carbon black (parts per 100 rubber)	50	50	50	50	0
Liquid: Hexane	16	13	9	2	5
Petrol	—	21	—	5	9
Cyclohexane	34	26	14	—	—
Benzene	190	180–230	150	125	210
Carbon tetrachloride	—	100	—	37	64
Ethyl alcohol	3	15	14	9	15
Ethyl ether	—	47	—	—	—
Acetone	—	120	—	175	345
Methyl ethyl ketone	160	175	155	—	—
n-Butyl acetate	145	125	105	—	—
Carbon disulphide	—	65	—	—	—
Aniline	—	215–400	—	275	—

Note

BS 2751 specifies that for vulcanised nitrile rubbers (25–33% acrylonitrile) the swelling in 70/30 *iso*-octane/toluene mixture after 24 h at 40°C must not exceed 30% (for rubbers up to 60 IRHD hardness) or 25% (above 60 IRHD).

35.23 IDENTIFICATION

N present (5–13%); Cl, P, S nominally absent (a little Cl may be present from a third monomer; *see* §35.21); the presence of sulphur indicates the vulcanised material. Nitrile, polyurethane and vinylpyridine (or methylvinylpyridine) copolymers are the only important rubbers containing substantial amounts of nitrogen; they are identified by the infra-red absorption spectra of the pyrolysates. Nitrile rubber is hydrolysed by boiling 80% (w/w) sulphuric acid, but not attacked by gentle boiling with the 30% acid. **Combustion** Burns with a smoky flame and distinctive persistent odour. **Pyrolysis** (beginning *c*. 300°C) becomes exothermic at *c*. 380°C, giving products of which 14% is volatile at 25°C, including saturated and unsaturated hydrocarbons, hydrogen cyanide, cyanogen and ammonia. The liquid pyrolysate is acid usually of dark colour, responds to colour tests 1 and 2 in §10.23 and has a characteristic infra-red absorption spectrum with diagnostic bands at 4·5, 10·4, 6·2, 6·3 and 11·0 μm (the first, due to —C≡N, affords distinction from other rubbers). **Other tests** Colour reactions of pyrolysate, *see* ASTM D297.

Quantitative estimation of acrylonitrile—(i) From the nitrogen content; vulcanised samples must first be extracted with isopropanol. Acrylonitrile content = 3·79 × nitrogen content (details as in BS 903, Part B12). (ii) From the infra-red absorption spectrum (raw rubber only).

35.3 PHYSICS

35.31 GENERAL PROPERTIES

Specific gravity Raw rubber, increases, with acrylonitrile content, from *c*. 0·95 (20% by wt.) to 1·02 (45%). **Refractive index** Raw (25–40% acrylonitrile), $n_D^{20-25} = 1\cdot519–1\cdot521$. **Water absorption** (equilibrium) Vulcanisates with 50 parts FT carbon black (per 100 rubber), finely divided and exposed to air at 86% r.h. and 25°C, (%) nitrile rubber (30% acrylonitrile), 0·38–0·42; natural rubber, 0·54–0·62. Immersion of nitrile rubbers in water may result in slight loss in weight by extraction of soluble matter. **Permeability and Diffusivity** Unfilled vulcanisate (27% acrylonitrile except where otherwise shown) at 25°C

	H_2	N_2	O_2	CO_2	CH_4	He
Permeability (10^{-10} cm²)/(s cmHg)	16	1·05–1·2	*c*. 4	30	3·2	12
do., 20% acrylonitrile	—	2·5	8	64	—	—
do., 39% acrylonitrile	—	0·24	0·96	7·5	—	—
Diffusivity (10^{-6} cm²)/s	4·2	0·23	0·36	0·17	—	—

35.32 THERMAL PROPERTIES

Specific heat Raw rubber (40% acrylonitrile) *c*. 0·47; in general, $C_p = 0\cdot27 + (0\cdot00068 \times \text{absolute temperature})$. **Conductivity** Unfilled

vulcanisates (18–35% acrylonitrile) at 60°C, c. 6 $(10^{-4}$ cal$)/($cm s °C$)$.
Coefficient of linear expansion $(10^{-4}/°C)$ Unfilled vulcanisates:

Acrylonitrile content (%)	18	26	40
Below glass temperature	0·80	0·70	0·65–0·73
Above glass temperature	2·4	2·3	2·2

Physical effects of temperature

Raw rubbers—Harden and become brittle when progressively cooled; glass temperature (°C) = approx. $(-85 + 1·40\,A)$ where A = percentage acrylonitrile (between 20 and 40%). Unlike natural, isoprene, butadiene and chloroprene rubbers, they do not crystallise on prolonged cooling. On heating, they soften; pyrolysis begins at c. 300°C.

Vulcanised rubbers—Glass temperature is 6–13°C above that of the corresponding raw polymer; brittleness temperature rises with acrylonitrile content similarly to glass temperature, but is lowered by plasticisers, e.g. by 45°C by 50 parts (per 100 rubber) of dibenzyl sebacate. On cooling, resilience decreases more than with natural rubber, and the more the higher the acrylonitrile content ($= A$), e.g. temperature of minimum rebound resilience (vulcanisates with c. 50 parts carbon black per 100 rubber) for $A = 0$, 22, 26, 36–40 and 51% is -20, -8, 0, $+20$, and $+33$°C respectively (these figures, obtained with a falling ball, are not comparable with those for butadiene rubber, §33.32). At elevated temperatures tensile strength and breaking elongation decrease markedly, usually more so than with natural rubber (*see* §35.34). BS 2751 for vulcanised nitrile rubbers requires that the ratio of elastic modulus at -20°C to that at $+20$°C must not exceed 3 or 4 (depending on hardness of rubber).

35.33 ELECTRICAL PROPERTIES

Volume resistivity 10^9–10^{12} ohm cm; BS 3222 for a nitrile rubber vulcanisate containing carbon black specifies a minimum of 2×10^9 ohm cm; also minimum **surface resistivity** 5×10^9 ohms. **Dielectric strength** Raw rubbers (27 and 35% acrylonitrile) 16 and 21 kV/mm; vulcanised (containing carbon black), BS 3222 specifies a minimum of 2 kV/mm. **Dielectric constant** Passes through a maximum with rise of temperature, e.g. unfilled vulcanisates tested at 1 MHz.

Temperature (°C)	0	20	max. (temp.)	100
27% acrylonitrile	4·0	5·5	11 (50°C)	9
40% acrylonitrile	4·0	4·8	14 (65°C)	12

Other data Raw rubbers at room temperature, frequency 50 and 800 Hz, 11 and 10 (27% acrylonitrile); 17 and 15 (35%). **Power factor** With increasing temperature this passes through a maximum, then a minimum, then rises; maxima and minima occur at higher temperatures the greater the frequency. *See* Table 35.T2.

Table 35.T2. MAXIMUM AND MINIMUM VALUES OF POWER FACTOR

	Acrylonitrile content (%)	Power factor (at °C given in brackets) Max.	Min.
Raw rubbers; 50 Hz	27	0·03 (−20)	0·001 (−5)
	35	0·03 (−10)	0·002 (+5)
Vulcanisates*†; 50 Hz	27	0·02–0·03 (−20)	0·003 (0 to 5)
	35	0·02 (0)	0·006 (20)
Vulcanisates‡§; 10⁶ Hz	27	0·37 (23)	0·02 (*c.* 100)
	40	0·42 (32)	0·03 (*c.* 100)

*Mixes without or with filler (talc and kaolin, 70 parts each per 100 rubber)
†Values at 40°C, 27% acrylonitrile, 0·02–0·04; 35% do., 0·01
‡Unfilled mixes, conditioned at 75% r.h.
§Values at 0°C, 27% acrylonitrile, 0·20; 40% do., 0·15

35.34 MECHANICAL PROPERTIES

The properties of vulcanisates depend on the proportion of acrylonitrile in the rubber, and on the other ingredients of the mix, especially the nature and amount of plasticiser. Increasing acrylonitrile lowers the resilience and increases stiffening and loss of resilience at low temperatures; plasticisers have the opposite effects and also give softer, more extensible vulcanisates.

The tensile strength in unfilled vulcanisates is low, but those containing reinforcing carbon blacks may reach values near those for similar natural and styrene/butadiene rubbers. Resilience (rebound) is low, especially in unplasticised vulcanisates, and hysteresis (heat build-up) correspondingly high. Resistance to abrasion may be better, but to flex-cracking is worse, than for natural rubber, e.g. in laboratory tests on mixes with 45 parts channel black (per 100 rubber):

	Nitrile rubbers	Natural rubber
Abrasion loss, ml/kWh	65–80	270
Flexing life, minutes to failure	55–70	130

Some typical mechanical properties of vulcanisates are given in Table 35.T3; additional data appear below.

Table 35.T3. TYPICAL PROPERTIES OF NITRILE RUBBER VULCANISATES, *c.* 20°C UNLESS OTHERWISE STATED

Acrylonitrile content of initial rubber (%)	*Unfilled vulcanisates*		*Vulcanisates with 50 parts carbon black (per 100 rubber)**	
	26–27	40	26–27	40
Elastic modulus, Young's (kgf/mm²)				
static	—	—	0·7	0·8–0·9‖
dynamic, 60 Hz	—	—	1·5	1·4
Tensile strength (kgf/mm²)	0·4–0·7	0·45–0·9	1·0–3·0	1·1–3·2
100°C	0·15	0·15	—	—
Breaking elongation (%)	350–800	470–650	350–800	310–700
100°C	80	150	—	—
Stress at 300% elongation (kgf/mm²)	0·1–0·3	0·1–0·25	0·3–2·0	0·35–1·5
Tear strength (kgf/mm)	0·7–2·0	1·3–2·4	5·4	5·2
Hardness (IRHD)	40–45	45–53	45–75	45–78
−30°C	—	—	54–95†	—
Compression set (%)	—	22‡	14–33§	22‡
Resilience, rebound (%)	50–55	25–30	30–55	*c.* 20
Internal friction, in compression, 60 Hz (kP)	—	—	60–70	87

*Various types
†Rubbers without (95) and with various plasticisers; hardness at 20°C, 50–73
‡Compressed by 40% for 22 h at 70°C; recovery 30 min at room temperature; set expressed as percentage of compression
§ As (‡) but 30% compression
‖With 35% acrylonitrile

Other data

(1) Vulcanisates with 15 parts zinc oxide and 5 parts reinforcing carbon black (per 100 rubber), tested at 20°C unless otherwise stated:

	Acrylonitrile (%)	
	27	35
Elastic modulus, Young's (kgf/mm²), static	0·40	0·40
dynamic, 17 Hz	3·0	4·0
Damping*, 17 Hz (%)	25	30
at −20/0/40–100°C	50†/30/25	0‡/55†/25
Resilience, rebound (%)	50	25
at −20/0/40/60–100°C	10/20/55/54	14/3/48/55

*Loss tangent (tan δ) = % damping ÷ 50 π (approx.)
†Peak of temperature/damping curve
‡Rigid at this temperature

(2) BS 2751, for vulcanised 25–33% acrylonitrile rubbers containing carbon black, specifies the following values (bracketed compression set figure is from BS 3222 for low compression set vulcanisates):

Hardness (IRHD)	41–50	51–60	61–70	71–80	81–88
Tensile strength, min. (kgf/mm^2)	0·77	0·77	0·84	1·26	1·26
Breaking elongation, min. (%)	600	450	400	250	150
Compression set, max. (%)*	30	30 (17)	30	20	20

*Compressed by 25% for 24 h at 70°C; recovery 10 min at room temperature; set expressed at percentage of compression

35.4 FABRICATION

Nitrile rubbers are compounded and processed like natural rubber, but with the following differences. (i) Except for easy-processing (low Mooney viscosity) grades, preliminary breakdown is even more important; it is best effected on a cold, tight mill, as chemical peptisers are relatively ineffective (*see* §35.22). (ii) Large proportions of plasticisers, especially esters and aromatic petroleum oils, are commonly incorporated to help processing and improve resilience and low-temperature performance; use of an insoluble polymeric plasticiser avoids extraction by contact of the vulcanisate with liquids. (iii) Unvulcanised mixes are deficient in tack (self-adhesion) unless they contain appropriate additives, e.g. pitch, coal tar, wood rosin, coumarone resin, liquid butadiene/acrylonitrile copolymers. (iv) Less sulphur is needed for vulcanisation, namely: 1–2 parts per 100 rubber (the less the higher the acrylonitrile content); good dispersion is essential owing to the low solubility of sulphur in nitrile rubber, hence sulphur coated with, e.g. magnesium carbonate is often used. (v) Sulphurless vulcanisation, e.g. by dicumyl peroxide, tetramethylthiuram disulphide, or tetrachloro-*p*-benzoquinone, is commonly used. (vi) Reinforcing carbon blacks are needed to get the best mechanical strength, but as this may be less important than swelling resistance, 'softer' (less reinforcing) blacks in large proportions are often used as diluents.

Nitrile rubber can be bonded to metals.

35.5 SERVICEABILITY

The vulcanisates resemble those of styrene/butadiene rubbers (Section 34) in respect of atmospheric oxidation and of cracking (when stretched) in air containing ozone, and in the effects of lowered or raised temperature; however, especially with high-acrylonitrile rubbers, the vulcanisates generally show lower resilience and greater stiffening at low temperatures. Nitrile rubbers give vulcanisates with greater resistance to hydrocarbon liquids (oils, petrols) than styrene/

butadiene or natural rubber, an advantage that increases with the acrylonitrile content. They also resist heat ageing and light better than natural or styrene/butadiene rubbers.

35.6 UTILISATION

Nitrile rubbers are used primarily for products used in contact with hydrocarbon liquids or with oils or greases, e.g. petrol and oil hose, gaskets, oil seals, tank linings, oil-well parts, fuel-cell liners, belts handling oily or greasy products. Other applications include (i) adhesives (with or without addition of phenol/formaldehyde resin) for leather, wood, upholstery, metals, porcelain, glass, brake shoes and linings; (ii) blends with polyvinyl chloride (Section 11), where it acts as a vulcanisable plasticiser giving low flammability and good resistance to ozone, heat, moisture, abrasion and aromatic hydrocarbons; used for cable covers (because readily coloured), car sealing strips, and footwear.

35.7 HISTORY

In parallel with their development of styrene/butadiene copolymers (*see* §34.7) W. Bock and E. Tschunkur in 1929 made acrylonitrile/butadiene copolymers, the first 'nitrile' rubbers. Commercial development in Germany (1935 onwards) produced Buna N (Perbunan) and Buna NN (Perbunan Extra). Following German–U.S. negotiations, started in 1932, nitrile rubbers were produced in U.S.A. from 1939 by the major rubber companies and Standard Oil Development Co. 'Cold' polymerisation, using redox catalysts (*see* §34.21), was applied to the production of nitrile rubbers from *c.* 1950. Nitrile rubbers are now made in several countries in Europe and N. America, Japan and Russia.

35.8 ADDITIONAL NOTES

35.81 NITRILE RUBBER LATEX

(Breon, Butakon AL, Chemigum, Ciago AB, Elaprim, Europrene N, FR-N, Hycar, Nilac, Nitrex, Perbunan, Polysar, SKN-10P and 40P, Sto-Chem, Tylac.)

Preparation Emulsion copolymerisation of acrylonitrile and butadiene by either the 'hot' or 'cold' process (*see* §35.21) yields an aq. suspension (latex) with rubber content and particle size too low for most uses;

it is therefore usually subjected to freezing, to increase particle size, and then concentrated (*see* §34.81). Commercial latices have the following properties ('hot' and 'cold' latices differ much less than corresponding styrene/butadiene rubber latices): Total solids (%), 33–64 (mostly 45–55); acrylonitrile content (% in polymer), 20–45; Mooney viscosity of polymer (100°C, 4 min reading), 65–200; average particle size (μm), 0·05–0·18 (0·005–0·01) for impregnating fabrics, etc.); pH, 6·5–11·0 (mostly *c.* 9–11); viscosity (P), 0·12 (40% total solids) to 18 (60%); surface tension (dyn/cm), 40–55. *Compounding and vulcanisation* are basically as for natural rubber latex (§29.81), using sulphur (1·5–2·0 parts per 100 rubber), zinc oxide and an organic accelerator, e.g. a zinc dialkyldithiocarbamate.

Properties of Vulcanisates Most important is resistance to hydrocarbon and some other liquids (cf. §35.22).

Uses Dipped articles (e.g. gloves) and foam products to resist organic solvents; impregnation of fabrics and paper; binders for non-woven fabrics (sometimes blended with polyvinyl chloride latex).

35.82 NITRILE TERPOLYMER RUBBER LATICES

Most important of these are (i) carboxylated nitrile latices (Breon 1571, Chemigum 520, Ciago AB 71, Elaprim, Hycar, Perbunan Latex 4M, SKN-40-0·5, -40-3 and -40 1GP, Tylac) made by including 0·5–10% of methacrylic acid in the acrylonitrile/butadiene monomer mixture, the whole being copolymerised in emulsion, generally as for nitrile rubber latex (§35.81); (ii) latices of acrylonitrile/butadiene/styrene terpolymers containing 10–30% styrene (Nitrex 2612 and 2625; for the solid 'ABS' terpolymers *see* §4.81). Commercial latices have the following properties: total solids 35–50%; rubber particle size, 0·04–0·12 μm; pH, 6–11; viscosity (40–50% solids), 15–150 cP. Carboxylated nitrile latices have greater mechanical stability than nitrile latices. Acrylonitrile/butadiene/styrene latices give unvulcanised films with tensile strength as high as vulcanised films of normal latices (2·0 kgf/mm^2 or above, with 500% breaking elongation).

Vulcanisation (carboxylated nitrile latices). The —COOH groups permit cross-linking (vulcanisation) by water-soluble substances, thus avoiding the need to make aq. dispersions of sulphur, zinc oxide and accelerators, and difficulties due to these 'filtering off' when impregnating textiles, etc.; commonly used are sodium aluminate, ammonium zincate, urea- and melamine-formaldehyde resins, these being added to the latex and the dried film heated; other vulcanising agents are zinc oxide (active at room temperature), epoxy and phenol-formaldehyde resins, polyhydroxy compounds, polyamines, polyimines. The vulcanisates generally resemble those from nitrile latices (§35.81), but give high tensile strengths ($\geqslant 2$ kgf/mm^2) even when unfilled.

Uses (carboxylated nitrile latices) Oil-resistant coatings and dipped articles (gloves), binders for non-woven fabrics, leather finishes, impregnation of paper to improve tear strength; (acrylonitrile/butadiene/styrene latices) leather finishes, impregnation of paper.

35.9 FURTHER LITERATURE

BLACKLEY, D. C., *High Polymer Latices*, London and New York (1966)

BOSTRÖM, S. (Ed), *Kautschuk-Handbuch*, Vols 1 and 2, Stuttgart (1959–60)

BREUERS, W. and LUTTROPP, H., *Buna: Herstellung, Prüfung, Eigenschaften*, Berlin (1954)

GÉNIN, G. and MORISSON, B. (Ed), *Encyclopédie Technologique de l'Industrie du Caoutchouc*, Vol. I, Titre II (Paris 1958)

WHITBY, G. S. (Ed), *Synthetic Rubber*, New York and London (1954)

STERN, H. J., *Rubber, Natural and Synthetic*, 2nd edn, London and New York (1967)

Specifications
BS 1673: Part 7 (chemical analysis of acrylonitrile/butadiene rubber (NBR))
BS 2751: Vulcanised butadiene/acrylonitrile rubber compounds
BS 3222: Low compression set butadiene/acrylonitrile vulcanised rubber compounds

SECTION 36

CHLOROPENE RUBBERS

SYNONYMS AND TRADE NAMES

Poly (2-chloro-1,3-butadiene), polychloroprene PCP, GR-M, CR. Baypren, Butzclor, Denka Chloroprene, Mustone, Nairit, Neoprene, Sovprene, Svitpren.

GENERAL CHARACTERISTICS

Raw polychloroprene rubber varies in colour from off-white to dark brown, depending on type. Polymers having sulphur linkages in the chain ('sulphur-modified' types, e.g. Neoprene G, Butaclor S) are generally amber while mercaptan-modified types (e.g. Neoprene W, Butaclor M) are off-white to grey. The latter have better storage stability. All types exhibit crystallisation to a greater or less degree. This is a gradual process which reveals itself as stiffening or hardening of the polymer, particularly at low temperatures. Latices and special purpose grades with specific crystallisation characteristics are available.

Unvulcanised compounded mixes of sulphur-modified rubbers have less 'nerve' and better flow, and are more tacky than those based on mercaptan-modified types. The latter, however, mix more quickly with less generation of heat, and are more stable dimensionally, e.g. during extrusion.

Vulcanisates resemble those of natural rubber but have superior resistance to heat, oils, and ozone. Sulphur-modified types have better resilience and tear strength than mercaptan-modified types, which, however, are superior in heat resistance and compression set.

36.1 STRUCTURE

Simplest Fundamental Unit

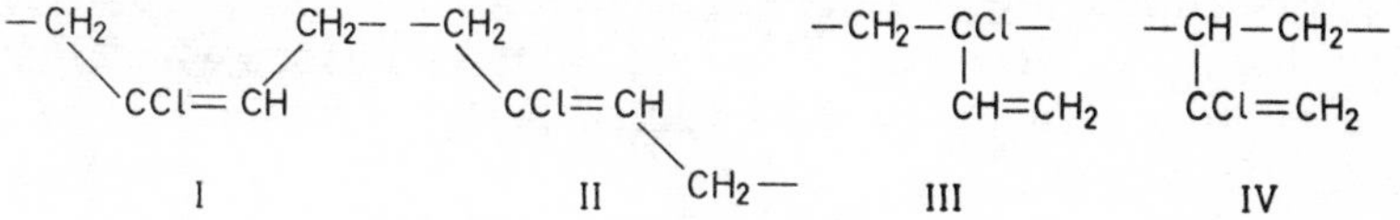

$$C_4H_5Cl, \text{ mol. wt.} = 88.5$$

Polychloroprene contains approximately 85% *trans*-1,4 units (II), 10% *cis*-1,4 units (I), 1·5% 1,2 units (III), and 1% 3,4 units (IV). Sulphur-modified polymers contain about 1 atom of sulphur for every 100 chloroprene units in the polymer chain.

Molecular weight Sulphur-modified types: 20000–950000 (greatest frequency about 100000). Mercaptan-modified types: 180000–200000.

Degree of polymerisation Approximately 1000–3500.

X-ray data Amorphous, but crystallises readily on account of high content of *trans*-1,4 monomer units. Because microcrystalline on stretching, with a diagram similar to β-gutta percha. Unit cell dimensions: $a = 8\cdot8$–9 Å; $b = 8\cdot2$–$10\cdot2$ Å; c (identity period) $= 4\cdot3$–$4\cdot8$ Å.

36.2 CHEMISTRY

36.21 PREPARATION

Normally from acetylene, via monovinylacetylene:

$$CH{=}CH \xrightarrow[\text{(Catalyst Cu}_2\text{Cl}_2 + \text{H}_2\text{O)}]{\text{Dimerisation}} CH{\equiv}C{-}CH{=}CH_2 \xrightarrow{\text{HCl}}$$

$$CH_2{=}\underset{\underset{Cl}{|}}{C}{-}CH{=}CH_2 \xrightarrow[\text{(Catalyst, potassium persulphate)}]{\text{Polymerisation}}$$

$$\left[{-}CH_2{-}\underset{\underset{Cl}{|}}{C}{=}CH{-}CH_2{-} \right]_n$$

Chloroprene is an unstable liquid (b.p. $59\cdot4°$C, density $0\cdot958$) which polymerises spontaneously to give, in the completely polymerised form, μ-polychloroprene which is a resilient, non-dissolving, non-millable material resembling a soft vulcanised rubber; at intermediate stages of polymerisation plastic millable materials similar to un-vulcanised rubber are produced and these are the basis of commercial materials. The polymers are obtained as aq. emulsions which may be freeze coagulated with the aid of acetic acid, formed into rope, and chipped. In sulphur-modified types polymerisation is carried out in the presence of sulphur, and a thiuram disulphide is used as a short-stop, plasticity controller, and stabiliser. Copolymers have been prepared commercially, e.g. Neoprene ILA (copolymer of chloroprene and acrylonitrile) which has exceptional oil resistance. Latices are available having solids contents in the range 35–60%, particle size 110–160 nm, pH usually above 10; a carboxylated type is manufactured in Germany (Baypren latex 4R).

A more recent process starts with chlorination of C_4 petroleum fractions to obtain 3,4-dichlorobut-1-ene (together with 1,4-dichlorobut-2-ene), followed by dehydrochlorination to give the required monomer.

36.22 PROPERTIES

Soluble (milled raw polymer) in benzene, toluene, xylene, carbon tetrachloride, chlorobenzene, *o*-dichlorobenzene, methylene chloride, methyl ethyl ketone, cyclohexanone, butyl acetate, ethyl propionate; blends often used in preparation of adhesives, e.g. acetone/paraffin oil, ethyl acetate/hexane. Polymers with high gel contents (Neoprenes S and WB) and copolymers (Neoprene ILA) have different solubility characteristics. **Plasticised by** petroleum oils (especially naphthenic and aromatic types); esters (e.g. dioctyl sebacate, butyl oleate, trioctyl phosphate); chlorinated diphenyl; unsaturated vegetable oils; resins and polymeric plasticisers. High viscosity polymers may be extended with mineral oils to give cheap compositions. Peptisers (e.g. mercaptans, thiurams, guanidines) are used to assist processing. **Swollen by** aromatic and chlorinated hydrocarbons. Swelling of vulcanised polymer is reduced by incorporation of fillers. Volume swelling of typical composition containing 50 phr* SRF black, after 7 days at 25°C (%): toluene, 185; carbon tetrachloride, 170; ethyl alcohol, 0; methyl ethyl ketone, 85; dibutyl phthalate, 110. One of the most important features of polychloroprene is its low swelling in commercial hydrocarbon oils: for this purpose compositions containing high proportions of carbon black (up to 300 phr*) and aromatic oils are used (the latter are leached out in service and thus contribute to low net swelling). Resistance to oils not as good as in nitrile or polysulphide rubbers. **Relatively unaffected by** (as vulcanised polymer) aliphatic hydrocarbons, alcohols, lubricating oils, commercial fuels and greases, hydraulic fluids (except ester and chlorinated diphenyl types), vegetable and animal oils and fats, inorganic acids, bases, and salts, water (*see* §36.31), ozone. **Severely attacked by** conc. sulphuric acid, chlorosulphonic acid, chromic acid, nitric acid, pickling solutions (nitric acid/hydrofluoric acid), sodium hypochlorite, liquid chlorine, chlorine dioxide, alcoholic alkalis, thionyl chloride, acetyl chloride, titanium tetrachloride.

36.23 IDENTIFICATION

Thermoplastic (unvulcanised state); Cl and S (usually) present; N and P nominally absent.

Combustion Basically non-inflammable but burns with sooty flame if kept hot (e.g. in contact with bunsen flame) leaving swollen black residue; extinguishes on removal from flame; odour of hydrogen chloride. **Pyrolysis** Swells and darkens, yields strongly acid distillate, leaves charred residue; free monomer not formed. **Other tests** Polychloroprene (pure polymer, up to 40% Cl) is distinguished from

*phr = parts (by weight) per 100 parts of raw rubber.

polyvinyl chloride (plasticiser-free, 56–57% Cl) as follows: 1. By decomposition in concentrated nitric acid or in boiling 80% sulphuric acid; PVC is unaffected. 2. By fairly rapid decolorisation of a dil. solution of iodine (e.g. 0·02% in carbon tetrachloride); PVC, subject to type of plasticiser present, remains ineffective. In the infra-red spectrum (vulcanised film) an absorption band centred on 833 cm^{-1} distinguishes natural rubber and polychloroprene from all other rubbers, while natural rubber differs from polychloroprene in exhibiting a band centred on 1380 cm^{-1}. The pyrolysate may give a variable spectrum lacking in characteristic features. However, the following diagnostic absorptions may be useful: 12·2 μm (820 cm^{-1}); 13·4 μm (747 cm^{-1}); 13·0 μm (769 cm^{-1}); 11·3 μm (885 cm^{-1}); 14·3 μm (699 cm^{-1}).

36.3 PHYSICS

36.31 GENERAL PROPERTIES

Specific gravity 1·20–1·25 (raw polymer; for calculation of sp. gr. of vulcanisates *see* Section 58). **Refractive index** Raw polymer, 1·558 (at 25°C). **Water absorption** Compositions formulated for optimum water resistance may be cured with lead oxide but this is suitable only for dark coloured products. Finely divided silica fillers with a small proportion of diethylene glycol can be used with the normal curing systems for both dark and light coloured stocks to give comparable results. A high state of cure is essential. Absorption is not linear with time, e.g. in silica filled compositions swelling is rapid at first, then levels off. Typical values for three different compositions (absorption given as weight increase related to surface area of the sample, after 80 days' immersion at 70°C):

Filler	Curing system	Absorption (mg/cm^2)
Carbon black	Red lead	10
Carbon black/silica	Magnesium oxide/zinc oxide	10
Carbon black	Magnesium oxide/zinc oxide	55

In terms of volume increase, swellings of less than 10% can be obtained with suitable formulations. Effect on physical properties (carbon black/silica filled, MgO/ZnO cured composition):

Property	Immersed in water*	
	Before	*After*
Volume increase (%)	—	10
Tensile strength (kgf/mm^2)	1·72	1·2
Extension at break (%)	490	320
Hardness, IRHD	68	75

*90 days at 100°C

Permeability Vulcanisates of chloroprene rubber are relatively impermeable to gases, being better than natural rubber but not as good as those of butyl rubber in this respect.

Permeability of Neoprene G (unfilled vulcanisate) to various gases:

Method of expression	Hydrogen	Oxygen	Nitrogen	Carbon dioxide	Air
(i) As percentage relative to natural rubber, at 25°C	27	17	14	20	15
(ii) Permeability at 25°C	13·6	3·9	1·2	25·7	—
at 50°C (10^{-10} cm^2)/(s cmHg)	37·5	13·3	4·7	74·3	—

Lower permeability is obtained in filled compositions especially when fillers having a plate-like structure, e.g. mica, are used. Water vapour permeability of vulcanisate at 20–30°C (10^{-10} g)/(cm s cmHg), 0·7–1·45.

36.32 THERMAL PROPERTIES

Specific heat Raw, 0·52; unfilled vulcanisate, 0·49–0·52; vulcanisate containing carbon black, 0·40. **Coefficient of linear expansion** (at 25°C) Raw, 2×10^{-4}/°C; unfilled vulcanisate, $2·3–2·5 \times 10^{-4}$/°C. **Conductivity** cal/(cm s °C) Raw polymer or unfilled vulcanisate, $4·6 \times 10^{-4}$; vulcanisate containing carbon black, $5·0 \times 10^{-4}$. **Physical effects of temperature** *Raw polymer*: Stiffens and hardens progressively as temperature falls, crystallisation occurs most rapidly at about −5°C; glass transition temperature, *c.* −45°C. Crystalline m.p. *c.* 45°C (higher for more crystalline polymer, prepared at lower temperature). Softens on heating and begins to decompose at 150°C. Suffers extensive degradation at 250°C; at 300–310°C decomposition accelerates sharply and is accompanied by a spontaneous rise in

411

temperature to 420–430°C. *Vulcanised polymer* (filled): Crystallisation is most rapid at $c.$ -12°C and is slower than in raw polymers. It is retarded by use of sulphur containing curing agents, resinous or petroleum plasticisers, and the presence of a large proportion of fillers; accelerated by ester plasticisers. Brittleness temperature, -40 to -70°C; glass transition temperature, -43 to -49°C (unfilled vulcanisate, -44°C). Flexibility at low temperatures is improved by incorporation of suitable plasticisers, e.g. butyl oleate. Thaw temperature, i.e. melting point of the crystalline phase, varies with polymerisation temperature, crystallisation temperature, and strain in specimen. Heat resistance varies with both grade of polymer and type of additives used. Mercaptan-modified types are superior in this respect; antioxidants are especially important (secondary amines suitable); calcium carbonate or soft carbon black should be used as fillers, rapeseed oil as plasticiser for best resistance to heat ageing. Curing system should include high proportion of zinc oxide. Performance of typical compositions:

Property	Exposure to dry heat*		Exposure to moist heat†	
	Before	After	Before	After
Tensile strength (kgf/mm^2)	1·05	0·63	2·4	0·9
Extension at break (%)	500	100	680	50
Hardness, IRHD	65	88	59	79

*Aged 5 days at 150°C
†Aged 30 days in steam at 194°C)

Resistance to ageing in the absence of air is considerably better, properties being retained 15 times longer. Maximum service temperature is about 120°C. **Flame resistance** Self extinguishing owing to high chlorine content; improved by use of mineral fillers (e.g. hydrated alumina), zinc borate, antimony trioxide, and, as plasticisers, tritolyl phosphate or chlorinated paraffins.

36.33 ELECTRICAL PROPERTIES

Polychloroprene is generally inferior in this respect to natural rubber, SBR, butyl and silicone rubbers; however, it is extensively used in the sheathing of cables because of its good fire and weather resistance. Mercaptan-modified vulcanisates have slightly higher resistances than sulphur-modified types; mineral fillers improve insulation resistance and dielectric strength; carbon blacks vary in their effects; most plasticisers have adverse effect.

Properties of typical compositions based on Neoprene W (mercaptan-modified) with zinc oxide/magnesium oxide curing system; at room temperature:

(i) Filler (ii) Softener (parts per 100 rubber)	D.C. Resistivity (ohm.cm)	Dielectric strength (kv/mm)	Dielectric constant at 1 kHz	Power factor at 1 kHz
(i) SAF carbon black, 40 (ii) Process oil, 12	10^8	1·2	32	0·058
(i) Clay, 35; FEF c. black 25; SRF c. black 25 (ii) Process oil, 12	10^8	4·9	9	0·04
(i) Clay, 90; titania, 5 (ii) Process oil, 10	10^{12}	29·6	6	0·02
(i) Clay, 100; whiting, 45; FEF c. black, 20 (ii) Hydrocarbon resin, 30	2.10^{13}	23·6	7	0·035
Unfilled vulcanisate	33.10^{10}	—	7·3	0·047

Compositions formulated for optimum water resistance (see §36.31) also have satisfactory electrical properties.

Conductive compositions can be prepared with suitable carbon black fillers, and values of AC resistivity as low as $1·2 \times 10^2$ ohm cm are obtained using up to 100 parts of acetylene black per 100 rubber.

36.34 MECHANICAL PROPERTIES

As with other rubbers, the mechanical properties of polychloroprene can be varied over a very wide range of values depending upon the type and quantity of fillers, softeners, and curing agents used. There are also differences between the various grades of the polymer. Sulphur-modified rubbers are usually superior in flex-cracking resistance, tear strength, and resilience characteristics, whereas mercaptan-modified types have better tensile strength and compression set, and are characterised by lower hardness and modulus for a given filler loading.

Table 36.T1. TYPICAL PROPERTIES OF VULCANISATES AT 20–25°C

Property	Unfilled vulcanisates	Vulcanisates with 50–60 parts carbon black per 100 rubber†	
		Sulphur modified	Mercaptan modified
Elastic modulus (kgf/mm^2)			
(Young's static)	0·1–0·3	0·3–0·5 (type unspecified)	
(Young's, dynamic):			
3–4 Hz*	0·35–0·42		
460 Hz	0·43	—	—
1775 Hz	0·46		
1070 Hz‖	0·10		
(Shear, dynamic):			
50–100 Hz*	—	—	0·42
1500 Hz*	—	—	0·88
Tensile strength (kgf/mm^2)	1·3–2·2	2·3–2·5	1·3–2·4
Extension at break (%)	–800–1000	200–450	200–600
Stress at 300% extension (kgf/mm^2)	1·2	1·9–2·5	0·6–2·3
Tear strength (kgf/mm)	1	4·3–6·5	2·5
Hardness (IRHD)	43	70–88	60–84
Compression set (%)‡	14	20–24	9–18
Resilience (rebound) (%)	58	55–68	29–66
Loss tangent (tan δ)			
460 Hz	0·67	—	—
1775 Hz	1·18	—	—
1070 Hz‖	14·8	—	—
50–100 Hz*	—	—	0·25
1500 Hz*	—	—	0·40
Abrasion loss (ml/hp h)	—	410–550	430–575
(mm^3)§	274	—	61–156

*Temperature not stated
†Various types
‡After 22 h at 70°C. (ASTM D395 Method B)
§DIN 53516
‖At –1°C

Other data (1) Effects of various fillers in compounds of comparable hardness based on Neoprene W (mercaptan-modified), with zinc oxide/magnesium oxide curing system, at 25°C:

Filler(parts per 100 rubber)	Hardness (IRHD)	Tensile strength (kgf/mm²)	Extension at break (%)	Compression set (%) 70 h 100°C	Yerzley resilience† (%)
SRF carbon black (44)	68	2·04	350	34	73
MT carbon black (88)	70	1·34	400	33	71
EPC carbon black (29)	65	1·83	380	44	69
Hard clay (84)	73	1·72	680	74	69
Calcium carbonate (88)	75	1·16	690	47	74
Pptd. silica (47)	74	2·25	750	69	57
Zinc oxide (136)	64	1·97	800	62	70

*ASTM D395 Method B
†ASTM D945

(2) BS 2752, for vulcanised chloroprene rubber compounds containing carbon black, specifies the following values:

Hardness (IRHD)	41–50	51–60	61–70	71–80	81–88
Suggested carbon black content (phr)	1	25	45	55	80
Tensile strength (kgf/mm²), minimum	1·12	1·41	1·41	1·69	1·55
Extension at break (%), minimum	550	450	300	200	100
Compression set, % of compression applied, maximum (at 70°C)	30	30	25	25	25

36.4 FABRICATION

Can be masticated, compounded, and processed (extrusion, calendering, moulding, etc.) in much the same way as natural and other general purpose synthetic rubbers. Vulcanised with metal oxides, generally magnesium oxide used in conjunction with zinc oxide, but red lead and litharge often used in water-resistant compounds; the latter may tend to scorch during processing. Small amounts of sulphur are sometimes used to improve modulus and inhibit crystallisation. Accelerators, not needed with sulphur-modified rubbers, are essential with mercaptan-modified types, ethylenethiourea being most common. MBTS, thiuram sulphides, and sodium acetate used as retarders, i.e. to reduce risk of scorch. Antioxidants, plasticisers, fillers, and other additives as for natural rubber and SBR (*see* Sections 29 and 34).

36.5 SERVICEABILITY

Inherently extremely resistant to ozone; discolours on prolonged exposure to light but a phenolic antioxidant system minimises this;

compounds incorporating antioxidants have excellent stability at room temperature; heat resistance better than in natural rubber. Crystallisation limits low-temperature performance unless special plasticisers used. Characterised by low inflammability and good resistance to chemical attack and swelling in oils. Certain plasticised compounds susceptible to microbiological attack. Particularly sensitive to high energy radiation which results in cross-linking, insolubility, higher modulus, and increased resistance to compression set.

36.6 UTILISATION

Extremely varied but fall into two main groups: those in which oil or solvent resistance is important, e.g. seals, components for the motor industry, hoses and accessories; and those in which resistance to weathering or corrosion is required, e.g. cable jackets, motor vehicle glazing strip, chemical plant coatings, marine applications. Used in conveyer belting and industrial clothing. Very widely used in impact adhesives (special purpose fast-crystallising types); fastest crystallising types used in golf-ball covers because of similarity to gutta percha. Small amount used in tyres in U.S.A. (solid tyres and coloured sidewalls). Latices used for dipped products (e.g. surgical and industrial gloves), foams, adhesives, binders for non-woven fabrics, paper-, leather-, and textile-treatments, and modifiers for concrete and asphalt. Adheres to metals (especially if a priming coat of chlorinated rubber is used, or if compounded with phenolic resins). Used as a protective interlayer between copper conductors and vulcanised natural rubber-insulation.

36.7 HISTORY

The first step towards the production of polychloroprene was made in the U.S.A. in 1925 with J. A. Nieuwland's preparation of monovinylacetylene by passing acetylene through aq. cuprous chloride. This was taken up by the Du Pont Company as a starting material for synthetic rubber and much of the subsequent development was carried out by W. H. Carothers. By 1931 chloroprene had been prepared, polymerised, and the molecular structure of the μ polymer examined. The commercial product came on the market in 1932 under the trade name Du Prene, later changed to Neoprene*, and a latex version also became available about the same time. Early work was also carried out in the U.S.S.R resulting in the production of Sovprene about 1935. In Germany, too, several patents on chloroprene rubbers were taken out by IG Farbenindustrie in the 1930s but production does not appear to have begun until 1957 very likely because an oil-resistant rubber was

---

*Now, like nylon, used as generic name *neoprene*, without a capital N.

already available as Buna N (§35.7). Work on a synthesis of chloroprene in Japan was described in 1939. During the Second World War production of neoprene came under the control of the U.S. Government and it was designated GR-M (M = monovinylacetylene): production rose from 1738 long tons in 1939 to 58 102 tons in 1944. Post-war production was at a rather lower rate until 1951, since when it has risen rapidly. World consumption in 1965 reached 165 000 tons. Currently polychloroprene is being produced on a commercial scale in U.K., France, West Germany, U.S.S.R., Czechoslovakia, and Japan as well as U.S.A.

36.9 FURTHER LITERATURE

MARK, H. and WHITBY, G. S. (Eds), *High Polymers*, Vol. I, W. H. Carothers: Collected Papers Pt. 2, Interscience Publishers, New York (1940)

WHITBY, G. S. (Ed), *Synthetic Rubber*, Chapter 22 by NEAL, A. M. and MAYO, L. R., John Wiley & Sons Inc., New York (Chapman & Hall Ltd., London) (1954)

PENN, W. S., *Synthetic Rubber Technology*, Vol. I, Chapters 20 to 27, MacLaren, London (1960)

MURRAY, R. M. and THOMPSON, D. C., *The Neoprenes*, E. I. du Pont de Nemours & Co. (Inc.), Elastomer Chemicals Dept., Wilmington, Delaware (1964)

HOLLIS, L. E., *Chem. and Ind.,* 1030 (1969)

Specification
BS 1673: Part 8 (Chemical Analysis of Chloroprene Rubber)
BS 2752 (Vulcanised Chloroprene Rubber Compounds)

SECTION 37

ETHYLENE/PROPYLENE RUBBERS*

SYNONYMS AND TRADE NAMES
EPM, EPR, EPDM, EPT; C 23, Dutral, Intalon, Keltan, Nordel, Royalene, Vistalon.

GENERAL CHARACTERISTICS

Two main types are produced commercially: saturated copolymers vulcanisable with organic peroxides; and terpolymers containing a minor proportion of a diene to confer unsaturation and make sulphur vulcanisation possible. There appears to be no significant difference between these two types in processing behaviour or vulcanisate properties.

Raw rubbers are light grey to amber coloured and virtually without odour. They have good storage stability, do not crystallise at low temperatures or on stretching, and are tough and 'nervy'.

Unvulcanised compounded mixes have a degree of thermoplasticity which confers good 'green strength' (cohesion) for processing, but have poor tack. Vulcanisates are characterised by good resistance to attack by heat, polar liquids, and ozone, good low-temperature flexibility and abrasion resistance.

37.1 STRUCTURE

Simplest Fundamental Unit

Copolymer rubbers consist of randomly alternating blocks of from 10–20 monomer units, the basic structure being thus represented as follows:

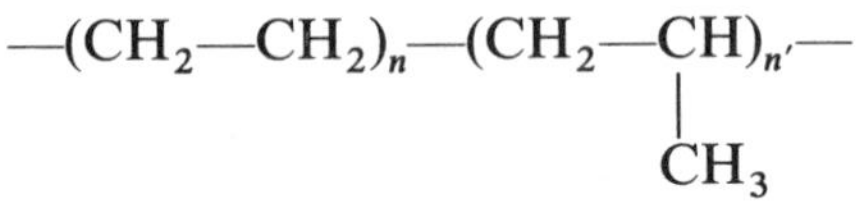

$$-(CH_2-CH_2)_n-(CH_2-CH)_{n'}-$$
$$\overset{|}{CH_3}$$

where n and $n' = 10$–20.

A high content of either monomer in the copolymer increases the probability of the existence of long sequences of the monomer in question; the greatest dispersion in the lengths of sequences occurs in copolymers containing equimolar amounts of ethylene and propylene. Usual ethylene content is 40–60%.

Inclusion of a small proportion of a diene, such as cyclopentadiene, cyclo-octadiene or norbornadiene, yields a terpolymer that can be vulcanised with sulphur, e.g. basic structure,

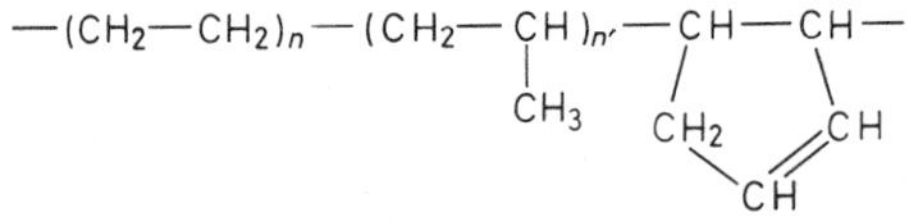

$$-(CH_2-CH_2)_n-(CH_2-\underset{CH_3}{\overset{|}{CH}})_{n'}-CH-CH-$$

For fluorinated ethylene/propylene eopolymers *see* §3.81.

Molecular weight 100 000–200 000.

X-ray data Amorphous. (During polymerisation care is taken to avoid the formation of long sequences of ethylene units which give rise to crystallisation; but see terpolymers §37.32.)

37.2 CHEMISTRY

37.21 PREPARATION

Copolymerisation of ethylene and propylene, with or without the addition of a small proportion (*c.* 2 %) or a diene, depends on the use of stereospecific catalysts (Ziegler-Natta type) and involves a complex anionic-type chain propagation mechanism (*see* Section 54).

The main constituents of these catalysts are halides of transition metals (commonly vanadium or titanium), and alkyls or alkyl halides usually of aluminium. There are many patents or patent applications covering polymerisations of this type, and the preparation of halogenated varieties of the copolymers.

Batch-wise processes can be employed with or without the use of hydrocarbon solvents. The critical factors are the provision of a continuous feed of ethylene to the reactor and the maintenance of a constant ratio of ethylene to propylene in the liquid phase, to avoid preferential polymerisation of the ethylene and ensure homogeneity of the copolymer. Liquid propylene is introduced into an autoclave at -10 to $-20°C$ followed by ethylene at a pressure of a few atmospheres, together with the catalyst which is added intermittently or continuously, in solution form if desired. The pressure in the autoclave is regulated by controlling the flow of ethylene. Chain transfer agents, e.g. zinc alkyls, are used to control molecular weight.

One continuous commercial process (U.S. Rubber Co.) places emphasis on ensuring the purity of the reactants, especially the catalysts themselves, since traces of water, sulphur, carbonyl compounds, or other impurities can seriously disturb the catalyst system. In this process, a terpolymer incorporating (probably) dicyclopentadiene is prepared by means of vanadium based catalysts, polymerisation taking place in solution (hexane). The reaction temperature is maintained at about 38°C and a pressure of 200–250 lbf/in^2 is applied to keep the ethylene in solution. The reaction products are stripped of unreacted monomers and carefully washed with steam and de-aerated water to remove catalyst residues which adversely affect the properties of the rubber. Coagulation takes place at the same time and the washed crumb is finally separated and dried.

37.22 PROPERTIES

Soluble (milled raw polymer) in benzene, hexane, heptane, cyclo-hexane, carbon tetrachloride, perchloroethylene. **Plasticised by** mineral oils, adipate, phthalate and sebacate plasticisers. **Swollen by** non-polar liquids (incl. mineral oils), tetralin, decalin, turpentine, chlorinated hydrocarbons, ethers (incl. tetra hydrofurum). Volume swelling of typical composition (sulphur-cured terpolymer) containing 60 parts FEF black per 100 polymer after 28 days at 25°C (%): benzene, 103; carbon tetrachloride, 175; ethyl alcohol, 0; acetone, 4; dibutyl phthalate, 4. Peroxide-cured copolymers show a similar pattern of swelling behaviour. **Relatively unaffected by** alcohols, glycols, ketones, phosphate plasticisers, alkalis, phosphoric acid, water, ozone. **Decomposed** chlorosulphonic acid (98%), nitric acid (70%).

37.23 IDENTIFICATION REACTIONS

Thermoplastic in unvulcanised state; S and N may be present in vulcanisates; P and Cl nominally absent.

Combustion Burns readily without distinctive odour; leaves little carbonised residue. **Pyrolysis** Complex mixtures of hydrocarbon monomers are formed. **Other tests** There are no simple chemical tests specific to ethylene-propylene rubbers. They are most likely to be confused with butyl rubbers, and possibly some polyacrylates. An ester odour during combustion, however, characterises the polyacrylates, and butyl rubbers can usually be distinguished by the tests given in §38.23 (provided that reference standards are used since ethylene-propylene rubbers may give a slight positive reaction). Infra-red diagnostic absorption bands (for pyrolysates): 7·3 μm (1370 cm^{-1}); 11·0 μm (909 cm^{-1}); 11·3 μm (885 cm^{-1}); 10·4 μm (962 cm^{-1}); 13·8 μm (725 cm^{-1}).

37.3 PHYSICS

37.31 GENERAL PROPERTIES

Specific gravity 0·85–0·87 (raw polymer; for calculation of sp. gr. of vulcanisates *see* Section 58). **Refractive index** 1·48. **Water absorption** Comparable with that of butyl rubber (lower than natural rubber or SBR). Weight increase of vulcanisate (copolymer type) containing 60 parts FEF black per 100 polymer after 30 days' immersion at 100°C: approx. 5 mg/cm^2 surface area (cf. similar natural rubber and SBR vulcanisates under same conditions: approx. 17 mg/cm^2). Effect of immersion in distilled water on physical properties of typical carbon black filled vulcanisates:

Property	Copolymer		Terpolymer	
	3 days at 100°C	365 days at 25°C	3 days at 100°C	365 days at 25°C
Volume increase (%)	0·7	0·6	1·1	2
Retained tensile strength (%)	106	108	105	104
Retained extension at break (%)	102	109	85	44
Hardness change (IRHD)	+2	+3	0	+3

Permeability to gases This is generally comparable with that of natural rubber, e.g. at 30°C (10^{-10} cm^2)/(s cmHg), O_2, 25; N_2, 8·5; CO_2, 108. Permeability of Dutral (copolymer type, gum vulcanisate) to various gases:

Method of expression	Hydrogen	Oxygen	Nitrogen
(i) As percentage relative to natural rubber, at 30°C	82	160	133
(ii) Permeability (10^{-10} cm^2)/(s cmHg) at 30°C	47·5	42·5	15·2
at 50°C	121·4	108·1	44·9

37.32 THERMAL PROPERTIES

Specific heat Raw (copolymer), 0·52; filled vulcanisate (terpolymer), 0·56. **Coefficient of linear expansion** Raw (copolymer) $1·8 \times 10^{-4}$/°C; terpolymer, $2·3–2·4 \times 10^{-4}$/°C. **Thermal conductivity** Raw (copolymer), $8·5 \times 10^{-4}$ cal/(cm s °C). **Physical effects of temperature** *Raw polymer* Although essentially amorphous, some crystallisation has been detected in terpolymers at temperatures between room temperature and the glass transition point. Significant stiffening begins at *c.* 50°C; glass transition temperatures, *c.* −50 to −58°C. Brittleness temperature, *c.* −90°C. *Vulcanised polymer* (*filled*) Low temperature flexibility is good and can be still further enhanced by the use of ester-type plasticisers and paraffinic oils. Glass transition temperatures, *c.* −52 to −56°C (unfilled vulcanisates, *c.* −54 to −59°C). Brittleness temperature, *c.* −70°C.

Heat resistance is better than that of natural rubber but depends on the type of curing system and fillers used. High filler and plasticiser loadings and the presence of sulphur have an adverse effect whereas MBT and zinc oxide are beneficial. The normal maximum service temperature is *c.* 150°C (dry heat) while resistance to moist heat is still better. Heat resistance of typical terpolymer vulcanisates (curing system: accelerators and sulphur; fillers: FEF black and, in the case of the compounds exposed to dry heat, zinc oxide):

Property	Exposure to dry heat*		Exposure to moist heat†	
	Before	After	Before	After
Tensile strength (kgf/mm²)	1·37	0·35	1·69	1·54
Extension at break (%)	620	100	430	370
Hardness (IRHD)	70	85	68	69

*Oven aged 5 days at 177°C
†Aged 6 weeks in steam at 194°C

37.33 ELECTRICAL PROPERTIES

Ethylene/propylene rubbers have excellent electrical properties, especially when compounded with clay or calcium carbonate fillers; in view of their additional advantage of high ozone-resistance they are thus useful for cable covering and allied applications.

Properties of typical vulcanisates at 20°C:

Rubber type	Curing system	Filler (parts per 100 rubber)	D.C. resistivity (ohm.cm)	Dielectric strength (kV/mm)	Dielectric constant at 1 kHz	Power factor at 1 kHz
Copolymer	Uncured	None	5×10^{16}	28	2·2	0·0015
Copolymer	Peroxide and sulphur	clay (125); EPC black (10)	$1·56 \times 10^{14}$	40	3·34	0·0079
Terpolymer	Sulphur and accelerators	clay (120); process oil (20)	—	—	3·36*	0·0297*

*After soaking for 1 month in water at 75°C

37.34 MECHANICAL PROPERTIES

Like other non-crystallising polymers unfilled vulcanisates of ethylene/propylene rubbers have low tensile strength (*c.* 0·14 kgf/mm²) and reinforcing fillers are required to produce practical compounds. There is no significant difference between copolymer and terpolymer type vulcanisates in mechanical properties. Dynamic properties are similar to those of SBR.

Properties of typical vulcanisates at 20–25°C:

Property	Unfilled vulcanisates	Vulcanisates with 50–60 parts carbon black (various types) per 100 rubber
Dynamic Young's modulus (kgf/mm^2)	0·14–0·56	0·5
Tensile strength (kgf/mm^2)	0·12	1·0–1·6
Extension at break (%)	400	250–750
Stress at 300% extension (kgf/mm^2)	—	0·9–1·2
Tear strength kgf/mm^2)	—	2·5–4
Hardness (IRHD)	—	40–75
Compression set (22 h at 70°C)	—	5–20
Resilience (rebound) (%)	65–80	40–55

Properties of typical vulcanisates of approximately the same hardness:

Rubber type	Curing system	Filler (parts per 100 rubber)	Hardness (IRHD)	Tensile strength (kgf/mm^2)	Extension at break (%)	Compression set* (%) 22 h, 70°C	Resilience† (%)
Copolymer	Peroxide and sulphur	HAF black (100); EPC black (50)	65	1·8	380	13	—
Copolymer	Peroxide and sulphur	Silica (50)	64	1·7	450	—	—
Terpolymer	Sulphur and accelerators	MT black (130); process oil (20)	66	0·54	520	13	72
Terpolymer	Sulphur and accelerators	SAF black (50); process oil (20)	63	2·34	530	15	67
Terpolymer	Sulphur and accelerators	Calcium carbonate (200); process oil (20)	60	0·33	580	45	60
Terpolymer	Sulphur and accelerators	Hard clay (120); process oil (20)	61	1·26	820	47	64

*ASTM D395 Method B
†ASTM D945

37.4 FABRICATION

Ethylene/propylene rubbers are compounded and processed in much the same way as natural rubber and general purpose synthetic rubbers but differ in the curing systems used. Saturated *copolymers* (*see* §37.1 and 37.2) require organic peroxides to effect cross-linking and these may interfere with other compounding ingredients or give off a bad odour; an acid environment reduces the cross-linking efficiency of peroxides. A small proportion of sulphur is often added to regulate the reaction. Diene-containing *terpolymers* are cured with sulphur, zinc oxide, and accelerator systems of the conventional type. No breakdown (i.e. reduction of molecular weight) occurs during milling; hot-mixing with fillers in the presence of a cross-linking promoter, e.g. quinone dioxime, is sometimes carried out to improve mechanical properties. Tackifiers (e.g. phenolic resins) are required to produce building tack. Antioxidants are only used if exceptional heat resistance is required. Ethylene/propylene rubbers may be blended with other rubbers or thermoplastics and may be bonded to metals during vulcanisation with the aid of special proprietary cements (e.g. Thixon).

37.5 SERVICEABILITY

The outstanding characteristics of these rubbers are excellent resistance to ozone, weathering, and heat- and light-ageing, in which respects they are superior to natural rubber and most other unsaturated rubbers. Resistance to chemicals, other than hydrocarbons (e.g. oils) is also good. Because of their amorphous structure, good flexibility is retained at low temperatures, unlike crystallising elastomers such as polychloroprene.

37.6 UTILISATION

Ethylene/propylene rubbers have been suggested for such a wide range of applications that they can be regarded as general purpose materials. However, largely because of the patents situation (*see below*), their application has tended to be restricted. Potentially important applications include: tyres, insulation and sheathing for electrical cables, footwear, cellular products, proofed fabric, belting and sealing strip. When blended with polypropylene they improve the impact strength.

37.7 HISTORY

The invention of ethylene/propylene rubbers was but one aspect of the intensive research in stereochemistry which followed the discovery of a new class of polymerisation catalysts by K. Ziegler during the period 1952–55. Within a few years workers in Italy, Germany, and the United States succeeded in producing commercially exploitable elastomers, but first success must be accorded to the group led by G. Natta and sponsored by the Montecatini company, who brought out a saturated rubber in 1959. However, so many patents, differing in details of production were applied for within a short space of time that no single company could establish a clear lead; the development of unsaturated rubbers (involving chlorosulphonation or introduction of dienes) to overcome some of the disadvantages of the earlier types, and attempts to obtain patents rights for the use of ethylene/propylene rubbers in certain applications still further complicated the situation. In the U.S.A. consumption of ethylene/propylene rubbers (mainly terpolymers) rose from 4000 tons in 1964 to 35000 tons in 1967, with an estimated increase to over 100000 tons in 1972. In Europe consumption was estimated at 7500 tons in 1967, and a demand for up to 80000 tons is forecast for 1973. World production capacity was estimated to be 109000 tons in 1968, with a forecast of 208000 tons for 1971.

37.9 FURTHER LITERATURE

NATTA, G., CRESPI, G., VALVASSORI, A. and SARTORI, G., *Rubb. Chem. Technol.* **36**, No. 5, 1583 (1963)

HAM, G. E. (Ed), *High Polymers,* Vol. XVIII, Copolymerisation, Chapter IV C, by CRESPI, G., VALVASSORI, A. and SARTORI, G., Interscience Publishers, New York (1964)

BUTYL RUBBER

With notes on latex, halogenated rubbers and polyisobutylene

SYNONYMS AND TRADE NAMES

Isobutylene/isoprene rubber; isobutene/isoprene rubber; GR-I; IIR. Bucar; Socabutyl.

GENERAL CHARACTERISTICS

Raw butyl rubber, a soft yellowish translucent substance that exhibits a somewhat tacky surface along with considerable cold flow; resembles in appearance and consistency crepe rubber (*see* §29.83) that has been massed and softened by milling. It is also available as an aq. dispersion or latex (*see* §38.81).

Butyl rubber is not generally blended with other sulphur-vulcanisable rubbers (*see* §38.4(ii)).

Vulcanisates resemble natural rubber vulcanisates but show greater resistance to heat ageing, oxidation, ozone-cracking, and general chemical attack, along with lower permeability to gases. A low isoprene content favours oxidation and ozone resistance, but a higher proportion increases the rate of vulcanisation and the resistance of the product to heat. As it is never sufficiently unsaturated to produce a high degree of cross-linkage (*see* §38.1), butyl rubber does not give hard vulcanisates or ebonites.

38.1 STRUCTURE

Simplest Fundamental Unit

$$-\left[-CH_2 - \underset{\underset{CH_3}{|}}{\overset{\overset{CH_3}{|}}{C}} -\right]_n -CH_2 - \underset{\underset{}{}}{\overset{\overset{CH_3}{|}}{C}} = CH - CH_2 -$$

isobutylene, C_4H_8, mol. wt. = 56; isoprene, C_5H_8, mol. wt. = 68. The average value of *n* ranges from 30–200.

The isoprene units (which are randomly distributed and mostly in the 1,4 configuration) provide the unsaturation that permits the copolymer to be vulcanised; the small number of double bonds remaining in the vulcanisate means that it resists oxidative and similar forms of chemical attack.

Molecular weight (by viscosity): Range, $<50000-10^6$; peak of molecular weight–weight percentage distribution curve, *c.* 300000. Viscosity-average, 300000–450000.

X-ray data Amorphous, but crystalline on stretching or cooling. The pattern produced on stretching is indistinguishable from that given by polyisobutylene, i.e. identity period (*c*), 18·6 Å; other dimensions of rectangular unit cell, $a = 6.94$ Å, $b = 11.96$ Å. The isobutylene units are arranged head-to-tail, and the chain has a helical configuration.

38.2 CHEMISTRY

38.21 PREPARATION

Isobutylene ($CH_2 = C(CH_3)_2$, b.p. $-6.9°C$) is extracted from a C_4 petroleum fraction by selective absorption in sulphuric acid of a concentration (65%) such that other components (e.g. other butylenes and butanes) are not appreciably taken up. It is then steam-distilled from the acid, washed with water and aq. alkali and finally distilled to yield a product over 99% pure.

For the preparation of butyl rubber, isobutylene mixed with a small proportion of a diene (usually isoprene, $CH_2 = C(CH_3) - CH = CH_2$, *see* §32.21) is dissolved in an inert solvent (methyl or methylene chloride, or liquefied ethylene) and treated with a cationic initiator (e.g. a Friedel-Crafts catalyst such as $AlCl_3$ or BF_3, together with a trace of a co-catalyst such as H_2O) usually also dissolved in an inert solvent. Polymerisation takes place extremely rapidly, and to obtain polymers of high molecular weight it is necessary to introduce the catalyst to the monomer solution at a very low temperature (-95 to $-100°C$). However, when suitable provision is made to remove the heat of the reaction the reactants may be fed together in such a way that the process can be made continuous. The polymer separates as a precipitate, and the resulting slurry is passed into hot water to flash off residual solvent and unreacted monomer. An antioxidant (preventing subsequent breakdown) and zinc stearate (preventing cohesion of the crumb) are then added and the product is filtered off, and the moist crumb hot-extruded, milled, sheeted and cooled before packing.

Commercial rubbers are sometimes described as of *low, medium* or *high* unsaturation, i.e. containing approx. 0·6–1·0, 1·0–1·8, or 2–2·5 mol.% isoprene. Other variables are (i) Mooney viscosity (normally *c.* 45, but in some grades 70–89) and (ii) the type of antioxidant; *regular* grades contain one of the staining type (e.g. phenyl-β-naphthylamine) though non-staining grades are available (e.g. containing an alkylaminophenol).

Polymerisation can also be effected in a solution in pentane, giving a high molecular weight product, but as the polymer remains in solution, resulting in a very high viscosity, conversion cannot exceed *c.* 5%, as compared with *c.* 50% in the process described above.

427

38.22 PROPERTIES

Soluble (raw rubber) in hydrocarbons, halogenated hydrocarbons and carbon disulphide. Vulcanisates are insoluble except under conditions causing breakdown of the polymer. **Plasticised by** peptisers such as β-naphthyl or xylyl mercaptan, pentachlorothiophenol or its zinc salt and dicumyl peroxide, but butyl rubber, having a low content of double bonds, does not undergo chemical plasticisation (peptisation) so readily as natural rubber. Physical plasticisation is effected by substantially saturated substances of low polarity, such as hydro-carbon (petroleum) oils and waxes, petroleum jelly, polyisobutylene, bitumens, aromatic and higher aliphatic esters (e.g. dioctyl phthalate and sebacate), dibenzyl ether, metal soaps, resins, e.g. hydrogenated ester gum. **Swollen** (vulcanisates) by hydrocarbons (especially paraffins and cycloparaffins), most halogenated hydrocarbons, carbon disulphide, higher ketones, diethyl ether, oleic acid and some esters (*see* swelling data below). **Relatively unaffected** (raw and vulcanised) by oxygen, ozone, water, salt solutions, inorganic acids (except conc. HNO_3, H_2SO_4 and H_2CrO_4 and hot conc. HCl), alkalis (including conc. NH_4OH and NaOH), hydrogen peroxide (up to 30 %), alcohols, glycols, glycerol, phenols, nitrobenzene, aniline, xylidine, animal and vegetable oils and some ketones, ethers, esters and fatty acids. **Decomposed by** conc. nitric, sulphuric and chromic acids, and 10 % chlorosulphonic acid.

Swelling of vulcanisates (as volume increase, %)

(i) Equilibrium swelling of unfilled vulcanisate (in brackets: vulcanisate with 45 parts SRF carbon black per 100 rubber) at 25°C: cyclohexane 540 (350), carbon tetrachloride 470, *n*-heptane 313 (240), carbon disulphide (220), benzene 188 (140), ethyl ether (85), ethylene dichloride (35), ethyl acetate 21 (22), methyl ethyl ketone 17, acetone (7), aniline (5).

(ii) Swelling of vulcanisate with 75 parts FEF carbon black, after immersion for 70 h at 100°C: ASTM oil No. 3, 296 and 122 for 'low' and 'high' unsaturation respectively (*see* §38.21).

38.23 IDENTIFICATION

Cl, N, P and S nominally absent in raw butyl rubber (N present in antioxidant, *see* §38.21) but one or more of these elements present in vulcanisates.

Combustion Burns with a clear flame (distinction from natural, synthetic isoprene, butadiene and styrene/butadiene rubbers, which give smoky flames). **Pyrolysis** Melts but does not readily char;

depolymerises giving white fumes that condense to a clear neutral liquid, mainly isobutylene. The latter may be identified by (i) yellow coloration on filter paper freshly dipped in a solution made by dissolving 1 g yellow mercuric oxide in 20 ml boiling 5N sulphuric acid; or (ii) formation of methoxy*iso*butyl mercuriacetate, m.p. *c*. 55°C, by passing the pyrolytic vapours through an ice-cold trap and leading the non-condensed portion into a 3·5–5% methanol solution of mercuric acetate, evaporating the alcohol, boiling the residue with 40–60°C petroleum ether, filtering, concentrating the filtrate to small volume, allowing the mercury derivative to crystallise, and drying it at 30–40°C (for details *see* WAKE, W. C., *The Analysis of Rubber and Rubber-like Polymers,* 2nd edn, London (1969), and ASTM D297, which also describes other colour tests).

Other tests Inert to boiling 80% sulphuric acid (distinction from silicone and ester rubbers). Infra-red absorption spectrum has a strong doublet band at 7·2 and 7·3 μm (distinction from the single 7·3 μm band of natural rubber), a strong band at 11·3 μm and a medium doublet at 8·0–8·2 μm. (Note: the spectrum does not distinguish butyl rubber from polyisobutylene or from chlorinated or brominated butyl rubber).

38.3 PHYSICS

38.31 GENERAL PROPERTIES

Specific gravity Raw rubber, 0·91–0·96; unfilled vulcanisates, 0·92–0·98. **Refractive index** n_D^{25} Raw, 1·5078 (2% isoprene); 1·5081 (3% do.). **Water absorption** Much less than with natural rubber, e.g. unfilled vulcanisates, after immersion for 7 days at 100°C (%): butyl, 2·9; natural, 11–12·6. Absorption by vulcanisate containing 140 parts calcined clay (per 100 rubber), after immersion for 28 days at 75°C, *c*. 2 mg/cm^2. **Permeability** Much lower than that of natural rubber, e.g. pressure loss in inner tubes (butyl/natural), 0·07–0·125. Permeability to gases at 25°C (10^{-10} cm^2)/(s cmHg), H_2, 7·3; O_2, 1·3; N_2, 0·32; CO_2, 5·2; He, 8·5; CH_4, 0·8; C_2H_2, 1·7. Permeability to water vapour, 0·03–0·16 (10^{-10} g)/(cm s cmHg).

38.32 THERMAL PROPERTIES

(Raw rubber or unfilled vulcanisates unless otherwise stated.)
Specific heat 0·44–0·46. **Conductivity** (10^{-4} cal)/(cm s °C), 3·1; higher for filled vulcanisates. **Coefficient of linear expansion** (10^{-4}/°C) *c*. 1·8; vulcanisate with 50 parts carbon black (per 100 rubber), *c*. 1.1. **Physical effects of temperature** *Raw rubber* At low temperatures,

becomes hard and finally brittle; glass temperature $-75°C$ to $-67°C$. At elevated temperatures becomes softer and weaker and decomposes slowly above 130°C. *Vulcanisates* Glass temperature (unfilled rubber), $-63°C$; brittleness temperature (ASTM D746) $-40°C$, or down to $-62°C$ by incorporation of plasticisers. The resilience or rebound of unplasticised vulcanisates is low at room temperature, and they remain sluggish up to *c*. 40°C, but above this, the resilience increases and becomes similar to that of natural rubber (*see* §38.34).

38.33 ELECTRICAL PROPERTIES

	Raw rubber	Unfilled vulcanisate	Filled vulcanisates		
			A	B	C
Volume resistivity (ohm cm)	—	*c*. 10^{17}	$1·2.10^{15}$ (8.10^{13})	4.10^{15} ($3·2.10^{15}$)	—
Dielectric strength (kV/mm)	—	24	—	—	22–35
Dielectric constant					
1 kHz	2·3–2·35	2·11–2·42	—	—	3·0 (3·0)
1300 MHz	2·12–2·35	2·38	—	—	—
frequency not given	—	—	2·9 (4·2)	3·2 (3·5)	—
Power factor*					
1 kHz	0·0005–0·001	0·0054	—	—	0·0054 (0·007)
1300 MHz	0·0004–0·0008	0·0012	—	—	—
frequency not given	—	—	0·015 (0·058)	0·024 (0·030)	—

*Depends greatly on moisture content
A 140 parts calcined clay (per 100 rubber), vulcanised with sulphur and tetramethylthiuram disulphide (bracketed values: after 28 days' immersion in water). **B** As **A**, but vulcanised with *p*-quinone dioxime dibenzoate (Dibenzo GMF). **C** Insulation mix, 35–40 % rubber content (bracketed values: after 24 h immersion in water).

38.34 MECHANICAL PROPERTIES

Like those of natural rubber, unfilled vulcanisates of butyl rubber exhibit high tensile strength; reinforcing fillers produce little, if any, increase, but they increase tear strength.

Stiffness (stress at fixed elongation) is generally similar to that of natural rubber vulcanisates; it is greater—and the breaking elongation lower—the greater the unsaturation of the copolymer (and hence

430

the extent of cross-linking produced by vulcanisation, *see* §38.1).

Resilience is normally low, but increases with rise of temperature, e.g. Lüpke rebound, (%):

Temperature, °C	−40	0	20	60	100
Unfilled vulcanisate	10	8	12	30	50
Filled vulcanisate (50 parts EPC black per 100 rubber)	7	6	13	48	76

Resilience also increases with plasticisation (*see* §38.22), and with 'heat treatment' (*see* §38.4iv), which can reduce the internal viscosity of the vulcanisate to one-half or one-third.

Abrasion Butyl vulcanisates show good abrasion resistance (equal to or better than that of a styrene/butadiene rubber in a tyre tread mix). Flexing life (cut-growth resistance) of filled vulcanisates, cycles/mm, 750 and 16 respectively, for 'low' and 'high' unsaturation (*see* §38.21).

Some other mechanical properties are shown in Table 38.T1.

Table 38.T1. TYPICAL MECHANICAL PROPERTIES OF BUTYL RUBBER VULCANISATES

	Unfilled vulcanisates	Filled vulcanisates*	Filled vulcanisates†				
Hardness (IRHD)	30–35	—	41–50	51–60	61–70	71–80	81–88
Tensile strength (kgf/mm)	1·7–2·1	0·9–2·1	1·05	1·05	1·05	0·88	0·88
Breaking elongation (%)	750–950	300–700	600	500	400	300	250
Stress at 300% elongation (kgf/mm^2)	0·07–0·13	low, 0·3; high, 1·1–1·3	—	—	—	—	—
Tear strength (kgf/mm)	0·9–1·2	4·5–6·0	—	—	—	—	—
Compression set‡ (%)	—	—	40	40	40	40	40

*'Low' and 'high' refer to degree of unsaturation (*see* §38.21)
†From BS 3227; values (minima for tensile strength and elongation; maxima for compression set) based on medium-unsaturation butyl
‡Compressed 25% for 24 h at 70°C; recovery 10 min at room temperature; set expressed as percentage of compression

38.4 FABRICATION

Butyl rubber is generally processed like natural rubber, except for the following differences.

(i) It does not readily plasticise when worked on hot rolls, but responds to certain peptisers or softeners (*see* §38.22).

(ii) It vulcanises only slowly with sulphur, needing ultra-accelerators and high temperatures (150–230°C) to achieve reasonably short (10–20 min) cures; hence sulphur-vulcanising mixes must not contain (or even be contaminated with) other rubbers, softeners, etc., that react more readily. Accordingly, special vulcanising systems have been devised, notably with *p*-quinone dioxime*('GMF') or its derivatives, e.g. *p*-quinone dioxime dibenzoate ('Dibenzo GMF'), which make possible vulcanisation in as little as 1 min at 200°C, and also confer good age- and heat-resistance.

(iii) It can be vulcanised by certain alkylphenol/formaldehyde resins (e.g. 'methylolphenol' resins) plus a halogen-containing activator (e.g. $SnCl_2 \cdot 2H_2O$, or zinc oxide plus either PVC or chlorosulphonated polyethylene). This confers low compression set and exceptional age- and heat-resistance.

(iv) The normally low room-temperature resilience of butyl rubber vulcanisates containing filler (carbon black, clay, silica) can be improved by 'heat treatment' or 'thermal interaction', which is effected either by heating the polymer/filler masterbatch for 2–4 h in steam at *c*. 160°C, or by working it hot on a roll mill (e.g. 10–20 min at 180–205°C) or in an internal mixer (e.g. 7–10 min at 135°C). The second process is facilitated by inclusion of 'promoters', e.g. N-methyl-N,4-dinitrosoaniline, *p*-quinone dioxime, *p*-dinitrosobenzene.

Butyl rubber can be bonded to metals during vulcanisation.

38.5 SERVICEABILITY

Butyl rubber vulcanisates have exceptional resistance to deterioration (ageing) by oxidation, ozone and heat, also to the action of corrosive chemicals in general, though many organic liquids (hydrocarbons, chlorinated hydrocarbons) have a swelling action similar to that on other hydrocarbon rubbers. The vulcanisates have relatively low permeability to gases, and good electrical properties; they are serviceable from *c*. $-50°C$ to $+125°C$ (or higher for short periods). Normally, at room temperature they are less resilient than vulcanisates of natural or other synthetic rubbers, and hence can have a useful damping and sound- or shock-absorbing action.

38.6 UTILISATION

In the tyre field the low air permeability of butyl rubber has made it largely replace natural rubber for inner tubes, and for the same reason it is used in tubeless tyres. Butyl tyre covers have been the subject of many trials, showing improved ('softer', quieter) riding qualities—these being attributed to its high damping capacity—no 'squealing' on turns, good skid-resistance, tread wear-resistance equal to styrene/butadiene rubber, and less tread-groove and sidewall cracking.

*The true vulcanising agent is the corresponding dinitroso compound (e.g. *p*-dinitrosobenzene) formed by oxidation of the quinone dioxime by (e.g.) red lead or dibenzthiazyl disulphide included in the mix.

Nevertheless, though commercially available since *c.* 1960, butyl tyre covers have not become widely used.

Other uses include: proofed fabrics (e.g. for raincoats, convertible car tops, tank and pool linings), motor-car components (radiator hose, pedal pads, seals, dampers), weather seals and gaskets in buildings, tyre curing bags, conveyer belts for use at elevated temperatures, steam and acid hose, chemical plant linings, air cushions and bellows, pneumatic springs and accumulator bags, ozone-resistant cable insulation, and as a constituent of adhesives.

38.7 HISTORY

Isobutylene was discovered by Faraday in 1825; A. M. Butlerov and Gorianov prepared liquid polymers in 1873, and the rubbery solid polymers (§38.83), discovered by M. Otto and M. Müller-Cunradi in the early 1930s, were first manufactured by I. G. Farbenindustrie A.G. (Oppanol B) and by the Standard Oil Company, *c.* 1935.

Butyl rubber has been developed from polyisobutylene. The parent substance is unvulcanisable, but R. M. Thomas and W. J. Sparks in 1937 made a vulcanisable copolymer by introducing *c.* 2% of isoprene; the proportion was kept low to provide sufficient unsaturation for vulcanisation but little unsaturation in the final product.

38.8 ADDITIONAL NOTES

38.81 BUTYL RUBBER LATEX

Aq. dispersions (latices) containing 60–70% total solids, are made either by emulsifying a hydrocarbon solution of butyl rubber with water containing an emulsifier and removing the solvent, or by mechanically dispersing the rubber, plus a mineral filler (e.g. kaolin), in water with or without a dispersing agent. They are used for coating and impregnating leather, paper (especially air-barrier and food-wrapping papers) and textiles (e.g. tyre cord, for bonding to butyl rubber mixes), also as adhesives (including binders for non-woven fabrics) and in emulsion paints. Foam rubber can be made from butyl rubber latex plus 5–15% of styrene/butadiene rubber latex.

38.82 HALOGENATED BUTYL RUBBERS

Butyl rubbers containing a small proportion of combined chlorine or bromine vulcanise more quickly than normal butyl rubber because the halogen atoms provide additional sites for cross-linking. For this reason they can be blended with more unsaturated rubbers, e.g. natural, styrene/butadiene, chloroprene, and also give stronger bonds to these rubbers.

Brominated butyl rubber is available commercially (Hycar 2202), or may be prepared as required by milling butyl rubber with either charcoal containing adsorbed bromine (to the extent of *c.* 7% on the rubber), or a substance that liberates bromine on heating, e.g. 10% of N-bromosuccinimide or 7·5% of dibromomethylhydantoin; this method is used to make 'tie-gum' mixes for bonding rubber to metal.

Chlorinated butyl rubber (Butyl HT) contains 1·1–1·3% combined chlorine and 1–2 mol.% of unsaturation; **molecular weight,** 350000–400000; **specific gravity,** 0·92. It shows outstanding resistance to heat (up to 200°C), low compression set and low gas permeability, with good compression flexing, tear strength and resistance to oxygen, ozone and chemicals. It can be co-vulcanised with natural or styrene/butadiene rubber. Uses include weather-excluding gaskets and seals in buildings, inner tubes and tyre curing bags.

38.83 POLYISOBUTYLENE

(Poly-2-methylpropene, polyisobutene, PIB. Isolene, Oppanol B, Vistanex.)

This precursor of butyl rubber (for history, *see* §38.7) is prepared by low-temperature cationic polymerisation of isobutylene (as in §38.21, but with the diene omitted); having the following repeat unit,

$$-CH_2-C(CH_3)_2- \qquad C_4H_8, \text{ mol. wt.} = 56$$

it is not vulcanisable by normal means and cannot be converted to a practicable elastomer.

Polyisobutylenes of moderate molecular weight are viscous liquids but the more important polymers (100000–400000 or more in molecular weight*) are colourless, highly extensible, tough rubber-like materials (remaining so over a wide temperature range, but at the same time showing considerable cold flow). Their general behaviour is similar to that of raw butyl rubber (*see* structure, §38.1; solvents, §38.22; identification, §38.23) and some other of their properties are as follows. **Specific gravity** 0·913–0·92. **Refractive index** 1·51. **Water absorption** Practically zero. Outstanding features: (i) *resistance* to ozone, and to chemical and biological attack, (ii) low *permeability* to water vapour and gases, e.g. water vapour, 010 $(10^{-10} \text{ g})/(\text{cm s cmHg})$ at 39·5°C, and (iii) excellent *electrical proper-*

---

*The molecular weight decreases on vigorous agitation of a solution.

ties, namely **volume resistivity,** *c.* 10^{18} ohm cm; **dielectric strength,** 23–24 kV/mm; **dielectric constant,** *c.* 2·3 (50 Hz to 1 MHz); **power factor,** 0·000 15 (1 kHz), 0·000 35 (1 MHz).

Uses Liquid low molecular weight polyisobutylene, used as an impregnant, has been employed to function as a condenser dielectric. The high molecular weight polyisobutylenes are used as inert linings for metal tanks and coatings on balloon fabrics; also in unvulcanised adhesive rubber compositions (e.g. on pressure-sensitive tapes), in caulking and sealing composition, to plasticise and tackify polyethylene (for melt coating and heat sealing applications), as an additive to asphalts and waxes, and in vulcanisable rubber mixes (e.g. to impart heat- and ozone-resistance in cable insulation).

For note on polybutene-1, *see* §2.81.

38.9 FURTHER LITERATURE

GÜTERBOCK, H., *Polyisobutylen und Isobutylen-Mischpolymerisate,* Berlin (1959)
Polymer Corporation Ltd., *Polysar Butyl Handbook*, Sarnia, 1966, p 496.

Specifications
BS 3227 (vulcanised butyl rubbers)
BS 4470 (raw butyl rubber)

POLYSULPHIDE RUBBERS

With notes on Polysulphones

SYNONYMS AND TRADE NAMES

Elastothiomers, thioplasts, GR-P. Carbogum, Elastron, Ethanite, Hikatol, Hydrite, Lastoprene, Novoplas A, Perduren, Resinit, Stamikol, Thiogomme, Thiogutt, Thiokol, Thionite, Thioprene, Thiorubber, Vulcaplas.

GENERAL CHARACTERISTICS

Commercial forms include solid and liquid rubbers, pre-vulcanised moulding powders, and latices. The solid rubbers range from brown to white in colour and some types (e.g. Thiokol A) have an unpleasant mercaptan-like odour. Vulcanised rubbers have poor mechanical properties but excellent solvent resistance. Liquid rubbers are clear amber fluids. Latices have solids contents of over 50% and a particle size of 2–15 μm.

39.1 STRUCTURE

Simplest Fundamental Unit

$$-R-S-S-$$
$$\downarrow \quad \downarrow$$
$$S \quad S$$

where R = an alkylene or related bivalent radical, e.g.

$$-CH_2-CH_2-$$

(*see also* §39.21). Disulphide (—S—S—) groups, lacking the lateral sulphur atoms of the tetrasulphide structure, may also be present.

Molecular weight Solid types: not established but presumed to be very high. Liquid types: 500–7500.

X-ray data Amorphous but some types exhibit crystallinity on stretching. Identity period along fibre axis: polyethylene tetrasulphide, 4·3 Å; polyethylene disulphide, 8·8 Å.

39.2 CHEMISTRY

39.21 PREPARATION

The basic reaction is that of condensation between an aliphatic dihalide and sodium polysulphide (obtained by boiling aq. sodium hydroxide or sodium sulphide with sulphur), the simplest product (e.g. Thiokol A) being derived from ethylene dichloride and sodium tetrasulphide,

$$ClCH_2CH_2Cl + Na_2S_4 \rightarrow \left[-CH_2-CH_2-\underset{\underset{S}{\downarrow}}{S}-\underset{\underset{S}{\downarrow}}{S}- \right] + 2NaCl$$

In commercial production a dihalide is stirred into an excess of an aq. solution of sodium polysulphide at 70°C, in the presence of magnesium hydroxide to assist dispersion and facilitate the reaction. The latex so obtained may be coagulated with acid, washed, dried, and pressed into blocks.

Heating the tetrasulphide polymer with conc. aq. alkali removes the labile sulphur atoms to give a hard, non-rubbery disulphide polymer, which may also be obtained directly by reaction of ethylene dichloride with sodium disulphide or by oxidation of ethylene dimercaptan. The disulphide form can be readily converted to the rubbery tetrasulphide (e.g. on milling with sulphur and a vulcanisation accelerator).

Several types of polysulphide rubbers have been produced commercially, by varying the dihalide, using mixed dihalides, varying the type of sodium polysulphide, or incorporating modifiers to produce either cross-linkage (e.g. 1,2,3-trichloropropane) or control of chain length (e.g. sodium hydrosulphide and sodium sulphide). Liquid polymers are usually based on bis(chloroethyl) formal and incorporate a small proportion of a trifunctional modifier (as above), which renders them cross-linkable. Latices are treated with chain scission promotors which have a softening effect and confer film-forming properties.

Table 39.T1. CHEMICAL BASIS OF SOME POLYSULPHIDE RUBBERS THAT HAVE BEEN PRODUCED COMMERCIALLY

Proprietary name	Dihalide component	Bifunctional group (R- see §39.1)
Thiokol A, Ethanite	Ethylene dichloride	$-(CH_2)_2-$ (I)
Thiokols B and D, Perduren G, Novoplas A	Bis (2-chloroethyl) ether	$-(CH_2)_2O(CH_2)_2-$ (II)
Thiokols F and DX	Bis (2-chloroethyl) ether, ethylene dichloride	Mixed: I and II
Thiokol N, Thiokol latices MX and WD-6	Ethylene dichloride, 1,2-propylene dichloride	Mixed: I and $-CH_2-\underset{\underset{CH_3}{\mid}}{CH}-$
*Thiokols LP-2, LP-3, ST, Perduren H	Bis (2-chloroethyl) formal	$-(CH_2)_2OCH_2O(CH_2)_2-$ (III)
Thiokol FA, Thiokol latex MF, *Thiokol PR-1	Bis (2-chloroethyl) formal, ethylene dichloride	Mixed: I and III
Vulcaplas	1,3-Glycerol dichlorohydrin	$-CH_2CH(OH)CH_2-$

*Contain cross-linking agent

39.22 PROPERTIES

Solubility (unvulcanised polymers). Ethylene dichloride (Thiokol A) type, insoluble; other solid rubbers are very resistant to solvents but may be slightly soluble in chlorinated hydrocarbons. Uncured liquid polymers dissolve in cyclic hydrocarbons (toluene, cyclohexane), chlorinated hydrocarbons (ethylene dichloride, carbon tetrachloride), phenols, dioxan, ketones, esters, and nitro-paraffins. Glycerol dichlorohydrin type (Vulcaplas) soluble in ethylene glycol monoethyl ether (Cellosolve). **Plasticised by** MBT, MBTS, DPG (*see* §29.4 Fabrication), liquid polysulphide rubbers. **Relatively unaffected by** (as vulcanisates) aliphatic hydrocarbons (petrol, oils, greases), water, alcohols, dilute acids, ozone. **Swollen by** aromatic and chlorinated hydrocarbons; the degree of swelling varies with type but is small compared with that of other elastomers.

Table 39.T2. VOLUME SWELLING OF VULCANISATES CONTAINING 60 PARTS (PER 100 OF RUBBER) OF SRF CARBON BLACK, AFTER 1 MONTH AT 25°C

Polymer	Toluene	Carbon tetra-chloride	Methyl ethyl ketone	Dibutyl phthalate	Ethyl alcohol
Thiokol PR-1	45	29	23	5	1
Thiokol FA	55	36	28	7	2
Thiokol ST	79	48	49	8	5

Decomposed by conc. sulphuric, hydrochloric, and nitric acids; strong alkalis remove lateral sulphur atoms from tetrasulphides.

39.23 IDENTIFICATION

Thermoplastic; N, Cl absent, S present. Sulphur content of various types (%): Thiokol A, 84; FA, 47; ST, 37 (cf. Ebonite, 20–35%).

Combustion The rubbers burn fairly readily, with evolution of sulphur dioxide and other pungent products but relatively little soot, and leave a charred residue. **Pyrolysis** Evolve yellow acid fumes and hydrogen sulphide; charred residue. **Other tests** 1. The polymers are almost always dark in colour, and a mercaptan-like odour is often present. The high sulphur content—*see above*—is also characteristic; the polymers are distinguished from ebonite by their physical properties and the Storch-Morawski test (§5.23) in which polysulphide rubbers give a red-violet to red-brown colour (vulcanised natural rubber gives no colour). 2. On decomposition by alkali, and reduction of the disulphide product with sodium sulphite, followed by acidification, a solution is obtained that yields a dithiol compound $R(SH)_2$, and hydrogen sulphide. For estimation of the labile sulphur in tetra-sulphide rubbers, by absorption in sodium sulphite solution and titration of the thiosulphate formed, *see* PARKER, L. F. C., *India Rubb. J.*, **108**, 387 (1945). 3. Polysulphide rubbers are rapidly attacked when gently warmed with a mixture composed of equal volumes of conc. sulphuric and nitric acids (other rubbers react, but less rapidly). 4. Polysulphide rubbers based on a formal yield formaldehyde on warming with 6 N sulphuric acid (*see* chromotropic acid test for formaldehyde, §26.23). 5. Many polysulphide rubbers yield dithian (m.p. 111–112°C) when distilled with hydriodic acid. 6. In the tri-chloroacetic acid test (*see* §29.23), simple unvulcanised polysulphide rubbers impart a pink colour, and produce a bright ruby-red solution on boiling, with a salmon pink precipitate on dilution. *See also* distinction between natural and synthetic isoprene rubbers, 7. The main diagnostic features in the infra-red spectrum of polysulphide rubber pyrolysates are the high aliphatic ether absorption in the region of 9 μm and the relatively low intensity of CH absorptions. 8. For ultra-violet absorption spectra of polysulphide rubbers *see* BAER, J. E. and CARMACK, M., *J. Amer. chem. Soc.*, **71**, 1215 (1949).

39.3 PHYSICS

39.31 GENERAL PROPERTIES

Specific gravity Liquid types, 1·13–1·31; ethylene dichloride (Thiokol A) type, 1·55–1·6. **Refractive index** Liquid types, 1·56–1·57; solid types, 1·6–1·7. **Water absorption** is affected by polymer type (generally decreases as sulphur content increases) and compounding (decreased by blending with other elastomers, e.g. SBR or butyl rubber). All types swell in hot water.

Table 39.T3. VOLUME SWELLING OF VULCANISATES WITH 60 PARTS (PER 100 OF RUBBER) OF SRF CARBON BLACK AFTER IMMERSION IN WATER FOR 1 MONTH

Polysulphide type	Sulphur content (%)	Volume swelling (%) at 25°C	at 60°C
Thiokol PR-1	50	5	26
Thiokol FA	47	5	56
Thiokol ST	37	5	41

Permeability This property, like several others, is related to sulphur content but is generally much lower than that of natural rubber and comparable to that of butyl rubber.

Table 39.T4. PERMEABILITY OF THIOKOL B (UNFILLED VULCANISATE) TO
VARIOUS GASES AT 25°C

Method of expression	Hydrogen	Oxygen	Carbon dioxide
As percentage relative to vulcanised natural rubber	3·2	1·2	2·4
Permeability coefficient, $(10^{-10}\ cm^2)/(s\ cmHg)$	1·6	0·29	3·2

Permeability to water vapour (at 25°C): $c.\ 0·1\ .\ 10^{-10}$ g/(cm s cmHg).
Permeability to solvent vapours (e.g. methyl alcohol, carbon tetrachloride, benzene) is lower than that of nitrile or polychloroprene
rubbers.

39.32 THERMAL PROPERTIES

Specific heat 0·31. **Conductivity** $7·2 \times 10^{-4}$ cal/(cm s °C). **Physical
effects of temperature** Polysulphide rubbers generally are not resistant
to high temperatures, e.g. the ethylene dichloride (Thiokol A) type
softens at $c.$ 70°C and has a brittle point of 7°C. Other types have
rather wider service temperature ranges (e.g. Thiokol FA, −45 to
120°C; ST, −50 to 100°C) with second order transition temperatures
of −40 to −60°C and brittle points of −15 to −50°C.

Curing systems are important in determining resistance to heat
ageing, e.g. p-quinone dioxime causes chain-scission and reversion
at high temperatures; tellurium dioxide and manganese dioxide are
more satisfactory.

Flame resistance Polysulphide rubbers burn rapidly and, despite
their high sulphur content, have sufficiently high heats of combustion
to be used as fuels for military rocket motors. Liquid polymers have
flash points of 180–230°C (open cup method).

39.33 ELECTRICAL PROPERTIES

Solid polysulphide rubbers have limited electrical applications.
Liquid types are more important, having some use in encapsulating
electrical components. The properties of the cured end-products of
both types are similar.

Table 39.T5. PROPERTIES OF TYPICAL COMPOSITIONS (VULCANISED UNLESS STATED OTHERWISE) AT 25°C

Type	Filler	Volume resistivity (ohm cm)	Dielectric constant at 1 kHz	Power factor at 1 kHz
Solid (Thiokol ST)	unfilled unvulcanised	7.10^{12}	8·95	0·004
	SRF black	$8·4.10^6$	—	—
	Calcium carbonate	$1·7.10^{10}$	6·7	0·031
Cured liquid (Thiokol LP-32)	SRF black	$3·3.10^9$	28	0·07
	Calcium carbonate	$4·4.10^{11}$	7·5	0·005

39.34 MECHANICAL PROPERTIES

Uncured unfilled polysulphide rubbers have very low strength and it is essential to use reinforcing fillers, e.g. SRF or FEF carbon blacks; however, even reinforced cured rubbers never attain strength and resilience comparable with natural rubber or the general purpose synthetic rubbers. Compression set and abrasion resistance are also poor.

Typical properties of vulcanisates (containing 50–60 parts per 100 rubber of carbon black, unless otherwise stated) at 20–25°C are as follows:

Property	Solid types	Liquid types (cured)
Tensile strength (kgf/mm^2) (Unfilled)	0·4–1·1 0·14–1·1	0·14–0·6 —
Extension at break (%) (Unfilled)	300–450 450–700	450–950 —
Stress at 300% extension (kgf/mm^2) (Unfilled)	0·35–0·93 0·1–0·4	— —
Hardness (IRHD)	70–78	45–60

Compression set, as percentage of compression applied for 22 h at 70°C (ASTM D395 Method B). Solid types, 37–100. **Abrasion resistance,** as ml/(Lp. h) (Du Pont abrader). Solid types, 600. **Rebound resilience,** percentage (Lüpke pendulum). Solid types, 15–25. For the *dynamic shear properties* of an unfilled polysulphide vulcanisate, *see* FLETCHER, W.P. and GENT, A.N., *J. Sci. Instrum.*, **29**, 186 (1952); *Brit. J. Appl. Phys.*, **8**, 194 (1957).

39.4 FABRICATION

Solid polysulphide rubbers can be processed on conventional machinery although they are rather less tractable than natural rubber or general purpose synthetic rubbers, and they often give off an offensive odour. Breakdown by milling or softening, e.g. with MBTS or liquid polysulphides, is often necessary before proceeding to extrusion and other fabrication methods. The rubbers are vulcanised with zinc oxide and a small proportion of sulphur, but the formal types (Thiokol ST) require an oxidising system, e.g. *p*-quinone dioxime. Because of low intrinsic mechanical properties it is essential to use a reinforcing filler (commonly SRF carbon black) but processing difficulties limit the range of types that may be employed; however, blending with small proportions of natural rubber, polychloroprene, or nitrile rubber improves processing characteristics without seriously impairing solvent resistance. Tackifiers such as natural rubber or coumarone resins are essential when stocks are to be frictioned on to fabric. The rubbers can be bonded to brass-plated metal during vulcanisation.

The liquid polymers are converted into solid rubbers by addition of oxidising agents, e.g. lead dioxide, cumene hydroperoxide, or *p*-quinone dioxime; curing generally takes place in one or two days at room temperature, although heat is sometimes required. The fillers used are as for the solid polymers but non-black fillers, such as silica or titanium dioxide, are common.

Aq. dispersions can be used in much the same way as other rubber latices but the particles are uncharged and can be easily re-dispersed after settling out. Polysulphide rubbers have also been produced in powder form suitable for moulding or for application as protective coatings by a fluidised bed technique or flame-spraying.

39.5 SERVICEABILITY

Having a saturated structure, all polysulphide rubbers are extremely resistant to ozone attack and outdoor exposure; the high-sulphur types perform especially well and are comparable to butyl rubber in this respect. Heat resistance and low temperature behaviour are relatively poor, and in mechanical properties generally polysulphides are among the poorest of rubbers. The outstanding advantages are exceptional resistance to oils and solvents, and low permeability to gases. Unless specially compounded, polysulphides, like other rubbers, may be attacked by micro-organisms and insects. High energy radiation appears to cleave disulphide and polysulphide linkages, but the physical results vary: some polysulphide rubbers

442

undergo progressive softening almost to the liquid state, others show marked increases in hardness and modulus, and ultimately become brittle.

39.6 UTILISATION

These have been largely determined by, on the one hand, outstanding solvent- and oil-resistance and, on the other, poor strength and handling characteristics; the development of other oil-resistant rubbers with better mechanical properties has also been a limiting factor. Thus, whereas in the years immediately following their introduction, and during the Second World War, many different grades of solid polysulphide rubbers were tried in most of the applications pertaining to natural rubber, many of these have since been discontinued and a more limited range of uses remains. These include linings for oil and paint hoses, printing rolls, moulded seals, coated fabrics, and some cable coverings. Liquid polymers are extensively used in sealants for building, civil engineering, and boat-building; also for cast printing plates and rollers, impregnated leather gaskets, paints and adhesives, flexible moulds, and cast solid fuel units for missiles. Liquid polymers can also be used to impart flexibility and toughness to thermosetting resins, e.g. in pheno-formaldehyde adhesives and epoxy resin casting compounds. Certain liquid types (e.g. Thiokol VA-7, which is structurally related to Thiokol B) are used as vulcanising agents for other sulphur-cured rubbers. Latices, sometimes blended with those of PVC or polyvinylidene chloride, have found use as protective coatings and linings, e.g. in concrete fuel-storage tanks. Flame-sprayed polysulphide powders have been used to form anti-cavitation coatings on external propeller shaft casings and rudder struts of ships.

39.7 HISTORY

The polysulphides were among the earliest synthetic rubbers to be developed. In 1839 C. Löwig and S. Weidmann reacted 1,2-dichloro-ethane with potassium pentasulphide to obtain a rubbery material; similar results were reported by J. M. Crafts, O. Masson, V. Meyer and other nineteenth-century experimenters. Patents covering the industrial application of thioplasts were taken cut in 1926 by J. Baer (in Switzerland) and in 1927 by J. C. Patrick (U.S.A.). Subsequently, both the Thiokol Corpn. in U.S.A. and IG Farbenindustrie in Germany brought out commercial polymers, the first American material (Thiokol A) appearing in 1930. Commercial forms also appeared before the Second World War or in its early years, in Russia, U.K. (Dunlop), Belgium, Holland, France, Italy, Canada, and Japan.

During this period a polysulphide based on Thiokol N was re-designated GR-P and produced, under the synthetic rubber production programme of the U.S. Government, at a rate of about 2000 tons per year, mainly for use as retreading material (from 1938 to 1949 all Thiokols were manufactured by Dow Chemicals for Thiokol Corpn.). Production in Germany during the war is believed to have reached 10 000 tons per year.

Liquid polymers, introduced in 1943, have since been extensively developed, a plant being opened for this purpose by the Thiokol Corpn. in 1955.

39.8 ADDITIONAL NOTES

39.81 POLYSULPHONES

Like polysulphides in having sulphur in the main chain, but differing in other characteristics, polysulphones are represented by the following repeat unit.

$$-R-\underset{\underset{O}{\downarrow}}{\overset{\overset{O}{\uparrow}}{S}}-$$

where R = either an alkene or arylene type of structure (*see* complex example below).

Polymers of this type can be obtained in various ways, chiefly by (i) a radical-initiated reaction of an olefin or related monomer (e.g. propylene or styrene) with sulphur dioxide, which at moderate or low temperature yields an alternating hydrocarbon/SO$_2$ copolymer, or (ii) self-condensation of a phenylene-type monomer having a chlorosulphonyl (—SO$_2$Cl) group, or—as industrially developed (Union Carbide Corp., 1965)—condensation of a bisphenol with a dichloroarylene sulphone, e.g. diphenylolpropane (*see* §25.21) with 4,4'-dichlorodiphenyl sulphone.

Products of the last kind are in the same class as polycarbonates (Section 24C) and polyphenylene oxide (§26.83), being tough, heat-resistant, high-strength thermoplastics serviceable to at least 150°C. They are soluble in polar or chlorinated solvents (and dissolve or swell in aromatic hydrocarbons) and are subject to environmental stress-cracking in some alcohols and aliphatic hydrocarbons, but they resist oxidation and exhibit thermoplasticity only at high temperatures, e.g. while some polysulphones soften at 200–300°C, the more complex kinds are moulded or extruded at *c.* 350°C and decompose only above 500°C.

Further data Generally amorphous in structure, transparent, and light amber in colour; **specific gravity,** 1·24–1·25; **refractive index,** 1·63; **water absorption** (disc, 24 h), 0·2%; **combustion,** self-extinguishing; **glass temperature,** 190°C; **volume resistivity,** $> 10^{16}$ ohm cm (10^{13} at 175°C); **dielectric constant,** *c.* 3; **power factor,** up to 0·003 (up to 0·006 at 175°C); **elastic modulus,** Young's, 250 kgf/mm²; **tensile strength** at yield, 7 kgf/mm²; **extension at break,** 50–100%, with yield at 5–7%; **impact strength** (Izod), 1·3 ft lbf/in of notch (1·2 at −40°C); **creep under load,** very low (e.g. at room temperature at 2 kgf/mm², under 1% p.a.).

Applications Engineering uses; tough, heat-resistant parts (mouldings, extrusions, also films; can be melt-bonded to steel) including electrical components.

39.9 FURTHER LITERATURE

GAYLORD, N. G. (Ed), *Polyethers: Part 3 Polyalkylene Sulfides and other Polythioethers* (*High Polymers,* Vol. XIII), New York and London (1962)

BARRON, H., *Modern Synthetic Rubbers* (Chapter 19), London (1949)

PENN, W. S., *Synthetic Rubber Technology* (Volume I, chapters 35 and 36), London (1960) *Polyarylene sulphones:*

ROSE, J. B., *Chem. and Ind.,* 461 (1968)

POLYURETHANES

SYNONYMS AND TRADE NAMES

Polyester or polyether intermediates Caradol, Daltocel, Daltoflex, Daltolac, Desmophen, Multron, Niax, Teracol.
Isocyanate intermediates* Caradate, Desmodur, Hylene, Mondur, Suprasec.
Rubbers Adiprene (B, C, L, LD), Chemigum SL (= Neothane) and XSL, Duthane, Elastothane, Estane, Genthane S, Prescollan, Solthane, Texin, Vulcaprene A, Vulkollan.
Fibres (a) Perlon U. (b) Elastomeric (spandex) type: Dorlastan, Elura, Glospan, Lycra, Numa, Spanzelle, Vyrene.

GENERAL CHARACTERISTICS

A wide class of materials, ranging from durable oil-resistant rubbers and fibres (either relatively inextensible, or extensible and elastic) to surface coatings, adhesives, and flexible or rigid foams. They are *not* direct polymers of urethanes, but are derived from the reaction of polyesters or polyethers with di- or poly-isocyanates, to produce complex structures containing urethane linkages.

40.1 STRUCTURE

Simplest Fundamental Unit
There are two classes of polymers containing urethane

$$\left(-\underset{\underset{H}{|}}{N}-\underset{\underset{O}{\|}}{C}-O-\right)$$

groups.

I. Linear polymers—Of these, now relatively unimportant, the best known is a nylon-like substance (Perlon U), the repeat unit of which is represented by

$$-CONH(CH_2)_6NHCO \cdot O(CH_2)_4O-$$

$$C_{12}H_{22}O_4N_2, \text{ mol. wt.} = 258$$

II. Complex polymers—The major component of these is a prepolymer, composed of a linear polyester or polyether that has been extended several-fold in chain length by coupling through urethane linkages. Further treatment of the prepolymer (denoted by —[PP]—,

---

*Common abbreviations for di-isocyanates are as follows (d.i. = di-isocyanate): DADI, dianisidine d.i. (3,3'-dimethoxy-4,4'-biphenyl d.i.); HDI, 1,6-hexamethylene d.i.; MDI, 4,4'-diphenylmethane d.i.; NDI, 1,5-naphthalene d.i.; PDI, phenylene d.i.; TODI, tolidine d.i. (3,3'-dimethyl 4,4'-biphenyl d.i.); TDI, mixed 2,4- and 2,6-tolylene d.i.; XDI, xenylene d.i. (4,4'-biphenyl d.i.). TDI is the most used.

see also §40.21II) can further extend the length of its molecules, and can make it at intervals cross-linked (denoted by —(X)—; hence an

ultimate structure can be represented by:

$$—[PP]—(X)—[PP]'—(X)—$$
$$[PP]''$$

A denser cross-linked structure is obtained if a *branched* prepolymer is used.

Molecular weight Linear polymers (Perlon U), *c.* 10000. In complex structures the molecular weight of the initial polyester or polyether may range from 500–6000; but the value rises several-fold by extension through urethane linkages, and in the final network is indefinitely large.

X-ray data The earlier linear polymers were crystalline; identity period (Perlon U monofil), 18·8 Å. Diagrams from rubbers and foams are random and diffuse, although not completely amorphous in type: the pattern from elastomeric fibres is similar, but when highly stretched they may show orientation and partial crystallinity.

40.2 CHEMISTRY

40.21 PREPARATION

Basic reactions
Production of polyurethanes depends on the reactivity of isocyanate (—NCO) groups, compounds containing which are obtained by condensation of primary amines with phosgene, i.e. $RNH_2 + COCl_2 \rightarrow RNCO + 2HCl$.

The isocyanate group readily reacts (catalysed especially by bases) with hydroxyl (—OH) or amino (—NH$_2$) groups:

(i) —NCO + —OH → —NHCO·O—   (urethane linkage)

(ii) —NCO + —NH$_2$ → —NHCO·NH—   (urea linkage)

reaction (ii) being particularly rapid.

In the presence of excess isocyanate the above products undergo further reaction, yielding branched structures that lead to cross-linkage (the —NHCO— group, *vide infra*, reacts similarly):

$$—NCO·O—$$
(iii) —NHCO·O—   + —NCO → | (allophanate linkage)
$$CONH—$$

(iv) $-NHCO \cdot NH- + -NCO \rightarrow$
$$\begin{array}{c} -NCO \cdot NH- \\ | \\ CONH- \end{array}$$
(biuret linkage)

(v) $-NHCO- \quad + -NCO \rightarrow$
$$\begin{array}{c} -NCO- \\ | \\ CONH- \end{array}$$
(acylurea linkage)

It is to be noted that reactions (i)–(v) involve transfer of a H atom to the N of the isocyanate without elimination of volatile matter. However, with the hydroxyl groups of water or carboxylic acids reaction (i) leads to unstable products, which undergo decomposition to amines or amides with evolution of carbon dioxide:

(vi) $-NCO + H_2O \quad \rightarrow [-NHCO \cdot OH] \quad \rightarrow -NH_2 \quad + CO_2$

(vii) $-NCO + -COOH \rightarrow [-NHCO \cdot OCO-] \rightarrow -NHCO- + CO_2$

Production of Polymers

I. Linear polymers

The best known of these (Perlon U) derives from direct reaction of 1,6-hexamethylene di-isocyanate with 1,4-butanediol in an inert solvent such as *o*-dichlorobenzene:

$$OCN(CH_2)_6NCO + HO(CH_2)_4OH \rightarrow$$
$$-CONH(CH_2)_6NHCO \cdot O(CH_2)_4O-$$

Alternatively, simple linear polymers are obtained when a diamine reacts with a bischloroformate (*see also* §54.13v).

II. Complex polymers

The structure of these materials is indicated under §40.1 II. A first step in their production is the preparation of a 'segmented' prepolymer, by coupling a hydroxyl-terminated polyester or polyether with a polyfunctional isocyanate; this extends the molecular size through urethane linkages (as in (i)). A typical polyester is a poly-(diethyleneglycol) adipate having a molecular weight 2000–3000; a typical polyether is obtained by polymerisation of propylene oxide, usually in the presence of a small proportion of glycerol or sorbitol to provide branched structures. The isocyanate is commonly a difunctional one, such as TDI (*see* footnote p. 446) which is chosen for its reactivity, cheapness and relatively low volatility (hence low toxicity).

The resulting prepolymer has —OH terminal groups when a deficiency of isocyanate is used, but —NCO terminal groups if an excess is employed, e.g.:

$$(n-1) \text{ R(NCO)}_2 + n\text{HO[PE]OH} \rightarrow$$

$$\text{HO[PE]O}\left[\text{CONHRNHCO·O[PE]O}\right]_{n-2} \text{CONHRNHCO·O[PE]OH}$$

or

$$(n+1) \text{ R(NCO)}_2 + n\text{HO[PE]OH} \rightarrow$$

$$\text{OCNRNHCO·O[PE]O}\left[\text{CONHRNHCO·O[PE]O}\right]_{n-1} \text{CONHRNCO}$$

where HO[PE]OH represents the initial linear polyester or polyether, and the urethane linkages are shown as —NHCO·O— (or —O·CONH—).

Prepolymers of the above kind, of which the formula may be abbreviated to HO[PP]OH or OCN[PP]NCO, also those of the non-linear (branched) kind, provide intermediates for the cross-linked products described below.

Rubbers

There are three main types:

(1) *Millable rubbers* (first to be developed, and processable on conventional rubber machinery). These are based on isocyanate-terminated ester- or ether-type prepolymers that can be chain-extended by reaction with water, glycols, amines or amine-alcohols. Cross-linkage (vulcanisation) is effected through the free isocyanate groups by incorporation of catalysts (e.g. *tertiary* amines) and heating. Rubbers with better storage stability contain no free isocyanate groups but are made vulcanisable by addition of di-isocyanates during milling; aliphatic amino-groups included in a prepolymer enhance the structural irregularity, elasticity and reactivity. *Vulcaprene A*, an early storage-stable polyesteramide rubber of this kind (based on glycol, ethanolamine, adipic acid and a di-isocyanate) was curable by isocyanates, formaldehyde derivatives, or chromates. *Adiprene C*, a polyether-urethane based on poly(1,4-oxybutylene) glycol and TDI, can be vulcanised with peroxides or sulphur.

(2) *Liquid casting rubbers,* usually prepared in the following sequence: (a) prepolymer, (b) mixing with chain-extender in liquid state, (c) chain-extension, with some cross-linkage, in a mould, (d) further cross-linkage by heating after removal from mould. The prepolymers are similar to those used in (1), but castor oil or polypropylene glycol-TDI types have also been developed.

(3) *Thermoplastic processing rubbers* (processable on plastics equipment). There are two types, e.g. (a) *Estane 5740 XI* (based on adipic acid, 1,4-butanediol and MDI), which although thermoplastic is virtually cross-linked—needing no vulcanising additives—and has good extensibility and elasticity, and (b) *Texin,* heat-cured rubbers, prepared from hydroxy-terminated polyesters, MDI, and a diol, giving materials that undergo allophanate-type cross-linkage during processing or post-moulding treatment.

ELASTOMERIC THREADS

These may be obtained from an isocyanate-ended prepolymer—e.g. derived from poly(ethylene/propylene adipate) and MDI—which is extended by reaction with an equivalent amount of a dihydroxy compound—e.g. 1,4-butanediol—to give a product, composed of very long flexible chains, that can be extruded in solution (e.g. in dimethyl-formamide). Alternatively, a solution of a prepolymer is either (a) treated with a diamine (e.g. ethylenediamine) and dry-spun before reaction takes place, or (b) wet-spun into a coagulating bath containing a diamine, followed by heating to complete the polymerisation. Considerable variation is obviously possible; while some fibrous elastomers appear to be composed of linear but hydrogen-bonded molecules, others are almost certainly covalently cross-linked.

FOAMS

In the preparation of polyurethane foams a polyfunctional isocyanate is made to react with a prepolymer and with water, yielding respectively a rubber and carbon dioxide sufficient to effect the expansion (*see* vi). Originally it was necessary to make a viscous isocyanate-ended prepolymer, as described earlier, and then to react it with a small proportion of water in the presence of an amine activator; however, polyethers made from propylene oxide, for which the reaction can be catalyst-activated, have led to a 'one-shot' method—needing no prepolymer—where both the polymerisation and the evolution of gas occur simultaneously at the correct relative rates.

Flexible foams produced by this technique are usually made from TDI and a branched polyether of mol. wt. 2000–5000. Polymerisation is catalysed by an organo-metallic compound (e.g. stannous 2-ethylhexoate) and a tertiary amine catalyst (e.g. triethylenediamine, or 1,4-diaza[2,2,2]-bicyclo-octane, often abbreviated to DABCO), which acts at room temperature and also promotes the evolution of carbon dioxide. A surface-active agent (e.g. a water-soluble silicone derivative) to promote production of a uniformly-textured foam is also included, together with—if required—an auxiliary volatile blowing agent (e.g. trichlorofluoromethane).

Rigid foams require more cross-linkage than the flexible kinds, and consequently are made from branched polyols of relatively low molecular weight (*c*. 500). In a 'one-shot' two-component system, a branched polyether, containing a volatile blowing agent, is treated with a di-isocyanate such as MDI (which yields stiffer products than TDI), and the heat of the reaction effects the 'rise' of the mixture.

Flexibility or rigidity depends upon both the chemical and the physical structure; in flexible foams the cells are open and inter-connected, while rigid foams have mostly closed cells.

COATINGS AND ADHESIVES

(1) *Two-component systems* For one form of coating, a solution of a hydroxyl-terminated polyester (e.g. in an inert ester or ketone) is mixed with one of a non-volatile di-isocyanate just before application. The hardness of the ultimate insoluble films depends on the complexity of their composition and the degree of branching; rubbery prepoly-mers, solvent-based and curable with di-isocyanates at room tempera-ture or by heat, adhere well to metals and textile fabrics. Alternatively, isocyanate-ended polymers can be cured by intermediates of epoxy resins containing hydroxyl groups (§25.21) or fatty acid polyamides (§23.21.6).

(2) *One-component systems* A branched isocyanate-ended polymer, applied from solution, will cure to a hard coating by reaction with atmospheric moisture (see basic reactions vi and ii); thus, castor oil (largely glycerol ricinoleate, with three —OH groups per molecule), when treated with an excess of a di-isocyanate, becomes a drying oil. Any toxicity hazards associated with isocyanate-curing stoving enamels and lacquers are reduced by using involatile adducts (e.g. those with phenol, which react with OH— groups only above 150°C, or with 1-pentanol, which reacts with OH— and NH_2— groups only when baked), the isocyanate being consumed in the curing reaction as soon as liberated. This technique also permits the formulation of one-can stoving lacquers, i.e. comprising two potentially reactive components in one solution; adducts of isocyanate-ended pre-polymers, reactive only at an elevated temperature, have also been used as adhesives for nylon*. A different kind of coating medium, known as a *urethane oil* or *urethane alkyd*, is made by treating a drying oil (a poly-unsaturated glyceride, such as linseed oil) with glycerol, to yield a mixture of mono- and di-glycerides, an excess of which is then caused to react with a di-isocyanate; this gives a product that is isocyanate-free but highly unsaturated, thus it can dry rapidly by atmospheric oxidation. (For drying oils and 'drying', *see also* §24A.81.)

---

*The adhesion of rubbers to polar surfaces (metals, textiles) can be enhanced by simple treat-ment with 4,4′, 4″-triphenylmethane tri-isocyanate, which probably reacts with surface layers of hydroxide or moisture.

40.22 PROPERTIES

Solvents The preparation of prepolymers from polyols and iso-cyanates is carried out in inert media (esters, ketones, or chlorinated aromatic hydrocarbons, with aromatic hydrocarbons as diluents). Linear and lightly cross-linked products dissolve in tetrahydrofuran, pyridine, dimethylformamide, dimethyl sulphoxide, formic acid, or phenols (suitable diluents: methylene chloride, acetone, toluene, alcohol). However, many polyurethanes dissolve (or disintegrate) only with decomposition. For example, ester-type polyurethanes are attacked by alcoholic alkali, hot aq. alkali or moderately conc. mineral acids; ether-type polyurethanes are more resistant but tend to disperse in hot alcoholic alkali, or in 95% sulphuric acid. **Plasticised by** phthalates and long-chain fatty acid esters, sulphonamides, and some synthetic rubbers (for polyurethane rubbers); but, in general, hardness can be regulated by composition, hence additional plasticisers are employed only where very high flexibility is required. Castor oil softens polyurethane surface coatings (and improves their resistance to water). Cellulose acetate butyrate improves flow properties in foam manufacture. **Swollen by** aromatic hydrocarbons (rubbers, up to 100–200% increase in volume swelling in benzene), chlorinated hydrocarbons (but usually undamaged by dry cleaning), ketones, esters; some products swell (without dissolution) in formic acid or phenols. **Relatively unaffected by** water (but see below), aliphatic hydrocarbons (petrol, mineral oils), vegetable oils, dry-cleaning solvents (may cause temporary swelling), dil. alkalis, oxygen, ozone. **Decomposed by** (*see also* under Solvents, above) prolonged contact with water, dil. acids, or moist heat (causes swelling and slow hydrolysis, particularly in some ester-type polyurethanes, *see also* footnote to Table 40.T1 on p. 455); chlorine bleach solutions (may cause yellowing and decomposition); prolonged exposure to light (discoloration of derivatives of aromatic isocyanates).

40.23 IDENTIFICATION

N present, sometimes in small proportion. Cl, S, P nominally absent.
Combustion Softens, may melt (drip), chars, burns; combustion may cease when unassisted, but foams not containing retardants continue burning, with candle-like flame. **Pyrolysis** Melts, chars, evolves mildly acid or alkaline vapours with sharp but non-characteristic odour (isocyanates—*toxic*, inhalation dangerous—amines and olefins are among the products). Polyurethanes derived from adipate esters respond to the ONB colour test (§23.23). **Other tests** (1) *p*-Dimethylaminobenzaldehyde, applied in solution or as a powder to a sample dissolved in or swollen with glacial acetic acid, produces a

bright yellow colour. (2) In conc. nitric acid at room temperature, elastomeric threads turn pink to dark red in colour, followed by dispersion (cf. no immediate change in natural rubber threads). (3) Colours are obtained if the product from hydrolysis in alcoholic acid is diazotised and coupled with alkaline β-naphthol (see test for aniline, §27.83). (4) Distinction between ester- and ether-type polyurethanes (e.g. as foams): (i) Ester-types dissolve or disintegrate when boiled with 2N sodium hydroxide; ether-types are inert. (ii) Add to a small sample, dissolved or suspended in 2N alcoholic sodium hydroxide, a few drops of saturated alcoholic hydroxylamine hydrochloride; warm gently, and after 1 min acidify with dil. hydrochloric acid, then add 1 drop of 1% ferric chloride; red to violet colours result from ester-type polyurethanes, ether types are inert. (iii) Ester-types derived from adipic acid respond to the ONB test (see **Pyrolysis**, above); ether types produce only a yellow colour. Various di-isocyanates yield colours with 4-nitrobenzene diazofluoroborate (see OSTROMOW, H., Adhasion, **8**, 453 (1964). (iv) Ether-type foams usually feel more 'greasy' than the ester types.

Infra-red spectroscopy can assist in identification of polyurethanes*, as can hydrolysis followed by paper chromatography, and pyrolysis with gas-liquid chromatography (See CORISH, P. J., Anal. Chem., **31**, 1298 (1959); SCHRÖDER, ELIZ., Plaste u. Kaut., **9**, 121, 186 (1962), **10**, 25 (1963); HASLAM, J. and WILLIS, H. A., Identification and Analysis of Plastics, London (1965).

40.3 PHYSICS

40.31 GENERAL PROPERTIES

Specific gravity Linear polymers (Perlon U), 1·2. Rubbers, 1·05–1·31. Elastomeric fibres, 1·1–1·28. Flexible foams, according to application, 0·015–0·08 (15–80 kg/m^3 or 1–5 lb/ft^3), commonly 0·03 (30 kg/m^3 or 1·9 lb/ft^3). Rigid foams, commonly 0·030–0·032 (30–32 kg/m^3 or 2 lb/ft^3), high density type up to 0·13 (130 kg/m^3 or 8 lb/ft^3). **Refractive index** Of the order of 1·50–1·55. **Water absorption** (%): Rubbers, dependent on compounding, 1–5. Rigid foams (largely closed cells), low, e.g. 1–3 after one month's immersion; flexible foams (open cells), spongelike. **Water retention** (%) Elastomeric yarns, 8–11. **Moisture regain** (%, 65% r.h.) Linear polymers (Perlon U monofil), 1·2. Elastomeric fibres, 1·2–1·3. **Permeability** Rubbers Low per-

---

*For instance, pyrolysates from polyurethane rubbers show a strong broad band at 8·2 μm; those from polyesters usually show also a band at 8·5 μm (sometimes stronger than the absorption band at 8·2 μm) and a medium-strong broad band at 9·4 μm; see BS 4181, Addendum.

meability to gases and to petrol vapour. Typical values at room temperature $(10^{-10}$ $cm^2)/(s\ cmHg)$:

Type of polyurethane	N_2	O_2	CO_2
Polyester	0·027	—	0·4
Polyether (Adiprene)	0·49	1·5–4·8	14–40
Polyesteramide (Vulcaprene)	0·49	—	—

Permeability to water vapour $(10^{-10}$ $g)/(cm\ s\ cmHg)$ 0·3–10.

Films In general, low; moisture-cured urethane alkyds (§40.21 Coatings) show relatively high permeability to water vapour. *Foams* Rigid kinds, low permeability to gases and vapours; flexible kinds (open cells), high permeability.

40.32 THERMAL PROPERTIES

Specific heat Polyurethane rubber, 0·45. **Conductivity** $(10^{-4}$ cal)/ (cm s °C) Rubbers (Adiprene), 2·8–4·2; rigid foams (largely closed cells), 0·8–0·85; do. but blown with trichlorofluoromethane, 0·4–0·55; flexible foam (sp. gr. 0·03, 30°C), 0·85. Foams used for textile laminates contain approximately 98 % air. **Coefficient of linear expansion** (/°C) Rubbers, *c.* 2.10^{-4}; rigid foam, $0·5–1.10^{-4}$. **Physical effects of temperature** Rubbers are serviceable down to -30 or $-35°C$; but some show no embrittlement until -60 or $-70°C$ (although all increase in stiffness, e.g. modulus of rigidity approximately doubled at $-25°C$); also up to 80–100°C, but with declining tear strength and abrasion resistance; the service limit in mineral oil is higher (110–120°C) than in air. Flexible foams do not embrittle at $-50°C$, but decline in rate of recovery from deformation; they are serviceable up to 110–150°C (ester types usually superior to ethers). Rigid foam can be used at 110–120°C, though with reduction in load-bearing capacity. Linear polymer (Perlon U monofil), softens *c.* 170°C then melts. Elastomeric fibres commonly soften (become sticky) at 175–200°C, melt at 210–260°C, thereafter may suffer oxidative loss, ultimately with decomposition (above 300°C) and depolymerisation (400°C). Fabrics containing these threads need to be stabilised or set, e.g. baked for a few minutes at 150°C, but oversetting with excessive loss of elastic modulus and recovery must be avoided. All polyurethanes undergo slow degradation at *c.* 200°C (isocyanates—toxic —being released) and further decomposition at 200–300°C (amines and olefins).

40.33 ELECTRICAL PROPERTIES (polyurethane rubbers)

Volume resistivity Of the order of $10^{10}–10^{12}$ ohm cm $(10^9–10^{11}$ at high humidity). *Surface resistivity* 10^9–10^{11} ohm. **Dielectric strength** about 20 kV/mm. **Dielectric constant and Power factor** (at 1 MHz) *c.* 6–7, and 0·006–0·08 respectively.

40.34 MECHANICAL PROPERTIES

Elastic modulus (Young's) (kgf/mm^2) Linear polymer (Perlon U monofilament), 275 (25 gf/den.). Elastomeric threads, initial modulus greater than that of threads of natural rubber, e.g. *c.* 1 (0·1 gf/den.). Stress (kgf/mm^2) at given deformation: Rubbers, at 20% extension, 0·1–1·5; at 300–500%, 0·5–2·5. Flexible foams, at 40% compression, 0·02–0·05; ester types up to 0·1. **Elastic recovery** Linear polymer (Perlon U monofilament), *c.* 70% from 10% extension. Recovery of elastomeric threads is less rapid and complete than with those of natural rubber (e.g. 10% increase in length after 500% extension) but original length is slowly regained. Flexible foams, *compressed* by 50–75% for 24 h at 70°C, recovery in 30 min at room temperature to within 5–10% of initial thickness; fatigue resistance in repeated cycling is low. Rigid foams, elastic up to *c.* 5% compression (crush beyond 10%).

Tensile strength and Extension at break

	Tensile strength (kgf/mm^2)	Extension at break (%)
Linear polymer (Perlon U):		
monofilaments	24 (2·2 gf/den.)	50
fibres	up to 80 (7·5 gf/den.)	*c.* 20
Elastomeric fibres	*c.* 5–9 (0·5–0·8 gf/den.)	up to 600
Rubbers	2–5*	*c.* 400–700
Flexible foams	*c.* 0·015–0·02	up to 500

*Some reduced in strength if stored under humid conditions

Flexural strength Rigid foams, 0·035–0·04 kgf/mm^2. Compression stress at 10% compression: rigid foams, 0·015–0·035 kgf/mm^2; stiffness is greater parallel to, than across, direction of foam rise. **Hardness, Friction and Abrasion** Rubbers available soft to horn-like (Shore A 60–98); marked hysteresis in dynamic stressing causes good damping. Tear strength is high, and abrasion resistance is at least twice that of natural rubber. Coefficient of friction: rubber/steel, 0·15–0·4 (much reduced by lubrication); do./concrete, *c.* 1 (independent of lubrication).

40.4 FABRICATION

Rubbers may be extruded, compression- or injection-moulded, or spread-, spray-, or dip-coated from solution; vulcanisable (optional) with peroxides, isocyanates, epoxy resins and amines. Generally no reinforcing fillers are required. Flexible foam, cut by bandknife or hot wire, can be laminated to textiles, etc. by adhesives or by flame fusion; production of rigid foam *in situ* offers advantages in filling large moulds and cavities without leaving air gaps. For spinning of elastomeric threads, *see* §40.21 II.

40.5 SERVICEABILITY

Polyurethanes resist degradation by micro-organisms (mildew, etc.); they can also be sterilised, and (with care) dry-cleaned; they are oil-, petrol- and ozone-resistant. The rubbers can have a high modulus and are superior to most others in abrasion and tear resistance. In general, ether-type polyurethanes are more inert than the ester types, e.g. to slow hydrolysis in hot water or steam. They are relatively resistant to high energy radiation. Serviceable at temperatures down to $-30°C$ or up to 100 or 150°C (maximum *c.* 80°C under dynamic conditions), though respectively with some stiffening or some decline in mechanical properties. The foams can be used at higher temperatures than those of expanded thermoplastics. Coatings and surfaces of foams may yellow on exposure to light or elevated temperatures but not necessarily with further detrimental effects. Elastomeric fibres resist oxidation and the effects of sunlight more than those of natural rubber.

40.6 UTILISATION

Rubbers Soft printing rollers, vibration supports, buffers and bumpers, oil seals, footwear components; abrasion resistant solid tyres, driving belts, hard rollers; constructional material for stamping dies (for sheet metal); sprayable abrasion-resistant coatings for metal, textiles, etc. *Elastomeric fibres* Light-weight stretch fabrics (foundation garments, swimwear, etc.). *Flexible foams* Padding, as employed in upholstery, bedding, garments, crash panels (in cars), drop mats (for unloading goods), and carpet underlays; synthetic sponges; thermal insulation, including textile laminates; acoustic insulation, especially at high frequencies; vibration absorption. *Rigid foams* Packaging, light-weight sandwich-type constructional panels; acoustic and thermal insulation (*in situ* cavity filling eliminates convection currents). *Coatings* Abrasion-resistant thermal insulation, including textile laminates; acoustic insulation, especially at high frequencies; vibration absorption.

40.7 HISTORY

In the nineteenth century, isocyanates received much attention from chemists such as A. Wurtz (discovered the reaction with hydroxy- and amino-compounds, 1848), A. W. Hofmann, and W. Hentschel (discovered high yield route through phosgenation of primary amines, 1884); but they did not become commercially important until 1937, when O. Bayer applied them, in Germany, to the synthesis of polymeric materials (e.g. Perlon U fibres, 1941). This was extended to rubbers, adhesives, and rigid foams in 1940–45, and later further developments took place in Germany, the U.K. and the U.S.A.; e.g. Vulcaprene rubbers and coatings in 1949, Valkollan rubbers in 1950, and flexible foams in 1952 (1956 in U.K.). Relatively inexpensive ether-type intermediates appeared in 1957, and the first elastomeric fibre (Lycra) was introduced in 1958.

40.9 FURTHER LITERATURE

SAUNDERS, J. H. and FRISCH, K. C., *Polyurethanes. Chemistry and Technology (High Polymers,* Vol. XVI), New York and London (1962)

GAYLORD, N. G. (Ed), *Polyethers (High Polymers,* Vol. XIII, Part I, pp. 315–408, FRISCH, K. C. and DAVIS, S., polyurethanes from polyalkylene oxides), New York and London (1963)

PHILLIPS, L. N. and PARKER, D. B. V., *Polyurethanes. Chemistry, Technology and Properties,* London (1964)

HEALY, T. T. (Ed), *Polyurethane Foams,* London (1964)

BUIST, J. M. and GUDGEON, H. (Eds), *Advances in Polyurethane Technology,* London (1968)

WRIGHT, P. and CUMMINGS, A. P. C., *Solid Polyurethane Elastomers,* London (1969)

Specifications
BS 3379 (polyether type flexible foam components)
BS 3667 (indentation hardness, flexible foams)
BS 4021 (foam for laminates)
BS 4443 (test methods for flexible foams)
ASTM 2341 (specification for rigid foams)
ASTM 2406 (test methods for flexible foams)

SILICONES
Rubbers and Resins

With notes on fluids, dispersions, organochlorosilanes, aminosilanes, silicon esters and 'bouncing putty'

SYNONYMS AND TRADE NAMES

Organo-silicon oxide polymers, Polysiloxanes. *Rubbers* Adrub, LS-53, LS-63, NSR, Polysil, Silastic, Silastomer, Silcoset, Silicol, Sil-O-Flex, SKT (and variants denoted by extra letters); DP, E, K, KW, MS, S, SE and W followed by numbers. *Resins* Dri-film, Sylgard; DC, DP, R, SR followed by numbers.

GENERAL CHARACTERISTICS

Raw silicone rubbers are transparent, colourless, 'limp', virtually fluid materials (viscosity 10^7 to 4×10^7 cSt), not generally miscible with other rubbers. They are marketed (*a*) as the raw 'gums', (*b*) compounded mixes containing fillers, vulcanising agents, etc. (*see* §41.4), (*c*) solutions, and (*d*) spreading pastes, based on low molecular weight 'gums'.

Vulcanisates vary little in properties over a wide temperature range, and have exceptional resistance to deterioration during prolonged heating and to oxidation and ozone attack. Chemical resistance is good, and electrical properties are excellent. At normal temperatures their strength is poorer than for organic rubbers, but equal to these at elevated temperatures.

Uncured silicone resins are marketed as: (*a*) powder or flake, (*b*) moulding compositions (containing filler, etc.), (*c*) fluids 3500–9000 cP, and (*d*) solutions. They are not miscible with organic resins unless containing a high proportion of phenyl groups (*see* §41.1).

Cured resins are very stable to prolonged heating in air (apart from a gradual loss in weight, (*see* §41.32)), resistant to chemical attack (except organic liquids and strong acids), and have good electrical properties and outstanding water-repellance. Their strength is generally inferior to that of organic resins.

41.1 STRUCTURE

Simplest Fundamental Unit

$$
\begin{array}{c}
\text{R} \\
| \\
-\text{Si}-\text{O}- \\
| \\
\text{R}
\end{array}
$$

where R is an organic substituent (usually CH_3—).

Additional to the above difunctional unit (**D**), there are monofunctional (**M**) and trifunctional (**T**) units, i.e.

$$
\begin{array}{c}
\text{R} \\
| \\
\text{R}-\text{Si}-\text{O}- \\
| \\
\text{R}
\end{array}
\qquad\qquad
\begin{array}{c}
\text{R} \\
| \\
-\text{Si}-\text{O}- \\
| \\
\text{O} \\
|
\end{array}
$$

Note. The units shown above must be so linked together that there is always *one* oxygen atom between two adjacent silicon atoms; to ensure this condition, the units are sometimes written as follows, the $O_{\frac{1}{2}}$ indicating that an oxygen atom is shared between the two silicon atoms.

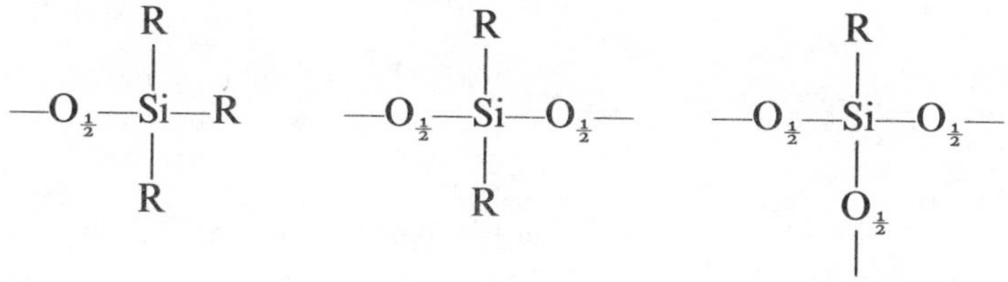

$$
-O_{\frac{1}{2}}-\underset{\underset{R}{|}}{\overset{\overset{R}{|}}{Si}}-R
\qquad
-O_{\frac{1}{2}}-\underset{\underset{R}{|}}{\overset{\overset{R}{|}}{Si}}-O_{\frac{1}{2}}-
\qquad
-O_{\frac{1}{2}}-\underset{\underset{O_{\frac{1}{2}}}{|}}{\overset{\overset{R}{|}}{Si}}-O_{\frac{1}{2}}-
$$

    monofunctional   difunctional   trifunctional

In *silicone rubbers* the molecule is a substantially linear chain of **D** units with **M** end-units, and R is usually a hydrocarbon radical. The main types are: (i) general-purpose dimethyl or 'methyl' silicone rubber, where R is CH_3—; (ii) cold-resistant methylphenyl silicone rubber, with 5–15 mol.% CH_3— replaced by C_6H_5—; (iii) readily cross-linkable methylvinyl silicone rubber, with 0·1–4·5 mol.% CH_3— replaced by CH_2=CH—; (iv) room-temperature vulcanising or 'RTV' rubber, containing H, OH, alkoxy or acyloxy groups; (v) swelling-resistant silicone rubber, containing cyanoalkyl or fluoro-groups, e.g. $CF_3CH_2CH_2$—.

In *silicone resins* there are more 'T' units, which form branching points or cross-links (by condensation from the —OH groups formed by hydrolysis of a trichlorosilane, *see* §41.21), and thus make possible a higher degree of 'curing'. The R groups are generally alkyl (usually methyl) and phenyl. Methyl groups alone give hard but relatively weak resins; longer alkyl groups make the resin softer, less heat-resistant and more soluble. Phenyl groups increase strength and heat-resistance, but give a brittle resin, hence the best result is obtained with phenyl plus methyl groups.

Molecular weight Raw rubbers: mean, normally 4×10^5 to $1·5 \times 10^6$; extreme values: 2500, $2·8 \times 10^6$. Degree of polymerisation: mean, 4000–20 000.

X-ray data Polydimethylsiloxane is amorphous but shows crystallinity when cooled or stretched; the monoclinic cell contains 6 monomer units: $\beta = 60°$, $a = 13·0$ Å, b (identity period) = 8·3 Å; $c = 7·75$ Å.

41.2 CHEMISTRY

41.21 PREPARATION

Starting materials. These are normally organosilicon chlorides or organochlorosilanes, R_3SiCl, R_2SiCl_2, $RSiCl_3$, which by hydrolysis (replacement of Cl by OH) and polymerisation (condensation by loss of H_2O) give **M**, **D** and **T** units respectively in the polymer (*see* §41.1). The most important organochlorosilanes are:

			b.p., °C
I.	$(CH_3)_3SiCl$,	trimethylchlorosilane,	57·5
II.	$(CH_3)_2SiCl_2$,	dimethyldichlorosilane,	70
III.	CH_3SiCl_3,	methyltrichlorosilane,	66
IV.	$(C_6H_5)_3SiCl$,	triphenylchlorosilane,	365–378 (m.p. 111)
V.	$(C_6H_5)_2SiCl_2$,	diphenyldichlorosilane,	303–305
VI.	$C_6H_5SiCl_3$,	phenyltrichlorosilane,	199–201
VII.	$(CH_3)(C_6H_5)SiCl_2$,	methylphenyldichlorosilane,	205
VIII.	$(CH_3)(CH_2 = CH)SiCl_2$,	methylvinyldichlorosilane,	93

Organochlorosilanes are made by:

(i) *'Direct' process* Silicon is reacted with methyl chloride, at 250–280°C in presence of copper, giving mostly II plus some I and III; IV, V and VI (from chlorobenzene) are made similarly, or at 375–425°C using silver.

(ii) *Grignard process* This uses, e.g. methyl magnesium chloride (CH_3MgCl) which with silicon tetrachloride ($SiCl_4$) gives mostly II plus some I and III; III plus phenyl magnesium chloride gives VII.

(iii) *Olefin addition processes*, e.g. ethylene ($CH_2{=}CH_2$) plus trichlorosilane ($HSiCl_3$) give ethyltrichlorosilane, $C_2H_5SiCl_3$; acetylene ($CH{\equiv}CH$) plus trichlorosilane gives vinyltrichlorosilane $(CH_2{=}CH)SiCl_3$.

(iv) *Aromatisation* Benzene (C_6H_6) heated with trichlorosilane ($+ BCl_3$ catalyst) gives mostly V plus some IV and VI.

Heating with aluminium trichloride causes 'redistribution', e.g. a mixture of I and III is converted into the more useful II.

Rubbers are obtained when a difunctional organochlorosilane (e.g. II for dimethyl silicone rubber) is hydrolysed (e.g. with dil. HCl) giving mainly liquid cyclic polysiloxanes, $O[Si(CH_3)_2O]_nSi(CH_3)_2$, where $n = 2$ to 7, mostly 3. These are purified by vacuum distillation, then polymerised at 150–200°C with a (usually alkaline) catalyst to give a rubber or 'gum'. In this process (called 'equilibration') molecular weight is controlled by adding a very small proportion of a monofunctional 'chain stopper', e.g. I, which on hydrolysis yields **M** end-units (*see* §41.1). Trifunctional substances (e.g. III) must be absent, since they produce chain branching and cross-linking by hydrolysing to **T** units.

For methylphenyl and methylvinyl silicone rubbers, II is partly replaced by V or VII and by VIII respectively.

Resins are obtained when organochlorosilanes (usually II, III, V, VI, VII and sometimes alkoxysilanes), mixed with an organic solvent, are hydrolysed at a carefully controlled temperature; the organic layer is washed (e.g. to remove alkaline catalyst) and part of the solvent distilled off. The resulting solution of low-molecular weight liquid polysiloxane is 'bodied' by heating with a polymerisation catalyst, e.g. at 120–200°C with zinc octoate (2-ethylhexoate).

41.22 PROPERTIES

Note. The action of liquids varies according to the nature of the silicone polymer. The following data relate essentially to general-purpose ('methyl') silicone polymers. Those containing fluorine or cyano-groups (§41.1) are more resistant to hydrocarbons (*see* Table 41.T1).

Soluble in (raw rubber) hydrocarbons, chlorinated hydrocarbons, and some ketones, esters and ethers; (uncured resins) organic liquids, e.g. toluene, xylene, petroleum spirit, *n*-butyl acetate. **Swollen by** (vulcanised rubber) non-polar liquids, e.g. hydrocarbons (some mineral oils have little effect), chlorinated hydrocarbons, esters, methyl ethyl ketone, some silicone fluids, carbon disulphide, water at 70°C and above; (cured resins) toluene and some other hydrocarbons, carbon tetrachloride, methyl chloride, acetone, methyl ethyl ketone, liquid ammonia, liquid sulphur dioxide, glacial acetic acid. **Relatively unaffected by** (vulcanised rubber) water (cold), hydrogen peroxide (to 90%), some mineral lubricating oils, animal and vegetable oils, glycol, glycerol, acetone, some silicone fluids, most hydraulic fluids, most cold dil. acids (*c.* 10% HCl, HNO_3, H_2SO_4, H_3PO_4), cold glacial acetic acid, cold 50% aq. caustic alkalis, conc. aq. ammonia, liquid ammonia, aniline; (cured resins) water (to 40°C), hydrogen peroxide (3%), petroleum hydrocarbons, linseed oil, alcohols, phenol, cold dil. acids (10% HCl, HNO_3, H_2SO_4), conc. phosphoric acid, 20% acetic acid, 50% aq. caustic alkalis, conc. aq. ammonia, gaseous chlorine and sulphur dioxide, and ozone. **Decomposed by** (rubbers) cold conc. nitric, sulphuric and (with some grades) hydrochloric acids, also by hot dil. acids and alkalis, and high-pressure steam; (resins) may be attacked by conc. hydrochloric and sulphuric acids.

Table 41.T1. SWELLING OF SILICONE RUBBER VULCANISATES
Swelling after immersion for 72–168 h at 15–40°C (% by vol.)

	General purpose rubbers	Rubbers containing fluorine (F) or cyano (CN) groups
n-Octane	190	—
70:30 *iso*-Octane-toluene	180–200	18–25 (F); 20 (CN)
Petrol	120–270	50 (CN); jet fuel: 10 (F), 15 (CN)
Cyclohexane	200	—
Benzene	110–150	—
Toluene	100–200	26 (F)
Carbon tetrachloride	110–190	20–25 (F)
Methyl ethyl ketone	50–130	—
Ethyl alcohol	1·5–8	5 (F)
Hydraulic fluids	2–60	2–25 (F)

41.23 IDENTIFICATION

N, Cl, S, P nominally absent in dimethyl-, methylphenyl- and methylvinyl-siloxanes; F and N present in fluoro- and cyano-silicone rubbers respectively; the last are distinguished from other nitrogen-containing rubbers by behaviour on burning and by characteristic infra-red absorption bands. Ignition of silicone rubbers and resins (or wet ashing with conc. H_2SO_4 and HNO_3 or with conc. H_2SO_4 and SeO_2) leaves a residue of silica, identifiable by (*a*) giving a blue coloration with ammonium molybdate, (*b*) dissolving completely in hydrofluoric acid, giving a solution that leaves no residue on evaporation.

N.B. This test will only distinguish silicone materials if they do not contain silica fillers.

If a silicone-treated fabric is heated with conc. sulphuric acid in a test tube, the glass wall becomes hydrophobic (to acid and to water). Rubbers and resins show characteristic infra-red absorption bands at 7·9 and 12·8–13·1 μm ($>Si(CH_3)$—), 7·9 and 12·5 μm (—Si-$(CH_3)_2$—), 4·6 μm ($>SiH$—) and 9·2 and 9·8 μm (—Si—O—Si—). (*See* HASLAM, J. and WILLIS, H. A., *Identification and Analysis of Plastics,* London and Princeton (1965).) Rubbers are hydrolysed and dissolved by boiling 80% (w/w) sulphuric acid. **Combustion** Silicones burn with a more or less smoky flame according to the aromatic group content (*see* §41.1); a glass surface held in the flame is coated white with silica (may initially be blackish if phenyl groups are present). **Pyrolysis** Pyrolysates show the main infra-red absorption peaks of the polymer, substantially unchanged (*see* BS 4181, Addendum).

See also KLINE, G. M., *Analytical Chemistry of Polymers,* Vols. I and III, New York and London (1959, 1962); SMITH, J. C. B., *Analyst,* **85**, 465 (1960) (review on analysis of silanes, siloxanes and other organosilicon compounds).

41.3 PHYSICS

41.31 GENERAL PROPERTIES

Specific gravity Raw rubbers, 0·96–0·98 (fluorine-containing, 1·0). Resins, 1·0–1·2 (glass-filled, 1·11–1·23; moulded compositions, 1·6–1·7). **Refractive index** Rubbers, 1·404; resins, 1·405–1·49. **Water absorption** *Rubbers* Depends on type of filler (some silicas give high values), e.g. 1–6 mg/cm^2 after 7 days at room temperature; 0·4–10% after 30 days (thickness and temperature not stated). *Resins* Moulding 3 mm thick, 0·2% after 24 h; resin–glass fibre laminates, 3·2 mm thick, 0·03–0·2% after 24 h at room temperature. **Permeability** Silicone rubber is more permeable to gases than most other rubbers, e.g. at 20–30°C $(10^{-10}$ cm$^2)$/(s cmHg) H$_2$, 530–720; O$_2$, 100–650; N$_2$, 260–320; CO$_2$, 180–3000; He, 380; water vapour $(10^{-10}$ g)/ (cm s cmHg), 0·5–8·5.

41.32 THERMAL PROPERTIES

Specific heat Resins, 0·36–0·37. **Conductivity** $(10^{-4}$ cal)/(cm s °C) Rubbers (generally greater than for organic rubbers), 4–10. Resins, 3·5. **Coefficient of linear expansion** $(10^{-4}$/°C) Rubbers, 2·5–4·0. Resins, $\alpha = 1·13$, $\beta = 0·038$. **Physical effects of temperature**

N.B. The *thermal stability of silicone polymers* is due to the high bond-energy of the main structure (—Si—O— compared with —C—C—; 90 and 60 kcal/mol respectively). The Si—C and C—H substituent bonds also are very stable.

Silicone rubbers harden and become brittle when progressively cooled, but generally remain flexible to lower temperatures than organic rubbers. Dimethylsiloxane polymers, and those containing up to *c.* 5% or above *c.* 30% methylphenylsiloxane units (§41.1), crystallise on cooling; crystallisation is generally quicker than with other synthetic rubbers, and is most rapid at temperatures such that the rubber may become hard and brittle well above the second-order (glass) transition temperature:

Polymer	Second-order transition (°C)	Crystallisation temperature‡ (°C)	Brittleness* temperature‡ (°C)	Temp. (°C) at which Young's modulus = 7 kgf/mm^2
Dimethylsiloxane	-125 ± 7	-52 to -54	-60 to -65	-50
Copolymer of above with:				
Methylphenylsiloxane (10 mol. %)	-112†	—	*c.* -90	-100
γ-Cyanopropylmethylsiloxane	—	—	—	-120
γ-Trifluoropropylmethylsiloxane	—	—	-68	-60

*Other (unspecified) rubbers; to -100°C and below
†Rises by *c.* 0·9 or 1·9°C per 1 mol. % methylphenylsiloxane or diphenylsiloxane respectively up to *c.* 60%
‡Defined as 'temperature of maximum crystallisation'

(*see* POLMANTEER, K. E., SERVAIS, P. C. and KONKLE, M. S., *Ind. Engng. Chem.,* **44**, 1576 (1952); POLMANTEER, K. E. and HUNTER, M. J., *J. appl. pol. Sci.,* **1**, 3 (1959); BORISOV, M. F., *Soviet Rubber Technology,* No. 7, 5 (1966).)

With rise of temperature, general-purpose silicone rubbers show an increase in compression set above *c.* 100°C; up to *c.* 200°C stiffness and hardness remain practically unchanged but tensile strength and breaking elongation gradually decrease. At higher temperatures chain scission causes softening.

Silicone resins that are flexible at room temperatures have brittleness temperatures of -70°C or lower. The resins soften at temperatures ranging from *c.* 30°C to above 200°C according to the degree of cure (cross-linkage). Prolonged heating causes gradual loss of weight by breakdown to volatile products e.g. benzene and cyclic siloxanes from methylphenylsiloxanes at 135–240°C.

41.33 ELECTRICAL PROPERTIES

Rubbers Typical data are shown in Table 41.T2.

Table 41.T2. ELECTRICAL PROPERTIES OF FILLED SILICONE RUBBER VULCANISATES
(unfilled rubbers in parentheses)

	Room temperature (usually 20 or 25°C)	c. 100°C	c. 200°C
Volume resistivity (ohm.cm)	10^{13}–10^{16} ($3 \cdot 4 . 10^{14}$)*†	$2 . 10^{14}$–$3 . 10^{14}$	$2 . 10^{11}$–10^{12}
Dielectric strength (kV/mm)	16–22¶	17–20	17–20
Dielectric constant			
60–100 Hz	3·0–3·6 (2·67)*‡	2·8–3·3	2·4–4·7
10^6 Hz	2·9–3·8‡	2·7–3·2	2·4–3·0
Power factor§			
60–100 Hz	0·001–0·008*‖	0·0025–0·04	0·013–0·3
	(0·00024)		
10^6 Hz	0·001–0·003‖	0·0015–0·007	0·002–0·01
10^{10} Hz	0·017	0·01	0·008

*Values for an electrical grade rubber are as follows

	Conditioned to 50% r.h.	2 mm sheet immersed 14 days in water at 70°C
Resistivity (ohm.cm)	$1 \cdot 5 . 10^{15}$	$2 . 10^{14}$
Dielectric constant, 60 Hz	3·6	4·3
Power factor, 60 Hz	0·004	0·009

†Mostly above 10^{15}; fluorine-containing rubber, 10^{13}.
‡For most grades; some are as high as 9 (60 Hz) or 8 (10^6 Hz); a grade giving 2·95 at 10^6 Hz gave 2·85 at 10^{10} Hz
§At room temperature power factor passes through a maximum at a frequency ranging from < 100 Hz to $3 \cdot 10^9$ Hz according to the nature of the rubber
‖For most grades; some are as high as 0·067 (100 Hz) or 0·045 (10^6 Hz)
¶Up to 40 in thin layers

Table 41.T3. ELECTRICAL PROPERTIES OF SILICONE RESINS

	Resins for surface coating, bonding, casting, impregnating and potting	Moulded filled composition
Volume resistivity (ohm.cm)	$3 . 10^{13}$–$3 . 10^{15}$*	10^{11}
Surface resistivity (ohm)	—	10^{10}
Dielectric strength (kV/mm)	20–120†	6–12
Dielectric constant		
60–10^6 Hz	2·75–2·85‡	3·8–3·9
Power factor		
60–1000 Hz	0·001–0·008§	c. 0·04
10^5–10^6 Hz	0·004–0·02§	

*Values about the same day or after 40 h immersion in water
†Dry, at room temperature; c. 50% lower at 100°C; 20–30% lower for film tested wet
‡Little change with frequency, or whether dry or wet
§Little change whether dry or wet; up to 0·10 at 200°C

Silicone rubbers resist tracking and corona discharge, and retain insulating properties even when burnt, because they have a siliceous (not carbonaceous) 'skeleton'.

Conductive (anti-static) rubbers can be made with a volume resistivity down to a few hundred ohm cm.

Resins Typical data appear in Table 41.T3. Arc (tracking) resistance of silicone resins is greater than with organic resins.

41.34 MECHANICAL PROPERTIES

Silicone rubber vulcanisates are generally inferior to those of organic rubbers in tensile and tear strengths and extensibility, also in abrasion resistance (e.g. only 15–45% that of a carbon black reinforced natural rubber). Strength is especially low in absence of reinforcing fillers. Compression set at room temperatures is comparable with that for organic rubbers, but there is less increase in set on raising or lowering the temperature. Typical properties are shown in Table 41.T4.

Table 41.T4. MECHANICAL PROPERTIES OF SILICONE RUBBER VULCANISATES
(at room temperature except where otherwise stated)

	Unfilled vulcanisates	*Dimethylsiloxane rubber vulcanisates containing 50 parts silica (various types) per 100 rubber*	*Other vulcanisates containing fillers*
Elastic modulus, Young's, static (kgf/mm^2)	—	—	0·10–0·28
Tensile strength (kgf/mm^2)	0·035; 0·5*	0·5–0·65	0·35–1·5†
Breaking elongation (%)	200*	120–250	40–800†
Stress at 300% elongation (kgf/mm^2)	—	—	0·45
Tear strength (kgf/mm)	—	—	0·5–4·0
Hardness (IRHD)	50–60*	52–80	35–90
Compression set (%)‡	—	—	10§–70
Resilience, rebound by Tripsometer (%)	—	—	46–54

*From rubber containing vinyl groups
†The high values are for rubber reinforced with silica (8–10 nm particle size) coated with a butyl ester ('Valron Estersil')
‡Compressed for 22 h at 150°C; recovery for 30 min at room temperature; set expressed as percentage of compression
§'Low set' rubber

Properties at various temperatures (not same rubbers for all properties)

Temperature (°C)	−60	−20	+20	100	200	300
Tensile strength (kgf/mm^2)	c. 1·0	—	0·7	0·5	0·3	0·2
Hardness (IRHD)	55–90	47–62	46–60	46–58	45–55	—
Compression set (%)*	70–100	13–25	4–10	4–12	27–45	—

*Lower figure is for 'low set' rubber

Silicone resins vary, according to the polymer type and fillers and the degree of cure (cross-linkage), from soft and rubbery (Young's modulus, c. 0·35 kgf/mm^2; Shore A hardness, 35–40; tensile strength,

0·55–0·7 kgf/mm^2; breaking elongation, 100–150%) to hard (Young's modulus, *c*. 40 kgf/mm^2; tensile strength, 2·2–4·2 kgf/mm^2; breaking elongation, 5–10%; compression strength, *c*. 80 kgf/mm^2; flexural strength, 55–70 kgf/mm^2; impact strength, 4–7ft lbf/in notch).

41.4 FABRICATION

Silicone rubbers are mixed with vulcanising agents, fillers, etc., using roll mills or internal mixers generally as for other rubbers.

Vulcanising agents are normally organic peroxides (up to 6 parts per 100 rubber, usually *c*. 1 part per 100 rubber) as follows: (i) for dimethyl- and methylphenyl-siloxane (i.e. saturated) rubbers: highly active peroxides are used, e.g. benzoyl peroxide, 2,4-dichlorobenzoyl peroxide, *tert*.butyl perbenzoate; (ii) for vinyl-containing (unsaturated) rubbers: less active peroxides are suitable, e.g. di-*tert*.butyl peroxide, dicumyl peroxide, *tert*.butyl peracetate.

N.B. With the highly active peroxides these rubbers vulcanise at room temperature, giving 'RTV' mixes.

Rubbers with relatively large vinyl content (4–5 mol %) can be hot-vulcanised with sulphur and organic accelerators; rubbers with —OH or alkoxy groups can be vulcanised at room temperature by bases, acids and metal salts, and those with acyloxy groups by moisture*.

To obtain good mechanical strength, reinforcing fillers must be used; these are normally fine silicas (particle size 8–50 μm) sometimes surface-treated to improve wetting and dispersion. Reinforcing carbon blacks inhibit vulcanisation by the highly active peroxides but can be used with the less active peroxides. Semi-reinforcing fillers are diatomite, kaolin and precipitated calcium carbonate. Other ingredients are used as follows: (i) to retard thermal degradation and embrittlement: ferric oxide, ferric 2-ethylhexoate, ferric phosphate, titanium dioxide, fine thermal carbon black; (ii) to reduce compression set: cadmium oxide; (iii) as pigments: inorganic oxides. Unless a silicone (or similar acting) fluid is included, unvulcanised mixes containing fine silicas become stiff and elastic (non-plastic or 'set up'), and hence unworkable, on keeping. Compounded mixes that have been stored usually require pre-milling on a roll mill before further processing.

Vulcanisation is normally in two stages: (1) under pressure, e.g. in a mould or in steam, at a temperature between 110° and 170°C depending on the activity of the vulcanising agent; (2) ('post cure')

*The basic reaction is $2\left(-\overset{|}{\underset{|}{Si}}-OCOR\right) + H_2O \rightarrow -\overset{|}{\underset{|}{Si}}-O-\overset{|}{\underset{|}{Si}}- + 2RCOOH.$

in air at 200–250°C (i.e. at or above the service temperature) for up to 24 h, to increase strength, reduce set and remove volatile residues from the peroxide (grades are available that do not need 'post cure'). Silicone rubbers show considerable shrinkage (2–7% linear) during hot vulcanisation, but only 0·2–0·6% in room temperature vulcanisation.

The rubbers can be moulded (compression, injection or transfer), extruded or calendered like other rubber mixes (keeping the temperature low in the last two processes) or spread on fabric as solutions in organic liquids; they are also made in cellular form by incorporating a blowing agent, and can be bonded to most metals.

Silicone resins are cured by mixing with cross-linking agents, fillers (for moulding compositions), etc., normally with the resin dissolved in an organic liquid. Typical catalysts are ferric, cobalt and zinc 2-ethylhexoates or naphthenates, trialkylamines, triethanolamine (with or without benzoyl peroxide), choline 2-ethylhexoate, or (for resins containing vinyl groups) organic peroxides. Fillers are commonly heat-resistant fibres (glass, asbestos), mica or diatomite; pigments are inorganic. When a solution is used for coating, dipping, impregnating and laminating, the solvent is removed before curing, which is normally carried out in stages at increasing temperatures, the final one being at above the service temperature, e.g. 30 min to 24 h at 150·250°C.

For mouldings, the solvent is removed, the solid partially cured, then comminuted and used for compression or transfer moulding under heat and pressure. In making laminates the impregnated material (e.g. glass-fibre fabric) is partially cured, then plied up and given a consolidating cure under pressure. To develop the best mechanical and electrical properties, laminates and mouldings are given a final or 'after'-cure.

Solventless liquid resins (used as casting, potting, embedding and encapsulating compositions) are reactive polysiloxanes to which a catalyst is added immediately before use; cure then takes place in, e.g. 4 h at 65°C or 3–7 days at room temperature.

Silicone resins can be made in cellular form by incorporating a blowing agent.

41.5 SERVICEABILITY

Silicone rubber vulcanisates have a long serviceable life at elevated temperatures (up to 80 times that of organic rubbers), e.g. 10–20 years at 120°C, 2–5 years at 200°C, 3–24 months at 250°C, 1–8 weeks at 300°C, 6–170 h at 350°C, 10–120 min at 400°C; maximum service temperature is normally 250–300°C. Fire risks are less than with organic rubbers because burning leaves a heat- and electrically-insulating residue of silica. Silicone rubbers, especially those containing phenyl groups (§41.1), resist low temperatures better than organic rubbers, e.g. minimum service temperature is $-100°C$ or $-70°C$ according as use involves static or dynamic stressing. They resist oxidation, chemical attack, high-energy radiation (similarly to polyvinyl chloride, chloroprene and polysulphide rubbers, *see* Sections 11, 36, 39), and when properly vulcanised are not subject to fungal or bacterial attack; they are tasteless, odourless and physiologically inert.

Silicone resins cover a service temperature range from $-100°C$ to $250°C$ (life up to 6 months at latter temperature). They have exceptional water-repellence, have high resistance to oxidation and chemical attack, and are not subject to fungal or bacterial attack.

41.6 UTILISATION

Silicone rubbers are used where their exceptional properties justify the relatively high cost, e.g. seals, gaskets, rollers, etc., to operate at low or high temperature or out-of-doors; room-temperature-vulcanising adhesives* and casting and potting compositions; coated glass-fibre fabric for heater ducts and diaphragms; aircraft dampers, de-icing equipment and instrument tubing; electrical insulation and cable covers to withstand high temperature or even flame; non-stick rollers and belts; articles for contact with foodstuffs, beverages and medicinal preparations; surgically implanted body parts.

Silicone resins are used for protective heat-resistant and/or water-repellent coatings and paints; impregnation of paper, fabrics, tapes, and electrical equipment; lamination of fabrics, especially glass-fibre and asbestos; bonding of mica and asbestos; mouldings, e.g. electrical insulation; room-temperature-curing sealing, potting and encapsulating compositions; medical equipment; release and non-stick agents, e.g. for moulds, release papers, bread-baking tins, frying pans.

41.7 HISTORY

J. B. A. Dumas (1840) predicted the existence of organosilicon compounds, and the first of these (tetraethylsilicon) was prepared by C. Friedel and J. M. Crafts (1863) though in 1857 H. Buff and F. Wöhler had prepared the first chlorosilane ($SiHCl_3$). From *c.* 1900 to 1944 F. S. Kipping and co-workers made an exhaustive study of organosilicon chemistry and obtained polysiloxanes (silicones).

---

*Some are vulcanised by exposure to moist air, *see* §41.4.

Further research and development by E. G. Rochow (c. 1930 on-wards), J. F. Hyde (from 1931) and R. R. McGregor (from 1938) led to commercial production of fluids and resins in the U.S.A.; the researches of K. A. Andrianov (1938 onwards) led to silicone production in Russia (1947). Silicone rubber first appeared in 1942, and in 1943 the Dow Corning Corporation was formed to exploit them, leading to full-scale production in U.S.A. (1946) and the U.K. (1954; Albright & Wilson Ltd., now transferred to Midland Silicones Ltd.). Until 1957 only compounded rubbers were sold, but subsequently uncompounded 'gums' have also been available. World output of all silicones (excluding Russia) is estimated at 10000–15000 tons per annum, about one-third being rubber.

41.8 ADDITIONAL NOTES

41.81 SILICONE FLUIDS (Drisil, Releasil, Silcodyne, Viscasil; DC, S, SF followed by numbers)

These are polyorganosiloxanes analogous in structure and preparation to silicone rubbers (§§41.1, 41.2) but of lower molecular weight. The molecule may be linear (usually comprised of dimethylsiloxane units, and represented by $(CH_3)_3SiO[Si(CH_3)_2O]_nSi(CH_3)_3$ or branched, as made by including in the reaction mixture a small amount of a trifunctional substance (e.g. methyltrichlorosilane).

Properties depend on molecular size; data for linear dimethylsiloxane fluids are as follows:

n (in formula above)	0	6	c. 14	c. 350	—
Sp. gr. 25°/25°C	0·76	0·91	0·94	0·97	0·97
Freezing point (°C)	−68	−64	−65	−50	−46
Boiling point (°C)	100	153/5 mm	>200	*	*
Viscosity at 25°C (c St)†	0·65	3·9	10	1000	10000
Viscosity-temperature coefficient‡	0·31	0·53	0·57	0·62	0·58
Refractive index, n_D^{25}	1·375	1·395	1·399	1·404	1·404
Surface tension at 25°C, dyn/cm	14·8	18·0	20	21	21

*Not appreciably volatile at 200°C
†Available with viscosity up to c. 10^6 c St
‡Proportional drop in viscosity between 100° and 210°F

Other properties Thermal conductivity c. 3·7 $(10^{-4}$ cal)/(cm s °C); coefficient of cubical expansion, 0·001/°C; dielectric strength, 10–20 kV/mm; dielectric constant, 2·2–2·8; power factor, 0·0001–0·0004 (between 100 and 10^6 Hz); volume resistivity, c. 10^{15} ohm cm.

In absence of air the fluids are stable up to 300–400°C, depending on the type; above this, thermal degradation causes reduction in viscosity. In the presence of air they are stable almost indefinitely up to 150°C, but at 170–250°C oxidation commences, forming water and formaldehyde and finally causing gelling. Viscosity changes little with rate of shear, and is less affected by temperature than with other types of liquid. The fluids are colourless, odourless, non-toxic, chemically inert (but decomposed by concentrated mineral acids), and combustible only if external heat is applied continuously; they are miscible with hydrocarbon and chlorinated hydrocarbon liquids but not with water, lower alcohols, glycol or acetone. Other useful properties are low freezing point, good dielectric characteristics, and low surface tension.

Silicone fluids are used as mould release agents, paint and polish additives (reduce stickiness in wax polishes), water-repellent (also non-adhesive and lubricating or softening) finishes for paper, textiles, leather, glass, concrete and masonry; lubricants (alone or thickened with finely divided silica, lithium soaps or carbon black to form non-melting greases); damping, hydraulic and dielectric fluids; diffusion pump fluids; non-flam heating fluids; foam stabilising agents (with miscible liquids); or anti-foaming agents (with non-miscible liquids).

41.82 SILICONE DISPERSIONS

Aq. dispersions of silicone fluids, including a cross-linking catalyst, are applied to textile fabrics (to make them water-repellent and stain-repellent, to paper (water-repellant, non-adhesive) and to leather (water-repellent). The silicone is finally cured to the resin stage, usually by heating, e.g. for 5 min at 150°C.

41.83 ORGANOCHLOROSILANES, AMINOSILANES AND SILICON ESTERS

Organochlorosilanes (*see also* §41.21) are hydrolysed by water and can react with moisture adsorbed on a solid substrate, to give adherent films of polyorganosiloxanes. Treatment with mixed methylchlorosilanes ('Dri-film') is used thus to impart water-repellance to paper, textiles, leather, wood, glass and ceramics (especially steatite electrical insulators). Treatment of glass fibre with vinyltrichlorosilane ($CH_2{=}CH{\cdot}SiCl_3$, b.p. 92°C) improves its adhesion to polyester laminating resins (*see also* §42.4).

Aminosilanes (Dow Z6030, Z6040; MS 2730, 2733). Compounds such as γ-aminopropyltriethoxysilane ($NH_2(CH_2)_3Si(OC_2H_5)_3$) and δ-aminobutyltriethoxysilane ($NH_2(CH_2)_4Si(OC_2H_5)_3$) are used as finishes for glass-fibre fabric to improve the bond to epoxy, melamine, phenolic and polyester resins and hence increase the water-resistance of laminates made with these, also to enable the fabric to be dyed.

Silicon esters are hydrolysed by moisture to form silicic acid or silica. Ethyl orthosilicate (a mixture of condensed tetraethoxysilanes, b.p. $>110°C$), in a volatile solvent, is used to make concrete, stone, etc., water-repellent. Other uses are as a bonding agent in refractory and ceramic materials, including sand moulds, as paint additives and (in aq. emulsion) to give an anti-slip finish to textiles.

41.84 'BOUNCING PUTTY'

Is made by reacting a dimethylsilicone (§41.1) with ferric chloride and a boron compound (e.g. B_2O_3), and incorporating fillers, etc. It exhibits an apparently contradictory combination of *plasticity* (flows like a viscous liquid; can be drawn into threads) with rubber-like *elasticity* (80% rebound) and even *brittle fracture* under a sudden blow. It has been used for golf ball cores, levelling devices and exercisers for injured muscles.

41.9 FURTHER LITERATURE

ANDRIANOV, K. A., *Organic Silicon Compounds* (transl. from Russian), Washington, D.C. (1958)

EABORN, C., *Organosilicon Compounds*, London (1960)

FREEMAN, G. G., *Silicones: an Introduction to their Chemistry and Applications*, London (1962)

FORDHAM, S. (Ed), *Silicones*, London (1960)

MCGREGOR, R. R., *Silicones and their Uses*, New York, London, Toronto (1954)

MEALS, R. M. and LEWIS, F. M., *Silicones*, New York (1959)

NOLL, W., *Chemistry and Technology of Silicones*, New York and London (1968)

Standards
BS 2899 (silicone rubber insulation)
BS 3826 (silicone-based water repellents for masonry)
ASTM D2526 (ozone-resistant silicone rubber insulation)

SECTION 42

GLASS
Principally as fibres

With notes on asbestos and other inorganic fibres, and polyphosphonitrile chloride

TRADE NAMES
Fibres: Fibreglass, Duraglas, Versil.

GENERAL CHARACTERISTICS
Glass, in bulk, and in the form of ceramic glazes, is an extremely hard and durable material, too well known to require further description. Coarse glass fibres (glass silk or glass wool) are employed in filtration and for thermal insulation; they irritate the human skin, though it has been claimed that fibres below 0·013 mm in diameter are innocuous. Fine glass fibres (as used for textiles) resemble silk or rayon; fabrics usually exhibit a harsh 'handle' on sudden crumpling, but a smoother one on gentler treatment, when lubricated yarns and fibres can slide over each other. Glass fibres for textiles are of very small diameter (down to 0·005 mm; 15 km of fibre weigh less than 1 g) and usually circular in cross-section.

42.1 STRUCTURE

Simplest Fundamental Unit
Glass, an inorganic substance regarded as a super-cooled liquid, is comprised of an amorphous network having a limited degree of order but no regular repeat unit, the general structure being a random one and partly uncertain. In the crystalline form of *silica,* the main component of glass, four oxygen atoms are disposed (tetrahedrally) around each silicon atom, and each O atom is shared by two Si atoms (spacing, *see* under **x-ray data**), thus making a regular network (*see* I, representing two of the simplest structural units of silica). In *glass*, which softens at a lower temperature than silica, the continuity of the network is interrupted by the presence of additional oxygen and ions of various metals, i.e. silica is largely converted to silicates by

$$-O-\underset{\underset{O}{|}}{\overset{\overset{O}{|}}{Si}}-O-\underset{\underset{|}{|}}{\overset{\overset{O}{|}}{Si}}- \quad \xrightarrow[\text{(or CaO)}]{Na_2O} \quad \left[-O-\underset{\underset{O}{|}}{\overset{\overset{O}{|}}{Si}}-O\right]^{-} + \left[O-\underset{|}{\overset{\overset{O}{|}}{Si}}-\right]^{-} \quad 2Na^+ \text{(or } Ca^{2+})$$

$$\text{I} \qquad\qquad\qquad\qquad \text{II}$$

reaction with suitable oxides (as in common crown or soda glass, *see* II). Other metals incorporated include potassium (hard glass) and lead (flint glass) while barium is usually present in the denser crown and flint glasses. Inclusion of aluminium raises the softening

472

temperature. Acid borosilicate glasses, having higher thermal resistance and lower expansion than ordinary glass, contain a high proportion of silica to which boric acid (as B_2O_3) is added. *See also* §42.21.

Molecular weight Indeterminate.

X-ray data In a spatial network of SiO_4 tetrahedra (as in crystalline *silica*), the Si—O distance is 1·62 Å, the O—O distance 2·65 Å, and the Si—Si distance 3·2 Å; but *glass* as a whole is substantially amorphous and even the drawn fibres are without appreciable fine structure.

42.2 CHEMISTRY

42.21 PREPARATION

The mineral known as obsidian represents a natural form of glass but is of no economic importance. Instead, glass is made artificially by fusing refractory acid oxides (e.g. SiO_2, B_2O_3) with alkalis, alkaline earths, or other metallic oxides (added as such or as carbonates, etc.). The various kinds of glass are briefly indicated in §42.1; below are shown some typical compositions (%):

Kind of glass	SiO_2	B_2O_3	CaO	Na_2O	K_2O	MgO	Al_2O_3
Common (window)	71	—	13	14	0·2	—	1·5
Borosilicate (Pyrex-type)	81	13	—	3·7	0·3	—	2·2
'A' (alkali-type)	73	—	8	└———14———┘		3·5	1·5
'E' (electrical grade)	54	8·5	17·5	└—under 1—┘		4·5	15

For spinning fine filaments, a melt made from selected marbles of glass flows through tiny holes in the bases of platinum crucibles (bushings) which are heated electrically to c. 1100°C. It is then either withdrawn continuously on fast-rotating drums, or is blown into staple fibre by jets of air or high-pressure steam. To prevent chafing, the fibres are lubricated immediately after spinning; a textile size may be added at the same time. For fabrication of resin-bonded products, *see* §42.4.

42.22 PROPERTIES

There are no direct solvents for glass, but it is attacked by hydro-fluoric acid (vapour, liquid, and solution), by hot phosphoric acid, and by strong alkalis. Fine fibres are attacked by hot dil. acids ('E' being less resistant than 'A' type glass); 'A' and borosilicate (Pyrex-type) glasses, relatively acid-resistant, are attacked by hot dil. alkalis.

42.23 IDENTIFICATION

Combustion and pyrolysis Glass fibres do not burn in a Bunsen flame, but melt into a bead when approaching red heat (lubricant and size chars and may discolour the bead). Glass fibres can thus be distinguished from asbestos and silica fibres, which do not melt in a Bunsen flame. On pyrolysis, lubricant may distil from glass fibres; further reactions, nil.

42.3 PHYSICS

42.31 GENERAL PROPERTIES

Specific gravity (an additive function of the component oxides) Common glass, 'A', 2·50; 'E', 2·58; dense crown or barium, *c.* 3·5; dense flint, 3·0–6·0; Pyrex-type, 2·23 (vitreous silica, 2·20). Fibres: 'A', 2·46; 'E', 2·55; fabric laminates (50% glass), 1·7–1·8; bonded mat (40% glass), 1·5–1·6. **Refractive index** Common and optical glasses, 1·49–1·52; dense crown or barium, up to 1·6; flint, 1·55–1·6; dense optical flint, 1·6–1·9; tallurium oxide glass, over 2; Pyrex-type, 1·48. Fibres: 'A', 1·52; 'E', 1·55. **Birefringence** (of fibres) Nil. **Light transmission** Pyrex-type (2 mm thick, %): 400–700 nm, 90; 340 nm, 80; 300 nm, 20. Common glass transmits the shorter infra-red radiation but is practically opaque to wavelengths beyond 4 μm (greenhouses function largely on this account). **Water absorption** Practically zero. **Water retention of fibres** *c.* 7%. **Moisture regain of fibres** (surface adsorption only, at 65% r.h., 25°C), *c.* 0·1%. **Permeability** Glass is practically impermeable to gases and water vapour, even when very thin; but it is permeable under electrolytic conditions, e.g. to sodium ions.

42.32 THERMAL PROPERTIES

Specific heat Common glass, *c.* 0·2; Pyrex-type, 0·23. **Conductivity** (10^{-4} cal)/(cm s °C) at room temperature; values slowly rise with temperature rise: common glass, 'A', *c.* 20; 'E', 25; Pyrex-type, 27; fused silica, *c.* 30. Fibres: lightly felted insulating quilting, 0·75; fabric laminates, 5–10; bonded mat, 5–6·5. **Coefficient of linear expansion** (10^{-4}/°C) At room temperature (values rise at high temperature): common glass, 0·09–0·11; flint glass, 0·09; 'A', 0·07; 'E'. 0·05; Pyrex-type, 0·035; fused silica, 0·0055. Fabric laminates, *c.* 0·1; bonded mat, 0·2–0·3. **Physical effects of temperature** Stable over a wide range, but acquires a permanent set when deformed at high temperature. Fibre strength declines above 360°C. Softening temperature: over 700°C (common soda-lime glass begins to soften $c.$ 570°C); Pyrex-type and 'E' glass fibres (viscosity 10^8 P), $c.$ 800°C, fabrication temperature (viscosity 10^4 P), $c.$ 1200°C; specially refractory fibres, e.g. aluminium silicate, retain characteristics up to 1250°C (silica softens 1500°C, melts above 1700°C).

42.33 ELECTRICAL PROPERTIES

Volume resistivity (ohm cm) Common (soda-lime) glasses, 10^{11}–10^{13} at room temperature; rises with silica content (Pyrex-type glass, 10^{15}) but falls with temperature rise, e.g. 10^8 at 200°C, 10^4 at annealing temperature (400–500°C), very low when melted. Resistivity of fibres varies according to surface contamination; experimental values, not necessarily representative: commercial yarn, 65% r.h., 10^7; 90% r.h., $5 . 10^4$; purified yarn, 65% r.h., $2 . 10^{10}$; 90% r.h., 10^6. *Surface resistivity* (ohm) Common glass, dry, $c.$ 10^{13}; Pyrex-type glass, 35% r.h., 10^{14}; 85% r.h., $c.$ 10^8. **Dielectric strength** (kV/mm) Common (window) glass, 8–12; Pyrex-type, over 12 (e.g. 3 mm plate under oil, breakdown at 40 kV). Values are higher for thin samples, e.g. zinc silicate crown, 18 (2·3 mm), 43 (0·4 mm). Fabric laminates, 6–8 (and higher). **Dielectric constant** Common glass, $c.$ 7 (wide frequency range); 'E', 6·4 (6·1 at 10^{10} Hz); Pyrex-type, 4·6; vitreous silica, 3·6–3·8; tellurium oxide glass, $c.$ 25. Value tends to rise with temperature rise. Fabric laminates, 3·5–5. **Power factor** Common glass, and 'E', of the order of 0·005 (wide frequency range); Pyrex-type, 0·003 (50 Hz), 0·005 (10 kHz), 0·0046 (1 MHz). Fabric laminates, 0·001–0·05.

42.34 MECHANICAL PROPERTIES (mainly of fibres)

Elastic modulus (Young's) (kgf/mm^2) Common glasses, 5000–8000 (220–350 gf/den.); Pyrex-type, 6000–7000 (vitreous silica, over 7500); fabric laminates, up to 4000. *Bulk modulus*: 'A' glass, 4900; 'E', 3500; vitreous silica, 3800. **Elastic recovery** Extensible elastically almost to breaking point; no yield, obeys Hooke's law. **Extension at break** $c.$ 3%. **Tensile strength** (kgf/mm^2) Glass in bulk, up to 35 (may fail at considerably lower stress because of surface flaws; safe design load, under 1 kgf/mm^2). Fibres, 0·01–0·005 mm diameter, 125–250 (5–11 gf/den.); ultra-fine experimental, up to 1000 (40 gf/den.). An apparent increase of tensile strength with fibre fineness may be regarded as spurious; fibres produced under very uniform conditions gave a value of about 350 (14 gf/den.), independent of diameter (THOMAS, W. F., *Phys. and Chem. Glasses,* **1**, 4 (1960)); surface flaws initiate premature breakdown. Fibres of 'E' glass are some 10% stronger than those of 'A'; e.g. 280 and 245 respectively. Fabric laminates,

28–56; bonded mat, 18. **Compression strength** (kgf/mm²) Glass in bulk, 70–100. Fabric laminates, 20–30; bonded mat, 18. **Flexural strength** (kgf/mm²) Fabric laminates, 35–60; bonded mat, 22, 'E' glass is superior to 'A'. **Hardness** Glass is harder than any organic polymer. Bulk glass: Mohs scale, 5–6 ('A', 6; 'E', 6·5). Shore sclero-scope rebound, 120 (max. possible = 140). Pyrex-type, kgf/mm², 50 gf load: Vickers DPH, 580; Knoop hardness number, 550. **Friction and abrasion** Approximate values for coefficient of friction, μ_{stat} : Glass/glass, lubricated, 0·1–0·6; do., lubricant free, 0·9–1·0; glass/metal, lubricated, 0·2–0·3; do., lubricant free, 0·5–0·7. For coefficient of friction of various fibres/glass, *see* §13.34 (cellulose), §14.34 (cellulose acetate), §20.34 (wool), and §23.34 (nylon). Although flexibility increases with fibre fineness, the dynamic flexural resistance of glass fibre yarns is poor, because of self-abrasion, unless a lubricant is present. Abrasion resistance in friction against materials softer than glass is high.

42.4 FABRICATION

Glass can be cast, moulded, blown, ground, and highly polished. For expanded glass and devitrified glass *see* §42.8. For fabrication of fibres, *see* §42.21.

Although fibres are strong, knot-strength is very low; they are best joined, therefore, with an adhesive such as polyvinyl acetate. In the preparation of articles from glass-fibre cloth or mat and polyester bonding resins, it is customary to treat the surface of the glass, freed from size or lubricant, with a reagent that promotes adhesion, e.g. (i) methacrylato-chromic chloride (Volan finish) and (ii) vinyl-trichlorosilane (Bjorksten finish); these react with —OH groups and moisture on the surface of the glass, and the unsaturated groups thus attached (*see* III and IV, respectively) can then co-polymerise with the resin. *See also* §41.8.

III   IV

42.5 SERVICEABILITY

Glass is hardly affected by sunlight or exterior weathering, is extremely permanent if not subjected to abrasion, and normally devitrifies only in the course of centuries (but *see* §42.82). Fibres suffer chemical attack more readily than glass in bulk, but under normal circumstances may be considered permanent; lubricant applied to reduce self-abrasion may supply nutrient matter for the growth of moulds.

42.6 UTILISATION

(i) *Textile fibres* High-temperature electrical insulation (cloth, tape, braid); fireproof fabrics; resin reinforcement (cloth, mat). (ii) *Coarser staple* Acoustic and/or thermal insulation; resin reinforcement; air filters; battery plate separators.

Glass-fibre fabric laminates or mat, bonded with polyester or epoxy resins are important constructional materials, being used, for example, for vehicle bodies, aircraft and marine components (including boat hulls), protective helmets, vats, and battery boxes.

42.7 HISTORY

Knives, arrow-heads, and other tools of natural glass (obsidian) have been found on Stone Age sites. Man-made glass, as a glaze on pottery, dates back prior to 4000 B.C., and it appears as a free substance, employed for jewellery (e.g. small beads, shaped on wires), in Egypt and Mesopotamia from about 2500 B.C.

An improved standard of production rendered it more plentiful around 1500 B.C., when glass vessels of fine workmanship were first produced; these were fabricated by grinding and polishing after removal of a sand core, around which they were fused (e.g. the beautiful tableware surviving from the time of Thotmose III). Recipes for making glass are given in Assyrian texts of the seventh century B.C., and at this period some vessels were moulded between dies, but the art of 'blowing' glass was unknown until late in the first century B.C.; invented probably in Egypt or Syria, the technique spread rapidly, and from the first until the fourth century A.D. Roman glassware became a common article of commerce in Europe (including Britain) and elsewhere. Production, continued by the Arabs, was later brought to a high state of development in Venice during the thirteenth to seventeenth centuries. English glass, first used for church windows, came into prominent production from the sixteenth century onwards. There is no doubt that the transparency of glass (e.g. put to use in barometers, thermometers, spectrometers, telescopes, chemical glasswear, etc.) greatly facilitated the growth of the scientific outlook in its development from this period onwards.

Glass fibres, known in antiquity, were made sufficiently fine for fabrics by Florentine and Venetian workers in the fifteenth to sixteenth century, but were not pursued commercially; a woven glass fabric was produced by Riva, in Venice, in 1713. An attempt to re-introduce glass for textile purposes was made by Schwabe, in Manchester, in 1840; about this era bunches of relatively coarse filaments were employed for decorative purposes, and later 'glass wool' came into common use, notably as an asbestos substitute in Germany in 1914–18. Fine, spinnable, fibres—and fabrics therefrom—were not properly developed until *c.* 1936; more refractory fibres, consisting largely of silica, for insulation at very high temperatures, appeared in 1948.

Today the romantic craft of glassworking is represented largely by a mechanised industry. Among many investigators of the structure of glass, W. H. Zachariasen and B. E. Warren are prominent names.

42.8 ADDITIONAL NOTES

42.81 EXPANDED GLASS (Foamglass)

This is a low-density material, obtained from compacted masses of powdered glass and finely divided carbon reacted at high temperature. It is a rigid product with closed cells, possessing a specific gravity of approximately 0·16 (10 lb/ft^3), and is especially useful for thermal insulation at low temperatures. The conductivity, (10^{-4} cal)/(cm s °C), at -40°C is *c.* 1 rising linearly to *c.* 1·5 at 10°C, and 1·9 at 150°C.

42.82 DEVITRIFIED GLASS (Pyroceram)

Glass that has been seeded with a nucleating agent, such as titania, may be shaped in the normal way as desired and can then, by controlled re-heating, be converted to a devitrified densely-polycrystalline ceramic-like product that is very hard and resistant to thermal shock. Material of this kind has been used for wear-resistant bearings, insulators, and rocket nose cones. *See* McMILLAN, §42.9.

42.83 ASBESTOS (mineral silk, rock wool)

Asbestos occurs in the serpentine and amphibole groups of minerals and consists of hydrated metal silicates, being thus partly related to glass; but it is unique in being naturally fibrous and thereby achieves economic significance.

The most important source is *chrysotile* or white asbestos, $Mg_3Si_2O_5(OH)_4$ (or $Mg_6Si_4O_{11}(OH)_6 \cdot H_2O$), a fibrous form of serpentine found in Canada, S. Africa, and the Urals; amphibole kinds, represented by a coarser, less refractory but more acid-resisting form, *crocidolite*, [Na, Mg]$_2$Fe$_5$(Si$_4$O$_{11}$)$_2$(OH)$_2$ (the blue asbestos of Cape Colony), and a fawn-coloured variety known as *amosite*, [Ca, Fe, Mg]$_7$([Si, Al]$_4$O$_{11}$)$_2$(OH)$_2$, are also worked commercially. The fibres are usually less than 50 mm long, and can be reduced by repeated cleavage to extremely small width; ultimate fibres of chrysotile have been shown, by electron microscopy, to be less than 5.10^{-5} mm in diameter. The x-ray diagram, distinctive for the two kinds of asbestos, is of the fibre type, sharp, and highly characteristic (SiO$_2$ chains oriented parallel to the fibre axis). Identity period (Å): chrysotile, 14·6; amphibole types, 5·3.

Properties

Asbestos is unaffected by organic liquids, water, alkalis and moderately concentrated acids, but is decomposed by hot acids and is slowly rotted by sea-water. Chrysotile asbestos is rapidly decomposed by strong acids (losing 58% in weight); the amphibole types are more resistant, but all forms are decomposed by hydrofluoric acid. For chemical analysis of asbestos products, *see* FRIESER, E., *Z. ges. Textil-Industrie,* **60**, 704, 751 (1958); also §42.9. **Specific gravity** Ranges from 2·2–2·5 (chrysotile) and up to 3·5 for the varieties containing iron. **Refractive index** Chrysotile, 1·5–1·57 (sample from S. Africa: $n_{||}$, 1·549; $n_\perp$, 1·534. *Birefringence* ($n_{||} - n_\perp$), 0·015); varieties containing iron, *c.* 1·6–1·7. **Moisture regain** (65% r.h.): *c.* 1%. **Water retention** of fibres: *c.* 20%. **Specific heat** 0·2–0·25. **Thermal conductivity**, as $(10^{-4}$ cal)/(cm s °C). Asbestos cloth, 3–4; asbestos-cement sheet, 7; amosite fibres (approx. values, 100–200°C),

Packing density (g/ml)	0·2	0·4	0·6
Conductivity	1·5	2·65	5·1

Melting point (after loss of water above 500°C) Chrysotile, 1550°C; others varieties, 1200–1400°C.

Electrical properties are not outstanding, but are maintained at high temperature. **Volume resistivity** (dry state, ohm cm) Chrysotile, up to 10^9; other varieties, up to 10^{11} (but decreases considerably with rise of moisture regain, especially in chrysotile, e.g. at 65% r.h., 10^5 or less). **Dielectric strength** (asbestos-covered wire), 2·8 kV/mm.

Asbestos fibres are exceptionally strong, and extensibility is very low. **Tensile strength** (kgf/mm^2) usually quoted as over 200 (tested sample S. African chrysotile, up to 175), but under special conditions a tensile strength of over 500—and **elastic modulus** of 15 000 kgf/mm^2— have been observed. (ZUKOWSKI and GAZE, *Nature*, 183, 35 (1959).) Amosite is similar in mechanical properties to chrysotile; crocidolite is slightly stronger. However, the fibres are very smooth and readily slip over each other; hence, when spinning them into yarns, cotton fibres are usually incorporated to increase friction and the strength of the product.

Utilisation

Asbestos fibres find uses as follows. (i) *Heat resistance* High-tempera-ture electrical insulation; brake and clutch linings; fireproof clothing, blankets, industrial fabrics; packing for steam valves and pumps. (ii) *Thermal insulation* Space- and joint-packing material, woven cloths and tapes, millboards; a mixture of 85 pt magnesia and 15 pt asbestos fibre, applied initially in a wet state, is commonly used for heat-resisting insulation. Asbestos is also employed for high-strength plastics reinforcement (as fibre, roving, mat, etc.), for acoustic insulation (sound absorbing tiles), in abrasion-resistance resin-bonded floor tiles, and (shortest fibres) for reinforcing a range of cement and bitumen products.

History

The name of the material derives from a-$\sigma\beta\epsilon\sigma\tau o\varsigma$ (= unquenchable). In antiquity, when ropes or fabrics of asbestos fibres were occasionally produced, they were regarded as almost magical curiosities, because of their resistance to burning. Thus, asbestos, then known also as amianthus ($a\mu\iota a\nu\tau o\varsigma$ = undefiled), was employed for the wicks of perpetual lamps in Greek temples, and Charlemagne (A.D. 742–814) is said to have been credited with supernatural powers because he possessed a cloth that could be cleaned by fire. The sixteenth-century naturalist, Morcati, classed *linum asbestinium*, or asbestos, as associated with the legendary fire-resisting creatures called sala-manders.

Modern production of asbestos yarns and fabrics dates only from the nineteenth century, when spinning of the fibres was developed on a commercial basis, first in Italy, and later (from Canadian sources of chrysotile) in Canada and Great Britain.

Further literature *See* §42.9.

42.84 OTHER INORGANIC FIBRES

Among inorganic fibres developed mainly for their heat resistance and low thermal expansion, are those of silica itself; also fibres of high silica content, serviceable at 1000°C or more, obtained by leaching out metallic oxides or sodium borate from glass fibres, followed by dehydration (Refrasil, Vycor). Fibres of aluminium silicate, blown from an arc-furnace melt containing a small proportion of soda glass, can withstand similar temperatures and find uses in flexible jointing for furnace lining and in space capsules (Fiberfrax, Triton). Potassium titanate, as felted short-length fibres melting at 1400°C, has been employed for furnace insulation, and highly refractory fibrous alumina and fibrous boron have also been produced.

For further reference, *see* CARROLL-PORCZYNSKI, C. Z. *Advanced Materials,* Guildford (1962).

42.85 POLYPHOSPHONITRILIC CHLORIDE

This inorganic polymer, not otherwise related to glass or asbestos, is mainly of academic interest but has potential technical applications, e.g. in heat resistance. It possesses the fundamental unit:

$$\begin{array}{c} \text{Cl} \\ | \\ -\text{N}{=}\text{P}- \\ | \\ \text{Cl} \end{array}$$

and forms both linear and cyclic (tri- and tetra-meric) polymers termed phosphazenes. Liebig isolated compounds of this series in 1834, and in 1895 Stokes prepared the crude monomer by heating phosphorus pentachloride with ammonium chloride in an inert solvent. Schenck and Römer (1924) found that the complex crystalline product polymerises when heated in a sealed tube at 250–300°C, to yield a colourless rubber-like material, which is amorphous but (like natural rubber) exhibits a sharp x-ray diffraction pattern when held in tension, the identity period being 5·7 Å. The polymer swells, but does not disperse, in benzene; it slowly hydrolyses and hardens in moist air, depolymerises when heated above 250–350°C under reduced pressure, but gives a black horny mass when heated in air.

Many derivatives, in which the —N=P— structure is maintained but the chlorine is replaced by other radicals (e.g. F, Br, CH_3, C_6H_5), have been prepared. For reviews of phosphonitrilic compounds *see* EMELEUS, H. J. and SHARPE, A. G. (Eds), *Advanc. Inorg. Chem.,* Vol. 1 (PADDOCK, N. L. and SEARLE, H. T., pp. 347–383), New York and London (1959); and SHAW, R. A., *et al., Chem. Rev.,* 62, 247 (1962). *See also Chem. Soc. Special Publication No. 15, Inorganic Polymers* (HABER, C. P., pp. 115–146), London (1961), LAPPERT, M. F. and LEIGH, G. J. (Eds), *Developments in Inorganic Polymer Chemistry,* London and New York (1962); STONE, F. G. A. and GRAHAM, W. A. G. (Eds), *Inorganic Polymers,* New York and London (1962); GIMBLETT, F. G. R., *Inorganic Polymer Chemistry,* London (1963). HUNTER, D. N., *Inorganic Polymers,* Oxford (1963); ANDRIANOV, K. A. (trans. from Russian by BRADLEY, D. C.), *Metalorganic Polymers,* London and New York (1965).

42.9 FURTHER LITERATURE

GLASS

MOREY, G. W., *The Properties of Glass*, 2nd edn, New York (1954) (American Chem. Soc. Monograph)

JONES, G. O., *Glass*, London (1956) (Methuen 'Monographs on Physical Subjects')

MACKENZIE, J. D. (Ed), *Modern Aspects of the Vitreous State*, London, Vol. I (1960), Vol. 2 (1962), Vol. 3 (1964)

MCMILLAN, P. W., *Glass-ceramics*, London (1964)

CARROLL-PORCZYNSKI, C. Z., *Inorganic Fibres*, London (1958) (includes glass asbestos, etc.)

Specifications

As reinforcing fibres: BSS *3396* (fabrics), *3496* (mat), *3691* and *3749* (rovings); *ASTM D579* (fabrics), *D2150* (rovings), *D2343* (fibres, yarns, rovings). As fabrics for lamination: *ASTM D2408, D2409, D2410* and *D2660* (respectively with aminosilane, vinylsilane, chrome complex and acrylic-silane finishes). As woven tapes, *BS 3779, ASTM D580*. Analysis: *ASTM C169* (soda-lime glass)

ASBESTOS

Asbestos Textile Institute, *Handbook of Asbestos Textiles*, New Brunswick, N.J. (1953)

SINCLAIR, W. E., *Asbestos (Origin, Production, Utilisation)*, London (1955)

BOWLES, O. (Bureau of Mines), *The Asbestos Industry (Bulletin 552)*, Washington (1955). Also: *Asbestos, Materials Survey (Information Circular 7880)*, Washington (1959)

CARROLL-PORCZYNSKI, C. Z., *Asbestos (Rock to Fabric)*, Manchester (1956)

BERGER, H. (transl. from German by R. E. Oesper), *Asbestos Fundamentals*, New York (1963)

BERGER, H. (transl. by OESPER, R. E.), *Asbestos with Plastics and Rubber*, New York (1966)

HODGSON, A. A., *Fibrous Silicates (Roy. Inst. Chem. Lecture Series (1965))*, London (1966)

Specifications

BS 3057 (insulating paper)

BSS 3249, 3258, 3387 (roved cables)

ASTM D315 (tapes), *D1571* (cloth), *D1918* (content in textiles)

PART 2

SIMPLEST FUNDAMENTAL UNIT

The formula given at the beginning of each of Sections 1 to 42 represents the simplest repeating unit—sometimes called a *mer* (from μερος, a part)—of which, together with at least two terminal units*, a molecule of a poly*mer* is composed. According to the number of external chemical bonds associated with a fundamental unit, there are two main types of polymers, as described below.

(i) *Linear polymers*

The simplest of these, derived from monomers of the alkene or bi-functional type, are composed of repeat units joined end-to-end to form very strong (primary-bonded†) chain molecules, which—by reason of residual interchain forces—cohere to form weaker (secondary-bonded‡) aggregates. Hence, since the molecules are relatively easily separated, polymers having this structure are usually both *soluble* and *fusible**. Most fibres, unsupported films, thermoplastics and unvulcanised rubbers are linear polymers . Note, however, that if a simple two-ended repeat unit is modified (e.g. by loss of a hydrogen atom), polymers with *branched* molecules can be formed; it is also possible to have chains radiating from a central multi-functional unit, i.e. polymers with *star-type* molecules.

*The terminal units consist of atoms or groups derived from a polymerisation initiator, a monomer (e.g. by disproportionation) or a solvent (e.g. by chain transfer), or can arise from a substance specially added (e.g. monofunctional substance, to control molecular weight). They necessarily differ from the repeating units but usually represent only a small fraction of each macromolecule (*see* use of end-groups for determination of molecular weight. Section 52).
†That is chemical, covalent bonds, c. 1–2 Å long, with energies up to c. 100 kcal/mol.
‡The following items are recognised: dispersion (London), dipole-induction (Debye), and dipole-dipole (Keesom) forces, producing physical bonds c. 3–5 Å long, with energies from c. 0·5–5 kcal/mole; also *hydrogen bonds*, c. 2·5 Å long, with energies up to c. 10 kcal/mol.
§However, when strong interchain bonds (notably H-bonds) are present and/or the chains have a sufficiently regular structure to crystallise, solubility and fusibility become restricted or even prevented. An extreme example of this is provided by *cellulose,* a linear polymer that, nevertheless, dissolves only in a very limited range of solvents and does not melt; note, however, that substitution of the —OH groups (i.e. reduction of H-bonding) returns it to the general type, a cellulose ester being both soluble and fusible.
‖Their differing properties arise mainly from their secondary structure being one of three kinds, i.e. (a) partially crystalline and thereby tending to be rigid, or (b) amorphous and highly extensible (rubber-like), or (c) amorphous and rigid (glass-like). The difference between (b) and (c) lies in whether or not the segments of the molecules can be moved freely, by rotation about primary bonds (a circumstance depending principally on the temperature; *see* T_g, Section 68).

(ii) *Cross-linked or network polymers*

In a trifunctional system (yielding repeat units with more than two external bonds) branching leads to bridging and the structure of the polymer becomes one of strong chains joined by cross-chains, or, in the ultimate, a fully three-dimensional primary-bonded network— of which, however, it is still possible to devise an average idealised repeat unit. Polymers of this kind are necessarily insoluble and infusible, and therefore much less tractable than linear polymers*. It is therefore usual to fabricate them in an intermediate state and to complete the cross-linkage *in situ*. Thus, the formation of cross-linked or network polymers is represented, for instance, by the resin- or reactive-finishing of cellulose fibres†, the 'drying' or curing of surface coatings, the setting of certain casting resins and adhesives, the insolubilisation of gelatin, the curing of thermosetting (and room-temperature-setting) plastics including the production of laminated boards, and the vulcanisation of rubbers.

It should be understood that regular structural features larger than the simplest repeating unit occur; for instance, in olefin or vinyl-type polymers with large side groups the units cannot be accommodated in straight (i.e. carbon zig-zag) lines, but the chains crystallise instead in a twisted conformation giving rise to a long-period repeat unit, e.g. stretched polyisobutylene displays a period (identity period, *see* Section 53) of 18·6 Å, possibly corresponding to eight repeat units coiled in (one turn of) a helix.

See also Section 53 (X-ray data) and Section 54 (Preparation).

---

*However, particularly if the cross-links are relatively few in number or fairly long, such a polymer can *swell* in an appropriate liquid and can be *softened*—but not melted—by heat (*see* Sections 55 and 68 respectively).

†Treated in fabric form to impart crease resistance, crease retention, smooth drying, etc. (*see* §13.86).

MOLECULAR WEIGHT AND DEGREE OF POLYMERISATION

DEFINITIONS

Molecular weight (*mol. wt.*) The sum of the weights of the atoms in an average molecule (*see below*) expressed relative to the oxygen atom $O = 16 \cdot 00$.

Degree of polymerisation (*D.P.*) The number of repeat units (*see* previous section) in an average molecule of a polymer. Thus, mol. wt./(repeat unit weight) = D.P.

METHODS OF MEASUREMENT

Introduction

Ordinarily, a chemical compound consists of relatively small molecules each of which contains the same number and arrangement of atoms, i.e. it is *homogeneous* in composition. A linear high polymer, however, consists of molecules that are not only very long but also variable in length by reason of the random extent of their growth, i.e. it is *heterogeneous* in composition*, some molecules having more—and some less—than the average number of repeat units.

The heterogeneity means that, when determined experimentally, the value obtained for the 'average' mol. wt. or D.P. depends on the method of measurement. The principal averages are as follows:

NUMBER-AVERAGE MOLECULAR WEIGHT (M_n)

This is an ordinary arithmetic mean or average molecular weight as obtained from methods that more or less 'count' molecules directly, and thus permit the weight to be divided by the *number* of molecules present (irrespective of their length). It is defined by:

$$\overline{M}_n = \frac{\sum n_i m_i}{\sum n_i}$$

---

*Physical differences arise mostly from the variation of molecular length. Chemical differences arise because of the associated variation of the fraction of each molecule represented by the end-groups; this fraction, however, is usually so small that for many purposes the end-groups can be ignored (but *see* §52.1).

where the sample consists of n_1 molecules each of mass m_1, n_2 molecules of mass m_2, etc., and the summation extends over all species present. It is an average sensitive to species of low molecular weight.

WEIGHT-AVERAGE MOLECULAR WEIGHT ($\bar{M}_w$)

This is obtained from methods that take into account the *weight* (or length if unbranched) of the molecules. It is defined by:

$$\bar{M}_w = \frac{\sum w_i m_i}{\sum w_i} = \frac{\sum n_i m_i^2}{\sum n_i m_i}$$

where n and m have the same meanings as before (and $w = nm$). Since it involves (mass)^2, this average gives bias to species of high molecular weight.

Z-AVERAGE MOLECULAR WEIGHT ($\bar{M}_z$)

This is an average involving weight to a still greater extent than $\bar{M}_w$, but is less important. It is defined by:

$$\bar{M}_z = \frac{\sum w_i m_i^2}{\sum w_i m_i} = \frac{\sum n_i m_i^3}{\sum n_i m_i^2}$$

where m, n, and w have the same meanings as before.

VISCOSITY-AVERAGE MOLECULAR WEIGHT ($\bar{M}_v$)

This is obtained from measurement of the viscosities of dil. solutions. It is defined by:

$$\bar{M}_v = \left[\frac{\sum w_i m_i^a}{\sum w_i} \right]^{1/a} = \left[\frac{\sum n_i m_i^{a+1}}{\sum n_i m_i} \right]^{1/a}$$

where n, m, and w have the same meanings as before, and a is a constant lying between 0·5 and 1·0 (*see also* §52.6).

In a polymer that is homogeneous in composition (as distinct from heterogeneous, *see* earlier) all the above molecular weight averages are alike, but normally $\bar{M}_w$ is always higher than $\bar{M}_n$, and the ratio $\bar{M}_w/\bar{M}_n$ affords a measure of the breadth of the distribution of molecular weight. Usually it ranges from *c.* 1·5 (narrow distribution) up to *c.* 10 (broad). The viscosity-average molecular weight lies between $\bar{M}_n$ and $\bar{M}_w$ and closer to $\bar{M}_w$ than $\bar{M}_n$; in the upper limit (a = unity), $\bar{M}_v \equiv \bar{M}_w$. It is seen that the various averages lie in the order:

$$\bar{M}_n < \bar{M}_v < \bar{M}_w < \bar{M}_z.$$

Some of the methods employed for the determination of the molecular weight of soluble high polymers are outlined below. Except in certain circumstances, there is as yet no technique for assessing the molecular weight of insoluble polymers (including cross-linked and net-work structures, which, indeed, may not have separate molecules).

CHEMICAL METHODS

52.1 END-GROUP DETERMINATION

The terminal units of a linear polymer are different from the repeat units (*see* previous section) and chemical reactions peculiar to them enable analytical methods to be used to count the ends—and hence the number—of molecules in a weighed sample. Thus, terminal hydroxyl (—OH) or carboxyl (—COOH) groups, or unsaturation, may be estimated by standard analytical methods. Physical methods of estimation may also be used, e.g. infra-red absorption spectrometry, and determination of the residues of radioactive initiators.

End-group analysis is usually suitable only for molecular weights up to 20000–25000, beyond which end-groups tend to represent a fraction too small for accurate estimation. The result from any end-group determination is a *number-average* molecular weight.

PHYSICAL METHODS

N.B. In the following methods, because of the deviation of polymer solutions from ideality, the basis for the calculation of molecular weight depends on the use of very dil. solutions and extrapolation of the measurements to infinite dilution.

52.2 OSMOTIC PRESSURE

In very dil. solutions each polymer molecule (whatever its size) makes the same contribution to osmotic pressure, and the molecular weight, M, can be obtained from the equation

$$\left(\frac{\pi}{c}\right)_{c=0} = \frac{RT}{M}$$

where π is the osmotic pressure (in dynes/cm^2), and R, T, and c are respectively the gas constant ($8 \cdot 314 . 10^7$ erg/mol K), absolute temperature (K), and concentration of the solution (g/cm^3). Naturally to avoid a spurious rise in osmotic pressure, trace impurities of low molecular weight must be absent.

As an alternative to determination of osmotic pressure, the change in another colligative property can be measured, e.g. lowering of vapour pressure, elevation of boiling point, or lowering of freezing point, but in these instances the effects observed are relatively much smaller. The result from osmotic pressure and related methods is a *number-average* molecular weight.

52.3 LIGHT SCATTERING

For molecules with linear dimensions less than *c*. 0·05 of the wavelength of the light, the amount of Rayleigh scattering by a dust-free solution is proportional to the concentration by weight and to the molecular weight. The intensity of the scattered light (i) and of the primary beam (I_0, which exceeds 1000-fold i), is measured photoelectrically to obtain the *turbidity* ($\tau = i/I_0 t$, where t is the path length or thickness), which expresses the total scattering and is related to the molecular weight of a dissolved polymer by the equation:

$$\tau = kcM$$

where c and M have the previous meanings, and k is a constant involving the refractive index of the solution and the wavelength of the light. The treatment can be extended to the determination of molecular weight over a very wide range, but it is essential that traces of dust and undissolved polymer be absent to avoid spuriously large readings.

Since each polymer molecule contributes to the scattering according to the square of its mass, the result is a *weight-average* molecular weight, $\bar{M}_w$.

52.4 SEDIMENTATION (EQUILIBRIUM METHOD)

When an ultracentrifuge is run at a steady low speed for a long time (1–2 weeks), polymer molecules in solution tend to sediment outwards in the centrifugal field, but Brownian movement tends to cause diffusion in the reverse direction. From the resulting distribution at equilibrium, molecular weight is calculated by means of the equation:

$$M = \frac{2RT \ln (c_2/c_1)}{(1 - V\rho)\omega^2(r_2^2 - r_1^2)}$$

where ω is the angular velocity of rotation in radians/second, $(1 - V\rho)$ is a buoyancy factor (V = specific volume of the polymer; ρ = density of the solvent), and c_1 and c_2 are the weight concentrations (g/cm^3) of the polymer—as determined from refractive index measurements—at the radii of rotation r_1 and r_2.

The result, which depends on the detailed treatment employed, is a *weight-average* molecular weight, e.g. $\bar{M}_w$ or $\bar{M}_z$ or an intermediate average.

52.5 SEDIMENTATION AND DIFFUSION (RATE METHOD)

The velocity at which polymer molecules *sediment* through a solvent is measured while an ultracentrifuge is run for 2–3 h at a high centrifugal field, e.g. 200 000 g. The velocity at which the molecules *diffuse* is assessed separately, in stationary apparatus in which a sharp plane boundary is formed between the polymer solution and pure solvent above it, and the blurring of the solvent/solution boundary is followed as the polymer diffuses.

From the two sets of observations the molecular weight is calculated by means of the equation:

$$M = \frac{RT}{(1 - V\rho)} \cdot \frac{S}{D}$$

where R, T, V, and ρ are as previously, S is the sedimentation constant (= velocity of sedimentation/acceleration due to centrifugal field), and D is the diffusion coefficient (= a proportionality constant, relating (i) the mass, m, of solute diffusing per second through and normal to a plane area 1 cm^2, and (ii) the concentration, c, at a radial distance, x, from the axis of rotation; thus $dm/dt = D \cdot dc/dx$).

Since sedimentation velocity depends very steeply on concentration, it is necessary to work with highly dil. solutions in order to extrapolate to infinite dilution. The result from the rate method is a *complex* average molecular weight lying between $\bar{M}_n$ and $\bar{M}_w$.

52.6 SOLUTION VISCOSITY

Viscosity measurements, requiring relatively simple apparatus, provide a rapid and convenient molecular characterisation of a polymer, but do not by themselves give an absolute determination. For most linear polymers the *intrinsic viscosity* or *limiting viscosity number*, denoted by $[\eta]$, is related to molecular weight by the Mark-Houwink equation:

$$[\eta] = KM^a$$

where K and a are constants for a given polymer/solvent system, that can be found from absolute molecular weight determinations (e.g. from osmotic pressure or light scattering), and $[\eta]$ is equal to

$$\left(\frac{\eta_{sp}}{c} \right)_{c=0}$$

c being as previously, while the specific viscosity, η_{sp}, is given by

$$\eta_{sp} = \frac{\eta_s - \eta_0}{\eta_0}$$

$$= \frac{\eta_s}{\eta_0} - 1$$

where η_0 and η_s are the viscosity of the solvent and solution respectively; which, ignoring the slight difference between the densities,

$$= \frac{\text{solution flow time}}{\text{solvent flow time}} - 1.$$

The result is a *viscosity-average* molecular weight (*see also* earlier in this section).

COMPARISON TABLE

Table 52.T1. MOLECULAR WEIGHT OF COMMON POLYMERS

Order of average molecular weight	Polymer and approx. mol. wt. (number-average unless otherwise stated). Approx. Degree of Polymerisation enclosed thus [ ].
400–15 000 (and upwards)	Epoxy resin intermediates, 400–6000. Phenolic resins, novolacs, 500 [2–13]. Polyurethane intermediates, 500–6000. Polysulphide rubbers, liquid types, 500–7500. Coumarone/indene resins, 1000 [10]. Polyvinyl acetate, 5000–500 000 [60–6000]. Unsaturated polyesters, uncured, 7000–40 000. Polyvinyl alcohol, 9000–120 000 [200–3000]. Polyurethanes, linear, *c*. 10 000.
15 000–50 000 (and upwards)	Polyethylene terephthalate, 15 000–20 000 [80–105]. Polyformaldehyde, 15 000–30 000 [500–1000]. Alkyd resins, 17 000–39 000. Polyamides, 20 000–100 000. Polyethylene, high density, 20 000–200 000 [1500–15 000]; do., low density, up to 50 000. Polyvinylidene chloride, 20 000–200 000 [>200]. Acrylonitrile/butadiene rubbers, *c*. 20 000–10⁶ [*c*. 400–20 000]. Polycarbonates, *c*. 30 000 [*c*. 120]. Regenerated proteins, 30 000–50 000 [*c*. 400]. Alginic acid, regenerated, 30 000–100 000 [200–600]. Styrene/butadiene rubbers, 'hot' types, 30 000–100 000.
50 000–100 000 (and upwards)	Polyacrylonitrile, 50 000–100 000 [1000–2000]. Polyvinyl chloride, 50 000–120 000 [800–2000]. Cellulose, regenerated, 50 000–150 000 [350–1000]. Cellulose nitrate, 50 000–150 000 [250–1000]. Polyvinyl acetals, 50 000–500 000. Wool, 60 000–80 000. Polyvinyl fluoride, 60 000–180 000. Isoprene rubber, synthetic, 77 000–2·5 . 10⁶. Polypropylene, >80 000 [>2000]. Cellulose acetate, *c*. 80 000–100 000 [300–400]. Silk, 84 000–150 000 [>1000].
Over 100 000	Polybutene, 10⁵–10⁶. Chloroprene rubber, sulphur-modified types, *c*. 100 000; do., mercaptan-modified types, 180 000–200 000; [1000–3500], Ethylene/propylene rubbers, 100 000–200 000. Polychlorotrifluoroethylene, 100 000–200 000 [850–1700]. Chlorinated rubber, 100 000–400 000. Styrene/butadiene rubbers, 'cold' types, 110 000–260 000. Alginic acid, native >150 000 [*c*. 1000]. Polystyrene, moulding grade, 200 000–300 000 [2000–3000]. Butadiene rubber, 'solution' types, 260 000–430 000; 'emulsion types, *c*. 450000. Butyl rubber, 300 000–450 000*. Cellulose, native, 300 000–>10⁶ [2000–>6000]. Silicone rubbers, 400 000–1·5 . 10⁶ [4000–10 000]. Acrylate plastics, 500 000–10⁶ [5000–10 000]. Polytetrafluoroethylene, 500 000–5 . 10⁶. Natural rubber, unmilled, 680 000–840 000; milled, 120 000–140 000. Cross-linked polymers, including cured amino, epoxy, phenolic and polyester resins, polyurethanes, vulcanised rubbers, and ebonite.

*Viscosity average

Table 52.T1 shows the *approximate* molecular weight (usually implying $\bar{M}_n$) of common commercially-available polymers. Some synthetic polymers are produced in more than one grade (i.e. soft or hard, etc.) depending on the degree of polymerisation*. It is often possible, experimentally, to obtain polymers with much higher molecular weights, but without technical advantage; indeed the high molecular weight of some natural polymers sometimes has to be lowered.

FURTHER LITERATURE

ALLEN, P. W. (Ed), *Techniques of Polymer Characterisation*, London (1959) (includes end-group determination, osmometry, light-scattering and viscometry)

RAFIKOV, S. R., PAVLOVA, S. A. and TVERDOKHLEBOVA, I. I., *Determination of Molecular Weights and Polydispersity of High Polymers,* Moscow (1963); English translation, Jerusalem (1964)

MEYER, K. H., *Natural and Synthetic High Polymers,* 2nd edn (Vol. IV of *High Polymers*), New York and London, 15–42 (1950)

Specification
BS 1673: Part 6 (determination of limiting viscosity number of raw rubbers)

*Useful mechanical properties appear only above a certain molecular weight. The following values represent roughly the *minimum* degree of polymerisation required: polyesters, 30; hydrocarbon polymers, 80; cellulose derivatives, over 150. Fibre-forming polymers require an average chain length of at least 1000 Å.

SECTION 53

X-RAY DATA

GENERAL INTRODUCTION

When a narrow pencil of x-rays impinges on a small sample of crystalline matter much of the beam passes straight through, but some of it may be scattered in a characteristic manner, as revealed by the use of an x-ray spectrometer or the pattern produced on a suitably disposed fluorescent screen or photographic plate.

The phenomenon is caused by diffraction of the x-rays by the ordered spacings of the atoms constituting the crystals. It is akin to the diffraction of light by a finely-ruled grating; except that (a) since atomic and molecular spacings are involved it is necessary to use radiation of a comparable wavelength (i.e. x-rays), and (b) a crystal does not act merely as a two-dimensional grating but as a three-dimensional lattice possessing various sets of parallel planes (reflection from which causes partial or complete extinction, save at specific angles of incidence where reinforcement produces a diffraction pattern*).

To interpret the pattern it is usual to employ monochromatic x-rays, such as the K_α rays of copper ($\lambda = 1.54$ Å); then, from measurement of the pattern (including its sharpness, intensity, and other details) and from the geometry of the equipment, it is possible to deduce much information concerning the internal structure of crystalline matter (see also **Further literature** at the end of this Section).

As an elementary approach to the subject, in connection with high polymers, one may consider first a single crystal, which when placed with one of its principal axes perpendicular to a monochromatic x-ray beam, and rotated or oscillated about the axis (to ensure that the rays strike at the correct angles), produces—on a flat plate or screen normal to the beam—a pattern consisting of series of small spots lying on *layer lines* (e.g. Figure 53.1a). In a similar manner, the passage of an x-ray beam through a crystalline powder yields a series of concentric rings (*Powder diagram,* Figure 53.1b) since each

---

*Reinforcement occurs, according to Bragg's law, when

$$n\lambda = 2d\sin\theta$$

where n = an integer (the *order* of the reflection), λ = wavelength, d = spacing of the lattice planes, and θ = the *glancing angle* at which the beam falls on, and is reflected from, each plane.

particle acts as a tiny crystal (and this time without needing rotation or oscillation, since, being randomly disposed, some of the particles will receive the x-ray beam at a suitable glancing angle). Similarly, randomly arranged crystallites in a crystalline polymer tend to give a powder-type x-ray diagram. However, when crystallites acquire a unidirectional orientation (e.g. as in a drawn fibre, so that macro-anisotropy tending towards that in a single crystal is present), the ring pattern reduces to arcs or spots again (*Fibre diagram,* Figure 53.1c).

A fibre diagram yields the *identity period,* i.e. the smallest repeat distance along the fibre axis (which frequently corresponds to the length of the simplest fundamental unit, *see* Section 51), and the dimensions and angles of the *unit cell,* i.e. the simplest spatial repeat unit of the crystal structure. Scattering by the non-crystalline portions of polymers, or by non-crystalline polymers, yields much less information than that afforded by crystalline regions; while in some instances an 'average spacing' of oriented chains may be determined, unoriented amorphous polymers resemble liquids in giving only diffuse haloes (e.g. Figure 53.1d).

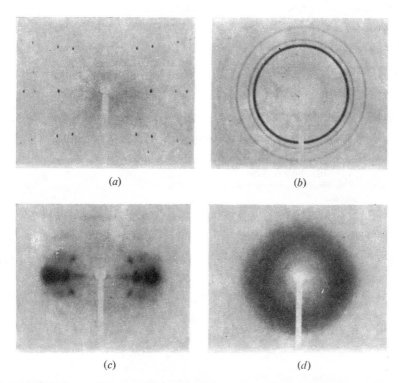

(a)   *(b)*

(c)   *(d)*

Figure 53.1. *Diffraction of x-rays. (a) Rotation diagram, from a single crystal (urea). (b) Powder diagram, randomly-disposed crystallites (powdered urea). (c) Fibre diagram, oriented crystallites (Terylene fibres). (d) Liquid-type diagram (amorphous Terylene).*

UNITS

Measurement of atomic and molecular spacings is expressed in tenth-metres or Ångström units (Å or Å.U.)

$$1 \text{ Å} = 10^{-10} \text{ m} = 10^{-8} \text{ cm} = 10^{-4} \text{ } \mu\text{m}.$$

COMPARISON TABLE

Table 53.T1 gives an indication of the internal structure of common polymers as revealed by x-rays. The results are for room temperature (*c.* 15–25°C); at a lower or higher temperature the internal state may be different (see Section 68).

Table 53.T1. THE INTERNAL STRUCTURE OF COMMON POLYMERS AT ROOM TEMPERATURE

Abbreviations: A = Substantially amorphous in any solid date (unless otherwise indicated).
A_C = Amorphous but with slight crystallinity.
A_Q = Amorphous when *quenched* (shock cooled from melt).
C = Largely crystalline in any solid state (unless otherwise indicated).
C_O = Crystalline in the oriented state (as fibres or films).
C_S = Crystalline after slow cooling or long storage.
C_T = Crystalline when held in tension.

Section No.	Polymer	Internal state	Identity period (Å)
1	Polyethylene	C	2·5
	Chlorosulphonated polyethylene	$A^{\cdot}$	—
	Ethylene/vinyl acetate copolymers	A	—
2	Polypropylene	C	6·5
	Polybutene-1	C	—
	Poly-4-methylpentene-1	A_C or C	—
3	Polytetrafluoroethylene	C	2·5–2·6
	Tetrafluoroethylene/hexafluoropropylene copolymer	A_C or C	—
	Polychlorotrifluoroethylene	C or A_Q	—
	Polyvinyl fluoride	A_C, C_O	—
	Polyvinylidene fluoride	C	—
	Fluoro-rubbers	A	—
4	Polystyrene	A	—
	ABS	A	—
5	Coumarone/indene resins	A	—
6	Polyvinyl acetate	A	—
7	Polyvinyl alcohol	C or A_C, C_O	2·6
8	Polyvinyl formal, etc.	A	—
9	Acrylic resins	A	—
	Acrylate rubbers	A	—
10	Acrylic fibres	A, C_O^*	*c.* 4·5
11	Polyvinyl chloride	A or A_C	*c.* 5
12	Polyvinylidene chloride	C, A_Q	4·7
13	Cellulose	C	10·3
14	Cellulose secondary acetate	A_C, A_Q	10·3
	Cellulose triacetate	C, A_Q	10·4
	Cellulose acetate butyrate	A	—
15	Cellulose nitrate	A or A_C, C_O	25·6, 38·3
16	Cellulose ethers (water-soluble)	A or A_C	—
17	Cellulose ethers (water-insoluble)	A or C†	—

Section No.	Polymer	Internal state	Identity period (Å)
18	Alginic acid, etc.	C	5·0‡, 8·7
19	Starch	A§, C	10·5
20	Silk	C	3·5 or 7
	Wool	C	3·3 or 6·7
21	Regenerated proteins	A‖	c. 4·5
23	Polyamides	C, A_Q	17·2¶
24A	Polyesters (alkyds)	C	—
24B	Polyesters (PET)	C, A_Q	10·8
24C	Polyesters (polycarbonates)	A_C	21·5
24D	Polyesters (unsaturated)	A or A_C	—
25	Epoxy resins	A	—
26	Polyformaldehyde	C	17·3
27	Amino resins	A	—
28	Phenolic resins	A	—
29	Natural rubber	A, C_S, C_T	8·1 or 25·1
	Gutta percha	C	8·8, 4·8 or 9·2
	Cyclised rubber	A	—
30	Chlorinated rubber	A, C_O	—
	Rubber hydrochloride	C	10·4
31	Ebonite	A	—
32	Isoprene rubber (synthetic)	A, C_S, C_T	8·1 or 25·1
33	Butadiene rubber	A, C_S, C_T	8·6
34	Styrene/butadiene rubbers	A	—
35	Nitrile rubbers	A	—
36	Chloroprene rubbers	A, C_S, C_T	4·3–4·8
37	Ethylene/propylene rubbers	A	—
38	Butyl rubber	A, C_S, C_T	18·6
39	Polysulphide rubbers	A, C_T**	4·3,†† 8·8‡‡
	Polysulphones	A	—
40	Polyurethanes	A	18·8§§
41	Silicones	A, C_S, C_T	8·3‖‖
42	Glass	A	—
	Asbestos	C	5·3¶¶
	Polyphosphonitrilic chloride	A, C_T	5·7
—	Liquid or molten polymers	A	—

*Only partially crystalline; copolymers usually amorphous
†Crystalline when fully etherified or nearly so
‡Sodium alginate fibres
§Dried starch
‖Fibres, A_C
¶Nylon 6.6 Nylon 6, 7.6 or 17.2; nylon 11, 14.9; nylon 6.10, 22.4
**Some types
††Polyethylene tetrasulphide
‡‡Polyethylene disulphide
§§Perlon U fibre
‖‖Polydimethylsiloxane
¶¶Amphibole type; chrysotile, 14·6 Å

FURTHER LITERATURE

BUNN, C. W., *Chemical Crystallography*, 2nd edn, London (1961)

VAINSHTEIN, B. K., *Diffraction of X-rays by Chain Molecules*, Moscow (1963); English version, Amsterdam (1966)

HERMANS, P. H., *Experimentia*, **19**, 553 (1963) (X-ray determination of crystallinity of high polymers)

KLUG, H. P. and ALEXANDER, L. E., *X-Ray Diffraction Procedures*, New York, 621 (1954)

MEYER, K. H., *Natural and Synthetic High Polymers*, 2nd edn (Vol. IV of *High Polymers*), New York and London, 48 (1950)

SECTION 54

PREPARATION OF HIGH POLYMERS

At least one method of preparation is indicated, where possible, for each polymer in Sections 1 to 42. In commercial practice, however, the precise techniques employed may show considerable variation, depending for instance on the availability of the raw materials or the type of product required. This is particularly so in the extraction, regeneration, or modification of natural organic polymers, such as cellulose, rubber, and the proteins; materials of this kind appear, in the first instance, to be synthesised by the action of specific enzymes, but the complex mechanisms involved are as yet imperfectly understood.

The synthesis of man-made organic polymers is a simpler matter: the molecules of a *monomer* of low molecular weight are made to polymerise—i.e. to join in a line or network—to yield a *polymer* of high molecular weight and a repetitive structure (*see also* Simplest Fundamental Unit, Section 51). Products of this type can be built up either by *addition polymerisation* or *condensation polymerisation*, notes on which appear below.

54.1 LINEAR POLYMERS*

54.11 ADDITION POLYMERISATION

This is a process applied to an unsaturated monomer—represented in the simplest form by $CH_2{=}CHX$—a molecule of which, once activated, can *add* a second molecule; and since the activity is then automatically transferred to the added molecule it, in turn, can add a further monomer molecule, and so on.

The initial activation (and subsequent polymerisation) may be effected by free radicals or ions, and the process of addition then proceeds, by a radical or ionic mechanism, as a *chain reaction*—monomer successively adding to (and in turn becoming) the active end of a growing chain molecule—until the radical or ion responsible' for the activity is neutralised or transferred elsewhere.

---

*Including branched-chain polymers and graft copolymers.

498

RADICAL POLYMERISATION

Free radicals can be produced by the action of heat or ultra-violet light on an unsaturated monomer, or by the decomposition of a catalyst—more correctly termed an *initiator* (e.g. benzoyl peroxide or azobisisobutyronitrile)—which itself needs to be activated (decomposed into radicals) by heat or ultra-violet light or, in some instances, by the presence of an *accelerator* (e.g. polyvalent metal naphthenates, tertiary amines, or mercaptans). Reduction-oxidation reactions (redox systems) provide a convenient source of free radicals in aq. solution.

The process of radical addition polymerisation can be represented as follows, where R—R (or R—R') = an initiator, R' = a free radical derived from R—R (or R—R'), and M = an unsaturated monomer such as styrene or vinyl acetate).

Initiation (impartation of initial activity)
(i) The initiator is activated, yielding free radicals,

e.g. $$R-R \longrightarrow 2R\cdot$$

then (ii) a free radical combines with a monomer molecule (to which the activity of an unpaired electron, denoted by ', is transferred),

$$R\cdot + M \longrightarrow RM\cdot$$

Propagation (successive addition of monomer).
As a polymer radical increases in length, it retains its activity at the growing end,

$$RM\cdot + M \longrightarrow RMM\cdot$$

$$RMM\cdot + M \longrightarrow RMMM\cdot, \text{ etc.}$$

$$(\text{or } RM\cdot + (n-1)M \longrightarrow RM_n\cdot).$$

Termination (loss of activity)
Occurs by (i) transfer,

$$RM_n\cdot + TH \text{ (transfer agent)} \longrightarrow RM_nH + T\cdot \text{ (new radical)}$$

or (ii) combination,

$$2 \sim CH_2C\cdot HX \longrightarrow \sim CH_2CHX - CHXCH_2 \sim$$

or (iii) disproportionation,

$$2 \sim CH_2C\cdot HX \longrightarrow \sim CH{=}CHX + \sim CH_2CH_2X$$

It can be shown that the rate of polymerisation is proportional to, and the average chain length (molecular weight) is inversely proportional to, the square root of the initiator concentration; also, since the initial production of radicals requires energy, the rate rises exponentially with temperature.

Radical polymerisation can be brought about in a liquid monomer (*bulk polymerisation,* which tends to give a low average molecular weight), or with the monomer dissolved in an inert solvent (*solution polymerisation*), or suspended or emulsified in water (*suspension* and *emulsion polymerisation** respectively, the last yielding a polymer of

*Emulsion polymerisation permits the use of a water soluble initiator, or a redox system (combination of reducing and oxidising agents) whereby polymerisation can be brought about at room temperature (e.g. *see* 'cold' rubbers, §34.21).

high molecular weight but contaminated with the other components of the emulsifying system). In the first two methods viscosity increases as the polymerisation proceeds, making transfer of the heat of polymerisation difficult; this, however, is avoided in the last two methods where each tiny drop of monomer is surrounded by water, and the product is eventually obtained—from the suspension technique—as a free-running 'powder' (minute spheres), or—from emulsion—as a high-solids low-viscosity latex (the extremely small particles of which can, if required, be coagulated or precipitated).

It is to be noted that in suspension polymerisation the initiator is dissolved in the monomer, where it subsequently produces free radicals; whereas in emulsion polymerisation free radicals enter the emulsified monomer after first being generated in the aq. phase.

IONIC POLYMERISATION

The ease with which many monomers can be activated by free radicals accounts for the popularity of radical-initiated addition polymerisation. However, initiation, propagation, and termination reactions similar to those described above apply in the two forms of ionic polymerisation, except that the mechanism is necessarily somewhat different, e.g. the rate of polymerisation is relatively independent of temperature (since no initial absorption of energy is required) and termination does not occur by combination or disproportionation (since each chain bears a charge of like sign).

In ionic polymerisation initiation is effected by addition of suitable catalysts, which, being ionic, are of two main kinds. For example, (i) the initiation of *cationic polymerisation* proceeds by conversion of the monomer to a *carbonium ion* (C^+) which is created by addition of a proton-donating substance, such as a strong acid (H_2SO_4, $HClO_4$), or a Lewis acid ($AlCl_3$, BF_3, $SnCl_4$) usually along with a trace of a co-catalyst,* and (ii) in initiation of *anionic polymerisation* the monomer is converted to a *carbanion* (C^-) by addition of a proton-abstracting or electron-donating substance, such as an alkali metal alkyl (e.g. LiBu) or amide (KNH_2) or alkoxide (NaOEt), or an alkali metal or alkali metal complex (e.g. $Na \cdot C_{10}H_8$). Monomers with electron-donating substituents, which supply a high electron-density at the double bond, favour polymerisation by electron-accepting initiators, i.e. cations (isobutylene, for instance, polymerises only in this way); whereas monomers with electron-abstracting groups (e.g. vinyl and vinylidene chlorides) respond best to electron-donating initiators, i.e. anions.

Ionic polymerisation is usually carried out in an inert non-aqueous

---

*Thus, aluminium trichloride with a trace of water or hydrogen chloride, is converted to a proton-donating strong acid, e.g.

$$AlCl_3 + H_2O \rightarrow H^+[AlCl_3OH]^-$$

medium*, where the catalyst not only effects transfer of charge to the monomer but also supplies a *counter ion* which becomes associated with the C^+ (or C^-) ion and moves along with it during the propagation of the polymer—polymer ion and counter ion forming an *ion-pair* (bound tightly together in a medium of low dielectric constant but less so in a polar solvent). Since little or no activation is needed, the rate of ionic polymerisation is usually very high, but the degree of polymerisation attained depends inversely on the temperature, since a low temperature—while having little effect on the propagation —suppresses the termination (which, occurring by transfer reactions —chiefly proton transfer—involves an intake of energy and thus increases with temperature rise); consequently, in order to obtain a product of high molecular weight, some ionic polymerisations are conducted at a very low temperature, e.g. isobutene (§§38.21 and 38.83) when polymerised at $-100°C$ very rapidly yields a polymer with a molecular weight of several millions.

It should be noted that in both radical and ionic systems the rate at which a polymer is propagated can be very high[†], consequently, soon after commencement of polymerisation it is possible to have fully-formed polymer along with as yet uninitiated monomer.

For examples of the preparation of some typical addition polymers *see*, for instance, §§6.21, 9.21, 11.21 (plastics), and §§34.21, 35.21, 36.21 (rubbers); *see also* (i), (ii), (iii), (v) and (vi) of §54.13 below.

54.12 CONDENSATION POLYMERISATION

As the name implies, this process proceeds by a series of *condensation* reactions, such as that between a hydroxyl (—OH) and a carboxyl (—COOH) group. A bifunctional monomer—or monomers—leads to a linear polymer. This has a repeat unit *lower* in weight than the monomer unit, since condensation involves elimination of a substance of low molecular weight, commonly water.

Other points of difference from addition polymerisation are as follows: (*a*) An initiating catalyst—in this instance an acid or base— though useful is not always essential. (*b*) Polymer is *not* built up by successive condensation of monomer on the end of a growing chain, but any species may condense together, i.e. reaction can occur between monomer/monomer, monomer/polymer, or polymer/polymer; hence, in this form of polymerisation the molecular weight

---

*A hydrocarbon, such as toluene, or a more polar liquid such as ether, N,N-dimethylformamide, or liquid ammonia; the presence of water, or other ionisable solvent, by supplying ions of opposite sign, terminates ionic polymerisation or prevents it from starting.
† However, in certain thermally dependent systems it is possible for the overall rate of production of polymer to decline to zero and even to become negative, the rate of propagation being then overtaken by that of depropagation; see ceiling temperature, Section 68.

reaches high values only towards the end of the reaction (cf. penultimate para §54.11 above) and, in order to attain it, it may be necessary to remove the by-product, e.g. by carrying out the condensation in a melt under vacuum. A further distinction from addition polymerisation is that, since the degree of condensation can be fairly readily controlled, it is possible to obtain intermediate products the condensation of which can be continued at a later stage (*see also* §54.2).

The preparation of a condensation polymer, effected in solution or a melt, can be represented (for a polyester) as follows:

SELF-CONDENSATION

$$n \, \text{HORCOH} \rightarrow \text{HORC} \left[\text{ORC} \right]_{n-2} - \text{ORCOH} + (n-1)\text{H}_2\text{O}$$

CONDENSATION BETWEEN TWO DIFFERENT MONOMERS

$$n \, \text{HOROH} + n \, \text{HOCR'COH} \rightarrow \text{HORO} \left[\text{CR'CORO} \right]_{n-1} - \text{CR'COH}$$
$$+ (2n-1)\text{H}_2\text{O}$$

It will be seen that to attain high molecular weight it is necessary to have closely equivalent quantities of the reactive groups, otherwise chain-ends become terminated by the group in excess; this equivalence is automatically present in a self-condensing monomer, such as a hydroxy acid, and also in the product represented by the condensation of one molecule of a diol (or diamine) with one molecule of a dibasic acid*.

For examples of the preparation of some typical condensation polymers *see*, for instance, §§23.21, 24B.21, 24C.21, 27.21, 28.21 and 39.21; *see also* (iii), (iv), (v) and (vi) of §54.13 below.

54.13 SPECIAL FORMS OF POLYMERISATION

(i) STEREOSPECIFIC POLYMERISATION

In 1953 K. Ziegler used a special organo-metallic catalyst to polymerise ethylene (normally requiring high pressure and an elevated temperature) at ordinary pressures and temperatures, and the resulting polyethylene proved to be less branched and thereby more crystalline and denser than usual. This important discovery of a new method of polymerisation and a new form of the polymer led to a further discovery, made by G. Natta in 1954, which may be described as follows.

*'Nylon salt' is an important example of a compound of this type (§23.21(3)).

The symmetry of the ethylene molecule allows it to produce only one kind of linear—unbranched—polymer; however, α-olefin molecules (and those of vinyl monomers) are asymmetrical and thus on polymerisation yield not only two possible ways of addition (i.e. head-to-tail or head-to-head) but also each repeat unit can exist in enantiomorphic configurations (i.e. mirror-images in *d-* or *l-*forms). Normally, in vinyl polymers derived via radical or ionic initiation the order of addition is mainly head-to-tail, but *d-* and *l-*forms of the repeat units occur quite randomly. A similar consideration applies if α-olefins are polymerised in the same way—although these reach only a moderate molecular weight. However, Natta made the discovery that when an α-olefin is polymerised with a Ziegler catalyst it yields a polymer of high molecular weight which is sterically ordered, i.e. in a given chain the configuration of each repeat unit tends always to be the same, hence the polymer is highly regular in structure and highly crystalline.

A Ziegler catalyst consists of a metal alkyl dissolved in a hydrocarbon solvent and reacted with a halide of a transition metal to form an insoluble complex (e.g. typically $Al(C_2H_5)_3$ complexed with $TiCl_4$), and since it directs a monomer molecule to enter a growing chain in a particular manner it is described as *stereospecific*. If in the resulting polymer each repeat unit has the same configuration it is said to be *isotactic**; a regularly alternating sequence of *d-* and *l-*repeat units is called *syndiotactic*, and molecules displaying a random arrangement† are *atactic‡*.

Other stereospecific catalysts, developed since Natta's discovery, consist of polyvalent metals or their oxides supported on carbon or silica, etc., and in some instances it is possible to obtain a highly stereospecific polymer from a soluble catalyst; e.g. polymerisation of isoprene with a dispersion of lithium, or lithium butyl dissolved in a non-polar solvent, yields largely the *cis*-1,4-polymer.

For examples of stereospecific polymerisation see §§ 32.21, 32.82, 33.21, and 37.21; for further literature *see* §54.3.

(ii) SOLID-STATE POLYMERISATION

Certain monomers have been found to polymerise in the solid state, e.g. crystalline acrylamide (CH_2=$CHCONH_2$, m.p. 85°C), after initiation by exposure to γ-rays, slowly polymerises to a crystalline water-soluble polymer of high molecular weight (termination reactions being suppressed by reason of the restricted mobility).

---

*For example, commercial polypropylene is over 70% isotactic; polymerised with a non-stereospecific catalyst it yields only irregular (atactic) polymers of low molecular weight.
†Still, however, predominantly head-to-tail in sequence (as with the atactic and syndiotactic arrangements).
‡Polystyrene, polyvinyl acetate, polyvinyl chloride, and polymethyl methacrylate as commercially produced are typical.

(iii) RING-OPENING POLYMERISATION

In this process molecules of a heterocyclic monomer are opened and become joined to each other, end to end, by what would appear— although proceeding in two possible ways—to be *addition polymerisation*, but the products closely resemble (and are often identical with) those derived from *condensation polymerisation**. In some forms of the ring-opening reaction, where a trace of an ionic catalyst—e.g. water—is required, the ring probably undergoes hydrolytic scission to form a monomer of the condensable type which thereafter undergoes condensation polymerisation (with regeneration of the catalyst); but other forms of ring opening require an ionic initiator and then proceed as a true addition polymerisation. Free radicals, in general, do not open ring molecules.

Ring-opening polymerisation has important industrial applications; *see*, for instance, §§23.21(i), 23.81 (Nylon 12), 26.84, and 41.21 (rubbers). The curing of epoxy resins, §25.21, also depends on ring opening. Formation of stable ring molecules† as an alternative to condensation polymerisation is usually undesirable and to be avoided.

(iv) INTERFACIAL POLYMERISATION

A typical example illustrating this process‡ is mentioned in §23.21(5); the appropriate condensation reaction between hexamethylene-diamine and adipyl chloride is represented as follows:

$$n\ H_2N(CH_2)_6NH_2 + n\ ClC(CH_2)_4CCl \xrightarrow{NaOH}$$

$$\left[-NH(CH_2)_6NHC(CH_2)_4C- \right]_n + 2n\ HCl$$

If the components are separately dissolved in immiscible liquids of different density, which are then brought into contact, reaction takes place at the interface so rapidly—even at *room temperature*—that the product may be withdrawn continuously as a film until the reagents are exhausted. A modified form of the process has been applied to fibre production, and it has been employed for coating

---

*One way of looking at this feature is to regard a condensation reaction as having taken place earlier when the ring monomer was formed. Thus, a lactone ring opens to produce a polyester by an addition reaction, but the lactone itself might have been obtained by condensation of a hydroxy-acid. However, if lactone formation were avoided, the same polyester might be derived directly from the hydroxy-acid by a condensation reaction. The end-products, identical from either path, would have the structure of a *condensation-type polymer*.

†For example certain 5- or 6-membered cyclic structures, which are nearly strain free.

‡An extension of the Schotten–Baumann reaction between an acid chloride and a substance with labile hydrogen atoms.

purposes, e.g. a yarn or textile is impregnated with one component and then passed through a bath containing the other. The cost and corrosiveness of the dicarboxylic acid chloride restricts its use.

(v) URETHANE-TYPE POLYMERISATION

This process can proceed as a true condensation polymerisation by reaction between a diamine and a bischloroformate and in this form can be used for interfacial polymerisation, see (iv) above), e.g.

$$n\,H_2NRNH_2 + n\,ClCOR'OCCl \rightarrow \left[-NHRNHCOR'OC-\right]_n$$
$$\qquad\qquad\;\; \underset{O}{\|}\quad\;\underset{O}{\|} \qquad\qquad\qquad \underset{O}{\|}\qquad\underset{O}{\|}$$

$$+2n\,HCl$$

But more usually it takes place, like ring-opening polymerisation (see (iii) above), by *addition* rather than condensation reactions, although the product has a structure identical with that obtained by condensation polymerisation.

This particular form of the process depends on the reactivity of the. $N{=}C$: double bond of the isocyanate group, $-NCO$, which reacts at room temperature with any substance containing labile H atoms, e.g. with a hydroxyl group,

$$RN{=}C + R'OH \rightarrow RNHCOR'$$
$$\quad\underset{O}{\|}\qquad\qquad\qquad\underset{O}{\|}$$

The
$$\begin{array}{c}-N-C-O-\\ \;\,|\quad\| \\ \;\,H\quad O\end{array}$$
group, obtained by either method, is known as a *urethane linkage** and reactions of the second kind are important in the preparation of *polyurethanes* (on which see information summarised in §40.21).

(vi) COPOLYMERISATION

A copolymer contains mixed repeat units derived from mixed monomers, and in radical *addition copolymerisation* the structure of the product obtained from two monomers depends on their relative polarisability. This is expressed by the *monomer reactivity ratios*; i.e. in a binary system, the rate at which monomer A adds to polymer radical $A^{\cdot}$ compared with the rate of addition of the other monomer, B, to $A^{\cdot}$ (and, correspondingly, the rate at which B adds to $B^{\cdot}$ compared with the rate of addition of A to $B^{\cdot}$).

A monomer reactivity ratio of <1 means that monomer favours addition to the opposite kind of radical, and if both monomers do this

---

*Named from *urethane* or ethyl carbamate, $NH_2COOC_2H_5$, though polyurethanes are *not* prepared from this compound.

exclusively (both ratios = 0)* the result is a regularly alternating copolymer (—ABABABAB—). If the chain end has no influence on the addition—i.e. a monomer adds equally well to its own or the opposite kind of radical (product of the ratios = 1)—the result is a random copolymer.† If the ratio >1, monomer favours addition to its own kind, giving sequences of the same repeat unit, and if both monomers did this exclusively (a condition not experienced in practice) the result would be two homopolymers produced simultaneously. It will be seen that, in general, if two monomers are dissimilar in electronic structure they will naturally tend to alternate in a copolymer, whereas if they are alike in structure a random copolymer is produced (the composition depending on the relative reactivities and concentrations of the monomers). Ionic copolymerisation proceeds at faster rates (and different monomer reactivity ratios) than the radical type; it is important in certain industrial applications, e.g. in the preparation of butyl rubber, §38.21.

In *condensation copolymerisation* there is usually no selective reactivity, each reactive group being uninfluenced by the main part of the molecule to which it is attached; hence the repeat units in a condensation copolymer tend to be randomly distributed in the same proportion as in the initial monomer mixture.

The preparation of some important copolymers is detailed in §§10.21, 34.21, 35.21, 37.21, 38.21 and 41.21, and some special forms of copolymerisation are described below.

Block copolymerisation yields runs of one kind of repeat unit alternating with runs of another kind, e.g.

$$—AAAABBBBAAAA—$$

Products of this type can be obtained by addition polymerisation in several ways, e.g. (i) by addition of a different monomer to a system in which the first monomer has been exhausted but polymer ions are still active ('living'), polymerisation then recommences when the new monomer is supplied; (ii) by masticating a rubber (or treating it with ultrasonic waves) in the presence of a vinyl monomer, the process first producing free radicals by fracturing rubber molecules and subsequently extending them by initiating polymerisation of the monomer on the ends (*see also* §29.85). Long chain molecules of

---

* Such monomers cannot be polymerised by themselves, but nevertheless can be copolymerised.
† In these circumstances,

$$\frac{k_{A\cdot A}}{k_{B\cdot A}} = \frac{k_{A\cdot B}}{k_{B\cdot B}} = 1$$

$$\therefore \frac{k_{A\cdot A}}{k_{A\cdot B}} = \frac{k_{B\cdot A}}{k_{B\cdot B}}$$

or $\quad r_A = 1/r_B \quad$ (thus $r_A r_B = 1$)

where $k_{A\cdot A}$, $k_{B\cdot A}$ etc. are the appropriate rate constants, and r_A and r_B are the respective monomer reactivity ratios.

urethane-linkages chain-extending a polyester or polyether (as in §40.21) constitute an extreme example of block copolymerisation; a related instance is provided by cured epoxy resins (§25.21), though cross-linkage is also present.

Block copolymers can also be obtained by condensation copolymerisation of mixed suitably prior-condensed homopolymers of low molecular weight.

Graft copolymerisation yields a 'backbone' of units of one type on to which are grafted chains of units of a different type, e.g.

$$—\text{AAAA AAAAAA}—\quad \text{etc.}$$
$$|$$
$$\text{BBBBB}—\quad \text{etc.}$$

Thus, if monomer B is addition-polymerised in the presence of preformed polymer A, which may be in solution or dispersed (as latex), by chain transfer reactions some of B becomes grafted on to A and grows as a pendant chain. Sites for growth of B on A can also be produced by photolysis of appropriate groups, or inclusion of peroxy-groups, on A.

The compatibility of styrene/butadiene rubber with polystyrene is much improved—thus raising impact strength—when the SBR is grafted with polystyrene chains (*see* §4.81). Inclusion of double bonds in the main chain of a polyester permits its subsequent cross-linkage by polystyrene chains (*see* §24D.1).

For further literature on copolymerisation *see* §54.3.

54.2 THREE-DIMENSIONAL POLYMERS

The synthesis of a non-linear polymer involves reactions and mechanisms similar to those described for linear polymers, except that a three-dimensional structure requires inclusion of at least one component having a functionality* greater than 2. The polymerisation may be carried out directly in one stage, but since an end-product consisting of a cross-linked or network structure is fairly intractable it is more often convenient to prepare it via a moderately stable but reactive intermediate. This is first fabricated† to a desired shape, or applied in a desired position, and may then be reacted to form the final stage *in situ*.

Processes of this kind depend on the preparation of intermediates possessing unsaturation or reactive groups, which can subsequently yield the required cross-linkage or network, e.g. on heating, or on addition of a catalyst or further component to the system. Examples are seen in moulding and laminating with phenol-formaldehyde and

See Section 51.
†Usually after compounding with fillers, pigments, extenders, or fibre reinforcement, etc.

other thermosetting resins (§§27.21, 28.21), formalisation of poly-vinyl alcohol and regenerated proteins (§§7.21, 21.21), vulcanisation of rubbers (§29.4, Fabrication) and the curing or hardening of epoxy, polyester and other bonding resins (which may sometimes be done at room temperature); the 'drying' of paint and stoving enamel films, and the crease-resisting and crease-setting of resin-finished cellulosic textiles (§13.86) also depend on the formation of complex molecular networks. A further method of cross-linkage which may be mentioned is that brought about when certain linear polymers are exposed to high-energy radiation; cross-linkage of polyethylene by γ-rays provides an example.

54.3 FURTHER LITERATURE

Theory

FRITH, ELIZABETH M. and TUCKETT, R. F., *Linear Polymers,* London, 44 (1951)

D'ALLELIO, G. F., *Fundamental Principles of Polymerisation,* New York (1952)

FLORY, P. J., *Principles of Polymer Chemistry,* Ithaca, N.Y. (1953)

PLESCH, P. H. (Ed), *Cationic Polymerisation and Related Complexes,* Cambridge (1953)

PLESCH, P. H. (Ed), *Chemistry of Cationic Polymerisation,* Oxford (1963)

BOVEY, F. A., KOLTHOFF, I. M., MEDALIA, A. I. and MEEHAN, E. J., *Emulsion Polymerisation (High Polymers,* Vol. IX), New York and London (1955)

BAMFORD, C. H., BARB, W. G., JENKINS, A. D. and ONYON, P. F., *The Kinetics of Vinyl Polymerisation by Radical Mechanisms,* London and New York (1958) (includes copolymerisation)

SMITH, D. A. (Ed), *Addition Polymers,* London (1968)

Synthesis and manufacture

SORENSON, W. R. and CAMPBELL, T. W., *Preparative Methods of Polymer Chemistry,* 2nd edn, New York (1968)

PINNER, S. H., *A Practical Course in Polymer Chemistry,* Oxford (1961)

SMITH, W. M., *Manufacture of Plastics,* New York and London (1964)

LONG, R. (Ed), *Production of Polymer and Plastics Intermediates from Petroleum,* London (1967)

OVERBERGER, C. G. (Ed), *Macromolecular Synthesis: a Periodic Publication of Methods,* New York and London, Vol. I (1963)

Stereospecific polymerisation

NATTA, G. and DANUSSO, F. (Eds), *Stereoregular Polymers and Stereospecific Polymerisation,* Oxford (1967) (two vols., translated from the original Italian papers)

PETERLIN, A., *et al.* (Eds), *Macromolecular Reviews,* Vol. 2 (pp. 115–268, BOOR, J., *The Nature of the active site in the Ziegler catalyst*), New York and London (1967)

KETLEY, A. D. (Ed), *The Stereochemistry of Macromolecules* (3 vols.), London and New York (1967–1968)

Copolymerisation

BURLANT, W. J. and HOFFMAN, A. S., *Block and Graft Copolymers,* London (1960)

CERESA, J. R., *Block and Graft Copolymers,* London and New York (1962)

HAM, G. E., *Copolymerisation (High Polymers,* Vol. XVIII), New York and London (1964)

FETTES, E. M. (Ed), *Chemical Reactions of Polymers (High Polymers,* Vol. XIX), New York and London, 1085 (1964). (WATSON, W. F., 'Mechanical Reactions', including block copolymerisation)

SECTION 55

SOLVENTS

GENERAL PRINCIPLES

It is commonly found that liquids of a similar kind may be mixed in all proportions, but those very different in composition are often immiscible. The same generalisation* applies to the dissolution of a solid in a liquid, including—with certain reservations—the dissolution of a linear polymer in a solvent.

Swelling (penetration of liquid into polymer) and *solution* (dispersion of polymer into solvent) require that the intermolecular forces† be approximately the same in magnitude in the solid and liquid—otherwise the greater cohesion in one phase tends to exclude molecules of the other phase. The processes, depending on thermal motion, are assisted by interaction (*see* below) between the solid and liquid, and solution is also assisted by mechanical agitation.

Crystallinity in a linear polymer restricts solubility, but sometimes this can be improved by copolymerisation, which by reducing the regularity of the molecular structure renders crystallisation difficult or impossible. The presence of covalent cross-links prevents solubility; however, when the cross-link density is low, solvent penetrates accessible regions until equilibrium is reached, with the polymer in a highly swollen state (e.g. as with lightly cross-linked rubber)—nevertheless, when the structure is largely that of a covalently-bonded spatial network (as in a fully-cured phenol-formaldehyde resin) there may be virtually no swelling.

Examples of linear polymers with weak and relatively strong intermolecular cohesion are to be seen, respectively, in polystyrene (which dissolves in hydrocarbons but leaves water—a more cohesive liquid—untouched) and polyvinyl alcohol (which is untouched by hydrocarbons but dissolves in water). Note, however, that polyethylene, though a hydrocarbon like polystyrene, is in addition polycrystalline and fails to dissolve in hydrocarbons until approaching its crystalline m.p.; similarly, cellulose, with —OH groups like polyvinyl alcohol, readily swells in but is not dissolved by water

---

*The alchemists had a phrase for it: *similia similibus solvuntur* (like dissolves like).
†Measured by the 'internal pressure' or *cohesive energy density*, i.e. the energy needed to separate the molecules of 1 ml of a substance (units: cal/ml). For a liquid this is equivalent to the latent heat of vaporisation per millilitre at constant volume. *See also* footnote ‡, Section 51.

since it contains H-bonded crystalline regions that need the presence of highly polar complex ions for their dissolution.

THEORETICAL ASPECTS

Notes are given below on two criteria of solvent efficiency, namely *solubility parameter* and *polymer/solvent interaction parameter*, and on the *theta temperature*, and *swelling*.

SOLUBILITY PARAMETER

When a polymer and a solvent mix (i.e. when solution occurs) there is a decrease in free energy, ΔF, which is represented by

$$\Delta F = \Delta H - T\Delta S$$

where ΔH is the heat of mixing (i.e. the change in cohesive energy or heat content), T is the absolute temperature, and ΔS is the change in entropy.

The separation of solute molecules by a solvent increases their freedom, and the entropy change is positive; therefore, if the free energy is to decrease, i.e. if ΔF is to be negative, ΔH needs to be algebraically less than $T\Delta S$.

The heat of mixing, when endothermic, is given by

$$\Delta H = V_m(\delta_1 - \delta_2)^2 V_1 V_2$$

where V_m is the molar volume of the mixture, δ_1 and δ_2 are the *solubility parameters* of solvent and solute respectively, and V_1, V_2 are the corresponding volume fractions of each component. The solubility parameters are the square roots of the respective cohesive energy densities (*see* earlier footnote), and may be estimated experimentally.

Since ΔH depends on δ_1 and δ_2, and needs to be small for a decrease in free energy, solution is most likely to take place when $\delta_1 = \delta_2$, i.e. when $\Delta H = 0$ (solvent and polymer then having the same cohesive energy density). These theoretical conditions often apply in practice, where it is found that—although the value of the solubility parameter rises, with increasing polarity, roughly from 5 up to 24 $(\text{cal/ml})^{\frac{1}{2}}$ —the parameter of a satisfactory solvent is not more than about 1 removed from that of the polymer it dissolves.

When there is interaction between the solute and solvent, energy is released and ΔH is negative, so that there is a greater decrease in free energy, i.e. solution is further assisted.

For values of solubility parameters *see*, for instance, pp. IV 341– IV 368 of *Polymer Handbook* (*see* **Further Literature,** end of this section). The values show some variation with different solvent systems; for additional parameters, relating to dipole moment and hydrogen bonding, *see* CROWLEY, J. D., TEAGUE, G.S. and LOWE, J. W., *J. Paint Technology*, **38**, 269–280 (1966), and for subdivision of the parameter, *see* HANSEN, C. H., *J. Paint Technology*, **39**, 104–117 (1967).

POLYMER/SOLVENT INTERACTION PARAMETER

The heat of mixing can be defined as

$$\Delta H = kT\chi_1 n_1 V_2$$

where k is Boltzmann's constant ($1\cdot38.10^{-16}$ erg/deg), T is the absolute temperature, n_1 is the number of solvent molecules, V_2 is the volume fraction of the solute and χ_1 —sometimes written as μ— is a *polymer/solvent interaction parameter* ($kT\chi_1$ being the energy difference between a solvent molecule surrounded by polymer molecules and surrounded by pure solvent).

It can be shown that χ_1 bears a complex relation to the change in free energy, ΔF; and for ΔF to be negative, so that solution takes place, the value of χ_1 cannot much exceed $0\cdot5$. Lower values of χ_1, representing appreciable decrease in free energy, are indicative of good solvents, but above the value of c. $0\cdot5$ the system consists of swollen polymer and virtually pure solvent. The critical value, below which solution occurs and above which a polymer only swells in a solvent (the swelling decreasing as the value of χ_1 increases), is given by

$$\chi_c = 0\cdot5\,(1+1/x^{\frac{1}{2}})^2$$

where x is the degree of polymerisation.

For values of the interaction parameter *see*, for instance, pp. IV 157–IV 162 of *Polymer Handbook* (*see* **Further Literature,** end of this section).

THETA TEMPERATURE

Solubility improves, poor solvents becoming better ones, with rise of temperature. At the *theta temperature* polymer/solvent interaction is zero (i.e. $\Delta H = 0$; polymer molecules assume the unperturbed conformation in solution and the solvent is said to be a poor one). Above the θ temperature interaction causes the polymer molecules to become more extended (i.e. the solvent is a good one); below the θ temperature the polymer molecules are contracted, and ultimately precipitate from solution.

SWELLING

Cross-linkage prevents dissolution, but in a liquid that would otherwise act as a solvent a lightly cross-linked polymer (such as vulcanised rubber) swells until attainment of *equilibrium sorption.* This is inversely related to *cross-link density* (measured as mol cross-links per gramme polymer) in the following way,

$$-\ln(1 - v_r) - v_r - \chi_1 v_r^2 = \rho V_0 \left(\frac{1}{M_c} - \frac{2}{M} \right) (v_r^{\frac{1}{3}} - v_r/2)$$

where v_r = volume fraction of polymer in the swollen system
        χ_1 = polymer/solvent interaction parameter
        ρ = density of unswollen polymer
        V_0 = molar volume of solvent
        M_c = average mol. wt. of polymer between cross-links
        M = average mol. wt. of polymer before cross-linking.

The above relationship is often used, especially with vulcanised rubbers, to determine M_c and hence the cross-link density, which is $1/(2M_c)$. For an approximate result, if M is much greater than M_c and if the swelling is considerable (as is usual, notably in rubbers), the negative terms on the right-hand side of the equation can be neglected, since then $2/M$ is small compared with $1/M_c$, and $v_r/2$ is small compared with $v_r^{\frac{1}{3}}$.

See also NAUNTON, W. J. S. (Ed), *The Applied Science of Rubber,* London, 992 (1961); BATEMAN, L. (Ed), *The Chemistry and Physics of Rubber-like Substances,* London and New York, 465 (1963).

PRACTICAL ASPECTS

The solubility, or otherwise, of high polymers is of great importance in many industries. The main organic solvents are coal-tar and petroleum hydrocarbons (and their derivatives) together with alcohols (and their ester and ether derivatives) and ketones, some produced by microbiological means. Halogenation of hydrocarbon solvents usually increases their effectiveness and reduces inflammability, but increases the cost and in some instances the toxicity. Mixed solvents may offer advantages, for example, a polymer dissolved in an expensive solvent may tolerate considerable dilution with a cheap non-solvent*, and even non-solvents when mixed may exhibit effective solvent action, i.e. if the solubility parameter of the blend approaches that of the polymer concerned. The rate of evaporation of solvents can

---

*For a solution of a given concentration, at the point of maximum tolerance (where haziness commences), the *dilution ratio* = (vol. of non-solvent)/(vol. of solvent).

influence the properties of films, and use of the correct blends is an important item in the surface coatings industry.

A practical disadvantage of polymer solutions, especially with polymers of high molecular weight, is the high viscosity even at relatively low concentrations. This can often be avoided, however, by employing aq. latices (sometimes called 'emulsions'), such systems remaining freely fluid at concentrations of up to, say, 50% (vol/vol) solids; hence, latices are important in impregnating and surface coating applications.

Solutions normally become thinner with rise of temperature, but many other factors affect their viscosity. For instance, addition of rosin to a solution of natural rubber causes contraction of the molecules of the hydrocarbon polymer with consequent reduction in solution viscosity. Similarly, the properties of aq. solutions of polymers having ionizable substituents ($-COOH$, $-NH_2$, etc.) can depend on the pH of the system; e.g. the viscosity of a solution of polyacrylic acid rises markedly on neutralisation and is at a maximum somewhat beyond pH 7, above which the viscosity then declines. Alginic acid behaves in a comparable way, swelling in water but dissolving only as an alkali metal or ammonium salt. The interaction between a solvent and a polar polymer tends to be reduced by the presence of electrolytes, addition of which may cause precipitation of an otherwise water-soluble material.

Block and graft copolymers in which the components differ appreciably in character may sometimes be dissolved by solvents having greater attraction for one kind of repeat unit than for the other. Two types of films can be cast from such solutions, differing in properties according to whether the one component or the other was more extended in the solution. (*See*, for example, MERRETT, F. M., *J. Pol. Sci.,* **24**, 467 (1957).)

Fibres are usually required to be resistant to the action of water, wash liquors, and dry-cleaning solvents, but sometimes water- or alkali-soluble fibres are used, e.g. for soluble laundry bags and for temporary scaffolding yarns employed in special forms of weaving. Differential effects of liquids on fibres (and other forms of polymers) are of assistance in identification and in the analysis of mixtures.

Vulcanised rubbers are often required to resist the swelling action of hydrocarbons, e.g. petrol and lubricating oils. This can be achieved by: (i) selecting a polar rubber, such as a nitrile rubber (Section 35), chloroprene rubber (Section 36), a polysulphide (Section 39) or a fluorinated rubber (§3.85); (ii) ensuring adequate vulcanisation—see formula earlier, under *Swelling*, for cross-link density; (iii) incorporation of non-swelling fillers; and (iv) incorporation of extractable softeners that counterbalance the absorption of the hydrocarbon. However, since excessive cross-linkage by sulphur may impair ageing-resistance, while increase of cross-linkage or filler-content makes the product harder and stiffer, there are limits to the application of (ii) and (iii).

FURTHER LITERATURE

Practical

RIDDICK, J. A. and TOOPS, E. E., *Organic Solvents* (A. WEISSBERGER (Ed), *Technique of Organic Chemistry*, Vol. VII), New York and London (1955)

DURRANS, T. H., *Solvents*, 7th edn, London (1957)

MELLAN, I., *Source Book of Industrial Solvents*, Vols. 1–3, New York (1957–9)

MARSDEN, C. and MANN, S., *Solvents Guide*, London (1963)

BRANDRUP, J. and IMMERGUT, E. H. (Eds), *Polymer Handbook*, New York and London (1966) (pp. IV 185–IV 234, solvents and non-solvents)

BROWNING, ETHEL, *Toxicity of Industrial Organic Solvents*, London (H.M.S.O.) (1953)

Theoretical

MINISTRY OF SUPPLY, *Services Rubber Investigations, Rubber in Engineering*, London (1945)

HILDEBRAND, J. H. and SCOTT, R. L., *The Solubility of Non-electrolytes*, New York (1950)

FLORY, P. J., *Principles of Polymer Chemistry*, New York (1953)

TOMPA, H., *Polymer Solutions*, London (1956)

HUGGINS, M. L., *Physical Chemistry of High Polymers*, New York (1958)

MORAWETZ, H., *Macromolecules in Solution* (*High Polymers*, Vol. XXI), New York and London (1966)

Specifications

BS 903: Part A16 (Resistance of vulcanised rubbers to liquids)

ASTM D471 (Resistance of vulcanised rubbers to liquids)

ASTM D543 (Resistance of plastics to reagents)

ASTM D1239 (Resistance of films to extraction by chemicals)

See also BS and ASTM specifications for various organic solvents.

SECTION 56

PLASTICISERS

A plasticiser is a substance added to a high polymer to reduce brittleness or impart flexibility, or to increase flexibility especially at low
temperature*. It acts like a solvent†, penetrating the polymer and
lowering the intermolecular cohesion. In this its action is similar
to that of heat; in fact, plasticisers lower the glass temperature, T_g
(*see* Section 68). Thus, amorphous linear polymers tend to be readily
plasticised by appropriate compounds, but crystalline polymers are
more difficult, and cross-linked polymers—except for lightly cross-
linked rubbers—can hardly be plasticised at all, though some degree
of plasticisation is possible with ebonite (Section 31).

A plasticiser needs to fulfil certain conditions, chief of which are
that it shall be *compatible* and *efficient*, i.e. it should not exude from
the plasticised mass and it should impart reasonable flexibility.
Usually, too, its effects should be *permanent*; therefore, it should be
involatile (unlike many solvents), it should resist extraction by liquids
with which the mass may come into contact (e.g. water, detergent
solutions, petrol and mineral oils, or vegetable oils and fats), and in
the course of time it should neither change physically or chemically
nor undergo migration to neighbouring materials. It may also need to
withstand heat and heat-ageing and/or to retain its plasticising action
down to low temperatures. Other considerations may be: non-
toxicity (e.g. in packaging of foods), fire-retardancy, maintenance of
good electrical properties, and low cost. Plasticisers almost invariably
lower tensile strength, an effect that often needs to be kept to a
minimum.

Substances largely fulfilling the above requirements usually have a
moderate molecular weight—not so low as to render them volatile,
not so high as to restrict compatibility and processing—and most of
them are high-boiling liquids (a few are solid at room temperature).
A *primary plasticiser* is one compatible in all proportions with a given
polymer, a *secondary plasticiser* is more limited in compatibility but
may often be blended with a primary one with advantage. Some

---

*The term sometimes implies a substance that facilitates the processing of a polymer, making it
softer and more plastic when hot, though the final product may be hard at room temperature.
Camphor, which plasticises cellulose nitrate by forming a loose compound with it, may be
considered to act in this way.
†The parameter χ_1 (*see* Section 55) measures plasticising action; the lower the value, the
greater the polymer/plasticiser compatibility.

plasticisers are initially incompatible at room temperature but become operative above a certain temperature and then remain incorporated on cooling to room temperature (e.g. *see* the PVC-paste technique, §11.4 iv).

Table 56.T1. SOME COMMON PLASTICISERS AND THEIR PROPERTIES*

Plasticiser	Viscosity at 25°C (cSt)	Freezing point (°C)	Boiling point, at 760 mmHg (°C)	Specific gravity at 20–25°C	Refractive index, at 20–25°C, 5893Å
Ethylene glycol	16	−12	198	1·115	1·43
Glycerol	c. 10³	0§	290‖	1·26	1·475
Triacetin	14	−40	260	1·16	1·43
Camphor	—	+176	204	1·0	1·53
Castor oil	680	−12	>300‖	0·96	1·48
Butyl acetyl ricinoleate	c. 25	c. −50	235¶	0·93	1·46
Dibutyl phthalate	16	−35	335	1·04	1·49
Dioctyl phthalate†	55	−50	385	0·98	1·485
Ditridecyl phthalate	200	−30	285¶	0·95	1·48
Dioctyl adipate†	12	−70	220¶	0·93	1·45
Dioctyl sebacate†	20	−55	256¶	0·91	1·445
Triphenyl phosphate	—	+48	410	1·19	1·56
Tritolyl phosphate‡	75	<−35	430	1·16	1·56
Trixylyl phosphate	105	—	243–265**	1·14	1·55
Chlorinated biphenyls	Oily liquids to resinous solids		275 to >400	c. 1·2–1·7	>1·6
Epoxidised soya-bean oil	c. 350	—	>150¶	c. 0·99	1·47
Polyesters	10–10⁴	—	—	c. 1·0–1·1	c. 1·45–1·50
Mineral oils and petroleum fractions	20–1000††	−50 to +7	c. 300–500	0·86–1·05	c. 1·56
Pine tar	1200–4300	—	—	1·03–1·09	—

Solubility and *compatibility*. Ethylene glycol and glycerol are miscible in all proportions with water, triacetin is c. 7% soluble (hydrolyses in hot water) and camphor c. 0·2%. The rest of the plasticisers listed are largely insoluble in water—under 0·2% and often very much less—and water is largely insoluble in them. Most of these lyophobic plasticisers are compatible with cellulose nitrate and polyvinyl chloride but only some are suitable for cellulose acetate and certain other polymers
†Di-2-ethylhexyl ester
‡ = Tricresyl phosphate
§Melts at 17–20°C
‖Boils with partial decomposition
¶At 4–5 mmHg
**At 10 mmHg
††At 60°C, 7–500 c St; at 100°C, 3–60 c St

Some of the more prominent plasticisers and their properties are indicated in Table 56.T1. Water has a marked plasticising action on hydrophilic polymers (e.g. cellulose, gelatin, polyvinyl alcohol) but usually it is preferable to employ less volatile H-bonded substances— such as glycol, glycerol, pentaerythritol, sugars (e.g. sorbitol), urea and sulphonamides—some of which act as *humectants*, assisting retention of moisture. For plasticisation of hydrophobic polymers, and notably for PVC, numerous plasticisers have been developed, many of which are high-boiling esters. The principal classes of these are: castor oil and its derivatives; tri-aryl phosphates; di-alkyl phthalates, adipates and sebacates; and (for rubbers) mineral oils, naphthenic and aromatic fractions from petroleum, coal tar fractions and pine tar. Other hydrophobic plasticisers are chlorinated bi- and poly-phenyls, epoxidised oils (act as both stabilisers and plasticisers) and certain *polymeric plasticisers* (having the advantage of minimum or zero migration) such as liquid polyesters, polyisobutene, and nitrile rubbers. Polymerisable plasticisers (allyl, acrylic, or maleic esters) offer interesting possibilities of incorporation of a plasticiser followed by cross-linkage or formation of a network.

For service at low temperatures the low-viscosity adipate and sebacate esters are useful; usually low-temperature plasticisers also effect least lowering of the electrical resistivity of a polymer. For service at high temperatures, also for extreme resistance to water, very long-chain esters (e.g. di-tridecyl phthalate) are suitable, though their viscosity at room temperature and below is high; phosphate esters offer fire-retardancy (especially if halogenated) and some heat resistance but give poor low-temperature flexibility. Economic considerations may, of course, dictate the plasticiser ultimately chosen.

Internal plasticisation. One of the best ways of permanently plasticising a polymer is either partially to modify it chemically, or to copolymerise the monomer with a small proportion of another monomer, which yields a product softer and more readily fabricated than the homopolymer. Among the synthetic rubbers, ethylene/propylene copolymers are an excellent example of internal plasticisation.

FURTHER LITERATURE

CURME, G. O. (Ed), *Glycols,* New York (1952) (includes use as plasticisers)

MINER, C. S. and DALTON, N. N., *Glycerol,* New York (1953)

DOOLITTLE, A. K., *Technology of Solvents and Plasticisers,* New York and London (1954)

BUTTREY, D. N., *Plasticisers,* London (1960)

MELLAN, I., *The Behaviour of Plasticisers,* Oxford (1961)

MELLAN, I., *Industrial Plasticisers,* Oxford (1963)

MELLAN, I., *Plasticiser Evaluation and Performance,* Park Ridge, N.J. (1967)

MELLAN, I., *Compatibility and Solubility: Polymers and Resins, Plasticisers and Esters,* Park Ridge, N.J. (1968)

PARKYN, B., LAMB, F. and CLIFTON, B. V., *Polyesters* Vol. 2. *Unsaturated Polyesters and Polyester Plasticisers,* London (1967)

WEINER, J. and ROTH, Ł., *Plasticisers for Paper (Inst. Paper Chem.* Biblio. Series No. 202), Wisconsin (1963) (abstracts of 165 refs.)

BRUINS, P. F. (Ed), *Plasticiser Technology,* Vol. 1, London (1965)

GOULD, R. (Ed), *Plasticisation and Plasticiser Processes* (Amer. Chem. Soc., *Advances in Chemistry* series 48), Washington (1965)

RITCHIE, P. D. (Ed), *Physics of Plastics (Plastics Inst. Monograph),* London, 323 (1965); LANNON, D. A. and HOSKINS, E. J. (effects of plasticisers and fillers, mainly in PVC and cellulose derivatives)

Specifications
ASTM D1045 (Testing of plasticisers)
See also BS and ASTM specifications for various plasticisers (by name)
See also Further Literature, Section 55. For removal of plasticisers, *see under* Identification, Section 57.

IDENTIFICATION

INTRODUCTION

When the probable composition is known, a few simple tests—sometimes even a single one—will generally suffice to confirm the identity of a high polymer; accordingly, numerous tests are given in various parts of this book. Additionally, when the composition of a polymer is unknown, employment of such tests (often in combination with, for instance, solvent and non-solvent information) makes it possible to classify the material as belonging to a particular group.

However, while simple tests and identification schemes may supply useful clues, it is better, if possible, to employ one (or more) of the more complex analytical techniques, since these, although mostly requiring sophisticated equipment, usually afford a much more complete and unequivocal characterisation of the material examined.

The main part of the present section deals with certain preliminary procedures and simple tests, described immediately below; the more complex analytical techniques are briefly surveyed later.

PRELIMINARY PROCEDURES AND TESTS

When a substance of unknown composition is to be examined, the following procedures may be used in conjunction with the confirmatory tests given in §§1.23 to 42.23 and elsewhere.

1   ESTABLISHMENT OF GENERAL CHARACTER

1.1 HOMOGENEITY OF SAMPLE

If a sample contains more than one polymer and/or non-polymeric material it is sometimes possible, by using an optical microscope, to carry out characterising reactions individually on the mixed components, but usually it is necessary first to separate them. Various methods are available for this, e.g. hand-sorting, sifting, flotation, selective extraction (preferably in a Soxhlet apparatus, with a sample in a finely-divided state), selective precipitation from a dissolved sample, or isolation of one component by selective chemical destruction of the rest. Ester-type and related plasticisers can be largely removed by distillation under reduced pressure, or can be extracted with solvents such as petroleum ether or chloroform (provided that the polymer is insoluble); dyes of the direct class can often be stripped by steeping at room temperature in 25% pyridine in water, and some other dyes respond to warmer and stronger solutions (a solution of polyvinylpyrrolidone may also be useful here, *see* §7.82). The solubility of a polymer as such is, of course, an indication that it is not a cross-linked structure or network; these may swell in liquids but cannot be dissolved without chemical modification.

1.2 RECOGNITION OF POLYMERIC NATURE

Since, among organic compounds, the attribute of appreciable physical strength is unique to high polymers, if a sample is initially in the form of a fibre or film, or has been moulded or extruded, it is very likely to be polymeric in nature. Similarly, a high degree of elastic deformability (i.e. of an elastomer) is an alternative indication of high molecular weight. However, some polymers are brittle and as those behaving in this way are mostly soluble they are deposited as brittle films on evaporation of the solvent. In any instance, therefore, when a film has been obtained it may be useful to judge its general character in the following manner.

Needle test—Probe a deposited film with a mounted needle (or the tip of a penknife blade): if the film is polymeric it will be either *tough* or *rubbery*—and may perhaps come freely away from the substrate—or it will be *brittle*, tending to be shattered (or scratched locally) as the needle is pushed through it—or, again, it may lift from the substrate as a film which in this instance is easily broken.

N.B. In contrast, evaporation of a solution of a solid of low molecular weight yields a brittle *crystalline* deposit, while a grease or wax gives a film in which the needle can be fairly freely pushed about, without lifting or shattering it.

1.3 SIMPLE CHARACTERISATION

It cannot be over-emphasised that careful observation should be the rule throughout an examination; in this respect valuable clues may be forthcoming from initial combustion and pyrolysis tests, which may be conducted as here outlined.

Combustion test A small sample held in tweezers is slowly approached towards the side (and near the base) of a clear bunsen flame, preferably from a micro-burner, and the behaviour noted as below.

(i) *Before the sample ignites*—e.g. whether it shrinks or melts (indicating a thermoplastic substance or otherwise).

(ii) *When touched into the flame and withdrawn*—e.g. combustible or not (burning with a very sooty flame indicates unsaturation; burning only when *in* the flame often indicates the presence of chlorine).

(iii) *When extinguished*—odour may be characteristic, e.g. that of monomer or oligomers, rubber, sulphur, formaldehyde, celery (nylon), burnt paper or protein, etc.

(iv) *When allowed to burn freely*—may burn away completely, or leave a carbonaceous or inorganic residue, e.g. white or pale grey from silicone polymers and certain fillers.

For detailed results of this test, *see* §§1.23 to 42.23.

*Pyrolysis test** Heat a small sample—a few milligrammes in the bottom of a 50 mm × 10 mm ignition tube is sufficient—and test the pyrolytic vapours with a moistened indicator paper.

Neutral vapours are evolved from hydrocarbon polymers, silicones, and some polyesters.

Acid vapours may come from carbohydrate polymers and their derivatives; *highly acid vapours* often indicate the presence of chlorine, e.g. in polyvinyl chloride or rubber hydrochloride.

Alkaline vapours almost always indicate the presence of nitrogen, e.g. in polyamides, polyurethanes, proteins, and amino-formaldehyde resins.

The odour of the fumes and the nature of any residue should be noted as in the previous test. For detailed results, *see* §§1.23 to 42.23.

2. ESTABLISHMENT OF ELEMENTS PRESENT
(other than C, H or O)

Unless the material to be examined is in very short supply it is usually worth conducting a simple qualitative analysis to find which additional elements are present. These may be detected in the following ways:

2.1 CHLORINE (BEILSTEIN COPPER WIRE TEST)

Heat one end of a stout copper wire—size 18–22 SWG—in a clear bunsen flame until it ceases to impart any colour, and while the wire is

---

*Polymers may be identified through the characterisation of their pyrolysates by gas-liquid chromatography or infra-red spectroscopy (*see* later in this section).

still hot push it into, or rub it on, the sample to be tested, then return it to the flame. A distinct green coloration indicates chlorine (or bromine or iodine).

For polymers responding to this test, see the chlorine sections of Table 57.T1. For detection of *fluorine*, which does not behave in the above way, *see* §3.23.

2.2 NITROGEN (SODA-LIME TEST)

A small sample in the bottom of an ignition tube (e.g. size 50 mm × 10 mm) is well covered with granular soda-lime, which is packed down in a thick layer above it. The sample-end of the tube is then heated and the volatile products that diffuse through the soda-lime plug are tested with a moistened indicator paper. An *alkaline reaction* indicates the presence of NITROGEN.

N.B. A few ester-type polymers containing nitrogen—e.g. cellulose nitrate and some derived from acrylonitrile (vinyl cyanide)— yield acid products, and thus do not respond; otherwise this is a very sensitive test.

For polymers containing nitrogen *see* Table 57.T1.

2.3 PHOSPHORUS

Reagents Solution **I**: dissolve 0·5 g ammonium molybdate in 10 ml water and add 1 ml conc. nitric acid. Solution **II**: dissolve 0·05 g benzidine (or b.hydrochloride) in 1 ml glacial acetic acid and dilute to 10 ml with water.

Procedure—A few milligrammes of the sample to be examined is mixed with phosphate-free calcium oxide or fusion mixture (2 parts by weight $Na_2CO_3 + 1$ part KNO_3) and heated to redness in a small glass tube. The tube is cracked by touching it while hot into a few drops of water in a watchglass, and the contents thus exposed are dissolved by addition of a few drops of conc. nitric acid.

If PHOSPHORUS is present in appreciable amount, addition of a few drops of Solution **I** followed by warming produces a yellow precipitate (best seen against a black ground). To detect smaller proportions of the element, the same mixture is cooled and treated with one drop of Solution **II** followed by a few drops of a saturated sodium acetate solution; trace amounts of PHOSPHORUS produce a blue coloration (best seen against a white ground).

For polymers responding to this test, see the phosphorus sections of Table 57.T1. Precipitation as quinoline phosphomolybdate, an alternative to the ammonium salt, offers advantages in quantitative work.

2.4 SULPHUR

(i) Sulphur combined in organic compounds effects catalytic decomposition—yielding bubbles of nitrogen—in a solution consisting of 1 g sodium azide, 1 g potassium iodide, and a trace of iodine in a little water. It depends on the reaction,

$$2NaN_3 + I_2 \xrightarrow[\text{catalyst}]{\text{S as}} 2NaI + 3N_2$$

and is suitable for use, for example, when fibres are examined under a low-power microscope.

(ii) Free sulphur (e.g. as present in vulcanised rubbers) can be detected with thallium sulphide paper, which is made by impregnating filter paper with a 0·5 % solution of thallium carbonate, drying the product in warm air and exposing it over warm ammonium sulphide solution until black. The sample to be tested is extracted with sulphur-free pyridine and the extract is spotted on to the prepared paper, which is next dried and then bleached in very dilute nitric acid. A red-brown stain is left by SULPHUR.

(iii) For detection of sulphur in any form see under sodium fusion (2.5 (iv) below) and for polymers responding see the sulphur sections of Table 57.T1.

2.5 CHLORINE, NITROGEN, PHOSPHORUS AND SULPHUR (SODIUM FUSION OR LASSAIGNE TEST)

Heat a small sample with a pellet of sodium in a small ignition tube (e.g. 50 mm × 10 mm) and when the reaction is complete—and free sodium has burnt away—drop the hot mass into a test-tube half filled with water. Boil the mixture, or grind it in a mortar, filter the product and apply the following tests to separate portions of the filtrate.

(i) *Chlorine* Boil with dil. nitric acid and add silver nitrate solution. A white precipitate, soluble in ammonium hydroxide and re-precipitated by nitric acid indicates CHLORINE (or bromine or iodine).

(ii) *Nitrogen* Boil with fresh ferrous sulphate solution, clarify by addition of dil. hydrochloric acid, and add ferric chloride solution; a blue precipitate, or a permanent blue colour imparted to a filter paper when the solution is put through it, indicates NITROGEN.

(iii) *Phosphorus* Acidify a portion of the filtrate with conc. nitric acid and test with Solutions **I** and **II** as described in 2.3 above.

(iv) *Sulphur* To a portion of the filtrate add 1 or 2 drops of a fresh solution of sodium nitroprusside; production of a true violet colour indicates SULPHUR. A filtrate containing sulphur will blacken metallic silver, and with a solution of lead acetate yields a black precipitate.

The responses of polymers to the above tests are summarised in Table 57.T1.

An alternative procedure to sodium fusion depends on combustion of a small sample in oxygen, followed by colorimetric detection of the elements. *See*, for instance, HASLAM, J., HAMILTON, J. B. and SQUIRREL, D. C. M., *Analyst*, **86**, 239 (1961) (uses reagents different from those described above for detection of Cl, N and S).

Table 57.T1. CLASSIFICATION BY DETECTION OF ELEMENTS

Elements found	Principal polymers and related materials indicated
Cl only	Chlorinated polyethylene, polychlorotrifluoroethylene*, polyvinyl and polyvinylidene chlorides and copolymer modifications, chlorinated polyethers, chlorinated and hydrochlorinated rubber, chloroprene rubber; chlorinated butyl rubber. *Chlorinated forms of paraffin wax, diphenyl, and naphthalene.*
N only	Polyvinylcarbazole and polyvinylpyrrolidone, acrylic fibres, cellulose nitrate, chitin, silk and some regenerated proteins, polyamides (nylons), epoxy resins (amine-cured), amino- and phenol-formaldehyde resins†; nitrile rubbers (unvulcanised), polyurethanes. *Nitrogen is present in certain dyes, catalysts, accelerators,. antioxidants, antiozonants, blowing agents, and textile finishes; traces are detectable in raw cotton, natural rubber, and like biological products.*
P only	Polyphosphonic esters.‡ *Phosphate plasticisers. Traces of phosphorus are detectable in biological products such as raw cotton, flax, and natural rubber, and in some forms of starch and whiting (filler).*
S only	Hydrocarbon rubbers (natural and synthetic§, vulcanised‖), polysulphide rubbers (traces of chlorine), polysulphones. *Sulphonated and sulphated surface-active agents.*
Cl, N	Modacrylic fibres (vinyl or vinylidene chloride/acrylonitrile copolymers). *Halogenated dyes; chlorine occasionally present in gelatin.*
Cl, N, P	Polyphosphonitrilic chloride. *Some flame-resistant textile finishes.*¶
Cl, S	Chlorosulphonated polyethylene, vulcanisates of chloroprene rubber or chlorinated butyl rubber. *Chlorine detectable in rubber vulcanised with sulphur chloride.*
N, P	Some proteins, amino-formaldehyde resins (phosphoric acid catalysed).
N, P, S	Casein (phosphorus- and sulphur-contents each under 1 %).
N, S	Wool (traces of phosphorus), gelatin (chlorine sometimes present), some regenerated proteins; sulphonamide resins; nitrile rubbers (vulcanised).** *Sulphur dyes.*
None, or trace amounts only as indicated earlier	Polyethylene, polypropylene, polybutene, polymethylpentene, ethylene/propylene and ethylene/vinyl acetate copolymers, polytetrafluoroethylene*, polyvinyl and polyvinylidene fluorides and copolymers*, fluoro-rubbers* (not containing nitrogen), polyvinyl-alcohol (do. esters, ethers, and acetals), acrylic resins (not containing nitrogen), cellulose (and most cellulose esters and ethers), starch, alginates, polysters (most types), polyformaldehyde, polyglycols; phenol-formaldehyde resins†; hydrocarbon rubbers (natural and synthetic§, unvulcanised), silicones, glass, asbestos. *Some plasticisers; wax, bitumen, natural rosins and gums.*

*Fluorine present
†Nitrogen present in phenolic resins prepared via ammonia or hexamethylenetetramine, and in cashew nutshell liquid resin
‡ *See* mention in §24C.81
§Including synthetic isoprene, butadiene, styrene/butadiene, butyl, and ethylene/propylene rubbers
‖Compounded without nitrogenous components
¶E.g. Tetrakis(hydroxymethyl)phosphonium chloride in conjunction with melamine-formaldehyde
**Other vulcanised rubbers often contain a *small* proportion of nitrogen

3. ADDITIONAL TESTS

3.1 SAPONIFICATION NUMBER

The determination of the saponification number of an ester, such as a fat, can be extended to certain ester-type polymers. The method (detailed in textbooks of organic analysis) consists in boiling under reflux a dissolved sample of known weight with a known excess of standard alcoholic potassium hydroxide, following which the unconsumed alkali is titrated and the saponification number is expressed as milligrammes of potassium hydroxide per gramme polymer.

Saponification numbers of 200–650 mg KOH/g (approximately equal to the theoretical values) are obtained from the following materials:

polyvinyl acetate, polyacrylate esters and the free acids (glycolic alkali, boiling at a high temperature, is needed for polymethacrylates), cellulose esters, linear polyesters (e.g. alkyd resins, polyethylene terephthalate, drying oils; also ester-type plasticisers, and other animal and vegetable oils) and sulphonamide resins.

The following cyano-, chloro- and sulphide-type polymers give less satisfactory results, being saponified only in part and/or yielding coloured insoluble intermediates:

acrylonitrile polymers (acrylic and modacrylic fibres), vinyl and vinylidene chloride polymers and copolymers, chlorinated and hydrochlorinated rubber, chloroprene rubber and polysulphide rubbers.

Materials of the following types yield zero or very low saponification numbers (0–20 mg KOH/g):

hydrocarbon polymers (all types), polyvinyl alcohol, cellulose and most cellulose ethers, starch, proteins (partial decomp. on prolonged reflux), polyamides, polyethers, silicones, non-saponifiable plasticisers and fillers.

When carbon dioxide is excluded, the presence of potassium carbonate in the saponification residues indicates urea, melamine, or a carbonate in the original sample; similarly, potassium nitrate results from saponification of cellulose nitrate. Addition of sulphuric acid to the residues precipitates insoluble acids, e.g. terephthalic acid from saponification of ethylene terephthalate (*see* §24B.23).

3.2 IODINE NUMBER

In like manner to saponification, estimation of the iodine number of an unsaturated compound, such as a fat, can be extended to certain unsaturated polymers. The usual method employed consists in the addition of iodine monochloride across the double bonds, followed by titration of the unconsumed reagent, but the process is complicated by side-reactions and entails a lengthy procedure (*see* WAKE, under *Rubbers* in **Further Literature** at end of this section).

Iodine number is expressed as g iodine per 100 g polymer (e.g. up to *c.* 200 g I_2/100 g in drying oils, and above 400 in a highly unsaturated rubber) or may be converted to mol double bonds per 100 g polymer.

3.3 MISCELLANEOUS TESTS

Tests relating to characteristic *cross-sectional shape* and *staining behaviour* have been developed chiefly in connection with examination of textile fibres under the low powered optical microscope; *see,* for instance, §13.23 (cross-section of fibres) and reference to *Identification of Textile Materials* at the end of this section. Staining tests can sometimes be applied to fabrics and films, and even to polymers in bulk, and they may prove useful in differentiating between materials allied in composition, e.g. between various acrylic fibres, §10.23(5), and between cotton and viscose rayon, §13.23 (iodine test).

Finally it may be mentioned that, additional to the use of simple chemical and physico-chemical tests, polymer identification may often be facilitated or confirmed by the determination of simple physical properties such as specific gravity (*see* Section 58), refractive index (59), and birefringence (60). Examination under ultra-violet light in a dark room reveals whether a fluorescent substance is present, and shows up irregularities or contamination, but usually fluorescence is insufficiently characteristic to be of much use in polymer identification.

MORE COMPLEX ANALYTICAL TECHNIQUES

See also **Further Literature** at the end of this section.

4. OPTICAL AND ELECTRON MICROSCOPY

Both the optical microscope and the electron microscope may be used in polymer identification; the former for general purposes particularly in connection with fibres (as mentioned in 3.3 above) and—more especially in the form of a polarising microscope—for the examination of anisotropic materials, such as oriented fibres or films and spherulitic structures, while the electron microscope (though it mostly requires indirect observation by means of replicas) with its very high resolving power permits examination of the microfibrillar state of associated chain molecules and of the morphology of polymer single crystals.

5. END-GROUP ANALYSIS

See Section 52.

6. X-RAY DIFFRACTION ANALYSIS

See Section 53.

7. DIFFERENTIAL THERMAL ANALYSIS (DTA)

The basic requirements for this technique are as follows. In the top of a metal block, which can be enclosed in an inert atmosphere and heated at a uniform rate (e.g. $1-2°C/min$), are two small holes of the same diameter and depth. One hole is filled with an inert reference powder, and the other with the same powder to which has been added some of the material to be tested. Thermocouples inserted in the powders give a balanced output as the block is heated up, except that any thermal change undergone by the test material—and *not*, of course, appearing in the inert reference substance—produces a differential potential that can be amplified and recorded. In this way one obtains a characteristic 'thermal spectrum' or *thermogram*— in the form of a plot of temperature differential (ΔT)/temperature (T), showing peaks or troughs representing change in the specific heat of the sample, or other exothermic or endothermic changes, e.g. at the glass transition (T_g) or the crystalline melting point (T_m), or where the sample undergoes dehydration, decomposition or allied reactions.

8. CHROMATOGRAPHIC ANALYSIS

The several forms of this versatile technique are based on the following principles: the polymer to be tested is split into simpler components (e.g. by hydrolysis or pyrolysis) and a small sample of the mixed components is introduced into a system where a *mobile phase* moves in a uniform way past or through a *stationary phase*; the components (dissolved or dispersed in the mobile phase) may then become differentially separated, i.e. in a given time those with least affinity for the mobile phase and most readily taken up by the stationary phase (by adsorption, partition, ion-exchange, or other mechanisms) will show least movement, and vice versa. Thus, under controlled experimental conditions, the components of a mixture may be separated in a characteristic sequence in what is termed a *chromatogram*.

The technique is a very flexible one since the mobile phase may be liquid or gaseous, while the stationary phase may be solid or liquid (commonly a liquid supported on a solid); in addition, it is capable of high precision and usually requires only a very small test sample, e.g. less than 1 mg of solid or 0·05 ml of a solution.

The commonest forms of chromatography are briefly described below.

8.1 LIQUID–SOLID CHROMATOGRAPHY

This is well exemplified in paper chromatography, which is carried out as follows:

Paper chromatography
A single spot (sometimes a narrow line)* of the test solution is applied to a stationary phase consisting of a sheet of filter paper, or the like, and dried; a mobile phase, consisting of an appropriate solvent mixture which moves through the paper by capillary action, is then allowed to ascend (or descend) past the dried spot in order to separate the components as indicated earlier. Elution (i.e. washing out of the components of the spot) is continued until the 'front' of the liquid phase has traversed a suitable distance—e.g. 150–300 mm—and its position is then marked.

If the resolved components are coloured their separation may at once be evident†, but usually their respective positions need to be established by drying the paper and spraying it with appropriate reagents, or by examination for fluorescence under ultra-violet light, or by other methods. For a specified system, operated under controlled conditions, the distance traversed by a particular substance is characteristic, i.e. the ratio of the distance traversed to the maximum possible distance (represented by the liquid front) is constant, and is called the R_F value.

In *ring paper chromatography*, a variation of the above procedure, the initial spot of the test solution is applied centrally to a circular filter paper, and after drying is fitted with a small wick which dips into the eluting liquid; the liquid then spreads radially from the central spot, distributing the components in concentric rings.

Allied to paper methods is *thin layer chromatography* (*TLC*) where the sorption medium (the stationary phase) consists of a specially prepared thin layer of a fine powder, such as starch, silica, alumina, or an ion-exchange resin. The advantages of this method lie in the speed of development of a chromatogram, the many kinds of sub-

---

*For comparison purposes it is usual to include spots of known materials adjacent to that of the test solution.
†Hence the name chromatogram.

strates available, and that reagents that would be unsuitable for paper may be employed. Less used for identification but important in quantitative analysis is *column chromatography*, where the stationary phase is contained in a vertical tube, into the top of which is introduced the solution to be examined and down which an eluting liquid is caused to flow; the separated 'fractions' of the sample are then successively collected as they emerge at the bottom and are subsequently analysed.

8.2 GAS–LIQUID CHROMATOGRAPHY (GLC)

This extremely powerful method of analysis needs only a very small test sample, which is injected into a stream of an inert gas (the mobile phase) and thereby carried, in the gaseous or vaporised state, past an involatile liquid (the stationary phase). The liquid is distributed on an inert solid of large surface area contained in a 'column'; e.g. a high-boiling ester or polyglycol is supported on kieselguhr or glass ballotini in a long tube, which is coiled into a helix for ease of maintenance at a constant temperature (commonly several hundred °C).

For examination of the composition of solvent mixtures, the purity of monomers, etc., a sample may be injected directly above the hot column, but for polymer analysis a tiny sample is usually depolymerised *in situ* (by being enclosed in, or coated on to, a platinum filament which is inserted in the gas stream and then heated electrically). In this way a 'puff' of pyrolytic vapour enters the gas stream and passes into the hot column, where by the process of partition it is separated into fast or slow-moving components*. Detection of the components as they emerge from the column is effected physically, by katharometric or ionisation methods, and the resulting characteristic *'chromatogram'* is recorded automatically as a series of peaks plotted against time (the time taken by reference substances to traverse the column under identical conditions being separately determined).

9. INFRA-RED SPECTROSCOPY

Vibrational and rotational movements in the segments and substituent atoms or groups of chain molecules commonly fall—like those of substances of lower molecular weight—in the same range of frequency as that of infra-red radiation, i.e. from approximately $3 . 10^{12}$ to $3 . 10^{14}$ Hz, corresponding (inversely) to a wavelength range of 1–100 μm*. Hence, radiant energy pertaining to this region can be selectively

---

*It is sometimes possible to use an adsorbing *solid*, free from liquid; the method then becomes one of *gas-solid chromatography*.

*1 micrometre (1μm), formerly known as a micron (μ) = 10^{-6} metre. Thus 1 to 100 μm = 10^{-4} to 10^{-2} cm.

absorbed by a high polymer and an infra-red absorption spectrum may be obtained, which—being highly characteristic for a given structure (and likened to a fingerprint)—provides both a powerful and rapid method of polymer identification.

A typical commercial infra-red spectrometer utilises the shorter wavelengths and commonly operates over the range 2·5–15 μm (4000–667 wave numbers, i.e. wavelengths per centimetre). It consists essentially of a source of IR radiation (e.g. a Nernst filament), the rays from which are focused on the sample (by means of mirrors), and then pass through it to be dispersed by a prism or diffraction grating into a spectrum which in turn is focused on—and made to traverse past—a detector (e.g. a bolometer or thermocouple); ancillary electronic equipment then automatically records the spectrum obtained.

Since glass is opaque to all but the shortest of IR wavelengths, the prism is made from a crystal of a metallic halide (e.g. LiF, NaCl, or CaF_2); similarly, as water is absorbent, its presence must be excluded and the test specimen is examined either dissolved in a relatively non-absorbing liquid (e.g. $CHCl_3$, CS_2, in a cell with walls of NaCl or AgCl) or as a very thin unsupported film; alternatively the sample can be finely-divided and dispersed as a 'mull' in paraffin oil, or dispersed in potassium bromide (high pressure in a die causing the salt to cohere as a thin plate). A reference beam, which is derived from the same IR-source and traverses the same path except for passage through the sample, provides a datum line for the resulting absorption spectrum, which is recorded as percentage transmission (sometimes as percentage absorption) against wavelength and/or wave number.

The spectrum is compared with those from polymers of known composition, of which typical spectra are available (*see* **Further Literature** at end of this section). By placing a sample of a known polymer in the reference beam an exact comparison may be made with an unknown sample, no absorption bands arising if the two samples are identical. Sometimes it is more convenient to obtain a spectrum of a liquid pyrolysate, rather than that of a polymer itself; the two spectra are, of course, quite different. Working from model compounds, it has been possible to assign specific molecular structures or groups to numerous absorption bands, e.g. to relate the following groups to the given wavelength (μm): $-C{\equiv}N$ at 4·5; $>CO$, between 5·7 and 5·9; $-CH_3$, 7·3; *trans* $-CH{=}CH-$, 10·4; $-CH{=}CH_2$, 10·1 and 11·0; aliphatic hydrocarbon chain $-(CH_2)_n-$, 13·9; $\equiv C-Cl$, c. 14·5 (broad); and $RR'C{=}CH_2$ (from pyrolysates) at 11.3.

Infra-red absorption spectroscopy is highly specific, and since simple determinations are readily carried out—and require only very small samples—they are important in the qualitative characterisation of many polymeric materials. Extension to quantitative estimation of particular groups permits examination of such items as: the purity of a monomer, amount of residual monomer, the course of a polymerisation (by following loss of unsaturation or reactive groups), also determination of number-average molecular weight (*see* §52.1) or the ratio of the components in a copolymer, or such structural information as *cis-trans* configuration and the comparative index of tacticity (§54.13).

For publications containing IR-spectra *see under* **Further Literature** at the end of this section.

10. NUCLEAR MAGNETIC RESONANCE (NMR) SPECTROSCOPY, AND ELECTRON SPIN RESONANCE OR ELECTRON PARAMAGNETIC RESONANCE (ESR OR EPR) SPECTROSCOPY

These are two separate subjects which nevertheless run parallel in certain aspects. Involving elaborate equipment, neither would be used for simple identification, but the first gives a powerful insight into structural details in polymers while the second can provide similar insight into some of the states they assume.

NMR is associated with atomic nuclei having charge and appropriate spin, and thereby possessing magnetic moment. The hydrogen nucleus (proton) is the chief example in polymeric materials. A nucleus of this type, when in a powerful magnetic field, aligns itself either with or in opposition to the field and thus may resonate (at high frequency) between two energy levels, the frequency at which resonance occurs being characteristic for a given strength of magnetic field. **ESR** (or **EPR**) is also associated with high frequency transitions —not, however, of nuclei but of unpaired electrons (free radicals)— and, again, the frequency (this time extremely high) at which resonance occurs is characteristic for a given magnetic field strength.

In NMR spectroscopy the test sample, dissolved in a suitable solvent in a small container, is placed in a powerful magnetic field, across which is generated a fixed HF-field (usually 6.10^7 Hz, i.e. a low-energy RF-field). If the magnetic field strength is then varied, sharp resonance occurs at a particular value and is detected by absorption of energy from the HF-field. In *ESR spectroscopy* a powerful magnetic field is again utilised but the test sample placed in it is contained in a cavity resonator connected to a klystron that generates microwaves of *c*. 3 cm wavelength (*c*. 10^{10} Hz, i.e. a high-energy RF-field) normal to the magnetic field. As with NMR spectroscopy, a particular value of magnetic field strength causes resonance, which is detected as a reduction in intensity of the emergent microwaves. Thus, although the exciting frequencies are different (and differently generated) in the two instances, both techniques are conveniently carried out by variation of the field strength of a powerful d.c. electromagnet, and the spectra obtained are plotted (automatically) as HF-field strength against magnetic field strength. Certain forms of NMR equipment use a powerful permanent magnet and then vary the frequency of the exciting radiation.

Since differently-positioned protons resonate differently, NMR spectroscopy readily reveals structural features such as those arising from the various kinds of tacticity. ESR spectroscopy, depending on the presence of free radicals and their generation by bond fission, detects the progress of such varied processes as polymerisation, oxidation and related chemical reactions, irradiation, and mechanical stress (as in compression, extension, and creep under load).

FURTHER LITERATURE

General

FEIGL, F. and ANGER, V., *Spot Tests in Organic Analysis* (7th edn), London (1966)
KE, B., *Newer Methods of Polymer Characterisation* (*Polymer Reviews* No. 6), New York and London (1964) (includes IR, NMR, DTA)
KLINE, G. M. (Ed), *Analytical Chemistry of Polymers* (*High Polymers,* Vol. XII), New York and London. Part I *Analysis of Monomers and Polymeric Materials* (1959), *Part III Identification Procedures and Chemical Analysis* (1962)
FLETT, M. ST. C., *Physical Aids to the Organic Chemist*, Amsterdam and New York (1962) (chromatography, infra-red spectrometry, NMR and ESR, etc.)

Fibres

TEXTILE INSTITUTE, *Identification of Textile Materials* (5th edn), Manchester (1965)
GARNER, W., *Textile Laboratory Manual*. Vol. 2 *Resins and Finishes*, 3rd edn, London and New York (1966). Vol. 5, *Fibres*, 3rd edn (1967)
GILES, C. H. and WATERS, E., *J. Text. Inst.*, **42**, p. 909 (1951)
    Supplemented by DANDEKAR, S. D. and GILES, C. H., *J. Text. Inst.*, **53**, p. 430 (1962) (identification of textile finishes)
ASTM D276. Identification of Fibres in Textiles.

Films

KAPPELMEIER, C. P. A. (Ed), *Chemical Analysis of Resin-based Coating Materials*, New York (1959)

Plastics

HASLAM, J. and WILLIS, H. A., *Identification and Analysis of Plastics*, London (1965)
SAUNDERS, K. J., *The Identification of Plastics and Rubbers*, London (1966)
KRAUSE, A. and LANGE, A., *Introduction to the Chemical Analysis of Plastics*, London (1969)

Rubbers (*see also* SAUNDERS above)

WAKE, W. C., *The Analysis of Rubber and Rubber-like Polymers*, 2nd edn, London (1969)

DTA

SLADE, P. E., Jr. and JENKINS, L. T. (Eds), *Techniques and Methods of Polymer Evaluation*, Vol. 1, *Thermal Analysis*, London (1966)
SCHWENKER, R. F. (Ed), *Thermoanalysis of Fibres and Fibre-forming Polymers'* New York (1966)

Chromatography   See §22e.

Infra-red Spectroscopy

BELLAMY, L. J., *The Infrared Spectra of Complex Molecules*, London and New York (1954)
CROSS, A. D., *Practical IR Spectroscopy*, London (1960)
ELLIOTT, A., *Infrared Spectra and Structure of Organic Long-chain Polymers*, London, Ed. Arnold & Co. (1969)

531

HARNS, D. L., *Identification of Complex Organic Materials by Infra Red Spectra of their Pyrolysates, Analyt. Chem.*, **25**, No. 8, 1140 (1953)

HUMMEL, D., *Kunststoffe- Lack- und Gummi-Analyse, Chemische und Infrarot Spektroskopische Methoden*, Munich (1958)

HUMMEL, D. O., *Infrared Spectra of Polymers in the Medium and Long Wavelength Regions*, New York and London (1966)

KLINE, G. M. (Ed), *Analytical Chemistry of Polymers. Part II Analysis of Molecular Structure and Chemical Groups* (Chapter VIII, TRYON, M. and HOROWITZ, E.: infrared spectrophotometry), New York (1962)

MILLER, R. G. J. (Ed), *Laboratory Methods of Infrared Spectroscopy*, London (1965)

NYQUIST, R. A., *Infrared Spectra of Plastics and Resins*, Midland, Michigan (1961)

OFFIC. DIG. FEDRN. SOC. PAINT TECHNOL, **32**, (427) 517 (1960)

SADTLER RES. LABS, *25 Most Useful IR Spectra*, Philadelphia, Pa. (1960)

SADTLER RES. LABS, *Monomers and Polymers*, Vol. IV. *Spectra of Monomers and Polymers with Classification Indices*, Philadelphia (1962)

SADTLER RES. LABS, *Infrared Spectra of Commercial Products*, Philadelphia (1968) (includes polymers and pyrolysates)

SZYMANSKI, H. A., *Theory and Practice of IR Spectroscopy*, New York (1963)

ZBINDEN, R., *Infrared Spectroscopy of High Polymers*, New York and London (1964)

U.S. DEPT. OF THE ARMY, *Identification of Elastomers by IR Spectra of Pyrolysates*, Illinois (1958)

U.S. DEPT. OF COMM., OFFICE OF TECH. SERVICES, *Infrared Spectra of Plastics and Resins* (1954)

AATCC TECH. MANUAL, *Section on IR Spectra of Textile Fibres*

BS 4181: 1967. *Method for Identification of Rubbers* (by infrared absorption spectra of pyrolysates); plus *Addendum*, in the press

NMR and ESR

BOVEY, F. A., *Nuclear Magnetic Resonance Spectroscopy*, New York and London (1969)

CHAPMAN, D. and MAGNUS, P. D., *Introduction to Practical High-resolution NMR Spectroscopy*, London and New York (1966)

JACKMAN, L. M., *Applications of NMR in Organic Chemistry*, Oxford (1964)

JONES, R. A. Y., *et al.*, *The Techniques of NMR and ESR*, London (1965)

ROBERTS, J. D., *Nuclear Magnetic Resonance*, London and New York (1959)

SPECIFIC GRAVITY AND DENSITY

DEFINITIONS

Specific gravity (or *relative density*). The mass of a given portion of matter compared with the mass of an equal volume of water. The value relates to a stated temperature (e.g. 20 or 25°C) and sometimes is expressed with respect to water at its maximum density (4°C).

Density. The mass per unit volume of a substance, measured at a stated temperature.

The ratio of the true density of a powder, foam, etc., to its apparent density is called the *bulk factor*; it is the volume of any weight of uncompressed material divided by the volume of the same weight of material after compression to eliminate voids or cells, e.g. when moulded under pressure.

UNITS

Specific gravity, as a ratio of similar units, is without dimensions.

Density is usually expressed in *grammes per millilitre* (*or per cubic centimetre*).

$$1 \text{ g/ml} \equiv 0.036 \text{ lb/in}^3 \equiv 62.43 \text{ lb/ft}^3$$

Note. Since the mass of 1 ml of water is 1.0000 g at 4°C, and not much different at room temperature (e.g. 0.9971 g at 25°C), for most practical purposes specific gravity and density (expressed in g/ml) may be taken as numerically the same. *See also* **Note on SI Units,** page 659.

NOTES

Common polymers are all comparatively 'light-weight' materials. At the lower end of the range polymethylpentene (§2.82)—with a specific gravity of 0.83—is one of the least dense solids known. Its ability to float in water is shared by several other polymers—e.g. polyethylene, polypropylene, natural rubber, and polyisobutylene—but even at the upper end of the range only a few polymers in the unfilled state are twice as dense as water, e.g. polychlorotrifluoroethylene and polytetrafluoroethylene—(the last, with a specific gravity of 2.2, being one of the densest).

The apparent density can often be reduced, by the use of foaming or blowing techniques, to a very low value*, while by inclusion of a mineral filler or metal powder the specific gravity can be raised to double or treble that of an unfilled polymer. Since the volume of a composite material is the sum of the volumes of its constituents, the specific gravity of such a material (e.g. of a vulcanised rubber) is represented by

$$S = \frac{100}{(W_a/S_a)+(W_b/S_b)+(W_c/S_c)+\ldots}$$

where W_a, W_b, W_c, etc., are the respective percentages (by weight) of the constituents with specific gravities S_a, S_b, S_c, etc.

Specific gravity and density may be measured by Archimedean displacement, pyknometers (and specific gravity bottles), and 'sink or swim' methods including the use of a *density gradient tube*. The last, with density increasing from the top to the bottom of a liquid column, may be set up with an aq. solution of a salt, or with miscible liquids of different densities (such as water and glycerol, or petroleum ether and carbon tetrachloride), and preferably should be maintained at a constant temperature. Density gradient tubes are calibrated by introduction of tiny pieces of substances of which the density has been previously determined, and they are specially suited to rapid determinations on fibres, powders, and other small samples; they may also provide an automatic separation of loosely mixed components. An alternative method, in which the density of a small solid sample is matched with that of a liquid by adjustment of the temperature, depends on the higher coefficient of expansion of a liquid, e.g. carbon tetrachloride (specific gravity at 20°C, 1·59; at 70°C, 1·50).

Some liquids covering a range of specific gravity, suitable for use in its measurement, are given in Table 58.T1.

Table 58.T1. SPECIFIC GRAVITY OF LIQUIDS*
(approximate values at 20–25°C)

n-heptane	0·75	ethyl iodide	1·93
toluene	0·87	bromoform	2·85
water	1·00	acetylene tetrabromide	2·95
o-dichlorobenzene	1·30	Sonstadt's solution†	3·19
chloroform	1·499	methylene iodide	3·32
carbon tetrachloride	1·595	Rohrbach's solution‡	3·58
pentachloroethane	1·67	Clerici's solution§	4·9

*Note: although fairly inert chemically, some of the liquids listed may cause the swelling or dissolution of certain polymers. Aq. solutions of inorganic salts are sometimes useful, e.g. $ZnCl_2$ (at 70% concn), 1·93; ZnI_2 (at 75% concn), 2·39; calcium borotungstate (Kleins solution), up to 3·28. The specific gravity of various plasticisers is shown in Table 56.T1
†Also known as Thoulet's solution, this is saturated aq. potassium iodide saturated with mercuric iodide
‡Saturated aq. barium iodide saturated with mercuric iodide
§Saturated aq. thallium formate saturated with thallium malonate (poisonous)

*E.g. To an apparent density of as little as 0·015 g/ml (about 1 lb/ft^3). For typical values, *see:* expanded polystyrene (§4.6), cellular rubber (§29.88), microporous ebonite (§31.4), polyurethane foams (§40.31).

COMPARISON TABLE

Table 58.T2 lists the specific gravity of polymers and related materials; more detailed information on individual polymers will be found in §§1.31–42.31.

Table 58.T2. SPECIFIC GRAVITY OF COMMON POLYMERS AND RELATED SUBSTANCES
AT 20–25°C

Range	Material (unfilled unless otherwise stated)
Under 0·8	Expanded polymers, *balsa wood*, under 0·1. *Cork*, 0·25. *Spruce*, 0·5. *Oak, beech (plywood)*, 0·75.
0·8–1·0	Polymethylpentene, 0·83. Ethylene/propylene rubbers, raw, 0·85–0·87. Polybutene-1, 0·85–0·95. Butadiene rubber, raw, 0·88–0·91. Polypropylene, 0·89–0·92. Leather, 0·9. Butyl rubber, raw, 0·91–0·96; do., vulcanised, 0·92–0·98. Isoprene rubber, raw, 0·92. Polyethylene, low density, 0·92–0·94. Natural rubber, raw, *c.* 0·93; do., vulcanised, 0·92–1·0. Styrene/butadiene rubbers, raw, 0·93; do., vulcanised, 0·94–1·0. Ethylene/vinyl acetate copolymers, 0·93–0·95. Polyethylene, high density, 0·94–0·97. Acrylonitrile/butadiene rubbers, raw, 0·95–1·02. Silicone rubbers, raw, 0·96–0·98.
1·0–1·2	Silicone resins, 1·0–1·2; do., glass-filled, 1·11–1·23. Alkyd resins, 1·0–1·4. Polyamides, fibres, plastics, 1·04–1·14. Polystyrene, 1·05–1·06. Ebonite, 1·08–1·25. Coumarone/indene resins, 1·1–1·14. Polyurethanes, fibres, 1·1–1·28; do., rubbers, 1·17–1·31. Polyvinyl chloride, plasticised, 1·1–1·7. Polyvinyl acetals, 1·11–1·2. Rubber hydrochloride, 1·11–1·27. Chlorosulphonated polyethylene, 1·11–1·28. Ethylcellulose, 1·13–1·15. Polysulphide rubbers (liquid types), 1·13–1·31. Epoxy resins, 1·15–1·2. Polyacrylonitrile, 1·17–1·18. Polyvinyl acetate, 1·17–1·19. Polymethyl methacrylate, 1·18–1·19. Polycarbonates, 1·2.
1·2–1·4	Chloroprene rubber, raw, 1·2–1·25. Polysulphones, 1·24–1·25. Polyester, unsaturated, *cotton flock, wood flour*, 1·25. Polyvinyl alcohol, 1·25–1·35. Cellulose acetate, 1·28–1·32. Casein-formaldehyde, 1·35. Wool, 1·30–1·32. Phenolic resins, 1·3–1·34; do., cellulose filled, 1·36–1·46. Silk, raw, 1·36. Cellulose nitrate, polyethylene terephthalate, polyvinyl fluoride, 1·38. Polyvinyl chloride, rigid, 1·39.
1·4–1·6	Polyformaldehyde, 1·42. Hemp, jute, 1·48–1·49. Flax, starch, amino/formaldehyde resins (cellulose-filled), 1·5. Regenerated cellulose: fibres, 1·51–1·52; films, 1·53. Cotton, ramie, 1·53–1·55. Phenolic resins, mineral- or glass-filled, 1·54–1·92. Chlorinated rubber, 1·58–1·69. Sodium carboxymethylcellulose, 1·59. Alginic acid, polysulphide rubbers (Thiokol A), *c.* 1·6.
1·6–1·8	Polyesters (glass-reinforced), epoxy resins (silica-filled), 1·6–2·0. Polyvinylidene chloride, 1·67–1·71. Calcium alginate, 1·75–1·78. Ebonite (filled), *carbon black, c.* 1·8. Amino/formaldehyde resins (mineral- or glass-filled), 1·8–2·0.
Above 1·8	Fluorubbers, 1·80–1·86. *Sulphur, rhombic*, 2·07. Polytetrafluoroethylene, 2·1–2·3. Polychlorotrifluoroethylene, 2·11–2·13. Tetrafluoroethylene/hexafluoropropylene copolymers, 2·15–2·16. Glass, various types, 2·2–6·0; do., fibres, 2·46–2·55. Asbestos, 2·2–3·5. *Clays*, 2·6. *Calcium carbonate, aluminium*, 2·7. *Slate*, 2·8. *Mica*, 2·95. *Titanium dioxide*, 3·88. *Barytes*, 4·45. *Zinc oxide*, 5·57. *Zinc*, 7·0. *Iron, steels*, 7·0–7·9. *Brass*, 8·4–8·7. *Copper*, 8·5. *Silver*, 10·3–10·5. *Lead*, 11·3–11·4. *Mercury*, 13·5. *Tungsten*, 18·6. *Gold*, 19·3. *Platinum*, 21·5. *Osmium*, 22·5.

FURTHER LITERATURE

REILLY, J. and RAE, W. N., *Physico-chemical Methods,* 5th edn, Vol. I, London, 577 (1954)

British Standards
BS 903: Part A1 (method of test, vulcanised rubber)
BS 903: Part E1 (apparent density of cellular ebonite)
BS 4443 (apparent density of flexible cellular materials)
BS 903: Part H3 (density of rubber thread)
BS 2782: Method 501A (apparent density and bulk factor of moulding materials)
BS 2782: Methods 509 B and C (density of polyethylene film)
BS 2782: Method 509A (density of plastics)

ASTM Specifications
D792 (methods of test, plastics)
D954 (apparent density, non-pouring powders)
D1182 (apparent density, granular powders)
D297 (methods of test, rubber products)
D1505 (density of plastics by density-gradient technique)
D1564 (method of test, slab flexible urethane foam)
D2406 (method of test, moulded flexible urethane foam)
D1622 (method of test, rigid cellular plastics)
D1895 (method of test, apparent density, bulk factor and pourability of plastics materials)

REFRACTIVE INDEX

DEFINITIONS

The *refractive index* of a substance is the ratio of the velocity of electromagnetic radiation in a vacuum (i.e. 3.10^{10} cm/s) to the velocity in the given medium. Thus it determines the extent of the *lowering of the velocity of light* on passing from a vacuum into a given transparent and optically isotropic medium (or the *retardation* of the light, which—expressed with respect to unit thickness of the retarding medium—is related to the amount by which the wavelength is decreased in that medium). As a consequence, refractive index also determines the extent of the *refraction*, or deviation, that occurs when a ray of light passes at a slanting angle from a vacuum (or other medium*) into the stated medium; it is independent of the angle of incidence† and, as a constant denoted by n or μ, can be expressed trigonometrically as

$$n = \text{sine (angle of incidence)}/\text{sine (angle of refraction)}$$

UNITS

A ratio, without dimensions. It is necessary to state the wavelength of the light used and the temperature of the substance measured (*see below*), also—with a hygroscopic polymer—the moisture content or humidity to which it is conditioned.†

---

*When, instead of a vacuum, a second medium is present both media should be named (as, for instance, water/glass); this, however, can usually be ignored if the second medium is air, for which the refractive index is very small (only 1·0003 at s.t.p.; cf. vac/vac. = 1·0000), i.e. refractive index relative to air and relative to a vacuum are practically the same.

†The angle of incidence being measured as that between the incident ray and the *normal* to the surface (or interface) at the point of incidence. The angle of refraction is measured similarly, with respect to the refracted ray and the normal.

†For example (HERMANS, P. H., *Physics and Chemistry of Cellulose Fibres*, Amsterdam, 1949), the refractive index of dry cellulose (n_0) can be obtained from that at 65% r.h. (n_{65}) from the relation $n_0 = 1 + 1·17 (n_{65} - 1) - 0·0715$.

NOTES

Refractive index, being dependent on wavelength* and temperature†, is commonly measured with monochromatic radiation (e.g. sodium light, $\lambda = 5893$ Å, or mercury yellow light $\lambda = 5791$ Å) and at a fixed temperature (e.g. 20 or 25°C).

On suitably shaped polished samples refractive index can be measured directly, using the sample as part of a refractometer of the Abbe or Pulfrich type (several forms of which are available; *see*, for instance ASTM D542) along with a liquid having a higher refractive index than that of the sample; alternatively, if the material can be fabricated as a prism, a spectrometer can be used. Another method, providing a rough assessment, depends on observation of the disappearance of the outline of an optically isotropic sample when its refractive index is matched by that of a liquid in which it is immersed.

Two methods involving microscopy, which can be employed to determine the refractive index of very small samples (such as fibres), are as follows. (i) *Becke-line method*. This requires a microscope of moderate power (*c.* 150-times, without a condenser) and a selection of inert liquids with known refractive indices near that of the sample. Tests are carried out by introducing a suitable liquid between a cover slip and a slide carrying a few grains or fibres of the material to be examined. A bright line (Becke-line) can be seen close to the margin of the sample; when the focus is *lowered* the line moves—inwards or outwards—towards the phase of *lower* refractive index, and vice versa. When no movement is observable—or at an interpolated point of reversal—liquid and sample have the same refractive index. In a modification of this method the temperature is varied until a match is

---

*Refractive index increases from the red to the violet end of the spectrum. An indication of the *dispersion* so produced is given by $n_F - n_C$, where n_F and n_C are refractive indices for light corresponding to the F and C Fraunhofer lines, i.e. for greenish blue ($\lambda = 4861$ Å) and red ($\lambda = 6563$ Å) light respectively. Some typical values for dispersion expressed in this way are given below; also shown are v values ($\equiv$ reciprocal of the *dispersive power* of the given substance) expressed as $v = (n_D - 1)/(n_F - n_C)$; ($n_D$ is for sodium light, $\lambda = 5893$ Å)

	n_D	$n_F - n_C$	$(n_D - 1)/(n_F - n_C)$
Fused silica	1·4584	0·007	65
Common glass (soda lime)	1·5171	0·0087	59
Polymethyl methacrylate	1·49	0·009	54
Rock salt (NaCl)	1·5442	0·013	43
Light flint glass	1·5780	0·0139	42
Potassium bromide	1·5590	0·017	33
Polystyrene	1·59	0·019	31
Dense flint glass	1·7170	0·0243	29
Densest flint glass	1·9626	0·0488	20

---

†Usually refractive index decreases with rise of temperature. Approx. temperature coefficient (/°C): water, -0.0001; alcohol, -0.0004; carbon disulphide, -0.0008; the coefficient for organic polymers is smaller, e.g. *c.* -0.00005.

obtained; in another, the wavelength of the light is varied. Anisotropic materials may need to be examined in polarised light (*see*
Section 60). (ii) *Retardation method*. An interference microscope can
be used to measure the retardation or path-length difference between
light that has passed through a sample and that which, at the same
time, has passed through a liquid medium of known refractive index;
as with the rough method of assessment mentioned earlier, when no
retardation is observed, sample and liquid have the same refractive
index. Alternatively, conversion of the measured retardation (relative
to the liquid of known refractive index) to retardation relative to a
vacuum r_0, and measurement of the sample thickness t, enables the
refractive index to be calculated, i.e. $n =$ (path in vacuum)/(path in
the sample) $= (t + r_0)/t$. As above, for anisotropic materials, *see also*
Section 60.

The refractive index of an isotropic polymer is related to its density
(ρ) through the empirical rule of Gladstone and Dale that $(n-1)/\rho$
is constant, whence the molar refraction (R) is defined as

$$R = m(n-1)/\rho$$

where m is the molecular weight of the repeat unit. A more precise
relation (Lorentz and Lorenz) is that

$$R = m(n^2 - 1)/\rho(n^2 + 2)$$

The value of R is constant for a given polymer and independent of its
molecular weight. For certain polymers R is an additive function of
the refractivities of the atoms and bonds constituting the repeat unit,
and values of n derived from R calculated in this way have been found
to agree well with observed values*. With anisopropic fibres the
'isotropic' refractive index should be used, i.e.

$$n_{ISO} = (n_{||} + 2n_{\perp})/3$$

where $n_{||}$ and $n_{\perp}$ have the meanings given in Section 60. For a relation
between refractive index and dielectric constant, *see under* Notes,
Section 71.

Refractive index is important in the design of lenses and achro-
matic combinations, and in light-carrying probes, also (with birefringence, *see* next Section) it can be useful in establishing, non-
destructively, the identity of very small samples, e.g. single fibres.
Low values of refractive index are provided by fluorocarbons, and
high values are commonly associated with dense substances (e.g. flint
glass, methylene iodide) though density is not an essential requisite.

The refractive indices of various plasticisers are shown in Table
56.T1, and some liquids covering a range of values are given in
Table 59.T1 below.

---

*See, for instance, NOSE, S., *J. Polym. Sci.*, B2, 1127 (1964)

Table 59.T1. REFRACTIVE INDEX OF LIQUIDS*
(approximate values at 20–25°C)

Perfluoromethylcyclohexane	1·275	Nitrobenzene, o-dichlorobenzene	1·55
Water	1·333	Benzyl benzoate‡	1·57
Silicone fluids†	from 1·37	Aniline	1·585
Glycerol, 50% aq.	1·40	Bromoform	1·595
n-Decane	1·41	Quinoline	1·62
Chloroform	1·445	Carbon disulphide	1·63
Carbon tetrachloride	1·46	Acetylene tetrabromide	1·64
Olive oil	c. 1·46	1-Bromonaphthalene	1·66
Liquid paraffin, castor oil	c. 1·47	Clerici's solution§	1·685
Benzene, toluene	1·50	Methylene iodide	1·74
Monochlorobenzene	1·525	Do., sat. with sulphur	1·78
Benzyl alcohol	1·54	West's solution‖	2·05

*For sodium light, relative to air. A range of values can be obtained by mixing appropriate liquids, e.g. dilution of methylene iodide or 1-bromonaphthalene (or a chlorinated biphenyl, Table 56.T1) with n-decane or liquid paraffin, or dilution of Clerici's solution with water.
†Up to 1·52 with rising viscosity
‡M.p. 21°C
§Aq. thallium formate and malonate, having a sp. gr. of 4·15 (poisonous)
‖Phosphorus (8 parts) and sulphur (1 part) dissolved in methylene iodide (1 part); needs care in preparation and use

COMPARISON TABLE

Table 59.T2 lists the refractive index of polymers and related materials.

Table 59.T2. REFRACTIVE INDEX OF COMMON POLYMERS AND RELATED MATERIALS AT 20–25°C

Range	Material (unfilled)
Under 1·4	Vacuum or air, 1·00. Water, 1·33. Tetrafluoroethylene/hexafluoropropylene copolymers, 1·34. Polytetrafluoroethylene, 1·37–1·38. Methylcellulose, 1·4.
1·4–1·50	Silicones, rubbers, 1·404; do., resins, 1·405–1·49. Polyformaldehyde, 1·41. Polychlorotrifluoroethylene, 1·43. Polyvinyl acetals, 1·45–1·50. Silica, as fused quartz, 1·459. Polyvinyl fluoride, 1·46. Polyvinyl acetate, 1·46–1·47. Polymethylpentene, 1·465. Ethylcellulose, 1·47. Cellulose acetate, 1·47–1·5. Alkyd resins, 1·47–1·57. Ethylene/propylene rubbers, 1·48. Glass, various types, 1·48–>2. Ethylene/vinyl acetate copolymers, 1·482–1·485. Polypropylene, polymethyl methacrylate, 1·49. Cellulose nitrate, sodium carboxymethylcellulose, polyamides (non-fibrous), 1·50.
1·50–1·525	Glass, as fibres, 1·50–1·55. Polyvinyl alcohol, 1·50–1·53. Butyl rubber, raw, 1·5078–1·5081. Polyethylene, 1·51–1·52. Polyacrylonitrile, 1·511–1·515. Butadiene rubber, raw, 1·5158–1·5175. Natural rubber, raw, 1·519; do., vulcanised, 1·526. Acrylonitrile/butadiene rubbers, 1·519–1·521. Polyamides (nylons 11 and 6.10), 1·52. Calcium alginate, 1·525.
1·525–1·55	Starch, 1·53. Casein-formaldehyde, 1·53–1·54. Polyvinyl chloride, rigid, 1·53–1·56. Regenerated cellulose, fibres, 1·532–1·534; do., films, 1·53. Rubber hydrochloride, 1·533. Styrene/butadiene rubbers, 1·5345. Jute, 1·536. Asbestos, polyamides (nylons 6 and 6.6), 1·54. Urea-formaldehyde resins, phenolic resins, c. 1·55.
1·55–1·60	Polyvinyl chloride (plasticised), chlorinated rubber, 1·55–1·60. Cotton, 1·555. Wool, 1·556. Chloroprene rubber, raw, 1·5578–1·5580. Epoxy resins, cast, c. 1·56. Polysulphide rubbers, liquid types, 1·56–1·57. Flax, ramie, 1·563. Polycarbonates, 1·585. Polystyrene, 1·59. Silk, 1·591–1·595.
Above 1·60	Polyvinylidene chloride, 1·60–1·63. Coumarone/indene resins, 1·60–1·65. Ebonite, 1·60–1·65. Polysulphide rubbers, solid types, 1·6–1·7. Melamine-formaldehyde, polysulphones, 1·65. Polyethylene terephthalate, fibres, films, 1·65–1·66. Polyvinylcarbazole, 1·69. Diamond, 2·42. Titanium dioxide (rutile), 2·6–2·9.

FURTHER LITERATURE

PARTINGTON, J. R., *An Advanced Treatise on Physical Chemistry,* Vol. 4, *Physico-Chemical Optics,* London (1953)

REILLY, J. and RAE, W. N., *Physico-chemical Methods,* 5th edn, Vol. 2, London, 334 (1954)

MEREDITH, R. and HEARLE, J. W. S. (Eds), *Physical Methods of Investigating Textiles* (pp. 320–345 FAUST, R. C., Optical properties of fibres), New York (1959)

MORTON, W. E. and HEARLE, J. W. S., *Physical Properties of Textile Fibres,* London (1962)

Specifications
ASTM D542 (measurement of ref. ind. of transparent plastics)

SECTION 60

BIREFRINGENCE

DEFINITIONS

When drawn or strained, most high polymers acquire an oriented structure and become—permanently or temporarily—optically anisotropic, i.e. a ray of light entering such a material tends to be split into two components, plane-polarised at right angles to each other and travelling at different velocities, and the material is said to exhibit the phenomenon of double refraction or *birefringence*.

In a substance of this kind birefringence (Δn) is represented quantitatively as

$$\Delta n = n_{\parallel} - n_{\perp}$$

where $n_{\parallel}$ and $n_{\perp}$ (sometimes written n_{γ} and n_{α}) are, respectively, the refractive indices for light vibrating parallel to and transversely to the *optic axis* of the material, which in general lies parallel to the direction of orientation.

The Becke-line technique (Section 59), employed with a polarising microscope, is suitable for the determination of $n_{\parallel}$ and $n_{\perp}$. However, as an alternative, birefringence can be obtained without directly involving refractive indices by determining (with a polarising microscope and compensator) the *retardation, r*, of the slower component—relative to the faster—after passing through the sample, together with the sample thickness, t, at the point of measurement. Then, when r and t are measured in the same units (generally micrometres, μm), birefringence is represented by

$$\Delta n \text{ (or } n_{\parallel} - n_{\perp}) = r/t$$

i.e. it is the retardation per unit thickness*.

UNITS

As a difference of similar ratios ($n_{\parallel} - n_{\perp}$), or the ratio of two path lengths (r/t), birefringence is without units.

*Proof: For thickness t the retardation, compared with the path in vacuum, for the $\parallel$ ray $= t(n_{\parallel} - 1)$ and for the $\perp$ ray $= t(n_{\perp} - 1)$; hence the relative retardation between the $\parallel$ and $\perp$ rays $= t(n_{\parallel} - n_{\perp}) = r$; i.e. $r/t = (n_{\parallel} - n_{\perp})$.

542

NOTES

In transparent and optically anisotropic polymers, it is usual to find that light polarised in a plane parallel to that of the molecular orientation has a lower velocity than when it is polarised in a plane transverse to it. This means that the birefringence, $n_{\parallel} - n_{\perp}$, is usually positive. However, in a few instances, where long oriented molecules possess side groups conferring strong interchain attraction, conditions contrary to those just described apply; the velocity of light is then more restricted—and the refractive index is higher—in the transverse rather than the parallel direction. Polymers in this category thus exhibit *negative birefringence* (*see* examples in Table 60.T1). Adjustment of the orientation may give such materials negative, zero, or positive birefringence.

Birefringence—a property of many crystalline substances including many minerals—is evident in many fibres, especially if they are both highly oriented and crystalline (though the last is not essential, e.g. highly oriented polystyrene is birefringent but not crystalline). Birefringence is also found in unidirectionally-strained films (and in non-uniformly biaxially-stretched films) and it often appears when normally amorphous rubber-like polymers are subjected to mechanical strain; similarly a polymer melt or solution can show birefringence when made to flow.

As a measure of molecular alignment, birefringence has been employed to check the state of orientation during the manufacture of certain fibres, notably those of polyethylene terephthalate. It is useful in fibre identification, being especially valuable for the characterisation of fibres of a given chemical composition but different physical states, e.g. low or high tenacity fibres. It has applications in the detection of residual strain in transparent mouldings, etc., and in the use of transparent models (e.g. of celluloid or rubber) for analysis of the distribution of stress in strained structures.

COMPARISON TABLE

Table 60.T1 shows the approximate birefringence of some common fibres, and includes three minerals widely spaced on the scale.

Table 60.T1. BIREFRINGENCE OF COMMON FIBRES AND SOME MATERIALS*

$n_{\parallel} - n_{\perp}$ (approx.)	Fibre or mineral
	Calcite ($-0\cdot16$).
Negative	Polystyrene (down to $-0\cdot03$), polyvinylidene chloride ($-0\cdot008$), poly-acrylonitrile ($-0\cdot003$ to $-0\cdot004$†), cellulose nitrate.
Zero or small (up to 0·005)	PVC, VC-vinyl acetate copolymers, cellulose triacetate‡, calcium alginate (0·001), regenerated proteins§, glass.
0·005	VC-acrylonitrile copolymers (c. 0·003), sec. cellulose acetate, zein (up to 0·005).
0·01	Quartz (0·009). Wool (0·009–0·013).
0·015	Asbestos (S. African sample).
0·026	Viscose rayon (some forms higher; polynosics, 0·038).
0·022–0·037	Polyvinyl alcohol, formalised‖.
0·04	Cotton (0·045), jute, polyethylene (low density).
0·05	Polyethylene (high density), silk (0·053–0·057), polyamides¶.
0·06	Flax (0·064), ramie.
Up to 0·18 or 0·20	Polyethylene terephthalate.
0·35	Crocoite (PbCrO$_4$).

*In these, n_γ ($\equiv n_{\parallel}$) is the refractive index of the extraordinary ray vibrating in a plane through the optic axis, while n_α ($\equiv n_{\perp}$) is that of the ordinary ray vibrating transversely to the optic axis
†Some copolymers positive
‡Sometimes negative
§Zein higher (see below)
‖Highly oriented, up to 0·04
¶Some up to 0·06

FURTHER LITERATURE

(*See also* Section 59, Further literature.)

TRELOAR, L. R. G., *The Physics of Rubber Elasticity,* 2nd edn, Oxford, 197 (1958) (photoelastic properties of rubber)
THETFORD, A. and SIMMENS, S. C., *J. Microscopy,* **89,** 143 (1969) (birefringence phenomena in cylindrical fibres)

WATER ABSORPTION
(*PLASTICS AND RUBBERS*)

DEFINITION

The amount absorbed (units, *see below*) when a test piece of specified dimensions is immersed in water at a specified temperature for a specified time.

UNITS

The water absorbed by a polymer can be expressed as: (i) a weight-percentage of the initial dry weight, or volume-percentage of the initial volume; or (ii) the weight (in g or mg), or volume (in ml), absorbed by a test piece of given dimensions; or (iii) the weight or volume absorbed per unit area of the test-piece surface (e.g. mg/cm^2 or g/m^2). It is necessary to state the temperature and time of immersion.

NOTES

In many polymers the rate of absorption of water is very slow, and unless the test material is thin or finely divided, absorption within a convenient test period is often incomplete, i.e. the test gives only an arbitrary result instead of the equilibrium (maximum) absorption value. In such cases expressing the result as in (i), above, has the disadvantage that it depends on the test piece dimensions (indeed, unless these are stated the result has little significance), whilst expressing it as in (ii) makes comparisons between results from different test pieces difficult; hence (iii) is the preferred method because the result is substantially independent of test piece dimensions.

The following general features of water absorption by polymers should be noted: (a) absorption may be due largely to small and variable amounts of water-soluble or water-absorbent minor constituents, e.g. proteins and sugars in natural rubber (Section 29) or residues of emulsifiers, etc., in synthetic rubbers (Sections 32–38); (b) in the initial stages of absorption—i.e. whilst this is essentially superficial—the amount absorbed is roughly proportional to the square root of the immersion time; (c) the time required to reach equilibrium (maximum) absorption, or any given fraction of this, is proportional to the square of the linear dimensions for similarly shaped test pieces—i.e. (approximately) the square of the thickness for sheets; (d) absorption increases—in rate and amount—with increase of temperature.

For many purposes a useful quantity is the *equilibrium water vapour absorption*, determined by exposing finely divided material to air of controlled (usually high) relative humidity at a controlled temperature.

For standard methods of testing plastics and rubbers, *see* references below; for equivalent data on fibres, *see* under Moisture Regain and Water Retention, Sections 62 and 63.

COMPARISON TABLE

It has not proved possible to draw up a meaningful complete table of water absorption values, because many of the data given in the literature do not quote the test conditions, and those quoted vary so widely as to make comparisons difficult. However, the following brief list (Table 61.T1) indicates the order of magnitude of absorption effects for some selected polymers.

Table 61.T1. WATER ABSORPTION (%) OF POLYMERS AT 23–28°C

Range	Material (test conditions in brackets)
0–0.03	Polytetrafluoroethylene, 0.00 ('prolonged' immersion). Polypropylene, 0.01–0.03 (3.2 mm thick disc, 24 h). Polyethylene, 0.024 (3 × 6 mm strip, 90 days immersion).
0.03–0.4	Polystyrene (conventional), 0.03–0.05; do. (toughened), 0.05–0.3; polysulphones, 0.2; styrene/acrylonitrile copolymer (SAN), 0.2–0.3; acrylonitrile/butadiene/ styrene polymer blends (ABS plastics), 0.1–0.35 (all 3.2 mm sheet, 24 h).
0.25–1.6	Ebonite (1 mm sheet, 315 days).
c. 1	Polyvinyl chloride, unplasticised (0.4 mm sheet, 32 days).
1–6	Hydrocarbon rubbers (vulcanised): natural, styrene/butadiene, ethylene/ propylene, butyl (2 mm sheet, 365 days).
9	Polyamide (Durethan BK 28F) (1 mm sheet, 365 days).
7–19	Polar rubbers (vulcanised): acrylonitrile/butadiene, chloroprene, chlorosulphonated polyethylene (2 mm sheets, 365 days).
Over 20	Cellulose: vulcanised fibre (§13.82), 25 or more; regenerated film (air-dried), over 100; do. (never-dried), c. 300.
(Soluble)	Certain cellulose ethers; alkali and ammonium alginates; starch*; gelatin; polyvinyl alcohol; polyvinylpyrrolidone; alkali metal salts of polyacrylic and polymethacrylic acids; polyacrylamide; low mol. wt. condensates of urea-, melamine- and phenol-formaldehyde resins.

*See §19.22

FURTHER LITERATURE

DAYNES, H. A., *Trans. Faraday Soc.*, **33**, 531 (1937) (absorption of water by rubber)

Specifications
BS 903: Part A16 (liquid immersion tests on vulcanised rubber)
BS 903: Part A18 (equilibrium water vapour absorption of vulcanised rubber)
BS 2782: Methods 502 A–G (water absorption of plastics)
ASTM D471 (liquid immersion tests on vulcanised rubber)
ASTM D570 (water absorption of plastics)

Water absorption tests are included in:
BS 771 (phenolic moulding materials)
BS 1322 (amino plastics)
BS 1540 (moulded insulating materials)

MOISTURE REGAIN
FIBRES

DEFINITION

Moisture regain represents the equilibrium value of the adsorption of water vapour that occurs when dry fibres are exposed to an atmosphere of specified humidity, at a specified temperature (*see also* (i) and (ii) below).

UNITS

The weight of the moisture adsorbed is expressed as a *percentage* of the weight of the initial dry sample; thus, for example, the water vapour adsorbed from an atmosphere at—as commonly—65% r.h. and 25°C is determined in grammes per 100 g dry fibres.

NOTES

(i) *Moisture content*

Sorbed water vapour is sometimes expressed as *moisture content*. The difference between this term and moisture regain is as follows.

$$\text{Moisture regain}\,(\%) = \frac{\text{Wt. of moisture}}{\text{Wt. of } \textit{dry} \text{ sample}} \times 100$$

$$\text{Moisture content}\,(\%) = \frac{\text{Wt. of moisture}}{\text{Wt. of } \textit{moist} \text{ sample}} \times 100$$

whence

$$\text{Moisture content} = \frac{\text{Moisture regain}}{(\text{Moisture regain}/100) + 1}$$

(ii) *Adsorption and desorption moisture regain*

Instead of weighing a *dry* sample, and finding its increase in weight when 'conditioned' to a given relative humidity and temperature, it is usually more convenient to weigh a *conditioned* sample and then to find its dry weight. However, when this course is followed, the value obtained for the moisture regain (calculated as previously, with respect to the *dry* weight) may depend on the history of the sample, i.e. a sample that—prior to being conditioned to the given humidity—has been exposed to a higher humidity (e.g. 100% r.h.) may exhibit hysteresis and retain slightly more moisture than it would gain if merely conditioned to the given humidity from the dry state. This gives rise to what is termed, somewhat confusingly, *desorption moisture regain*, while the more literal regain of moisture by a dry sample is described as *adsorption moisture regain*; however, for many practical purposes and especially when the values are small the result from either procedure may be considered the same.

(iii) *Implications and measurement*

The sorption of water vapour by fibrous high polymers, being primarily a phenomenon associated with hydrogen bonding, is most pronounced in substances having molecules of a hydrophilic type (e.g. with —OH or >NH groups) and is virtually absent from hydrophobic polymers such as polyethylene and polyvinyl chloride (though a hydrophilic dressing or finish may cause uptake of a small proportion of water vapour). The dimensional changes produced by the sorption or desorption are sometimes put to good use, as in the employment of protein fibres in hair hygrometers that record the relative humidity of the atmosphere, but more often the effects need to be avoided or guarded against*. Control of the moisture regain (or moisture content) is thus important in many technical operations, notably during the processing of textile fibres and fabrics, in paper-making, in bonding rubber to textiles (e.g. belting fabrics, which should be dry) and in bonding wood with synthetic resin glues; there is also the economic aspect in that one does not want to buy water when paying for fibres (nor to sell them dry if water can legally be included—see reference below to commercial standards).

The water taken up may be determined in several ways, probably the most used being gravimetric methods depending on weighing a sample in both a conditioned state and the dry state†; physical entrainment methods are also employed, e.g. measurement of the volume of water distilled as an azeotrope when a weighed sample is boiled with toluene (*see* ref. below to Dean and Stark method), and chemical determination of water present can be made (*see* refs. below to Karl Fischer reagent). Where a continuous indication, or continuous control, of adsorbed moisture is required, moisture meters

---

*They lead, too, to the need to re-tune a gut-stringed musical instrument when the relative humidity changes in a concert hall.

†Included in these methods is the delicate 'spring balance' technique of hanging a small sample on the end of a helical spring of quartz fibre and, on changing the humidity of the ambient atmosphere, following the uptake or loss in weight by the corresponding change in the extension of the spring.

(direct reading and non-destructive) are of great use. They are calibrated in moisture regain or moisture content but mostly depend on measurement of electrical resistivity or dielectric loss (*see* Sections 69 and 72); one commercial instrument is of a nuclear kind, measuring the moderating effect of hydrogen—particularly that in water—in slowing down fast neutrons.

COMPARISON TABLE

Table 62.T1 shows the range of the moisture regain of common fibres at 65% r.h. and 25°C. More detailed information, including moisture regain at other humidities, is given in §§1.31–42.31 and elsewhere.

Table 62.T1. MOISTURE REGAIN OF COMMON FIBRES,
AT 65% r.h. AND 25°C

Approximate M.R., %	Fibres (M.R., %; desorption values enclosed thus [ ])
Under 0·1	Polyvinylidene chloride, polypropylene (up to 0·15).
0·1–0·5	Glass, c. 0·1. Polyvinyl chloride, 0·10–0·15. Polyethylene terephthalate, 0·45.
0·5–1·0	Polyamide (nylon 11), polyvinyl acetate (films), c. 1.
1–2	Polyurethane, 1·2–1·3. Polyacrylonitrile, 1·2–2·0 [1·6–3·0]. Polyamide (nylon 6.10), c. 1·7.
2–5	Cellulose triacetate, 2·5–4·5 [3·3–3·9]. Polyamide (nylon 6.6), 3·9–4·2; (nylon 6), c. 4. Polyvinyl alcohol (formalised), 4·5–5·0.
5–10	Cellulose sec. acetate, 5·9–6·0 [7·5–8·0]. Cotton, 6·8–8·5 [8·2–10·2]. Flax, 9–12. Regenerated proteins, 9·7–12·3. Hemp, 9·8–10·3 [9·9–11·3].
10–20	Silk, 10 [11]. Jute, 10–12 [12–14]. Viscose rayon (normal), 12–13·5 [14–16]; (high tenacity), 11·5–14·5 [13–15]; (polynosic), 12 [14]. Wool, 13·9 [15·7].
Over 20	Calcium alginate, 20–30.

FURTHER LITERATURE

URQUHART, A. R. and ECKERSALL, N., *J. Text. Inst.*, **21**, T499 (1930) (moisture relations of cotton); **23**, T163 (1932) (moisture relations of viscose rayon)

HUTTON, E. A. and GARSIDE, Miss J., *J. Text. Inst.*, **40**, T161 (1949) (moisture regain of silk); T170 (moisture regain of nylon)

GIBBONS, G. C., *J. Text. Inst.*, **44**, T201 (1953) (moisture regain of methylceullulose and cellulose acetate)

JEFFRIES, R., *J. Text. Inst.*, **51**, T339 (1960) (sorption of water vapour by cellulose at 0–100% r.h. and 30–90°C); T399 (sorption of water vapour by cellulose esters, wool, nylon, etc., at 0–100% r.h. and 30–90°C); T441 (sorption of water vapour by cellulose and cellulose acetate at 0–100% r.h. and 120–150°C)

HEARLE, J. W. S. and PETERS, R. H. (Eds), *Moisture in Textiles*, London (1960)

MORTON, W. E. and HEARLE, J. W. S., *Physical Properties of Textile Fibres*, London and Manchester (1962) (includes theories of moisture sorption and discussion of sorption phenomena)

PANDE, A., *Lab. Pract.*, **12**, 741 (1963) (measurement based on the Karl Fischer reagent)

Specifications
BS 756 (apparatus for determination of moisture by Dean and Stark method)
BS 1051 (moisture in textiles, includes commercial standards of regain)

BS 2511 (determination of moisture by Karl Fischer method)
ASTM D179 and *D2495* (cotton)
ASTM D540 (man-made staple fibres)
ASTM D629 (quantitative analysis of textiles)
ASTM D1576, D2118 and *D2462* (wool)
ASTM D1909 (commercial moisture regains)

SECTION 63

WATER RETENTION
FIBRES

DEFINITION

The water retained by wet fibres after centrifuging under conditions in which surplus liquid is removed.

UNITS

The weight of water retained is expressed as a percentage with respect to the dry weight of the sample.

NOTES

The absorption or imbibition of water by fibres and textiles, and its proper control, has important domestic and industrial implications; *see also* Section 62. Water retention or water imbibition is, however, an arbitrary measure and the values found depend somewhat upon the experimental conditions. The results listed in this book were obtained after soaking small samples (*c.* 0·5 g) in water and then centrifuging them (for 10–15 min at 1000- to 1200-times the force of gravity) in glass tubes (12·5 mm in internal diameter) stoppered at the top and fitted with a coarse sintered glass plate at the bottom. Small variations in the speed or time of centrifuging, and variations in the state of the sample (loose fibre, cut yarn, or fabric) prove to be unimportant; wax, grease, and other hydrophobic dressings or contaminants cause low values of water retention.

Cross-linkage can bring about a reduction in absorption—i.e. in *swelling* (*see also* Section 55)—but does not necessarily do so if effected while fibres are in a swollen state (*see* resin-treatment of cellulose, §13.86).

COMPARISON TABLE

Table 63.T1 indicates the water retention of some common fibres, free from grease, under the conditions specified above.

Table 63.T1. WATER RETENTION OF COMMON FIBRES

Range (%)	Fibres (approx. water retention*, %)
Under 5	Polyethylene, polypropylene (up to 5).
5–10	Polystyrene (5), polyvinylidene chloride (Saran, 5); vinyl chloride/acrylonitrile copolymers (6); glass (7);polyvinyl chloride, unplast. (8).
10–15	Polyacrylonitrile (10–20); polyester (Terylene, 4–12), vinyl chloride/vinyl acetate copolymers (Vinyon, 12); nylon 6 (Perlon, 15); cellulose triacetate (13–18).
20–30	Asbestos (20); nylon 6.6 (25); cellulose acetate, sec. (30), polyvinyl alcohol, formalised (Vinylon, 30).†
c. 40	Ramie, scoured wool (40); regenerated proteins (40–50).
c. 50	Cotton, bleached or mercerised (45–50); degummed silk (50).
c. 70	Hemp (65); flax, jute, raw silk (70).
c. 100	Regenerated cellulose (viscose rayon), 80–100‡; calcium alginate (over 100).

*For experimental conditions, *see* under **Notes**
†Another type (Kanebian), *c.* 55%
‡Rayon treated with an amine-formaldehyde finish, 30% or lower; high tenacity and polynosic rayons, 50–80%

FURTHER LITERATURE

MORTON, W. E. and HEARLE, J. W. S. (*see* Section 62) includes a chapter on retention of liquid water

SECTION 64

PERMEABILITY

DEFINITION

Permeability, as here discussed, refers to the passage of a gas or vapour through a solid barrier, which is often in the form of a thin film.

The quantity of gas (Q) passing in an interval of time (t) through a barrier of a measured area (a), with a difference of pressure—or of partial pressure—(δp), across the thickness (δx), is given by

$$Q = Pat\, \delta p / \delta x$$

where P is a proportionality constant known as the *permeability**.

Permeability may thus be regarded as the volume (at s.t.p.)—or the mass—of the permeating substance passing through a unit cube of the barrier material in unit time under unit pressure difference, under steady-state conditions†.

N.B. Since the permeability of a *gas* is largely independent of film thickness, it is usual to express the property as above, with respect to the pressure difference per unit thickness. With a vapour, however, since it is found that permeability may show considerable variation with thickness, it is better to state the rate of transmission, the barrier thickness and the difference of partial pressure across it (or, for water, the difference in relative humidity and the temperature). Vapour transmission rates expressed in this last way are related to, but not equal to P; nevertheless they are commonly spoken of as permeability.

UNITS

Permeability measurements are quoted in the literature in various ways, and occasionally are incompletely defined. Bearing in mind the

---

*Also called *permeability coefficient* and *permeability constant*, though these are not necessarily the same as permeability; permeability is sometimes used to mean the property, and the other terms its numerical value.

†The reciprocal of permeability represents the resistance offered by a unit cube; i.e. it is the pressure difference needed to transport unit volume or mass through unit area in unit time.

remarks above, and that $P = Q/(A.t.dp/dx)$, the units appearing most suited to scientific work are as follows:

For a gas—as $cm^3/(cm^2 \text{ s cmHg/cm})$*, which can be rendered as $(cm^3 \text{ cm})/(cm^2 \text{ s cmHg})$ or simplified to $cm^2/(\text{s cmHg})$. The last is the preferred form, but the unit is too large in practice; hence, in order to avoid very small numbers, *in the present book* the permeability of gases is expressed in terms of $(10^{-10} \text{ cm}^2)/(\text{s cmHg})$.

For a vapour—as $g/(cm^2 \text{ s cmHg})$, stating also the film thickness. Values given in these units are called 'transmittance'. Again the unit is too large in practice, and *in the present book* the transmission of vapours is expressed, wherever possible, in terms of $(10^{-10} \text{ g})/(cm^2 \text{ s cmHg})$. In other cases (and in Table 64.T2) permeability is given for 1 cm thickness, i.e. as $(10^{-10} \text{ g})/(\text{cm s cmHg})$.

Units of a different type may be preferred industrially, as providing 'transmission rates' in the range of 1 to 100000. For example, the permeability of a *gas* may be quoted in terms of $cm^3/(m^2.24 \text{ h})$ and that of a *vapour* in terms of $g/(m^2.24 \text{ h})$, stating in each instance the pressure difference and the film thickness. Areas may be expressed in terms of 100 in^2, and pressures in Torr (i.e. mmHg). As defined earlier, all volumes are corrected to standard temperature and pressure.

See also **Note on SI units,** page 659.

CONVERSION FACTORS

$(10^{-10} \text{ cm}^2)/(\text{s cmHg})$—or $(10^{-10} \text{ cm}^3)/(cm^2 \text{ s cmHg/cm})$ or $(10^{-10} cm^3.\text{cm})/(cm^2 \text{ s cmHg})$—has the following values in other units:

(i) $76.10^{-10} \text{ cm}^3/(cm^2 \text{ s atm/cm})$, or $76.10^{-10} \text{ cm}^2/(\text{s atm})$.

(ii) $66 \text{ cm}^3/(m^2.24 \text{ h.atm/mm})$ or $66.10^{-5} \text{ cm}^2/(24 \text{ h.atm})$.

(iii) $45.10^{-16} M \text{ g}/(cm^2 \text{ s cmHg/cm})$, or $45.10^{-16} M \text{ g}/(\text{cm s cmHg})$.

(iv) $29.10^{-4} M \text{ g}/(m^2 \text{ 24 h atm/mm})$, or $29.10^{-9} M \text{ g}/(\text{cm}.24 \text{ h.}$ atm) where M is the mol. wt. of the permeating gas or vapour. A table including additional conversion factors relating to volumes is given by W. M. Smith on pp. 172–3 of *The Manufacture of Plastics*, New York and London, 1964.

NOTES

High polymers show a small but measurable permeability to gases and vapours, the transport of which takes place (i) interstitially, through permanent or transient pores in the barrier, and (ii) by a process of sorption or dissolution on one side of the barrier, followed

---

*The preferred SI volume unit is millilitre (ml), not cubic centimetre (cm^3), but the latter is used here for consistency with cm and cm^2.

by diffusion through it, and subsequent desorption or evaporation on the far side. Permeability, represented for many gases by the product of the diffusion constant of the permeant/polymer system $(D)^*$ and the solubility coefficient $(S)^*$—i.e. $P = DS$—thus differs from simple porosity; likewise, at the other extreme, permeable films differ from those of thin metal or glass, which except in certain instances are completely impermeable. The property tends to increase with rise of temperature, to an extent depending on the polymer and gas (usually between c 2·5 and 8% per °C), and is greater above than below the glass temperature (i.e. greater in rubbers than in rigid plastics; *see* T_g, Section 68); permeability is also much increased by the presence of plasticisers, while on the other hand it is decreased by crystallinity, by cross-linkage (e.g. vulcanisation) and by inert fillers, especially those of a laminar plane-orientable kind (e.g. mica, graphite, aluminium leaf powders).

An understanding of permeability is particularly important where films or thin-walled containers are used, in various forms of packaging, in connection with either the exclusion or retention of gases or vapours (especially water vapour), also in relation to rubber products such as tyre inner tubes and gas-retaining rubberised fabrics. One of the simplest methods of examining the property with respect to vapours depends on sealing the open top of a shallow vessel, containing the appropriate liquid or an absorbent, with a thin sheet or film of the material to be tested, then exposing the weighed assembly to a specified atmosphere and re-weighing at intervals; for example, the vessel could contain water which permeates outwards into an atmosphere of known humidity, or a desiccant that absorbs water permeating inwards from such an atmosphere.

COMPARISON TABLES

It is difficult to correlate results obtained by different techniques, and the tables that follow must be considered approximate only; Table 64.T1 shows the permeability of various polymers to nitrogen and carbon dioxide, and Table 64.T2 shows the permeability to water vapour. More detailed data for individual polymers will be found in §§1.31–42.31.

Solubility coefficient is the concentration of dissolved gas (i.e. vols.—at s.t.p.—per unit vol. of barrier material) when the material is in equilibrium with the gas at unit pressure, e.g. 1 cmHg. *Diffusion coefficient* is the volume of gas diffusing in unit time through unit area of the barrier material when the concentration gradient of the gas dissolved in the material is unity, i.e. the concentration changes by 1 ml/ml per unit thickness (e.g. 1 cm) of the material; also called *diffusivity*.

Table 64.T1. PERMEABILITY OF COMMON POLYMERS TO NITROGEN AND CARBON DIOXIDE*

Range	*Material, at 20–30°C. Permeability expressed as* $(10^{-10} \text{ cm}^2)/(\text{s cmHg})$	
	Nitrogen	*Carbon dioxide*
<0·001	Polyvinylidene chloride 0·0009	—
0·001–0·01	Polychlorotrifluoroethylene, 0·003–0·13. Cellulose (regenerated), 0·0032. Polyvinyl fluoride, 0·004. Polyester (packaging film), 0·005–0·006. Rubber hydrochloride, 0·008†.	Cellulose (regenerated), 0·0047.
0·01–0·1	Nylon,0·01–0·02; (nylon 6), 0·01. Polyformaldehyde (copolymer film), 0·012; do., acetal (Delrin), 0·022. Ebonite‡, 0·025. Polyurethane (polyester), 0·027. Polyvinyl chloride (rigid), 0·12§; Styrene/acrylonitrile copolymer, 0·046. Phenol/formaldehyde resin, 0·095. Polystyrene, <0·1–0·4; do. (oriented), 0·7. Polymethyl methacrylate, 0·1.	Polyvinylidene chloride, 0·03, Polychlorotrifluoroethylene, 0·05–1·25. Polyvinyl fluoride, 0·09. Polyester (packaging film), 0·09–0·15. Epoxy resin, 0·09–1·5.
0·1–1·0	Cellulose nitrate, 0·12. Cellulose acetate, 0·16.‖ Polyethylene (high density), 0·18–0·3. Polypropylene, 0·2 (oriented) to 0·5. Polycarbonate, 0·23–0·3. Acrylonitrile/butadiene (39/61) rubber, 0·24. Butyl rubber, 0·32. Fluoro-rubber (Viton A), 0·44. Polyurethane rubbers (Adiprene, Vulcaprene), 0·49.	Nylon, 0·16–0·20; (nylon 6), 0·09. Rubber hydrochloride, 0·17¶. Polyformaldehyde, acetal (Delrin), 0·19–0·30; do. (copolymer), 0·04. Polyurethane (polyester), 0·4.
1–10	Acrylonitrile/butadiene (27/73) rubber, 1·05–1·2; do. (20/80), 2·5. Chlorosulphonated polyethylene, 1·16. Chloroprene rubber, 1·2. Polyethylene (low density), 1·2–2. Styrene/butadiene rubbers, 6·3. Butadiene rubber, 6·5. Polymethylpentene, 6·5. Natural rubber, 7–9. Ethylcellulose, 8·4. Ethylene/propylene rubber, 8·5–15.	Styrene/acrylonitrile copolymer, 1·08. Polypropylene, 1·2 (blown film) to 9 (cast film). Polyvinyl chloride (rigid), 1·6**. Cellulose nitrate, 2·12. Cellulose acetate, 2·4 to (plasticised) 18. Polyethylene (high density), 3–4·3. Polysulphide rubbers, 3·2. Butyl rubber, 5·2. Acrylonitrile/butadiene (39/61) rubber, 7·5. Fluororubber (Viton A), 7·8. Polycarbonate 8·5–11·2. Ionomers, 6.
10–100	—	Polystyrene, 12. Polyurethane rubber (Adiprene), 14–40. Polyethylene (low density), 17–35. Chlorosulphonated polyethylene, 21. Chloroprene rubber, 26. Acrylonitrile/butadiene (27/73) rubber, 30; do. (20/80), 64. Ethylcellulose, 41.
>100	Silicone rubbers, 260–320	Ethylene/propylene rubbers, 108. Styrene/butadiene rubbers, 125. Natural rubber, 130–135. Butadiene rubber, 140. Silicone rubbers, 180–3000.

*Values for hydrogen and oxygen are generally between those for nitrogen and carbon dioxide. Values for rubbers relate to unfilled vulcanisates (except butyl and silicone rubbers, not defined)
†Up to 0·62 when plasticised
‡At 67°C
§Up to 0·5 when plasticised
‖Up to 0·5 when plasticised
¶Up to 1·8 when plasticised
**Up to 12 when plasticised; 20 % plasticiser, 7·1

Table 64.T2. PERMEABILITY OF COMMON POLYMERS TO WATER VAPOUR*

Range	Material, at 20–30°C. Permeability expressed as $(10^{-10}$ g$)/($cm s cmHg$)$
<0·01	Polyvinylidene chloride, 0·0004–0·08**. Polychlorotrifluoroethylene, 0·00023–0·03**.
0·01–0·1	Polyethylene (high density), 0·010–0·012†. Rubber hydrochloride, 0·013–1·5**. Regenerated cellulose (nitrocellulose-coated), 0·025–0·032. Polytetrafluoroethylene, 0·03. Polypropylene, 0·03–0·07††. Butyl rubber, 0·03–0·16. Nylon, 0·055–1·4. Polyethylene (low density), 0·065–0·17**. Chlorinated polyether (Penton), 0·09†. Ethylene/vinyl acetate copolymer, 0·04–0·06. Ebonite, 0·08–0·42.
0·1–1·0	Polyisobutylene, 0·10†. Polyester (packaging film), 0·1–0·18**. Polyvinyl chloride (rigid), 0·2‡. Chlorinated rubber, 0·23§. Polyurethane rubber (Adiprene), 0·27. Polyester (unsaturated, cast film), c. 0·4. Polyformaldehyde, 0·4–0·8. Silicone rubbers, 0·5–8·5. Polycarbonate, 0·55–0·8. Polymethylpentene, 0·6†. Styrene/acrylonitrile copolymer, 0·7–1·0**. Chloroprene rubber, 0·7–1·45**. Polystyrene, acrylonitrile/butadiene rubbers, 0·8. Polyacrylonitrile†, chlorosulphonated polyethylene, 1·0.
>1	Ethylcellulose, 1·1–10. Cellulose acetate, 1·2–4·4‖. Natural rubber (raw, or unfilled vulcanisate), 1·5–3·8. Styrene/butadiene rubbers, 1·9. Polyvinyl alcohol, 1·9–11. Cellulose nitrate, 2·1–5·0. Polyurethane rubber (Estane), 2·3; do. (Vorite), 10. Polymethyl methacrylate, 3·0†. Butadiene rubber, 3·8. Cellulose, 11.¶

*The tabulated values have been converted to absolute units in order to render those for different materials comparable. This involves the assumption that transmission rate is inversely proportional to thickness. As this is not always true, and the thickness tested varies (at least from 0·013 to 0·35 mm), comparisons of different materials are only approximate
†At 38–40°C
‡Up to 1·1 when plasticised
§Up to 0·7 when plasticised
‖Up to 8·5 when plasticised
¶At 35°C
**Values at 40°C also fall in range shown
††Values at 40°C higher than those shown.

FURTHER LITERATURE

AMERONGEN, G. J. VAN, *J. appl. Phys.*, **17**, 972 (1946) (permeability of rubbers to gases)
HENNESSY, B. J., MEAD, J. A. and STENING, T. C., *Permeability of Plastics Films*, London (1966) (includes theoretical and practical considerations)
LEBOVITS, A., *Mod. Plas.*, **43**, No. 7, 139 (1966) (review and examples of applications)
OSWIN, C. R. and PRESTON, L. N., *Protective Wrappings*, London (1966) (lists resistance of films to transmission of O_2 and H_2O vap.)
PINNER, S. H. (Ed), *Modern Packaging Films*, London (1967) (includes references to permeability)
GLOVER, D., *New Scientist*, **35**, 678 (1967) (versatility of packaging films; plot of transmission of O_2 and H_2O vap.)
CRANK, J. and PARK, G. S. (Eds), *Diffusion in Polymers*, London and New York (1968)
SMITH, T. G., *Review of Diffusion in Polymer-Penetrant Systems*, US National Aeronautics and Space Administration; Maryland Univ., Dept. of Chem. Engng, Washington D.C. (1968) (theory; diffusion coefficient data)

Standards
BS 903: Part A17 (for vulcanised rubber and gases)
BS 3177 (flexible sheet materials and water vapour; also appears as *Method 513* in *BS 2782*)
ASTM D697 (plastic sheets and water vapour)
    D813 (rubbers and volatile liquids)
    D815 (rubber-coated fabrics and hydrogen)
    D1434 (gas transmission of plastic sheeting)

SPECIFIC HEAT

DEFINITION

The energy required to raise a unit mass through unit interval of temperature (which quantity, so expressed, is sometimes referred to as *thermal capacity**). Alternatively, for practical purposes specific heat may be considered as a *ratio* representing (the energy required to raise a given mass through unit temperature interval)/(the energy required to raise an equal mass of water through the same temperature interval). It is necessary to state the temperature at which the measurement is made.

UNITS

The units commonly used are cal/(g °C), or Btu/(lb °F), but if considered as a ratio—*see above*—specific heat is without dimensions. In all three instances, however, the values may be taken as numerically the same. (*See also* **Note on SI Units,** page 659.)

NOTES

For many solid chemical elements the product of specific heat and atomic weight is roughly constant, being approximately 6·4; i.e. the *atomic heat* or *thermal capacity of each atom* is the same (Dulong and Petit's law), although at room temperature carbon is an exception to this rule. For many compounds specific heat is approximately the sum of the specific heats of the component elements accounted proportionately; in high polymers, the atoms of which are usually of low atomic weight, the value of the specific heat at room temperature lies within the range 0·2–0·55.

Specific heat rises slightly with increase of temperature. It shows an increase in value beyond T_g, and in a crystalline polymer the heat of fusion causes a large increase at T_m (*see* Section 68) though beyond that temperature it returns to lower values.

*However, thermal capacity sometimes refers to the energy required to raise the whole of a given body through unit temperature interval. It then represents mass × specific heat; also, since the specific heat of water can be assumed to be unity, it represents the *water equivalent* of the given body, i.e. the mass of water that the same energy would raise through unit temperature interval.

COMPARISON TABLE

Table 65.T1 shows the specific heat of common polymers and some associated substances.

Table 65.T1. SPECIFIC HEAT OF POLYMERS AND RELATED SUBSTANCES

Range	Material* (specific heat at 20°C)
Under 0·1	*Gold, lead, mercury, platinum*, 0·03; copper†, *zinc, brass*, 0·09.
0·1–0·2	*Iron, steel, nickel*, 0·11; *zinc oxide*, 0·12; *graphite, sulphur (rhombic), titanium dioxide*, 0·17; *silica (fused* quartz*), steatite*, 0·18–0·2; *carbon black, whiting* (CaCO₃), 0·2.
0·2–0·3	Glass, asbestos, 0·2–0·25; polytetrafluoroethylene, 0·25; polycarbonates, 0·28–0·3; *Mica, talc*, 0·21; *aluminium, china clay (air dry powder)*, 0·22; *air*, 0·24‡; polyvinyl chloride (unplasticised), 0·25; *magnesium*, 0·25; fluorinated ethylene/ propylene polychlorotrifluoroethylene, 0·28; *sodium*, 0·29; (*molten*, at 100– 250°C), 0·32; phenol-formaldehyde resin (mineral/glass filled), 0·28–0·32.
0·3–0·4	Wool, silk, polyethylene terephthalate, polysulphide rubbers, c. 0·3; polyvinyl chloride (plasticised), 0·3–0·5; polyvinylidene chloride, 0·32; polystyrene, cellulose, 0·32–0·35 (cell. derivatives up to 0·4); ebonite (natural rubber, un- loaded), 0·33–0·34; ABS plastics, 0·38; polymethyl methacrylate, polyformalde- hyde, 0·35; polyacrylonitrile, casein plastics, leather, 0·36; phenolformaldehyde resin (wood flour/cotton flock filled), 0·35–0·40; silicone resins, 0·36–0·37; phenol formaldehyde resin (unfilled), 0·38–0·42.
0·4–0·5	Amino-resins (cell. filled), 0·4; natural rubber, 0·45; polyurethanes, styrene/ butadiene rubbers, butyl rubber, c. 0·45; nitrile rubber§, c. 0·47; polyesters (unsaturated), c. 0·5; polyamides, 0·4–0·55; *Wood*, c. 0·42.
Over 0·5	Ethylene/propylene rubbers, chloroprene rubber, 0·52; polyethylene, 0·55; *Alcohol, paraffin oil (varies)*, 0·55; *water*, 1·0; *hydrogen*, 3·4‖.

*Values for rubbers are for the *raw* material; unfilled vulcanisates generally give closely similar values. For filled vulcanisates, see **Notes** below
†Reference material: electrical grade (electrolytic) copper, 0·0931 (0–100°C)
‡At atmospheric pressure; at constant volume, 0·17
§40% acrylonitrile
‖At atmospheric pressure; at constant volume, 2·4

FURTHER LITERATURE

REILLY, J. and RAE, W. N., *Physico-chemical Methods*, Vol. I, London, 541 (1954)
PARTINGTON, J. R., *An Advanced Treatise on Physical Chemistry*, Vol. III, London, 264 (1952)

Specification
ASTM C351 (determination of specific heat of thermal insulation)

THERMAL CONDUCTIVITY

DEFINITION

The amount of heat Q passing in time t through a slab of a substance, of transverse area a and thickness b, with a temperature difference $d\theta$ across it, may be defined as

$$Q = Kat\,d\theta/b \qquad \text{(Fourier's law)}$$

where K is the *thermal conductivity*.

Since $K = Qb/(at\,d\theta)$, it may be regarded as the amount of heat transferred in unit time across a unit cube, when the temperature gradient—measured between opposite faces in the direction of the heat flow—is unity.

Thermal diffusivity The ratio of thermal conductivity to (specific heat × density), known as the thermal diffusivity, is used in problems relating to heat flow.

UNITS

Thermal conductivity is measured in the c.g.s. system as (cal cam)/(cm² s °C), which is more conveniently expressed as cal/(cm s °C).

To convert cal/(cm s °C) to Btu/(ft² h °F per inch), multiply by 2903. To convert the same units to Btu(ft² h °F per ft)—or Btu/(ft h °F)—multiply by 241·9. See also **Note on SI Units,** page 659.

NOTES

The value varies slightly with temperature, and changes at T_g and T_m (*see* Section 68). In an oriented material it is greater in the direction of orientation than perpendicularly to it.

Organic polymers are relatively poor conductors of heat, and many can be expanded to solid foams or cellular structures in which thermal conductivity diminishes to a very low value. In expanded material of a given ratio of cell volume to solid the conductivity declines with decrease in cell size and with increase in the proportion of closed cells. Polyurethane foams prepared with an auxiliary blowing agent such as trichlorofluoromethane have a lower thermal conductivity than those blown with carbon dioxide. The warmth of clothing is a matter of thermal insulation and depends to a large extent on the low conductivity of the still air extrapped in the interstices of the cloth; on a weight for weight basis the insulation is usually increased when a fabric is combined with a low-density foam (giving warm but light-weight garments).

Incorporation of non-polymeric substances, such as finely-divided aluminium or quartz powder, in cast or moulded plastics and rubbers increases their thermal conductivity. Aluminium wool, which provides long paths of relatively high conductivity, can be used to assist the dissipation of the exotherm in cast resins and to improve the conductivity of the final product.

COMPARISON TABLE

Table 66.T1 indicates the order of the thermal conductivity of common polymers and some associated substances. More details are given under individual polymers in §§1.32–42.32.

Table 66.T1. THERMAL CONDUCTIVITY OF POLYMERS AND RELATED SUBSTANCES

Range	Material, conductivity at 20°C, expressed as $(10^{-4}$ cal)/(cm s °C)
Under 1	Glass fibre quilt, expanded plastics and rubbers, 0·7–0·8 (polyurethane foams, 0·4–0·85). *Still air,* 0·6; *balsa wood,* 0·75.
1–2	Loosely packed fibres (cotton, wool, silk, etc., also average textile fabric), *c.* 1. *Felt,* 1·1; *charcoal, cork, sand,* 1·3. Expanded glass, loosely-packed asbestos fibres, *c.* 1·5. Polystyrene, 1·9–3·3; do., high-impact, 1–3.
2–3	Polyvinylidene chloride, 2·5. Cyclised rubber, 2·75; chlorinated rubber, 3. *Wood (transverse to grain), c.* 2; *wood (parallel to grain), paper, mineral and vegetable oils, c.* 3.
3–4	Polyvinyl chloride, butyl rubber, polyurethane rubbers, 3–4; cellulose nitrate, epoxy resins, 3–5; phenol-formaldehyde resins (unfilled), 3–6; natural rubber (raw), 3·2; do. (vulcanised, unfilled), 3·4–3·6; ebonite (natural rubber, unloaded), 3·7–4·4; casein plastics, polyvinyl acetal and acetate, polyethylene terephthalate, polymethylpentene *asbestos sheet* (soft), *hydrogen* (NTP), *c.* 4.
4–5	ABS plastics, 4–7. Polymethyl methacrylate, 4·5; polycarbonates, chloroprene rubber, 4·6; phenol formaldehyde resin (wood flour/cotton flock filled), 4·7; alkyds, unsat. polyesters, polyvinyl alcohol, 5. Silicone rubber, 4–10; cell. acetate, polyamides, 4·5–7·5; styrene/butadiene rubbers, 4·6–5·9.
5–6	Cotton (compressed), polyformaldehyde, 5·5; polytetrafluoroethylene, 5·8; phenol-formaldehyde resin laminates (normal to plies), 5–8. Fluorinated ethylene/propylene, 6. Nitrile rubbers (60°C), *c.* 6.
6–10	Polyesters (glass reinforced), 6–10. Natural rubber (vulcanised, with 50 parts HAF black per 100 rubber), 6·8; polysulphides, 7·2; amino-resins (cellulose mineral filled), 8/*c.* 15, polyethylene, 8–10; ethylene/propylene rubbers, 8·5; *sulphur (rhombic),* 6·5; *carbon black,* 6·7; *concrete, c.* 7; *whiting* ($CaCO_3$), *c.* 8.

Range	Material, conductivity at 20°C, expressed as $(10^{-4}$ cal)/(cm s °C)
10–50	Epoxy resins (silica-filled), 15; glass, 20–27; *still water* (80°C), *brickwork, asbestos-cement sheet (hard), zinc oxide, c.* 16; silica (fused), *c.* 30; *ceramics, c.* 35.
50–1000	*Marble*, 50–70; *quartz*, 150–300; *mercury* (50°C), 190; *lead*, 830.
Over 1000	*Iron, steel*, 1100; *sodium* (*molten*, at 250°C), 1900; *brass, zinc, magnesium alloys*, 2500; *aluminium*, 5000.
Nearly 10^4	*Copper*, 9200; *silver*, 9700.

FURTHER LITERATURE

ANDERSON, D. R., *Chem. Rev.*, **66**, 677 (1966) (thermal conductivity of polymers)
REES, W. H., *Text. Inst. and Ind.*, 29 (1964) (light-weight clothing)

Specifications
BS 874 (heat insulation; determination of thermal conductivity)
ASTM D1518 (thermal transmittance of textile fabric and batting)
ASTM C177, C518 (determination of thermal conductivity of materials)
ASTM D2326 (determination of thermal conductivity of cellular plastics)

COEFFICIENT OF LINEAR EXPANSION

DEFINITION

The reversible increase in length of a unit length of a material, per unit rise of temperature.

For an isotropic substance the coefficient of linear expansion is approximately one-third of the coefficient of cubical expansion.

UNITS

The coefficient of linear expansion, α, is normally expressed as length increase/(length °C), i.e. /°C.

To convert /°C to /°F, divide by 1·8.

NOTES

The rate of expansion tends to increase with rise of temperature, but over a short range the coefficient of linear expansion of a given substance may be assumed constant provided that no change of state (e.g. from polycrystalline to amorphous) occurs within the temperature interval. In crystalline polymers, however, the coefficient may be different in the directions of the three principal axes; thus, interchain forces of the van der Waals type give, in the direction of the bonding, a relatively high coefficient of expansion (e.g. $c.\ 2.10^{-4}$/°C in polyethylene) but the value is lowered by hydrogen bonding, while along a fibre or stress axis the value may be negative, i.e. an extended primary-bonded chain molecule tends to contract with rise of temperature.

For a number of polymers the *coefficient of expansion of the free volume*—i.e. the difference between the volume expansion coefficients in the rubbery and glassy states—tends to be constant, with a value of $4·8.10^{-4}$/°C.

The thermal expansion of organic polymers is high in relation to other constructional materials but may be reduced by inclusion of mineral fillers, e.g. quartz powder and other fillers. When one material is to be bonded to another it is advisable that they should have similar coefficients of expansion.

COMPARISON TABLE

Table 67.T1 shows the coefficient of linear expansion of common polymers and some associated substances. More details appear under individual polymers in §1.32 to 42.32.

Table 67.T1. COEFFICIENT OF LINEAR EXPANSION OF POLYMERS AND RELATED SUBSTANCES

Range	Material, with coefficient at 20°C expressed as $10^{-4}/°C$
Under 0·1	Fused silica, 0·005. *Low-expansion steel (Invar)*, 0·009. Glasses, 0·04–0·11. *Wood (along grain)*, 0·05. *Ceramics, 0·05–0·09. Platinum, 0·09.*
0·1–0·5	*Iron, steel, c.* 0·11. Polyesters, unsaturated (glass-reinforced), 0·15–0·3. Phenol-formaldehyde resins (filled), 0·15–0·45. *Copper, brass,* 0·18. Amino-resins (filled), 0·2–0·6. *Aluminium,* 0·25. *Zinc alloys,* 0·27. *Lead,* 0·29. Epoxy resins (silica-filled), *c.* 0·3. Epoxy resins (cast), casein-formaldehyde, polyethylene terephthalate, *wood (transverse to grain), c.* 0·5.
0·5–1·0	Polyurethane foams (rigid), polyesters (unsaturated, unfilled), 0·5–1. Poly-carbonates, 0·6–0·7. Polystyrene, 0·7. *Sulphur (rhombic),* 0·65–0·7. Ebonite, 0·65–0·8. Polyvinyl acetal, 0·7. Polyvinyl chloride (rigid), 0·7–0·8; do. (plasti-cised), up to *c.* 2·5. Polymethyl methacrylate, 0·75. Cyclised rubber, 0·75–0·8. ABS plastics, polyformaldehyde, polyvinyl formal, phenol-formaldehyde resins (unfilled), Penton 0·8. Cellulose sec. acetate, 0·8–1·6. Polyvinyl acetate, 0·85. Alkyd resins, polyvinyl alcohol, *c.* 1.
1·0–2·0	Polyamides, 1–1·5. Cellulose nitrate, 1–1·6. Polypropylene, 1.1. Butyl rubber, 1·1–1·8. PTFE, 1·2. Chlorinated rubber, 1·25. Polyethylene, 1·3–2·2. Poly-butene-1, cellulose triacetate, polyvinyl butyral, 1·5. Polyvinylidene chloride, 0·8–1·75. Ethylene/propylene rubbers, 1·8–2·4. Polyurethane rubber, *c.* 2.
Over 2·0	The following rubbers, as unfilled vulcanisates: natural, 2·2–2·3; nitrile, 2·2–2·4; styrene/butadiene, 2·2–2·5; chloroprene, 2·3–2·5; butadiene, 2·37–2·45; silicone, 2·5–4.

FURTHER LITERATURE

BS 903: Part E9 (coefficient of expansion of cellular ebonite)
ASTM E228, E289 (measurement of expansion of rigid solids)
ASTM D696 (measurement of linear thermal expansion of plastics)
ASTM D864 (measurement of cubical thermal expansion of plastics)

SECTION 68

PHYSICAL EFFECTS OF TEMPERATURE

DEFINITIONS

THE GLASS TEMPERATURE, T_g

An amorphous linear polymer may be relatively hard and inextensible, or highly extensible with good short-term elastic recovery*; i.e. it may be 'glass-like' as in polystyrene, or 'rubber-like' as in a raw rubber.

The physical state exhibited depends mainly on the temperature, and the point at which the change from one state to the other occurs is called *the glass/rubber transition temperature* of the polymer, or simply *the glass temperature, T_g*.

The glass temperature is associated with amorphous linear polymers† and with the amorphous regions of partially-crystalline polymers. A rubber is composed of very long chain molecules in an amorphous state, and the characteristic elastic extensibility arises from elongation of randomly coiled (kinked) chains which subsequently return to the coiled form. T_g is the temperature below which this kind of mobility is lost. The introduction of crystallinity also restricts rubberiness, and in a wholly-crystalline polymer T_g would be absent.

THE MELTING TEMPERATURE, T_m

Whereas the molecules of a rubber are randomly disposed, those of a hard polymer can be either randomly disposed—as in the 'glass-like' state above—or arranged in a more ordered manner, forming where possible crystalline regions (crystallites and spherulites) embedded in a non-crystalline or amorphous phase.

Like other crystalline substances, crystallites cannot exist above a certain temperature, i.e. they melt, and the polymer then changes to a rubber (or to a viscous melt if of only moderate molecular weight).

---

*Also with good long-term recovery when lightly cross-linked, as in lightly vulcanised rubber, to prevent viscous flow.

†Thus, below T_g an amorphous linear polymer is hard and rigid, but above it soft and flexible. Hence, the state assumed by such a polymer at *room temperature* depends on its T_g; if T_g is above room temperature the polymer would be classed as a *thermoplastic material*, if below, a *rubber*.

The temperature at which the change occurs is called *the crystalline melting temperature, T_m.*

NOTES

RELATIONS BETWEEN T_g AND T_m

T_g—as the temperature below which long-range elasticity ceases to exist, represents the point where the mobility of chain molecules (especially translation of chain segments, which occurs largely by *rotation* about primary bonds) becomes markedly restricted by interchain cohesion. Nevertheless, the change of state at T_g—i.e. from glass to rubber, or vice-versa on cooling—involves no major changes in basic physical properties and is therefore called a **second-order transition.**

Thus, at T_g there is no sudden change in volume or energy content, but there is a change in the first derivative of such a property. For example, a change occurs in the rate at which specific volume (or density, or refractive index) changes with temperature, and in corresponding relations with pressure; similarly, there is a change in the rate at which heat content varies with temperature, i.e. at T_g a change occurs in specific heat or thermal capacity. The temperature coefficient of thermal conductivity also alters (and can reverse its sign). In particular there is a very marked change—as much as 1000-fold— in elastic modulus, from *high* below T_g (and insensitive to temperature) to very *low* above T_g (and sensitive to temperature). This can be observed directly by measurement, for instance, of Young's modulus or the modulus of torsional rigidity, and is evident in related properties, e.g. as a change in the coefficient of restitution (elastic recovery) or in the penetrability (hardness).

However, the transition temperature is not well-defined, but varies with the *rate* of measurement, due to the time-dependence of segmental relaxation in chain molecules (*see also* footnote*, page 569). Thus, a test conducted at a high rate of deformation may yield a relatively high result—e.g. T_b, the brittleness temperature measured by an impact test can be somewhat higher than the normal value of T_g—while measurements made at a very slow rate may reveal no particular point of transition, one state then blending into the other. Sometimes, however, additional second-order transitions are observed, generally but not always below T_g; for instance, in polystyrene, below the temperature where the main chains lose their freedom, secondary values arise due to loss of rotation of side groups. In the presence of crystallinity the prominence of the effects associated with amorphous systems becomes reduced or obscured.

T_m—differing from T_g—involves major changes in basic physical properties; thus, there is a change in volume and in energy content (due to the heat of fusion). A change of state occurring at T_m is therefore called a **first-order transition.**

In rubbers, since crystallinity necessarily increases stiffness, T_m is preferably absent or very low. However, some rubbers crystallise on stretching (though above the normal T_m) or on long standing (if the temperature is below T_m). In fibres and films a high degree of crystallinity is usually desired, to provide the required strength, and T_m should be high to give the maximum upper limit of serviceability.

As already explained, T_g is characteristic of the amorphous state, and T_m is associated with crystallinity. There is, however, an approximate relation between the two: T_m, concerned with greater energy changes than T_g, is always the higher, and when both temperatures are expressed on the absolute scale (i.e. as K) the ratio T_g/T_m usually lies between 0·5 and 0·8.

PRACTICAL CONSIDERATIONS

T_g can be important in practice in several ways*. For instance, below T_g rubbers are completely unserviceable (ceasing to exist as such, and even before reaching T_g they may become very stiff); but above T_g moulding and extrusion of amorphous thermoplastics becomes possible, as does the drawing of fibres and films. Similarly, for the formation of a continuous film by air-drying of a polymer dispersion (as in an 'emulsion' paint), the T_g of the polymer or copolymer concerned needs to be below room temperature. Thermal bulking (texturising) and setting of textile yarns, and the thermal setting or creasing of fabrics, also require them to be taken above T_g (and, for crystalline polymers, near to T_m) although the presence of water, as steam, lowers the temperature needed. Moreover, it is only between T_g and T_m that crystallisation from the molten state can be effected. A melt of a crystallisable polymer when rapidly cooled (shock-cooled or quenched) below T_m yields a transparent and amorphous— or very finely crystalline—product; this condition in some materials may remain stable as such, but others may slowly become crystalline and opaque.

Both T_g and T_m are lowered by the presence of plasticisers (including, as above, water) and to a less extent by stabilising additives, residual monomers, oligomers, etc.; T_g rises with increasing molecular weight, at least up to a certain degree of polymerisation, but for materials in the practical polymeric range T_m is virtually independent of molecular weight. Blends of polymers show the individual T_g values, but copolymers only one value which, however, does not necessarily

*For relations with impact strength, *see* Section 78 (*see also* earlier and in footnote * on page 569).

follow a 'mixture law'. Fillers do not usually influence T_g. The tendency to crystallise is lowered by an irregular structure, as in a copolymer such as styrene/butadiene rubber, but it is enhanced by a high order of regularity of structure such as is produced by stereo-specific catalysts, e.g. in commercial polypropylene, which is largely isotactic.

T_g is determinable by methods that depend on obtaining a change in slope in a plot of a basic property, such as density or refractive index, against temperature, or obtaining a displacement in the graph of a first derivative, such as a coefficient of expansion, plotted against temperature; also by observation of the sudden change that occurs in the elastic modulus and in the nuclear magnetic resonance line width (however, N.M.R. gives higher T_g values owing to the more rapid time scale, and the same may apply to modulus measurements*).

T_m can be determined by standard methods including the disappearance of birefringence and x-ray crystal pattern. Both T_g and T_m can be obtained by dilatometry (change in specific volume/temperature relation at T_g, and in specific volume at T_m) and by differential thermal analysis (change in specific heat at T_g, and in the heat content at T_m). More arbitrary, but practical, tests include the determination of brittleness temperature, heat distortion temperature, softening temperature, etc., usually involving loading of bar specimens or indentation tests (*see* **Further Literature**).

ULTIMATE EFFECTS OF TEMPERATURE

Linear polymers heated beyond certain limits undergo decomposition reactions that can greatly impair their physical properties, e.g. drastic reduction in strength, due—if precautions are not taken against it—to oxidative degradation or other form of chemical breakdown; *see*, for instance, heat ageing of rubbers (§§29, General characteristics, 29.4 and 29.5), various pyrolytic products (in §§1.23–42.23), and general thermal effects (in §§1.32–42.32). For reversion to monomer *see below*.

Ceiling temperature
At a sufficiently elevated temperature certain linear .polymers† undergo reversal of the polymerisation process, i.e. they depolymerise, by reverting largely to monomer (and if this

---

*This last point serves to illustrate the fact—mentioned earlier—that T_g is a *time-dependent* phenomenon, temperature and time in this connection being complementary in their effects. For example: if an amorphous polymer exhibiting the rubbery state (i.e. above its T_g) is submitted to an alternating stress, to produce an alternating deformation, at a sufficiently high frequency (short period) the segments of the chain molecules may be unable to follow in step; thus, what is normally a rubber can, without change of temperature, at a sufficiently high rate of deformation become rigid (i.e. behave as if it were below its T_g). See *also* footnote page 568 and Section 74.

†Often those derived from asymmetrical monomers, particularly polymethyl methacrylate.

is removed, or free to escape, depropagation may continue until no polymer is left). Consequently, in a closed system that yields a polymer of this kind there is a *ceiling temperature*, T_c, at and above which polymer ceases to be produced.

This phenomenon comes about because the activation energy of depropagation, E_d, exceeds that of propagation, E_p; hence, since the rates of both processes rise exponentially with temperature, the rate of depropagation—though usually negligible at room temperature—eventually equals and then exceeds that of propagation. Thus, with regard to the *overall* rate of production of polymer from a monomer (i.e. $E_p - E_d$) it can be said: *just below T_c* the rate—earlier rising with temperature—begins to decline; *at T_c* the rate is zero (i.e. monomer molecules are in equilibrium with polymer radicals); and *above T_c* the rate becomes negative (i.e. polymer tends to depolymerise).

For a system comprising a liquid monomer in which the polymer is insoluble, T_c is fairly sharp and the polymer cannot exist above it*; but if the monomer dissolves the polymer there is a succession of values of T_c, depending on the extent of conversion of the monomer†. When a common solvent is present, it can be shown that

$$T_c = (E_d - E_p)/R \ln (A_d/A_p[M])$$

where E is the respective activation energy, A the respective collision frequency factor, and $[M]$ the monomer concentration. Thus, monomer held at a particular value of T_c would polymerise only so far as to leave the appropriate concentration of it unpolymerised‡ (while polymer, correspondingly, would depolymerise until providing the same concentration of monomer).

Lightly or moderately cross-linked polymers cannot revert to the original monomer(s) and cannot be melted or made truly thermoplastic, but they (or their amorphous portions) may be softened and they may then, before their decomposition temperature is reached, be deformed (as with ebonite, Section 31). However, highly cross-linked or net-work polymers, such as fully cured amino- or phenol-formaldehyde resins, neither soften nor melt with rise of temperature, but ultimately char and evolve volatile products. A few linear polymers behave similarly if they are highly hydrogen-bonded, e.g. cellulose, which shows neither T_g nor T_m before undergoing thermal decomposition.

Since a temperature as high as, say, 250°C can be withstood by relatively few polymers—e.g. fluoropolymers (Section 3), silicones (Section 41), and special polyamides (§23.82c)—and no organic polymer approaches the thermal resistance of glass (softens without decomposition, 500–800°C) or asbestos (1200–1500°C), there is considerable interest in increasing the upper operating limit.

For chemical degradation at moderate to high temperatures, *see* under *depolymerisation* in Section 54 and *pyrolysis* in Section 57.

---

*In much the same way that a liquid cannot exist above its critical temperature.

†Thus, a given monomer-polymer mixture will be in equilibrium at a particular T_c, while a mixture containing a higher proportion of monomer will be in equilibrium at a higher T_c.

‡*N.B.* If the initial concentration of monomer were the same as the equilibrium concentration, polymerisation would not take place. Similarly, if monomer is held at a temperature *above* T_c polymerisation will not take place, e.g. some monomers have been observed to polymerise only below room temperature.

FURTHER LITERATURE

LEE, W. A., *Glass Transition Temperatures in Homopolymers (RAE Technical Report 65151)*, Farnborough (1965) (bibliography and assessment of available data)

BOYER, R. F., *Rubber Chem. and Tech.*, **36**, 1303 (1963) (relation of transition temperatures to chemical structure)

BOYER, R. F. (Ed), *Transitions and Relaxations in Polymers*, New York and London (1966)

BOYER, R. F., *Polymer Engng and Sci.*, **8**, 161 (1968) (dependence of mechanical properties on molecular motion)

REISS, H. (Ed), *Prog. in Solid-state Chem.*, **3**, 407 (1957) (M. C. SHEN and A. EISENBERG: glass transitions)

MANDELKERN, L., *Polymer Engng and Sci.*, **7**, 232 (1967) (Crystallisation-melting process; relates mainly to polyethylene)

MARVEL, C. S., *Pure and Appl. Chem.*, **16**, Nos. 2–3 (1968) (thermally-stable polymers)

WRIGHT, W. W. and LEE, W. A., *Progress in High Polymers*, **2**, 189 (1968) (thermally-stable polymers)

FRAZER, A. H., *High Temperature Resistant Polymers*, New York and London (1968)

Specifications
  Softening tests
    BS 771 (phenolic resins; heat resistance by loaded bar)
    BP 903: Parts D1 and D2 (plastic yield of ebonite); *E2 and E3* (do. of cellular ebonite)
    BS 1493 (polystyrene, softening point)
    BS 2782: Methods 102A–H (deformation at elevated temperatures); *Methods 103A–C* (softening and melting points of resins and nylon)
    ASTM D530 (hard vulcanised rubber; includes test D648)
    ASTM D648 (heat distortion by loaded bar)
    ASTM D789, D1457, D2116, D2117 and D2133 (melting point of plastics)
    ASTM D1525 (Vicat test: temperature of penetration of loaded flattened needle)
    ASTM E28 (bitumens and rosins; ball and ring softening)
  Brittleness tests
    BS 903: Part A25 (brittleness temperature of rubber, by impact test)
    BS 2782: Method 104A ('cold bend' temperature of PVC)
    ASTM D746 (brittleness temperature, by impact test, of plastics and elastomers)
    ASTM D758 (impact resistance of plastics above and below room temperature)
  Low-temperature stiffening
    BS 903: Part A13 (rigidity modulus test on rubber)
    BS 2782: Methods 104 B and C (cold flex and extensibility tests on PVC)
    ASTM D1053 (Gehman stiffening test on rubber)

ELECTRICAL RESISTIVITY

DEFINITIONS

The ohmic resistance (R) of an electrical conductor or resistor, which is proportional to the length (l) and inversely proportional to the cross-sectional area (a, measured normal to the direction of the current and assumed to be uniform), is given by

$$R = \rho l/a$$

where ρ is a constant known as the *volume resistivity* or internal resistivity, or the *specific resistance*, which in theory is independent of the length (or thickness) or cross-sectional area of the given material.

In practice, especially where conduction is greater over a surface than in the bulk, distinction needs to be made between (i) *volume resistivity*—which, if l and a are reduced to unity, can be regarded as *the resistance between opposite faces of a cube of unit edge**, and (ii) *surface resistivity*, which can be regarded as *the resistance between opposite edges* of a square of the surface†.

> *N.B.* Whether (i) or (ii) is more important depends on the geometry of the system considered. For instance, when measurements are made between opposite sides of a small central portion of a large sheet the result depends largely on *volume resistivity*, and surface resistivity can usually be neglected; when measurements are made along a filament, or through a fibre assembly, the result may depend more on *surface resistivity* than on internal conduction.

It is convenient to express the resistivity of fibres or yarns (where a is difficult to measure) as *mass resistivity* or *mass specific resistance* (R_s), which is the resistance between the ends of a sample of mass 1 g, uniform in cross-sectional area and 1 cm long. The value of R_s is equal to $\rho \times density$, and since the density of most fibres is under 2 g/ml, R_s is somewhat less than twice the value of ρ.

---

*Normally, as in this book, a 1 cm cube; but in SI units (*see* Note, page 659) a 1 m cube.
†It is unnecessary to specify the *size* of the square. Increasing the size of a square of surface increases the width of the conductive path but lengthens it proportionately; hence a square of any size should have the same surface resistance.

UNITS*

Volume resistivity is measured in ohms cm²/cm, or *ohm centimetres*. Surface resistivity is measured in *ohms*. The units of mass resistivity are g ohms/cm². In all instances *temperature* and *humidity* should be specified, also the test-piece thickness and its conditioning, the material and dimensions of the electrodes, and the voltage—including whether direct or alternating (and the frequency) and the time for which it is applied. *See also* **Note on SI Units,** page 659.

NOTES

Most high polymers are classed as non-conductors, and non-polar ones—such as polyethylene and polytetrafluoroethylene—are among the best electrical insulators known.†

Polymers showing appreciable *moisture absorption*—such as cellulose (e.g. as cotton, or as vulcanised fibre, *see* §13.82 iii) and formalised casein—are more limited in electrical applications; their resistivity usually falls exponentially with rise of humidity, thus a small increase in moisture content may cause a several-fold decrease in electrical resistance. The relation between resistance and *temperature* is also an inverse exponential one, contrary to that for metals‡; thus, the logarithm of the resistivity is a linear function of the absolute temperature, and a rise of 10°C may cause a several-fold drop in resistivity.

The resistivity of a polymer may be affected adversely, or with advantage, in various ways, e.g. by the presence of initiator fragments (especially if ionic) or other impurities, and even if these are absent, and the polymer is not much affected by high humidity, contamination by mould growth or a conductive substance (e.g. H_2SO_4 formed on ebonite, *see* §§31.33 and 31.5) may severely lower surface resistivity. Plasticisation and application of antistatic dressings (*see later*) similarly cause lowering of resistivity. On the other hand, cross-linkage often raises it, and so does any treatment or modification that lowers the hygroscopicity of a polymer; thus, the resistivity of

*Sometimes the units of volume resistivity are incorrectly expressed as *ohm/cm³* and those of surface resistivity, also incorrectly, as *ohms/cm²*. Certain standard specifications express resistivity as a logarithm; the units are then log(*ohm.cm*) and log *ohms* for volume and surface resistivity respectively.

† No insulator, however, is a perfect non-conductor, even though when a d.c. voltage is applied to a polymer the resistivity tends to show an asymptotic increase with time; this comes about because a polymer with conducting electrodes attached constitutes a condenser, and application of a potential difference produces—following the initial rush of charge—a small current composed of (i) an *absorption current*, declining with time as the dielectric polarises, and (ii) a permanent *leakage current*.

‡And for glass, which when sufficiently heated (to near redness) loses its normal resistivity to become an ionically-conducting melt.

cotton is improved by partial acetylation (*see* §§14.21 and 14.81) while silicone compounds, conferring water repellence, are used to improve the surface resistivity of glass.

High resistivity is occasionally an undesirable feature. For instance, a spark arising from accumulation of electric charge may initiate an explosive chain reaction. This is prevented, if the contributing factor is a rubber sheet, a driving or conveyer belt, a rubber tyred vehicle, or even rubber footwear, by compounding with special carbon blacks, to render the rubber conductive (*see* §29.33). The production of electricity and its retention (or insufficiently fast disappearance) can prove troublesome during the processing of various films and textile fibres; it is also responsible for 'fog-marking' of newly-woven fabrics (soiling of charged threads, by attraction of dirt particles, when a loom is left idle). Similarly, air-suspended dirt can produce dust patterns on certain moulded plastics (e.g. polyethylene) and is a cause of the 'greying' of certain fabrics (e.g. nylon). These troubles can be largely prevented, where appropriate, by raising the relative humidity (for the more conductive fibres), by the use of a conductive dressing (e.g. glycerol, magnesium oleate, or an 'antistatic oil'— often consisting of a conductive liquid, to dissipate a charge, and a lubricant to lessen the chance of its formation), or by engendering localised ionisation (e.g. by means of a 'static eliminator', depending on a high voltage silent discharge, or a β-ray source). More permanent results are obtained (for mouldings) by incorporation of a conductive additive that slowly migrates to the surface, and (for high resistivity fibres) by graft copolymerisation of a more hygroscopic material on to the surface. In fibres of medium to low resistivity, moisture regain (Section 62) can be determined by measurement of the electrical conductance, and this has important applications in the textile industry.

Volume resistivity is usually measured on thin sheets or films of uniform thickness, using circular electrodes of graphite, conductive paint, thin metal foil, or mercury; employment of a concentric guard ring eliminates effects of surface conduction. *Surface resistivity* is measured between electrodes in the form of a disc and concentric ring; employment of a guard plate on the far side of the sample eliminates effects of volume conduction. *See* Specifications at the end of this section—also included in these are practical tests for *insulation resistance* as measured between plug or screw electrodes inserted in sheet insulating materials. In most determinations readings are taken 1 min after application of a potential difference of 500 volts d.c., the resistance being obtained by measurement of the resulting current or by means of a bridge circuit using a d.c. amplifier. For some purposes tests are made after a sample has been immersed for several hours in water and then wiped dry.

For electrically conductive and 'antistatic' polymers (plastics, rubbers) special methods of measurement have to be employed because of the relatively high contact resistance between such materials and the usual electrodes (*see* NORMAN, R. H. and *BSS* 2044 and 2050 under **Further Literature** below); the result is usually calculated as volume resistivity on the assumption that surface conduction can be neglected.

COMPARISON TABLE

Table 69.T1 shows the volume resistivity of various polymers and related materials. Values for surface resistivity, which can be much dependent on moisture and contamination, are to be found in §§1.33–42.33.

Table 69.T1. VOLUME RESISTIVITY OF COMMON POLYMERS AND RELATED SUBSTANCES

Note. Temperature of test assumed 20–25°C, relative humidity 65% unless otherwise stated.

Range for lower limit of resistivity	Material (approximate resistivity in ohm cm)
10^{-6}–10^{-5}	*Good conductors*: silver, copper, gold, aluminium.
10^{-5}–10^{-3}	*Resistance alloys*: Constantan, Manganin, c. 5.10^{-5}; Nichrome, $1\cdot1.10^{-4}$.
10^{-3}–10^{-2}	*Carbon, as graphite, c.* 5.10^{-3}.
1–10^2	Conducting rubbers and polyethylene, carbon black filled, *c.* 1–10^2 or more*.
10^6–10^7	Cellulose, natural or regenerated; as fibres or films†.
10^7–10^9	Alginates†, regenerated proteins (casein-formaldehyde)†, polyvinyl alcohol†, asbestos. Phenol-formaldehyde resins, cellulose filled, 10^8–10^{11}.
10^9–10^{11}	Wool, 10^9–10^{10}. Polysulphide rubbers, vulcanised, filled, *c.* 10^9–10^{11}; do., unvulcanised, 7.10^{12}. Acrylonitrile/butadiene rubbers, 10^9–10^{12}. Silk; cyano-ethylcellulose; glass fibre, purified, 10^{10}. Cellulose nitrate, 10^{10}–10^{11}. Poly-urethanes, 10^{10}–10^{12}. Cellulose acetate, plastics, 10^{10}–10^{13}; do., sec. acet. fibres, 10^{11}–10^{12}; do., triacet. fibres, $>10^{14}$.
10^{11}–10^{13}	Amino-formaldehyde resins; chloroprene rubber, unvulcanised or vulcanised; glass, common, soft, 10^{11}–10^{13}. Polyacrylonitrile, fibres, 10^{12}–10^{14}.
10^{13}–10^{15}	Polyvinyl acetate, cast film, 10^{13}–10^{14}. Ethyl cellulose; chlorinated rubber; styrene/butadiene rubbers, unvulcanised or vulcanised; silicone resins, 10^{13}–10^{15}. Epoxy resins; vinyl chloride/vinyl acetate copolymer; silicone rubbers, vulcanised, 10^{13}–10^{16}. Polyamides; alkyds; polyester resins, unsat., glass-filled; polyvinyl acetals; chlorosulphonated polyethylene; ethylene/vinyl acetate copolymers, *c.* 10^{14}. Polyvinyl chloride, rigid‡; polyformaldehyde; isoprene rubber (synthetic), vulcanised, unfilled, $>10^{14}$. Ebonite, loaded, 10^{14}–10^{15}; do., unloaded, 10^{16}–10^{17}. Polyvinylidene chloride, 10^{14}–10^{16}. ABS plastics, 2.10^{14}–10^{15}. Polyesters, unsat., unfilled; natural rubber, unvulcanised or vulcanised; do., hydrochloride; gutta percha, commercial (do., purified, $>10^{17}$); butyl rubber, vulcanised, filled; silicone fluids; borosilicate glass (Pyrex type); *porcelain silica (fused quartz), c.* 10^{15}.

Range for lower limit of resistivity	Material (approximate resistivity in ohm cm)
$10^{15}-10^{17}$	Butadiene rubber, unvulcanised or vulcanised, $3-5.10^{15}$. *Sulphur, rhombic,* 5.10^{15}. Chlorinated polyvinyl chloride, 6.10^{15}. Polyethylene, high density, $10^{15}-10^{16}$. Polymethyl methacrylate, $10^{15}-10^{17}$. Polystyrene, $10^{18}-10^{19}$; do., high impact, $c. 10^{16}$. Polycarbonates, chlorinated polyethers, polyvinylcarbazole, 10^{16}. Polypropylene, polymethylpentene, polysulphones $>10^{16}$. Ethylene/propylene rubbers, unvulcanised, 5.10^{16}; do., vulcanised, filled, $1·5.10^{14}$. Aniline-formaldehyde resins, $10^{16}-10^{17}$. Cyclised rubber, $10^{16}-6.10^{17}$. Poly-chlorotrifluoroethylene; polyphenylene oxide; butyl rubber, vulcanised, un-filled (do., filled, $c. 10^{15}$), 10^{17}.
Over 10^{17}	Polytetrafluoroethylene; natural rubber, purified, vulcanised; polyethylene, low density; polyisobutylene; polyethylene terephthalate. *Amber, top grade. Paraffin wax.*

*Flexible polymeric articles with resistance below 5.10^{14} ohms are classed as 'conducting' (BS 2050)
†Much affected by change of humidity; when quite dry, resistivity can be high, but at 90% r.h. it may be 100–10000 times lower than 65% r.h.
‡For plasticised PVC, resistivity depends greatly on amount and nature of plasticiser

FURTHER LITERATURE

MEREDITH, R. and HEARLE, J. W. S. (Eds), *Physical Methods of Investigating Textiles*, New York (1959) (HEARLE, J. W. S.: chapter 13)

MORTON, W. E. and HEARLE, J. W. S., *Physical Properties of Textile Fibres*, Manchester and London (1962) (resistance, pp. 457–486; static, pp. 487–523)

LEVER, A. E. and RHYS, J. A., *Properties and Testing of Plastics Materials*, 3rd edn, London, 208 (1968) (electrical properties and testing)

SCHMITZ, J. V. (Ed), *Testing of Polymers*, Vol. 1, (NewYork-London-Sydney), pp. 213–35 (SCOTT, A. H., Direct current dielectric conductance measurements); pp. 271–95 (WARFIELD, R. W., Characterisation of polymers by electrical resistivity techniques)

NORMAN, R. H., *Conductive Rubber*, London (1957) (properties, applications, test methods)

KATON, J. E. (Ed), *Organic Semi-conducting Polymers*, New York (1968)

Specifications
BS 771 (phenolic mouldings)
BS 903: Parts C1 and C2 (surface and volume resistivity tests on rubber and ebonite)
BS 1322 (aminoplasts)
BS 1540 (moulded insulators for use at radio frequencies; includes surface resistivity)
BS 2782: Part 2, Methods 202, 203 and 204 (testing of plastics)
BS 2044 (resistivity tests on antistatic and conductive rubbers)
BS 2050 (electrical resistance of conductive and antistatic products made from flexible polymers)
ASTM D257 (electrical resistance tests of insulating materials)
ASTM D991 (volume resistivity tests of electrically conducting rubber and rubber-like materials)

DIELECTRIC STRENGTH

DEFINITION

Dielectric strength is the lowest electric stress, expressed as a potential gradient, that causes breakdown of an insulating material.

UNITS

Dielectric strength, as a quotient of (dielectric breakdown voltage*)/ (test-piece thickness), is expressed as *kilovolts per millimetre*; or, alternatively, as kilovolts per centimetre, or as *volts per mil* (1 mil = 0·001 in); 1 kV/mm = 25·4 V/mil.

It is not, however, constant for a given substance but decreases with increasing *thickness*†; hence, it is necessary to quote the thickness of the material tested. Also of importance are temperature, ambient humidity, the method of application of the voltage (e.g. rate of increase, period of dwell, type of electrode system) and whether or not the test is conducted with the test-piece and electrodes immersed in oil. *See also* **Note on SI Units,** page 659.

NOTES

Dielectric strength is, naturally, an important element in many forms of electrical insulation, e.g. in capacitors, high-voltage cables, switch-gear and transformers. The strength observed is always much less than the theoretical value, largely because local inhomogeneities and electrode effects (discharges and mechanical damage) produce regions of high stress that lead to premature breakdown. It is thus obvious that a good insulator should possess a non-ionic and non-hygroscopic structure; impurities, too, should be absent, e.g. as in a polymer prepared from a pure monomer without addition of initiator or solvent.

*Dielectric breakdown voltage is the power frequency RMS alternating voltage, near sinusoidal in wave form (e.g. with peak approx. $\sqrt{2}$-times the RMS value), at which under specified conditions of test a dielectric fails.

†For a sheet of thickness t, the *dielectric breakdown voltage* at power frequencies is often approximately proportional to $\sqrt{t}$; therefore, the *dielectric strength* (= breakdown voltage/t) is approximately proportional to $1/\sqrt{t}$.

Dielectric strength is usually measured on sheets or films of known thickness, using specified electrodes (e.g. disc and plate) and increasing the applied voltage either at a steady rate (e.g. rising at 500 or 1000 V/s) or in stages (e.g. rising by 50 or 100 V in 20 s steps) until electrical breakdown takes place; alternatively, tests are made at various voltages to determine the one-minute value, i.e. the voltage at which breakdown occurs in 1 min. The electrodes and test piece are submerged in oil if the breakdown voltage is high (to avoid a flash-over) or if the insulating material is to be used in oil.

Tracking. When a spark or arc discharge strikes along the surface of an organic polymer it may leave behind a charred path, over which subsequent discharges take place more and more readily. A minute leakage current arising from surface contamination may produce a similar effect. The phenomenon is especially apparent with phenol-formaldehyde resins but much less so with aniline-formaldehyde resins while melamine-formaldehyde resins are virtually non-tracking. Polymers that revert to monomer or other volatiles on pyrolysis (e.g. *see* §§2.33, 3.33, 4.33, 9.33 and 26.33; *see also* ceiling temperature, Section 68) are usually non-tracking and arc-resistant though they may be ignited in air by an arc*. Susceptibility to tracking is examined without and with deliberate contamination of the surface with an aq. solution of an electrolyte (*see*, for instance, IVES and RILEY—below—for test method).

COMPARISON TABLE

Table 70.T1 indicates the approximate value of the dielectric strength of common polymers and related materials.

Table 70.T1. DIELECTRIC STRENGTH OF COMMON POLYMERS AND RELATED SUBSTANCES

Notes: (1) Temperature of test assumed 20–30°C, relative humidity 65%, unless otherwise stated.
(2) Dielectric strength depends on test piece thickness (*see* under **Units** earlier) and as this varies between 0·1 and 3·2 mm for the different materials, the data are not all strictly comparable. Values for 'thin films' are enclosed thus [ ].

Range	Material (*approximate dielectric strength in* kV/mm)
Under 5	*Air, narrow gaps; high voltage, c.* 0·8. Asbestos, covering on wire, *c.* 3. Cellulose, cotton covering on wire, 4. Phenol-formaldehyde resins, cellulose filled, 4–12.
5 and above	Cellulose, vulcanised fibre, <8; do., do., electrical grade, up to 12. *Porcelain*; *silica* (*fused quartz*); shellac, 8–10. Phenol-formaldehyde resins, cast; glass, common, soft, 8–12. Regenerated proteins (casein-formaldehyde*p. 579), 8–28.

*Glass, showing no carbonisation, is non-tracking and non-combustible but conducts as an ionic melt when sufficiently heated.

Range	Material (approximate dielectric strength in kV/mm)
10 and above	Natural rubber, vulcanised, filled, 10–20; do., unvulcanised, 16–18. Glass, borosilicate, Pyrex-type, >12 [>40]. Alkyd resins, 12–16. Polyesters, unsaturated; amino-formaldehyde resins, 12–18. Cellulose acetate, >14 [80]. Polyvinyl chloride, unplasticised (do., plasticised, 20–28), 14–16). Epoxy resins, cast, 14–18. Silk, 14–24. Polypropylene, 14–32.
15 and above	Polyvinyl acetate [16–32]. Phenol-formaldehyde resins, mineral-filled, low loss, 15. Styrene/butadiene rubbers, vulcanised, filled, 15–25; do., unvulcanised, 24–36. Chlorinated polyethers; polymethyl methacrylate (75% r.h.), [20]; polyvinyl butyral, 16; do., formal and acetal [40]. Polyamides [120], polycarbonates [120], polyvinylidene chloride [120–200], 16. Polyphenylene oxide, 16–20. Acrylonitrile/butadiene rubbers, unvulcanised, 16–21. Silicone rubbers, vulcanised, mineral filled, 16–22. Vinyl chloride/vinyl acetate copolymers, plasticised, 16–24. Chlorinated rubber, 16–80 [up to 100]. Phenol-formaldehyde resins, paper laminates, up to 20.
20 and above	Chlorosulphonated polyethylene; ethylene/vinyl acetate copolymers; aniline-formaldehyde resins; polyurethanes; coumarone resins [55]; ethylcellulose [up to 60]; polyformaldehyde [>200], 20. Polytetrafluoroethylene, >20 [80]. Silicone resins, 20–120. Polyethylene, 20–160 [up to 320]. Butyl rubber, vulcanised, 22–35. Isoprene rubber (synthetic), vulcanised, unfilled; polyisobutylene, 23–24. Cellulose nitrate [up to 48]; polychlorotrifluoroethylene [100], 24.
25 and over	Chloroprene rubber, vulcanised, filled†, 25–30. Polymethylpentene; ethylene/propylene rubbers, unvulcanised (do., vulcanised, filled, 40), 28. Polystyrene, >30 [120–160]; do., high impact, 18–27. Natural rubber, vulcanised, unfilled, 34–100‡. Polypropylene, 32. ABS plastics, 35. Chlorinated polyvinyl chloride, 39. Polyvinyl alcohol, 40§. Ebonite, loaded, 50–115; do., unloaded, 90–150. Vinyl chloride/vinyl acetate copolymer (unplasticised), 55. Polyvinylcarbazole [40–50]. Polyethylene terephthalate [100–160]. Mica.

N.B. In electrolyte condensers films of aluminium oxide, c. 10^{-4} mm thick, can withstand 500–700 V; i.e. they have a dielectric strength of over 5000 kV/mm.

*Varies greatly with humidity
†Carbon black filling gives lower values
‡Range shown includes variation in time of application and nature (a.c. or d.c.) of voltage
§For dry material; under humid conditions, as low as 0·4 kV/mm

FURTHER LITERATURE

Soc. Chem. Ind., *Monograph No. 5,* London (1959) (pp. 121–137: IVES, G. C. and RILEY, M. M., electrical tracking testing)

SCHMITZ, J. V. (Ed), *Testing of Plastics,* Vol. 1, New York, London, Sydney, 297 (1965) (DAKIN, T. W., high voltage testing of polymers)

Specifications
BS 903: Part C4 (tests on vulcanised rubber and ebonite)
BS 1137 (laminated paper sheets)
BS 1313 (laminated paper tubes)
BS 1322 (aminoplastic mouldings)
BS 1540 (moulded insulating materials at radio frequencies)
BS 2782: Part 2, Method 201 (tests on plastics)
ASTM D149 (tests on insulating materials at power frequencies)
ASTM D1389 (non-destructive proof voltage tests on thin insulating materials)
ASTM D495 (arc resistance tests of plastics)

DIELECTRIC CONSTANT

DEFINITIONS

The dielectric constant of an insulating material is the factor by which the capacitance of a condenser is increased when the given material is substituted for a vacuum (or air) between the plates.

Alternative practical considerations relating to dielectric constant are as follows. When an insulating material (of dielectric constant, K) is introduced so as to fill the space between two conductors or electrodes bearing fixed charges, polarised and polarisable portions of its molecules tend to become oriented in the direction of the electric field and to oppose it, i.e. the field strength *decreases* (to $1/K$ of the initial value). Conversely, if the material is introduced between conductors maintained at a fixed potential difference, the amount of charge on them rises, i.e. the capacity of the system *increases* (to K-times the original value*). The value of the dielectric constant is thus a measure of the ability of a non-conductor to store energy when subjected to an electric field.

UNITS

Dielectric constant, also known as *permittivity*† and *specific inductive capacity*, is a ratio without dimensions.

The value depends on the *frequency, temperature,* and *moisture content* (if any) at which measurement is made, and these factors should be stated (*see below*).

NOTES

Application of a steady potential difference to an ideal condenser (with a vacuum as dielectric) produces an initial (and practically instantaneous) surge current due to the flow of electrons to or from the plates. Application of an alternating potential difference produces a corresponding alternating current, which leads the applied voltage by a phase angle of 90 degrees. Under the last conditions, though a

---

*The capacitance of any condenser is thus proportional to the dielectric constant of the material between the plates.

†Strictly speaking *permittivity* is an absolute quantity (unit: F/m); *relative permittivity* or *dielectric constant* is its value for a material relative to that for a vacuum.

current continues to flow alternatingly into and out of the condenser, there is no loss of energy, i.e. the currect is 'watt-less' (*see* Section 72).

If, as in all practical instances, the condenser incorporates a material dielectric, an alternating voltage produces (i) *electronic displacements,* (ii) *atomic and molecular displacements* (due to dipoles and induced dipoles, the extent of the associated current depending on the frequency and temperature*) and (iii) a small *leakage current* (which, following in phase with the voltage, represents a constant loss of energy†).

These items are reflected in the dielectric constant, the value of which, in the absence of dipoles and with a minimum leakage current, remains small, e.g. between 2 and 3 in pure hydrocarbon polymers compared with a vacuum as unity. Polymers with more polar or polarisable components tend, as described, to possess a higher dielectric constant, except at very high frequencies to which the dipoles cannot respond. Water has a high dielectric constant (*c.* 80) and in hygroscopic polymers the dielectric constant rises with increasing humidity. At very high frequencies, to which only the bound electrons respond (and at lower frequencies in those materials where the atomic or molecular polarisation is small) the dielectric constant tends towards a true constant, the value of which has been shown to be equal to n^2, where n is the refractive index of the dielectric concerned.

COMPARISON TABLE

Table 71.T1 indicates the dielectric constant of common polymers and related substances at 20–25°C for frequencies of *c.* 50 Hz and 1 MHz; values at intermediate and higher frequencies, and at higher temperatures, are in some instances to be found in §§1.33–42.33.

Table 71.T1. DIELECTRIC CONSTANT (K)

Note: Temperature of test assumed 20–25°C, relative humidity 65%, unless otherwise stated. Test frequency indicated as follows: 50–100 Hz, without brackets; *c.* 1 MHz, enclosed thus [ ]; other frequencies, 'over wide range', or not stated, enclosed thus ( ).

Range for lower limit of K	Material (approximate value of K)
1–2	*Vacuum,* 1·0000. *Hydrogen, at atmospheric pressure,* 1·0003; *air,* 1·0006. Poly-*p*-xylylene, *c.* 1·7. *Petrol, c.* 2.

*For example, at very low frequencies dipoles may follow an applied voltage in a manner similar to electronic displacements, but at very high frequencies they do not move at all, while intermediately (near the resonance frequency where there is a fall in dielectric constant as frequency increases) there is a maximum dipole-movement and maximum loss of energy from internal friction (*see also* Section 72).

†The leakage loss is equivalent in effect to a high resistance in parallel with a capacitance, while the dipole loss (*see* note * above) is equivalent to a low resistance in series with a capacitance.

Range for lower limit of K	Material (approximate value of K)
2–3	Polytetrafluoroethylene (2·0–2·1). Polymethylpentene, 2·1 [2·1]. Ethylene/propylene rubbers, unvulcanised (2·2), do., vulcanised, filled (c. 3). Polypropylene, 2·25 [2·25]. Silicone fluids (2·2–2·8). Polyethylene (2·28). Polyisobutylene; *paraffin wax*; butadiene rubbers, raw, 2·3 [do.]. Butyl rubber, unvulcanised or vulcanised unfilled (2·1–2·4); do., do. filled, (c. 3). Natural rubber, unvulcanised (2·4–2·6); do., vulcanised, unfilled (2·5–3·3); do., do., mineral filled, 2·8–8. Polychlorotrifluoroethylene (2·4–2·7). Shellac, c. 2·5. Polystyrene, 2·5–2·6 [2·4–2·5]; do., high impact, 2·5–3·5 [c. 2·6]. Chlorinated natural rubber, ethylcellulose [2·5–3·5]. Polyphenylene oxide [2·55–2·7]. Isoprene rubber (synthetic), vulcanised, unfilled; gutta percha, purified (do., commercial (3·2)), (2·6). Styrene/butadiene rubbers, unvulcanised, 2·6 [2·35]; do., vulcanised, unfilled, 2·9 [2·7]; do., do., filled, [2·4–3·8]. Cyclised rubber, [2·6–2·7]. Silicone rubbers, vulcanised, unfilled, 2·67; do., mineral filled, 3·0–3·6 [2·9–3·8]. Polycarbonates, 2·7–3·1. Ebonite, [2·7–3·9]*. Silicone resins, 2·75–2·85 [do.]. Ethylene-vinyl acetate copolymer, [2·8]. Rubber hydrochloride, (2·8–3·5). Asbestos, (2·9). ABS plastics, 2·9–4·9 [3·7–4·1]. Chlorinated polyvinyl chloride, 2·93 [2·90]. Coumarone resins, polyvinyl acetals, (3·0). Polyvinylidene chloride, c. 3 [do.]. Polyvinylcarbazole, (c. 3) polysulphones.
3–4	Polyethylene terephthalate, 3·0–3·2 [do.]. Polyvinyl chloride (rigid)†, (3·0–3·3). Vinyl chloride/vinyl acetate copolymer, unplasticised, 3·0–3·5 [do.]. Polyvinyl acetate‡, (3·0–3·5). Aniline-formaldehyde resins, 3·4 [do.]. Alkyd resins, [3–5]. Polymethyl methacrylate, 3·3 [2·75]. Polyester resins, unsaturated, unfilled or glass filled, [3·5–5]. Polyvinyl alcohol, (3·5 to >10)§. Silica, vitreous, (3·5–3·8). Polyformaldehyde, c. 3·7 [do.]. *Sulphur, rhombic,* 4·0. Epoxy resins, c. 4. Polyamides, nylon 6.6, air dry, c. 4 [c. 3]. Wool (dry‖), [c. 4].
4–6	Silk (dry), [4·2]. Cellulose acetate, [4–5]; *oak, dry*, 4–5. Cellulose, vulcanised fibre, [4·5]; do., cotton, 65% r.h., (6–18)¶; do., regenerated films, dry, 7·7 [6·7]. Glass, borosilicate, Pyrex type, (4·6). Acrylonitrile/butadiene rubbers, vulcanised, unfilled, [4·8–5·5]; do., unvulcanised, 11–17. Phenol-formaldehyde resins, novolak, (c. 5); do., cast, (4–8); do., laminates, (4–20); do., filled mouldings, (5–15). Porcelain, c. 5–6 [4–5]. Steatite, 5·4.
6–9	Polyurethanes, [6–7]. *Mica*, slate, 6–7. Chlorosulphonated polyethylene (6–7). Chloroprene rubber, vulcanised, mineral filled, (6–7); do., do., carbon black filled, (up to 32); do., unfilled, (7·3). Polyacrylonitrile, 6·5 [4·2]. Regenerated proteins, casein-formaldehyde, c. 7 [do.]. Polysulphide rubbers, unvulcanised, or vulcanised mineral filled, (7–9); do., carbon black filled, (up to 28). Aminoformaldehyde resins, cellulose filled, 7·5 [6·5]; do., silica or glass filled, up to 14 [7·5]; cellulose nitrate 7 [6]; glass (common) (c. 7).
10 and above	Cyanoethylcellulose, (13) [11]. *Ethyl alcohol, c.* 25. Natural rubber, vulcanised, 45 parts furnace black per 100 rubber, (28). *Water, c.* 80. *Titanium dioxide, c.* 90. *High-permittivity ceramics,* over 1000.

*The higher value is for acrylonitrile/butadiene rubber ebonite
†Values for plasticised PVC vary widely
‡For dry films; at high humidity: up to 10–10·5
§The higher value is at 65% r.h.
‖Value rises steeply above 10% regain
¶Varies widely with test frequency; for *dry* cotton (3·0–3·2)

FURTHER LITERATURE

BÖTTCHER, C. J. F., *Theory of Electric Polarisation* (1952)
MORTON, W. E. and HEARLE, J. W. S., *Physical Properties of Textile Fibres*, Manchester and London (1962) (Chapter 19, dielectric properties)
LEVER, A. E. and RHYS, J. A., *Properties and Testing of Plastics Materials*, 3rd edn, London,

208 (1968) (electrical properties and testing)

SCHMITZ, J. V. (Ed), *Testing of Polymers,* Vol. 1, New York, London, Sydney, 237 (1965) (TUCKER, R. W., dielectric constant and loss)

Specifications
BS 771 (phenolic mouldings)
BS 903 : Part C3 (tests on vulcanised rubber and ebonite)
BS 1137 (phenolic paper laminates, at power frequencies).
BS 1493 (polystyrene moulding materials*)
BS 1540 (moulded insulating materials, at radio frequencies)
BS 2067 (dielectric constant and power factor at radio frequencies; Hartshorn and Ward method)
BS 2076 (phenolic paper laminates, at radio frequencies)
BS 2782 : Part 2 (electrical properties of plastics: *method 205*, dielectric constant and power factor at 50 Hz; *method 206*, do. at 800–1600 Hz; *method 207**, do. at 10^4–10^8 Hz)
ASTM D150 (dielectric constant and power factor tests)
ASTM D669 (dielectric constant and dissipation factor tests on laminates)

*Uses method of BS 2067.

POWER FACTOR

DEFINITION

In an electric circuit in which current alternates out-of-phase with an applied alternating voltage, the power* is less than the apparent power, i.e. the wattage is less than the product of the voltage and the current. The proportionality constant relating the true wattage to the apparent wattage (or volt-amperage) is called the *power factor*. Thus,

$$\text{true watts} (= \text{power}) = \text{power factor} \times \text{apparent watts}$$

or,

$$\text{power factor} = \text{watts/volt-amperes.}$$

UNITS

Power factor is a ratio without dimensions. The practice of expressing it as a percentage is to be deprecated. Like dielectric constant (Section 71) the value depends on the *frequency, temperature,* and *moisture content* (if any) at which measurement is made, and these factors should be stated.

In most insulating materials very little of the potential energy of an alternating electric field is lost (as heat), hence the power factor is small, and in some instances very small. The minimum value (zero) is associated with the application of an alternating voltage to an ideal condenser, which—*see also* under Notes, Section 71—produces a displacement current but occasions no loss of energy and is therefore described as 'watt-less' (i.e. no true watts, power factor = 0).

NOTES

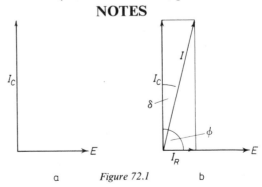

a   *Figure 72.1*   b

*Power = rate at which energy is absorbed or dissipated. It is measured in *watts* (1 watt = 1 joule/s = 0·24 cal/s) and in a direct current circuit—or in an (ideal) alternating current circuit devoid of capacitance or inductance—is equal to *volts × amperes.*

The above-mentioned 'watt-less' state, entailing no dissipation of energy, is represented vectorially in Fig. 72a, where a pure capacitance current, I_c, leads the applied voltage, E, by 90 degrees. In all material dielectrics, however, this ideal quadrature is destroyed because in addition to the capacitance current, a small current flows in phase with the applied voltage. The conditions are then represented as shown in Fig. 72b, where the bulk of the current (I_c) is still 90 degrees ahead of the voltage but the presence of the small in-phase current, I_R (for components *see* ii and iii, Section 71), means that the total current, I, leads the voltage by an angle less than a right angle, i.e. by the *phase angle, ϕ*.

For this it follows that

$$\text{power factor} \ (= \text{watts/volt-amperes} = I_R/I) = \cos \phi$$

The phase angle is also equal to $(90 - \delta)$ degrees, where δ is the *loss angle** (or dielectric loss angle), the tangent of which is called the loss tangent or *dissipation factor,* i.e.

$$\text{dissipation factor} \ (= I_R/I_C) = \tan \delta$$

N.B. When I_R is very small—as normally, in an insulator—I_c is nearly equal to I. Thus, *under these conditions* it can be said that

$$I_R/I_c \equiv I_R/I$$

or
$$\tan \delta \equiv \cos \phi \ (\text{or} \ \sin \delta)$$

i.e. for most dielectrics dissipation factor and power factor are identical[†].

Now, since in a condenser the in-phase current, I_R, is equal to $I_c \tan \delta$, and the capacitance current, I_c, is equal to $E \, 2\pi f C_0 K$ (where f is the frequency, C_0 is the capacity in the absence of a dielectric, and K is the dielectric constant), it follows that the power loss—or heat produced—in the dielectric can be expressed as,

$$\text{power loss} = EI_R$$
$$= EI_c \tan \delta$$
$$= E^2 \, 2\pi f C_0 K \tan \delta.$$

Thus the power loss is proportional to f, K, and $\tan \delta$, and if it is to be restricted (especially at high frequencies) it is necessary to have low values of K and $\tan \delta$. The product of these last items is called the dielectric loss index or *loss factor*, i.e.

---

*Compare *Fig. 73* in Section 73. However, the electrical and mechanical cases differ because in a perfect *dielectric* an alternating voltage produces a current 90 degrees *out*-of-phase with it, whereas in a perfectly elastic material an alternating stress produces a strain *in*-phase with it. In both instances departure from the ideal behaviour is measured by a loss angle δ.

[†] The difference is very small even when $\delta = 5°$ and $\phi \ (= 90° - \delta) = 85°$; namely, $\tan 5° = 0\cdot0875$, $\cos 85° = 0\cdot0872$.

$$\text{loss factor} = \text{dielectric constant} \times \text{dissipation factor}$$

As dielectric constant extends over only a limited range, the value of the loss factor is governed in particular by the dissipation factor (or, as above, the power factor).

Polymers such as polyethylene, polytetrafluoroethylene, polypropylene and polystyrene possess low dielectric constant and exceptionally low power factor ($\equiv$ dissipation factor) and as they also exhibit high dielectric strength and exceptionally high resistivity, these materials provide excellent low-loss insulation even at very high frequencies and high voltages, e.g. in television, radar, and microwave circuits. On the other hand, practically the reverse of these characteristics is required if a low loss factor is undesirable, i.e. when dielectric heating—readily apparent, for example, in polyvinyl chloride, polyamides, and urea-formaldehyde intermediates—is used for such purposes as high-frequency welding and glue setting, preheating of plastics before moulding, heat sealing of films (including electronic 'sewing' of garments), and the drying and twist-setting of textiles.

COMPARISON TABLE

Table 72.T1 shows the power factor of common polymers and related substances at 20–25°C for frequencies of c. 50 and 10^6 Hz; values at other temperatures and frequencies are in some instances to be found under individual polymers in §§1.33–42.33.

Table 72.T1. POWER FACTOR OF COMMON POLYMERS AND RELATED SUBSTANCES

Note: Temperature of test assumed 20–25°C, relative humidity 65% unless otherwise stated. Test frequency indicated as follows: 50–100 Hz, without brackets; c. 1 MHz, enclosed thus [ ]; other frequencies, 'over wide range', or not stated, enclosed thus ( ).

Range for lower limit of power factor	Material (approximate value of power factor)
0	*Vacuum (ideal dielectric).*
0·0001–0·0006	*Paraffin wax.* Silicone fluids, 0·0001–0·0004 [do.]. Polystyrene, 0·0001–0·0005 [0·0001–0·0004]; do., high impact, 0·003–0·005 [0·0009–0·001]. Polyethylene, low-density, (<0·0002). Polytetrafluoroethylene, (0·0002 or lower). Isoprene rubber (synthetic), vulcanised, unfilled, (dry, 0·0002; wet, 0·0004). Silicone rubber, vulcanised, unfilled, 0·00024; do., do., mineral filled, 0·001–0·08 [0·001–0·003]. Polyisobutylene, [0·00035]. Polyethylene, high density, (up to 0·0005). Polypropylene, 0·0005 [do.]. Butyl rubber, unvulcanised, (0·0005–0·003); do., vulcanised, unfilled, (0·0054); do., do., mineral filled, 0·005–0·007. Polyphenylene oxide, (0·0002) [0·0006].
0·001–0·005	*Silica, fused quartz; mica, best (poorest grades up to* 0·05); polyvinylcarbazole, c. 0·001. Polycarbonates, 0·001 [0·01]. Styrene/butadiene rubbers, unvulcanised, 0·001–0·004 [0·005]; do., vulcanised, 0·004–0·022 [0·010–0·016]. Silicone resins, 0·001–0·008 [0·004–0·02]. Natural rubber, unvulcanised, (0·0014–0·0029); do., do., purified, (0·0008–0·0026); do., vulcanised, unfilled, 0·002 [0·017–0·05]; do., do., filled, (0·006–0·014). Ethylene/propylene rubbers, unvulcanised, (0·0015); do., vulcanised filled, (c. 0·008). Cyanoethylcellulose,

Range for lower limit of power factor	Material (approximate value of power factor)
	(0·002) [0·01]. Gutta percha, commercial, (0·002–0·005); do., purified, (<0·002). Epoxy resins, 0·002–0·03. Chlorinated rubber, unplasticised, [0·006]; polyethylene terephthalate, [0·014], 0·003. Polyformaldehyde, 0·003–0·005 [do.]. Polysulphide rubbers, unvulcanised, (0·004); do., vulcanised, (0·005–0·03); do., carbon black filled, (up to 0·07). Glass, common, soft, (c. 0·005); do., borosilicate, Pyrex-type, 0·003 [0·0046].
0·005–0·01	Coumarone resins, 0·005 [0·0005]. Ebonite, [0·005–0·035*]. ABS plastics, 0·005–0·007 [0·07–0·08]. Cyclised rubber, 0·006 [0·002]. Acrylonitrile/butadiene rubbers, vulcanised, 0·006 [0·37]. Polyurethanes, [0·006–0·08]. Butadiene rubber, unvulcanised (do., vulcanished, unfilled, 0·002, mineral filled, 0·014); *porcelain,* 0·008.
0·01–0·02	Polyvinyl acetals, (c. 0·01). Aniline-formaldehyde resin, 0·01 [0·001–0·003]; do., filled, c. 0·05. Vinyl chloride/vinyl acetate copolymer, 0·01–0·02 [do.]. Polyamides, nylon 6.6, air dry, 0·01–0·02 [c. 0·03]. Polychlorotrifluoroethylene, (0·01–0·02). Cellulose acetate, 0·01–0·05 [do.]. Chlorinated polyvinylchloride 55% r.h.), 0·011 [0·009]. Alkyd resins, [0·012–0·03]. Chlorinated polyethers, 0·016 [0·01]. Polyvinyl chloride (rigid)†, (0·02).
0·02–0·05	Phenol-formaldehyde resins, novolak, (c. 0·02); do., cast, (0·01–0·05); do., filled mouldings, (0·02–0·2); do., laminates, (0·02–0·4). Polyester resins, unsaturated, unfilled, [0·02–0·04]; do., glass filled, [0·015]. Chloroprene rubber, vulcanised, (0·02–0·04); do., do., carbon black filled, (up to 0·058); do., unfilled, (0·047). Ethylcellulose, [0·03 or less]. Amino-formaldehyde resins, 0·03 [do.]; do., glass filled, up to 0·3 [up to 0·5]. Chlorosulphonated polyethylene, polyvinyl acetate, (0·03). Polyvinylidene chloride, 0·03–0·1 [do.]. Polyvinyl alcohol, dry, (0·03); do., 65% r.h., (>0·1). Regenerated protein, casein-formaldehyde, 0·05 [do.]. Rubber hydrochloride, (c. 0·05).
0·05–0·1	Polymethyl methacrylate, 0·06 [0·02]. Cellulose, cotton, 40% r.h., 0·1 [0·03]; do., regenerated film, dry, 0·009 [0·06]; do., vulcanised fibre, [0·05]. Cellulose nitrate, 0·1 [c. 0·09]. Polyacrylonitrile, 0·11 [0·03].
1·0	*Pure resistors.*

*The higher figure is for acrylonitrile/butadiene rubber ebonite
†Values for plasticised PVC vary widely

FURTHER LITERATURE

See under this title in Section 71.

ELASTIC MODULUS

DEFINITION

The ratio of stress to strain, measured within the range where deformation is reversible and proportional to the stress. The measure listed in §§1.34–42.34 is generally *Young's modulus, E,* which is the ratio of the tensile (or compressive) stress to the extension (or compression) strain, i.e.

$$E = (F/a)/(l/L)$$
$$= (FL)/(al)$$

where F/a is the ratio of the tensile (or compressive) force to the initial cross-sectional area and l/L is the ratio of the length increase (or decrease) to the initial length.

For materials subjected to shear deformation—i.e. relative displacement of two parallel planes in a direction parallel to them—the *shear* or *rigidity modulus, G,* applies. This is represented by

G = shear stress/shear strain

= (force/area of one face)/(displacement/distance between faces)

This modulus also determines the resistance of, e.g., a cylinder or fibre to torsion. It is less than Young's modulus, being related to the latter thus: $E/2G = (1 + \text{Poisson's ratio})^*$; for rubbers, Poisson's ratio for small strains is *c.* 0·5, and G is thus about one-third of E.

UNITS

Elastic modulus (Young's or shear) has the same dimensions as stress and is listed in kgf/mm^2; for fibres it can be expressed in gf/denier.

For other units, and conversion factors between units, *see* under Units, Section 75.

---

*Poisson's ratio is lateral strain divided by longitudinal strain in a material subject to stress; thus, in a stretched tensile test piece it equals (the proportional decrease in width or thickness)/ (the proportional increase in length).

NOTES

The value of Young's modulus indicates the resistance of a material to reversible longitudinal deformation. It can be considered as the theoretical stress required to double the length of a specimen, but this is not realised in practice because either the material breaks short (e.g. glass, most metals and hard plastics), or the stress/strain relationship is not linear.*

For most plastics materials Young's modulus is less than 1/10 that of metals, whilst for rubbers it is only 1/10000 or less†; some fibres, however, have a modulus approaching that for metals. For materials in the rubber-like (high elastic) state, elastic modulus increases with increase in degree of cross-linking.

Young's modulus can be derived from measurements of extension under load, or of the bending of a rod or beam (*see* Section 76) or of deformation under compressive load. The strain should be kept small so as to maintain a linear stress-strain relationship; this applies particularly to compression, where the linear range is especially small if the shape factor (ratio of cross-sectional dimensions to height) is large, and if the end faces of the test piece cannot slip freely over the compressing surfaces (the test piece then 'barrels' and the stress increases much more rapidly than strain). Deformation in shear gives a bigger linear range, so that measurements of shear modulus are often advantageous.

The elastic modulus as normally measured is the *complex modulus,* so called because it can be analysed into an *in-phase* or *storage modulus* and an *out-of-phase* or *loss modulus,* corresponding respectively to components of the stress *in* phase and 90 degrees *out* of phase with the applied strain (*see* Fig. 73.1); the out-of-phase modulus determines the energy lost in cyclic deformations (cf. Section 74).

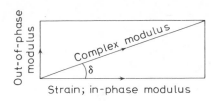

Figure 73.1

---

*For example, with rubbers up to moderate extensions or compressions, stress is approximately proportional to $(\lambda - \lambda^{-2})$, where λ is the ratio of strained to initial (unstrained) length; thus the force/extension curve is concave to the extension axis, but above c. 100–200% extension the curvature is reversed because the molecular network is approaching its limiting extension; *see also* Section 75, Fig. 75.2.

†This applies to normal temperatures; on cooling to the neighbourhood of the glass temperature —*see* Section 68—the modulus of rubbers increases greatly.

The ratio of out-of-phase modulus to in-phase modulus is the *loss tangent* (tan δ; δ being the *loss angle*)‡. A related quantity is the *internal friction*, equal to the out-of-phase modulus divided by the angular frequency of the deformation cycles.

Both the in-phase and out-of-phase moduli, but especially the latter, increase with increasing rate of deformation. Hence modulus measured at high deformation rates, notably under 'dynamic' conditions (e.g. impact, vibration, rapidly repeated deformation cycles), is higher than the 'static' modulus measured under equilibrium conditions or by slow deformation*.

Other elastic moduli

Modulus of volume compression (bulk modulus). This is the ratio of the change in external pressure to the change in volume for reversible conditions, and indicates the resistance of a substance to volume compression. It is high in organic polymers; materials appear in about the same order as for Young's modulus, but the values are higher, sometimes greatly so. Thus, the bulk modulus for soft rubbers is similar to that for water and the less compressible organic liquids, i.e. initially *c.* 200 kgf/mm^2, as compared with Young's modulus *c.* 0·2 kgf/mm^2.

Rubber technologist's 'modulus'. This is the stress (calculated on the initial cross-section) at a stated elongation, usually a multiple of 100%. It is not a modulus in the strict sense, and is better called 'stress at $x\%$ elongation', or simply '$x\%$ stress value'.

Hardness (Section 79) as measured on vulcanised rubber is essentially a function of elastic modulus.

COMPARISON TABLES

The approximate values of Young's modulus of common polymers and related substances, in the massive state, are given in Table 73.T1; values for fibres are given in Table 73.T2; for more detailed information *see* §§1.34–42.34. Values of shear (rigidity) modulus of certain materials are given in Table 73.T3:

Table 73.T1. ELASTIC MODULUS OF COMMON PLASTICS, RUBBERS AND RELATED MATERIALS, AT 20–25°C

Range for lower limit of modulus	Material in bulk form unless otherwise stated. Approximate value of Young's modulus in kgf/mm^2, and—where important—at 65% r.h.
< 10	Vulcanised rubbers (unfilled), 0·1–0·3; do., (*c.* 50 parts carbon black per 100 rubber), 0·35–0·9. Silicone resins (soft), 0·35. Polyvinyl butyral (plasticised), down to 0·5–1·5. Alkyd resins (oil-modified), 4–8.

*At a sufficiently high *rate of deformation* (or cyclic deformation frequency) an elastomer passes through the equivalent of the *glass temperature* (Section 68).

‡There is an analogy here, though not an exact parallel, with energy loss in dielectrics under alternating electrical stress (compare Figure 72b and associated footnote, in Section 72).

Range for lower limit of modulus	Material in bulk form unless otherwise stated. Approximate value of Young's modulus in kgf/mm², and—where important—at 65% r.h.
10–100	Polyethylene (low density), 15–35. Polybutene-1, 18. Polyvinylidene chloride, 20–60. Polytetrafluoroethylene, silicone resins (hard), 40. Fluorinated ethylene/propylene copolymers, Nylon 6, 70. Nylon 11, polyethylene (high density), 100.
100–200	Cellulose acetate (plastics), 100–200. Chlorinated natural rubber, 100–400. Chlorinated polyether (Penton), 110. Polypropylene, 120. Polychlorotri-fluoroethylene, 130. Polyethylene (cross-linked high density), 170. Ebonite (unloaded), 170–320. Nylon 6.10, 180–200. ABS plastics, 190–270..
200–300	Polyvinyl formal, do. acetal, do. butyral, >200. Cellulose nitrate, Nylon 6.6, 200–300. Epoxy resins (cast), 200–450. Cyanoethylcellulose, 225–240. Polyphenylene oxide, 230. Polycarbonates, polysulphones, 250. Phenol-formaldehyde resins (cast), polyvinyl chloride (rigid), 280. Polystyrene, 280–350; do., high impact, 270–320. Polyacrylates, polymethacrylates, aniline-formaldehyde resins, 300.
300–500	Polyformaldehyde, 300+. Cellulose ethers (insoluble), regenerated proteins, unsaturated polyesters (cast), 350. Ebonite (loaded), 400–600. Phenol-formaldehyde resin laminates (asbestos), 400–1400.
500–1000	Phenol-formaldehyde resin laminates (cotton), 500–1000; do. (paper), 700–1500. Phenol-formaldehyde resins (wood flour or cotton filled), 600–1000; do. (mineral filled), 1000–2000. Amino-formaldehyde resins (MF, UF), 700–1000. Polycarbonates (glass-reinforced), 900. Aniline-formaldehyde resins (reinforced), 1000.
1000–2000	Epoxy resins (filled), up to 1750. *Concrete, c.* 2000.
2000–5000	*Ceramics, c.* 3000. *Magnesium, c.* 4000.
5000–10 000	Glass, 5000–8000. *Aluminium, silica (fused quartz),* 7000. *Brass, iron (cast),* 10 000.
10 000–50 000	*Copper (hard drawn),* 12 500. *Iron (wrought), steel,* 20 000. *Tungsten,* 35 000.
c. 100 000	*Diamond.*

Table 73.T2. ELASTIC MODULUS OF COMMON FIBRES AND RELATED MATERIALS AT 20–25°C

Range for lower limit of modulus		Material in form of filaments, threads or wires. Approximate value of Young's modulus in kgf/mm², and—where important— at 65% r.h. Values in gf/den. are enclosed thus [ ].
kgf/mm²	gf/den.	
<200	<20	Rubbers (vulcanised, soft; thread), *c.* 0·15 [0·02]. Polyurethane (elastomeric thread), 1 [0·1]. Polyethylene (low density), 80 [10]. Polyvinylidene chloride, 100–150 [7–10]. Wool, 100–300 [9–25].
200–400	20–40	Nylon 6, 200–300 [20–30 (higher figures = cont. fil.)]; do. 6.6, 200–250 [20–25]; do. 6.10, 400 [40]. Regenerated proteins, 250 [20]. Cellulose acetate, 300–450 [25–40]. Polyvinylalcohol, 300–1200 [25–100]; do. (high tenacity), 2000 [180]. Polypropylene (monofil), 325 [40]. Polyacrylonitrile, 400–700 [40–70]. Polyester (Terylene), 400–1450 [30–115]; do. (high-tenacity), 1400–1600 [110–130]; do. (film), 350.
400–1000	40–100	Polyethylene (high density), 400 [50]. Polyvinyl chloride (cont. fil.), 500 [40]; (staple relaxed), *c.* 100 [8]. Nylon 11, 500 [50]. Polynonamethyleneurea (Urylon), up to 600. Viscose rayon, 600–900; do. (high tenacity), 1400 [100]. Cotton, 600–1100 [45–80]. Polypropylene (cont. fil. yarn), 650 [80]. Silk, 700–1000 [60–80].
1000–5000	100–250	Calcium alginate, 2000 [125]. *Brass,* [130]. *Copper (hard-drawn),* [155]. Flax, hemp, jute, 2500–5000 [180–360]. Glass, 5000–8000 [220–350].
—	>250	*Aluminium, steel,* [290]. *Silica (fused quartz),* [350]. Asbestos, 15 000 [700]. *Carbon fibres,* 20 000–40 000 [1600–3000].

Table 73.T3. MODULUS OF RIGIDITY (SHEAR MODULUS) OF COMMON FIBRES AND
RELATED SUBSTANCES

Range for lower limit of modulus kgf/mm²	Material in the form of filaments or wires. Approximate value of shear modulus, in kgf/mm², and—where important—at 65% r.h. Values in gf/den. enclosed thus [ ].
Under 10	Rubber* (natural vulcanised, unfilled), 0·03–0·07; do*. (vulcanised, 50 parts HAF black per 100 rubber), 0·14–0·18. Polyethylene, 5 [0·5].
10–100	Polyvinylidene chloride, 40–60 [2·5–4]. Polyamides (nylon), 50–60 [5–6]. Cellulose acetate, 60–80 [5–7]. Polyesters (Terylene), polyvinyl chloride, c. 90 [7]. Viscose rayon (including high-tenacity type), [6–7·5], wool, [8–9], c. 100.
100–1000	Polyvinyl alcohol (formalised fibres), [10], vinyl chloride/do. acetate copolymers [10–11], polyacrylonitrile [12], 110–120. Flax [10], regenerated proteins (formalised casein and ground-nut fibres) [12], c. 130. Cotton [15–18], silk [16–20], 200–250. Kapok, 300–600 [20–40].
Over 1000	Glass, 2000–4000 [90–180]. *Aluminium*, 2700 [110]. *Brass, copper*, 3500–4000 [45–50]. *Silica (fused quartz)*, 4750 [240]. *Steel*, 8000 [115].

*In massive form

FURTHER LITERATURE

(*See also* further literature under Sections 74 and 75)

MEREDITH, R., *The Mechanical Properties of Textile Fibres*, New York (1956)

MEREDITH, R. and HEARLE, J. W. S. (Eds), *Physical Methods of Investigating Textiles*, New York (1959) (Chapter 8, mechanical properties, by MEREDITH, R.)

PAYNE, A. R. and SCOTT, J. R., *Engineering Design with Rubber*, London, 69 (1960) (measurement of elastic moduli by dynamic tests)

RITCHIE, P. D. (Ed), *Physics of Plastics*, London and Princeton, 24 (1965) (VINCENT, P. I., deformation of high polymers)

Standards

BS 903: Part A13 (rigidity or shear modulus of rubber at normal and low temperatures)

BS 903: Part A14 (shear modulus of rubber)

BS 903: Part A24 (in-phase and out-of-phase moduli of rubber)

BS 2782: Methods 302A and D (Young's modulus of rigid plastics in tension and in bend)

ASTM D638 (tensile properties—including modulus—of plastics)

ASTM D695 (compressive properties—including modulus—of plastics)

ASTM D797 (Young's modulus of elastomers at and below room temperature)

ASTM D882 (tensile properties—including modulus—of thin plastic sheeting)

ASTM D1043 (stiffness properties of non-rigid plastics by torsion test)

VISCO-ELASTIC BEHAVIOUR

DEFINITIONS

The term visco-elastic covers several properties (detailed below) connected with the facts that (i) most polymers are not perfectly elastic (i.e. recovery after deformation is incomplete), and (ii) the response to an applied force, and recovery after removal of the force, are time-dependent, not instantaneous. In this connection the following items are of interest:

Recovery (*elastic recovery*)—The extent to which the original dimensions of a material are recovered upon removal of a deforming force. It can have time-independent and time-dependent components.

Set—The deformation remaining after a material has been deformed and then allowed to recover.

Creep or cold flow*—The gradual increase in deformation in a material subjected to a constant force.

Stress relaxation—The gradual decay of stress in a material held at a constant deformation.

Resilience—The proportion of the applied energy returned during a cycle of deformation and recovery. It depends on the rate of deformation, i.e. the cycle time.

Hysteresis†—The energy lost during a given cycle of deformation and recovery.

Heat build-up—The heat generated in a succession of cycles of deformation and recovery, due to the conversion of hysteresis energy into thermal energy; it is usually measured by the rise of temperature of the test specimen.

UNITS

Recovery

For fibres, *recovery* is expressed as the percentage recovery from a given tensile stress or strain. The value depends on the time and/or

---

*Not necessarily in the cold (e.g. the ASTM D530 test is at 49°C).

†Strictly speaking, hysteresis means 'lagging' (of strain relative to stress) but is commonly used to mean the resulting energy loss.

rate of application of the stress (or strain) and the time of recovery.

Resilience For rubbers it is usual to measure *rebound resilience,* which is the height of rebound expressed as a percentage of the height of fall of a striker, i.e. as the proportion of the impact energy returned.

Set is determined by applying a given strain (or stress) for a given time and measuring the deformation remaining after a given time of recovery; it is commonly expressed as the percentage of a given compression that remains.

Creep, cold flow and the other terms listed above are not generally included in Sections 1 to 42 since the data are as yet insufficiently consistent.

NOTES

The deformation of a polymeric material is a complex process involving the following phenomena.

(i) *Short-range (high modulus) elasticity*

This involves *stretching* (or *compression*) of bonds between atoms—also of bonds between molecules—and deformation of bond angles. It is important in rigid materials, e.g. polymers that are crystalline or below their glass temperature (Section 68). It extends over only a very short range of deformation, and the modulus (Young's) is high and only slightly affected by temperature; Hooke's law is obeyed, and the strain is in phase with the stress, i.e. produced (or disappearing) instantaneously on applying (or removing) the deforming force. This kind of elasticity can be likened to (is, in fact, nearly identical with) that of a helical steel spring.

(ii) *Long-range (low modulus, rubber-like, or 'high') elasticity*

This property, unique to high polymers, is due to extension (or compression) of randomly-coiled or kinked long chain molecules, the extension (or compression) occurring largely by *rotation* of chain segments about primary bonds*. The forces are many thousandfold smaller than those involved in short-range elasticity; hence a rubber—in which there is little restriction on chain uncoiling and coiling—can be easily and extensively deformed, and on removal of the deforming force the original state (which, being less ordered, has the higher entropy) is quickly restored by thermal motion of the chain segments. Thus the modulus is low, and markedly affected by temperature.

---

*It depends also on the angular disposition of the primary bonds associated with each atom in the main chain. For carbon chains the bonds lie at the valency angle of *c.* 109·5 degrees (if the angle were 180 degrees the chains would have only a rod-like configuration).

Long-range deformation and recovery from it are, however, delayed in action, i.e. time-dependent, so that the strain is more or less out-of-phase with the stress (cf. Section 73), causing *hysteresis* and *heat build-up*. Though normally rapid in a rubber showing good 'snap', deformation and recovery become slow at low temperatures, so that stiffness and hysteresis increase, as they do also if deformation (e.g. in repeated cycles of alternating stress) is rapid—indeed, at sufficiently high deformation rates (or cycle frequency) a normally soft rubber can appear quite rigid, showing short-range elasticity as in (i) above. (The effects of lowered temperature and increased deformation rate are mathematically related; *see* PAYNE, A. R. and SCOTT, J. R., *Engineering Design with Rubber,* pp. 22–41 and 109–114; FERRY, J. D., *Viscoelastic Properties of Polymers,* Chapter 11). This kind of elasticity can be likened to that of a system consisting of a helical spring and an oil dash-pot in parallel (Kelvin or Voigt element)*.

(iii) *Non-elastic (viscous or plastic) deformation*

Stress applied to a linear polymer, or indeed to any one in which the molecules are not firmly and permanently cross-linked, may bring about relative displacement of the molecules which is *irrecoverable*. Creep, cold flow, and stress relaxation—also set, at least in part—are associated with such rheological behaviour, which can be likened to that of a dash-pot† (movement in such being irrecoverable in absence of a spring).

Short-range (high modulus) elasticity, though often masked by the effects of high elasticity, is a component of the deformation of any polymeric material. It is in part responsible for the behaviour of bulked ('stretch') textile yarns; and, though it may also be delayed in action in a complex system, it is important in the crease recovery, smooth-drying, and crease retention of textile fabrics (notably in cellulosic fabrics, where these properties are enhanced when plasticity is reduced by cross-linkage, e.g. *see* §13.86).

Long range (low modulus or 'high') elasticity of a very evident kind is particularly manifest in unfilled rubbers that have been lightly vulcanised (cross-linked) to prevent non-elastic flow. These, at normal or moderately elevated temperatures, show recovery that is substantially complete and instantaneous ('snap'); however, below the glass temperature (Section 68) such materials become rigid, i.e. lose their long-range elasticity. Conversely, raising the temperature

---

*Increasing the rate of application of the force increases the phase lag between force and deformation (and hence the hysteresis), due to increased damping by the dash-pot.

†Plastic deformation is time-dependent but the *initial* response of materials exhibiting it is *elastic,* i.e. their behaviour can be likened to that of a helical spring and a dash-pot in series (Maxwell element).

For a more complete analogy of the complex visco-elastic behaviour of a polymeric material— i.e. immediate elasticity (i), delayed elasticity (ii), and permanent strain (iii)—we may picture a Maxwell element in series with a Kelvin element (*see* ii).

of certain normally hard and intextensible linear polymers causes them to become rubber-like, and also capable of non-elastic deformation. In any highly cross-linked or network polymer (e.g. ebonite, fully-cured phenolic resins), however, at room temperature such long-range elasticity—and also viscous flow—are largely absent, the materials being capable of only short-range deformation.

The possibility of plastic deformation is generally detrimental in finished products and may render them unsuitable for applications involving prolonged periods of stress, particularly at elevated temperatures; on the other hand, the phenomenon is of great value in the fabrication of polymers in the plastic state, i.e. drawing of fibres or films and extrusion or moulding of plastics or rubbers (*see* Section 81).

COMPARISON TABLE

It is not feasible to give a comparison table for the properties discussed above, as the data for different materials are often not comparable; information on recovery, set, resilience and related phenomena for individual polymers is given in §§1.34–42.34.

FURTHER LITERATURE

(*See also* under Further Literature, Sections 73 and 75)

MARK, H. and WHITBY, G. S. (Eds), *Advances in Colloid Science,* Vol. II (GUTH, E., JAMES, H. M. and MARK, H., pp. 253–298, kinetic theory of rubber elasticity), New York (1946)

ALFREY, T., Jr., *Mechanical Behaviour of High Polymers* (*High Polymers,* Vol. VI), New York and London (1948) (includes recovery and creep of plastics and rubbers)

MEREDITH, R., *The Mechanical Properties of Textile Fibres,* New York (1956)

TRELOAR, L. R. G., *The Physics of Rubber Elasticity,* 2nd edn, Oxford (1958)

PAYNE, A. R. and SCOTT, J. R., *Engineering Design with Rubber,* London and New York (1960) (visco-elastic properties and their measurement)

FERRY, J. D., *Visco-elastic Properties of Polymers,* New York and London (1961)

NAUNTON, W. J. S. (Ed), *The Applied Science of Rubber,* London, 506 (1961) (EDWARDS, A. C. and FARRAND, G. N. S., elasticity and dynamic properties of rubber)

BATEMAN, L. (Ed), *The Chemistry and Physics of Rubber-Like Substances,* London and New York, 155 (1963) (MULLINS, L. and THOMAS, A. G., theory of rubber-like elasticity); 187 (GENT, A. N. and MASON, P., visco-elastic behaviour of rubber)

Standards

BS 903: *Part A5* (rubber, set after extension)
BS 903: *Part A6* (rubber, set after compression)
BS 903: *Part A8* (rubber, rebound resilience)
BS 903: *Part A15* (rubber, creep and stress relaxation)
BS 903: *Part A24* (rubber, hysteresis and other dynamic properties)
BS 903: *Part D6* (ebonite, recovery after indentation)
ASTM D395 and *1229* (rubber, set after compression)
ASTM D530 (ebonite, recovery and cold flow)
ASTM D621 (plastics, recovery after compression)
ASTM D945 (rubber, hysteresis and other dynamic properties)
ASTM D1054 (rubber, rebound resilience)
ASTM D1206 (rubber, creep)
ASTM D1329 (rubber, recovery)
ASTM D1390 (rubber, stress relaxation)
ASTM D2236 (plastics, hysteresis and other dynamic properties)

TENSILE STRENGTH AND EXTENSION AT BREAK

DEFINITIONS

Tensile strength, ultimate tensile strength, or *tenacity* is the tensile stress at the breaking point of a material.

Extension (*elongation*) *at break* is the tensile strain at the breaking point of a material.

UNITS*

Tensile strength. The stress is the ratio of the force at break to the *initial* cross-sectional area, and is here expressed in kgf/mm^2 (for conversion to other units, *see* below).

Sometimes it is convenient to indicate the strength of a filament (or composite thread or rope) as a *breaking load,* directly in gf or kgf.

The strength of polymeric materials, especially fibres, is also quoted as *specific strength,* which involves the density of the material, and is expressed as *breaking* (*failing*) *length* or as *gf/tex* or *gf/denier*† :

$$\text{(tensile strength)/density} = \text{breaking length (in km‡)}$$

$$= \text{gf/tex}$$

$$\text{(tensile strength)/(9} \times \text{density)} = \text{gf/denier.}$$

where tensile strength is in kgf/mm^2 and density in g/ml.

Breaking length can be regarded as the maximum length of (unextended) material that could be hung vertically, from its upper end, to bear its own weight under gravity without breaking.

Extension (*elongation*) *at break.* The extension or increase in length at break is expressed as a percentage of the initial length.

*See also Note on SI Units, page 659.

†Specific strength, as expressed in gf/den., is the maximum load that could be sustained by a fibre of such cross-section that a 9 km length would weigh 1 g; it is 9 times the specific strength expressed in kilometres. Specific strength expressed in gf/tex is the maximum load that could be sustained by a fibre of such cross-section that a 1 km length would weigh 1 g; it is thus equivalent to specific strength expressed in kilometres.

‡Strictly speaking the unit is length × acceleration, since tensile strength is a stress, the unit of which involves acceleration; however, since the acceleration is commonly understood to be the constant value due to gravity (9·81 m/s²), it is omitted. Breaking length would have no meaning in outer space.

597

Tensile strength and extension at break should be quoted with reference to the *temperature*, and—for polymers affected by moisture (especially fibres and films)—the *humidity* (e.g. 65% r.h.) at which the material was conditioned before test; for polymers showing marked plastic deformation (*see* Section 74) it is necessary to quote the *rate of strain* in the tensile test.

CONVERSION FACTORS

$$1 \text{ kgf/mm}^2 \equiv 981 . 10^5 \text{ dyn/cm}^2 \text{ (or approx. } 10^8 \text{ dyn/cm}^2\text{)}$$

$$\equiv 1422 \text{ lbf/in}^2$$

$$\equiv 0 \cdot 635 \text{ tonf/in}^2 \text{ or } 0 \cdot 711 \text{ U.S. tonf/in}^2$$

NOTES

TENSILE STRENGTH

The theoretical strength of polymers, calculated from bond energies, is some 10 times the highest tensile strength found in practice.* The discrepancy is ascribed to randomness in molecular configuration, and to minute faults, fissures, voids and inhomogeneities at which over-concentration of stress initiates breakdown.

In practice the tensile properties—strength, elongation, stress at fixed elongation—depend on many factors: type of polymer, degree of polymerisation (*see* Section 52, footnote page 492, re critical *minimum* chain length), degree of crystallinity, molecular orientation— (stretching or drawing fibres and films produces a parallel alignment that greatly increases strength in the direction of orientation), degree of cross-linkage (very marked with cellulose and vulcanised rubbers), the reinforcing effect of fine-particle fillers (*see*, e.g., §29.4—the action of carbon black in rubber is attributed to 'stick-slip' adhesion between the filler and polymer), the presence of plasticisers (Section 56; these—while increasing serviceability—usually reduce tensile strength), strain rate during testing (this can range between about 0·05 and 50000%/s) and whether the test uses constant *strain* rate or constant rate of *stressing*, and size of test piece (since the chance of occurrence of a weak spot increases with the volume of material tested).

It is of interest to note the order that different types of substances

---

*Thus, the calculated strength of cellulose is about 72000 kgf/mm² (OTT, E., *Cellulose and Cellulose Derivatives* (*High Polymers*, Vol. V), New York, 1943, p. 999).

assume when arranged in comparison tables according to strength. In this respect the appearance of vulcanised rubbers among the weak materials, when some of them are really quite strong, is apt to be misleading. It arises because strength is normally measured with respect to the *initial* cross-sectional area; if it were given in terms of the cross-section at break the order of the substances would be changed, since the values for vulcanised rubber would be much higher*, while the strength of relatively inextensible materials would be not much altered. Most of the fibres appear in the lower part of the table and are placed, correctly, among some of the strongest materials known.

WET STRENGTH (SWOLLEN STRENGTH)

A polymer that swells in a liquid without dissolving (*see also* Section 55) is thereby weakened, though the original strength may be largely regained if the liquid is dried out. This weakening can be a serious defect, e.g. in vulcanised rubber used in contact with petrol or mineral oils and in wet fibres of regenerated proteins or cellulose. However, cotton and other fibres of natural cellulose appear to increase in strength when wetted. This is partly a result of the way in which wet strength is expressed: because of the helical molecular configuration cotton swells in such a way that the cross-section increases but the fibre undergoes axial contraction, and to compensate for this, additional force is required when the fibre is elongated (if the wet strength were measured with respect to the initial cross-section when wet, the result might be lower than the dry strength).

STRESS–STRAIN RELATIONS

Tensile strength and extension at break indicate, respectively, the maximum stress and maximum strain that a material can withstand. Typical curves illustrating the wide range of stress–strain relations of some common polymers are shown in Figures 75.1 and 75.2. The curves for fibres (Figure 75.1) show average results from normal continuous filaments (except cotton). In high-tenacity fibres the extension at break is lower, and with cut-staple fibres the strength is usually reduced while the extension at break is increased.

Plastics, in general, follow curves of the same order as those of the weaker fibres, but the curves terminate at lower values. The curves,

---

*Thus, tensile strength as kgf per mm² of cross-section *at break* is approximately: rubber, 35; textile fibres, 20–90; piano wire, 205 (TRELOAR, L. R. G., *Endeavour*, 11, 95 (1952)).

though initially straight, later bend over towards the extension axis, and there may be a final portion (cold-drawing region, *see* below) in which extension increases with little or no increase in stress, or even a decrease (polyethylene curve, Figure 75.2).

A distinction must be made between *brittle* and *tough* plastics; at the usual rates of loading the former, even though having high strength, have very low extensibility (e.g. polystyrene, polymethyl methacrylate; Figure 75.2) whereas tough plastics have relatively high extensibility (e.g. plasticised PVC, polyethylene) and hence require much more energy to produce rupture, this energy being represented by the area under the stress/strain curve (*see also* Section 78).

Another important property of tough plastics is that they are capable of cold or hot 'drawing', i.e. permanent extension and consequent molecular orientation, giving greatly increased strength.

Rubbers give stress–strain curves which finally bend the opposite way to that typical of the more rigid materials, and there is no 'yield point' (*see* Figure 75.2; *see also* Section 73 regarding the form of the rubber stress/strain curve).

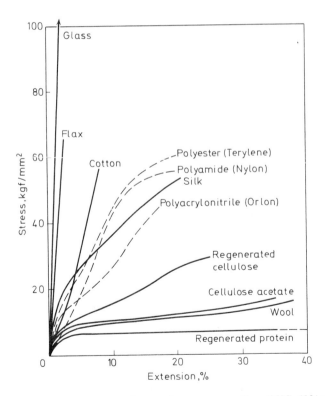

Figure 75.1. Stress-strain relations of some common fibres (25°C, 65% r.h.)

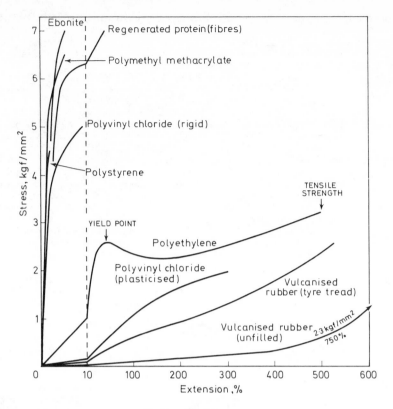

Figure 75.2. Stress-strain relations of some common plastics and rubbers (note change of extension scale at 10%)

Thus, the characteristics of rubbers and metals, in general, lie at extreme and opposite limits of the curves for fibres; thus, whereas a soft vulcanised rubber cord is extended by 50% of its length for a load of only 0·05 kgf/mm², a load of 50 kgf/mm² extends steel by only 0·25% of its initial length.

COMPARISON TABLES

The following are included in this Section:

Table 75.T1: Tensile strength, in kgf/mm².

Table 75.T2: Extension at break.

Table 75.T3: Specific strength of fibres, in gf/den.

Table 75.T4: Wet strength of fibres.

Table 75.T1. TENSILE STRENGTHS OF COMMON POLYMERS AND RELATED SUBSTANCES
AT 20–25°C

Range (for lower limit of T.S.)	Material: unfilled and unoriented unless otherwise stated; rubbers as vulcanisates; r.h., where important, 65%. Approximate tensile strength in kgf/mm².
Under 1	Silicone rubbers, 0·035–0·5; filled, 0·35–1·5. Styrene/butadiene rubbers, 0·14–0·28. Alkyd resins, 0·18–2·0. Butadiene rubber, 0·2–0·7. Nitrile rubbers, 0·4–0·9. Cyclised rubber, 0·45–3·5. Polysulphide rubbers, filled, 0·4–1·1. Ethylene/propylene rubbers, filled, 0·5–2·3. Silicone resins, soft, 0·55–0·7. Ethylene/vinyl acetate copolymers, 0·6–2·0. Butyl rubbers, filled, 0·9–2·1 Polyethylene (low density), 0·7–2·5. Chlorosulphonated polyethylene, 0·9–2·2.
1–5	Nitrile rubbers, filled, 1·0–3·2. Chloroprene rubber, filled, 1·2–2·3. Butadiene rubber, filled, 1·4–2·3. Styrene/butadiene rubbers, filled, 1·4–2·7. Chlorinated polyethylene, 1·5. Polytetrafluoroethylene, 1·5–2·5. Polyvinyl acetate, 1·5–3·5. Cross-linked polyethylene, 1·6–3·0. Polyacrylate rubbers, up to 1·7. Butyl rubber, 1·7–2·1. Natural rubber, 1·7–3·0; filled, 2·2–2·8; latex films or threads, up to 4·5. Polybutene-1, 1·7–3·2 (yield, 0·4–2·5). Isoprene rubber (synthetic), 1·7–3·8; filled, 2·3–2·9. Polyethylene (high density), 2·0–3·2. Polyvinyl chloride, plasticised, c. 2. Rubber hydrochloride, 2·0–3·6. *Porcelain, silica (fused quartz)*, 2–4 (*steatite, c.* 6). Polyurethane rubbers, 2–5. Cellulose acetate, 2–6. Ebonite, 2–8. Silicone resins, hard, 2·2–4·2. Fluorinated ethylene-propylene, 2·27. Ionomers, 2·5–3·9. Phenolic resins, filled, 2·5–6·5. Polypropylene, 2·8–10. Chlorinated rubber, 2·8–4·5. Polyesters (glass-reinforced mouldings), 2·8–5·3. Amino resins, filled, 3–8. Polyvinylidene chloride, 3–4. Polychlorotrifluoroethylene, 3·2–4·0. Polystyrene, 3·2–5·0; do., high impact, 2·1–3·5. Polyvinyl alcohol, 3·5; cast films, 0·4–12. Poly-3,3-dichloromethyloxacyclobutane, 3·5–4·2. Polyvinyl acetals, 3·5–7·0. ABS plastics*, 3·9–5·2. Polyvinyl chloride, rigid, 4·25–5·6. *Wood* (along grain): *pine, c.* 4; *beech, c.* 12 (*plywood, c.* 6; *densified plywood, c.* 20).
5–10	Polyvinylidene chloride films, 5–10. Vinyl chloride/vinyl acetate copolymers, 5·3–6·0. Regenerated proteins, 5–7. Ethylcellulose, up to 6. Chlorinated polyvinyl chloride, 6·0–6·7*. Polyamides: nylon 6, 6–7; nylons 6.10 and 11, 5–6; nylon 6.6, 7–8. Phenolic resins, cast, 6·5; laminates, 5–15. *Aluminium, cast*, 6–9; *rolled*, 9–15. Vinyl chloride/vinyl acetate copolymer fibres, staple, 8·5; cont. fil., 25–30. Polyformaldehyde, 6·5–7. Polycarbonates, cellulose nitrate, up to 7. Polysulphones, 7*. Polyesters (cast), polyphenylene oxide, polymethyl methacrylate, 7. Epoxy resins, up to 7 or 8.
10–25	*Iron, cast*, 10–20; *wrought*, 30–75. Polytetrafluoroethylene films, 10·5–17·5. Polycarbonates, glass-filled, 14. Wool, 15–20. Polyvinyl chloride fibres, staple, c. 15; cont. fil., 34. Cellulose acetate fibres, 16. Polyethylene terephthalate films, 16–18. Polypropylene films, 17·5. Polyvinyl fluoride films, up to 20. *Brass, gunmetal, c.* 20. *Aluminium wires*, 20–30; *Duralumin wires*, 40–55. Cuprammonium rayon, 23–32. Viscose rayon, 23–34.
25–50	Calcium alginate fibres, 25–30. Cotton, 25–80. Polyvinylidene chloride fibres, polyacrylonitrile fibres, 30–40. *Brass wires*, 30–40 and upwards. Glass, up to 35. Silk, 35–60. Polycarbonate fibres, up to 40. *Copper wires, hard drawn*, 40–45. *Steel, mild*, 40–50; *hard tempered*, up to 100. Polynosic rayon, 40–70. Jute, 40–80.
50–100	Polyethylene terephthalate fibres, 50–90. Flax, 50–100. Polyvinyl alcohol fibres, insolubilised, 50–105. Polyesters (glass-reinforced laminates), up to 85.
Over 100	*Silica fibres*, 100–120. Glass fibres, 125–1000. Asbestos fibres and *steel wires* (e.g. *piano wire*), up to 200. *Carbon fibres*, 210–300. *Tungsten wires* (e.g. *lamp filaments*), 400. *Sapphire whiskers*, over 4000.

*Yield stress

Table 75.T2. EXTENSION AT BREAK OF COMMON POLYMERS AT 20–25°C

Range (for lower limit of extension)	Material: unfilled and unoriented unless otherwise stated, rubbers as vulcanisates; r.h., where important, 65%. Approximate extension at break, in %.
Under 2	Amino-resins, 1. Phenol-formaldehyde resins, 1–5. Ebonite, 1–6. Jute, 1·8. Cyclised rubber, 1–30. Polyesters, cast or glass-reinforced, <2.
2–5	Flax, 2–2·6. Polystyrene, 2–3; high impact, 20–40. Hemp, 2·2. Regenerated proteins, 2·5. ABS plastics*, 3–10. Polyvinyl acetals, 3–10. Polypropylene glass, talc or asbestos, filled, 2–20. Chlorinated rubber, 3·5. Polymethyl methacrylate, 4. Epoxy resins, 4–7. Polyvinyl alcohol cast films, 2–550.
5–15	Silicone resins, hard, 5–10. Polyvinyl chloride, rigid, 5–25. Ethylcellulose, 5–50. Calcium alginate fibres, up to 6. Polyethylene terephthalate fibres, 6–50. Polynosic rayon, 7–10. Chlorinated polyvinyl chloride*, 8–9. Cotton, up to 10. Polyvinyl alcohol fibres, insolubilised, 10–26. Polyvinylidene chloride, 10–40.
15–30	Polyformaldehyde, 15–16. Regenerated cellulose films, 15–30. Viscose rayon fibres: staple, 18–20; cont. fil., 19–25. Silk, 20–25. Polyacrylonitrile fibres, 20–40. Polyvinylidene chloride fibres or films, 20–40. Polyvinyl chloride fibres: cont. fil., 20; staple, 180. Polyethylene (high density), 20–50. Polypropylene, moulded, 20–300 (yield elongation, c. 15). Vinyl chloride/vinyl acetate copolymer fibres: cont. fil., 20–30; staple, up to 100. Cellulose acetate, 25–50. Regenerated protein fibres, 25–60.
30–100	Cellulose nitrate, >30. Wool, 30–40. Silicone rubbers, filled, 40–800. Alkyd resins, polysulphones, 50–100. Polytetrafluoroethylene, 50–450. Rubber hydrochloride, 50–550. Polycarbonates, 60–100. Poly-3,3-dichloromethyloxacyclobutane, 60–160. Polyethylene terephthalate films, 70. Polypropylene film, biaxially oriented, 70; unoriented, 200–1000. Polyphenylene oxide, 80. Polyamides (nylon 6.6), up to 100.
100–500	Silicone resins, soft, 100–150. Polychlorotrifluoroethylene, 125. Cross-linked polyethylene, 150–250. Polyamides (nylon 6.10), up to 200. Silicone rubbers, 200. Polyamides (nylon 6), 200–300. Polyvinyl chloride (plasticised); vinyl chloride/vinyl acetate copolymers (plasticised), 200–450. Polyvinyl alcohol, 225. Ethylene/vinyl acetate copolymers filled, 230–560. Butyl rubber, filled, 300–700. Chlorinated rubber, plasticised, up to 250. Chlorosulphonated polyethylene, filled, 270–500. Nylon 11, up to 300. Polyethylene (low density), 300–900. Butadiene rubber, filled, 310–500. Fluorinated ethylene-propylene, 320. Polybutene-1, 350. Chloroprene rubber, filled, 350–800. Nitrile rubbers, 350–800; filled, 310–800. Ethylene/propylene rubbers, filled, 380–820. Styrene/butadiene rubbers, 400–600; filled, 400–650. Polysulphide rubbers, filled; polyurethane rubbers, 400–700. Isoprene rubber (synthetic), filled, 430–560. Natural rubber, filled, 450–600. Acrylate rubbers, 500.
500–1000	Elastomeric polyurethane fibres, up to 600. Butadiene rubber, 660–830. Natural rubber, 675–900; latex thread, 950. Isoprene rubber (synthetic), 720–1300. Butyl rubber, 750–950. Chloroprene rubber, 800–1000.

*Yield strain

Table 75.T3. SPECIFIC STRENGTH OF COMMON FIBRES AT 20–25°C

Range	*Fibres: r.h., where important, 65% (approximate specific strength in gf/den.)*
Under 1	Natural rubber thread, *c.* 0·5. Elastomeric polyurethanes, 0·5–0·8. Regenerated proteins, 0·6–1.
1–2	Polyethylene (low density), 1–1·5. Wool, 1·3–1·7. Polybutene-1, 1·5. Calcium alginate, 1·6–2·0. Rayon (cuprammonium and viscose), 1·7–2·5. Cotton, 1·9–5·9. Cellulose sec. acetate, 1·3.
2–4	Polyvinylidene chloride, 2–2·6. Vinyl chloride/vinyl acetate copolymers: cont. fil., 2·5; staple, 0·7. Polyvinyl chloride cont. fil., 2·7. Polyacrylonitrile, 3–4. Silk, 3–5. Jute, rayon (polynosic), 3–6. Polyamides (nylon 6), 3·5–7·5. Polycarbonates, 3·7. Flax, 3·7–7·4.
4–10	Polypropylene: monofil, 4; cont. fil yarn, 7·4. Polyamides (nylon 6.10), 4. Polyethylene terephthalate, 4–7. Polyvinyl alcohol (insolubilised), polyamides (nylon 6.6), 4–9. Polyethylene (high density), 4·5–6. Glass, 5–40. Polyamides (nylon 11), up to 5·5.

Table 75.T4. WET STRENGTH OF COMMON FIBRES

Range	*Fibres. (Wet strength expressed as percentage of tensile strength at 65% r.h.)*
Over 100	Flax, 105–110. Cotton, 110–120.
c. 100	Polypropylene, polyvinyl chloride, vinyl chloride/vinyl acetate copolymers, polyethylene terephthalate.
85–95	Polyacrylonitrile, 85. Polyamides, 85–90.
55–85	Cellulose acetate, polynosic and high-tenacity rayons, 55–80. Wool, 75. Silk, polyvinyl alcohol (insolubilised), 80.
Under 55	Regenerated proteins, viscose rayon, *c.* 50. Calcium alginate, *c.* 25.

FURTHER LITERATURE

ALFREY, T., Jr., *Mechanical Behaviour of High Polymers* (*High Polymers,* Vol. VI), New York and London, 476 (1948) (ultimate strength and related properties)

The following standards give details of tensile tests:
BS 903: Part A2 (soft vulcanised rubber)
BS 903: Part D5 (ebonite)
BS 903: Part E4 (cellular ebonite)
BS 903: Part F7 (soft cellular rubber)
BS 2782: Method 301 (plastics)
BS 3379 (soft polyurethane foam)
BS 4443 (flexible cellular materials)
ASTM D412 (soft vulcanised rubber)
ASTM D530 (hard vulcanised rubber; ebonite)
ASTM D638 (plastics)
ASTM D651 (moulded electrical insulation)
ASTM D759 (plastics, at sub- and super-normal temperatures)
ASTM D882 and *1923* (thin plastic sheeting)
ASTM D1564 (soft polyurethane foam)
ASTM D1623 (rigid cellular plastics)
ASTM D1708 (plastics, micro-test)
ASTM D2289 (plastics, high speed test)

SECTION 76

FLEXURAL STRENGTH

DEFINITION

Flexural strength, cross-breaking strength, or *modulus of rupture* is the maximum stress developed in the surface of a prescribed bar-shaped test piece subjected to a bending force, i.e. one acting in a direction perpendicular to the length of the bar.

UNITS*

Flexural strength is measured as a force per unit area, and is here expressed in kgf/mm². For conversion factors to other units, *see* under Units, Section 75.

NOTES

A bar-shaped test piece is held and loaded (i.e. the force is applied) by one of three methods:

(i) *Cantilever loading* Test-piece clamped at one end and loaded at the other; distance between clamping and loading points $= L_1$.

(ii) *3-Point loading* Test piece supported near its ends and loaded in the centre; distance between supports $= L_2$.

(iii) *4-Point loading* Test piece supported at two points towards the ends and loaded equally at the ends; distance between a support and a loading point $= L_3$. This last method has the advantage that stress does not vary along the test length (as it does in (i) and (ii) (test length is the part between clamping and loading points in (i) and between supports in (ii) and (iii)).

If the test piece is of rectangular cross-section, of width $= b$, and depth $= d$, and the force at break $= F$, the *flexural strength*

‘ in (i) $= 6\,FL_1/bd^2$

in (ii) $= 3\,FL_2/2bd^2$

and in (iii) $= 3\,FL_3/bd^2$.

See also Note on SI Units, page 659.

Flexural strength applies only to rigid materials, i.e. those that do not bend excessively under the load, and the formulae are derived on the assumptions (a) that the stress is proportional to strain right up to the breaking point or maximum stress, and (b) that shear stresses are absent, which is true only if d is very much less than L_1, L_2 or L_3. The results calculated as above are sometimes higher than the true maximum stress because assumption (a) is not valid, the upper part of the stress–strain curve being concave to the strain axis (cf. Section 75, Figure 75.2). Thus, flexural strength may be considerably greater than tensile strength* (see, e.g., data for ebonite in Section 31, Table 31.T4).

Young's modulus can be obtained from the slope of the *initial* part of the force/deflection (f/y) curve; e.g. for 3-point loading,

$$\text{Young's modulus} = \frac{fL_2^3}{4ybd^3}$$

REPEATED FLEXURE, OR BENDING FATIGUE

Dynamic flexure leads to lower breakdown values than those derived statically, the difference being attributed to internal friction, crystallisation and development of surface cracks. The *dynamic flexural strength*, or *fatigue strength*, of a plastics material is the maximum stress, developed alternately in tension and compression, that a specified test piece can withstand during a stated (large) number of bending cycles (e.g. at a frequency of 30 Hz); the *fatigue limit* is the corresponding value for a substantially infinite number of cycles; for details see ASTM D671 (test for repeated flexural stress (fatigue) of plastics).

In *flex-cracking* and *cut growth* tests on vulcanised rubbers, rapidly repeated bending causes development of surface cracks, or growth of a deliberately made cut, but the stresses set up are not measured; for details see BS 903: Parts A10 and A11. The so-called 'flexometer' fatigue tests do not involve bending, but only compression with or without shear.

COMPARISON TABLE

Table 76.T1 compares the flexural strengths of rigid polymers with those of some other rigid materials.

*Other factors may, however, contribute to this difference; see SCOTT, J. R., *Ebonite*, London, 146 (1958).

Table 76.T1. FLEXURAL STRENGTH OF SOME RIGID POLYMERS AT 20–25°C

Range for lower limit of Fl. Str.	Material: unfilled and unoriented unless otherwise stated (approximate flexural strength, in kgf/mm²)
Under 2	Polytetrafluoroethylene, low. Cyclised rubber, 1·1–6·5.
2–10	Ethylcellulose, 2–8·5. Cellulose acetate, 2·5–10. Polypropylene*, 2·9–4·9; glass filled, 5·6–7·75. Ebonite, 4–15. Cellulose nitrate, 5–10. Polyamides, 5–10. Polystyrene, 5–10·5; high-impact, 3·2–4·2. ABS plastics*, 5·7–8·0. *Wood, common kinds,* 6–10 (hickory, 13–14). Polycarbonates, 6–12. Chlorinated rubber, 7–10. Regenerated proteins, 7–12·5. Epoxy resins, 7–14. Polyvinyl butyral; amino-formaldehyde resins, *c.* 8. Polyformaldehyde, 8–12. Vinyl chloride/vinyl acetate copolymers, 8·5. Polyvinyl chloride (rigid); *porcelain*, 9·5. Cellulose (as vulcanised fibre); polyphenylene oxide, 10. Phenol-formaldehyde resins: cast, 10; filled mouldings, 6–10; laminates, 7–15.
Over 10	Polyacrylates, 10–14. Polyethylene: low density, 11–14; high density, 17–24. Unsaturated polyesters: cast, 11; glass-reinforced, 100. Polyvinylidene chloride, up to 12. Polyvinyl formal, *c.* 12. *Steatite*, 13. *Cast iron,* 30. Silicone resins, 55–70.

*Flexural yield strength

FURTHER LITERATURE

The following standards give details of flexural strength tests:
BS 771 (phenolic moulding materials)
BS 903: Part D4 (ebonite)
BS 903: Part E8 (cellular ebonite)
BS 2782: Method 304 (plastics)
BS 2572 (phenolic laminated sheet)
BS 2966 (phenolic resin/cotton sheet for electrical purposes)
ASTM D530 (hard vulcanised rubber; ebonite)
ASTM D790 (plastics)

COMPRESSION STRENGTH

DEFINITION

Compression strength or *crushing strength* is the maximum stress that a rigid material will withstand under longitudinal compression.

UNITS*

Compression strength is measured as force per unit area of the initial cross-section of the test piece, and is here expressed in kgf/mm². For conversion factors to other units, *see* under Units, Section 75.

NOTES

Compression strength is significant only in materials that shatter, crush or collapse under load; it is not necessarily the stress at the point where rupture begins. In compression, stress increases faster than the strain; however, the size and shape of the test piece, the rate of loading, and the presence or absence of lubrication at the end faces may profoundly influence the value obtained, which therefore is not a fundamental property of the material, but an arbitrary number specific to the test conditions used.

COMPARISON TABLE

Table 77.T1 compares the compression strengths of rigid polymers with those of some other rigid materials.

Table 77.T1. COMPRESSION STRENGTH OF SOME RIGID POLYMERS AT 20–25°C

Range for lower limit of Comp. Str.	Material: unfilled and unoriented unless otherwise stated (approximate compression strength, in kgf/mm²)
Under 2	Polytetrafluoroethylene, low. Polyethylene: low density, 1·1–1·7; high density, 1·8–3·0; high density cross-linked, 1·5–3·8.
2–10	Polyamides, 2–5. Cyclised rubber, 2·7–8. Polypropylene*, 3·9–4·6; glass-filled, 4·5–5·0. ABS plastics, 4·2–8. Ethylcellulose, 5–14. Cellulose acetate, 5–20. Polyvinyl chloride (rigid), 5·6–6·7. Polyvinylidene chloride, 6. Chlorinated rubber, 7. Ebonite, 7·5–8. Polycarbonates, 8. Polystyrene, 8–11; high impact, 5·6–11. Polyacrylates, 8–13. Chlorinated polyvinyl chloride*, 8–9.
10–20	Phenol-formaldehyde resins: cast, 10–20; filled mouldings, 10–25; laminates, 17–35. Polyformaldehyde, >12·6. Cellulose nitrate, 14–21. Cellulose (as vulcanised fibre), 15–20. Polyesters: cast, 15; moulded, 17; glass-reinforced, 50.
Over 20	Regenerated proteins, 20–35. *Aluminium and magnesium alloys*, 25–30. Amino-formaldehyde resins, up to 30. *Porcelain, cast iron, c.* 50. *Steatite*, 60–80. Glass, 70–100. Silicone resins, 80. *Steel*, >100.

*Yield stress

See also Notes on SI Units, page 659.

FURTHER LITERATURE

The following standards give details of test methods:
BS 771 (phenolic moulding materials)
BS 903: Part D3 (ebonite)
BS 903: Part E5 (cellular ebonite)
BS 2782: Methods 303 A, B, C (plastics)
ASTM D695 (rigid plastics)
ASTM D1621 (rigid cellular plastics)

SECTION 78

IMPACT STRENGTH

DEFINITION

Impact strength represents the 'toughness' or resistance of a rigid material to sudden application of a force, i.e. the strength at a very high rate of strain. It is conventionally assessed by determining the energy required to fracture a standard test piece under standard conditions.

UNITS*

Impact strength is generally expressed as either: (a) the energy of fracture (e.g. in kgf mm) of a test piece of stated dimensions; (b) the energy per unit *width* of the face of the test piece struck, e.g. as ft lbf/in width†, which is commonly used with *notched* test pieces (*see* Notes below), when impact strength is quoted as 'ft lbf/in of notch'—length of notch being, of course, equal to test piece width; or (c) the energy per unit *area* of the effective cross-section (i.e. between the notch and the opposite face in a notched test piece), e.g. kgf mm/mm².

Occasionally use is made of energy per unit *volume* of the stressed part of the test piece.

$$1 \text{ ft lbf} = 13{\cdot}825 \text{ kgf cm or } 138{\cdot}25 \text{ kgf mm}.$$

NOTES

The test piece may be either a plain (*unnotched*) bar, or *notched* by cutting a narrow groove across the middle of one face so as to concentrate the stress.

Shock loading, of either unnotched or notched test pieces, is obtained either by the swing of a bar pendulum or the vertical fall of a weight:

Pendulum tests In the *cantilever* or *Izod* test the bar-shaped test

---

See also Note on SI Units, page 659.

†The use of kgf mm/mm width would be more consistent with other metric units used in this book, but it is thought better to retain the more familiar ft lbf/in width; 1 ft lbf/in = 5·45 kgf mm/mm.

piece is clamped at one end as a cantilever (usually pointing vertically upwards), and the free portion is struck at a given distance from the clamp. In the *simple beam* or *Charpy* test the test piece is arranged horizontally, supported near each end, and struck at the centre. In both tests the pendulum incorporates a specified striking edge coincident with its centre of percussion, and is released from a fixed height so as to strike the test piece with a given velocity (e.g. 2·45 or 3·35 m/s) when at the bottom of its swing. The energy absorbed in fracturing the test piece is calculated from the difference between the initial down-swing of the pendulum and its over-swing; the result must be corrected for energy lost in friction, air-damping and the kinetic energy imparted to the broken portion ('flying fragment') of the test piece—these being measured by separate tests.

Falling-weight tests Here again either cantilever (this time horizontal) or simple beam tests can be made. However, in falling weight tests the only information obtained is that the test piece either breaks or does not, as the energy absorbed in breaking cannot readily be measured. 'Impact strength' is then defined as the impact blow that breaks just 50% of a large number of test pieces. As materials are never perfectly uniform, a blow *less* than this 'impact strength' will break some test pieces, and a blow *greater* will fail to break some; hence it is only possible to deduce approximately the 50%-breakage energy by tests with energies above and below this value (detailed procedures are given in the test method standards quoted below).

Variants of the falling weight test are used for sheet materials and films; the test piece is circular, supported round its edge, and struck in the centre by a falling weight (*see* ASTM D530 and D1709 in **Further Literature**).

The results of impact tests are only approximately similar for the pendulum and falling weight methods, or for the cantilever and simple beam forms of test. The results from notched test pieces are also dependent on the accuracy with which the stress-concentrating notch is cut, and on the shape of the cross-section of the notch, especially the radius at the bottom. Moreover, single-blow tests give no indication of the resistance of a specimen to repeated non-fracturing blows, by which some materials are soon rendered unserviceable.

Fibrous fillers, plasticisation, and molecular orientation tend to raise the impact strength, while cross-linkage and reduction of temperature usually lower it (polystyrene and specimens incorporating glass fibre increase in impact strength with fall of temperature). Edge-wise measurements on laminated materials give lower results than measurements normal to the plane of the laminations.

Pendulum machines, similar to those described for rigid bodies, are employed in testing rubbers, the rebound from the specimen providing a measure of the dynamic elasticity or rebound resilience (*see* Section 74).

COMPARISON TABLE

Table 78.T1 compares the impact strengths of rigid polymers with those of some other rigid materials; these values must be regarded as only approximate.

Table 78.T1. IMPACT STRENGTH OF SOME RIGID POLYMERS AT 20–25°C

Range for lower limit of Imp. Str.	Material: unfilled and unoriented unless otherwise stated (approximate impact strength, notched specimens, ft lbf/inch of notch)
Up to 0·5	Amino-formaldehyde resins, 0·2. Phenol-formaldehyde resins: cast, 0·25–0·5; filled mouldings, 0·1–2·5; laminates, 0·2–5. Polystyrene: 0·25–0·5; high impact, 0·8–1·7. Epoxy resins, 0·3–1. Polyesters: cast, 0·3; filled mouldings, 8; glass-reinforced, 16. Polymethyl methacrylate, 0·3–0·4. Poly-3,3-dichloromethyloxacyclobutane, 0·4. Polypropylene: 0·4–2·2; high impact, 1·5–12; glass filled, 0·5–2.
0·5–2	Ebonite, vinyl chloride/vinyl acetate copolymers, 0·5–1. Polyvinyl chloride (rigid): 0·8; 'impact modified', up to 15. Polyvinyl acetals; regenerated proteins, up to 1. Cellulose acetate, 1–7. Polysulphones, 1·3. Polyformaldehyde; chlorinated rubber, 1·5. *Porcelain, steatite, c.* 1·5. Polyphenylene oxide, 1·5–1·8. Polyamides, 1·5–4·5. Polyvinylidene chloride, c. 2.
2–3	Cellulose nitrate, ethylcellulose, 2–6. ABS plastics, 2·5–7·5. Polyethylene: high density, 2·5–20; low density, >20. Cyclised rubber, 3.
Over 3	Polytetrafluoroethylene, c. 4. Silicone resins, 4–7. Cellulose (as vulcanised fibre), up to 5. Polycarbonates, up to 16. *Aluminium, magnesium (cast)*,* 5–10. *Aluminium alloys*, c.* 20. *Steel (tough)*, c.* 100.

*Values not strictly comparable with those for organic polymers, but in general of high impact strength

FURTHER LITERATURE

ADAMS, C. H., *Bull. Amer. Soc. Test. Mat.,* No. 173, TP 102 (1951) (Izod impact tests on plastics)

HULSE, G., *SCI Monograph* No. 5, 157 (1959) (impact strength of tough plastics, and relation to other properties)

SPÄTH, W. and ROSNER, M. E., *Impact Testing of Materials,* London (1961)

ARENDS, C. B., *J. Appl. Polym. Sci.,* **9**, 3531 (1965) (mechanism of impact failure of plastics; impact testing)

SHOULBERG, R. H. and GOUZA, J. J., *SPE Journal,* **23**, 32 (Dec. 1967) (correlation and significance of various impact tests)

LEVER, A. E. and RHYS, J. A., *Properties and Testing of Plastics Materials,* 3rd edn, London, 65 (1968) (impact strength and testing)

MORRIS, A. C., *Plast. and Polym.,* **36**, 433 (1968) (relation of impact resistance of thermoplastics products to impact strength and processing and design factors)

KESKKULA, H. (Ed), *Polymer Modification of Rubbers and Plastics* (Applied Polymer Symposia No. 9), New York, 237 (1968) (TURLEY, S. G., effect of polymer structure on impact properties)

BOYER, R. F., *Pol. Engng and Sci.,* **8**, No. 3, 161 (1968) (complex relations of impact strength and glass temperature)

Standards
BS 903: Part D7 (ebonite)
BS 1322 (aminoplastic materials)
BS 2782: Methods 306 A–F (plastics; includes Izod and Charpy tests)
ASTM D256 (plastics; Izod and Charpy test procedures)
ASTM D530 (includes falling-ball test on ebonite sheet)
ASTM D758 (impact resistance of plastics above and below room temperature)
ASTM D1709 (falling dart test on polyethylene film)
ASTM D2444 (impact resistance of thermoplastics pipe by falling weight)

SECTION 79

HARDNESS

Hardness generally denotes the resistance of a material to local deformation, being measured as the resistance to penetration either by a loaded indentor (indentation hardness) or by a loaded sharp point moving over the surface (scratch hardness); these two forms of test are discussed below.

Since hardness in this sense is related to elastic modulus (*see below*), it increases with increasing cross-link density—so that fully cross-linked polymers (cured epoxy and melamine-formaldehyde resins; ebonite) are the hardest—and is reduced by plasticisers (cf. Table 79.T3 and Section 56).

'Rocker' hardness (*see below*) departs from this definition in that it depends largely on the damping properties of the test material. Measurement of 'hardness' by rebound resilience (*see* Section 74), as is done with metals, does not apply to high polymers, because their hardness and resilience are not directly related.

HARDNESS TESTS

Indentation hardness tests
Deformation under a loaded indentor may be *elastic* (reversible) and/or *plastic* (irreversible). Soft vulcanised rubbers and rubber-like plastics (e.g. plasticised polyvinyl chloride) are substantially elastic; hence the deformation (indentation) must be measured while the loaded indentor is pressing on the test piece. Hard plastics and ebonite show both types of deformation; hence ideally measurements should be made of the total (elastic + plastic) indentation under the load and the residual (plastic) indentation after removing the load, though the latter is not always done. The time of loading, test piece dimensions and temperature affect the result, as noted below.

The *indentor* is usually either (a) a ball or a hemispherical-ended plunger, e.g. in instruments measuring in International Rubber Hardness Degrees (IRHD; *see* ISO Recommendation R48), also in the Brinell, Rockwell, Pusey and Jones Plastometer, and Wallace Pocket instruments; (b) a truncated cone, e.g. Shore A Durometer; (c) a sharp-pointed cone, e.g. Shore D Durometer, or (d) a sharp pyramid, e.g. diamond pyramid and Knoop hardness.

The *indenting load** may be provided by either a dead weight (preferably) or a spring; the former is used in the 'IRHD' instruments and the Pusey and Jones Plastometer, and the latter in the Shore Durometers and the Wallace Pocket Meter. Dead-weight instruments sometimes apply a small 'minor' or 'contact' load, to give a definite zero position from which to measure the indentation produced by increasing the load to a much larger 'major' or 'total' load.

The *measurement* made is normally the depth of indentation or penetration of the indentor; for essentially plastic materials (*see* above) the measured quantity may be the diameter (or area) of the impression or permanent indentation left after removing the load or reducing it to the 'minor' value.

The test result is expressed in various ways as follows:

(*a*) Directly as depth of indentation, e.g. Pusey and Jones Plastometer, BS 903: Part D6 test on ebonite, BS 2782: Method 307A test on flexible PVC (there called 'softness number'), and 'ASTM hardness number' (ASTM D314). Here a high number denotes a soft material, and vice versa.

(*b*) As 'hardness degrees', such that 100 = infinitely hard and 0 = either extremely soft (Shore A Durometer) *or* 'infinitely' soft (IRHD). The IRHD value is derived from Young's modulus (E, kgf/mm^2) via a probit curve†, being defined as the percentage frequency corresponding to a probit value of $(5\cdot897 + 1\cdot428 \log E)$. The value of E is derived from its relation to the depth of indentation P (mm/100) by a ball of radius R (mm) under a load F (kgf), namely:

$$E = 263\ F/(P^{1\cdot35}\ R^{0\cdot65})$$

(*see* SCOTT, J. R., *Trans. Inst. Rubber Ind.*, **11**, 224 (1935)).
This relation applies for perfectly elastic isotropic materials; usually there is some departure from this ideal behaviour.

Except on very soft materials, IRHD and Shore A Durometer values are approximately the same.

(*c*) As force per unit area of the indentation (impression), e.g. Brinell, diamond pyramid, Knoop; sometimes the true (spherical or pyramidal) area is taken, or sometimes—as an adequate approximation—the projected area.

(*d*) On arbitrary scales derived from the depth of indentation, e.g. Rockwell, which can give even negative values of hardness!

Other features of indentation hardness tests. With relatively soft elastic materials, e.g. vulcanised rubbers, the result is influenced by

---

*'Force' is scientifically preferable, but 'load' has been used as a convenient and widely accepted term.

†The probit curve is derived from the 'normal' frequency distribution curve (representing the frequency y of occurrence of a value x) by integration with respect to x. It is a sigmoid curve extending from $y = 0$ at $x = -\infty$ to $y = 1$ (or 100%) at $x = +\infty$; the x co-ordinate is the 'probit', conventionally taken as = 5 at the mid-point of the curve.

the dimensions (especially thickness) of the test piece; the indentation usually increases somewhat with the time of loading, and may be influenced by temperature. Hence standard procedures define test piece dimensions, loading time, and temperature.

A *micro-test*, a scaled-down version of the normal IRHD test, has been developed for use on very small test pieces or finished articles.

Scratch hardness tests

These measure damage at fixed load, or load to cause a fixed amount of damage; lubricants, by reducing adhesion, can alter test values.

Mohs hardness is a qualitative scale of unequal intervals, based on increasing scratch resistance of minerals, with talc = 1 and diamond = 10 (*see* Table 79.T3). It is too broad for the present purpose, because all organic polymers are placed below 3 in Mohs hardness, i.e. they can be scratched with calcite; it serves, however, to place them in the same category as rock salt and 'finger nail', and shows that they are very much softer than glass or ceramics.

Koh-i-noor or *Faber-Castell pencil hardness*, obtained by using sharp-pointed graphite pencils of increasing hardness (6B, 5B, . . ., B, HB, F, H, 2H, . . ., 9H) until a definite scratch is produced, covers a much smaller range than the Mohs scale and is sometimes used to assess the hardness of surface coatings. The pencils must, of course, be reliably calibrated.

Martens hardness is obtained by measuring the width of a scratch produced by a loaded diamond point.

Rocker hardness tests

These tests, used mainly on organic coatings, measure 'hardness' by the damping of the oscillations of a pendulum supported on the test surface. The *Sward Hardness Rocker* is a circular pendulum, i.e. a ring resting on its edge and having its centre of gravity displaced from its geometrical centre so that it can rock to and fro; the test result is twice the number of oscillations required to reduce the amplitude from 22 degrees to 16 degrees, the instrument being adjusted so that the result for glass = 100 (*see* ASTM D2134; and BAKER, D. J. ELLEMAN, A. J. and MCKELVIE, A. N., *J. Oil Col. Chem. Assoc.*, **34**, 160–79 (1951), for theory of the rocker). The *Persoz* hardness pendulum rests on the test surface by two rigid balls, and rocks with a period of 1 second; the result is the time for the oscillation amplitude to decrease from 12 degrees to 4 degrees (PERSOZ, B., *Peint. Pig. Vernis*, **21**, No. 7, 194 (1945); French standard NF T30-016).

HARDNESS TEST APPARATUS

(The relation between the readings of some of these is shown in Table 79.T2.)

Dead-load Hardness Gauge*—Designed primarily for testing rubber, this is the best known apparatus working according to ISO Recommendation R48 and giving readings in IRHD. For the usual hardness range (30–95 IRHD) it uses a 2·5 mm diam. indenting ball, with dead loads of 30·5 gf (contact) and 580 gf (total). For very soft (10–35 IRHD) or very hard (85–100 IRHD)† rubbers the ball diameter is respectively 5 mm or 1 mm, but with the same loads.

Micro-hardness Gauge*—This scaled-down version of the above gauge has the ball diameter reduced to *c.* 1/6, i.e. 0·395 mm, and the loads to *c.* $(1/6)^2$, i.e. 0·85 gf (contact) and 15·7 gf (total). It is intended for the range 30–95 IRHD.

Pocket Hardness Meter*—This 'pocket' size instrument, with a hemispherical-ended plunger and approximately constant spring loading, reads in IRHD and covers the range 30–100 IRHD.

Shore Durometers‡—Several models—likewise 'pocket' size—are available for different hardness ranges. Most used are: Type A (or A2), with small frustoconical indentor and variable spring loading, i.e. decreasing as indentation increases; readings are in 'Shore A Degrees' (approx. = IRHD over the range 30–100); Type D, with sharp conical indentor and heavier (variable) spring loading, giving a more open scale (*c.* 30–100 corresponding to 80–100 Shore A) and hence suitable for semi-rigid materials.

Plastometer§— This uses a $\frac{1}{8}$ in (usually) or $\frac{1}{4}$ in diam. ball with dead-loads of 85 gf (contact) and 1085 gf (total); it reads directly as depth of indentation (mm/100). Initially designed to test rubber-covered rollers, it has ball-jointed feet to rest on curved surfaces.

Brinell Test—Brinell hardness is measured by impressing a hard steel ball into the test piece for a given time, the hardness (in the original form of the test) being expressed as the ratio of the applied load to the *true* area of the indentation, i.e.

$$\text{Brinell hardness} = \frac{2P}{\pi D\{D - \sqrt{(D^2 - d^2)}\}},$$

where P = load (kgf), D = diameter of ball (mm), and d = diameter (mm) of the impression measured immediately after removing the load. The units are kgf/mm^2, but the result is often quoted simply as a number; it increases with the true hardness of the material. Originally designed for metals, the Brinell test is suitable only for relatively rigid plastics and for ebonite. For these D is commonly 5 mm and P is 50 kgf, the result being expressed as ratio of load to the *projected* area of the impression, i.e. as P/Dh, where h = depth of the impression

*H. W. Wallace and Co. Ltd., Croydon, England.
†In the ranges of overlap (30–35 and 85–95) the modified tests for soft and hard rubbers respectively should be used.
‡Shore Instrument and Mfg. Co. Inc., Jamaica, N.Y., U.S.A.
§The Pusey and Jones Co., Wilmington, Del., U.S.A.

(mm) measured with the load acting (this is sometimes called *Meyer hardness*).

Rockwell Test—This measures hardness by the indentation produced by a loaded ball. The indentation is measured *either* (i) after increasing the load from a 'minor' to a 'major' one and then reducing it to the 'minor' value, thus measuring the residual (plastic) deformation (procedure for R, L, M, E and K scales, Table 79.T1); or (ii) after increasing the load from 'minor' to 'major', the reading being taken with the latter acting, so as to measure total deformation (α scale). In either case the sequence of operations is carried out to a fixed time-table, and the position of the ball under the 'minor' load (10 kgf) is taken as zero penetration.

The test is suitable only for relatively rigid materials; the Rockwell number, which increases with the true hardness, is designated by a letter denoting the ball size and 'major' load, and a number expressing hardness inversely as the depth of the indentation (h), i.e. in the form: (constant $- h$).

Table 79.T1. ROCKWELL HARDNESS*

Scale designation letter	Major load (kgf)	Ball diameter (in)	Time (s)
R	60	0·5	Between applying minor and major loads: 10 max.
L	60	0·25	Action of major load: 15
M	100	0·25	Between reducing load to minor and reading indentation:
E	100	0·125	15
K	150	0·125	
α	60	0·5	Between applying minor and major loads: 10 max.
			Between applying major load and reading indentation: 15

*ASTM D785, Test for Rockwell hardness of plastics and electrical insulating materials

Diamond pyramid hardness (DPH)—Obtained with a *Vickers* (or similar) machine, is measured as the ratio of the applied load (kgf) to the true surface area (mm^2) of the indentation made by a standard diamond pyramid indentor; it is employed when a Brinell steel ball is insufficiently hard for the material to be tested, but over the range covered by organic polymers the results are similar to those for Brinell hardness.

Knoop hardness number (KHN)—This is the ratio of the load (kgf) to the projected area (mm^2) of the rhomb-shaped indentation formed by a loaded diamond pyramid, the area being measured immediately after removing the load; it is particularly applicable to thin sheets and surface coatings, and is conveniently measured by the *Tukon Micro-hardness Tester*.

Pfund indentation hardness—This equals $1·27/d^2$, where d is the

617

diameter (mm) of the indentation produced by a 0·25 in diameter rigid ball, measured whilst the load (1 kgf) is still acting.

(For Knoop and Pfund tests *see* ASTM D1474.)

COMPARISON OF HARDNESS MEASUREMENTS

Relation between Rubber Hardness Scales—Table 79.T2 shows, on the same horizontal line, equivalent readings on the scales most used for vulcanised rubbers and rubber-like materials. The equivalence must be regarded as approximate, especially with very hard materials that show more or less plastic deformation.

Table 79.T2. RUBBER HARDNESS SCALES

	*IRHD and Shore A Durometer**	*Shore D Durometer*	*Pusey and Jones Plastometer*	
			$\frac{1}{8}$ inch *ball*	$\frac{1}{4}$ inch *ball*
↑	100	100	0	0
	98	60	—	—
hard	95	50	14	10
	90	40	27	20
	80	30	48	35
soft	70	22	68	50
	60	16	92	67
	50	12	125	90
	40	9	170	125
↓	30	7	260	185

*These are the same down to about 30; below this the Shore reading is slightly lower

COMPARISON TABLE

Table 79.T3 compares the hardness of common polymers at room temperature, and shows the relations between the Mohs, Brinell, Rockwell, Shore D Durometer and IRHD (approx. = Shore A) hardness values. The different materials and scales are not strictly comparable, and the values shown are to be considered only as approximate; more detailed values are given under individual polymers in Sections 1–42.

Table 79.T3. HARDNESS OF COMMON POLYMERS AND RELATED SUBSTANCES [†]

Mohs scale	Brinell hardness, kgf/mm²	Rockwell hardness		Shore Durometer D	IRHD	Material (grouped by average hardness)
		M scale	R scale			
					down to c. 10	Very soft vulcanised rubbers, e.g. printing rollers.
					30–45	Vulcanised rubbers, unfilled.
					40–50	Unsaturated polyester resins (flexible).
					55–70	Vulcanised rubbers, tyre-tread type; polyvinyl chloride (plasticised).
				20–50	65–95	Vulcanised rubbers, flooring and sole/heel types; polyvinyl chloride (plasticised).
			c. 25+	60+	98–100	'Hard rubbers' (pseudo-ebonites; Section 31).
	1–2		20–35	50–70	c. 97	Polyethylene, polybutene-1, polytetrafluoroethylene, fluorinated ethylene-propylene.
			30–95			Polypropylene (high impact).
		40–60				Polystyrene (high impact).
1. Talc	4–5					*Lead.*
	6–15	50–80	50–120			Cellulose (vulcanised fibre), cellulose acetate and nitrate, ethylcellulose, polyamides (nylon), coumarone/indene resins (hard), polyvinylidene chloride, phenol-formaldehyde resins (novolac), polypropylene (general-purpose, and filled), polystyrene (general purpose), ABS plastics, nylon 6.6, polycarbonates.
	10–20	80–90	110–120	75–95	c. 100	Ebonite, chlorinated rubber, polyvinyl formal, polyvinyl chloride (unplasticised), vinyl chloride/do. acetate copolymers.
	20–25	70–115	c. 125	80–85	c. 100	Regenerated protein (formalised casein), polyester and epoxy resins, polyformaldehyde, styrene/acrylonitrile copolymer (SAN).
2. Gypsum and rock salt	c. 25	80–120				Polymethyl methacrylate, polyacrylonitrile. *Aluminium* (alloys up to 150 Brinell).
Approx. 2–2·5 'Fingernail'	30–50	90–125				Aniline-, urea- and phenol-formaldehyde resins (fully cured); melamine-formaldehyde resins.
	c. 40					*Zinc.*
3. Calcite 3–3·5	c. 50	130				*Brass (soft to hard). Soft iron.*
	80–100					*Soft steel.*

Mohs scale	Brinell hardness kgf/mm²	Rockwell hardness		Shore Duro-meter D	IRHD	Material (grouped by average hardness)
		M scale	R scale			
4. Fluorspar	150–190					Cast iron (malleable). Wrought iron. Mild steel, stainless steel.
	170					Iridium.
5. Apatite	c. 200					
6. Felspar Approx. 6–6·5 'Penknife blade'	c. 340					Common glass‡, tool steels§. Borosilicate glass (Pyrex).
7. Quartz	c. 500					Porcelain, steatite. Agate, hard flint. Tool steels‖.
8. Topaz						Silicon carbide, tungsten carbide.
9. Corundum						
10. Diamond						Boron carbide.

†It is important to note that increasing hardness is *not* represented linearly in descending this table, e.g. the interval between 9 and 10 on the Mohs scale is much *larger* than that between 1 and 2, and not *smaller* as set out for the present purpose.
‡Common window glass, including plate glass.
§E.g. carbon steel woodworking tools.
‖E.g. high-speed steel milling cutters and hobs; 'file-hard' steel balls, as used for Brinell testing.

FURTHER LITERATURE

LYSAGHT, V. E., *Bull. Amer. Soc. Test. Mat.* No. 138, 39 (1946) (Knoop indentation of non-metallic materials)

BOOR, L., RYAN, J. D., MARKS, M. E. and BARTOE, W. F., *Proc. Amer. Soc. Test. Mat.* **47**, 1017 (1947), and *Bull. Amer. Soc. Test. Mat.* No. 145, 68 (1947) (hardness and abrasion resistance of plastics)

BERNHARDT, E. C., *Mod. Plast.* 26, 123, 174, 186 (Oct. 1948) (scratch resistance of plastics)

LYSAGHT, V. E., *Indentation Hardness Testing*, New York (1949) (pp. 219–234, hardness testing of plastics; the rest relates mainly to metals and machines)

SODEN, A. L., *A Practical Manual of Rubber Hardness Testing*, London (1951)

TABOR, D., *The Hardness of Metals*, Oxford (1951) (includes various hardness tests, but relates mainly to metals)

PARTINGTON, J. R., *An Advanced Treatise on Physical Chemistry*, Vol. 3, London, 231 (1952) (theory of hardness of solids)

SCOTT, J. R., *Trans. Inst. Rubber Ind.*, **11**, 224 (1935) (theory of indentation tests on rubber)

SCOTT, J. R., *J. Rubber Res.*, **17**, 145 (1948); **18**, 12 (1949) (basis of IRHD scale—since modified by I.S.O.—*see* below)

SCOTT, J. R. and SODEN, A. L., *Trans. Inst. Rubber Ind.*, **36**, 1 (1960) (micro-tests on rubber)

SCOTT, J. R., *Physical Testing of Rubbers*, London (1965), pp. 91–106 (tests on soft rubbers), pp. 213–4 (tests on ebonite), pp. 240–1 (tests on cellular rubbers)

FINK-JENSEN, P., *Hardness Testing of Organic Coatings*, London (1966)

LEVER, A. E. and RHYS, J. A., *The Properties and Testing of Plastics Materials*, 3rd edn, London, 89 (1968) (comprehensive list, with notes, of hardness tests)

SCHMITZ, J. V. and BROWN, W. E. (Eds), *Testing of Polymers*, Vol. 3, New York, London, Sydney, 111 (1967) (LIVINGSTON, D. I., indentation hardness testing)

Standards
BS 240 (Brinell hardness testing; relates mainly to metals)
BS 427 (Vickers hardness, mainly of metals)
BS 891 (Rockwell hardness testing; relates mainly to metals)
BS 860 (approximate correlation of hardness scales; relates mainly to metals)
BS 903: Part A26 (tests on soft vulcanised rubber)*
BS 903: Part D6 (indentation and recovery number of ebonite)
BS 2719 ('pocket' rubber hardness meters, methods of use and calibration)
BS 2782: Method 307A (softness number of flexible PVC)
ASTM D314 (ASTM hardness of rubber)
ASTM D530 (testing hard rubber; includes Rockwell hardness)
ASTM D785 (Rockwell hardness of plastics)
ASTM E18 (Rockwell hardness of metallic materials)
ASTM D531 (Pusey and Jones plastometer for indentation of rubber)
ASTM D1415 (international hardness of vulcanised rubbers)*
ASTM D1484 (penetration of hard rubber by type D durometer)
ASTM D2240 (durometer hardness of plastics and rubber)
ISO Recommendation R48, *Determination of Hardness of Vulcanised Rubbers,* 2nd edn, Geneva, 1968.

*Based on ISO Recommendation R48.

FRICTION AND ABRASION

FRICTION

Friction, or resistance to sliding motion, is an important property in several aspects of engineering, and, in relation to high polymers, features particularly with rubbers, certain plastics (polytetrafluoro-ethylene, polyamides), the 'blocking'* of thin films, and in a number of textile operations, e.g. spinning, weaving, raising, sewing, etc.

Friction is also a factor in the performance of textiles; thus, for instance, the crease recovery, etc., of cellulosic fabrics, which is improved by cross-linkage (*see* §13.86), is often further enhanced by the presence of a friction-reducing agent, e.g. polyethylene applied initially from aq. dispersion.

The frictional force, i.e. the force opposing sliding, is often found to be approximately independent of the area of contact but proportional to the applied load, it is therefore conventionally expressed as

$$F = \mu R$$

where F is the frictional force, R is the load, i.e. the normal component of the force between the two surfaces, and μ is a coefficient representing the frictional force per unit load.

From some systems the static coefficient of friction, μ_{stat} (or the friction at very low velocity of sliding) may be higher than the dynamic coefficient of friction, μ_{dyn}, which is effective when the relative velocity of the surfaces is high.

However, even for a given combination of materials, μ frequently shows large variations for comparatively small changes in such factors as: (i) the flatness and surface finish of the contacting bodies, (ii) the relative velocity between the bodies, (iii) the presence of substances that either increase the friction or act as lubricants (*see below*), (iv) the accumulation of wear debris between the surfaces, and (v) the temperature (*see below*), humidity, load, tension and twist (of yarns), and similar technological features. Therefore, the value of μ should not be quoted (as it often has been) without specifying the conditions under which the coefficient is measured.

---

*Cohesion, often due to electrostatic charges, but resembling friction in preventing easy sliding of one film over another.

Substances that *increase* friction, whether used as additives or superficially, include rosin, coumarone/indene resins, and finely divided silica; a rubber coating can be used for the same purpose. Friction is *reduced* by lubricants, e.g. oils, silicone fluids, certain waxes, molybdenum disulphide, French chalk, or certain surface coatings such as polyethylene (above) and polytetrafluoroethylene; liquid or waxy lubricants may be used as additives which exude to the surface. A reduction in hydrodynamic friction can be brought about by the presence of trace amounts of water-soluble polymers such as guar gum (§19.83) and polyethylene oxide (§26.82).

The effect of *temperature* depends on the particular material, and also the temperature range involved, as the coefficient of friction may pass through a maximum and/or minimum.

There are two basic theories of friction. Both depend on the assumption that the area of true contact is normally much smaller than the apparent area, and is then proportional to (load)^n, where n is usually between 0·6 and 1, depending on the geometry of the surfaces and whether they deform plastically or elastically. The first theory assumes that the frictional force arises from the hysteresis energy required to stretch small portions of one surface when they are temporarily attached to the other surface. With rubbers there is often a close correspondence between the friction/velocity curve and the hysteresis/frequency curve if it is assumed that a frequency of the order of 10^6 Hz corresponds to a velocity of the order of 10 mm/s.

The second theory, which is probably more appropriate to relatively rigid materials, assumes that when surfaces are brought together, the high pressure developed over the small areas that come into intimate contact causes them to adhere or weld together; friction is then interpreted as the force required to shear the interface or the component constituting the softer (i.e. weaker) side of it. Thus, the friction of relatively soft thermoplastics, sliding on glass or hard metal, is usually dependent on the cohesive or shear strength of the polymer, whereas when sliding on softer materials, friction is largely governed by the mechanical properties of the softer phase. On this basis it has been shown (by KING and TABOR, *Proc. Phys. Soc. Lond.,* 66B, 728 (1953)) that, for certain plastics, the coefficient of friction is approximately equivalent to the ratio of shear strength to hardness.

Although most plastics are unexceptional, with the value of μ lying approximately between 0·2 to 0·8 and not greatly affected by common lubricants, soft rubbers can exhibit exceptionally high coefficients of friction (μ up to 4 or more), and polymers such as nylon usually show low ones, e.g. μ against steel, 0·15. Polytetrafluoroethylene is unique in showing exceptionally low coefficients of friction in almost any combination (μ down to 0·02), irrespective of the presence or absence of possible lubricants, and this material has been successfully employed in low-friction bearings. Animal hair, and sheep's wool in particular, is remarkable because it possesses a lower coefficient of friction in a root-to-tip direction than from tip-to-root (for example, *see* §20.34).

A useful technique, known as friction welding, which is dependent upon the friction and poor thermal conductivity of certain thermoplastics, consists in first rubbing the faces or ends at high speed, until soft, and then bonding them together under moderate pressure.

The coefficient of friction is commonly determined by: (i) measuring the force required to drag a block of the one material over a plane made of (or covered with) the other, the two being pressed together by a force normal to the plane of contact (*see*, e.g. JAMES, D. I., *J. Sci. Instr.*, 38, 294 (1961)); or (ii) increasing the angle of tilt of the plane ti!! the block slides down it. Special instruments are used for textile yarns, etc. (e.g. 'Shirley' Yarn Friction Recorder and 'Shirley' General Purpose Yarn Friction Tester; Shirley Developments Ltd., Manchester).

'*Rolling friction*' is the resistance (equivalent to a frictional drag) to the rolling of a sphere or torus (e.g. tyre) along a plane; the mechanism of rolling friction differs from that described above, in being due primarily to hysteresis loss in the rolling member where it is deformed by contact with the plane. For a method of measurement *see* FLOM, D.G., *J. Appl. Phys.*, 31, 306–14 (1960).

As the coefficient of friction for a given material varies widely, according to the test conditions and other factors, it is not feasible to give a comparison table. Data for individual materials will be found in §§1.34–42.34.

ABRASION

Abrasion, the wearing away or 'wear' of surfaces, is related to— and is as important industrially as—friction. As a complex property, however, it proves difficult to analyse and measure, e.g. although several machines have been devised for the accelerated abrasion testing of textiles, plastics and rubbers, none gives results entirely in agreement with the performance observed in service. Therefore they can be used only for comparison purposes between similar materials under particular conditions of abrasion, rather than to obtain absolute values.

In textiles the abrasion resistance or wearing quality of viscose rayon and cellulose acetate fibres is, in general, inferior to that of cotton, which in turn is much inferior to nylon. In plastics, abrasion or wear resistance is important in such items as bearings and gear teeth. In this connection, polymers such as nylon and polyethylene (which show low frictional properties) exhibit high abrasion resistance against smooth metals and against abrasives of small particle size; against themselves, however, they are less commendable, since their low thermal conductivity can easily lead to local fusion and subsequent breakdown, cf. friction welding, mentioned earlier (running nylon rope through a loop of the same material is to be avoided for this reason). In glossy plastics, the ability of the surface to withstand the impact of grit is known as *mar resistance**; most thermoplastics, such as cellulose acetate and polystyrene, are poor in this respect, but in some of the cross-linked polyesters mar resistance has been found to have a high value (for standard test *see* **Further Literature**).

Abrasion of rubbers, notably that of tyre treads, necessarily receives considerable attention, yet no theory to account for the marked improvement effected by incorporation of carbon blacks is entirely acceptable. The observation that unvulcanised crepe rubber (§29.21) subjected to a stream of abrasive particles—e.g. when used as lining of conveyer shutes—resists abrasion better than the best carbon black reinforced vulcanisates is an example of the fact that under these conditions the highest wear resistance is shown by materials with low elastic modulus and high elasticity (i.e. recovery).

Several investigators have attempted to separate the effects of the various mechanisms of abrasion, such as (*a*) micro-cutting or scratching on sharp surfaces, (*b*) plastic deformation occurring progressively in the same place on blunt projections (not with rubbers), (*c*) fatigue failure caused by repeated elastic deformation on blunt projections (rubbers; this mechanism may be severe in an oxidative atmosphere), (*d*) strong adhesion between the surfaces, resulting in welding and pieces being torn out (with rubbers this mechanism is catastrophic, resulting in the formation of rolls of debris and an abrasion pattern of parallel ridges).

In laboratory *abrasion tests* for rubbers and plastics the test material is abraded by either a loose (granular) abradant, abrasive paper or cloth, a bonded abrasive wheel, or knives. However, even the most sophisticated of these has so far failed to correlate satisfactorily with service wear, if only because 'service' covers a great variety of conditions of wear, and the relative behaviour of different materials depends on these conditions. Hence, laboratory abrasion tests are useful chiefly for comparing similar materials intended for one type of service (e.g. tyre tread rubbers made from the same type of raw rubber), as a pointer to materials with improved abrasion resistance. For this reason no Comparison Table for abrasion resistance has been given.

---

*Important for instance in stoving enamel finishes on sheet metal (e.g. car bodies), also in aircraft windows to retain maximum visibility and contrast (as opposed to diffusion of light by minute scratches).

FURTHER LITERATURE

I. *Friction*

BUCKLE, H. and POLLITT, J., *J. Text. Inst.*, **39**, T199 (1948) (instrument for measuring friction of textile yarns against other materials)

BOWDEN, F. P. and TABOR, D., *The Friction and Lubrication of Solids*, Oxford, 164 (1954) (friction of polymers; the rest relates mainly to metals)

RUBENSTEIN, C., *Wear*, **2**, 296 (1958/9) (factors in friction of fibres, yarns and fabrics); *J. Appl. Phys.*, **32**, No. 8, 1445 (1961) (lubrication of polymers)

CONANT, F. S. and LISKA, J. W., *Rubber Chem. and Tech. (Rubber Reviews)*, **33**, No. 5, 1218 (1960) (friction of rubber-like materials)

HOWELL, H. G., MIESZKIS, K. W. and TABOR, D., *Friction in Textiles*, London (1959)

JAMES, D. I., NORMAN, R. H. and PAYNE, A. R., *SCI Monograph* No. 5, 233 (friction and dynamic properties of PVC)

HUFFINGTON, J. D., *Research*, **12**, No. 10/11, 443 (1959) (adhesion theory of friction of polymers)

BULGIN, D., HUBBARD, G. D. and WALTERS, M. H., *Proc. 4th Rubber Tech. Conf.*, London, 173 (1962) (relation of friction to hysteresis in rubbers)

PASCOE, M. W. and TABOR, D., *Proc. Roy. Soc.*, A235, April 24, 210 (1956) (friction and deformation of polymers)

ASTM E303 (measurement of frictional properties)

II. *Abrasion*

CLEGG, G. G., *J. Text. Inst.*, **40**, T449 (1949) (microscopic examination of worn textile articles)

JAMES, D. I. (Ed) and JOLLEY, M. E. (Transl.), *Abrasion of Rubber*, London (1967) (collected Russian papers)

BULGIN, D. and WALTERS, M. H., *Proc. Internat. Rubber Conf.*, Brighton, 445 (1967) (abrasion of elastomers under laboratory and service conditions)

DAVIES, G. R., *J. Ag. Eng. Res.*, **12**, 55 (1967) (abrasion tests on plastics, elastomers and ferrous metals)

HARPER, F. C., *Wear.* **4**, No. 6, 461 (1961) (wear of floorings; test methods)

BATEMAN, L. (Ed), *The Chemistry and Physics of Rubber-like Substances*, London, 355 (1963) (SCHALLAMACH, A., abrasion and tyre wear)

BS 903: Part A9 (vulcanised rubber: includes methods of testing with Akron, Dupont, and Dunlop (Lambourn) abrasion machines)

BS Handbook No. 11, *Methods of Test for Textiles* (includes determination of abrasion resistance by means of wear test machines of the Martindale and Linen Industry Research Association type)

ASTM D394 (vulcanised rubber; test with Dupont machine)

ASTM D1044 (surface abrasion resistance of transparent plastics)

ASTM D1242 (abrasion resistance of plastics)

ASTM D1630 (abrasion resistance of rubber soles and heels)

ASTM D673 (mar resistance of plastics surfaces)

III. *Friction and Abrasion*

BOWDEN, F. P. and TABOR, D., *Brit. J. Appl. Phys.*, **17**, 1527 (1966) (friction, lubrication and wear: survey of work during the last decade)

KRAGELSKII, I. V. (Transl., RONSON, L. and LANCASTER, J. K.), *Friction and Wear*, London (1965) (general discussion of mechanisms and theories)

SCHALLAMACH, A., *Rubber Chem. and Tech., (Rubber Reviews)*, **41**, No. 1, 209 (1968) (rubber friction and tyre wear); *Wear*, **1**, No. 5, 384 (1957/8) (friction and abrasion of rubber)

TABOR, D., *Wear*, **1**, No. 1, 5 (1957/8) (friction, lubrication and wear of synthetic fibres)

CHAPMAN, J. A., PASCOE, M. W. and TABOR, D., *J. Text. Inst.*, **46**, No. 1, p. 3 (1955) (friction and wear of fibres)

ASTM D2714 (friction and wear test machine)

FABRICATION

The numerous and diverse end-uses of high polymers must be nearly rivalled by the ways used to fashion them into the practical forms dictated by their properties and function. As a consequence, it is possible here to survey only briefly the many techniques employed— some of which have become highly sophisticated—and for elaboration of individual aspects reference must be made to more detailed works (e.g. as listed at the end of this section).

Below are summarised in turn processes for the production of *fibres* and *films*, and techniques used for the fabrication of *plastics* and *rubbers* (including moulding, extrusion and related processes, and manipulation of sheets, rods, and tubes). Additional data relating to fabrication (including brief notes on moulding requirements, etc.) appear in §§1.4–42.4 and elsewhere.

81.1 PRODUCTION OF FIBRES

Fibres are commonly *spun* by extrusion of a viscous liquid through a *spinneret* (a die-plate pierced with fine holes, cf. those of the silk moth caterpillar and spiders) with subsequent solidification and stretching of the continuous filaments so produced. Cutting to staple fibres, when required, is a later operation (*see below*); monofilaments, made in the same way as finer filaments, similarly may be cut for use as bristles.

Three main methods by which fibres are produced are as follows.

Wet spinning A viscous solution of a polymer of high molecular weight, or of an intermediate, is extruded through a spinneret into a bath that causes precipitation, thus coagulating the filaments, either because of dilution of the solvent or because of chemical reactions (reaction spinning).

Dry spinning A viscous solution of a polymer in a volatile solvent is extruded through a spinneret into a counter-current of warm gas (commonly air, but a hot inert gas if the solvent is of limited volatility), which removes the solvent and solidifies the filaments.

Melt spinning A thermoplastic polymer in the molten state is extruded through a spinneret into a cooler region (e.g. cold air) in which the filaments solidify.

In each method the extruded filaments are continuously withdrawn at a rate faster than that at which they are extruded, which causes them to become attenuated (in the still fluid region near the spinneret). They are then *drawn* (stretched, in the solid state) several times their length, either at room temperature (*cold drawing*) or above (i.e. above T_g, *see* Section 68), whereby the chain molecules become extended and highly oriented along the fibre axis, and tensile strength is correspondingly much increased*. Continuous filament yarns finally receive a lubricating finish and if necessary an antistatic treatment, respectively to prevent chafing (and to assist yarn cohesion) and electrostatic troubles (*see* Section 69). Rubber threads are either cut from thin sheet wrapped on a drum or turned from a block of laminated sheets (for square-section threads), or spun from compounded latex (round threads); threads of polyurethane rubbers have been produced by reaction spinning (*see* earlier).

Other methods of making fibres include: fibrillation of highly oriented thin films, notably those of polypropylene; sintering of an extruded dispersion of a refractory polymer, notably polytetrafluoroethylene; interfacial techniques (§54.13 v), a form of reaction spinning where the product obtained by reaction at an interface at room temperature may be continuously removed and stretched; melt-spinning from a perforated centrifuge; the action of jets of gas or steam on a molten polymer; mechanical separation of a viscous melt, e.g. as effected between rotating cylinders. Extrusion of two different melts through a compound spinneret can yield bi-component (or conjugate) fibres, e.g. with the components side by side (so that—at a later stage—permanent crimping is effected by differential thermal contraction) or with a non-soiling envelope over a stronger core.

Chemical treatment to modify certain physical properties—such as to improve elastic recovery—is usually carried out after spinning and weaving, i.e. as 'resin-finishing' of fabrics (*see* cross-linked cellulose, §13.86); fabrics finished in this way show, for instance, either better *crease recovery* (recovery to an uncreased state) or better *crease retention* (recovery of folded pleats or crease-lines) depending on whether treated when in the flat or folded state.

81.2 PRODUCTION OF FILMS

Thin self-supporting films can be made by methods related to those for fibres (*see* above), the fine holes of a spinneret being replaced by either a linear or a circular slit.

---

*Filaments gathered together in massive form as tow, which is subsequently cut, usually receive less stretching, and correspondingly the staple fibres derived from them are less strong but more extensible. Special techniques enable tapered bristles to be obtained from monofilaments.

Thus, by the extrusion of viscose through a slit into an appropriate coagulating bath, cellulose is regenerated in the form of a thin film (Cellophane). Similarly, a thermoplastic material—such as poly-ethylene—may be extruded through a linear slot die to yield flat films, while extrusion through an annular slot (tube die) yields tubular films. Extruded films may be employed in their initial state, but more often they are first stretched and oriented along their length (to become strong in that direction) or stretched both length-wise and width-wise, which produces *biaxial orientation* (films may then be both very thin but very strong along and across their length). With tubular films both operations can be carried out simultaneously and continuously, by introducing compressed air to *blow* the tube in the form of an elongated 'bubble'; the blown products are either slit and opened as flat films, or used directly for bags, sacks, etc.

Other means of obtaining thin films include extrusion on to a polished chill roller, or casting from solution on to a polished bed-plate or endless metal band, or precision calendering of a polymer (or compounded polymer) in the plastic state; any such films may subsequently be hot-drawn several times their length and then expanded laterally on a stenter (yielding products similar to those obtained by extrusion bubble-blowing). Not infrequently films are surface-coated, or different types are laminated together, to obtain particular properties. Incorporation of a blowing agent allows a polymer such as polystyrene to be extruded as films of the expanded (cellular) material, i.e. very thin and of low density.

Welding or sealing of films is important, e.g. for containers made from tubular films; for this, and for films made by paste techniques, also supported films (surface coatings), *see* later.

81.3 FABRICATION OF PLASTICS

I. MOULDING, EXTRUSION AND RELATED PROCESSES

Conversion factors:

To convert °C to °F multiply by 1·8 and add 32.

To convert lbf/in² to tonf/in² multiply by 0·00045.

To convert lbf/in² to kgf/mm² multiply by 0·0007.

See also **Note on SI Units,** page 659.

MOULDING (HIGH PRESSURE), EXTRUSION AND RELATED TECHNIQUES

Pressure moulding is effected (i) by direct compression of a charge in a heated mould, or (ii) by forcing a charge, already in a hot and plastic state, into a closed mould.

Compression moulding, the oldest process, is used mainly for thermosetting polymers. A mould (which may be hot initially, or is heated subsequently) is charged with a predetermined amount of moulding powder, which is first held under low pressure until it becomes plastic, then pressed at maximum pressure, so that the mould is completely filled and a slight excess ('flash') escapes, and both heat and pressure are maintained while the thermosetting reaction ('cure') takes place. Unless the mouldings are very thin, it is usual to pre-dry and pre-heat the moulding powder (commonly pelleted or pre-moulded), using an oven, infra-red radiation, or a high frequency electric field. In *transfer moulding*, a variation of compression moulding, a moulding powder is charged into a heated chamber adjacent to the mould proper, and when the charge becomes plastic it is forced through a narrow heated gate into the closed hot mould (or moulds); in this method the flow of the charge is better than in compression moulding, and for mouldings of thick cross-section the moulding cycle is shorter.

Injection moulding applies mainly to thermoplastic polymers. Granular moulding material is pre-dried and fed into a heated cylinder where it attains the plastic state; then by means of either a ram or a revolving (and usually reciprocating) screw producing a high pressure, it is rapidly injected through a nozzle into a closed mould, from which the moulding is removed after only a very short interval (thus, though usually heated to facilitate flow of the injected material, the mould needs to be cool enough to permit removal of the final product without causing its distortion). The moulding cycle is rapid, and automatic injection moulding machines can operate at a rate of several mouldings or sets of multi-impression mouldings (sprays) per minute. Unlike the 'flash' from compression moulding of thermosetting polymers*, the feed stalk ('sprue') and other waste portions, also imperfect mouldings, can be re-worked.

Compression and transfer moulding usually involve pressures of 2000–4000 lbf/in^2 (1·4–2·8 kgf/mm^2) of moulding cross-section, and temperatures of 125–175°C. The platens of the presses are heated electrically or by high pressure steam, and can usually be cooled by circulation of cold water. In injection moulding it is advantageous to use pressures of 5000–30000 lbf/in^2 (3·5–21 kgf/mm^2) with cylinder temperatures of 140–300°C† and mould temperatures of 20–135°C. The moulds, necessarily strong, are precision made in die-steel finished with a high polish or chromium surfaced, and though expensive initially are good for many thousands of mouldings.

Laminated boards are produced in high-pressure equipment allied to that used for moulding. They are composed of paper or fabric impregnated with a resin intermediate (e.g. of the phenol-formaldehyde type) then stacked in the requisite number of sheets and subjected to 'cure' between hot polished platens under high pressure; optimum

---

*Although thermoset flash cannot be re-worked, it is sometimes used as a filler.
†Up to 450°C for fluorinated ethylene/propylene copolymers (§3.81).

strength is obtained when the resin content of the board is such that all interstices ('voids') in the assembly are just filled.

Annealing and *sintering* resemble the corresponding metallurgical processes. The first consists in continued heating after moulding, usually out of the mould, in an oven or by high frequency induction, the purpose being to improve the physical characteristics; it is less important with plastics than as a postcure treatment of certain heat-resistant rubbers, to improve their properties. The second is used when a polymer shows insufficient flow to be pressure-moulded but, being initially in a finely-divided state, can be made to cohere under pressure and to increase in cohesion on continued heating, which as previously may be done out of the mould; it applies in particular to polytetrafluoroethylene.

For *blow-moulding* see immediately below, and for *low-pressure moulding* see later.

Extrusion

This resembles the injection process (above) except that the intermittent action of the ram, or plunger or screw in the injector cylinder is replaced by that of a continuously turning screw which is so arranged that chips or pellets of the moulding material are first heated to the plastic state then forced through a heated die-head. The extruded material is subsequently passed into water, or on to a chill roller, or some form of endless band ('haul-off') where it cools with retention of the extruded cross-section. Rods, tubes (pipes), thin sheets (also profiled shapes), films, and ribbons can thus be produced continuously; similarly a polymer can be extruded as a covering on wires, cables, cords, etc., by feeding them through the centre of the extrusion machine or through a T-head.

Depending on the polymer, the pressures used (which are lower than in injection moulding) range from 1000–3000 lbf/in^2 (0·7–2·1 kgf/mm^2) with temperatures ranging from 110–300°C.

Blow moulding, for fabrication of plastics bottles and similar articles, consists in intermittent extrusion of a thin-walled tube and automatically clamping round it a split mould, into which the tube is expanded by increasing the air pressure inside it.

Casting and low-pressure moulding

High polymers, when they can be melted, are usually too viscous to be cast under gravity (though some are satisfactory if the force is increased, as in rotational or centrifugal casting), but certain *monomers* or *intermediates* can be cast directly into moulds in which polymerisation is then completed; thus, though the rate of production

631

is generally lower, the expensive equipment associated with high-pressure moulding is avoided.

Polymethyl methacrylate may be cast in this way (as sheets and rods, and in blocks in which museum specimens may be embedded), polymerisation of the monomer being initiated by heat or ultra-violet light. To reduce the heat evolved during setting, especially in castings of thick cross-section, the methyl methacrylate may first be partially polymerised and/or 'diluted' by inclusion of fine-divided polymer.

A thick syrup of a phenol/formaldehyde intermediate, when cast hot (as rods, blocks, etc.) and slowly hardened by continuation of the condensation for several hours or days below 100°C, yields a homogeneous solid (Catalin) which is readily machineable, taking a good polish. Somewhat similarly, intermediate syrups of a kind hardening by addition-type polymerisation—consisting, for example, of styrene and an unsaturated polyester (*see* §24D.4), or an epoxy resin and its hardener (§25.21), or a reactive polyurethane (§40.21 II) or silicone (§41.21) blended with hardening agent or catalyst—may be cast and 'cured' at elevated or room temperature. Often inert fillers are included, with various advantages, e.g. reduction of the 'exotherm' of the reaction, mechanical reinforcement, or economy.

Since syrups of the polyester or epoxy kind flow fairly readily and evolve no volatile products during cure, and thus need only gentle force for their manipulation, they—or their end products—are often called *low-pressure* or '*contact-pressure*' resins. Such materials, incorporating reinforcing fillers (notably as glass-fibre reinforced plastics or GRP), are the basis of an important technique whereby, using relatively inexpensive moulds (e.g. of wood, plaster, or a casting resin), it is possible to fabricate large structures—such as tanks, car bodies and boats—economically, even if only a one-off product is required (though normally a mould would be used many times). Laminated boards and tubes, made by bonding glass-fibre fabrics with low-pressure resins—also GRP made from fibre mat or from yarns arranged parallel or criss-cross—are exceptionally strong yet relatively low in density.

PASTE TECHNIQUES

In some instances, particularly with polyvinyl chloride, very finely divided polymer can be compounded with a liquid plasticiser to provide a viscous dispersion or 'paste', which is stable as a mixture at room temperature but on heating to 150–160°C undergoes an irreversible physical change to become a homogeneous single-phase system, i.e. identical with plasticised polyvinyl chloride as prepared by other means. Pourable and spreadable pastes of this kind can be used for casting rods or small blocks, for slush mouldings, and for coating metals or—notably—fabrics, e.g. for leathercloth and for laminated assemblies (*see* §11.4).

Soft or rigid low-density foams find many and varied uses (as, for instance: absorbent sponges; resilient upholstery, carpet underlay, crash pads, and packaging; thermal and acoustic insulation; the central portions of light-weight non-buckling sandwich structures used in engineering). The expansion, which may be arranged to produce closed or interconnected cells, can be brought about in several ways, namely (i) by the setting of a whipped-up foam, (ii) by dissolving nitrogen under pressure in a molten polymer, followed by expansion under reduced pressure, (iii) by incorporation of a blowing agent, such as sodium bicarbonate, ammonium nitrite, or a nitroso- or azo-compound (e.g. azodicarbonamide), and (iv) by reactions combining the final stage of polymerisation and liberation of a gas such as carbon dioxide (*see*, particularly, polyurethane foams, §40.21).

IRRADIATION
Some high polymers when subjected to penetrating radiation (β-rays, γ-rays, x-rays) undergo degradation in molecular weight with consequent loss in their physical properties, others become cross-linked and thus insoluble and improved in thermal resistance, while some undergo complex changes (dependent also on the absence or presence of oxygen) ascribed to both chain scission and cross-linkage. These phenomena are of limited importance industrially, though graft copolymerisation, induced by radiation, has been used to change bulk and surface characteristics, e.g. of nylon or polyacrylonitrile with styrene or acrylic acid; perhaps the main commercial application is the cross-linkage of end-products following fabrication by standard methods, e.g. irradiation of polyethylene pipes and cable insulation.

II. MANIPULATION OF SHEETS, RODS AND TUBES

Apart from fabrication by moulding, casting, etc., various polymeric substances are available in the form of cast or extruded sheets, rods, tubes, laminates, etc., which—being constructional materials in their own right—can be fabricated by traditional engineering methods, and adaptations thereof, some of which receive mentioned below.

THERMOFORMING
This includes any process by which the geometry of rigid sheets, rods, etc., is changed to a desired configuration. One method is to soften a sheet (in an oven, water-bath, or oil bath, or by infra-red radiation or high frequency dielectric heating) and then mould it to curved shapes by pressing between dies, or forcing into a mould by vacuum and/or air pressure, until cold. In this way, even thermoset

phenolic laminates—rapidly heated to *c.* 160°C so as to confer limited plasticity—can have simple curvature imposed upon them (*post-forming*), but the greatest application is in the forming of sheets of thermoplastic polymers, as discussed below.

In *vacuum forming* a sheet of a thermoplastic polymer is secured in a frame that fits on top of a mould, and is rapidly softened by exposure to 'black heat' (infra-red radiation). The mould is then connected to a vacuum line and the softened sheet is thereby 'sucked' into the mould by atmospheric pressure. The now contoured sheet is removed when cool. Trays, doors, windows, bas-relief sculptures and maps, and also Braille literature, illustrate the varied applications of this technique, for which the moulds may be accurately but inexpensively constructed in wood, plaster, metal, etc. In *drape moulding*, which permits the production of deep-drawn articles, such as tanks and domes, part of the mould is made to rise into the softened sheet, which thus drapes itself smoothly over suitably-designed positive contours before the vacuum is applied.

WELDING

Thermoplastic tubes or sheets may be joined together if the ends, or edges, are first contacted with a hot-plate to soften them locally. Rubbing the ends vigorously together, to soften them by friction, is an alternative procedure. Larger sheets and tubing of rigid thermoplastic materials can be welded in the form of tanks, ducting, etc.—somewhat analogously to the process with metals—by means of filler rods and hot air (or nitrogen) torches operating at 250–400°C.

By application of a hot metal bit ('soldering iron') the edges of two thin thermoplastic sheets, lying page-wise one on the other, can be bonded together, but the low thermal conductivity of polymeric materials prevents more extensive use of this technique. However, thermoplastic packaging foils can be sealed together (for bags, etc.) by nipping them momentarily between electrically-heated metal jaws, adhesion to which is prevented by a thin coating of polytetra-fluoroethylene. With polymers having an appreciable dielectric loss (*see* power factor, Section 72)—notably plasticised polyvinyl chloride, as used for rainwear, etc.—welding can be rapidly effected by a combination of pressure and a high frequency electric field, i.e. by 'electronic sewing' (localised dielectric heating at 20–100 MHz). Where low dielectric loss in a plastic material makes the above process impracticable (e.g. with polyethylene) ultrasonic welding may be used, heat being created at an interface by compressional impulses (e.g. at 20 kHz).

USE OF ADHESIVES

While welding (above) generally relates to the cohesion of a material to itself, as effected by heat and pressure, *gluing* is usually concerned with bonding together two similar, or different, materials by means of an intermediate film of an *adhesive**.

All adhesives are initially applied in a liquid state (as solutions, dispersions, or melts), and many industries use various natural or synthetic polymers (proteins, gums, rubbers, etc.) in the form of solutions based on water or an organic solvent; generally, however, adhesives of this kind are not very satisfactory for constructional purposes, since—even if they can adhere (and water-soluble kinds will not readily do so to hydrophobic polymers)—the solvent must be free to escape, and the bond when formed must not deteriorate, e.g. at high humidity, high temperature, or as a result of biological attack. However, small pieces of thermoplastic materials can sometimes be stuck together quite satisfactorily by softening the surfaces with a solvent (or solvent containing dissolved polymer) and clamping them in contact until hard again, e.g. as with polymethyl methacrylate moistened with chloroform, or cellulose acetate with acetone.

When at least one of the surfaces to be stuck is permeable, it may be possible to use a solvent-based adhesive as above†, but more often superior results are obtained from condensation-type adhesives setting by loss of a volatile by-product, commonly water. Amino- and phenol-formaldehyde syrups are in this category, being 'curable' (cross-linkable) by either heat or incorporation of a 'hardener', e.g. an acid (sometimes disadvantageous) or paraformaldehyde (as with resorcinol-formaldehyde, *see* §28.81). In practice, the setting of wood glues of this kind can be accelerated by the use of (i) resistance heating, or (ii) high frequency dielectric heating (at 2–10 MHz), in which instance most heat is generated in the glue line. A phenolic resin in conjunction with a polyvinyl acetal is used for bonding light alloy components in aircraft construction (Redux).

A third class of adhesives, setting without loss of volatile matter, consists of (i) fusible polymers and blends thereof ('hot melt adhesives', offering—like solder—an advantage of setting as they cool) and (ii) cross-linkable syrups—notably those of epoxy resin intermediates, *see* §25.21—that can be cured by heat, or at room temperature by incorporation of a hardener. Such adhesives, when fully cured, provide very strong bonds (to various high polymers, wood, glass, metals, stone and concrete) which are relatively inert to chemical attack and not much affected by water or organic liquids, or heat; they do not, however, adhere to non-polar polymers such as polyethylene (they are often pre-mixed in small amounts and dispensed from polyethylene syringes or guns before setting). Natural and most

---

*Clearly, maximum strength requires that there shall be good contact between the adhesive and the adherends on either side; therefore, when a glued joint is being made it is essential to avoid contamination by any additional substance that might weaken it (grease, oils, etc.).
†Those that remain tacky after evaporation of the solvent from a spread film ('contact adhesives', based on natural or chloroprene rubber) are, of course, much used as adhesives for thin laminated boards (Formica, etc., e.g. for table tops and wall panels) and in re-soling footwear.

synthetic rubbers adhere strongly when vulcanised *in situ* against brass or metal that has been brass-plated. Adhesion to polyethylene, polypropylene, and—in particular—polytetrafluoroethylene, is always poor; the last material is therefore employed as a release agent (i.e. for 'non-stick' purposes) and can be bonded only after chemical etching (*see* §3.4) that virtually changes the composition of the surface.

MACHINING

High polymers are not difficult to cut and, when certain precautions are observed, most plastics, hard rubbers (ebonites), and filled or resin-bonded composites (e.g. laminated boards) can be sawn, drilled, turned, and in general machined with the conventional equipment employed for metal or wood.

It is most important, however, to recognise that cutting produces heat and that *the thermal conductivity of polymers is very much lower than that of metals*. Therefore, in the absence of lubricants a cutting edge can easily suffer a deleterious rise in temperature; moreover, if with a thermoplastic material the temperature in the cutting region exceeds the softening point, the working surfaces may become rubber-like. Thus, polystyrene and polyvinyl chloride, which soften below 100°C, are difficult to work unless both the cutter and surface being cut are adequately cooled, e.g. with compressed air or a stream of clear or soapy water; in the 'CeDe cut' technique, the cutting tool edge is cooled by carbon dioxide allowed to expand from a cylinder.

Other points of importance in machining plastics materials are: the tools (high speed steel, or carbide- or diamond-tipped) should be kept *sharp*, and the cutting speed should be high (up to 5 m/s) but usually the depth of cut should be shallow; drills should have a point angle of *c.* 30 degrees and a helix angle of *c.* 15 degrees, with ample clearance for removal of swarf. In cutting or tapping screw threads it is advisable to use a *coarse* pitch. In blanking or die-punching it is sometimes advisable to use warmed material or a hot table, although some plastics sheets can be punched or even cut with shears at room temperature. Thermoplasticity assists the production of a gloss finish, but most machinable polymers can be ground, sanded, and finally buffed to a high polish.

COATING

Though the wide subject of surface coating—the application of decorative and protective lacquers and paints—cannot even be summarised here, mention may be made of covering metal articles with a thick uniform coating of a high polymer by *powder coating*. This too has decorative and protective uses and can be achieved by pre-heating the object to be coated (say to 150–400°C) then bringing it into contact with a fine powder of a fusible polymer, followed if necessary by post-treatment in an oven or tunnel kiln.

Methods of bringing the heated article into contact with the fusible powder include simple dusting, but more particularly lowering the metal into a *fluidised bed** of the powder; also electrostatic spraying and flame spraying, the last needing no initial heating of the object to be coated and thus not being restricted to metal articles. Suitable powders include those of polyethylene, polyvinyl chloride, cellulose acetate-butyrate, nylon 11, and epoxy resin intermediates.

Plastics may be used to coat metals, as above, but can also themselves be subjected to *metallisation*, the object being to improve their appearance or reflectivity, to protect or strengthen them, or to increase the electrical conductivity; also to eliminate wiring in miniature electronic equipment, where metal circuits are 'printed' upon insulating sheets.

Transparent sheets or films, plain or embossed, will protect a highly reflective metal coating deposited on the reverse and can be used with economic advantage to simulate polished metal. Metallisation is effected by spray gun, by electro-plating (of surfaces rendered conductive), by vacuum evaporation or sputtering, and by chemical silvering.

81.4 FABRICATION OF RUBBERS

The mastication, compounding, and ultimate vulcanisation of natural rubber are dealt with in §§29.4 and 31.4 (ebonites). Synthetic rubbers are treated similarly, except that they may not need mastication and their composition may call for a special form of vulcanisation, e.g. chloroprene rubber is commonly vulcanised with metal oxides, saturated rubbers such as ethylene/propylene copolymers and silicone rubbers need organic peroxides (or, for fluoro-rubbers, polyamines) and polyurethane rubbers may need isocyanates (*see* §32.4–41.4).

The shaping of the rubber mix is effected either by moulding (compression, transfer, or injection), extrusion, or calendering, or articles may be hand-built (without or with textile reinforcement) to a required shape; alternatively, a mix may be (i) 'frictioned' on to a fabric, i.e. forced into the interstices by passage between calender rolls, or (ii) used in solution to spread on fabric or to make thin-walled articles by dipping of formers. Liquid rubbers (§29.8(10)) may be cast and then vulcanised in the mould. Natural rubber latex (concentrated to 60% solids) and synthetic rubber latices, compounded for subsequent vulcanisation, are employed in impregnating, coating, dipping and moulding techniques, also for making circular-section

---

*In which a finely divided powder, suspended in a current of air supplied through a porous base in a tank, behaves like a liquid.

thread by extrusion-coagulation, and for 'latex foam' rubber (used in upholstery, mattresses, carpet underlays, etc.).

Vulcanisation may be brought about by heat (usually between 120° and 160°C), simultaneously with moulding*, or—with non-moulded articles—by steam under pressure or by hot air, while with very active accelerators (§29.4) vulcanisation in hot water or even at room temperature is possible. Similarly, thin articles and proofings on fabrics can be 'cold-cured' with sulphur monochloride (§29.4). Long-length extruded articles are often vulcanised by a continuous process, e.g. by passage through a fluidised bed (for extruded strip) or through a chamber filled with high-temperature steam (for rubber-insulated cables). Tyres are made by assembling a 'carcass' of reinforcing plies of cord (high tenacity viscose rayon, polyamide, polyester, steel or glass fibres) coated with an adhesive rubber mix, applying an outer tread (of an abrasion-resistant rubber) and sidewalls, and vulcanising the whole while it is forced into a hot mould by the pressure of steam or superheated water in a flexible 'curing bag' (normally of vulcanised butyl rubber). Belting and reinforced hose are likewise assembled from a rubber mix and textile fabrics, cords or yarns, with subsequent vulcanisation.

The fully vulcanised hard product, ebonite, which behaves at room temperature as a rigid material rather than as a rubber, is either vulcanised in the shape required or is produced as sheets, rods or tubes that may be machined (sawn, drilled, turned, etc.) or bonded by means of adhesives. Both soft rubbers and ebonites may be made as low-density cellular products (*see Expansion techniques,* earlier).

In general, for processes used in fabricating rubbers see also earlier in this Section under **Fabrication of plastics,** and in the section that follows below.

FURTHER LITERATURE

Production of Fibres
HILL, R. (Ed), *Fibres from Synthetic Polymers,* London (Elsevier) (1953)
MARK, H. F., ATLAS, S. M. and CERNIA, E. (Eds), *Man-made Fibres: Science and Technology* (Vol. 1, principles of spinning processes), New York and London (1967)

Production of Films (including surface coatings†)
HARVEY, A. A. B., *Paint Finishing in Industry,* London (1958)
SEYMOUR, R. B., *Hot Organic Coatings,* New York (1959)
LEFAUX, R., *Emballages et Conditionnements Modernes,* Paris, 279 (1960) (plastics films)
NYLÉN, P. and SUNDERLAND, E., *Modern Surface Coatings,* London and New York (1965)
PARKER, D. H., *Principles of Surface Coating Technology,* New York (1965)
PINNER, S. H. (Ed), *Modern Packaging Films,* London (1967)
MARTENS, C. R. (Ed), *Technology of Paints, Varnishes, and Lacquers,* New York (1968)
SWEETING, O. J. (Ed), *The Science and Technology of Polymer Films,* Vol. 1, New York and London (1968)

*In injection moulding, the heat generated by friction during the rapid passage of the rubber through the narrow gate is sufficient to raise its temperature almost instantaneously to that required for vulcanisation.
† For surface coatings see also references in §24A.9, and under §81.3 Coating, above.

Fabrication of Plastics
I. *Moulding, extrusion, and related processes*
General
The Society of Plastics Industry, Inc., *Plastics Engineering Handbook*, New York and London, 3rd edn (1960) (materials, processes, design, testing)
BERNHARDT, E. C. (Ed), *Processing of Thermoplastic Materials*, New York (1959)
MILES, D. C. and BRISTON, J. H., *Polymer Technology*, London (1965)

Moulding (high pressure)
BUTLER, J., *Compression and Transfer Moulding of Plastics*, London (1959)
GROVES, W. R., *Plastics Moulding Plant*, Vol. 1, London (1963) (hydraulics, compression and transfer moulding equipment)
WALKER, J. S. and MARTIN, E. R., *Injection Moulding of Plastics*, London (1966)
Learning Systems Ltd., *The Elements of Injection Moulding of Thermoplastics, a Teaching Programme*, London (1968)
MINK, W., *Practical Injection Moulding of Plastics*, London (1964)
PENN, W. S., *Injection Moulding of Elastomers*, London (1969)
PYE, R. G. W., *Injection Mould Design*, London (1968)
MUNNS, M. G., *Plastics Moulding Plant*, Vol. 2, London (1964) (injection moulding equipment)
JONES, D. A. and MULLEN, T. W., *Blow Moulding*, New York and London (1961)

Extrusion
SIMONDS, H. R., WEITH, A. J. and SCHACK, W., *Extrusion of Plastics, Rubber and Metals*, New York (1952)
JACOBI, H. R. (Transl.) and EASTMAN, L. A. H., *Screw Extrusion of Plastics*, London (1963) (fundamental theory)
FISHER, E. G., *Extrusion of Plastics*, London (1964)
SCHENKEL, G., Transl. EASTMAN, L. A. H., *Plastics Extrusion Technology and Theory*, London and New York (1966)
GRIFF, A. L., *Plastics Extrusion Technology*, New York (1968)

Low-pressure moulding and GRP
BROWN, W. J., *Laminated Plastics*, London (1961)
OLEESKY, S. and MOHR, G., *Handbook of Reinforced Plastics*, New York and London (1964)
BEYER, W., *Glasverstärkte Kunststoffe*, 3rd edn, Munich (1963)
HOLISTER, G. S. and THOMAS, C., *Fibre-reinforced Materials*, Amsterdam, London and New York (1966)

Irradiation
CHARLESBY, A., *Atomic Radiation and Polymers*, Oxford, London, New York and Paris (1960)

II. *Manipulation of sheet, rods and tubes*
Thermoforming
ESTEVEZ, J. M. J. and POWELL, D. C., *Manipulation of Thermoplastic Sheet, Rod and Tube*, London (1960)
THIEL, A., *Principles of Vacuum Forming*, London (1965) (German edn 1957)

Welding
FARKAS, R. D., *Heat Sealing*, New York and London (1964)
HAIM, G., *Manual of Plastics Welding*, Vol. 1 (with ZADE, H. B.), London (1947); Vol. II, *Polyethylene* (with NEUMANN, J. A.); Vol. III, *Polyvinyl Chloride* (1959)
NEUMANN, J. A. and BOCKHOFF, F. J., *Welding of Plastics*, New York and London (1959)
The Institute of Welding, *Data on Welding of Thermoplastics*, London (1966)
FREDERICK, J. R., *Ultrasonic Engineering*, New York and London (1965)

Adhesives (and adhesion)
Society of Chemical Industry, *Adhesion and Adhesives—Fundamentals and Practice* (Conference papers), London (1954)
HURD, JOYCE, *Adhesives Guide* (BSIRA Res. Rpt. M39), London (1959)
GUTTMANN, W. H., *Concise Guide to Structural Adhesives*, New York and London (1961)
SKEIST, I. (Ed), *Handbook of Adhesives*, New York and London (1962)
SHIELDS, J., *Adhesives Handbook*, London (1970)

MCGUIRE, E. P., *Packaging and Paper Covering Adhesives*, New York (1963)

HOUWINK, R. and SALOMON, G. (Eds), *Adhesion and Adhesives*, 2nd edn, Amsterdam (Elsevier), Vol. 1, Adhesives (1965); vol. 2, Applications (1967)

PATRICK, R. L. (Ed), *Treatise on Adhesion and Adhesives*, New York, Vol. 1, Theory (1967); Vol. 2, Materials; (1969)

CAGLE, C. V., *Adhesive Bonding Techniques and Applications*, New York and London (1968)

Ministry of Technology, *Adhesion—Fundamentals and Practice* (Conference papers, London (1969)

ALNER, D. J. (Ed), *Aspects of Adhesion*, London (1965–1969) (proceedings of annual conferences)

Metallisation

HEPBURN, J. R. I., *The Metallisation of Plastics*, London (1947)

NARCUS, H., *Metallisation of Plastics*, New York (1960)

FURNESS, R. W., *The Practice of Plating on Plastics*, London (1968)

GOLDIE, W., *Metallic Coating of Plastics* (2 vols.), London (1968 and 1969)

Fabrication of Rubbers

PENN, W. S., *Synthetic Rubber Technology*, London (1960)

NAUNTON, W. J. S. (Ed), *The Applied Science of Rubber*, London (1961)

STERN, H. J., *Rubber, Natural and Synthetic*, 2nd edn, London (1967)

HOFMANN, W., *Vulcanisation and Vulcanising Agents*, London (1967)

VANDERBILT CO. INC., R. T., *The Vanderbilt Rubber Handbook*, New York (1968)

SCOTT, J. R. (Ed), *Progress of Rubber Technology*, Vol. 32, London, 57 (1968) (WHEELANS, M. A., injection moulding of rubber)

CRAIG, A. S., *Dictionary of Rubber Technology*, London (1969)

SERVICEABILITY

The narrower the limits within which a substance retains useful attributes, the fewer can be its applications; for example, if the properties of a polymeric substance are adversely affected by water it must necessarily be excluded from exterior use and its electrical applications will also be restricted. Even if all the immediate requirements are satisfied, however, including resistance to abrasion (when wear is an operative factor), the long-term time- and environment-dependent demands may be more difficult to satisfy. This is because—unlike, say, gold or ceramics—it is only under exceptional circumstances that high polymers survive unchanged indefinitely; normally they slowly lose strength and/or other essential properties, commonly becoming either unduly soft or hard and embrittled as the case may be, i.e. they undergo *ageing* (which arises from physical causes and/or chemical action), and sometimes they are subject to *biological attack*.

Notes on the serviceability of the individual polymers appear in Sections 1 to 42, and some general information on ageing and biological attack appears below. Other factors which influence serviceability, namely resistance to organic solvents, to water, and to temperature, also electrical and mechanical properties, are discussed in Sections 55, 61, 68, 69–72 and 73–80 respectively.

AGEING

When slow physico-chemical changes eventually prove detrimental to the proper functioning of a polymeric substance, the material is said to 'age' and ultimately to perish. Irreversible effects of this kind can be produced by the action of sunlight, oxygen, or ozone*, or by subjection to repeated changes in temperature and humidity, or to various forms of radiation, or to a severe combination of actions such as occurs in exposure to outdoor weathering. Repeated flexing or vibration can bring about related effects, such as development of

---

*Ozone is synthesised in air in the presence of ultraviolet radiation or a high voltage 'silent' discharge. In particular, it rapidly causes cracking on certain types of vulcanised rubbers when they are slightly stretched.

surface cracks, internal generation of heat, and loss of strength; continuous stress can also be inducive to premature ageing.

In some polymers the action of sunlight and atmospheric oxygen causes discoloration (e.g. of cellulose nitrate or polyvinyl chloride if unsatisfactorily stabilised) and often brings about scission of the main chain (cellulose, polyamides), the fall in molecular weight being apparent as a loss in strength. Some polymers soften with time, especially at elevated temperatures (e.g. butyl rubber vulcanisates), others become stiffer (e.g. vulcanisates of styrene/butadiene rubbers or nitrile rubbers), while some undergo slow syneresis—sometimes due to continued polymerisation or loss of volatile matter—that reveals itself as internal crazing (i.e. minute cracks, as may occur in unfilled polystyrene or amino-formaldehyde resins). Fillers, pigments, dyes, etc., may have a beneficial effect in lowering the rate at which a high polymer deteriorates (as with polyethylene, which very readily undergoes photo-oxidative degradation*, but—rather surprisingly—is rendered quite serviceable by incorporation of 2% of carbon black); but in other instances the additive may sensitise the polymer to attack (as with certain delustrants in polyamides, with tin-weighting of silk, and with metal naphthenates used for rot-proofing; and as do certain anthraquinonoid vat dyes in much accelerating the degradation of cellulose fabrics exposed—for example, as window curtains—to sunlight and a humid atmosphere). Similarly, traces of polyvalent metals or their ions can be detrimental, as in the decomposition of chlorine-containing polymers (polyvinyl chloride, polyvinylidene chloride) which can be accelerated by the presence of copper, zinc, or iron; in some instances, too, copper, manganese and (sometimes) iron can so adversely affect rubber that it is necessary to ensure their absence not only from a rubber mix but also from textile fibres used in close contact with the rubber, and (sometimes) even from equipment used for processing.

Substances added as *stabilisers* and/or *ultra-violet absorbers*† are important in these respects; their action—but imperfectly understood—involves termination of free-radical reactions or removal of a chemical product of the deterioration, an increasing concentration of which might have an autocatalytic effect. Antioxidants‡ (*see* §29.4) often bring about stabilisation by absorbing free radicals and thus inhibiting oxidative chain reactions, while stabilisers of a basic kind probably act by removing acid by-products favourable to detrimental chain reactions (e.g. *see* §11.22). In vulcanised rubbers antiozonants and anti-flex-cracking agents are used to retard cracking

---

*Whereby all the mechanical and electrical properties suffer.

†Commonly substituted *o*-hydroxybenzophenones or salicyclic esters.

‡Some, perhaps, preferentially oxidised and forming a protective layer, though this is not their main function.

642

caused, respectively, by ozone or repeated flexing. Waxes exert a protective action as surface sealants.

Numerous testing procedures intended to accelerate ageing (by artificial weathering, or exposure to light, to hot air, to hot oxygen under pressure, or to ozonised air) have been devised, especially for accelerated weathering of surface coatings and accelerated ageing of vulcanised rubbers, but the results from many of these are difficult to correlate with those of long-term service. This applies particularly when the samples are exposed to light from a carbon arc or mercury discharge lamp, although a xenon lamp—with a lower proportion of ultra-violet radiation—gives results more nearly corresponding to those from sunlight; the temperature, nature of atmosphere, and the humidity to which the samples are subjected during exposure are, of course, also important. Exposure to tropical sunlight* and tropical weathering represents a 'natural' way of accelerating ageing when determining the serviceability in temperate localities.

Some results illustrating the relative photodegradation of fibres are shown in Table 82. T1.

Table 82.T1. PHOTODEGRADATION OF TEXTILE FIBRES*

(i) Fibres as undyed yarns	Exposure-time† for degradation to half strength
Silk, degummed‡	2·9
Silk, raw	3·8
Nylon, delustred§	4·0
Cotton, bleached; viscose rayon, bright‖	8·8
Wool, bleached	11·8
Cellulose acetate	13·8

*It is emphasised that the results relate to the tests indicated and to particular samples of material; they are believed typical but not necessarily representative in general.
†Units are average summer months; samples exposed in air, at an inclination of 45 degrees to the vertical, behind a south-facing glass window 6 miles south of Manchester, England.
‡Resistance highest when silk is finished in an alkaline bath (pH 9–10).
§Resistance decreased two- to threefold in the presence of titanium dioxide delustrant.
‖Resistance much decreased at high humidity, and in the presence (in rayon) of a delustrant.

(ii) Comparison of two commercial yarns.

Exposure time in CPA lamp* (h)	Retention of tensile strength (%)	
	Polyethylene terephthalate	Polypropylene
128	99	81
308	96	42
424	95	28

*A water-cooled carbon arc (now largely replaced by the xenon lamp)

*In some instances enhanced by mirrors and with the test samples on an equatorial mounting to follow the sun.

BIOLOGICAL ATTACK

As is commonly observed, naturally-occurring polymers are subject to various forms of biological attack, which is necessary to preserve a balance in nature but in certain practical situations can be disastrous; for example, the rotting of wood provides an all too well-known instance of the complete loss of the characteristic high strength of the carbohydrate polymer, cellulose, while similarly vulnerable are any unprotected proteinaceous glue films.

Most of the synthetic polymers, differing from the natural ones, are less subject to attack, and some when pure are perhaps not susceptible at all (a statement that might also apply to some purified natural polymers, such as the hydrocarbon of natural rubber); but in practice such polymers may have associated with them a less immune material, e.g. another polymer (as in coated fabrics, laminated boards, or adhesive-bonded structures) or a secondary component (a filler, plasticisers, catalyst or emulsifier residues, surface dressing, etc.). Thus, for instance, even a seemingly inert polymer such as polystyrene may suffer from mould growth on its surface (which can so catastrophically lower its surface resistivity as to invalidate its application as an otherwise excellent electrical insulator); similarly, in lined fire hoses, acid produced by sulphur bacteria thriving on damp vulcanised rubber may cause chemical rotting of the adjacent cotton canvas, which is not itself attacked by the organisms.

Attack can arise from such differing causes as bacteria, fungi, insects, and rodents or larger animals. The loss from these causes, though guarded against, is enormous; and since the damage—especially from the first two causes, where it results from the action of enzymes or metabolic by-products—usually starts from and may be confined to a surface, the detrimental effects are exceptionally prominent where the surface/volume ratio is high, i.e. in *fibres* and *films*, and sometimes in open-celled solid foams, where a decrease in strength comes about much more rapidly than in bulk materials. Here again, however, at least two kinds of attack are discernible, one where the base material can quickly become degraded (with irrecoverable loss in strength) and one where attack on the other materials present (fillers, plasticisers, etc.) may cause different but equally detrimental effects, e.g. loss of surface resistivity (as instanced above) or patchy discoloration (sometimes apparent only on dyeing at a later date).

Most forms of attack, air-borne or soil-borne, are dependent on moisture and are accelerated in warm environments; so they are minimized in the cold and at their worst under tropical conditions. To combat them, the components employed should be carefully selected (and purified if necessary, e.g. natural gums and proteins can be removed from natural rubber and gutta percha), and sometimes it may be advisable to avoid such potential nutrient-supplying substances as: *fillers* consisting of wood dust, leather dust, or cellulosic fibres; *resins, plasticisers, lubricants, etc.,* based on vegetable fats or oils, glycerol, or ricinoleic, oleic, stearic, or sebacic acids; or *bonding agents* containing natural gums, starch, gelatin, etc.* The following steps may also be taken: (*a*) incorporation of bactericidal and/or fungicidal substances, in—say—1–2% concentration, and, (*b*) partial modification of the basic polymer.

Under (*a*) are included: *phenolic substances,* especially penta-chlorophenol and its sodium derivative or lauric ester, and *o*-phenyl-phenol; *organo-metallic compounds* such as copper 8-hydroxy-quinolinate, copper or zinc naphthenate, pyridylmercuric chloride, tributyltin oxide, and zinc dimethyldithiocarbamate; and *inorganic compounds* such as insoluble metallic oxides (e.g. those of iron and chromium, precipitated inside textile fibres as mineral khaki, etc., which also increase resistance to degradation by light), copper borate initially solubilised in a solution of a zirconium salt, and copper hydroxide initially solubilised in ammonium hydroxide. None of the above is wholly ideal, however, some of the additives being rather dangerously toxic, some deeply coloured, while others accelerate ageing (*see earlier*—for example, copper compounds are not generally permissible in rubbers, as they can promote rapid deterioration). Treatment with aq. formaldehyde, salicylanilide, borax, copper or zinc sulphate, etc., may be useful as a temporary measure, e.g. to inhibit growth of mildew, but being leached out by rain or on washing, these materials are unsatisfactory in a permanent capacity.

Under (*b*) are included: treatment with formaldehyde, e.g. *forma-lisation* of casein (§21.21); modification of cellulose by partial esterification or etherification, e.g. by *acetylation* (§14.81), *benzylation* (§17.81), *carboxymethylation* (§16.4), or *cyanoethylation* (§17.82); and certain kinds of 'resin treatment', usually entailing coating and/or internal deposition together with some cross-linkage to the base polymer, e.g. colloidal condensates of melamine-formaldehyde (§27.82) have been used to rot-proof cellulose.

Concerning attack by more visible forms of life, native and re-generated cellulose is devoured by silver fish (*Lepisma saccharina L.*) and termites, while the keratin of wool and natural bristles supports the larvae of the clothes moths and those of several species of beetles. It is to be observed that insects and rodents are not adverse to gnawing through non-assimilable polymers (e.g. cellulose acetate) in order to reach more digestible fare. Insects, however, are repelled by certain plasticisers, such as dimethyl phthalate, and most insects and their larvae are susceptible to the lethal—though controversial—effects of 2,2-bis(*p*-chlorophenyl)-1,1,1-trichloroethane (D.D.T) and the γ-isomer of 1,2,3,4,5,6-hexachlorocyclohexane (B.H.C., Gammexane).

---

*On the other hand, some ingredients added for other purposes (e.g. plasticisers, vulcanisation accelerators) may have a fungicidal or bactericidal—or fungistatic or bacteriostatic—action.

FURTHER LITERATURE

General

GREATHOUSE, G. A. and WESSEL, C. J. (Eds), *Deterioration of Materials,* New York (1954) (includes fibres, films, plastics and rubbers)

ROSATO, D. V. and SCHWARTZ, R. T. (Eds), *Environmental Effects in Polymeric Materials* (Vol. 1, *Environments*; Vol. 2, *Materials*), New York and London (1968)

Ageing

JELLINEK, H. H. G., *Degradation of Vinyl Polymers,* New York (1955)

BUIST, J. M., *Ageing and Weathering of Rubber,* London (1956)

GRASSIE, N., *The Chemistry of High Polymer Degradation Processes,* London (1956)

VALE, C. P. and TAYLOR, W. G. K., *Chem. and Ind.,* 268 (1961) (ultraviolet absorbers for prevention of photo-degradation)

BOLT, R. A. and CARROLL, J. G. (Eds), *Radiation Effects in Organic Materials,* New York and London (1963)

GORDON, G. Ya., *Stabilisation of Synthetic High Polymers,* (transl. from Russian), Jerusalem (1964)

MADORSKY, S. L., *Thermal Degradation of Organic Polymers,* New York and London (1964)

NEIMAN, M. B. (Ed), *Ageing and Stabilisation of Polymers,* Moscow (1964); English translation, New York (1965)

PINNER, S. H. (Ed), *Weathering and Degradation of Plastics,* (Symposium), Manchester (1966)

KAMAL, M. R. (Ed), *Weatherability of Plastic Materials,* (Applied Polymer Symposia No. 4), New York, London, Sydney (1967)

ALNER, D. A. (Ed), *Aspects of Adhesion,* London (1965–1969) (proceedings of annual conferences) New York–London–Sydney (1967)

Biological

SIU, R. G. U., *Microbial Decomposition of Cellulose,* New York (1951)

HEAP, W. M., *Microbiological Deterioration of Rubbers and Plastics,* RAPRA Information Circular 476, Shawbury (1965)

PACITTI, J., *Attack by Insects and Rodents on Rubbers and Plastics,* RAPRA Information Circular 475, Shawbury (1965)

Microbiological Deterioration in the Tropics, S.C.I. Monograph No. 23, London (1966)

Specifications

BS 903: Parts A10, A11 (rubber, flex-cracking and crack growth)

BS 903: Part A19 (rubber, air oven and oxygen-pressure ageing)

BS 903: Part A23 (rubber, ozone-cracking)

BS 903: Part G8 (rubberised fabric, accelerated ageing)

BS 903: Parts H8, H10 (rubber thread, accelerated ageing)

BS 1006 (fastness of textiles to light; applicable to certain plastics)

BS 2782: Method 104D (hot air ageing of PVC, applied before cold flex test)

BS 2782: Method 108A (ageing of polyethylene by hot milling)

BS 2782: Method 108B (ageing of polyethylene by hot air oven)

BS 3667: Parts 9 and 10 (flexible polyurethane foam, accelerated ageing)

ASTM D454 (accelerated ageing of rubber; hot air under pressure)

ASTM D518 (rubber, cracking by weathering)

ASTM D572 (accelerated ageing of rubber; hot oxygen under pressure)

ASTM D573 (accelerated ageing of rubber; hot air)

ASTM D756 (plastics, accelerated 'service' test)

ASTM D795 (plastics, mercury-arc light exposure)

ASTM D813 (rubber, crack growth)

ASTM D865 (rubber, 'test tube' heat ageing)

ASTM D1149 (rubber, ozone-cracking)

ASTM D1171 (rubber, weathering)

ASTM D1435 (plastics, weathering)

ASTM D1499 (plastics, light-plus-water exposure)

ASTM D1870 (tubular oven ageing)

ASTM D1672, D2309 (high-energy radiation test on polymers)

ASTM D2126 (rigid cellular plastics, simulated service test)

SECTION 83

HISTORY

INTRODUCTION

There is a certain fascination in the story of the utilisation of fibres, films, plastics and rubbers, since it begins in part before history and yet at the present time these materials command more (and increasing) attention than ever previously. It begins some 10 000 years ago, when man first put to his practical use the organic structural materials of the vegetable and animal worlds (i.e. cellulose and proteins, respectively); thus, with vegetable **fibres** (such as those of flax) things could be bound strongly together or a bow could be strung, while animal fibres (e.g. wool, either naturally on skins or spun and woven into fabrics) provided warmth and even decoration. Things could also be tied with strips of rawhide (a protein in which the fibrous structure is less immediately obvious), and where thin skins were used uncut (e.g. as sails) they may be regarded as the first crude unsupported **films***. Supported films also date far back into history, e.g. the craftsmen of ancient Egypt employed a variety of paints, varnishes, glazes and adhesives.

The story is somewhat different with regard to organic materials that need to be fabricated, e.g. by moulding or extrusion, because these—in contrast to the highly important inorganic substance, clay, which has been worked for many thousands of years†—remained undeveloped until well into the nineteenth century. Bitumen had been employed in small amounts by early peoples (e.g. in Egypt and Mesopotamia) and uses were found for such natural commodities as horn, tortoiseshell and whalebone, but the principal methods by which **plastics** are manipulated arose mainly out of those developed in the nineteenth century to deal with **rubbers**, i.e. with natural rubbers

*Though bearing little resemblance in flexibility and transparency to modern packaging films. It is of interest that in the Nile delta, still at an early stage of civilisation (c. 3500 B.C.), a different kind of fibrous sheet was made from a reed, Cyperus papyrus (flourishing today only in the regions of the Upper Nile); this was the writing base *papyrus*, forerunner of *paper*—the invention of the last, however, took place in China c. A.D. 100 or possibly earlier, although taking nearly 1000 years to reach Europe. It is of interest, too, that when (c. 170 B.C.) Pergamun, in Asia Minor, became cut off by the exigencies of war from the supply of papyrus, an animal-based substitute for it was devised, which—as *parchment* (Lat. *charta pergamena*) or vellum—proved to have superior properties of its own.

†And to *glass*, the fabrication of which has long been understood, the art of *blowing* glass being discovered and perfected in Roman times.

647

from various sources, for there were then no synthetic rubbers. Out of the early technological investigations of rubber came also much of the impetus for the production of man-made fibres, starting with those made from natural or derived polymers and later from synthetic polymers; and out of a blend of the technologies of rubbers, plastics and fibres came that of the production of unsupported films.

Below are briefly presented some principal events relating to fibres, films, plastics and rubbers; more detailed historical information on individual polymers will be found under §§1.7–42.7 and elsewhere.

FIBRES

Following on from what has been said above, as pre-history merged into history the cultivation of cotton and/or bast fibres developed in such countries as Egypt, India, and America*, the Mesopotamian lands of Assyria and Babylon produced wool, and China began the culture of silk. These and a few lesser fibres supplied man's needs for thousands of years, and it was not until 1664 that R. Hooke† (also R. A. F. de Reaumur‡, in 1734) suggested that textile fibres might be made artificially by copying the manner in which the silkworm 'wiredraws his clew' or thread of gum. Few attempts to do this were made, however, until J. W. Swan§ succeeded in producing filaments of cellulose nitrate (to be carbonised for use in electric lamps) and Count Hilaire de Chardonnet‖ adapted the process for spinning textile fibres, fabrics made from these artificial fibres being exhibited at the Paris Exposition of 1889.

Chardonnet's 'artificial silk' was followed by fibres of regenerated cellulose obtained by coagulation of viscose (first prepared by Cross and Bevan in 1892)¶ and by the cuprammonium process of L. H. Despeissis**. Fibres of regenerated cellulose were followed by early ones of regenerated proteins†† and then by those of acetone-soluble cellulose acetate‡‡.

Ultimately a wholly synthetic product (not initially dependent on a natural polymer) became available in 1934 when I. G. Farbenindustrie

---

* Fabrics dating back several thousand years have been unearthed from pre-historic tombs in Peru.

† *Micrographia*, London, 1665.

‡ *Mémoires pour servir à l'Histoire des Insectes*, Paris, 1734–42.

§ Brit. Pat. 5 978/1883.

‖ Fr. Pat. 165 349/1885.

¶ Brit. Pat. 8700/1892 (C. F. Cross, E. J. Bevan, and C. Beadle); the production of fibres was developed under Brit. Pat. 1020/1898 (C. Stearn).

** Fr. Pat. 203 741/1890; fibres first spun commercially in 1899.

†† E.g. Brit. Pat. 15 522/1894 (gelatin), Brit. Pat. 6700/1898 (proteins), both to A. Millar.

‡‡ This modification (secondary cellulose acetate) is due to G. W. Miles, Brit. Pat. 19 330/1905. Commercial production of fibres, using multi-holed spinnerets, was developed after the first World War by H. and C. Dreyfus, under Brit. Pat. 165 519/1921.

AG produced textile fibres of after-chlorinated polyvinyl chloride (PeCe).* The German fibre, spun from solution, was followed by the American melt-spun polyamide (nylon 6.6),† introduced commercially in 1938, and the solvent-spun vinyl chloride/vinyl acetate copolymer (Vinyon),‡ commercial production of which commenced shortly afterwards.

From these last comparatively recent developments in the long history of fibres have appeared—in association with parallel advances in theoretical aspects and the production of new polymers—the complex arrays of man-made fibres available today, of which may be instanced: cellulose triacetate, the more recent (but now mostly discontinued) regenerated proteins, polyamides (including nylon 6), polyesters (especially polyethylene terephthalate), acrylics and mod-acrylics (also the very strong carbon fibres obtained by pyrolysis of acrylic fibres), polypropylene, polyvinyl alcohol, polyvinyl chloride and glass fibres. An assembly of synthetic fibres, bonded in the form of a sheet to which is applied a permeable surface coating, now presents a challenge to natural leather for such purposes as shoe uppers.

FILMS

It is convenient to divide films into those of the supported and the unsupported kind.

SUPPORTED FILMS

As earlier mentioned, these featured in antiquity as surface coatings (e.g. based on egg-white or glue) and adhesives, and served special functions such as the sizing of textile warps. Oil painting, made possible by the use of a drying oil medium, appears to have been introduced by H. van Eyck early in the fifteenth century (the 'drying' being, in part, a form of polymerisation).

The nitration of cellulose—examined by H. Braconnot§, 1833, and developed by C. F. Schönbein‖, *c.* 1845—and dispersion of the product in organic solvents led to 'collodion' (used by F. Scott Archer for his wet-plate photographic process in 1851) and later, *c.* 1920, to low viscosity grades for metal lacquers and leathercloth. During

---

*Ger. Pat. 748 253/1938. PeCe fibre, made by Kunstseidenfabrik Wolfen, was initially called WK-Seide; fibres of PVC were made in Germany by F. Klatte in 1913, and for a time were produced commercially in 1931.

†E.g. Brit. Pat. 461 236/1935 and 461 237/1935; Brit. Pat. 533 307/1939 relates to melt-spinning of fibres.

‡Copolymers patented in Brit. Pat. 406 338/1928 (E. W. Reid) Brit. Pat. 518 555/1937, Brit. Pat. 518 700/1937 and U.S. Pat. 2 161 766/1937 cover use for fibres.

§*Liebigs Ann.*, 7 (1833) 243.

‖Brit. Pat. 11 407 (1846).

1917–18, when a less inflammable protective coating was needed for the fabric-covered fuselages of early military aircraft, production of acetone-soluble secondary cellulose acetate was developed on a large scale (and when the demand abated the plant producing the 'dope' was skilfully adapted to spin it into fibres).* Also in the 1914–18 period coumarone/indene resins were developed in Germany as rosin substitutes, and in 1917 polyvinyl acetate became available. Phenol-formaldehyde lacquers were marketed by Sir J. Swinburne in 1909; K. Albert and L. Berend made rosin-modified phenolic surface coatings in 1910, and oil-soluble (alkylphenol) polymers became available in 1928. Alkyd resins, first made c. 1910, were developed as surface coatings from c. 1920 onwards (notably as oil-modified types in 1927, following fundamental wotk by R. H. Kienle, and particularly after development of the catalytic oxidation of naphthalene to phthalic acid in 1930; also as styrenated alkyds in 1942).

Modern surface coatings include the above oil-modified alkyd resins; chlorinated rubber (c. 1915); butylated amino-resin stoving enamels (UF/cell.nitrate, 1926; MF/alkyd, 1938); aq.-based vinyl, acrylic, and styrene/butadiene latices or 'emulsion' paints (developed first in Germany in the late 1930s); polyurethanes of various kinds (from the 1940s); abrasion-resistant two-component epoxy resins (c. 1945); thixotropic (non-drip) polyamide-modified alkyd and other paints (1952–4) and thermosetting acrylic resins (late 1950s).

UNSUPPORTED FILMS

Apart from paper, noted earlier, man-made unsupported films—though now so commonplace—are relatively new. Plasticisation of cellulose nitrate, patented by A. Parkes† in 1864, led to Celluloid (1871) and Xylonite (1879) which could be produced as thin and transparent but rather stiff films. Plasticised cellulose nitrate was long employed as a photographic film base, films of secondary cellulose acetate not being developed until after 1903, while the triacetate was re-introduced (following earlier difficulties) only in the mid 1950s. Thin sheets of ethylcellulose also became available about this time.

Cellulose regenerated as transparent film was introduced in 1924 (Cellophane); rubber hydrochloride (Pliofilm, Tensolite) and poly-vinylidene chloride (Saran) came on the market as transparent and moisture-proof packaging films in 1937 and 1942 respectively. Alginate films were investigated c. 1940 but have not been pursued.

In addition to the above (including various forms of coated regenerated cellulose) modern packaging films are now based on polyethylene, polypropylene, polystyrene, polyethylene terephthalate, and unplasticised PVC (including in each instance highly stretched

---

* See previous footnote‡‡. p. 648.
† Brit. Pat. 2 675 (1864).

biaxially-oriented films); the many uses of thin sheet rubbers and plasticised PVC must also be considered. Films of polyvinyl fluoride, remarkable for exhibiting much greater stability to ultra-violet light than those of PVC or polyethylene, have been available since 1942.

PLASTICS

The first plastics (Gk. πλασσω to shape or mould) were those derived from natural polymers, following investigations conducted early in the nineteenth century. Thus, nitration of woody materials to obtain 'xyloidine' (by Braconnot, in 1833, see above) and nitration of paper or cotton (by J. Pelouze*, in 1838) to obtain 'pyroxylin', led to the industrial production of 'gun-cotton' by Schönbein (above) from 1845 onwards, and when it was found that the mechanical properties of cellulose nitrate were improved by plasticisation with camphor— as patented by A. Parkes† in Britain in 1864, and discovered independently by J. W. Hyatt‡ in America in 1869—the earliest plastics material (Celluloid, Xylonite) may be said to have made its debut. A discovery that the casein of milk—first used, somewhat unsuccessfully, in 1884 to prepare 'artificial horn'—could be hardened by treatment with formaldehyde to obtain tough horny products (Lactoform, Galalith) was made by A. Spitteler§ in 1897. Both 'nitrocellulose' and casein products—one derived from a vegetable source, the other from an animal source—are still in use. Cellulose acetate, also investigated in the nineteenth century‖, was not prominently employed as a plastics material until 1927.

The first fully synthetic plastics, although not immediately recognised as such, appear to have been polyvinyl chloride and polystyrene, since H. V. Regnault¶ prepared vinyl chloride in 1835 and observed its polymerisation, while styrene was prepared by J. Bonastre** in 1831 and polymers of it were obtained by E. Simon†† in 1839; however, neither polymer was developed to any extent until nearly a century later. The polymerisation of vinyl chloride was re-investigated by I. Ostromislenskii‡‡ in 1912, the use of copper catalysts§§ for the production of vinyl chloride was developed in 1933, and W. L.

---

* *C.R. Acad. Sci., Paris,* 7 (1838) 713.
† See † p. 650.
‡ U.S. Pat. 105 338/1870.
§ Brit. Pat. 24 742 (1897).
‖ P. Schützenberger, *Compt. Rend. Acad. Sci., Paris,* 61 (1865) 485; C. F. Cross and E. J. Bevan, Brit. Pat. 9676 (1894).
¶ *Liebigs Ann.,* 14 (1835) 22–38.
** *J. de Pharm.,* 17 (1831) 338.
†† *Liebigs Ann.,* 31 (1839) 258–277.
‡‡ Brit. Pat. 6299 (1912).
§§ U.S. Pat. 1 926 638 (1933); 1 934 324 (1934).

Semon* showed how the polymer could be plasticised to provide a rubber-like material, which was introduced commercially as 'Koroseal' c. 1934. Commencing c. 1929, the now classical investigations of H. Staudinger and his co-workers† into the mechanism of the polymerisation of styrene‡ led to commercial production of the polymer in 1936. Meanwhile, back in 1905–9 L. H. Baekeland§ had studied the chemistry of the phenol-formaldehyde reaction and had begun to manufacture 'Bakelite' thermosetting moulding resins; the lighter-coloured amino-formaldehyde resins, examined from c. 1894 onwards, made their first appearance commercially in 1926.

Approximately contemporarily with Staudinger's work, W. H. Carothers and his colleagues‖ conducted their equally classical researches on linear condensation¶ (leading to production of nylon 6.6 in 1938); another achievement that was to prove far-reaching was the high-pressure polymerisation of ethylene (Gibson and Fawcett, 1933). Hence, because of the expansion of the theoretical background and the commercial advent of several new polymers, the modern plastics industry is sometimes considered to date from this period, i.e. around 1930.

Since that time, however, there have been several further developments. A major one arose from a discovery, made by K. Ziegler** in 1953, that ethylene could be polymerised at room temperature and pressure, and in a very regular manner, by using a catalyst consisting of an ionic complex formed from a transition metal halide and a metal alkyl (e.g. from $TiCl_4 + AlEt_3$). This resulted not only in commercial production of high-density polyethylene in 1954, but also led G. Natta†† in the same year to find that catalysts of the above kind could yield polymers of α-olefins that were stereoregular, e.g. *isotactic*— each repeat unit along a chain molecule having the same spatial configuration. This most important discovery led in turn to commercial syntheses of polymers previously unknown, particularly such plastics as the crystalline—largely isotactic—form of polypropylene (available since 1959) and poly-4-methylpentene (1965) one of the least dense solids known, and future applications of Ziegler-Natta stereospecific catalysts seem likely to prove far-reaching.

Among other plastics or resins developed in relatively recent years are those formable from reactive intermediates, at or near room

*U.S. Pat. 1 929 453 (1933).

†See, for instance, *Trans. Farad. Soc.*, 32 (1936) 97–121.

‡Also polymerisation of formaldehyde to polyoxymethylenes.

§*Industr. Engng Chem.*, 1 (1909) 149–161.

‖ See: MARK, H. and WHITBY, G. S., *Collected Papers of W. H. Carothers (High Polymers*, Vol. 1), New York (1940).

¶Also on addition polymerisation (leading to production of chloroprene rubber in 1932).

Co-ordination catalysts for polymerisation of ethylene: K. Ziegler *et al.*, Bel. Pat. 533 362 (1955); *Angew. Chem.*, **67, 426, 541 (1955).

††Co-ordination catalysts for polymerisation of α-olefins to *stereoregular structures*: *J. Pol. Sci.*, 16, 143 (1955); *J. Amer. Chem. Soc.*, 77, 1708 (1955).

temperature and with minimum pressure—and often *in situ* where the final product may be required; examples of these are seen in unsaturated polyesters (introduced 1942–47), epoxy resins (1939–45) and polyurethanes (*c.* 1937 in Germany; developed there and elsewhere from *c.* 1949, notably as those that expand during formation and set as solid foams).

Still other plastics, developed because of their relatively high thermal resistance and necessarily covering a variety of chemical types, include polycarbonates (introduced 1958–60), polysulphones (1965), and materials of ingenious internal architecture, such as 'ladder' polymers (with molecules consisting basically of two long chains bridged by cross-links) investigated since *c.* 1959.

RUBBERS

Although rubber (*cis*-1,4-polyisoprene) occurs in several kinds of plants*, it was unknown to Europe before the second voyage of Columbus to the New World in 1493–96, but excavations at ancient Maya sites in Mexico† have produced balls of rubber that were used in a ceremonial ball-court game several centuries before Columbus's time. Reports of the sixteenth century mention rubber‡, particularly as featuring in W. Indian ball games§, and later there are records|| of early settlers copying the native practice of fashioning it into shoes and employing it to produce waterproof clothing; however, in Europe, native rubber—described as *gum-elastic* and *caoutchouc*— for many years after its first appearance remained little more than a curiosity.

The collection of latex from rubber trees was first described by C. F. Fresneau¶ in 1747, and a scientific account of rubber and its uses was compiled by C. B. de La Condamine** in 1745. J. Priestley††, in 1770, mentioned that small cubes of the substance could be purchased because of its property of 'wiping from paper the marks of a black-lead pencil', from which is derived the term *rubber* or (West) *India-rubber*. A patent for waterproofing fabrics by means of rubber

---

See, for instance, p. 162, NAUNTON, W. J. S., *Applied Science of Rubber*, London (1961), p. 152; G. Génin and B. Morisson (Eds) *Encyclopédie technologique de l'Industrie du Caoutchouc*, Paris, 1958, Vol. 1, pp. 63–82.

† Notably at Chichen-Itza, in the State of Yucatan.

‡E.g. P. M. d'Anghiera, *De Orbo Novo*, Alcala, 1530.

§E.g. de Oviedo, *Historia de los Indias*, Seville, 1535.

||E.g. F. J. de Torquemada, *De la Monarquia Indiana*, Seville, 1615.

¶*Sur une résine élastique . . . (Mém. Acad. Roy. Sci., Paris,* 1751); translation appears in *India Rubb. J.,* 121 (1951) 232–33, 732–36.

***Relation Abrégée d'un Voyage fait dans l'Intérieur de l'Amérique Méridionale . . .*, Paris, 1745; also in *Mém. Acad. Roy. Sci., Paris,* 1751.

††*Introduction to the Theory and Practice of Perspective*, London, 1770.

‡‡Brit. Pat. 1 801 (1791).

§§Brit. Pat. 4 804 (1823).

dissolved in turpentine was granted to S. Peal‡‡ in 1791; and in 1819 C. Macintosh§§, using the newly-available coal tar naphtha as solvent, developed a process of bonding together two plies of fabric with an interlayer of rubber ('"mackintosh" double textures').

In 1820 T. Hancock discovered the importance of the mechanical mastication of rubber*, which converts it into a more tractable form; and in 1842 he developed the even more important process that he termed *vulcanisation*†, i.e. heating rubber with sulphur (making it 'unadhesive, cold resisting, capable of enduring heat, greatly increased in elasticity and strength . . . and no longer soluble in essential oils'‡). F. W. Lüdersdorff§, in 1832, had first observed the effects of sulphur on rubber, but Hancock's vulcanisation process arose out of an examination of rubber that had been treated in a manner discovered semi-accidentally in 1839 by C. Goodyear‖; the last-named inventor is usually credited as first to observe the great improvement in physical properties brought about by reaction with sulphur, but he left his discovery unpatented until 1844¶. Hancock's investigation of vulcanisation included production of a hard substance by immersion of rubber in molten sulphur, but in 1851 N. Goodyear (brother of Charles) was granted a patent** for heating rubber with an excess of sulphur to obtain a similar result (Ebonite, Vulcanite); articles made from this product—including furniture—were shown at the Great Exhibition of 1851. 'Cold vulcanisation', by the action of a solution of sulphur monochloride, was discovered by A. Parkes in 1846††.

From this period (1842–50) onwards the rubber industry may be said to have become established‡‡, and in successive years much painstaking research was devoted to the academic aspects of the substance on which it was based. M. Faraday§§, in 1826, had shown rubber to be a hydrocarbon, and in 1860 C. G. Williams‖‖ distilled

*Elastic fabrics, etc. Brit. Pat. 4451 (1820); masticator, Brit. Pat. 7344 (1837). See also: *Personal Narrative of the Origin and Progress of the Caoutchouc or India-rubber Manufacture in England*, London, 1857 (reprinted 1920).
†Brit. Pat. 9952 (1843), see also Hancock's *Personal Narrative* above.
‡From an advertisement of 'Charles Macintosh and Co., sole manufacturer and patentees of the Vulcanised India Rubber', Manchester, 1861.
§*J. für tech. u. ökonom. Chem.*, 1832; *Auflösen u. Wiederherstellen des Federharzes, gennant 'Gummielasticum', zur Darstellung luft- u. wasserdichter Gegenstände . . .*, Berlin, 1832.
‖*Gum-elastic and its varieties*, New Haven, Conn., 1853–55.
¶U.S. Pat 3633 (1844), Brit. Pat. 10 027 (1844).
**U.S. Pat. 8075 (1851).
††Brit. Pat. 11 146 (1846).
‡‡Particularly among the numerous 'Gummifabriken' in Germany. In the pre-vulcanisation era, Hancock commenced business in London in 1820, and Macintosh established a factory in Glasgow in 1823 (moving to Manchester in 1824); the first American factory was that of Roxbury India Rubber, in 1828, and the first German factory was that established by F. Fonrobert in 1829, at Finsterwalde, in Brandenburg. J. B. Dunlop fitted his first rubber tyre to a wooden wheel in 1888 (leading to the vast commercial interest in pneumatic tyres of today).
§§*Quart. J. Sci.*, **21**, 19 (1826).
‖‖*Proc. Roy. Soc.*, **10**, 516 (1860).
¶¶Dipentene from isoprene: *Bull. Soc. Chim.*, Paris, **24**, 108 (1875); a rubber from isoprene: *C.R. Acad. Sci.*, Paris, **89**, 361, 1117 (1879).

rubber to obtain products that he termed isoprene and caoutchine (dipentene). In 1875, G. Bouchardat¶¶ converted isoprene into dipentene, and in 1878 obtained from it a form of rubber, of which he thereby concluded that isoprene might be a simple basis; in 1892 W. A. Tilden*, having observed that samples of isoprene derived from rubber and a non-rubber source (turpentine) slowly changed into similar rubbery products, may be considered to have established the feasibility of producing 'artificial rubber'. In 1900, I. L. Kondakov† obtained a rubber from 2,3-dimethyl-1,3-butadiene, which showed that physical structure rather than identical composition might provide properties hitherto unique to natural rubber.

With regard to early development of synthetic rubbers, the first sodium-polymerisation of butadiene was patented in 1910 by F. E. Matthews and E. H. Strange‡ following work in the U.K. and discoveries made (independently) by C. D. Harries§, in Germany, and S. V. Lebedev‖, in Russia. Polymers of this type (e.g. methyl rubber from dimethylbutadiene) were made by F. Hoffmann and others in production amounts in Germany during the First World War. These materials—which later (1930s) came to be termed Buna rubbers (from *butadien* and *natrium*)—were, however, poor in their performance. Development of superior butadiene rubbers, notably the emulsion-polymerised styrene/butadiene copolymer, had to await the needs of a later Germany (Buna S, 1934) and the U.S.A. during 1939–45 (GR-S, 1939, later termed SBR); the related copolymers with acrylonitrile, first examined in 1929, were produced in Germany in 1935 (Buna N) and in the U.S.A. in 1939 (GR-A, later termed NBR or nitrile rubber).

In 1931, extending a study on the reactions of acetylene that J. A. Nieuwland and colleagues¶ commenced in 1921, W. H. Carothers and his co-workers** polymerised 2-chloro-1,3-butadiene (chloroprene, derived from vinylacetylene) and announced the discovery of an entirely new synthetic rubber, polychloroprene, commercial production of which commenced in the following year (Duprene, later termed Neoprene). In 1926 a reaction first observed in 1839†† was developed by J. Baer‡‡, in Switzerland, to obtain a polysulphide rubber (Perduren), and in 1927 J. C. Patrick and colleagues§§, in

---

*Isoprene from turpentine: *J. chem. Soc.*, 411 (1884); a rubber from isoprene: *Chem. News*, **65**, 266 (1892).
†*J. prakt. Chem.*, **62**, 166 (1900).
‡Brit. Pat. 24 790 (1910).
§*Liebigs Ann.*, **383**, 157 (1911); claims in Untersuchung über die natürlichen u. künstlichen Kautschukarten, Berlin, 1919.
‖*J. Soc. phys.-chem. russe*, **42**, 919 (1910); **45**, 1313, 1377 (1913).
¶J. A. Nieuwland *et al.*, *J. Amer. chem. Soc.*, **53**, 4197 (1931).
W. H. Carothers *et al.*, *J. Amer. chem. Soc.*, **53, 4203 (1931); *see also* footnote ‖, page 652.
††*Poggendorffs Ann.*, **46**, 81 (1839); **49**, 128 (1840).
‡‡Brit. Pat. 279 406 (1926).
§§Brit. Pat. 302 270 (1928); U.S. Pat. 1 890 191 (1932).

America, initiated a parallel development (termed Thiokol rubbers in 1930).

Thus with butadiene rubbers (including later developments), chloroprene rubber, polysulphide rubbers, and commercial introduction of the ion-polymerised butyl rubber* in 1937, it was fully recognised that the remarkable properties of caoutchouc were no longer unique to that material. More recent developments†, including those in fluoro-, silicone, polyurethane and similar polymers, have brought additional special-purpose rubbers, and the major developments in Ziegler-Natta-type stereospecific catalysts‡ have, since 1955, brought the subject into a further (and vast) new era, e.g. *cis*-polyisoprene itself§—virtually *natural* rubber—was synthesised artificially in 1954, while ethylene/propylene rubbers‖, introduced in 1959, are made from components either of which polymerised separately yields a plastics material.

FURTHER LITERATURE

LUCAS, A., *Ancient Egyptian Materials and Industries*, 3rd edn, London (1948)

SENGUPTA, P., *Everyday Life in Ancient India*, 2nd edn, Oxford (1955)

FORBES, R. J., *Studies in Ancient Technology*, Vol. IV, *Textiles*, Leiden (1956)

FORBES, R. J., CIBA Review, 1968/2, *Textiles in Biblical Times*

FORBES, R. J., Imperial Chemical Industries Ltd., *Landmarks of the Plastics Industry*, London (1962)

SCHIDROWITZ, P. and DAWSON, T. R. (Eds), *History of the Rubber Industry*, Cambridge (1952)

*R. M. Thomas and W. J. Sparks, Brit. Pat. 513 521 (1937); *Ind. Engng Chem.*, **30**, 916 (1938). Butyl was the first example of a vulcanisable rubber obtained by introducing a small proportion of unsaturation into an otherwise non-crosslinkable polymer.

† *See*, for instance KENNEDY, J. P. and TÖRNQVIST, E. G. M. (Eds), *Polymer Chemistry of Synthetic Elastomers, Part 1*, New York and London, 21–94 (1968) (TÖRNQVIST, E. G. M.: historical background of synthetic elastomers); also DUCK, E. W. (and others), *Chem. and Ind.*, 219, 254, 286 (1969) (historical, present, and future aspects of synthetic rubbers).

‡ See references ** and ††, page 652, Alkyl-lithium catalysts are also used.

§ Introduced commercially in the U.S.A. by the Goodrich-Gulf, Firestone, and Goodyear companies in 1954–55, e.g. U.S. Pat. 3 114 743 (1954) and Brit. Pat. 827 365 (1955).

‖ Introduced by the Montecatini company in 1959, Ital. Pat. 554 803 (1957).

SECTION 84

ADDITIONAL REFERENCES

Some sources of reference relating to the main classes of polymeric materials, and to polymers in general, are given below. Suggestions for further reading on *specific polymers* are given in §§1.9–42.9, and references to *specific properties* (also serviceability and history) appear at the ends of Sections 51–83.

FIBRES

COOK, J. G., *Handbook of Textile Fibres, Vols. I and II*, London (1968)
KORNREICH, E., *Introduction to Fibres and Fabrics; Their Manufacture and Properties*, London and New York (1966)
MARK, H. F., ATLAS, S. M. and CERNIA, E. (Eds), *Man-Made Fibres: Science and Technology, Vols. I–III*, New York and London (1967–8)
MONCRIEFF, R. W., *Man-made Fibres*, 5th edn, London (1970)

Periodicals: *Journal of the Society of Dyers and Colourists, Journal of the Textile Institute, Melliand Textilberichte* (Engl. edn), *Modern Textiles Magazine, Textile Institute and Industry, Textile Manufacturer, Textile Month, Textile Research Journal, Textile World.*

FILMS*

PINNER, S. H. (Ed), *Modern Packaging Films*, London (1967)
PARK, W. R. R., *Plastics, Film Technology*, New York and London (1969)
SWEETING, O. J. (Ed), *The Science and Technology of Polymer Films, Vol. I*, New York and London (1968)

Periodicals: *Journal of the Oil and Colour Chemists' Association, Journal of Paint Technology, Modern Packaging, Packaging, Packaging Technology, Paper, Film and Foil Converter, TAPPI.*

PLASTICS

BRYDSON, J. A., *Plastics Materials*, London (1969)
DUBOIS, J. H. and JOHN, F. W., *Plastics*, New York and London (1967)
DUBOIS, P., *Plastiques Modernes, Vol. I. Plastophysicochimie*, Paris (1968)
LEVER, A. E. (Ed), *The Plastics Manual*, London (1968)
LEVER, A. E. and RHYS, J. A., *The Properties and Testing of Plastics Materials*, London (1968)
OGORKIEWICZ, R. M. (Ed), *Engineering Properties of Thermoplastics*, London and New York (1970)
RITCHIE, P. D. (Ed), *Physics of Plastics*, London and Princeton (1965)
SMITH, W. M. (Ed), *Manufacture of Plastics, Vol. I*, New York and London (1964)

Periodicals: *Applied Plastics and Reinforced Plastics Review, British Plastics, Kunststoffe, Modern Plastics, Plastics and Polymers, Plastics—Rubbers—Textiles, Plastics Technology, Reinforced Plastics, Society of Plastics Engineers Journal, Soviet Plastics.*

*For references to supported films (surface coatings) *see also* §24A.9 and under Further Literature, Section 81.

RUBBERS

BOSTRÖM, S. (Ed), *Kautschuk-Handbuch*, Vols. 1–5, Stuttgart (1959–62)

GÉNIN, B. and MORISSON, B. (Eds), *Encyclopédie Technologique de l'Industrie du Caoutchouc*, *Vols. I–IV*, Paris (1956–60)

NAUNTON, W. J. S. (Ed), *The Applied Science of Rubber*, London (1961)

PAYNE, A. R. and SCOTT, J. R., *Engineering Design with Rubber*, London and New York (1960)

PENN, W. S., *Synthetic Rubber Technology, Vol. I*, London (1960)

SCOTT, J. R., *Physical Testing of Rubbers*, London and New York (1965)

STERN, H. J., *Rubber: Natural and Synthetic*, London and New York (1967)

Periodicals: *Journal of the Institution of the Rubber Industry, Rubber Age, Rubber Chemistry and Technology* (includes *Rubber Reviews*), *Rubber Developments, Rubber Journal, Rubber World, Soviet Rubber Technology.*

POLYMERS IN GENERAL

BRANDRUP, J. and IMMERGUT, E. H. (Eds), *Polymer Handbook,* New York and London (1966)

MARK, H. F., GAYLORD, N. G. and BIKALES, N. M. (Eds), *Encyclopedia of Polymer Science and Technology; Plastics, Resins, Rubbers, Fibers*, Vol. 1– , New York and London (1964 onwards)

MILES, D. G. and BRISTON, J. H., *Polymer Technology*, London (1965)

Periodicals: *Advances in Polymer Science, Angewandte Chemie* (International edn), *British Polymer Journal, Chemical Abstracts* (macromolecular sections), *Chemistry and Industry, European Polymer Journal, Journal of Applied Polymer Science, Journal of Macromolecular Science, Journal of Polymer Science, Macromolecules, Makromolekulare Chemie, Materials Engineering, Polymer, Polymer Engineering and Science, Polymer Science USSR, Progress in High Polymers, RAPRA Abstracts.*

NOTE ON S I UNITS

The 'Systeme International des Unités' (SI for short), developed by the Conférence Générale des Poids et Mesures in 1954, is built up from six *basic* (or *base*) *units*:

Quantity	Name (and symbol) of unit
length	metre (m)
mass	kilogramme (kg)
time	second (s)
electric current	ampere (A)
thermodynamic temperature*	kelvin (K)
luminous intensity	candela (cd)

*Temperatures on the thermodynamic scale are proportional to the pressures (or to the volumes) of an ideal gas in a perfect constant-volume (or constant-pressure) gas thermometer.

From the above are formed numerous *derived units*, some of which are as follows:

Quantity	Name (and symbol)	Expressed in terms of basic units
frequency	hertz (Hz)	1 s^{-1} (or $1/\text{s}$)
force	newton (N)	1 kg m/s^2
energy*	joule (J)	$1 \text{ N m} = 1 \text{ kg m}^2/\text{s}^2$
power*	watt (W)	$1 \text{ J/s} = 1 \text{ kg m}^2/\text{s}^3$
electric potential	volt (V)	$1 \text{ W/A} = 1 \text{ kg m}^2/\text{As}^3$
electric resistance	ohm (Ω)	$1 \text{ V/A} = 1 \text{ kg m}^2/\text{A}^2 \text{ s}^3$

*The joule and watt are used respectively for *all* forms of energy and power—whether mechanical, thermal or electrical.

Essential features of SI units are:

(i) They are *consistent*, i.e. a derived unit involves only the basic unit(s), without any numerical factor, as will be seen from the above examples. Thus there is no need of factors for converting (for instance) ergs to calories, ft lbf to kWh, or so on. Also g (acceleration due to gravity), is not used.

(ii) The preferred multiples and sub-multiples of units are those in steps of 10^3, such as mega (M; 10^6), kilo (k; 10^3), milli (m; 10^{-3}) and micro (μ; 10^{-6}). However, 'in certain cases, for convenience' hecto (h; 10^2), deca (da; 10), deci (d; 10^{-1}) and centi (c; 10^{-2}), also litre, millilitre (ml), minute, hour and day 'may be used'. *N.B.* When m (metre) is multiplied by another unit it should preferably be placed second, because if it is placed first, *and if the multiplication dot is omitted,* it means milli; thus, Nm = newton × metre, but mN = milli newton.

(iii) It is recommended that in derived units any prefix denoting a multiple or sub-multiple should be attached *only* to the *numerator*.

Units used in this book

Below are listed certain quantities—other than simple ones like mass, length, time, area, volume—frequently used, showing (A) the units used in this book (if different from SI), (B) the SI unit, and (C) the factor by which values expressed in (A) units must be *multiplied* to give values in SI units, e.g. 2 g/ml $= 2 \times 10^3$ kg/m^3.

Quantity	A. Unit used in this book	B. SI Unit	C. Conversion factor
Density	g/ml	kg/m^3*	10^3
Permeability, gas	$(10^{-10}$ cm$^2)$/(s cmHg)	m^4/(s N)	7.5×10^{-18}
do., vapour	$\begin{cases} (10^{-10}$ g)/(cm^2 s cmHg)**	kg/(s N)	7.5×10^{-13}
(*see* Section 64)	$(10^{-10}$ g)/(cm s cmHg)	(kg m)/(s N)	7.5×10^{-15}
Surface tension	dyne/cm	N/m	10^{-3}
Viscosity, dynamic	P (poise); cP	N s/m^2	10^{-1}; 10^{-3}
do., kinematic	St (stoke); cSt	m^2/s	10^{-4}; 10^{-6}
Specific heat	cal/(g °C)†	J/(kg K)	4.19×10^3
Thermal conductivity	$(10^{-4}$ cal)/(cm s °C)	W/(m K)	4.19×10^{-2}
Electric resistivity (volume)	ohm cm (Ω cm)	Ω m	10^{-2}
Dielectric strength	kV/mm	V/m	10^6
Force	kgf	N	9.81
Stress‡	kgf/mm^2	N/m^2§	9.81×10^6
Pressure‖	lbf/in^2	N/m^2	6.89×10^3
Impact strength¶	$\begin{cases}$ kgf mm/mm^2	J/m^2	9.81×10^3
	ft lbf/in	J/m	53.3

*Mg/m^3 is suggested as a convenient alternative, since it gives the same numerical values as g/ml

†In the text, specific heat is given as *ratio* of the energy required to raise the temperature of a given mass by one degree, to that for an equal mass of water; this ratio is (nearly enough) numerically equal to the heat capacity of the material in cal/(g °C)

‡As in elastic modulus and tensile, compression and flexural strengths

§As this unit is inconveniently small, kN/m^2 or MN/m^2 is commonly used. Some authorities have provisionally adopted the *bar* ($= 10^5$ N/m^2), which is very nearly 1 (actually 1·02) kgf/cm^2

‖In moulding and similar processing operations

¶The kgf.mm/mm^2 unit applies when impact strength is expressed as energy/(cross-section of test piece), but ft lbf/in when expressed as energy per unit width, or length of notch when present (*see* Section 78)

**With thickness of test material stated

FURTHER LITERATURE

ISO Recommendation R1000, *Rules for the Use of Units of the International System of Units and a Selection of the Decimal Multiples and Sub-multiples of the SI Units,* 1st edn (1969)
BS 3763: 1964, *The International System (SI) Units.*
British Standards Institution, *The Use of SI Units,* publication PD 5686 (Jan. 1969)

AUTHOR INDEX*

Adams, C. H., 612
Adams, P. A., 310
Adriani, A., 344
Akin, R. B., 284
Albert, K., 309, 650
Alexander, J., 196
Alexander, L. E., 496
Alexander, P., 196, 208
Alfrey, T., Jr., 596, 602
Allen, P. W., 333, 492
Alliger, G., 326
Alner, D. J., 640, 646
Amerongen, G. J. van, 348, 558
Anderson, D. R., 563
Andrianov, K. A., 469, 471, 481
Anfinsen, C. B., 196
Anger, V., 531
Anson, M. L., 196, 204
Appleyard, H. M., 194
Archer, F. S., 163, 649
Arends, C. B., 612
Argana, C. P., 81
Atkinson, R. R., 148
Atlas, S. M., 638, 657
Audemars, 163

Baekeland, L. H., 309, 652
Baer, J., 443, 655
Baer, J. E., 439
Baeyer, A. von, 309
Bamberger, E., 14
Bamford, C. H., 508
Barb, W. G., 508
Barnett, G., 312
Barron, H., 445
Barson, C. A., 44
Bartoe, W. F., 619
Basdekis, C. H., 52
Bateman, L. C., 333, 341, 512, 596, 626
Baumann, E., 115, 128
Bayer, O., 457
Beadle, C., 140, 648
Bellamy, L. J., 531
Bemmelen, J. M., van, 234

Berend, L., 309, 650
Bergen, W. von, 196
Berger, H., 482
Bernhardt, E. C., 619, 639
Berthelot, M. M., 50, 234
Berzelius, J., 234
Bevan, E. J., 140, 156, 648, 651
Beyer, W., 639
Bikales, N. M., 358, 658
Binder, J. L., 377
Bing, M., 328
Biot, J. B., 185
Black, H., 168
Blackley, D. C., 341, 395, 406
Blais, J. F., 298
Block, R. J., 208
Bloomfield, G. F., 348
Blyth, J., 50
Bock, W., 392, 404
Bockhoff, F. J., 639
Boenig, H. V., 7, 20, 262
Bolt, R. A., 646
Bonastre, M., 50, 651
Boor, J. (Jr.), 508
Boor, L., 619
Borasky, R., 196
Bose, P. K., 239
Boström, S., 343, 362, 406, 658
Böttcher, C. J. F., 582
Bouchard, R., 81
Bouchardat, G., 370, 654
Boundy, R. H., 58
Bovey, F. A., 508, 532
Bowden, F. P., 626
Bowles, O., 482
Boyer, R. F., 58, 571, 612
Braconnot, H., 148, 163, 649, 651
Bradley, D. C., 481
Bradley, T. F., 261
Brandrup, J., 514, 658
Braun, J. von, 221
Breit, W., 335
Bretschneider, O., 37
Breuers, W., 382, 395, 406
Briston, J. H., 639, 658

TRADE NAME INDEX*

* Names of less importance have been omitted

SUBJECT INDEX*

---

* Page numbers in bold type indicate principal references. Items of less importance have been
  omitted and prefixes (*o, n, N, a* etc.) have been ignored in compiling this index.